JOYCE LEPETU

Uni  9.

MW01232976

from

BOTSWANA

ADVANCES IN AGROFORESTRY

Series editor:

P.K.R. Nair
School of Forest Resources and Conservation, University of Florida, Gainesville, Florida, USA.

Volume 1

NEW VISTAS IN AGROFORESTRY

A Compendium for the 1st World Congress of Agroforestry, 2004

Edited by P.K.R. Nair, M.R. Rao, and L.E. Buck

Reprinted from the journal *Agroforestry Systems*, Volumes 61 & 62, 2004

KLUWER ACADEMIC PUBLISHERS
Dordrecht / Boston / New York

A C.I.P. Catalogue record for this book is available from the Library of Congress

ISBN 1-4020-2412-6

Published by Kluwer Academic Publishers,
P.O. Box 17, 3300 AA Dordrecht, The Netherlands

printed on acid-free paper

Printed in The Netherlands

Aims and Scope

Agroforestry, the purposeful growing of trees and crops in interacting combinations,began to attain prominence in the late 1970s, when the international scientific community embraced its potentials in the tropics and recognized it as a practice in search of science. During the 1990s, the relevance of agroforestry for solving problems related to deterioration of family farms, increased soil erosion, surface and ground water pollution, and decreased biodiversity was recognized in the industrialized nations too. Thus, agroforestry is now receiving increasing attention as a sustainable land-management option the world over because of its ecological, economic, and social attributes. Consequently, the knowledge-base of agroforestry is being expanded at a rapid rate as illustrated by the increasing number and quality of scientific publications of various forms on different aspects of agroforestry.

Making full and efficient use of this upsurge in scientific agroforestry is both a challenge and an opportunity to the agroforestry scientific community. In order to help prepare themselves better for facing the challenge and seizing the opportunity, agoroforestry scientists need access to synthesized information on multi-dimensional aspects of scientific agroforesty.

The aim of this new book-series, *Advances in Agroforestry*, is to offer state-of-the art synthesis of research results and evaluations relating to different aspects of agroforestry. Its scope is broad enough to encompass any and all aspects of agroforestry research and development. Contributions are welcome as well as solicited from competent authors on any aspect of agroforestry. Volumes in the series will consist of reference books, subject-specific monographs, peer-reviewed publications out of conferences, comprehensive evaluations of specific projects, and other book-length compilations of scientific and professional merit and relevance to the science and practice of agroforestry worldwide.

ADVANCES IN AGROFORESTRY

Contents

Volume 1

Electronic journals at
KluwerOnline
WWW.KLUWERONLINE.NL
Contact your librarian for more information

Agroforestry Systems **61:** 1, 2004.

PREFACE

It was in late 2002 that the idea of preparing a collection of multi-authored chapters on different aspects of agroforestry as a compendium for the 1st World Congress of Agroforestry, June 2004, was tossed around. With the approval of the idea by the Congress Organizing Committee, serious efforts to make it a reality got under way in early 2003. The rigorously peer-reviewed and edited manuscripts were submitted to the publisher in December 2003. Considering the many different individuals involved in the task as authors and manuscript reviewers, we feel quite pleased that the task could be accomplished within this timeframe.

We are pleased also about the contents on several counts. First of all, the tropical-temperate mix of topics is a rare feature of a publication of this nature. In spite of the scientific commonalities between tropical and temperate practices of agroforestry, the differences between them are so enormous that it is often impossible to mesh them together in one publication. Secondly, several of the chapters are on topics that have not been discussed or described much in agroforestry literature. A third feature is that some of the authors, though well known in their own disciplinary areas, are somewhat new to agroforestry; the perceptions and outlooks of these scholars who are relatively uninfluenced by the past happenings in agroforestry gives a whole new dimension to agroforestry and broadens the scope of the subject. Finally, rather than just reviewing and summarizing past work, most chapters take the extra effort in attempting to outline the next steps. Agroforestry stands to gain enormously from the infusion of these new and different ideas and bold initiatives. Thus, we feel quite justified with the title of the volume: *New Vistas.*

The subject matter of various topics was primarily the choice of the editors, but the scope and contents of individual chapters were mostly of the respective authors. We recognize that there are several other topics that can be considered appropriate for such a volume; but space- and time constraints did not permit the luxury of including them. Moreover, widely described and well-published topics within agroforestry were kept out deliberately; preference was for relatively new subjects within the context of agroforestry.

Almost all who were identified and approached as potential chapter authors accepted our invitation enthusiastically. Even more importantly, they worked within the strict and tight time schedules to not only prepare the manuscripts, but also to respond to demands for repeated revisions and various other chores associated with the process. Equally gratifying was the prompt, thorough, and professional reviews provided by manuscript reviewers (see the list of reviewers at the end of the volume).

We sincerely thank all the authors and reviewers, who contributed enormously and cooperated so splendidly under strict and difficult time schedules. Special thanks go to Mr. Larry Schnell for copy editing the manuscripts.

P.K.R. Nair, Gainesville, Florida, USA
M.R. Rao, Secunderabad, India
L.E. Buck, Ithaca, New York, USA

February 2004

Introduction

Agroforestry Systems **61:** 5–17, 2004.

Agroforestry and the achievement of the Millennium Development Goals

D.P. Garrity
World Agroforestry Centre, United Nations Avenue, P.O. Box 30677–00100, Nairobi, Kenya; e-mail: d.garrity@cgiar.org

Key words: Commercialization, Domestication, Development goals, Markets, Nutrient replenishment, Policy, Tree products

Abstract

The Millennium Development Goals (MDGs) of the United Nations (UN) are at the heart of the global development agenda. This chapter examines the role of agroforestry research and development (R&D) in light of the MDGs. It reviews some of the ways in which agroforestry is substantively assisting to achieve the goals and discusses how the agenda can be realigned to further increase its effectiveness in helping developing countries to meet their MDG targets. Promising agroforestry pathways to increase on-farm food production and income contribute to the first MDG, which aims to cut the number of hungry and desperately poor by at least half by 2015. Such pathways include fertilizer tree systems for smallholders with limited access to adequate crop nutrients, and expanded tree cropping and improved tree product processing and marketing. These advances can also help address lack of enterprise opportunities on small-scale farms, inequitable returns to small-scale farmers (especially women), child malnutrition, and national tree-product deficits (especially timber). The rate of return to investment in research on tree crops is quite high (88%); but enterprise development and enhancement of tree-product marketing has been badly neglected. The products, processing, and marketing of tree products and services, through tree domestication and the commercialization of their products is a new frontier for agroforestry R&D. A major role for agroforestry also is emerging in the domain of environmental services. This entails the development of mechanisms to reward the rural poor for the environmental services such as watershed protection and carbon sequestration that they provide to society. Agroforestry R&D is contributing to virtually all of the MDGs. But recognition for that role must be won by ensuring that more developing countries have national agroforestry strategies, and that agroforestry is a recognized part of their programs to achieve the MDGs.

Introduction

At the United Nations Millennium Summit in September 2000 in New York, world leaders agreed to a set of time-bound and measurable goals for combating hunger, poverty, disease, illiteracy, environmental degradation, and discrimination against women. These Millennium Development Goals (MDGs) (see Table 1) are now at the heart of the global development agenda. The Summit's Millennium Declaration also outlined a plan for how to proceed to achieve the goals (www.un.org/millennium). Leaders from both developed and developing countries have started to match these commitments with resources and action, signaling the evolution of a global deal in which sustained political and economic reform by developing countries will be matched by greater support from the developed world in the form of aid, trade, debt relief, and investment. The MDGs provide a framework for all nations, and the entire development community, to work coherently together toward this common end.

In preparation for the August 2002 World Summit on Sustainable Development (WSSD) in Johannesburg, South Africa, UN Secretary-General Kofi Annan proposed the WEHAB initiative to provide focus and impetus to action in the five key thematic areas of Water, Energy, Health, Agriculture

Table 1. The United Nations Millennium Development Goals.

Goal 1. Eradicate extreme poverty and hunger	*Target for 2015: Halve the proportion of people living on less than a dollar a day and those who suffer from hunger.*
Goal 2. Achieve universal primary education	*Target for 2015: Ensure that all boys and girls complete primary school.*
Goal 3. Promote gender equality and empower women	*Targets for 2005 and 2015: Eliminate gender disparities in primary and secondary education preferably by 2005, and at all levels by 2015.*
Goal 4. Reduce child mortality	*Target for 2015: Reduce by two thirds the mortality rate among children under five*
Goal 5. Improve maternal health	*Target for 2015: Reduce by three-quarters the ratio of women dying in childbirth.*
Goal 6. Combat HIV/AIDS, malaria and other diseases	*Target for 2015: Halt and begin to reverse the spread of HIV/AIDS and the incidence of malaria and other major diseases.*
Goal 7. Ensure environmental sustainability	*Targets: Integrate the principles of sustainable development into country policies and programmes and reverse the loss of environmental resources; By 2015, reduce by half the proportion of people without access to safe drinking water; By 2020 achieve significant improvement in the lives of at least 100 million slum dwellers.*
Goal 8. Develop a global partnership for development	*Targets: Develop further an open trading and financial system that includes a commitment to good governance, development and poverty reduction – nationally and internationally; Address the least developed countries' special needs, and the special needs of landlocked and small island developing States; Deal comprehensively with developing countries' debt problems; Develop decent and productive work for youth; In cooperation with pharmaceutical companies, provide access to affordable essential drugs in developing countries; In cooperation with the private sector, make available the benefits of new technologies – especially information and communications technologies.*

Source: Human Development Report by UNDP 2003

and Biodiversity. The initiative provides a coherent international framework for the implementation of sustainable development surrounding these crucial issues (www.johannesburgsummit.org/html/documents/wehab_papers.html).

Advances in agroforestry can contribute significantly to the achievement of virtually all of the MDGs and the WEHAB initiative. Agroforestry focuses on the role of trees on farms and in agricultural landscapes to meet the triple bottom line of economic, social and ecological needs in today's world. Recognition of this role in overcoming key problems from local to global levels is growing. The World Agroforestry Centre (ICRAF) has identified seven key challenges related to the MDGs and WEHAB that agroforestry science and practice can materially address.

1. Help *eradicate hunger* through basic, pro-poor food production systems in disadvantaged areas based on agroforestry methods of *soil fertility and land regeneration*;
2. Lift more rural poor from *poverty* through market-driven, locally led *tree cultivation systems* that generate income and build assets;
3. Advance the *health and nutrition* of the rural poor through agroforestry systems;
4. Conserve *biodiversity* through integrated conservation-development solutions based on agroforestry technologies, innovative institutions, and better policies;
5. Protect *watershed services* through agroforestry-based solutions that enable the poor to be rewarded for their provision of these services;
6. Assist the rural poor to better adapt to *climate change*, and to benefit from emerging *carbon* markets, through tree cultivation; and
7. Build human and institutional *capacity* in agroforestry research and development.

This chapter examines these key components of the agenda for agroforestry research and development in the context of the MDGs. It reviews some of the ways in which agroforestry is demonstrably assisting to achieve the goals, and discusses how the agenda should be realigned to further increase effectiveness in helping developing countries to meet their MDG targets. We begin by focusing on the achievement of the first goal, eradication of extreme poverty and hunger, where the role of agroforestry is highly evident.

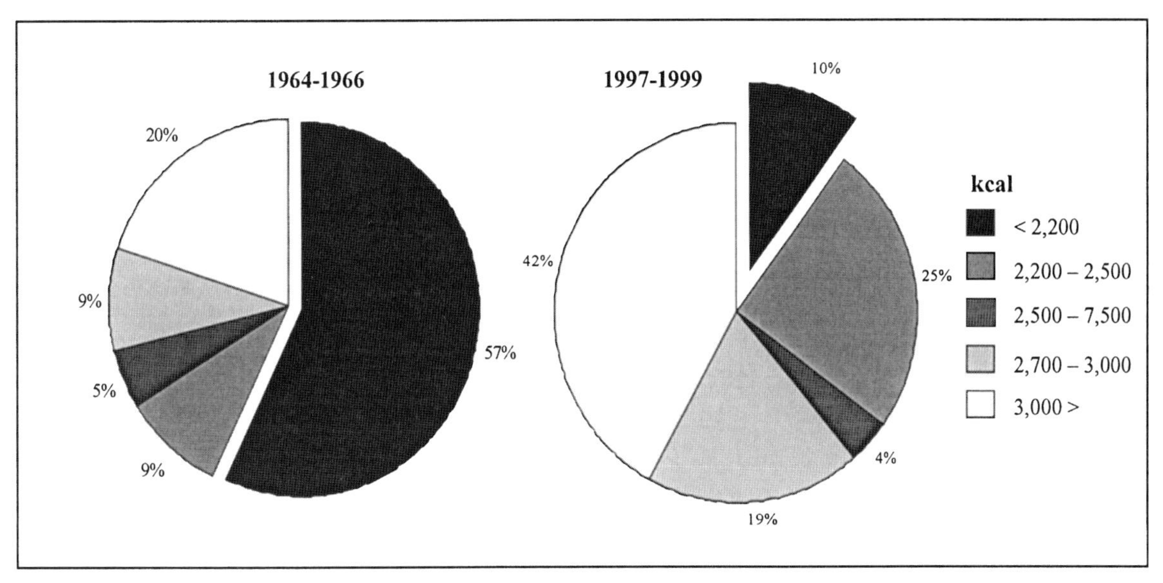

Figure 1. The world today produces enough food to feed everyone, yet hunger persists. Distribution of the world's population in terms of kilocalories per capita per day [1 calorie (cal) = 4.19 joules (J)]. Source: Bread for the World Institute (2003).

Eradicating extreme poverty and hunger

In a world of growing prosperity, there exists a massive and inexcusable failure of will to address the fundamental blight of hunger and desperate poverty that stunts the lives of the disadvantaged and excluded. The first MDG aims to cut at least in half by 2015 both the number of hungry, and the desperately poor who live on less than $1 a day. The world has made significant progress in reducing hunger and poverty in the recent past. During the last 30 years, the percentage of food-insecure people has declined by more than half, even though world population nearly doubled during that period. The percentage of world's population that is food insecure has fallen from 37% to 18%, and food availability has improved dramatically in the developing world in the past three decades. Daily per capita calorie availability increased from 2100 to 2700 (FAO 2001). Asia has gone from being a 'hopeless basket case' in the 1960s to being a 'food basket' in the 1990s, with huge declines in the numbers and proportions of the food insecure in East and Southeast Asia.

Currently, the three areas of the world with by far the greatest number of desperately poor are sub-Saharan Africa (291 million), South Asia (522 million), and China (213 million) (CGIAR 2000).

Alarmingly, the number of food-insecure people in sub-Saharan Africa during this period has more than doubled. Average per capita calorie availability in Africa is now below the minimum requirement for basic sustenance. Such trends must be reversed to achieve the MDGs. (See Figure 1)

Agroforestry and food security

Farmland in the developing world generally suffers from the continuous depletion of nutrients as farmers harvest without fertilizing adequately or fallowing the land. Nowhere is this more true than in Africa (Figure 2). Small-scale farmers have removed large quantities of nutrients from their soils without using sufficient amounts of manure or fertilizer to replenish fertility. This has resulted in high annual nutrient depletion rates of 22 kg nitrogen, 2.5 kg phosphorus, and 15 kg potassium per hectare of cultivated land over the past 30 years in 37 African countries – an annual loss equivalent to $4 billion worth of fertilizers (Sanchez 2002).

Commercial fertilizers cost two to six times as much in Africa as in Europe or Asia. Even at these prices, supplies are problematic due to poorly functioning markets and road infrastructure. Consequently, most African farmers have abandoned their use. This is having dire effects on smallholder food production. In the extensive maize (*Zea mays*)-producing belt of eastern and southern Africa, there are 12 million farm households that support some 60 million people, with farm sizes ranging from 0.3 to 3 hectares. Few farms

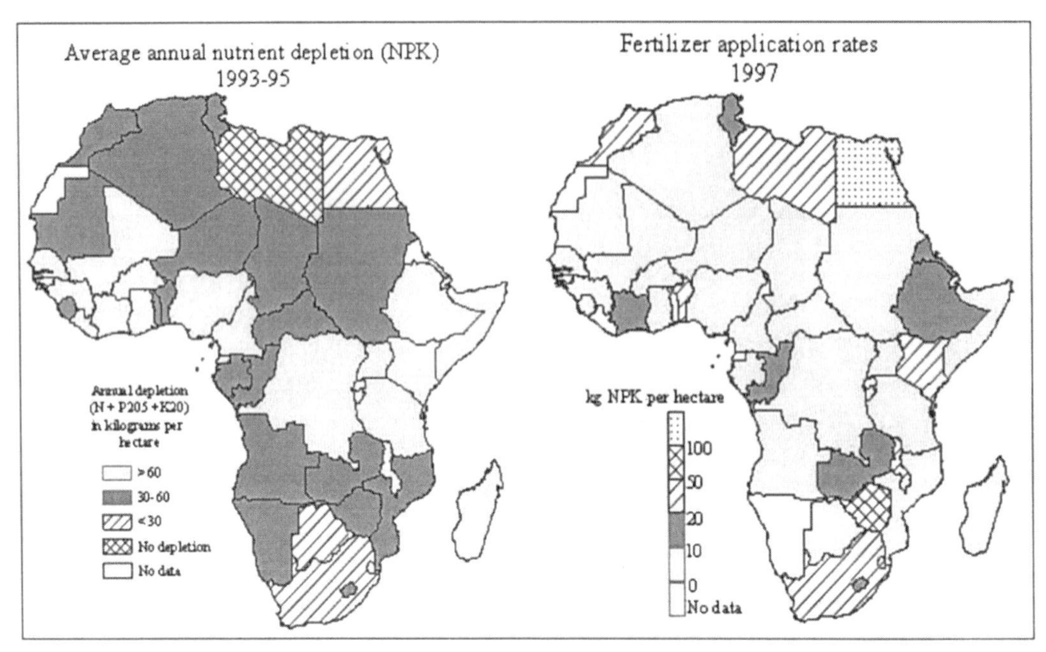

Figure 2. African soils are being depleted of nutrients due to very low average fertilizer application rates: Fertilizer tree systems are a practical option for integrated soil fertility replenishment that farmers can adopt with minimal cash cost. Sources: J. Henao and C. Baanante, 'Nutrient Depletion in the Agricultural Soils of Africa,' 2020 Vision Brief 62 (Washington, D.C: International Food Policy Research Institute, 1999) and FAOSTAT 1999. Compiled by Stan Wood, IFPRI.

Figure 3. Thousands of farmers in eastern and southern Africa are intercropping *Gliricidia sepium* and other nitrogen-fixing trees with their maize to provide crop nutrient needs and sustainably double or triple yields.

currently produce enough maize to feed the family, let alone provide a surplus for the market. For example, recent work shows that up to 90% of maize-growing families in eastern Zambia experience hunger for three to four months during normal years and there is severe famine in drought years (P.L. Mafongoya, pers. comm. 2003).

Many policy and infrastructural constraints have to be addressed to alleviate basic rural food insecurity in this region. But, in the meantime, one promising pathway is to enable smallholders to use fertilizer tree systems that increase on-farm food production. After years of experimentation with a wide range of soil fertility replenishment practices, three types of simple, practical fertilizer tree systems have been developed that are now achieving widespread adoption. These are: (1) improved fallows using trees and shrubs such as sesbania *(Sesbania sesban)* or tephrosia (*Tephrosia vogelii*), (2) mixed intercropping with gliricidia (*Gliricida sepium*), and (3) biomass transfer with wild sunflower (*Tithonia diversifolia*) or gliricidia (Place et al. 2002). They provide 50 to 200 kg N ha^{-1} to the associated cereal crops. Yield increases are typically two-to-three times that with current farmers' practices. These fertilizer tree systems have now reached over 100,000 households in the maize belt of eastern and southern Africa, and demand for tree seed and knowledge transfer is increasing exponentially (F. Kwesiga, pers. comm. 2002). These practices tend to be adopted to a greater extent by the poorest families in the villages, which is unusual for agricultural innovations (Place et al. 2002). Accelerated efforts are under way to further adapt these options and scale them up to reach the 12 million maize growing families of the region and beyond (Franzel et al. 2002; 2004). (See Figure 3)

Promising as these approaches may be, food security will not be achieved by focusing only on increasing crop productivity. There must also be vigorous agricultural market reforms. These reforms need to be based on an understanding of what is actually happening in the rural areas and on determining why previous efforts to ensure food security have failed. Integrated approaches, rather than sectoral approaches, will be key to achieving sustained food security in the future (Omamo and Lynam 2002).

Policy research must unravel the dynamic 'poverty traps' that reinforce the low demand for and low supply of improved technologies in smallholder agriculture (see, for example, Schreckenberg et al. 2002). Besides identifying policy prescriptions, we must have pragmatic solutions that Africa's cash-strapped governments can actually implement, and that empower farmers to solve their own problems.

Roughly one quarter of the farmland in the developing world has been degraded. Agricultural reform must address the risks of land degradation due to overcultivation, desertification, declining water supplies, and loss of biodiversity. The potential of agroforestry to rehabilitate degraded land, and to conserve soil and water on the working lands of the tropics, has long been recognized (Young 1997). But we cannot assume that conservation investments will be attractive to farmers simply because they are known to protect the resource base (Clay 1996). There is much work to be done to identify soil and water conservation practices that not only make sense, but also make money for smallholders. The challenge is to make them profitable to adopt. Tree-based soil conservation systems have real promise in this regard.

Collective action through community-level support for soil and water conservation is crucial to the timely application of agroforestry-based soil and water conservation over whole landscapes. Mobilizing community involvement in evolving locally sustainable conservation farming systems through the Landcare approach has shown great promise in Australia, the Philippines, and South Africa (Mercado et al. 2001; Franzel et al. 2004). There is bright scope for investigating its application more widely in developing countries, particularly in Africa.

Under the prevalent conditions of uncertainty and sparse information, risk-averse smallholders have always tended to self-insure by diversifying their enterprises. Market failure is very common in smallholder systems. Rural areas typically have markets with high transaction costs, and this makes production diversification a favourable choice (Omamo and Lynam 2002). In such situations, integrated agroforestry systems are a suitable pathway toward improved livelihoods. This is exemplified by the agroforests of Indonesia (Torquebiau 1992; Michon et al. 1995) and the tropical homegardens in general (Kumar and Nair 2004). Unfortunately, too little research attention has been given to how the successful agroforest systems observed in some parts of the tropics can be more widely expanded for the benefit of the smallholders living in remote areas with poor market infrastructure.

Agroforestry research and development must now seriously focus on land management interventions that reach the poorest and most vulnerable land users. This requires a deep understanding of poor land users, particularly women, and the problems they face. These constraints involve poor access to technology development processes, and to agricultural information; poor bargaining power in markets and for public services; and extreme household resource constraints. Participatory technology development processes will be fundamental to identifying successful pro-poor solutions that are more readily adopted. Accelerating the formation and functioning of effective farmers' associations that serve the poor is crucial in addressing these needs.

Agroforestry and poverty alleviation

Nearly three-quarters of the poor people who live on less than $1 a day are found in the rural areas of developing countries (Dixon et al. 2001). Agricultural development is therefore key to increasing poor people's access to both food and income. The evidence clearly indicates that in developing countries, agricultural growth nearly always benefits poor people, with greatest gains going to those most in need (Thirtle et al. 2001). Empirical work points to a strong relationship between agricultural productivity and poverty (Mellor 2001). It is only with rising farm incomes that

Figure 4. Creation of a cultivar: The African Plum, *Dacryodes edulis*, is one of many indigenous species of fruits now being domesticated from natural forests and improved for on-farm cultivation to create income opportunities by supplying expanding markets.

poverty can be reduced. A 1% yield increase is associated with a 1% drop in the number of people living on less than $1 a day. Although agriculture's poverty-fighting potential in developing countries is regaining wider recognition, national policies and funding trends have not reflected that promise. Agriculture continues to receive low levels of aid and investment from both donor organizations and developing country governments. Its share of total official development assistance averaged only 12% during the 1990s, while developing regions themselves spent only 4% of total expenditure on agriculture (FAO n.d.). Such trends must be reversed to achieve the MDGs.

Smallholder tree production contributes substantially to rural livelihoods and national economies, yet these contributions are not adequately quantified or appreciated. With less than half a hectare of natural forest remaining per person in the tropics, trees on farms in many countries are more important for tree product supply than trees in forests. This is true for both household and commercial purposes. Yet, obsolete policy objectives often act as a barrier to greater investment in trees on farms by farmers and entrepreneurs as discussed by Puri and Nair (2004) in the Indian context and by Scherr (2004) in the general context. Even where forest extraction gives way to tree cultivation, small-scale farmers are not sufficiently prepared to diversify and add value to their tree production. National planners are also ill-equipped to support agroforestry since few analyses have been carried out to identify the winners and losers in the cultivation and commercialization of tree products (as, for example, Shackleton et al. 2003). In many regions the enabling policies, species choice, tree husbandry skills, germplasm quality, and tree improvement lag far behind the overall demand for tree planting. Furthermore, the markets for tree products are often poorly organized and thus perform suboptimally (Russell and Franzel 2004). This causes spoilage of perishable tree products, lost income for producers, and restricted choices for consumers. The rural poor are further disadvantaged by lack of market price transparency, and the absence of processing techniques to add value to tree products.

Enhanced tree-based systems and improved tree product marketing have the potential to address key aspects of rural poverty, child malnutrition, poor access to conventional health care, national tree product deficits (especially timber), inequitable returns to small-scale farmers (especially women) from tree product marketing, and lack of enterprise opportunities on small-scale farms. Advances in these areas will contribute to seven of the eight UN Millennium Development Goals (MDG 1, 3, 4, 5, 6, 7, and 8). Work in this area focuses on the vulnerability and transformational opportunities of natural, financial, human and social capital assets of the rural poor and related stakeholders, within the sustainable livelihoods framework.

In Africa, agriculture has been growing at the rate of 3% per year since the mid-1980s. But because of increasingly competitive world markets, Africa's market share in most agricultural commodities is declining. This is combined with declining market prices for most commodities. Most agricultural development programs have focused on improving output, productivity, and exports. But as more countries adopt these programs the value of commodity exports continues to fall. Consequently, Africa is falling further behind.

A new approach is needed: a research and development strategy to reduce dependency on primary agricultural commodities, and to establish production of *added-value products* based on raw agricultural materials, with links to growing and emerging markets. For the African countries to compete successfully in the world economy, their agricultural research and development institutions must develop new skills in domestication of indigenous species (as discussed, for example, by Simons and Leakey 2004) and the processing/storage of their products, and in market analysis and market linkages. Research and extension services must adapt quickly to meet the challenges of globalization. There is an important role for 'bridging institutions' that catalyze improved integration of activity along the continuum from basic to applied, to adaptive research. They must convert technically workable innovations into commercially feasible ones.

Agroforestry research and development has traditionally focused on trees and tree production issues. Tree enterprise development and enhancement of tree product marketing has been neglected. It needs to be significantly reoriented. The frontier is to focus on the products, processing, and marketing of domesticated tree products and services (Russell and Franzel 2004).

The World Agroforestry Centre is now putting much greater attention on the development of tree products, and the expansion of their markets. With new analyses, and networking, we are building robust knowledge bases on these aspects. This complements our *Agroforestree* knowledge bases. The aim is to support new and existing tree product enterprises that favor small-scale farmers and entrepreneurs, and to improve the functioning and information of tree product markets. Increasingly, we see the science of agroforestry as a means to advance the business of smallholder agroforestry. Accelerated work on timber trees, fruit trees, fodder trees, medicinals, and fertilizer trees, is achieving greater diversity and productivity on the smallholder farm (Scherr 2004). This is opening up a range of business opportunities, not only for farmers but for the wider rural economy as a whole.

This effort will require attention to market chain and policy analysis for key tree product sub-sectors including beverage, extractives, fodder, fruit, fuelwood, medicine, and timber. Priorities concerning promising tree products will need to be made through the active involvement of producers, consumers and merchants. New products need to be identified for development with partners. The work must focus on the constraints and opportunities for pro-poor and gender-sensitive tree product enterprises and extension approaches, and building sustainable seed systems and better management approaches for the genetic resources of agroforestry trees.

Tree crops played a critical role as a springboard for economic growth in Southeast Asia during the past three decades (Tomich et al. 1994). Indeed, tree crops such as cacao (*Theobroma cacao*), coffee (*Coffea* spp.), and tea (*Camellia sinensis*) have also been mainstays of the economies of a number of African countries. But global overproduction of these few commodities has reduced their present and prospective profitability for smallholders. A new 'tree crops revolution' is needed (Leakey and Newton 1994) that greatly broadens the array of tree products that are produced, processed and delivered by developing countries to regional and global markets. These products could be a vehicle providing comparative advantage to poor land-locked countries, and countries that are not otherwise well-integrated into the global economy. These disadvantaged countries need to be more aware of the practical opportunities of tree products, and to build the market linkages that can capitalize on them.

We foresee much greater efforts to domesticate new and underutilized tree species, and to intensify their cultivation on smallholder farms (Simons 1996). Appropriate tree propagation and management methods for on-farm cultivation must be elucidated, and tree species improvement undertaken for priority taxa (Simons and Leakey 2004). Massive capacity building will be necessary to expand this work. Allied to this is the need to implement approaches to encourage sustainable tree seed and seedling systems, and the wise conservation and use of agroforestry tree genetic resources. (See Figure 4)

Does research on trees and tree crops for smallholders really pay off? Some might assume that it does not have a high rate of return. The evidence, however, shows otherwise. Alston and Pardey (2001) tabulated the rates of return to various types of agricultural research and development from thousands of studies. From the 108 studies they assembled that estimated the rate of return to investment in tree crops research, the average rate of return was 88%. The median rate of return was 33%. Such rates of return are outstanding, and they compare favorably with the rates of return on field crops research, which averaged 74% over 916 studies.

The flow of high-quality tree germplasm among countries, for the various types of fruit, timber, fodder, and medicinal trees of potential interest to smallholders, has to date been highly constrained compared with the flow of germplasm of the basic food crops. Serious calls for greater scientific and development attention to the many highly promising agroforestry species that are underexploited and underutilized have been with us for many years (NAS 1975; NAS 1979). But the response has been inadequate. More comprehensive and active international networks are needed in the coming decade to facilitate the flow of the best varieties and the flow of knowledge of how these species may contribute new or expanded enterprise opportunities.

Timber production on farm

Smallholder timber production is increasingly appreciated as an important source of many countries' wood supplies. This is particularly true for countries with low forest cover. Kenya has had one of the few comprehensive forest inventories ever done that included trees outside forests (Holmgren et al. 1994). The authors observed that two-thirds of the country's woody biomass occurs outside of conventional forests. In Bangladesh, it is estimated that 90% of the wood used is produced on agricultural land. In India, half of the timber now emanates from private farmlands. The smallholder may well be the timber producer of the future in many parts of the developing world, and increasingly a provider of high quality tropical timbers. But there is little understanding or documentation of these trends. Data for trees on farms are not collected, and are thus not included in national or global forest inventory data (Bellefontaine et al. 2002). This has to be redressed if the significance of agroforestry is to be better appreciated. The World Agroforestry Centre and the Forestry Department of the Food and Agriculture Organization of the United Nations (FAO) are working together to draw international attention to the need to estimate trends in farm-grown timber. That will provide a basis for greater focus on smallholder timber production systems.

Agroforestry for health and nutrition

Advances in agroforestry have many links with improving the health and nutrition of the rural poor. The expansion of fruit tree cultivation on farms can have a significant effect on the quality of child nutrition. This is particularly important as indigenous fruit tree resources in local forests are overexploited. Currently, eastern Africa has the lowest per capita fruit consumption of any region in the developing world. Work with national partners to domesticate a range of nutritious wild indigenous fruits that are popular in eastern, western, and southern Africa seeks to save these species from overexploitation, and develop them for local and regional markets. These efforts will contribute to the fourth Millennium Development Goal on reducing child mortality.

Recognition is growing that there are many complex linkages between agroforestry and the fight against HIV/AIDS (MDG 6). Forty million people are currently living with HIV. Some countries including Brazil, Senegal, Thailand and Uganda have shown that the spread of HIV can be stemmed. But most of sub-Saharan Africa is experiencing a disastrous AIDS epidemic. The global infection rate among adults is 1.2%, but it is 8.8% in sub-Saharan Africa. Sixty-eight percent of AIDS victims reside in this region. AIDS and associated diseases such as pulmonary tuberculosis are now the leading cause of death on the continent, accounting for up to 80% of all adult deaths in the parts of eastern and southern Africa that have prevalence rates of 20% to 30% (Mushati et al. 2003).

On one hand, HIV/AIDS poses significant threats to agroforestry. HIV/AIDS can reduce the economic incentive for farmers to undertake long-term investments. It undermines local knowledge about tree production, and makes property rights to land and trees less secure for the most vulnerable segments of the population. It dissipates labor and financial resources that would be needed to establish and maintain agroforestry practices. On the other hand, there is potential for agroforestry to generate much-needed income, improve nutrition, reduce labor demands, and stabilize the environment in AIDS-affected communities. The range of threats and the various opportunities have yet to be thoroughly explored, and incorporated into the research and development agenda. This is an urgent imperative.

Tree medicinals

Natural medicinal plants are the source of treatment for many diseases and ailments of the poor throughout the developing world (Rao et al. 2004). In Africa, for example, more than 80% of the population depends on medicinal plants for their medical needs. Two-thirds of the species from which medicinals are derived are trees. The vast majority of these tree products are obtained by extraction from natural forests. There is also increasing interest in natural medicines in the developed world, creating new or expanded markets for these products. This puts further extraction pressure on the forests. Many of the medicinal tree species are now overexploited. Some species are so depleted that their gene pools are greatly eroded (e.g., *Prunus africana*), and some are in danger of extinction. Meeting the expanding demand for medicinal trees can only be assured through greater efforts to domesticate them and promote their cultivation on farms. Vigorous research partnerships between agroforestry and the medical sciences will be crucial to ensure that the key tree medicinals are effectively developed for farm cultivation.

Agroforestry and the advancement of women

Sixty to eighty percent of the farmers in the developing world are women. In 1990, women subsistence farmers accounted for 62% of total female employment in low-income countries (Mehra and Gammage 1999). Rural women in developing countries grow and harvest most of the staple crops that feed their families. This is especially true in Africa. In sub-Saharan Africa, women account for 75% of household food production (UNDP 1999, as cited in Bread for the World Institute 2003). Food security throughout the developing world depends primarily on women. Yet they own only a small fraction of the world's farmland and receive less than 10% of agricultural extension delivery.

Because women are in charge of household and staple crops, female farmers often fail to gain from export-oriented agriculture. Women may have trouble diversifying their crops because they have difficulty obtaining the credit and land needed to shift to non-traditional exports. These realities have major implications for agroforestry research. Much more needs to be done to understand the kinds of traditional and non-traditional agroforestry products that are accessible to women, and to get research attention focused on them. This also applies to value-added processing activities and marketing. Greater attention to how women are affected by land and tree tenure practices is leading to awareness of the need to address these inequities. For example, women in Cameroon are very keen on cultivating *Dacryodes edulis* as its marketing season coincides with the need to pay school fees and buy school uniforms (Schreckenberg et al. 2002).

Trees are a medium for long-term investment on the farm. Thus, the propensity to cultivate them is particularly sensitive to property rights (Place and Otsuka 2002). Policy research in agroforestry must continue to strengthen our understanding of these linkages. We need to assist in identifying the means by which women's land rights can be made more secure to enhance the intensification of farming in general, and the acceleration of tree cultivation in particular (Place and Swallow 2002).

Building capacity and strengthening institutions

The second Millennium Development Goal focuses on achieving universal primary education. One in six children in the developing world do not attend school, and one in four start school but drop out before they can read or write. Education is a powerful weapon against poverty. It is therefore important to explore and implement practical and innovative measures in which the interrelated issues of education, food and health in rural areas can be tackled together. The *Farmers of the Future* initiative is an emerging global consortium spearheaded by the World Agroforestry Centre and FAO to promote the incorporation of agroforestry and natural resource management into basic education to enhance its quality and relevance for rural youth (Vandenbosch et al. 2002). The concepts of sustainable agriculture are taught through the lens of experiential learning that focuses on working trees for the farm. This can make a major impact, particularly on those that return to the farm after a few years of schooling.

Networking to strengthen agroforestry education at the tertiary level has been progressing for many years (Temu et al. 2003). There are now 123 member colleges and universities in the African Network for Agroforestry Education (ANAFE), and 35 member universities in the Southeast Asian Network for

Figure 5. Landscape mosaics that incorporate working trees enable smallholders living in upper watersheds to farm productively while preserving watershed functions and conserving natural biodiversity.

Agroforestry Education (SEANAFE). These networks assist the member institutions to incorporate agroforestry and multi-disciplinary approaches to land use in their curricula. The networks are fighting against the tendency for developing country agricultural universities to be marginalized, particularly in Africa (Omamo and Lynam 2002).

Ensuring environmental sustainability

The seventh MDG aims to ensure environmental sustainability, and to integrate the principles of sustainable development into country policies and programs to reverse the loss of environmental resources. Smallholder agroforestry systems generate environmental benefits of value to communities, national societies, and the global community. The environmental services that are of greatest relevance are watershed protection, biodiversity conservation, and climate change mitigation and adaptation. Thus, a major agroforestry research and development agenda is emerging in the domain of environmental services. The goal is to identify agroforestry systems and landscape mosaics that meet farmers' needs for food and income while enhancing these services. Policy reform and institutional innovations will enhance the adoption of effective technologies and resolve conflicts among stakeholders.

Pro-poor strategies to enhance watershed functions

There is a general impression that natural forests are crucial to protect watershed functions, particularly in upper catchment areas. This impression, which is part fact, and part fallacy, has led to severe conflicts between national governments and poor upland farming communities throughout the developing world. The real key to flood control and sustained water supply is maintaining the infiltration properties of the soil. This can be accomplished in working landscapes with an appropriate density of vegetative filters. Thus, agroforestry is in the center of the debate concerning how people can farm in watersheds while sustaining catchment functions that impact on downstream populations. New knowledge on suitable types of tree farming systems, their configuration in the landscape, and their location relative to lateral flows of water and soil is providing a foundation for options that enable upland communities to farm sustainably while protecting watershed services (Swallow et al. 2001). These management principles need to be refined for different spatial scales and contexts. Recent work by the World Agroforestry Centre's Southeast Asia Program, is identifying and applying policies, incentives, and institutional mechanisms that can enhance the adoption of win-win technologies and resolve persistent conflicts (van Noordwijk et al. 2003). (See Figure 5)

More effort is needed to link known tree-farming options to watershed level interventions, and to increase policy support for soil and water conservation efforts. For this we need more effective methods for wide area assessment of severity of land degradation. Here we can exploit the recent development of techniques for rapid monitoring of land quality across whole landscapes, such as the reflectance spectroscopy method that was recently developed (Shepherd and Walsh 2002).

Conserving biological diversity in working landscapes

The well being of the land is directly tied to the well being of its inhabitants. Only when rural people and poor farmers have a way to earn sustainable, stable livelihoods will the planet's biodiversity be safe. It is futile to attempt to conserve tropical forests without addressing the needs of poor local people, nor is it desirable (ASB 2002). As much as 90 percent of the biodiversity resources in the tropics are located in human-dominated or working landscapes. Agroforestry impinges on biodiversity in working landscapes in at least three ways. First, the intensification of agroforestry systems can reduce exploitation of nearby or even distant protected areas (Murniati et al. 2001; Garrity et al. 2003). Second, the expansion of agroforestry systems can increase biodiversity in working landscapes. And third, agroforestry development may increase the species and within-species diversity of trees in farming systems.

A new paradigm is emerging that integrates protected areas into their broader landscapes of human use and biodiversity conservation, particularly in agricultural areas that now constitute the principal land use in most of the developing world (Cunningham et al. 2002). The issue of how best to achieve a balance between production and biodiversity conservation is moving to the centre of much of ICRAF's work, particularly in Southeast Asia (van Noordwijk et al. 1997). It has become the basis for the concept of *ecoagriculture,* which refers to land-use systems managed for both agricultural production and wild biodiversity conservation (McNeely and Scherr 2003). Agroforestry is uniquely suited to provide ecoagriculture solutions (McNeely 2004). But much more must be done to understand and refine suitable options for widespread use. The global program on Alternatives to Slash-and-Burn has been on the forefront of identifying and applying such solutions in humid forest areas (Tomich et al. 1998).

Climate change mitigation and adaptation for rural development

Agroforestry will play a major role in the two key dimensions of climate change: mitigation of greenhouse gas emissions and adaptation to changing environmental conditions. Despite some efforts to reduce the impacts of climate change, the process will not be halted. Farmers will need to adapt to more extreme drought and flooding events, as well as the elevation in temperatures that are predicted to occur in coming decades. People differ in their vulnerability to such climate changes. The poorest rural populations in the regions with least responsibility for causing climate change are nevertheless likely to be most negatively affected. Agroforestry needs to play a role in increasing the resilience of smallholder farmers to climate change and other stresses. However, research on its prospective role in adaptation is only now getting under way.

Agroforestry was recognized by the Intergovernmental Panel on Climate Change as having a high potential for sequestering carbon as part of climate change mitigation strategies (Watson et al. 2000). Methods are now needed to determine the sequestration potential of specific agroforestry systems in particular agroclimatic conditions. Carbon offsets through tree farming will be a secondary product of smallholder agroforestry systems. A key question is how smallholders can benefit from carbon sequestration projects (Montagnini and Nair 2004). Methods are being developed to pursue carbon projects that will improve livelihoods and provide positive incentives to smallholder agroforesters.

Policies to harmonize environmental stewardship and rural development

Environmental policy related to agroforestry needs to address three main challenges: provide policy makers and civil society organizations with science-based evidence on the tradeoffs and complementarities between land use choices, the resulting environmental services, and the livelihoods of smallholder farmers; provide policy options for harmonizing environmental policy and concerns for sustainable rural development; and increase access to information about policy options. These challenges need to be tackled at international, national and local levels. Success depends on gathering field-based evidence in locations around the developing world, and synthesizing and interpreting it in ways that connect to the needs of national and international policy processes. Agroforestry can champion the perspective of smallholder farmers in these policy debates.

One growing aspect of this work is the development of mechanisms to reward the rural poor for the environmental services they provide to society (RUPES 2003). This will require methods to quantify these services in ways that are adaptable to rural smallholder-occupied landscapes, coupled with practical modalities for successful environmental service agreements.

Conclusion

Global targets are important mechanisms for the community of nations to come to grips with the complex development challenges of the early 21st century. This chapter has reviewed the ways that agroforestry research and development can contribute to achieving the Millennium Development Goals. National poverty-reduction-strategy papers (PRSPs) are one of the main tools by which individual countries articulate and monitor their plans to address the MDGs. To date, relatively few countries have incorporated an agroforestry perspective into their PRSPs, or have developed national agroforestry strategies. However, several developing countries have done this, and the results are promising. For instance, there are some notable examples from southern Africa. A growing number of donors are taking explicit steps to assist other countries to tackle this job. This creates a major opportunity for the world agroforestry community to be more proactive in drawing serious attention to the many benefits of greater investment in expanding the cultivation of trees on farms.

The role of agroforestry is also capturing ever-greater domestic recognition in the developed world, particularly in North America and Europe. This awareness is enhanced by the growing concerns in the north to evolve multifunctional rural landscapes, and to transform farm subsidies toward the conservation of soil, water and biodiversity in lieu of commodity payments. Growing appreciation of the role of agroforestry in the developed countries will enhance understanding and support for its expansion in the developing countries. This will better ensure that the needed investments are forthcoming to deliver on the promise of agroforestry to contribute substantively to achieving the Millennium Development Goals.

Acknowledgements

I am indebted to the many persons with whom I have interacted in the development of this paper. These include Brent Swallow, Frank Place, Tony Simons, Diane Russell, August Temu, Meine van Noordwijk, Tom Tomich, Bashir Jama, and many others.

References

Alston J.M. and Pardey P.G. 2001. Attribution and other problems in assessing the returns to agricultural R&D. Agr Econ 25: 141–162.

ASB. 2002. Balancing rainforest conservation and development. Alternatives to Slash-and-Burn, World Agroforestry Centre, Nairobi, Kenya.

Bellefontaine R., Petit S., Pain-Orcet M., Deleporte P. and Bertault J. 2002. Trees Outside Forests: Towards Better Awareness. Food and Agriculture Organization, Rome. 216 pp.

Bread for the World Institute. 2003. Agriculture in the Global Economy. Bread for the World Institute, Washington, DC. 164 pp.

CGIAR. 2000. A Food Secure World For All. Consultative Group on International Agricultural Research & Food and Agriculture Organization, Rome. 50 pp.

Clay D. 1996. Fighting an uphill battle: Population pressure and declining land productivity in Rwanda. Michigan State University International Development Working Paper 58: 2 pp.

Cunningham A.B., Scherr S.J. and McNeeley J.A. 2002. Matrix Matters: Biodiversity Research for Rural Landscape Mosaics. CIFOR and ICRAF, Bogor and Nairobi. 34 pp.

Dixon J., Gulliver A., Gibbon D. (eds) 2001. Farming Systems and Poverty. Food and Agricdulture Organization, Rome.

FAO. 2001. State of Food Insecurity in the World 2001. http://www.fao.org/docrep/x8200e/x8200e00.htm.

FAO. n.d. Mobilizing the Political Will and Resources to Banish World Hunger. Food and Agriculture Organization: Rome. 79 pp.

Franzel S., Cooper P., Denning G. and Eade D. 2002. Development and Agroforestry: Scaling Up the Impacts of Research. Oxfam, Oxford. 202 pp.

Franzel S., Denning G.L., Lilisøe J.P., and Mercado A.R. Jr. 2004. Scaling up the impact of agroforestry: Lessons from three sites in Africa and Asia (This volume).

Garrity D.P., Amoroso V.B., Koffa S., Catacutan D., Buenavista G., Fay P. and Dar W.D. 2003. Landcare on the poverty-protection interface in an Asian watershed. pp. 195–210. In: Campbell B.M. and Sayer J.A. (eds), Integrated Natural Resource Management: Linking Productivity, the Environment, and Development. CABI Publishing, Cambridge, MA, USA.

Holmgren P., Masakha E.J., and Sjoholm H. 1994. Not all African land is being degraded: a recent survey of trees on farms in Kenya reveals rapidly increasing forest resources. Ambio 23: 390–395.

Kumar B.M. and Nair P.K.R. 2004. The enigma of tropical homegardens (This volume).

Leakey R.R.B. and Newton A.C. 1994. Domestication of 'Cinderella' species as the start of a woody-plant revolution. pp. 3–5. In: Leakey R.R.B. and Newton A.C. (eds), Tropical Trees: Potential for Domestication and the Rebuilding of Forest Resources. HMSO, London.

McNeely J.A. 2004. Nature vs. Nurture: Managing Relationships Between Forests, Agroforestry and Wild Biodiversity (This volume).

McNeely J. and Scherr S. 2003. Ecoagriculture: Strategies to Feed the World and Save Wild Biodiversity. Island Press, Washington DC. 323 pp.

Mehra R. and Gammage S. 1999. Trends, countertrends and gaps in women's employment. World Dev 27(3): 538.

Mellor J. 2001. Reducing poverty, buffering economic shocks – Agriculture and the non-tradable economy. Background Paper for Roles of Agriculture Project. Rome: FAO. 2 pp.

Mercado A.R., Patindol M. and Garrity D.P. 2001. The Landcare experience in the Philippines: technical and institutional innovations for conservation farming. Dev in Practice 11: 495–508.

Michon G., de Foresta H., and Levang P. 1995. Strategies agroforestieres paysannes at developpement durable: les agroforets a dammar de Sumatra. Natures, Sciences, Societes 3: 207–221.

Montagnini F. and Nair P.K.R. 2004. Carbon sequestration: An underexploited environmental benefit of agroforest systems (This volume).

Murniati, Garrity D.P. and Gintings A.N. 2001. The contribution of Agroforestry systems to reducing farmers' dependence on the resources of adjacent national parks: a case study from Sumatra, Indonesia. Agroforest Syst 52: 171–184.

Mushati P., Gregson S., Mlilo M., Zvidzai C. and Nyamukapa C. 2003. Adult mortality and erosion of household viability in AIDS-afflicted towns, estates and villages in eastern Zimbabwe. Paper presented at the Scientific Meeting on Empirical Evidence for the Demographic and Socio-Economic Impact of AIDS. Durban, South Africa, 26–28 March 2003.

NAS. 1975. Underexploited Tropical Plants with Promising Economic Value. National Academy of Sciences, Washington, DC. 190 pp.

NAS. 1979. Tropical Legumes: Resources for the Future. National Academy of Sciences, Washington, DC. 332 pp.

Omamo S.W. and Lynam J.K. 2002. Agricultural science and technology policy in Africa. Discussion Paper 02-01. International Service for National Agricultural Research, The Hague, The Netherlands. 49 pp.

Place F. and Otsuka K. 2002. The role of tenure in the management of trees at the community level: Theoretical and empirical analyses from Uganda and Malawi. pp. 73–98. In: Meinzen-Dick, R., Knox, A., Place F. and Swallow B. (eds), Innovation in Natural Resource Management. Johns Hopkins University Press, Baltimore.

Place F. and Swallow B. 2002. Assessing the relationships between property rights and technology adoption in smallholder agriculture: Issues and empirical methods. pp. 45–71. In: Meinzen-Dick R., Knox A., Place F. and Swallow B. (eds), Innovation in Natural Resource Management. Johns Hopkins University Press, Baltimore, MD, USA.

Place F.S., Franzel S., De Wolf R., Rommelse R., Kwesiga F.R., Nianf A.I., and Jama, B.A. 2002. Agroforestry for soil fertility replenishment: Evidence on adoption processes in Kenya and Zambia. In: Barrett, C.B., Place F. and Aboud A.A. (eds), Natural Resource Management Practices in Sub-Saharan Africa. CABI Publishing and International Centre for Research in Agroforestry, Wallingford, UK. 335 pp.

Puri S. and Nair P.K.R. 2004. Agroforestry research for development in India: 25 years of experiences of a national program. (This volume).

Rao M.R., Palada M.C. and Becker B.N. 2004. Medicinal and aromatic plants in agroforestry systems (This volume).

RUPES. 2003. Program for Rewarding Upland Poor in Asia for the Environmental Services They Provide. www.worldagroforestrycentre.org/sea/Networks/RUPES/Index.htm.

Russell D. and Franzel S. 2004. Trees of prosperity: Agroforestry, markets and the African smallholder (This volume).

Sanchez P.A. 2002. Soil fertility and hunger in Africa. Science 295: 2019–2020.

Scherr S.J. 2004. Building Opportunities for small-farm agroforestry to supply domestic wood markets in developing countries (This volume).

Schreckenberg K., Degrande A., Mbosso Z., Baboule B., Boyd C., Enyong L., Kanmegne J. and Ngong C. 2002. The social and economic importance of *Dacroydes edulis* in Southern Cameroon. Forests, Trees and People: 12: 15–40.

Shackleton S., Wynberg R., Sullivan C., Shackleton C., Leakey R., Mander M., McHardy T., den Adel S., Botelle A., du Plessis P., Lombard C., Combrinck A., Cunningham A., O'Regan D. and Laird S. 2003. Marula commercialisation for sustainable and equitable livelihoods: Synthesis of a southern African case study, Winners and Losers – Final Technical Report to DFID (FRP Project R7795), Volume 4, Appendix 3.5. 57 pp.

Shepherd K. and Walsh M. 2002. Development of reflectance spectral libraries for characterization of soil properties. Soil Sci Soc Am J 66: 988–998.

Simons A.J. 1996. ICRAF's strategy for domestication of non-wood tree products. pp. 8–22. In: Leakey R.R.B., Temu A.B., Melnyk M. and Vantomme P. (eds), Domestication and Commercialization of Non-Timber Forest Products in Agroforestry Systems. Food and Agriculture Organization, Rome.

Simons A.J. and Leakey R.R.B. 2004. Tree domestication in tropical agroforestry (This volume).

Swallow B., Garrity D. and van Noordwijk M. 2001. The effects of scales, flows, and filters on property rights and collective action in watershed management. Water Policy 3: 457–274.

Temu A., Mwanje I. and Mogotsi K. 2003. Improving Agriculture and Natural Resources Education in Africa. World Agroforestry Centre, Nairobi, Kenya. 36 pp.

Thirtle C., Irz X., Lin L., McKenzie-Hill V. and Wiggins S. 2001. Relationship between changes in agricultural productivity and the incidence of poverty in developing countries. United Kingdom Department for International Development Report 7946. pp. 2.

Tomich T.P., Roemer M., and Vincent J. 1994. Development from a primary export base. pp. 151–194. In: Lindauer D. and Roemer M. (eds), Asia and Africa: Legacies and Opportunities in Development. Institute for Contemporary Studies, San Francisco.

Tomich T.P., van Noordwijk M., Vosti S. and Witcover J. 1998. Agricultural development with rainforest conservation: Methods for seeking best bet alternatives to slash-and-burn, with applications to Brazil and Indonesia. Agr Econ 19: 159–174.

Torquebiau E.F. 1992. Are tropical agroforestry homegardens sustainable? Agr Ecosyst Environ 41: 189–207.

UNDP. 2003. Human Development Report. United Nations Development Programme. New York. http://www.undp.org/hdr2003/.

van Noordwijk M., Tomich T.P., de Foresta H. and Michon G. 1997. To segregate – or to integrate? Agroforestry Today, Jan-March pp. 6–7.

van Noordwijk M., Tomich T.P. and Verbist B. 2003. Negotiation support models for integrated natural resource management in tropical forest margins. pp. 87–108. In: Campbell B.M. and Sayer J.A. (eds), Integrated Natural Resource Management: Linking Productivity, the Environment, and Development. CABI Publishing, Cambridge, MA, USA.

Vandenbosch T., Taylor P., Beniest J. and Bekele-Tesemma A. 2002. Farmers of the Future – a strategy for action. World Agroforestry Centre, Nairobi, Kenya. 72 pp.

Watson R., Noble I., Bolin B., Ravindranath N., Verardo D. and Dokken D. 2000. Land Use, Land-Use Change, and Forestry. Intergovernmental Panel on Climate Change & Cambridge University Press, Cambridge, UK. 377 pp.

Young A. 1997. Agroforestry for Soil Management. 2nd edition. CAB International & International Centre for Research in Agroforestry, Wallingford/Nairobi. 288 pp.

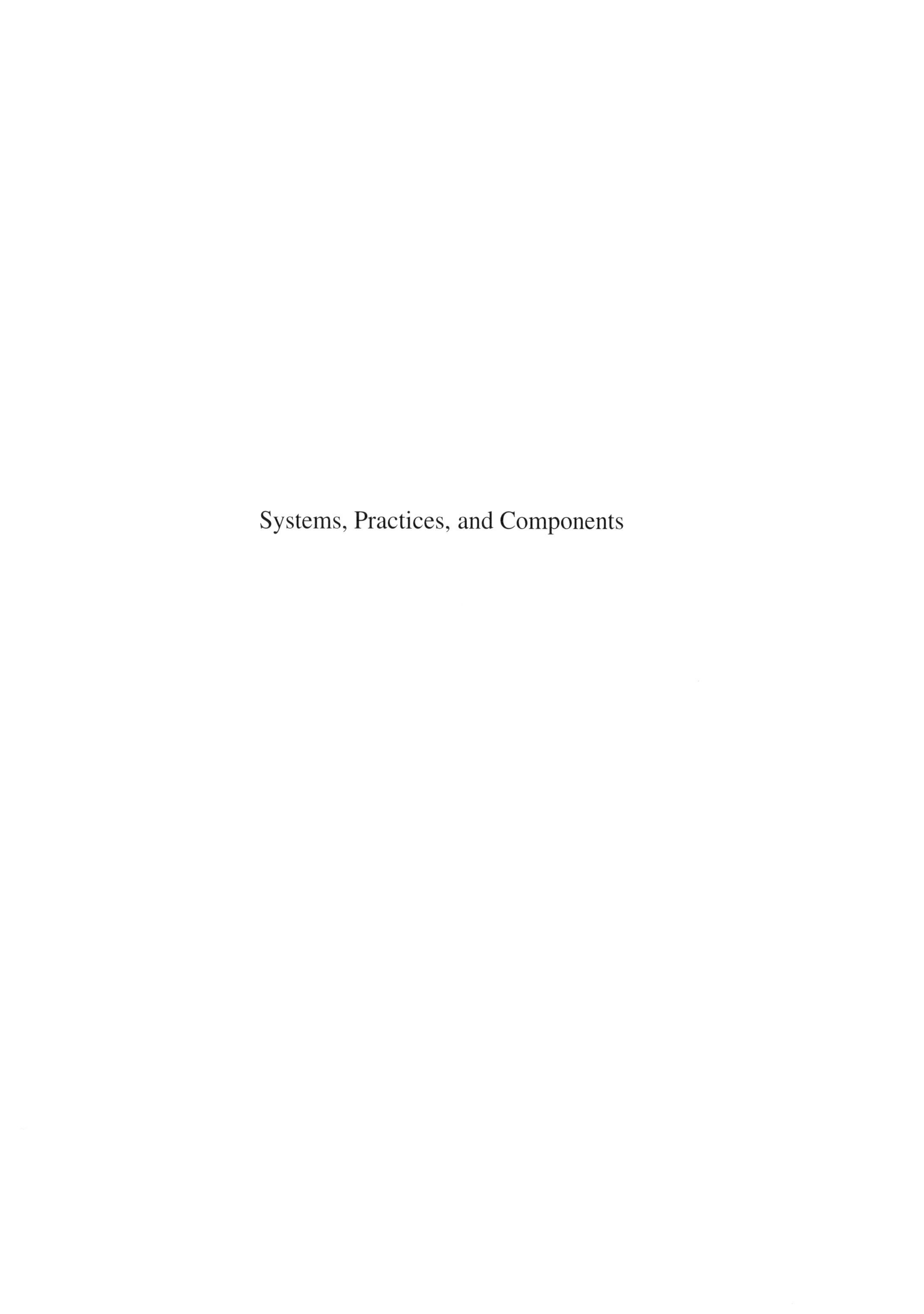

Systems, Practices, and Components

Agroforestry Systems **61:** 21–33, 2004.

Hardwood silvopasture management in North America

H.E. Garrett[1,*], M.S. Kerley[1], K.P. Ladyman[1], W.D. Walter[1], L.D. Godsey[1], J.W. Van Sambeek[2] and D.K. Brauer[3]
[1]*Center for Agroforestry, University of Missouri, Columbia, MO 65211, USA;* [2]*US Forest Service, North Central Research Station, Columbia, MO, USA;* [3]*Dale Bumpers, Small Family Farm Research Center, Booneville, AR, USA;* **Author for correspondence: e-mail: GarrettH@missouri.edu*

Key words: Animal forage, Cool-season forages, Light, Livestock, Shade, Hardwood regeneration

Abstract

Hardwood silvopasture management has great potential throughout the Central Hardwood Region in the United States, but has been little utilized due to the lack of available information on its application. However, more than one-third of farm woodlands within the region are being grazed without the benefit of the application of silvopasture principles. The University of Missouri Center for Agroforestry has undertaken a major research initiative to further develop and build upon the fragmented information that is available on hardwood silvopastoral management. Ten years of screening forage species (grasses and legumes) for shade tolerance has clearly demonstrated that many cool-season forages benefit from 40% to 60% shade when grown in Missouri – a finding that could likely be extrapolated to the entire region of the Midwestern United States. Grazing trials have proven to be successful in the short-term. Long-term research is currently underway to fully document the interactions between hardwood trees, cattle and forage.

Introduction

Land clearing for agricultural purposes began in the Central Hardwood Region of the United States with native Americans and continued under European occupation. Despite clearing practices, however, there was a strong emphasis by an agrarian society to intensively manage woodlands for multiple uses that included not only timber-based products, but also livestock grazing, harvesting of edibles, and hunting (Koch and Kennedy 1991). As the country moved from an agrarian to an industrialized society, the use of forests became more singular in purpose, with a strong emphasis placed on producing wood products. Throughout the 1930s, forest practices, such as high-grading, burning, and continuous grazing, were prevalent and continue to impact the overall productivity and quality of the Central Hardwood forests. Though these practices are not as broadly applied today, their influence is still visible in forests whose ages at harvest may be 80 to 120 years.

One of the challenges facing the forestry profession today is how to encourage hardwood landowners to invest in management when financial returns are futuristic at best. Agroforestry provides a means of encouraging good forest stewardship and does so with the outlook of positively influencing productivity and financial gain. In the course of integrating trees and shrubs into farming practices (such as with silvopasture management), multiple objectives are obtainable. Through the intentional integration of trees with livestock, silvopastoral practices strive to simultaneously optimize economic, environmental, and social benefits.

Silvopasture utilizes intensive management practices for growing trees with forage and livestock in agroecosystems. Grazed forests or woodlands extensively managed as 'natural ecosystems' similarly produce wood products and forages for livestock or wildlife, but without using intensive approaches and with the potential for dire consequences (Chandler 1940; Patric and Helvey 1986; Bezkorowajnyj et al. 1993). These practices are not classified as silvopasture or even agroforestry in general. Silvopasture management, like the other forms of agroforestry, requires shifting our thinking in both spatial and temporal domains, and demands skills in managing rather than reducing complexity. Essential to this is an understanding of hierarchical scalar relationships within ecosystems and recognition that defined ecosystem 'boundaries' exist primarily for managerial convenience (Garrett et al. 1994; Garrett and Buck 1997). The purpose of this chapter is to assemble in a single writing the fragmented information that is available on hardwood silvopasture in North America, and to develop a premise for its application.

Potential for hardwood silvopasture

While traditionally, trees and forest products have been employed on the farm for a multitude of purposes (timber, fence posts, windbreaks, livestock shelter, etc.), the utilization of farm forests and products has seldom been optimized. Hardwood silvopasture has great potential to optimize the use of these resources. Silvopasture can be the product of establishing trees in pastures, or grasses and legumes in managed forest stands. Because the practice endeavors to simultaneously produce forest products, high quality forage, and livestock on the same parcel of land, intensive management is required.

Although vast areas of forestland, pastureland, rangeland and even cropland that are suitable for the practice are known to exist, little data exist that would provide a reliable estimate of the area currently under silvopastoral management within the hardwood region. In the five hardwood states of Missouri, Illinois, Indiana, Iowa, and Ohio there are more than 7 million ha of cropland with an erodibility index (EI) > 10 (Noweg and Kurtz 1987) (For a definition of Erodibility Index, see End Note 1). Within the Central Hardwood Region (Michigan, Wisconsin, Minnesota, Missouri, Ohio, Indiana, Illinois, and Iowa) there is an estimated 34 million ha of forest; 6.6 million ha of this 'forest' is present on farms. Of this 6.6 million ha, 2.3 million (35%) is currently being pastured but without the benefit of intensive management (Table 1).

The use of forested land for passive grazing varies by regions reflecting diverse views held by landowners. In a survey of four rural Missouri counties conducted by the Missouri Department of Conservation, it was found that 68% of the woodlands were grazed (Hershey 1991). Passive grazing is detrimental to the forest and contributes little forage towards sustaining a cow-herd. In contrast, silvopasturing of woodlands that are available and currently being mismanaged, has the potential to increase the forested land that is under management for wood production while increasing available pasture area. In Missouri alone, there are 5.7 million ha of timberland of which 10% or less is under management. Missouri has the second highest number of cows of any state in the United States (www.mocattle.org/beef_facts.htm). National projections suggest that land area available for grazing is likely to be reduced due to conversion to other uses, and use of forages for grazing will decrease by 32% (Van Tassell et al. 2001). Silvopasturing would not only increase the pasture area available, but could also serve as an incentive for landowners to place currently unmanaged forestland under management.

Physical and biological relationships of hardwood silvopasture

Progression of silvopastoral research

The objective of employing silvopastoral practices is to integrate trees, forage, and grazing herbivores for a production benefit. The supposed advantage is that the sum of combined benefits is greater than the value of any component individually. The actual benefit realized from implementation of silvopastoral practices depends on understanding the production requirements of each component, and how synergism among the components can be maximized. The potential benefit of silvopastoral practices is not a new concept. Stephenson (1954) noted written reference to it in describing the 1783 backwoods of North Carolina. The idea that woodland grazing is inferior is, however, more recent. DenUyl and Day (1934) summarized that forage from farmwoods was inferior in nutritive quality and that grazing farmwoods would negatively impact timber stands due in large part to young tree damage. This is a view that is still widely held in forestry. However, further research (McIlvain and Shoop 1971; Lehmkuhler et al. 1999) has demonstrated that tree, forage, and livestock interactions can be manipulated to enhance forage production and animal growth without negative effects on tree performance. Today, research is focused on understanding the interaction dynamics of the silvopastoral practice. The expected outcome is that silvopastoral practices will be employable that improve the productivity of the grazing animal, the quality and diversity of forage available to the grazing animal and wildlife, and ef-

Table 1. Total forestland and forestland located on farms (pastured versus nonpastured) in the hardwood zone of the Midwestern United States in millions of hectares[a].

State	Total forestland[b]	Forestland on farms[c]		
		Total ha's	Pastured ha's	Nonpastured ha's
Michigan	7.8	0.46	0.06	0.40
Minnesota	6.8	0.81	0.30	0.51
Wisconsin	6.5	1.15	0.29	0.87
Missouri	5.7	1.85	0.92	0.94
Ohio	3.2	0.63	0.14	0.49
Indiana	1.8	0.52	0.11	0.41
Illinois	1.7	0.66	0.17	0.49
Iowa	0.8	0.53	0.26	0.28
Total hectares	34.3	6.61	2.25	4.39

[a]Inventory year 1997
[b]Smith, et al. 2001
[c]Glickman et al. 1999

fectively interpose timber stand improvement across a wide array of forested land.

Light availability in woodlots

Site selection (aspect and slope) and the application of management activities (including tree planting, pruning, thinning and logging) can have a dramatic influence on the light available for silvopasture practices. A direct relationship exists on south-facing slopes between degree of slope and percentage of direct solar radiation reaching the forest floor. Gaps up to two times the height of adjacent trees on south-facing slopes of 30 degrees have a greater percent direct solar radiation than similar gaps on slopes of 15 degrees (Fischer 1979). Furthermore, an inverse relationship exists between the same slope degrees (15% and 30%) and the percentage of direct solar radiation reaching the forest floor on north-facing aspects. The role of gap size has also been quantified in that gaps have little effect on available light when openings are smaller than 0.04 ha and larger than 0.4 ha in size (Dey and MacDonald 2001). Also, most studies of light intensity measured in gaps have emphasized light readings at the center of the gap, recognizing the decreased availability of light when moving from gap center towards the edge (Minckler 1961). For the silvopasture practice and growth of forages, the uneven distribution of light, such as may be created with gaps, is not desirable. But, understanding the effects of slope and aspect may play an important role in determining appropriate residual densities when thinning a forest stand. South-facing slopes that naturally receive greater solar exposure should, logically, have higher densities of trees than north-facing slopes that are predisposed to less direct sunlight – obviously the directional effects noted here refer to northern temperate latitudes.

Thinning studies of forested environments to create desirable levels of available light have been conducted. While these studies have primarily been designed in an attempt to create favorable light conditions for regeneration of desirable tree species, the results have direct application to creating light levels favorable for the growth of select forages. Two popular harvesting practices for opening mature forest canopies in hardwoods and allowing light to penetrate to the forest floor are group selection and the shelterwood methods. The group selection method creates patches of high light intensity, while the shelterwood method is designed to create a more even distribution of light throughout the forested understory. In young immature stands, release thinnings such as timber stand improvement (TSI), crop tree, and deferment cuts, will provide increased light levels for forage production while at the same time improving the growth of trees identified for retention.

Dey and Parker (1996) reported that removing 43% and 77% of the basal area with a shelterwood harvest in an oak (*Quercus* spp.) stand increased light intensities to 35% and 65%, respectively. Other studies have found that up to 50% of the basal area of hardwood forests may need to be removed to increase light levels to 35%–50% of that found in the open (Sander 1979; Marquis 1988; Dey and Parker 1996).

For many plants, full sunlight is not required to maximize growth. Many herbaceous plants need only about 10% of full sunlight to reach a state of growth where photosynthesis exceeds respiration, and will reach light saturation at 50% (C_3 plants) and 85% (C_4 plants) of full sunlight, respectively (Gardner et al. 1985). The light intensity within mature hardwood forests is typically $< 20\%$ and may be as low as 1% (Dey and MacDonald 2001). While the removal of some proportion of the overstory canopy can increase the intensity and duration of light reaching the forest floor, and thereby improve the growth of forage crops, the relationship of overstory removal to available light is not linear, with an estimated residual stocking density of 30% required to provide light levels of 50% of open values (Sander 1979).

Potential effects of converting forest stands to silvopasture on tree quality

Forestry practices that reduce stand stocking, such as thinning to accommodate silvopastoral management, can have both positive and negative effects on the residual trees. The quality of any tree is measured by its ability to meet the specifications/expectations that are defined by the product of end-use. Tree utilization is broad and may include products such as nuts and lumber, or services provided to society, such as filtering water (riparian buffer strips) and air (CO_2 sequestration). Tree species will vary in their ability to provide related end products. However, maintaining the health (growth and productivity) of individual trees is paramount to their utility. Any injury or inhibition of growth resulting from management (thinning, grazing, etc.) is undesirable.

Thinning or release of hardwood stands, especially young stands, normally benefits growth. In general, tree growth (especially diameter) has been shown to increase as stand density decreases (Dale 1968; Schlesinger 1978; Dale and Sonderman 1984). Even older stands may show strong growth responses to thinning. A study by Smith and Miller (1991) reported significant improvements in growth rates of released 75- to 80-year-old Appalachian hardwoods. Additionally, Dale (1975) observed improved growth of older overstory trees with the removal of the understory.

Stokes et al. (1995) identified a general trend of an increasing number of damaged residual trees per ha with a decreasing number of trees removed. Further, while the operations of a silvopasture practice might benefit from whole tree removal, the process of whole tree removal can result in significant damage to residual crop trees. In thinnings of hardwood stands, up to a third of the residual stand may be damaged (Ostrofsky et al. 1986). In most studies, a significant portion of the damage to residual trees has been linked with skidder activities and could have been avoided through better planning of logging activities (Ostrofsky et al. 1986). These wounds, while ultimately resulting in defect, may also become points of entry for pathogens that result in wood rot and/or tree death. Both decay and discoloration begin with the creation of a wound, and can ultimately lead to decreased value of residual timber (Shigo 1966). Tree injury is highly undesirable as it can decrease financial returns and it is one of the goals in silvopasture management to maximize the financial returns. Another effect of thinning, which especially must be considered in establishing a silvopasture program, is the development of epicormic branching (i.e., branching developing from dormant buds on the bole that are exposed to light). Many factors (age of the tree, crown position, vigor, evidence of previous sprouts, site aspect, and species) may be indicators of the potential for epicormic sprouting. Smith (1966) ranked several common hardwood species relative to their potential to develop epicormic branches. White oak (*Quercus alba*) had the highest likelihood followed by a grouping of black cherry (*Prunus serotina*), red oak (*Q. rubra*) and chestnut oak (*Q. prinus*). A grouping of hickory (*Carya* spp.), yellow poplar (*Liriodendron tulipifera*), red maple (*Acer rubrum*) and sugar maple (*Acer saccharum*) ranked third. White ash (*Fraxinus Americana*) showed the least likelihood. Trees are less likely to develop epicormic branching when they are vigorous dominants or co-dominants and have shown no previous evidence of epicormics (Lamson et al. 1990; Ward 1966). Brinkman (1955) noted that epicormic scars may result in lower log values. Thinnings designed to reduce stand density to 40% to 50% (similar to the requirement in silvopasture management), while leaving vigorous co-dominant crop trees that show no evidence of epicormic sprouts, has been shown to minimize epicormic development (Lamson et al. 1990).

Regeneration and sustainability of a silvopasture woodland

One of the criticisms of allowing livestock access to farm woodlands relates to their potential negative impact on forest reproduction and thus loss of sustainability. When nutritional forage is available, most livestock, especially cattle, will not seek out the less palatable woody browse (Hall et al. 1992; Hawley and Stickel 1948). However, any browsing of terminal shoots by domestic or wild animals will result in deformity and loss of tree growth (Den Uyl et al. 1938). To ensure sustainability, physical protection must be planned for, as well as some form of regeneration. Due to the inherent characteristics of hardwood silvopasture, planting of large seedlings with well-established root systems is desirable to help guarantee establishment and early fast growth.

Lehmkuhler et al. (2001) demonstrated the benefits of a single strand of high tensile electric-charged wire positioned along both sides of a row of trees in protecting them from livestock. A solar panel attached to batteries made the system cost effective. However, some landowners prefer a simpler means of protection. Wire cages or tree shelters might be better in such situations. Tree shelters (i.e., tubes within which seedlings begin growth) are becoming more popular but are expensive (Lantagne 1991; Minter et al. 1992; McCreary and Tecklin 2001). Individual wire cages are expensive too, but they have the advantage of being readily adaptable to protecting scattered 'strategically positioned' seedlings, whether planted in an understory or open pasture, not only from browse but also from hoof damage.

Studies on hardwood-silvopasture regeneration underway at the University of Missouri Center for Agroforestry utilize containerized oak seedlings protected by wire cages or tree shelters. In these studies, pole-size white oak stands have been thinned to approximate 50% crown cover. White oak seedlings, grown using a root production method (RPM) technology developed by the Forrest Keeling nursery in Ellsberry, MO, are strategically positioned in the understory in sufficient numbers per ha to represent the next generation stand. The RPM containerized seedlings which range in height from 1.2 to 2.0 m with a large root mass at time of planting, show remarkably fast juvenile growth and can be producing mast at age 3 (Lovelace 1998; Dey et al. 2003). The wire cages and 'tree shelters' provide excellent protection and the RPM seedlings provide greater assurance than bare root seedlings that the practice will be sustainable .

Grazing livestock

Cattle and sheep are primarily used in silvopastoral practices. The greatest difference between silvopastoral and 'open' management of cattle or sheep is the contrasting environmental conditions. In the open, such as a conventional pasture or range, radiant heat can be much more intense than in a shaded environment. Shade has been shown to improve animal performance, with primary emphasis placed upon heat stress amelioration. McDaniel and Roark (1956) conducted a shade experiment comparing artificial or natural shade to open pastures. The natural shade consisted of abundant, savanna-type tree spacing, and scanty shade, clusters of trees in the grazed pasture treatments. Cattle with shade gained more, as did their calves. During the daylight hours, the cows with the greatest and most uniformly distributed shade, spent the most time grazing, with grazing time decreasing concomitantly with decreasing shade. McIlvan and Shoop (1971) measured improved gains in yearling Hereford steers on rangeland given access to shade. Of particular interest in their findings was that shade could be used to create more uniform, or less spot grazing by cattle. Shade was noted to be nearly as effective as water placement or supplemental feeding location in promoting uniform grazing within a pasture. Silvopastoral practices could be extrapolated to encourage more uniform grazing and waste nutrient deposits within a pasture compared to open pasture or range. The natural shade, in McDaniel and Roark's work (1956), particularly the abundant shade treatment, resulted in superior weight gain of cattle compared to the artificial shade treatment. Planted or natural trees are usually an effective source of shade and potentially superior to manufactured shade structures because their radiosity is lower than flat-roofed structures, giving them a larger low temperature ground area (Kelly et al. 1950).

Heat and cold stress can adversely affect cattle throughout much of the temperate zone in North America. Protection from cold can be important for livestock in northern climates. Properly positioned trees and shrubs or natural forest stands can provide much needed protection for pastures, feedlots and calving areas. Reducing wind speed lowers animal stress, improves animal health and increases feeding efficiency of livestock. Canadian researchers have demonstrated that cattle on winter range require an additional 20% increase in feed energy, above maintenance, to offset the direct effects of exposure to a combination of cold temperatures and wind. Adequate wind protection has been found to reduce the direct effects of cold by more than half (Webster 1970). Similar findings have been reported for swine and dairy animals (Hintz 1983).

Worstell and Brody (1953) noted, however, that greater emphasis should be placed on prevention of heat stress than cold stress in cattle. When an animal is subjected to an environment that causes heat stress, the behavior of the animal adjusts quickly to alleviate the stress. The symptoms of heat stress begin to occur at 30 °C (Cartwright 1955). Two such behaviors are reduced intake, to lessen metabolic heat production, and seeking relief from the heat source, such as standing in water or under shade. In a study on the effects of heat on dairy animal milk production, University of Missouri animal science researchers found that increasing daily temperatures from 20 °C to 29 °C resulted in daily milk production decreasing from 39 kg to 26 kg and feed intake per cow decreasing from 38 to 32 kg (J. Spain, University of Missouri, 2003, pers. comm.). Changes were observed within 48 to 72 h following the change of temperature. Muller et al. (1994) and Spain et al. (1997) have also reported similar results. Alleviating heat stress by providing a shaded environment would potentially increase feed intake, which would increase animal growth and/or milk production. An animal in its thermoneutral zone would also be more likely to graze during daylight, increasing its ability to select high quality forage, and therefore a more nutrient dense diet. However, our understanding of the dynamics between shade and animal behavior is very limited. More research is needed to understand the extent of influence of shade and tree canopy structure on animal behavior under heat- and cold-stress environments.

Forage production

Forages grown in a silvopastoral practice need to be evaluated for the effect of canopy on production and nutrient quantity/quality. They also should be evaluated for the effect of grazing management strategies on productivity under shade. It is generally understood that forage yield in a silvopastoral practice decreases with increased tree basal area (Wolters 1973). Pearson (1990) noted that forage yields generally decrease as tree overstory increases. However, he also noted that integrating timber and grazing management can result in greater profit than is possible from either enterprise alone. The approximate 30 million ha of timberland in the north central region of the United States (Shifley and Sullivan 2002), much of which is not under management, indicates the scope and potential for silvopastoral practices. Therefore, the question that arises is, can forage species and tree–forage interactions be identified that have a positive influence on forage productivity?

Ehrenreich and Crosby (1960) determined that forage production increases when hardwood tree-crown cover is reduced below 50%. They found only small per unit reductions in grass production as crown cover increased from 50% to 80%, and that forage species composition under dense canopies differed from that under less dense canopies or in the open. As this research pointed out, forage species differ in their adaptability to shade stress. It has long been known that forage nutrient composition was altered by a shaded environment (Roberts 1926; Holechek et al. 1981). Forage grown under a forest canopy had higher crude protein and greater in vitro organic matter digestibility than forage grown in the open. Interestingly, Holechek et al. (1981) also found that forage species under a forest canopy were more diverse than in an open area, and that digestibility was influenced by shade and season – digestibility was greatest for grass grown under shade in the summer and for grass grown in the open in the fall, when adequate rainfall occurred. They suggested that the most efficacious grazing system would combine grazing under a forest canopy with open pastures at different periods of time during the grazing season. Frost and McDougald (1989) reported that herbaceous production was 115% to 200% greater under scattered oak than on open grassland. They attributed the response to more favorable physical and chemical soil properties and a more favorable soil temperature under the oak. Demonstrating the importance of tree–forage interactions, they measured differences in forage production under canopies of blue oak, (*Quercus douglasii*), interior live oak (*Quercus wislizenil*) and digger pine (*Pinus sabiniana*). They concluded that the increased forage production was in large part due to shading, in drought conditions, reducing moisture loss via evapotranspiration. Standiford and Howitt (1993) further defined the link between forage production under varying canopy and rainfall conditions and concluded that receiving less than 50 cm of rainfall will result in greater forage production under shade than in the open, with just the reverse result in areas receiving more than 50 cm of rainfall. Kay (1987) showed greater forage production in the open than under dense oak canopy woodlands with rainfall of 50 cm to 75 cm. Standiford and Howitt (1993) developed models in an attempt to quantify the various influences on forage production under canopy, and concluded that forage production was influenced by crown cover, rainfall, the interaction of rainfall and crown cover, and growing-degree days. Based upon their model, canopy had the greatest effect in depressing forage yield in higher rainfall areas.

Our assessment is that the productivity of forage in a silvopastoral practice in the temperate zone of North America is dependent upon tree and forage species in addition to canopy structure and climatic conditions (rainfall amount and seasonality). As noted by Clason and Sharrow (2000), the relationship between the overstory trees and the understory forage species must be compatible. Warm- and cool-season grasses respond differently to shade stress. Increased temperature increases cell wall content of cool-season grasses (Ford et al. 1979). Grasses respond to shade stress by increasing leaf-area and shoot-to-root ratio (Allard et al. 1991; Kephart et al. 1992) and by concentrating nitrogenous compounds (Kephart and Buxton 1993). Huck et al. (2001) evaluated several grass species under shade stress in Missouri and found that cool-season grass production was often increased under 45% sunlight compared to full sunlight, and that nitrogen and fiber digestibility was also improved in forages grown under shade. Interestingly, they also found that fiber concentration of the grasses increased under shade. These findings agreed with Garrett and Kurtz (1983) who measured higher in vitro digestibility of Kentucky 31 tall fescue (*Festuca arundinacea*) and orchardgrass (*Dactylis glomerata*) when grown under a walnut (*Juglans nigra*) canopy than when grown in open pastures. Lin et al. (1999, 2001) have recently identified several cool-season grasses and legumes for the Midwestern United States that perform better at 50% shade than in full sun. Not only were significant increases in yields observed, but the quality of the forage was also found to be superior under the shade. This research has demonstrated the possibility for selecting forage species that perform well under reduced sunlight conditions, even producing greater masses of dry matter with superior nutritional quality than when grown in full sunlight. By understanding the physiology of the response of different forage species to shade stress, hardwood silvopastoral practices can be more effectively designed to enhance grazing animal performance.

Integrating trees, forages, and animals

The four variables in a silvopastoral practice that can be subjected to management are tree species, tree density, forage species, and animal maintenance. The majority of research conducted has evaluated silvopastoral practices under conifers (mostly pine) with only limited evaluation of hardwood-based practices. Most hardwood research has been conducted with either oak species or nut-bearing species (e.g., black walnut). There is a strong correlation between light intensity and forage production under hardwoods as has also been demonstrated for pine (Pearson 1990). However, in certain instances under deciduous tree stands, forage production has been reported to be equal or even greater than in open exposure to sunlight. Fescue and orchardgrass production was greater under a 35-year-old walnut canopy than in open pastures (Garrett and Kurtz 1983). Forage production was found to be two-fold greater following walnut establishment into existing pastures than it was without trees (Neel 1939; Smith 1942). In both of these programs, the tree species was ideally suited for silvopasture due to the open characteristics of the crown and foliage (Garrett and Harper 1998). Furthermore, both grasses are known to be fairly shade tolerant (Lin et al. 1999). Spatial arrangement and management of trees can be designed such that tree damage from livestock is minimized. New or young plantings must be protected by electric fencing or by other means because tree species that are susceptible to predation from the grazing animal will sustain damage (Lehmkuhler et al. 2001). Older trees must be protected from soil compaction and physical damage to roots near the surface through the use of rotational grazing and livestock removal during wet periods. The general consensus of the reported literature is that tree performance does not have to be significantly altered in a silvopastoral practice once the tree is of sufficient size to prevent physical damage of the stem or browsing of the crown assuming proper management recommendations are followed (e.g., rotational grazing).

The quantity and quality of forage produced in a silvopastoral practice will be influenced by the forage species present. Typically, maximizing forage production will require that cool season grass and legume species be selected that are adaptable to shade. It is possible to use forage species that will increase dry matter yield and nutritional quality of forage compared to that produced in an open pasture. These forage species are well adapted and indigenous to the temperate zone of North America (Lin et al. 1999, 2001). Even with the proper selection of forage species (i.e., shade and drought tolerant), trees must be managed on a continuing basis to assure 40% to 60% full light for forage growth. Light availability is a function of tree spacing, tree crown diameter and tree crown density. Following the initial thinning of a forest stand, crown diameter increases and light availability decreases. Through a monitoring of changes in light availability, one can determine the need for repeated thinnings. In addition, additional light can be made available through the pruning of lower branches, which reduces the density of the crowns. Managed grazing practices, similar to open pastures, should be developed and implemented to maximize forage production in a silvopastoral practice. The increased forage production under a canopy would result in increased stocking rate potential and greater productivity per unit of land.

Animal performance can be enhanced via use of silvopastoral practices. This occurs from reduction of heat stress and improved forage availability and nutritional quality (Muller et al. 1994; Spain et al. 1997; Lin et al. 2001). Increased performance can be achieved without negative effects on tree growth if proper management practices are adopted. Stocking density should be designed to prevent over grazing of the forage production potential and damage to residual trees.

Economics of silvopasture

Economics is the study of 'choice.' Faced with a limited amount of resources, resource managers must make choices regarding use. From a supply and demand perspective, natural resources must be managed in such a way that maximum needs are met with a finite supply of resources. Benefits to good resource managers can be tangible, such as monetary rewards, or intangible, such as aesthetics or self-satisfaction. Regardless of the benefit to the resource manager, each benefit is the result of a choice, or trade-off that is made. The tradeoff often revolves around three main objectives: financial benefits, environmental benefits, and social benefits. From a silvopastoral perspective, a landowner must balance the use of land, animal and labor resources in a way that provides the greatest net benefit to him/her and society.

Ranchers who choose to graze their cattle on open pastures have made a choice to forego the cost of establishing trees in their pastures (a financial benefit of decreasing pasture cost). They are also foregoing the benefits of reducing heat stress on the animals, improving forage quality, sequestering carbon, perhaps mitigating nutrient-rich runoff, and the futuristic prospect of income from the trees. Although financial benefits can be derived in the future from current intangible benefits, decisions are often driven by immediate, measurable financial benefits.

Financial benefits of silvopasture

A rational decision-maker will not choose to adopt a land-use practice where costs outweigh benefits. However, because of difficulties in measuring all benefits, resource managers often choose to adopt a land-use practice that only maximizes financial returns in response to financial costs. As a result, the adopted land-use practice may not be the best. Benefits to the resource manager, society, and the environment may be lost due to lack of knowledge of all benefits and asymmetric decision information.

Although financial indicators may not reflect all benefits that could be attained through the adoption of a silvopasture practice, they do reflect benefits that are valued by the resource manager. Understand that value, price and cost are not synonymous terms. However, the terms benefit and value are very similar. The distinction between 'value' and 'price,' or 'value' and 'cost' is important when analyzing any natural resource practice. 'Price' is a quantitative indicator of the interaction of market forces. 'Value,' on the other hand, is a qualitative assessment of benefit, or utility. Financial indicators must be able to accurately reflect value while using monetary terms such as price.

Common financial indicators applied to silvopastoral practices include net present value (NPV), internal rate of return (IRR), annual equivalent value (AEV), or land expectation value (LEV) (Godsey 2000; Kurtz et al. 1996). These indicators measure the financial costs and benefits of silvopastoral practices for long-term, intermediate-term, or short-term analysis purposes. Although these indicators can be used for comparison purposes, resource managers understand that cash flow is the key to their sustainability.

Most financial analyses of silvopastoral practices come out of the southern pine area of the United States. Grado et al. (1998, 2001) focused on LEV to measure the success of silvopasture in Mississippi where they combined livestock with loblolly pine (*Pinus taeda*) and fee hunting. Six alternative land use treatments were analyzed over a period of eight years. Treatments consisted of loblolly pine only, loblolly mixed with livestock, and livestock only. Results showed that LEV for the silvopastoral practices was more than double that of pine only. The livestock only treatments yielded LEV's twice that of the silvopasture practices. However, in the process of determining

LEV, total costs and returns were calculated and average annual net income was derived. These indicators showed that both silvopasture treatments and only one of the livestock treatments had positive average annual net incomes per hectare.

Lundgren et al. (1983) calculated the real rate of return for four silvopasture practices with three rotation lengths (30, 40, and 60 years) and varying intensities of thinning (no thin, one thinning, two thinnings, and four thinnings). Their results showed that silvopasture managed under the described conditions had a very satisfactory real rate of return ranging from 0.5% to 4.5%. Husak and Grado (2001) built on these two studies to compare the economic performance of silvopasture with that of soybeans (*Glycine max*), rice (*Oryza sativa*), cattle, and a pine plantation. Using LEV, equivalent annual income (EAI), which is essentially the same as equivalent annual value, and rate of return (ROR) as the primary indicators for comparison, silvopasture practices were shown to outperform the other four land-use practices in all cases. This study found that silvopasture had an ROR of 14.6%, whereas cattle systems had an ROR of 12.9% and pine plantation systems an ROR of 13.4%. Husak and Grado (2001) further applied income from hunting leases ranging from \$7.41 ha^{-1} to \$19.76 ha^{-1} to the silvopasture system, showing that the hunting revenue could further increase the financial performance of silvopasture by between 5.1% and 26.4%, respectively.

Numerous other financial studies of silvopasture are available from the southern United States, yet the conclusions are the same. Silvopasture has been found to improve financial performance over pure forestry or pure livestock management practices. This financial benefit is measured by increased NPV (Harwell and Dangerfield 1990; Husak and Grado 2001), increased annual equivalents (whether it be annual equivalent income or land equivalent values) (Grado et al. 2001; Husak and Grado 2001), improved cash flow (Harwell and Dangerfield 1990), or higher rates of return (Lundgren et al. 1983; Harwell and Dangerfield 1990). Clason (1999) even showed that future net value (FNV) can be improved by adopting a silvopastoral practice with loblolly pine.

Few financial analyses of silvopasture in hardwood stands have been reported. Although there are silvopasture practices that incorporate cattle with mixed pine and hardwoods or just hardwoods in the southern United States, very little has been written regarding the economics of those practices (Zinkhan and Mercer 1997). Standiford and Howitt (1993) modeled resource dynamics and interrelationships by looking at the effects of 'multiple use management' (silvopasture) on firewood production, livestock production, and commercial hunting. Their study showed that managing for livestock, firewood and commercial hunting nearly doubled the expected NPV over managing for livestock alone. The potential for silvopasture in hardwood stands is relatively untapped. For states like Missouri, where oak-hickory forests are predominant, a silvopasture practice could provide a profitable means of inspiring landowners to manage underutilized forestland.

Currently, a study is being conducted by the University of Missouri Center for Agroforestry that is designed to evaluate the biology and economics of establishing a silvopasture practice in an existing pole-size, white oak stand, and measuring the financial implications of establishing silvopasture in existing stands versus establishing trees in existing pastures or grazing open pastures. Obviously, income generated from the initial commercial thinning will vary with the size, quantity and quality of the material harvested. However, regardless of the return from the initial thinnings, imposing a silvopasture practice on a previously unmanaged timber stand shortens the rotation of the timber income, creates faster tree growth and larger trees of greater value, and increases grazing land for cattle, thus improving the overall profitability of the land base.

Environmental and social benefits

Financial benefits are only one of many resulting from sound resource management. Zinkhan and Mercer (1997) noted that along with financial benefits (increasing economic returns, diversifying outputs and income, and shortening the wait for income), silvopasture practices also provide conservation and environmental benefits (Alavalapati et al. 2004). Although enhanced wildlife habitat is considered in many of the financial studies (Grado et al. 2001; Husak and Grado 2001), the value of this benefit is often measured only by hedonic pricing methods that look at revenue generated from hunting leases. However, there are also intangible benefits associated with improving wildlife habitat that are more difficult to measure.

Other benefits from silvopasture that are noted in the literature include the intangible benefits resulting from the synergies of tree – livestock interactions.

More specifically, the use of manure as fertilizer for the trees and the climate-stabilizing effects of the trees on the livestock that reduces animal stress are major benefits (Muller et al. 1994; Spain et al. 1997; Zinkhan and Mercer 1997). Water and air quality can also be significantly impacted as a result of hardwood silvopasture adoption. Agriculture-derived contaminants such as sediment, nutrients, and pesticides, constitute the largest diffuse source of water-quality degradation in the United States. Surface runoff and subsurface flow from pasture sites can cause significant nutrient loading of water sources. Excessive pollutants have deleterious ecological impacts on the receiving waters of streams and lakes. Employment of catchment strategies such as silvopasture management can help reduce pollutants before they enter aquatic systems (Nair and Graetz 2004).

Research has demonstrated that inclusion of trees within agricultural systems can improve water quality (Lowrance 1992). Nutrient uptake and removal by the soil and vegetation in a wooded ecosystem (either through tree plantings in pastures or grazing cattle in wooded settings) has been shown to minimize agricultural upland outputs from reaching stream channels (Nair and Graetz 2004). Forested areas function as bioassimilative transformers, changing the chemical composition of compounds. Under oxygenated soil conditions, resident bacteria and fungi mineralize runoff-derived nitrogen, which is then available for uptake by soil bacteria and plants. Livestock-created nutrients moving to streams and ground water are reduced due to absorption by roots. Greater infiltration of nutrient-transporting water occurs within forested areas than in cultivated soil. Processes involved include retention of sediment-bound nutrients in surface runoff, uptake of soluble nutrients by vegetation and microbes, and absorption of soluble nutrients by organic and inorganic soil particles (Garrett et al. 1994).

Tree planting in pastures or enhancement of the vigor of woodlots through thinning and incorporating silvopasture management, presents many opportunities to sequester atmospheric carbon dioxide and mitigate the impacts of global warming (Montagnini and Nair 2004). Fast-growing trees take up carbon at high rates. Trees, as long-lived perennials, are an effective storage system for sequestered carbon and carbon dioxide (Clinch 1999). Landowners should not overlook the potential financial value of silvopasture sites for carbon sequestration (Nair and Nair 2003). Silvopasture and all other agroforestry practices have the potential to generate income from carbon credits, with a global market ranging from \$2.45 to \$18.00 Mg^{-1} of sequestered carbon (Parcell 2000).

Clason and Sharrow (2000) also note intangible social benefits, such as aesthetics, social responsibility (being a good neighbor), and intergenerational responsibility (stewardship for future generations) that help determine the success of a silvopasture practice. Raedeke et al. (2003) found that in Missouri, intergenerational equity was the most important reason for planting trees in an agricultural landscape followed by erosion control and wind protection.

Conclusions

The historical recording of grazing under trees dates to the late 1600s. Initially, grazing animals were subjected to silvopastoral-type practices due to lack of management and fencing. Research then led to the conclusion that grazing under trees should be averted due to poor forage production and animal growth, and damage to trees and understory. Now, current research is demonstrating that managed grazing under tree canopies, can benefit a grazing-animal enterprise. Implementation of silvopastoral practices should be considered in open pastures for two reasons, to increase animal productivity and to generate additional value in timber and/or nut production. Silvopastoral practices should be considered in unimproved hardwood timber for two reasons as well, to improve animal productivity and to implement timber stand improvement practices and place millions of hectares of unmanaged forests under management. The optimum solution will most plausibly be a combination of open pastures and silvopastoral practices integrated into an animal production system.

Hardwood silvopasture is a practice with great potential for application. Within the Central Hardwood Region of the United States alone, there is an estimated 34 million ha of forest with 6.6 million ha occurring on farms. Much of this timber is not being managed due to the high cost of timber stand improvement and the long-term nature of the investment. Silvopasture provides landowners justification for managing their timber as a long-term investment while creating a cash flow from the forested land through the establishment of domestic forages and rotational grazing. While much of the science in support of silvopasture is still under development, enough is known to recommend adoption. Critical to the success of a silvopasture program is the availability of sufficient light to maintain the forage component. Research has shown that with the proper selection of forage species, light levels 40% to 60% of that in the open can result in yields equivalent to or greater than those when forages are grown in the open. Maintenance of the proper levels of light requires pruning and thinning of trees on a regular basis. Shade tolerant forages recommended for the Central United States could include some combination of the grasses Kentucky 31, tall fescue, smooth bromegrass (*Bromus inermis*), orchardgrass, or Kentucky bluegrass (*Poa pratensis*), mixed with a legume such as white clover (*Trifolium repens*) or, one of the lespedeza's (*Kummerowia* spp.). Forage should be grazed moderately so there is sufficient leaf surface area for the plants to respond when the livestock are rotated off the plot. The normal recommendation for cool-season grasses and legumes is to rotate the livestock when the forage has been grazed to a height of 8 cm to 10 cm. Furthermore, livestock must be removed from silvopastured areas during excessively wet periods to avoid soil compaction and tree damage.

Additional research is needed to fully develop silvopastoral practices. We need to better understand the ecological interplay between forage and tree species in silvopastoral practices, and to develop predictive models that allow us to gauge how grazing stress applied to the forage affects tree growth and productivity over time on a variety of soil types. It is important to determine, for example, how wood quality and/or nut production are affected in a silvopastoral practice. We also need to develop methods of establishing new plantings and ensuring the regeneration of tree species. Research can help optimize the benefits that hardwood silvopasture can provide to forestry, agriculture, wildlife, and the environment. In this way we are not repeating history by cycling back to the practice of silvopasture, but rather correcting the error in our ways of ever ceasing its practice.

End Note

1. Erodibility Index: Erodiblity Index (EI) is calculated based on the parameters used in the Universal Soil Loss Equation. In practical terms, soils with an EI < 5 are only slightly susceptible to erosion, and those with EI > 8 are highly susceptible. An EI between 8 and 15 indicates that crop land bare of cover could erode eight or more times the tolerance value and EI of 15 and higher suggests that erosion could occur at 15 or more times the tolerance value.

References

Alavalapati J., Shrestha R.K., Stainback G.A. and Matta J.R. 2004. Agroforestry development: An environmental economic perspective (This volume).

Allard G., Nelson C.J. and Pallardy S.G. 1991. Shade effects on growth of tall fescue: I. Leaf anatomy and dry matter partitioning. Crop Sci 31: 163–167.

Bezkorowajnyj P.G., Gordon A.M. and McBride R.A. 1993. The effect of cattle foot traffic on soil compaction in a silvopastoral system. Agroforest Syst 21: 1–10.

Brinkman K.A. 1955. Epicormic branching on oaks in sprout stands. Technical Paper 146. Research Paper, NE-47. U.S. Department of Agriculture, Forest Service, Columbus, Ohio. 8 pp.

Cartwright T.C. 1955. Responses of beef cattle to high ambient temperatures. J Anim Sci 14: 350–362.

Chandler R.F. Jr. 1940. The influence of grazing upon certain soil and climatic conditions in farm woodlands. J Amer Soc Agron 32: 216–230.

Clason T.R. 1999. Silvopastoral practices sustain timber and forage production in commercial loblolly pine plantations of northwest Louisiana, USA. Agroforest Syst 44: 293–303.

Clason T.R. and Sharrow S.H. 2000. Silvopastoral practices. pp. 119–147. In: Garrett H.E., Rietveld W.J. and Fisher R.F. (eds), North American Agroforestry: An Integrated Science and Practice. American Society of Agronomy, Madison, WI.

Clinch J.P. 1999. Economics of Irish forestry: Evaluating the returns to economy and society. The National Council for Forest Research and Development, University College of Dublin, Belfield, Dublin 4, Ireland, 276 pp.

Dale M.E. 1968. Growth response from thinning young even-aged white oak stands. Research Paper NE-112. USDA, Forest Service, Northeastern Forest Experiment Station, Upper Darby, PA. 19 pp.

Dale M.E. 1975. Effect of removing understory on growth of upland oak. Research Paper, NE-321. U.S. Department of Agriculture, Forest Service, Northeastern Forest Experiment Station, Upper Darby, PA. 10 pp.

Dale M.E. and Sonderman D.L. 1984. Effect of thinning on growth and potential quality of young white oak crop trees. USDA Forest service Research Paper NE-539. U.S. Dept. of Agriculture, Forest Service, Northeastern Forest Experiment Station, Upper Darby, PA.

DenUyl D. and Day R.K. 1934. Woodland carrying capacities and grazing injury studies. Purdue University Agr. Exp. Sta. Bulletin 391. Purdue University, Lafayette, IN.

DenUyl D., Diller O.D. and Day R.K. 1938. The development of natural reproduction in previously grazed farm woods. Purdue Univ. Agr. Exp. Sta. Bul. 431. Purdue University, Lafayette, IN.

Dey D.C., J. Kabrick J, Grabner J. and Gold M. 2003. Restoring oaks in the Missouri River floodplain. pp. 8–20. In: Proceedings 29th Annual Hardwood Sym. Hardwood silviculture and sustainability: 2001 and beyond. May 17–19, 2001. French Lick, IN. National Hardwood Lumber Association. Memphis, TN.

Dey D.C. and MacDonald G.B. 2001. Overstory Manipulation. pp. 157–175. In: Wagner R.G. and Colombo S.J. (eds), Regenerating the Canadian Forest: Principles and Practice for Ontario. Fitzhenry & Whiteside Limited, Markham, Ontario, Canada.

Dey D.C. and Parker W.C. 1996. Regeneration of Red Oak Using Shelterwood Systems: Ecophysiology, Silviculture and Management Recommendations. Ont. Min. Nat. Resour., Ont. For. Res. Inst. For. Res. Inf. Pap. 126. Ontario, Canada. 59 pp.

Ehrenreich J.H. and Crosby J.S. 1960. Herbage production is related to hardwood crown cover. J Forestry 56: 564–565.

Fischer B.C. 1979. Managing light in the selection method. pp. 43–53. In: Proceedings Regenerating Oaks in Upland Hardwood Forests, The 1979 J.S. Wright For. Conf., Purdue Univ., Lafayette, IN.

Ford C.W., Morrison I.M. and Wilson J.R. 1979. Temperature effects on lignin, hemicellulose, and cellulose in tropical and temperate grasses. Aust J Agric Res 30: 621–633.

Frost W.E. and McDougald N.K. 1989. Tree canopy effects on herbaceous production of annual rangeland during drought. J Range Manage 42: 281–283.

Gardner F.P., Pearce B.B. and Mitchell R.L. 1985. Physiology of crop plants. Iowa State University Press; Ames, IA.

Garrett H.E. and Buck L. 1997. Agroforestry practice and policy in the United States of America. Forest Ecol Manag 91: 5–15.

Garrett H.E. and Harper L.S. 1998. The science and practice of black walnut agroforestry in Missouri, USA: a temperate zone assessment. pp. 97–110. In: Buck L.E., Lassoie J.P. and Fernandes E.C.M. (eds), Agroforestry in Sustainable Agricultural Systems. CRC Press, New York, NY., 416 pp.

Garrett H.E., Buck L.E., Gold M.A., Hardesty L.H., Kurtz W.B., Lassoie J.P., Pearson H.A. and Slusher J.P. 1994. Agroforestry: An Integrated Land-Use Management System for Production and Farmland Conservation. USDA SCS, Washington, D.C., 59 pp.

Garrett H.E. and Kurtz W.B. 1983. Silvicultural and economic relationships of integrated forestry farming with black walnut. Agroforest Syst 1: 245–256.

Glickman D., Gonzalez M. and Bay D.M. 1999. 1997 Census of agriculture – State data. United States Department of Agriculture: National Agricultural Statistics Service. Vol. 1 Geographic Area Series, Part 51, USDA AC97-A-51. Washington, D.C., 530 pp.

Godsey L.D. 2000. Economic budgeting for agroforestry practices. Agroforestry in Action #3-2000, University of Missouri Center for Agroforestry, University of Missouri, Columbia, MO., 16 p.

Grado S.C., Hovermale C.H. and St. Louis D.G. 1998. A preliminary economic assessment of a silvopastoral system in south Mississippi. In: Proc., Southern Agroforestry Conference, Huntsville, AL. October 19–21, 1998.

Grado S.C., Hovermale C.H. and St. Louis D.G. 2001. A financial analysis of a silvopasture system in southern Mississippi. Agroforest Syst 53: 313–322.

Hall L.M., George M.R., McCreary D.D. and Adams T.E. 1992. Effects of cattle grazing on blue oak seedling damage and survival. J Range Manage 45: 503–506.

Harwell R.L. and Dangerfield C.W. Jr. 1990. Sustaining use on marginal land: Agroforestry in the Southeast. pp. 212–220. In: Williams P. (ed.), Agroforestry in North America: Proceedings of the First Conference on Agroforestry in North America, 13–16 August, 1989, Department of Environmental Biology, Ontario Agricultural College, University of Guelph, Guelph, Ontario, Canada.

Hawley R.C. and Stickel P.W. 1948. Protection against domestic animals: grazing. pp. 240–256. In: Forest Protection. John Wiley & Sons, Inc., New York.

Hershey F. A. 1991. The extent, nature and impact of domestic livestock grazing on Missouri woodlands. pp. 248–255. In: Garrett 'Gene' H.E. Garrett (ed.), Proceedings of the Second Conference on Agroforestry in North America; Springfield, MO.

Hintz D.L. 1983. Benefits associated with feedlot and livestock windbreaks. USDA Soil Conservation Service, Midwest National Technical Center, Technical Note 190-LI-1, 15 pp. Lincoln, NE.

Holechek J.L., Vavra M. and Skovlin J. 1981. Diet quality and performance of cattle on forest and grassland range. J Anim Sci 53: 291–298.

Huck M.B., Kerley M.S., Garrett H.E., McGraw R.L., Van Sambeek J.W. and Navarrete-Tindall N.E. 2001. Effect of shade on forage quality. pp. 125–135. In: Schroeder W. and Kort J. (eds), Proc of 7th North American Conf. on Agroforestry in North America. Saskatchewan, Canada.

Husak A.L. and Grado S.C. 2001. Monetary and wildlife benefits in a silvopastoral system. pp. 167–175. In: Proceedings of the 2000 Southern Forest Economics Workshop. Lexington, KY. March 28, 2000.

Kay B.L. 1987. Long-term effects of blue oak removal on forage production, forage quality, soil, and oak regeneration. pp. 351–357. In: Proc. Symp. Multiple Use of California's Hardwood Resources. USDA Forest Serv. Gen. Tech. Rep. PSW-100. USDA Forest Service, Pacific Southwest Forest and Range Expt. Station, Berkeley, CA.

Kelly C.F., Bond T.E. and Ittner N.R. 1950. Thermal design of livestock shades. Agr. Eng. 31: 601–606.

Kephart K.D. and Buxton D.R. 1993. Forage quality responses of cool and warm perennial grasses to shade. Crop Sci. 33: 831–837.

Kephart K.D., Buxton D.R. and Taylor S.E. 1992. Growth of cool and warm season perennial grasses in reduced irradiance. Crop Sci. 32: 1033–1038.

Koch N.E. and Kennedy J.J. 1991. Multiple-use forestry for social values. Ambio V. 20: 330–333.

Kurtz W.B., Garrett H.E., Slusher J.P. and Osburnj D.B. 1996. Economics of agroforestry. MU Guide, Forestry #G 5021, University of Missouri Extension, University of Missouri-Columbia, 2 pp.

Lamson N.I., Smith H.C., Perkey A.W. and Brock S.M. 1990. Crown release increases growth of crop trees. Research Paper, NE-635. U.S. Department of Agriculture, Forest Service, Northeastern Forest Experiment Station, Broomall, PA. 8 pp.

Lantagne D.O. 1991. Tree shelters increase heights of planted northern red oaks. Proceedings 8th Central Hardwood Conference, USDA, Forest Svc. GTR-NE-148. pp. 291–298. USDA Forest Service, Northeastern Forest Experiment Station, Radnor, PA.

Lehmkuhler J., Felton E.E.D., Schmidt D.A., Bader K.J., Moore A., Huck M.B., Garrett H.E. and Kerley M.S. 2003. Cattle performance and tree damage using various tree protection methods during tree establishment. Agroforest Syst 59: 35–42.

Lehmkuhler J.W., Kerley M.S., Garrett H.E., Cutter B.E. and McGraw R.L. 1999. Comparison of continuous and rotational silvopastoral systems for established walnut plantations in southwest Missouri, USA. Agroforest Syst 44: 267–279.

Lin C.H., McGraw R.L., George M.F. and Garrett H.E. 1999. Shade effects on forage crops with potential in temperate agroforestry practices. Agroforest Syst 44: 109–119.

Lin C.H., R.L. McGraw R.L., George M.F. and Garrett H.E. 2001. Nutritive quality and morphological development under partial shade of some forage species with agroforestry potential. Agroforest Syst 53: 1–13.

Lovelace W. 1998. The root protection method (RPM) system for producing container trees. Combined Proceedings International Plant Propagators' Society. 48: 556–557.

Lowrance R.R. 1992. Groundwater nitrate and denitrification in a coastal plain riparian forest. J Environ Qual 21: 401–405.

Lundgren G.K., Connor J.R. and Pearson H.A. 1983. An economic analysis of forest grazing on four timber management situations. South J Appl For 7: 119–124.

Marquis D.A. 1988. Guidelines for regenerating cherry-maple stands. pp. 167–188. In: Smith H.C., Perkey A.W. and Kidd W.E. Jr. (eds), Proceedings, Guidelines for Regenerating Appalachian Hardwood Stands, 24–26 May 1988, Morgantown, W.Va Soc. Am. Pub. 88-03.

McCreary D.D. and Tecklin J. 2001. The effects of different sizes of tree shelters on blue oak growth. West J Appl For 16: 153–158.
McDaniel A.H. and Roark C.B. 1956. Performance and grazing habits of Hereford and Aberdeen-angus cows and calves on improved pastures as related to types of shade. J Anim Sci 15: 59–63.
McIlvain E.H. and Shoop M.C. 1971. Shade for improving cattle gains and rangeland use. J Range Manage 24: 181–184.
Minckler Leon S. 1961. Measuring light in uneven-aged hardwood stands. Technical Paper 184. U.S. Department of Agriculture, Forest Service, Central States Forest Experiment Station. St. Paul, MN., 9 pp.
Minter W.F., Myers R.K. and Fischer B.C. 1992. Effects of tree shelters on northern red oak seedlings planted in harvested forest openings. North J Appl Forest 9: 58–63.
Montagnini F. and Nair P.K.R. 2004. Carbon sequestration: An under-exploited environmental benefit of agroforestry systems (This volume).
Muller, C.J.C., Botha J.A., Smith W.A. and Coetzer W.A. 1994. Production, physiological and behavioral responses of lactating friesian cows to a shade structure in a temperate climate. In: Budklin R. (ed.), Dairy Systems for the 21st Century. Proceedings, Third International Dairy Housing Conference. February 2–5, 1994, Orlando, FL. American Society of Agricultural Engineers. St. Joseph, MI.
Nair P.K.R. and Nair V.D. 2003. Carbon Storage in North American Agroforestry Systems. pp. 333–346. In: Kimble J., Heath L.S., Birdsey R.A. and Lal R. (eds), The Potential of U.S. Forest Soils to Sequester Carbon and Mitigate the Greenhouse Effect. CRC Press, Boca Raton, FL, USA.
Nair V.D. and Graetz D.A. 2004. Agroforestry as an approach to minimizing nutrient loss from heavily fertilized soils: The Florida experience (This volume).
Neel L.R. 1939. The effect of shade on pasture. Tenn. Agr. Exp. Sta. Circular #65, 2 p.
Noweg T.A. and Kurtz W.B. 1987. Eastern black walnut plantations: An economically viable option for conservation reserve lands within the corn belt. North J Appl Forest 4: 158–160.
Ostrofsky R.S., Seymour R.S. and Lemin Jr. R.C.. 1986. Damage to northern hardwoods from thinnings using whole-tree harvesting technology. Can. J Forest Res 16: 1238–1244.
Parcell J. 2000. Carbon sequestration and its value for Missouri producers. Online publication http://agebb.missouri.edu/mgt/carbon.htm [Last accessed November 20, 2001].
Patric J.H. and Helvey J.D. 1986. Some effects of grazing on soil and water in the eastern forest. NE-GTR-115. Bromall, PA. U.S. Department of Agriculture, Forest Service, Northeastern Forest Experiment Station, 25 pp.
Pearson H.A. 1990. Silvopasture: Forest grazing and agroforestry in the southern coastal plain. pp. 25–42. Proceedings of Practical Agroforestry in the Mid-South Conference. West Memphis, AR.
Raedeke A., Green J., Hodge S. and Valdivia C. 2003. Farmers, the practice of farming and the future of agroforestry: An application of Bourdieu's concept of field and habitus. Rural Sociol 68: 64–86.
Roberts E.N. 1926. Wyoming forage plants and their chemical composition studies. No. 7. Effects of altitude, seasonal variation, and shading experiments. Wyoming Agr. Exp. Sta. Bull. 146.
Sander I.L. 1979. Regenerating oaks with the shelterwood system. pp. 54–60. In: Proceedings Regenerating Oaks in Upland Hardwood Forests, The 1979 J.S. Wright For. Conf., Purdue Univ., Lafayette, IN.,
Schlesinger R.C. 1978. Increased growth of released white oak poles continues through two decades. J Forest 76: 726–727.
Shifley S.R. and Sullivan N.H. 2002. The status of timber resources in the North Central United States. Gen. Tech. Rep. NC-228. St. Paul, MN: USDA, Forest Service, N.C. Res. Sta. 47 pp.
Shigo A.L. 1966. Decay and discoloration following logging wounds on northern hardwoods. Research Paper, NE-47. Upper Darby, PA. U.S. Department of Agriculture, Forest Service, Northeastern Forest Experiment Station. 43 pp.
Smith W.B., Vissage J.S., Darr D.R., Sheffield R.M. 2001. Forest resources of the United States, 1997. Gen. Tech. Rep. NC-219. St. Paul, MN. U.S. Department of Agriculture, Forest Service, North Central Research Station. 190 pp.
Smith H.C. and Miller G.W. 1991. Releasing 75- to 80-year-old Appalachian hardwood sawtimber trees: 5-year d.b.h. response. pp. 402–413. In: McCormick L.H. and Gottschalk K.W. (eds), 8th Central Hardwood Forest Conference; Pennsylvania State University; University Park, PA.
Smith H.C. 1966. Epicormic branching on eight species of Appalachian hardwoods. Research Note, NE-53. U.S. Department of Agriculture, Forest Service, Northeastern Forest Experiment Station, Upper Darby, PA. 4 pp.
Smith R.M. 1942. Some effects of black locusts and black walnuts on southeastern Ohio pastures. Soil Sci 53: 385–398.
Spain J.N., Zulovich J. and Hardin D.K. 1997. Heat stress on a commercial dairy farm: An economic evaluation of cooling. pp. 936–941. In: Botcher R.W. and Hoff S.J. (eds), Livestock Environment V, Volume II, Proceedings of the Fifth International Symposium, May 29–31, 1997. Bloomington, MN.
Standiford R.B. and Howitt R.E. 1993. Multiple use management of California's hardwood rangelands. J Range Manage 46: 176–182.
Stephenson W.H. 1954. North Carolina's beef, pork and poultry trade. pp. 107–109. A Basic History of the Old South. D. Van Nostrand Co., Inc., Princeton, NJ.,
Stokes B.J., Kluender R.A., Klepac J.F. and Lortz D.A. 1995. Harvesting impacts as a function of removal intensity. pp. 207–216. In: XX IUFRO World Congress on Forest Operations and Environmental Protection. Tampere, Finland.
Van Tassell L.W., Bartlett E.T. and Mitchell J.E. 2001. Projected use of grazed forages in the United States: 2000 to 2050: A technical document supporting the 2000 USDA Forest Service RPA Assessment. Gen. Tech. Rep. RMRS-GTR-82. U.S. Department of Agriculture, Forest Service, Rocky Mountain Research Station, Fort Collins, CO, 73 pp.
Ward W.W. 1966. Epicormic branching of black and white oaks. Forest Sci 12: 290–296.
Webster A.J.F. 1970. Direct effects of cold weather on the energetic efficiency of beef production in different regions of Canada. Can J Anim Sci 50: 563–573.
Wolters G.L. 1973. Southern pine overstories influence herbage quality. J Range Manage 26: 423–425.
Worstell D.M. and Brody S. 1953. Environmental physiology and shelter engineering. Mo. Agr. Exp. Sta. Res. Bul. #515. University of Missouri, Columbia, MO.
Zinkhan F.C. and Mercer D.E. 1997. An assessment of agroforestry systems in the southern USA. Agroforest Syst 35: 303–332.

Agroforestry Systems **61:** 35–50, 2004.

Riparian forest buffers in agroecosystems – lessons learned from the Bear Creek Watershed, central Iowa, USA

R.C. Schultz[1,*], T.M. Isenhart [1], W.W. Simpkins[2] and J.P. Colletti[1]
[1]*Department of Natural Resource Ecology and Management, 253 Bessey, Iowa State University, Ames, IA 50011-1020;* [2]*Department of Geologic and Atmospheric Sciences, Iowa State University;* **Author for correspondence: e-mail: rschultz@iastate.edu*

Key words: Buffer design, Buffer maintenance, Filter strips, Surface runoff, Water quality

Abstract

Intensive agriculture can result in increased runoff of sediment and agricultural chemicals that pollute streams. Consensus is emerging that, despite our best efforts, it is unlikely that significant reductions in nutrient loading to surface waters will be achieved through traditional, in-field management alone. Riparian forest buffers can play an important role in the movement of water and NPS (non-point source) pollutants to surface water bodies and ground water. Riparian buffers are linear in nature and because of their position in the landscape provide effective connections between the upland and aquatic ecosystems. Present designs tend to use one model with a zone of unmanaged trees nearest the stream followed by a zone of managed trees with a zone of grasses adjacent to the crop field. Numerous variations of that design using trees, shrubs, native grasses and forbs or nonnative cool-season grasses may provide better function for riparian forest buffers in specific settings. Properly designed riparian buffers have been shown to effectively reduce surface NPS pollutant movement to streams and under the right geological riparian setting can also remove them from the groundwater. Flexibility in design can also be used to produce various market and nonmarket goods. Design flexibility should become more widely practiced in the application of this agroforestry practice.

Introduction

Intensive agriculture, as practiced in much of the temperate zone, is not environmentally friendly as evidenced by the major non-point source (NPS) pollution problems that have befallen many surface water bodies. The size of cultivated fields continues to increase to accommodate large equipment, and livestock production continues to be concentrated into larger facilities producing major amounts of waste that must be disposed of safely. In many landscapes of the Midwestern United States, more than 85% of the landscape is devoted to row crop agriculture or intensive grazing (Burkart et al. 1994). It is unlikely that these trends will slow in the near future as more food needs to be produced efficiently to feed a growing world population. Increased surface runoff laden with sediment and agrochemicals continue to provide higher and more frequent peak flows that result in more flooding, incision and widening of stream channels, reduction of base flow, and reduction in the quality of aquatic ecosystems (Menzel 1983).

Consensus is emerging that, despite our best efforts, it is unlikely that significant reductions in nutrient loading to surface waters will be achieved through traditional, in-field management alone (Dinnes et al. 2002). Recognizing the gravity of the problem, public agencies are looking to augment these traditional pollution-control efforts by utilizing landscape buffers as off-site sinks for contaminants (Mississippi River/Gulf of Mexico Watershed Nutrient Task Force 2001; Mitsch et al. 2001). One such effort in the United States is the Conservation Buffer Initiative, a public-private program with the goal to assist in the establishment of 3.2 million km of conservation buffers, mainly through the use of the continuous United States

Department of Agriculture (USDA) Conservation Reserve Program. A host of different buffers that reduce surface movement of soil and agricultural chemicals by wind and/or water have been developed. These include such practices as: alleycropping, constructed wetlands, cross-wind trap strips, field borders, filter strips, riparian forest buffers, vegetative barriers, and windbreaks and shelterbelts, to name a few. The effectiveness of most of these practices can be improved if they are used in combination throughout the landscape. However, while conservation buffers have great potential to assist in meeting our stewardship goals, Lowrance et al. (2002) concluded that key knowledge gaps still exist and that future research results need to be translated into up-to-date technical standards and guidelines for conservation buffer design, installation, and maintenance.

Infield conservation practices are designed to trap and slow the movement of materials near their source. Filter strips and riparian forest buffers are designed as ecotones between upland and aquatic ecosystems and provide the last conservation practice that slows movement and traps materials before they enter surface water bodies. These buffers may be the only major perennial plant conservation practice that is used along intermittent streams (channels that carry water only during the wet part of the year) and some first order streams (with no tributaries). Because of the small size of the stream, cultivation can be easily and relatively safely carried on right down to the channel bank putting the stream in direct contact with the cultivated field source of NPS pollutants. In higher order streams (streams with one or more tributaries; example: 4th order stream occurs when two 3rd order streams join) they should be supplemented with other perennial plant conservation practices to insure maximum phytoremediation of NPS pollutants before reaching the buffers.

Riparian forest buffers are a recognized agroforestry practice that not only provide phytoremediation for NPS pollutants but also increase biodiversity of both the terrestrial and aquatic ecosystems. They can also provide stream bank stabilization, moderate flooding damage, recharge groundwater, sequester carbon and provide recreational opportunities and fiber and nonfiber products to landowners.

This chapter will describe the structure, function, design and maintenance of riparian forest buffers and grass filters based on work that has been conducted in the Bear Creek Watershed in central Iowa, United States (42° 11′ N, 93° 30′ W). Because of the extensive research and demonstration activities that have been conducted for more than a decade, this watershed has been designated as a National Restoration Demonstration Watershed by the interagency team implementing the Clean Water Action Plan (1999) and as the Bear Creek Riparian Buffer National Research and Demonstration Area by the USDA (1998). Most of the discussion will address the establishment of new buffers along first to fourth order streams, where present cultivation extends to near the stream edge or riparian grazing is practiced by fencing in the narrow riparian corridor that may be difficult to cultivate with large machinery because of the tight meandering of the stream. Establishing buffers under these conditions usually requires complete restoration of a perennial plant community as none has existed for decades or has been very heavily impacted by grazing pressure.

What is a riparian forest buffer?

A riparian forest buffer is specifically defined as a three zone system consisting of an unmanaged woody zone adjacent to the water body followed upslope by a managed woody zone and bordered by a zone of grasses with or without forbs (Welsch 1991; Isenhart et al. 1997). The major objectives of a riparian buffer are:

1. To remove nutrients, sediment, organic matter, pesticides, and other pollutants from surface runoff and groundwater by deposition, absorption, plant uptake, denitrification, and other processes, and thereby reduce pollution and protect surface water and subsurface water quality while enhancing the ecosystem of the water body,
2. To create shade to lower water temperature to improve habitat for aquatic organisms, and
3. To provide a source of detritus and large woody debris for aquatic organisms and habitat for wildlife (USDA-NRCS 2003).

Riparian forest buffers can vary in design in response to management objectives. The vertical and horizontal structure of the woody and grass zones may differ from one location to the next, depending on:

1. Landowner objectives, concerns, experiences and biases,
2. Present condition of the site,
3. Major functions of the proposed buffer system, and
4. Short and long term management practices to be employed.

Table 1. Typical landowner concerns about riparian forest buffers.

1. How much can buffers reduce the movement of sediment and other pollutants to the stream?
2. Can buffers heal gullies?
3. What happens to water moving rapidly over grass waterways when it intersects a buffer?
4. Can buffers slow stream meandering?
5. What buffer vegetation produces the best wildlife habitat and fishery?
6. Can buffers reduce stream bank erosion?
7. If a riparian forest buffer is planted will trees fall into the stream and back up water into the fields or drain tiles?
8. Is the buffer a source of weed seeds?
9. Are cool-season grass filters just as effective as riparian forest buffers?
10. Will riparian forest buffers bring beavers that build dams that also back up water?
11. Will deer become a problem for crops?
12. How much maintenance is required and who will do it?
13. If fencing is used to keep livestock out of a buffer will it be damaged by floods?
14. How much land will be taken out of crop production or pasture?
15. Can specific products be harvested from the buffers to offset income losses?
16. Are there consultants who can design, plant and maintain the buffers?
17. Are there government programs to help pay for the practice?

They can consist of existing riparian forests or be established on previously cultivated or grazed land. While this chapter focuses primarily on riparian forest buffers in the agricultural landscape, they may also be found in forested, suburban and urban landscapes.

In a forest landscape riparian buffers are often referred to as streamside management zones and usually consist of only a forested zone. Management activities such as thinning, fertilizing and harvesting are usually dramatically restricted, if allowed at all. In suburban landscapes, riparian forest buffers usually consist of existing forests that border on residential or commercial developments. If these forests are protected and managed they can provide the natural functions of filtering and processing of suburban pollutants associated with runoff from roads, lawns and construction sites. They can also provide noise control and screening along with their wildlife habitat and aesthetic benefits. In the more urbanized environment, riparian forest buffers are often very narrow and fragmented to the point that they may not be completely functional forest ecosystems. However, with proper planning and zoning, these forests can play an important role in storm water management. These forests are often the largest and most continuous forests in the urban environment and therefore provide significant wildlife habitat and recreational opportunities. However, as use of these areas increase, their pollutant control ability is often reduced. Because these buffers often are located on highly prized land, the goal of suburban and urban planning is to protect them from development and overuse by the public (Palone and Todd 1997).

Riparian forest buffers in the agricultural landscape can take numerous forms. In the eastern United States, they may exist as narrow corridors of remnant forests along the stream or as irregular islands along portions of meandering streams. In many cases these forests have been grazed in the past. In much of the intensive agricultural belt of the Midwestern United States, riparian forest buffers have to be reestablished from scratch on land that was cropped or grazed for many decades. In the arid and semiarid west, riparian forest buffers consist of narrow tree and shrub zones in a vast expanse of grazed dry upland shrubs and grasses (Elmore 1992). Because riparian forest communities evolved in the most fertile and moist position of the landscape, they can often be easily reestablished. However, in many agricultural landscapes land uses have so dramatically changed the hydrology that these communities cannot be restored to their original condition. Stream channels have been incised and widened by higher discharge resulting from greater surface runoff from crop fields and heavily grazed pastures. Channelization of meandering streams, tiling of some landscapes and urbanization also have contributed to higher storm flow and lower baseflow (Menzel 1983). In many cases, water table depths have been lowered to the point that restored buffers require a different community structure to function properly. However, with proper planning and design the functions of a healthy riparian forest community can be reestablished.

As an agroforestry practice, riparian forest buffers play an important role in the movement of water through the agricultural landscape and in the movement of NPS pollutants to surface water bodies and ground water. They are linear in nature, and because of their position in the landscape, they provide effective connections between the upland and aquatic ecosystems in a watershed.

Assessing the need for riparian buffers

The design of a buffer system will strongly influence the expression of each of the buffer functions on a specific site. It is therefore imperative that before designing a riparian buffer an assessment of the site be conducted as well as the landowner's objectives and concerns identified. Concerns that landowners typically have about designing and adopting a riparian buffer are summarized in Table 1.

After landowner objectives have been identified and concerns have been satisfied it is time to assess the present condition of the stream and riparian zone. While a lot of information can be gained from aerial photos it is imperative that the entire site is walked and a detailed site map identifying problem areas is developed. Walking the site with the landowner is an excellent option. Items to look for and questions to answer while making the field assessment are indicated in Table 2.

Table 2. Field assessment guidelines for designing a riparian forest buffer.

Questions to address in a site inventory.
1. What is the order of the stream?
2. Is the stream listed as an impaired waterway?
3. Is it a naturally meandering or channelized reach?
4. Is the channel in contact with the flood plain or deeply incised?
5. What is the stage of channel evolution – is it deepening, widening or beginning to stabilize (Schumm et al. 1984)?
6. How high are the banks above the streambed and are they vertical or sloping?
7. Is bank erosion naturally slow or accelerated?
8. Are there major slumps or eroded sections and are these associated with gullies or areas of concentrated flow or on the outside of bends?
9. What is the dominant land use of the riparian zone?
10. What is the land-use immediately adjacent to the channel – does vegetation cover consist of annual row crops, a narrow strip of weedy vegetation, permanent forest or forage production, or grazed pasture?
11. Does the weedy vegetation consist of dense introduced grasses or annual weeds?
12. What is the condition of any forest vegetation?
13. Is livestock grazed rotationally, providing significant rest periods for re-growth or intensively grazed with complete access to the banks and channel?
14. Are there drainage tiles emptying into the stream?
15. Are there upslope areas of concentrated flow that should be in grassed waterways?
16. Is it a cold- or warm-water fishery?
17. Is there evidence of a healthy aquatic invertebrate community?

Once the landowner's questions have been addressed and the site assessment has been completed the design process may begin. If the adjacent land-use, including that of the riparian zone is row crop agriculture a riparian forest buffer or filter is probably needed (Figure 1). The choice between a riparian forest buffer and a filter strip will depend on landowner preferences and site conditions. Where steep vertical stream banks of ≥ 2 m exist, some kind of woody vegetation is recommended to provide strong perennial woody roots that strengthen stream bank soils. If the stream bank is gently sloping with a $\geq 3{:}1$ slope (for every 1 unit vertical rise there is a 3 unit horizontal setback), only a native grass/forb filter may be needed depending on the desires of the landowner and the long-term management of the buffer. These options will be discussed in detail in the following sections.

If a narrow strip of trees, 3 m to 10 m wide is already present along the stream bank only a filter may be needed to effectively intercept upslope runoff. If stream banks are vertical, and if roads, structures or other valuable property is being threatened by stream bank erosion, then stream bank bioengineering techniques may have to be used in addition to any riparian buffers (Figure 2). In-stream structures such as boulder weirs or riffle structures may also be used to help stabilize both the stream banks and the channel bed. If field drainage tiles pass below any buffer on their way to dumping into the stream, consider developing a wetland to intercept the flow before it enters the stream. Biological activity in the wetland can effectively remove a number of the primary chemical pollutants threatening surface waters (Crumpton 2001).

Figure 1. Pasture and crop sites that are in need of riparian buffers. Banks in A are gentle enough that native grasses and forbs could be planted at the edge of the bank. Banks in B are steep enough that their stability would be improved with woody roots.

Designing the buffer

The traditional riparian forest buffer consists of three zones (Figure 3). The first zone is unmanaged forest along the stream bank, left unmanaged to provide shading and large woody debris to the stream. The second zone consists of forest that is managed as a nutrient sink with systematic harvests to remove trees. The third zone consists of a grass and/or grass-forb filter strip that intercepts and slows the usually concentrated surface runoff from adjacent fields and spreads it over a wider front to move more slowly through the riparian buffer.

Experience with landowners in some regions suggests that zone 1 in the three-zone system is of concern to landowners who do not want trees falling into streams slowing water that they feel needs to be carried rapidly from the landscape. This concern resulted in the development of the two-zone multi-species riparian buffer consisting primarily of a woody zone and a grass or grass-forb filter zone (Figure 4). In this model, the whole woody zone is managed but may be planted to combinations of trees and shrubs.

Further experience with this model and landowner acceptance led to the development of a model with multiple combinations of woody and non-woody zones as discussed below (Figure 5). In this model the buffer is divided into three zones that may be occupied by trees, shrubs and grasses or grasses and forbs. Within each of these plant communities, a wide variety of species and zone widths can be used depending on design requirements.

Functions of grasses and forbs

Grasses and/or grasses and forbs provide excellent ground cover (Table 3). They can be designed to provide the frictional surface needed to slow surface runoff from the adjacent field allowing the sediment in the runoff to be dropped in the crop field and converting much of the concentrated flow into a wider front of slower moving water approaching sheet flow. Much of this slower moving water can infiltrate into the buffered soil where it will have extended contact time with the 'living filter' of the soil-plant system. Surface runoff that moves through the buffer continues to move slowly dropping further sediment and infiltrating more water and soluble nutrients.

Most non-native forage grasses such as *Bromus inermis*, *Phleum pratense* and *Poa pratensis* do not present stiff leaves to the on-coming surface runoff. Rather they are easily laid down, allowing water to move rapidly over them. This feature is very effective in the grass waterway conservation practice where the objective is to carry water rapidly down sloping depressions within row-crop fields while protecting the mineral soil surface. When these grasses are used in zone 3 of a riparian buffer they may be inundated with sediment in large runoff events because slopes are usually between 0% to 2%, slowing water enough to drop most of the sediment (Dillaha et al. 1989). When this occurs, the filter becomes ineffective in slowing further surface runoff. Even among the native warm-season grasses, differences exist in the presentation of a highly frictional surface to incoming runoff. While most native warm-season grasses are strong clump-forming types providing corridors for movement between the clumps, switchgrass (*Panicum*

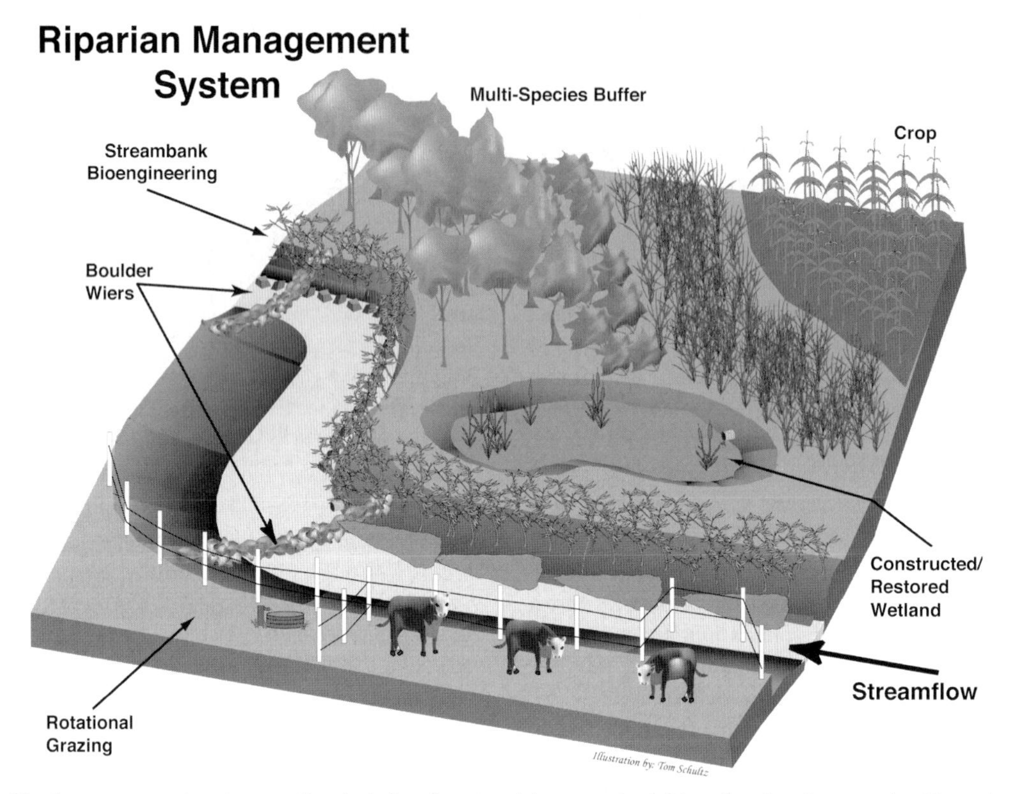

Figure 2. Riparian management system practices including from top right: streambank bioengineering, in-stream boulder weir structures, intensive rotational grazing, constructed wetlands and riparian buffer.
Source: Reprinted, with permission, from Schultz et al. (2000). © 2000 by American Society of Agronomy.

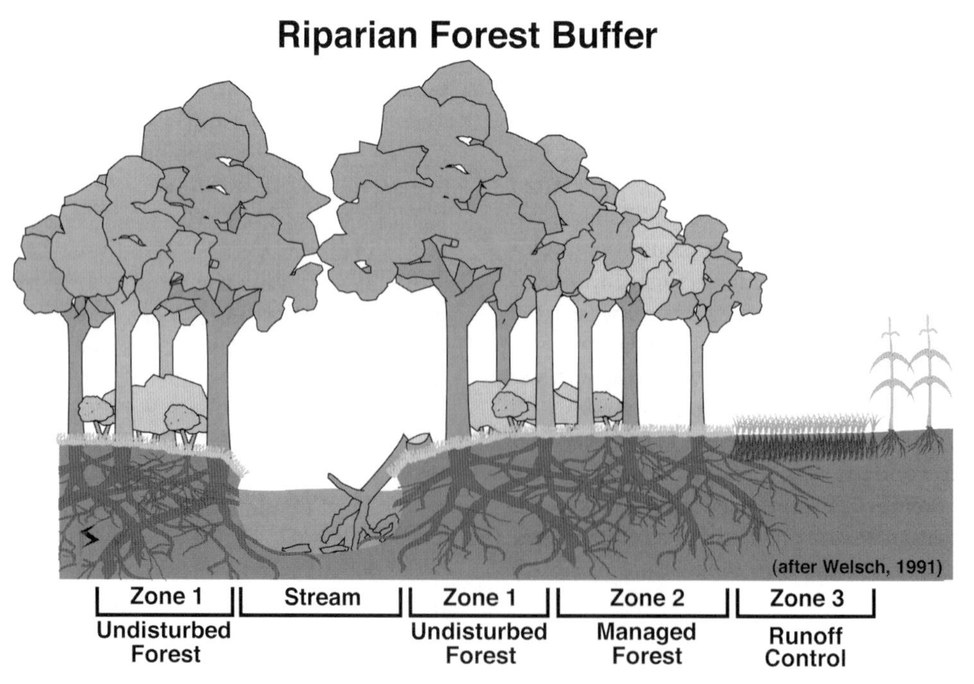

Figure 3. The traditional three zone riparian buffer.
Source: Reprinted, with permission, from Schultz et al. (2000). ©2000 by American Society of Agronomy.

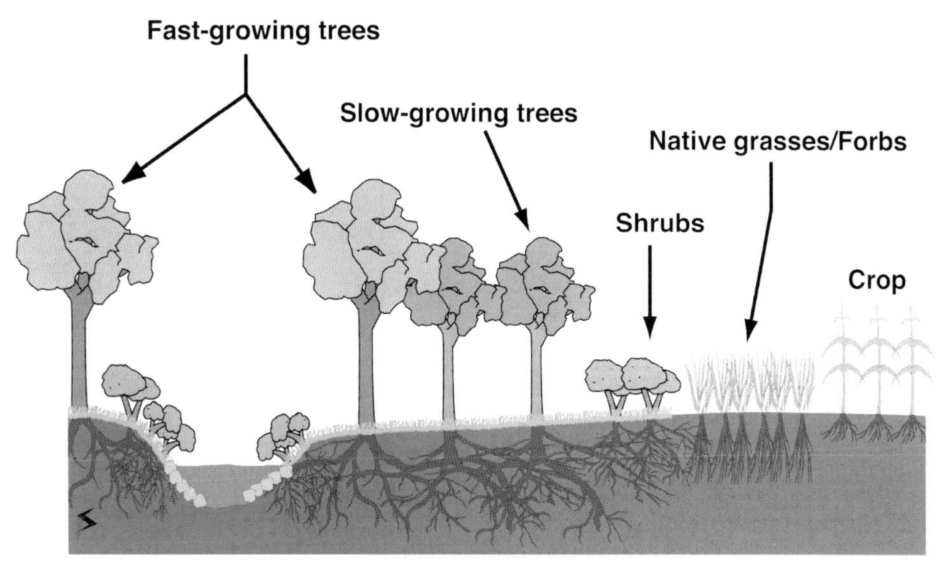

Figure 4. A two-zone multispecies riparian buffer with a woody and grass/forb zone. The whole woody zone is managed to keep large woody debris from falling in the stream. Shrubs are added to the system for a multitude of functions. With the added shrubs, a three-zone system provides numerous planting options as shown in Figure 5.
Source: Reprinted, with permission, from Schultz et al. (2000). ©2000 by American Society of Agronomy.

virgatum) does not do that and presents such an effective frictional surface to on-coming runoff that water slows, ponds and drops its sediment prior to entering the buffer (Dabney et al. 1994).

Grasses and forbs help to restore biological and physical soil quality by adding large amounts of carbon to the profile from rapid turnover of roots that contain more than 70% of the total biomass of the prairie plants. Litter additions from the aboveground biomass also add carbon to the surface soil. This carbon plays a key role in redeveloping soil macro-aggregate structure that helps facilitate the high infiltration rates needed to get surface runoff into the soil profile. The carbon also serves as a substrate for increased soil microbial activity that is important both in building soil structure and processing some agricultural chemicals that move in the surface and ground water.

Major differences in impacts on the soil ecosystem can occur depending on the dominance of C3 or C4 grasses in a buffer. C3 grasses tend to have lower lignin and C:N ratios than C4 grasses (Pearcy and Ehleringer 1984). This leads to more rapid litter decomposition, higher soil respiration, more active soil organic matter (OM), less soil C accumulation, less nitrogen immobilization and more net nitrogen mineralization. The result is faster restoration of soil's quality-related functions in filter systems planted to C3 grasses. However, these soil improvements usually are not noticed as deeply in the profile as under C4 grasses. The major Mollisol (prairie) soils of the world have developed under prairie plant communities dominated by C4 grasses. Given enough time, soils planted to C4 grass strips also demonstrate improved soil quality to greater depths.

While pure stands of switchgrass may be the most effective plant community at slowing surface runoff and trapping sediment, it is not the most effective in providing wildlife habitat. The dense nature of its stand of stems may provide good winter cover for upland game birds but it provides poor nesting cover. In landscapes where the slope of the adjacent fields providing the surface runoff is $>5\%$, a pure stand of switchgrass may be planted in a 5-m to 6-m wide strip at the leading edge of the filter with a mixture of other grasses and forbs planted on the down slope stream side of the switchgrass. The mixture of grasses and forbs provide more structural diversity for wildlife habitat. The result is that they can provide good nesting cover, because of the more open nature of the stand, and good winter cover, because they can resist being bent over by high winds and snow loads, for game bird species. Grasses and forbs do not shade the

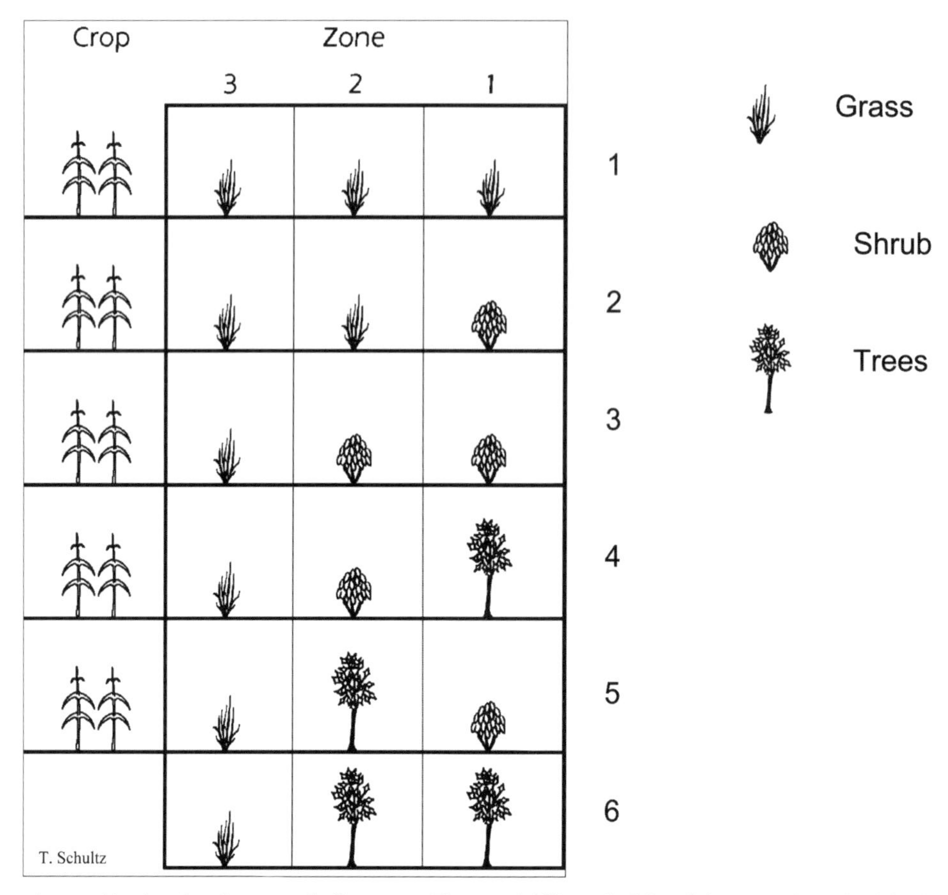

Figure 5. Vegetation combinations in a three-zone buffer system. The crop field is on the left and the stream is on the right. Vegetation zones can vary in width in addition to species and species of trees, shrubs and grasses can vary within a zone.

stream, which is important in warm water ecosystems, nor do they provide significant detrital inputs for the aquatic ecosystem.

Functions of shrubs

Shrubs add the longevity of a woody root system and a more permanent nutrient sink to the buffer (Table 3). While most shrubs are multi-stemmed they are able to sequester nutrients in their woody biomass. The woody root system provides strength to soil profiles. Annual leaf fall and fine-root turnover provide significant organic matter (OM) to the O and A horizons of the soil profile. OM additions are not usually found as deep in the profile as under C4 grass and forb ecosystems (Tufekcioglu et al. 1999) but C:N ratios of many native shrub leaves are conducive to rapid decomposition that helps to rapidly rebuild surface soil quality.

The multiple stem nature of shrub species provides an effective barrier that traps floodwater debris in the buffer rather than letting it move out into the adjacent fields. The multiple stem system also provides cover for wildlife and wind reductions near the ground providing thermal cover. Their size and shape provides another layer of vertical structure that fits between the grasses and trees. In most regions, a wide variety of shrub species can be planted in riparian buffer settings providing a diverse food source and cover for a wide variety of bird and wildlife species as well as providing the potential for ornamental woody cut production.

Because shrubs are usually limited to < 5 m in height, they do not provide much shade to the channel but provide perennial woody roots that can add strength to stream bank soils. While the larger woody roots of shrubs are not as massive as those on trees they can still add more effective strength to the stream bank profile than grasses and forbs. Because of their shorter

Table 3. Functions of the grass, shrub and tree components of riparian buffers.

Kind of Plant	Functions
Prairie grasses/forbs	1. Slow water entering the buffer
	2. Trap sediment and associated chemicals
	3. Add organic carbon to a range of soil depth
	4. Added carbon improves soil structure
	5. Improve infiltration capacity of the surface soil
	6. Above ground nutrient sink needs annual harvest
	7. Provide diverse wildlife habitat
	8. Do not significantly shade the stream channel
	9. Provide only fine organic matter input to stream
	10. Can provide forage and other products
Shrubs	1. Multiple stems act as a trap for flood debris
	2. Provide woody roots for bank stabilization
	3. Litter fall helps improve surface soil quality
	4. Above ground nutrient sink needs occasional harvest
	5. Adds vertical structure for wildlife habitat
	6. Do not significantly shade the stream channel
	7. Provide only fine organic matter input to stream
	8. Can provide ornamental products and berries
Trees	1. Strong, deep woody roots stabilize banks
	2. Litter fall helps improve surface soil quality
	3. Long-lived, large nutrient sink needs infrequent harvest
	4. Adds vertical structure for wildlife habitat
	5. Vertical structure may inhibit buffer use by grassland birds
	6. Shade stream, lowering temperature and stabilizing dissolved oxygen
	7. Provide both fine organic matter and large woody debris to the channel
	8. Can provide a wide variety of fiber products

stature, shrubs often are not major sources of in-stream detritus that can support aquatic invertebrates.

Functions of trees

The aboveground woody biomass of trees provides a large C and N sink that can be systematically removed to maintain the storage capacity of a buffer (Table 3). The large woody roots that often can be found extending 2 m to 3 m or more into the soil provide additional strength to stream banks. High transpiration rates and deep roots help to dewater moist stream banks improving their stability.

Native riparian species are always a good choice on most sites but many landowners have concerns with some of them because they tend to be short-lived and break up easily under natural storm events. Most of the native riparian species are prolific spring seeders and either root- or stump sprout; so, even though they may be short-lived and may succumb to storm events they rapidly re-establish themselves. Root sprouting species are ideal for planting directly adjacent or on stream banks because they can spread rapidly presenting a highly frictional surface to base and storm flows. Planting these species near the field edge of the buffer is not recommended, however, because these same root sprouting species often migrate out into portions of the crop field becoming a nuisance to farmers.

Depending on the depth of the aerated root zone, a wide variety of tree species can be planted. If channels are incised at least 2 m to 3 m in depth, many common upland species can be included in the design especially if care is taken to plant them on the microtopographic rises of the riparian zone. These species are often requested by landowners who see them as providing potential quality fiber products at some future date and hard mast for a variety of wildlife species. While conifers are usually not recommended for planting near the channel because of the damage that they can incur from flooding, they can provide effective winter cover if planted in double or triple rows on rises in the riparian zone. Species diversity in the tree component of any buffer is critical to the longevity of that component. Mixing species within and between rows is strongly recommended to reduce the potential of large holes in the buffer corridor resulting from insect or pathogen problems. Native species are also recommended over non-native species and hybrids as they are usually better adapted to the site over the long run.

The size and shape of trees provides yet another dimension to the vertical and horizontal structure of a buffer site. This addition to structure can provide habitat that will attract the largest diversity of wildlife and bird species to the buffer. However, if specific species of birds, such as grassland birds, are desired the added vertical structure can be detrimental because of the abrupt edges that are introduced in the narrow corridor. In addition, trees provide perches for major avian predators who can wreak havoc on other birds and mammals that are 'trapped' in the corridors created by the buffers.

Trees are effective in shading streams and providing both fine and large detritus to the channel. The fine detritus provides a readily available carbon source for many invertebrates while the large woody debris provides structure to the channel that can create pools and riffles that provide stability to channel discharge and additional habitat within the channel. Shading of many streams in agricultural landscapes can help control stream water temperature and reduce the wide fluctuations in dissolved oxygen that often degrades the quality of the fishery. (Wesche et al. 1987)

Questions may be raised about the appropriateness of planting trees in previously prairie landscapes if restoration is an issue. While native prairie grasses and forbs may be desirable trees may provide functions, such as stream bank stabilization, that are needed in the present riparian corridor. This is especially true in dominantly agricultural landscapes where the hydrology of the landscape has been drastically modified and many channels are no longer in contact with their riparian zone.

Examples of different combinations of plant communities in the three zone system

Filter strip

The combination depicted in Figure 5.1 is effective in narrow buffers along intermittent and first order streams in row-crop landscapes where shading of adjacent crops is a concern or where warm-water prairie fisheries are being managed. Species options include non-native cool-season forage grasses that could provide forage harvests, or, if fenced, could be flash-grazed several times per year. Native warm-season grasses and forbs provide another option if prairie restoration and grassland bird-habitat are desired or if serious surface runoff problems exist. Pure switchgrass could be planted if there is a biomass market or if crop field slopes are $\geq 5\%$. If the stream bank is gently sloped (slopes of 6:1) and the channel not deeply incised, then either of the grass species options would be effective. As the banks become steeper (slopes 4:1) and the channel more incised, the native prairie species would provide more bank-stability control because of deeper roots.

Filter strip with bank-side shrubs

The system depicted in Figures 5.2 and 5.3 can also be used in a headwaters reach (first order stream) along a warm-water fishery in a prairie or crop-field landscape. The advantage of shrubs over trees is that they do not shade the stream and do not provide large woody debris that some landowners may not want. The shrubs would be introduced primarily to provide more stream bank stability on channels that are incised with steep banks (slopes 4:1 to 3:1). While bioengineering of stream banks is an option, it is a costly one – and, realistically, not cost-effective in most settings. Depending on the severity of the incision, the shape of the banks and the dynamics of the channel both the stream-edge zone (zone 1) and the second zone could be planted to shrubs recognizing that, over time, some of the first rows of shrubs might not hold the eroding banks and may drop into the channel. Shrubs could also be planted along streams with flashy flood flow responses where large quantities of debris could be trapped between the shrubs and the channel instead of extending out over and beyond the whole buffer. This might be an important consideration in suburban buffers where trails are a part of the buffer.

Classic multi-species riparian forest buffer

Introduction of trees into the buffer is warranted when streams are deeply incised and in the widening morphological phase of the channel-evolution model (Schumm et al. 1984), when maximum aboveground nutrient storage is desired, cooler more stable stream temperatures are desired, and/or when the land owner wants maximal structural habitat diversity, a woody fiber crop, or wind protection. Figures 5.4 and 5.5 depict this option. Trees also provide large woody debris and finer detrital inputs to the channel. Addition of the shrubs in zone 2 not only provides more structure to the system but can effectively keep woody and other flood debris within the woody zone of the buffer (Schultz et al. 1995).

This buffer combination can be modified by reversing the shrub- and tree zones with the shrubs in zone 1 adjacent to the channel. This modification would be considered along warm-water prairie streams where the landowner wants woody products and other tree benefits but does not want the tree influence on the channel, including input of large woody debris. Similarly, this combination would be effective along deeply incised channels of prairie streams where woody vegetation is desired to help stabilize the stream banks. Multiple rows of shrubs and trees would be advised especially on outside bends that move continuously and where woody plants would provide effective control until they have reached critical sizes. This combination could also be effective in urban settings with widely spaced rows between which trails could be placed. The more open wider spacing would provide a more expansive visual experience and a safety factor for trail users.

Classic three-zone riparian forest buffer

The riparian forest buffer depicted in Figure 5.6 would be used in naturally forested landscapes and in cold-water fisheries especially where small streams have significant channel slope and can benefit from large woody debris inputs. In prairie landscapes, this model is also effective in deeply incised settings with actively meandering channels. In these settings making both the grass buffer and the woody zones as wide as possible can provide long-term bank control in lieu of stream bank bioengineering.

This model also provides the largest aboveground nutrient sink and can be managed as either a two or three zone system depending on the need to remove nutrients from the site and the desire for in-stream, large woody debris. By being sensitive to microsite conditions, a wide variety of species can be planted to provide a variety of fiber product options to the landowner.

All three zones can be planted to trees in riparian pastures. In this setting surface runoff laden with sediment is not a major concern but limiting access of livestock to the stream channel providing bank stability is. With the proper spacing, such a planting could be managed as a silvopastoral system with rotational grazing to control the time livestock spend on stream banks and in the channel.

Additional general design considerations

When working in agricultural landscapes, buffer edges should be as straight or gently curving as possible to accommodate farming practices (Figure 6). Buffers will be maintained and supported by farmers if they do not see them as hindering their farming operation. This often means spending time on site with the landowner to discuss his or her field layout. Differences in buffer width to accommodate field borders can be absorbed by either the woody or the grass-forb zone, depending on the site. In those landscapes where surface runoff is more problematic than stream bank stability, maintaining a wider grass-forb zone would be favored over a wider tree-shrub zone.

In tightly meandering reaches, the buffer should be placed outside of the channel meander belt. The areas within the meanders as well as other small odd-sized areas provide ideal opportunities for developing prime wildlife habitat. Along channelized streams, buffers should be as wide as possible to accommodate the future re-meandering of the channelized reach. The landowner is more likely to allow the re-meandering if the adjacent riparian area is no longer being used for crop production. If the channelized reach is part of an organized drainage district, only grasses or grasses and forbs should be planted across all three zones to allow access for channel maintenance.

The potential also exists in any buffer planting to include even more agroforestry options. These might include planting horticultural varieties of shrubs for making jams, selling nuts, or providing ornamental woody material for home decorating. Fruit or nut trees could also be planted. These options may provide unique opportunities for small-truck farmers or hobby farmers but may not suit grain and livestock farmers who often do not have the time to invest in such plantings. The potential also exists that these more ex-

Figure 6. Examples of buffers placed outside the meander belt of the stream and smooth buffer borders to accommodate efficient farming of the upland crop fields. Areas within the meander belt are also planted to a variety of species to provide diverse wildlife habitat.

pensive plants could be severely damaged during flood events and a substantial investment in time and money could be lost.

Establishing buffers

Establishing buffers is best accomplished by working with local natural resource professionals who understand the unique requirements of each site. The underlying assumption for most of the chapter has been that recommended buffers and filters would be restored on sites that previously were intensively managed for livestock grazing or row-crop agriculture. Some general establishment considerations follow.

One of the major differences between planting buffers and planting forest plantations to maximize timber production is that woody buffer plants must compete with ground vegetation that is designed to provide the frictional surface required to slow surface runoff. As a result, when planting a riparian forest buffer into previous row-crop field soils, it is advisable to plant a cover of grasses and/or forbs that provide minimal competition of young trees and shrubs. Density of the trees and shrub planting should be such that the surface cover community remains viable during the life of the buffer or at least until a major organic (O) horizon has been developed above the mineral soil.

Site preparation before planting is imperative if good establishment is to occur and should begin the fall prior to spring planting. Previously row-cropped sites may be shallow-disked several times, followed by broadcast or drilled seeding of undercover grasses. If planting into an abandoned pasture, using herbicide to create narrow strips or individual circles into which to plant the trees and shrubs is recommended. Herbicides can be used but great care must be taken because of the neighboring stream channel. Herbicide recommendations are best obtained from local natural resource professionals. Other weed-control options such as shade cloth or plastic or organic mulches can be effective in some circumstances but may be more costly to apply to the large plantings.

If planting native prairie grasses and forbs, disking should be followed by packing to allow proper depth control for the drilled small seeds of many prairie species. Broadcasting of prairie seeds followed by rolling can also be successful. Once again, it is important to contact local professionals for the best method of establishment. Planting of woody material can be done in rows or in a random broadcast manner. Rows allow seedlings to be planted more rapidly with machines and allow easier mechanical maintenance for the first few years. If using seedlings, select individuals with large root systems that include five or more large lateral roots to assure rapid establishment and growth. Replanting holes of missing woody plants during the first two to three years should be done to establish a continuous buffer barrier. Broadcast planting of tree seeds can also be done rapidly; but then there is no initial control of density and less control of species diversity. However, broadcast seeded plantings may require less early maintenance because germination produces very dense stands of seedlings that control competing vegetation. The dense sapling stands also thin themselves effectively over time.

Maintenance of planted riparian buffers

Mowing of both the woody and grass zones can be beneficial during the first few years of establishment. Native grasses and forbs often germinate and grow more slowly than local weeds or non-native grasses. Local weeds can pose a problem for the farmer. Mowing at 40 cm to 60 cm heights, twice a season, before major annual weed seeds mature, can minimize those problems. Using a flail chopper mower instead of a brush-hog will scatter the cut debris rather than windrowing it. Windrows of debris can choke some small young prairie plants. The last mowing may be planned for late summer or fall to allow enough regrowth to occur to provide fuel for prescribed spring burns.

Mowing between woody plant rows reduces shading of young seedlings and helps identify rows. However, during very hot dry windy weather shading by taller weeds may benefit young seedlings. Mowing between woody plant rows late in the season may concentrate rodents in the seedling rows where they may girdle plants, unless herbicides have kept the seedling rows clear of weeds. Careful use of pre-emergent herbicides in the woody plant rows may help rapid establishment of seedlings.

Prescribed fire is a useful tool to help establish prairie communities. Annual spring burns during the first three or four years can produce good stands of desired grasses and forbs. Burning regimes of three to five year frequency should be used once the desired species are established. Alternating fall burns may be used to stimulate forb species over grasses that are stimulated by spring burns. Care is required to conduct these prescribed burns in the narrow zones that can lie between woody species and extensive row crops. With good planning and careful attention to weather and site conditions, these burns can be successful. As with other planting and maintenance recommendations, it is always good to check with local professionals about the best burning regime to establish and maintain the filter zone planting.

Trees can be pruned to improve form if trees will be harvested for timber products in the future. This is especially true for the higher quality hardwood species that are planted on the higher microtopographic locations of the buffer. Systematic thinning and harvesting are required to maintain an active woody plant nutrient sink. Once a given number of stems in a buffer have reached the natural production limit of the buffer, no new nutrient storage biomass is added to the trees. Instead, the trees naturally thin themselves allowing the surviving trees to continue to expand. The trees that die naturally will add their nutrients back to the buffer soil making it available for transport to the stream. A management plan should be developed with the help of a local natural resource professional to schedule these removals and to assure regeneration of the next stand.

One of the major fallacies with buffers is that they can be planted and then left on their own. For them to remain effective buffers, extensive management is required. While not intensive during each year, annual site evaluation and long-term maintenance is a must. For example, if sediment berms are found to be building up along the filter edge between the crop field and the buffer, corrective disking and replanting may be required to reduce the chance that concentrated flow pathways do not develop through the buffer.

How well do buffers carry out their prescribed functions?

Extensive studies in the Bear Creek Watershed over the past decade have demonstrated the efficacy of buffers to remove non-point source pollution from surface and ground water as well as improve the biological integrity of both the terrestrial buffer and the adjacent aquatic stream ecosystem. However, these studies have also demonstrated that there is wide variability of functions related to the variability of alluvial landscapes and that buffers may not be effective at controlling pollution in some settings.

Process-oriented studies that have been completed as part of the Bear Creek Watershed Project include: soil carbon and aggregate dynamic studies (Marquez et al. 1999; Marquez 2000), denitrification and soil microbiological studies (D.M. Haake 2003. Land use effects on soil microbial carbon and nitrogen in riparian zones in Northeast Missouri. M.S. Thesis. Iowa State University. 56 pp.; J.L. Nelson. 2003. Denitrification in riparian soils in three NE Missouri watersheds.

M.S. Thesis. Iowa State University. 44 pp.; J.E. Pickle 1999. Microbial biomass and nitrate immobilization in a multi-species riparian buffer. M.S. Thesis. Iowa State University. 86 pp.); surface runoff and infiltration studies (Bharati et al. 2002; Lee et al. 1999, 2000, 2003); above- and below-ground biomass, carbon, nitrogen and soil respiration studies (Tufekcioglu et al. 1999; Tufekcioglu et al. 2000; Tufekcioglu et al. in press); fine root decay rates (M.E. Dornbush. 2001. Fine root decay: a comparison among three species. M.S. Thesis. Iowa State University. 109 pp.); streambank erosion studies (G.N. Zaimes 1999. Streambank erosion adjacent to riparian landuse practices and stream patterns along Bear Creek, in north central Iowa. M.S. Thesis. Iowa State University. 58 pp.; Zaimes et al. 2004); and surface flow path and spatio-temporal modeling studies (Hameed 1999; D. Webber. 2000. Comparing estimated surface flowpaths and sub-basins derived from digital elevation models of Bear Creek Watershed in central Iowa. M.S. Thesis. Iowa State University. 82 pp.). A series of studies have been conducted on stream substrate and fish diversity, small mammal diversity in native grass and forb vs. cool-season grass filter portions of the riparian forest buffer and bird species composition (unpublished data).

Hydrogeological studies have also been conducted to establish the connection between surface and groundwater processes. These studies have included: geology and hydrogeology of the 1990 planting site, groundwater interaction with Bear Creek, hydrogeology and groundwater quality below a cool-season grass buffer and two multi-species buffers, assessing groundwater velocity and denitrification potential beneath a multi-species buffer using natural-gradient tracer tests and application of geophysics and innovative groundwater sampling to optimize placement of future buffers in the watershed (Simpkins et al. 2002). Several socioeconomic analyses have also been conducted in the Bear Creek Watershed and in the Mark Twain Watershed in Missouri where companion studies have been conducted (Brewer 2002; Colletti et al. 1993). Major conclusions from these studies include those listed below.

1. A 7-m wide native-grass filter can reduce sediment loss by more than 95% and total nitrogen and phosphorus and nitrate and phosphate in the surface runoff by more than 60%. Adding a 9-m wide woody-buffer results in removal of 97% of the sediment and 80% of the nutrients. There also is a 20% increase in the removal of soluble nutrients with the added width.
2. Water can infiltrate up to five times faster in restored six-year old buffers than in row cropped fields or heavily grazed pastures.
3. Soils in riparian buffers contain up to 66% more total organic carbon in the top 50 cm than crop field soils. Poplar hybrids (*Populus* spp.) and switchgrass living and dead biomass sequester 3000 and 800 kg C ha^{-1} yr^{-1} and immobilize 37 and 16 kg N ha^{-1} yr^{-1}, respectively. Riparian buffers have more than eight times more below ground biomass than adjacent crop fields.
4. Buffers show a 2.5 fold increase in soil microbial biomass and a four-fold increase in denitrification in the surface 50 cm of soil when compared to crop field soils of the same mapping unit.
5. Tracer tests and isotope evidence shows that denitrification is the major groundwater nitrate removal mechanism in the buffers.
6. Stratigraphy below buffers can determine the effectiveness of nutrient removal from shallow groundwater. With a shallow confining layer of till below a loamy root zone buffers can remove up to 90% of the nitrate in groundwater. When the confining layer is found well below the rooting zone and porous sand and gravel are found between the till and the loam, residence time and contact with roots is dramatically reduced and buffers are unable to remove much nitrate from the groundwater. The difficulty in describing the stratigraphy below buffers makes it difficult to quantify the role that specific buffers might play in remediating agricultural chemicals in groundwater.
7. Buffered stream banks lose up to 80% less soil than row cropped or heavily grazed stream banks.
8. Riparian buffers can reach maximum efficiency for sediment removal in as little as 5 years and nutrient removal in as little as 10–15 years.
9. Streamside buffers cannot remove materials from field drainage tiles. But an acre of tile–intercepting wetland can remove 20–40 tons of N over a period of 60 years.
10. Stream segments with extended buffers exhibit greater stream substrate and fish species diversity. Vole and mouse species common to the region strongly prefer riparian forest buffers with prairie grass and forb zones instead of introduced cool-season grass zones. Riparian forest buffers support five times as many bird species as row-cropped or heavily grazed riparian areas.
11. To have a significant effect on stream water quality continuous riparian buffers should be placed high up in the watershed.

12. Eighty percent of farmers and town's people agree that buffers are an effective tool for improving stream water quality. These same persons believe that water quality in streams should be improved by 40%.
13. Ninety percent of financial agents who appraise agricultural land and lend money to farmers believe buffers are a net asset when considering market (financial) and nonmarket (conservation, aesthetic, environmental, etc.) benefits and government assistance. When market benefits exclusively are considered, only 46% think that buffers are 'a net asset.'
14. Buffers enhance recreational opportunities. Fishing, hunting and watching wildlife are popular uses.

Conclusion

Based on the studies that have been conducted in the Bear Creek watershed, riparian buffers are effective at reducing the impact of non-point source pollutants on stream and ground water in a small order watershed. These studies have also demonstrated, however, that there is great variability in rates of the important biological and physical processes that are responsible for these improved conditions. Statements about general buffer effectiveness over a wide range of landscapes must be made carefully and further research must be conducted to establish the range of process rates in other ecoregions around the country and the world.

It has been difficult to demonstrate the impact of over 6 km of buffer on the water quality in Bear Creek, itself, because about a third of the watershed lies above the buffered portion of the creek and that part of the watershed contains about twice as many kilometers of channel as the buffered reach. Yet at the field scale buffer efficacy has been demonstrated. This points to the need for buffering watersheds in a systematic manner beginning in the headwaters and moving down stream.

While this chapter has only dealt with the design and role of riparian buffers in improving stream water and ground water quality, it must be realized that they are just one of a number of conservation practices that may be needed to make up a riparian management system. Stream bank bioengineering, in-stream flow control structures, constructed wetlands and intensively managed grazing systems along with a host of upland conservation practices may all be needed to round out an effective pollution control system within a watershed.

References

Bharati L., Lee K-H., Isenhart T.M. and Schultz R.C. 2002. Riparian zone soil-water infiltration under crops, pasture and established buffers. Agroforest Syst 56: 249–257.

Brewer M.T. 2002. Financial agents, water quality and riparian forest buffers. M.S. Thesis. Iowa State University, 98 pp.

Burkhart M.R., Oberle S.L., Hewitt M.J. and Picklus J. 1994. A framework for regional agroecosystems characterization using the national resources inventory. J Environ Quality 23: 866–874.

Colletti J., Ball C., Premachandra W., Mize C., Schultz R., Rule L. and Gan J. 1993. A socio-economic assessment of the Bear Creek Watershed. pp. 295–302. In: Schultz R.C. and Colletti J.P. (eds), Opportunities for Agroforestry in the Temperate Zone Worldwide: Proceedings of the Third North American Agroforestry Conference. Ames, IA, USA.

Crumpton W.G. 2001. Using wetlands for water quality improvement in agricultural watersheds: The importance of a watershed scale approach. Water Sci Technol 44: 559–564.

Dabney S.M., McGregor K.C., Meyer L.D., Grissinger E.H. and Foster G.R. 1994. Vegetative barriers for runoff and sediment control. pp. 60–70. In: Mitchell J.K. (ed.), Integrated Resource Management and Landscape Modifications for Environmental Protection. American Society Agricultural Engineering, St. Joseph, MI. USA.

Dillaha T.A., Reneau R.B., Mostaghimi S. and Lee D. 1989. Vegetative filter strips for agricultural nonpoint source pollution control. Transactions American Society of Agricultural Engineers 32: 513–519.

Dinnes D.L., Karlen D.L., Jaynes D.B., Kaspar T.C., Hatfield J.L., Colvin T.S. and Cambardella C.A. 2002. Nitrogen management strategies to reduce nitrate leaching in tile-drained Midwestern soils. Agron J 94: 153–171.

Elmore W. 1992. Riparian responses to grazing practices. pp. 442–457. In: Naiman R.J. (ed.), Watershed Management. Springer-Verlag, New York, NY, USA.

Hameed S. 1999. Spatio-temporal Modeling in an Agricultural Watershed. Ph.D. Dissertation. Iowa State University, Ames, IA. 237 pp.

Isenhart T.M., Schultz R.C. and Colletti J.P. 1997. Watershed restoration and agricultural practices in the midwest: Bear Creek in Iowa. Chapter 15. In: Williams J.E., Dombeck M.P. and Woods C.A. (eds), Watershed Restoration: Principles and Practices. American Fisheries Society.

Lee K-H., Isenhart, T.M., Schultz R.C. and Mickelson S.K. 1999. Sediment and nutrient trapping abilities of switchgrass and bromegrass buffer strips. Agroforest Syst 44: 121–132.

Lee K-H., Isenhart, T.M., Schultz R.C. and Mickelson S.K. 2000. Multispecies riparian buffers trap sediment and nutrients during rainfall simulations. J Environ Quality 29: 1200–1205.

Lee K-H., Isenhart T.M. and Schultz R.C. 2003. Sediment and nutrient removal in an established multi-species riparian buffer. J Soil Water Conserv 58: 1–8.

Lowrance R., Dabney S. and Schultz R. 2002. Improving water and soil quality with conservation buffers. J Soil Water Conserv 57: 36A–43A.

Marquez C.O., Cambardella C.A., Isenhart T.M. and Schultz R.C. 1999. Assessing soil quality in a riparian buffer strip system by testing organic matter fractions. Agroforest Syst 44: 133–140.

Marquez C.O. 2000. Soil Aggregate Dynamics and Aggregate-associated Carbon under Different Vegetation Types in Riparian Soils. Ph.D. Dissertation. Iowa State University, Ames, IA. 214 pp.

Menzel B.W. 1983. Agricultural management practices and the integrity of instream biological habitat. pp. 305–329. In: Schaller F.W. and Bailey G.W. (eds), Agricultural Management and Water Quality. Iowa State University Press, Ames, IA.
Mississippi River/Gulf of Mexico Watershed Nutrient Task Force. 2001. Action Plan for Reducing, Mitigating, and Controlling Hypoxia in the Northern Gulf of Mexico. Washington, DC, USA.
Mitsch W.J., Day J.W., Gilliam J.W., Groffman P.M., Hey D.L., Randall G.L. and Wang N. 2001. Reducing nitrogen loading to the Gulf of Mexico from the Mississippi River Basin: Strategies to counter a persistent ecological problem. BioScience 51: 373–388.
Pearcy R.W. and Ehleringer J. 1984. Comparative ecophysiology of C3 and C4 plants. Plant Cell Environ. 7: 1–13.
Palone R.S. and Todd A.H. (eds) 1997. Chesapeake Bay riparian handbook: A Guide for Establishing and Maintaining Riparian Forest Buffers. USDA For. Serv. NA-TP-02-97. USDA-FS, Radnor, PA.
Schultz R.C., Colletti J.P., Isenhart T.M., Simpkins W.W., Mize C.W. and Thompson M.L. 1995. Design and placement of a multi-species riparian buffer strip system. Agroforest Syst 31: 117–132.
Schumm S.A., Harvey M.D. and Watson C.C. 1984. Incised channel morphology, dynamics and control. Water Resour. Publ., Littleton, CO.
Simpkins W.W., Wineland T.R., Andress R.J., Johnston D.A., Caron G.C., Isenhart T.M. and Schultz R.C. 2002. Hydrogeological constraints on riparian buffers for reduction of diffuse pollution: examples from the Bear Creek Watershed in Iowa, USA. Water Sci Technol 45: 61–68.
Tufekcioglu A., Raich J. W., Isenhart T.M. and Schultz R. C. 1999. Root biomass, soil respiration, and root distribution in crop fields and riparian buffer zones. Agroforest Syst 44: 163–174.
Tufekcioglu A., Raich J.W., Isenhart T.M. and Schultz R.C. 2000. Soil respiration in riparian buffers and adjacent croplands. Plant Soil. 229: 117–124.
Tufekcioglu A., Raich J.W., Isenhart T.M. and Schultz R.C. In press. Biomass, carbon and nitrogen dynamics of multi-species riparian buffer zones within an agricultural watershed in Iowa, USA. Agroforest Syst.
USDA-NRCS. 2003. National Handbook of Conservation Practices. United States Department of Agriculture, Washington, D.C.
Welsch D. J. 1991. Riparian Forest Buffers: Function and Design for Protection and Enhancement of Water Resources. NA-PR-07-91. USDA Forest Service, Radnor, Pennsylvania.
Wesche T.A., Goertler C.M. and Frye C.B. 1987. Contribution of riparian vegetation to trout cover in small streams. North Am J Fish Manage 7: 151–153.
Zaimes G.N., Schultz R.C. and Isenhart T.M. 2004. Stream bank erosion adjacent to riparian forest buffers, row-cropped fields and continuously-grazed pastures along Bear Creek in Central Iowa. J Soil Water Conserv 59: 19–27.

Agroforestry Systems **61:** 51–63, 2004.

Short-rotation woody crops and phytoremediation: Opportunities for agroforestry?

D.L. Rockwood[1,*], C.V. Naidu[1], D.R. Carter[1], M. Rahmani[2], T.A. Spriggs[3], C. Lin[4], G.R. Alker[5], J.G. Isebrands[6] and S.A. Segrest[7]
[1]*School of Forest Resources and Conservation, University of Florida, Gainesville, FL, 32611-0410, USA;* [2]*Food and Resource Economics Department, University of Florida, Gainesville, FL, USA;* [3]*CH2M Hill, Tampa, FL, USA;* [4]*Ecology & Environment, Tallahassee, FL, USA;* [5]*WRc, Swindon, UK;* [6]*Environmental Forestry Consultants, New London, WI, USA;* [7]*Common Purpose Institute, Temple Terrace, FL, USA;* **Author for correspondence: e-mail: dlrock@ufl.edu*

Key words: Eucalyptus, Fuelwood, *Populus*, Riparian buffers, Shelterbelts

Abstract

Worldwide, fuelwood demands, soil and groundwater contamination, and agriculture's impact on nature are growing concerns. Fast growing trees in short rotation woody crop (SRWC) systems may increasingly meet societal needs ranging from renewable energy to environmental mitigation and remediation. Phytoremediation, the use of plants for environmental cleanup, systems utilizing SRWCs have potential to remediate contaminated soil and groundwater. Non-hyperaccumulating, i.e., relatively low contaminant concentrating, species such as eucalypts (*Eucalypts* spp.), poplars (*Populus* spp.), and willows (*Salix* spp.) may phytoremediate while providing revenue from fuelwood and other timber products. Effective phytoremediation of contaminated sites by SRWCs depends on tree-contaminant interactions and on tree growth as influenced by silvicultural, genetic, and environmental factors. Locally adapted trees are essential for phytoremediation success. Among the different agroforestry practices, riparian buffers have the greatest opportunity for realizing the SRWC and phytoremediation potentials of fast growing trees. Agroforestry that combines SRWC and phytoremediation could be an emerging holistic approach for sustainable energy, agricultural development, and environmental mitigation globally.

Introduction

Short rotation woody crop (SRWC) systems involve the intensive culture of fast growing hardwoods at close spacing for rotations of 10 years or less, primarily for fuelwood, but also for some other traditional timber products. Phytoremediation broadly refers to the use of plants for environmental remediation. It has been suggested that SRWCs have potential for addressing the escalating demand for fuelwood as well as for phytoremediation (e.g., Glass 1999; Alker et al. 2002; Rockwood et al. 2004). For the perspective of examining this potential, this chapter will first present briefly the gravity of these two major problems, and then review research results on SRWCs and what we subsequently term dendroremediation, i.e., phytoremediation utilizing trees, in various agroforestry situations globally, with emphasis on Florida. Following that, we will discuss opportunities for incorporating agroforestry into SRWC and phytoremediation and consider prospects for applying dendroremediation in agroforestry systems.

At present, biomass is the world's fourth largest energy source, providing about 14% of the total. In industrialized countries, fuelwood averages only 3% of total energy consumption (Hall et al. 1999). Wood energy accounts for about 4% of the total primary energy in Europe and countries such as Australia, Canada, Japan, New Zealand, and the United States (R. Van den Broek pers. comm. 1997), although fuelwood use is 14% in Austria and more than 17% in Sweden and Finland. Household fuelwood use is of major importance in Europe, especially in rural Sweden, France, and Austria where small-scale biomass boilers are extensively used. Strong incentives in these countries for heat production have resulted in establishment of several large biomass fired heating plants and modification of fossil fuel plants for biomass use (Gustavsson et al. 1995; Gustavsson and Börjesson 1998).

In developing countries fuelwood constitutes 33% of total energy use (WEC-FAO 1999). More than one-third of the world's population depends on wood for cooking and heating. Of all the wood consumed annually in developing countries, 86% is used for fuel, and of that at least half is used for cooking. In many regions, wood is the only locally available energy source and requires little capital expenditure for acquisition or conversion to useful forms of energy. Africa relies heavily on wood for meeting basic energy needs, especially in tropical regions, with countries such as Angola, Chad, Ethiopia, and Tanzania at more than 90%. Fuelwood is an important source of energy in south and southeast Asia, with Bhutan, Cambodia, Myanmar, and Nepal supplying more than 80% of their energy from wood (RWEDP-FAO 1999). About 95% of families in rural Bangladesh use biomass for cooking purposes.

While environmental concerns such as CO_2 emissions and global warming have encouraged industrialized countries to use modern biomass energy systems, socioeconomic considerations, such as equity for low-income communities and their necessity for fuel, have encouraged developing countries to improve traditional biomass energy sources. The fuelwood crisis in developing countries has necessitated the need for SRWCs, not only to meet fuelwood demands but also to protect existing forests from further destruction and to maintain ecological balance (Pathak et al. 1981). In general, improvement of SRWC productivity depends on screening of species, density, fertilization, irrigation, stand management and harvesting techniques (S. Thaladi pers. comm. 1986). Hence, research and development efforts are underway for increasing fuelwood production of tree species that are traditionally used for this purpose and for selecting new ones. For example, farmers in India, Kenya, and Rwanda are turning to agroforestry for fuelwood and improving the environment.

Research on short-rotation woody crops

SRWCs are a carbon-neutral source of energy and do not contribute to CO_2 enrichment of the atmosphere (Tuskan and Walsh 2001). Use of SRWCs for cofiring, i.e., cocombustion with fossil fuels, may be a cost-effective, simple 'fuel switching strategy' to a more environmentally friendly fuel source in developed countries (Segrest et al. 2001).

World

More than 100 species identified as SRWCs for biofuels (El Bassam 1998) could be used for nearly 10 million ha worldwide. In India, recommended SRWCs, depending on rainfall, soils, and other factors, include *Acacia auriculiformis, A. catachu, A. nilotica, Albizzia lebbeck, A. procera, Azadirachta indica, Casuarina equisetifolia, Leucaena leucocephala, Melia azedarach, Parkinsonia aculeata, Prosopis cineraria, P. juliflora*, and *Terminalia arjuna* (C.V. Naidu and P.M. Swamy pers. comm. 1993). In Bangladesh, species such as *Acacia auriculiformis, A. mangium, Senna siamea, Eucalyptus camaldulensis* and *Leucaena leucocephala* grow well (S.N. Hossain and M. Hossain pers. comm. 1990; 1997). Worldwide, *Eucalyptus* spp. constitute about 38% of all SRWCs, and hardwoods make up about 63% of all plantations (FAO 2003).

In temperate regions, poplar (*Populus* spp.), willow (*Salix* spp.), and black locust (*Robinia pseudoacacia*) predominate as SRWCs. Poplars and willows have been planted for biomass production in Canada, the Netherlands, and Northern Ireland with up to 45 Mg ha^{-1} harvested at the end of the first three-year cycle. The climatic and edaphic conditions of southern Quebec, Canada, are conducive to SRWCs as two willow species produced between 15 and 20 dry Mg ha^{-1} yr^{-1} (Labrecque et al. 1993; 1994; 1997). Fertilization of willows with wastewater sludge significantly increased biomass productivity on treated plots and recycled an undesirable residue (Aronsson and Perttu 2001; Hytönen 1994; Labrecque et al. 1997). Willows have been grown as SRWCs in Northern Ireland since 1976. Black locust, which grows very rapidly, survives droughts and severe winters, tolerates infertile and acidic soils, and is widely used for erosion control and reforestation, may be well suited for SRWCs and commercial energy production in the United States, Europe and Asia (R.O. Miller et al. pers. comm. 1987). In Sweden, $\sim$16 000 ha of dedicated willow coppice are currently managed for fuelwood for district heating systems, accounting for 50% of the fuel supply.

Ratification of the Kyoto protocol in 2002 legally obligated the United Kingdom government to reduce CO_2 emissions 12.5% below the 1990 levels by 2010. Its implementation strategy and incentives include 1) Tradable Renewable Obligation Certificates (ROCs) for generators, Climate Change Levy tax exemptions for industrial and commercial users, Community and Household Capital Grants for nonprofit organizations and household implementers of renewable energy in general (wind, hydro, solar, biomass, etc), 2) Bioenergy Capital Grants for the installation of boilers and power plants, Bioenergy Infrastructure Scheme to harvest, store, and supply energy crops and forestry wood fuels to end-users of bioenergy (agricultural, forestry and wood wastes in addition to bioenergy crops), and 3) The Energy Crop Scheme of planting grants and funding to set up producer groups for energy crops. In order to minimize transportation costs associated with energy crops, decentralized power plants (small-scale electrical, thermal and combined heat and power) are encouraged. Several have already been commissioned and funded by the Bioenergy Capital Grants Scheme since 2001. In addition, current Renewable Obligations state that existing power stations powered by fossil fuel and biomass cofiring must source 75% of their biomass from dedicated energy crops starting April 2006. Currently 167 such stations generate 877 670 kW from biomass. If 75% of this capacity were generated from SRWCs, ~1 million dry Mg of wood chips would be required each year. As a result of these and other drivers, markets for renewable energy crops are developing rapidly.

In developed countries, energy from wood in direct costs is presently more expensive than from fossil fuels (Horgan 2002). The costs of SRWC fuelwood range from \$25 to 40 m^{-3}, although this cost might be reduced by at least 50% by technological improvements. For example, SRWC fuelwood in Brazil has been delivered for \$12 m^{-3}, or \$1.50 GJ^{-1}. Special subsidies and incentives are often needed to ensure energy independence and reduce negative externalities. Fuelwood is typically a site- and situation-specific energy option.

United States

Cofiring up to 5% SRWCs may be a very cost-effective means of creating renewable energy while using existing power plant infrastructure in the United States. Many leading policymakers believe that the model of future air quality federal regulations will be the Clean Air Act of 1990 (CAA) which introduced the concept of environmental capitalism. The CAA, instead of requiring that every power company put pollution control equipment on their smokestacks, instructs companies to reduce pollution in any way that works. This created market competition to find cheaper ways to reduce emissions and led to cheaper prices for coal and pollution control equipment. A small but significant segment of electricity customers may be willing to pay a small premium (5% to 10%) for 'Green Energy.'

The magnitude of CO_2 emissions in the United States, the world leader in C emissions, illustrates the opportunity for renewable energy sources (Segrest 2003). The 12 southern states of the United States collectively emit 39% of the United States' total carbon annual emissions of 1.36 Gg C. Florida is the eighth ranked state with 0.06 Gg C emissions per year (Segrest 2003). Of the fuels used by Florida's utilities (coal = 36.3%, natural gas = 23.0%, petroleum = 19.9%, nuclear = 16.9%, hydroelectric = 0.1%, other = 3.9%), 79.2% come from ~\$3 billion of fossil fuels, almost all of which are imported. The use of SRWCs, which contain almost no sulfur and about 50% of the nitrogen content of coal, would considerably reduce sulfur dioxide (SO_2) and nitrous oxides (NO_x) emissions, which cause acid rain and smog, respectively. In recent tests at a large power plant in central Florida, cofiring 2.5% SRWCs reduced NO_x emissions ~7% (Segrest et al. 2001).

Hybrid poplar, willow, and switch grass (*Panicum virgatum*) are now the most promising energy crops in the temperate United States, with *Eucalyptus* and sugarcane preeminent in semitropical and tropical regions. Several pulp and paper companies have commercialized SRWCs.

Willow SRWC research at the State University of New York (SUNY) is focused on optimizing commercial biomass and bioproduct production systems. SUNY has a willow-breeding program that produces new hybrids for biomass production and dendroremediation (Kopp et al. 2001). SRWC willow crops in the United States have, however, not yet reached the significance that they have in Sweden (Aronsson and Perttu 2001).

Hybrid poplar plantations in the Pacific Northwest and Minnesota have achieved considerable size and importance (Stanton et al. 2002). In the Pacific Northwest, ~20 000 ha are grown primarily on former agricultural lands in rotations of six to eight years for pulpwood, with ~25% of the total biomass used as fuel. Market opportunities exist for solid wood products; moreover, environmental applications such as pollution abatement have been implemented by many industries and communities. In Minnesota, electricity-generating companies are planting 20 000 ha to poplars.

An SRWC pilot test project is currently underway on a 50 ha clay settling area (CSA) near Lakeland, Florida, which is representative of ~160 000 ha of phosphate mined lands and unimproved pastures in central Florida. The project 1) documents 'real world' SRWC costs and yields, 2) develops guidelines for establishing and managing SRWCs on CSAs, and 3) evaluates genetic, cultural, and harvesting options for further improving the cost effectiveness of using cottonwood (*Populus deltoides*, *PD*), *Eucalyptus amplifolia* (*EA*), and *E. grandis* (*EG*) (Segrest et al. 2001). The cogongrass (*Imperata cylindrica*) infested site was cleared and treated with herbicides before bedding and planting. The site productivity has been influenced by site conditions, planting date, and planting method. Genotypes within species, fertilization, vegetation control, site preparation, planting stock quality, and method and time of planting have all been critical elements. *PD* was especially impacted by weed competition and drought. Clonal differences in storage/drought tolerance were observed as one clone failed completely. Surviving trees grew more than 2.5 m tall in less than five months. A clonal nursery planted under slightly better moisture conditions was less affected by post-planting drought. *EA* and *EG* in the studies, and particularly the over 500 000 *EA* and *EG* seedlings in the pilot tests, were much more vigorous than *PD* and competed more effectively with cogongrass.

The cost of *EA*, *EG*, and *PD* grown in central Florida, including delivery, has been estimated between \$1.67–2.52 GJ^{-1}. SRWC cost competitiveness will be very dependent on harvesting costs, as ~66% of the cost of delivered energywood may be due to harvesting with conventional feller-bunchers. Double-row harvesters such as the Claas, a high capacity forage harvester, have great promise for reducing harvesting costs. Double row planting, however, while important for yield enhancement and harvesting cost, is challenging on bedded CSAs.

Availability of land with low opportunity cost for high yield biomass crops is a major condition for establishing a biomass-based energy system. Florida, potentially one of the leading areas in the United States to produce biomass as a source for renewable energy, has such land for developing biomass crops, including more than 70 000 parcels of forest-, agricultural-, reclaimed-, or industrial land with a total area of 2.5 million ha in the peninsular region of the state (Rahmani et al. 1999). These parcels are used as crop- and pastureland, natural- or planted forest, reclaimed land, or for power plants, waste treatment facilities, food processing plants, other processing facilities, and transportation terminals (Table 1). Crop- and pastureland, with an area of 877 000 ha, has the highest potential for biomass production, estimated to be more than 5 Gg yr^{-1}. Potential biomass production for all types of land-use in peninsular Florida exceeds 13 Gg yr^{-1}. Some 81 000 ha of mined lands and more than 161 000 ha of pastureland in central Florida are 'marginal lands' with low opportunity cost and conversely high potential for biomass production (Stricker et al. 1995). Use of these lands for biomass energy production must be more enticing than existing uses in order for these lands to be allocated to SRWCs.

Successful demonstration of SRWC production and co-firing in Florida could lead to SRWC development in the Gulf Coast region and similar environments. Florida has several coal-fired power plants that can co-fire biomass at a fraction of the cost of other renewable energy options. Building a new, stand-alone biomass energy power plant does not currently compete with natural gas generation options. The total cost of growing, harvesting, transporting, and co-firing a SRWC must be at a cost reflecting a slight premium above the cost of coal, which, delivered to electric utilities in Florida, currently ranges from \$1.42 to \$1.66 GJ^{-1}. If Florida's electric utilities used SRWCs to generate just 2% of total electricity produced, a new farming industry with an economic impact of over \$100 million yr^{-1} would be created.

Review of dendroremediation

World

Many opportunities exist worldwide for linking dendroremediation with tangible biomass economic opportunities such as bioenergy, solid wood products, and reconstituted products. More intangible opportunities are soil erosion control, uptake of soil nutrients and agricultural and industrial wastes, carbon sequestration, and wildlife habitat (Licht and Isebrands 2003). Landfill-leachate management using land-treatment

Table 1. Potential biomass producing sites and production[a] in Florida by land use types.

Land-use types	Number of parcels larger than 5 ha	Area (1000 ha)	Potential biomass production (1000 Mg yr^{-1})
Crop & pasture	15 841	877	5 021
Pine forest	17 142	522	2 931
Hardwood plantations	8 332	462	2 796
Mixed hardwoods, dead trees	15 499	278	1 628
Oak, exotic forest	7 584	195	1 024
Reclaimed land	667	29	167
Industrial, other	5 200	94	NA
All Types	70 265	2 457	13 567

Source: Soil Conservation Service (1993), State Soil Geographic Data Base (Statsgo), Data Users Guide, SCS Misc. Pub. 1492.
[a]Estimates of biomass productivity based on soil types.

systems has been investigated and practiced in several countries including Sweden and Finland (Ettala 1987; K. Hasselgren pers. comm. 1989), Canada (Gordon et al. 1989), United States (Licht 1994), and Hong Kong (Wong and Leung 1989).

In the United Kingdom, a five-year project investigating the management of landfill leachate using short rotation coppice detailed the fate of leachate chemical components after application to willow coppice (Alker et al. 2002). Yields were statistically higher in leachate-irrigated areas in response to the improved nutritional status of the soil. Leachate-irrigated-, water-irrigated-, and control trees produced 9, 7, and 5 Mg ha^{-1} yr^{-1} drymatter (biomass), respectively. Based on mass budgets, 92% of the chlorine (Cl) applied in leachate leached from the soil and 41% of N applied was removed in harvested wood chips, but rainwater diluted Cl concentration in drainage water to ~21% of the applied leachate concentration and 21% of the applied N drained from the site. The concentrations of all chemical components, including heavy metals, in wood chips after leachate irrigation were not significantly different from those from controls. The project recommended suitable loading rates for a number of leachate chemical components. The leachates used were of relatively low strength, so the hydraulic loading rate could be high without exceeding these chemical loading recommendations. In order not to exceed chemical loading rates for extremely concentrated leachates, the hydraulic loading rate may need to be too low for the method to be economically or logistically feasible. In this case, land availability or establishment costs may result in the system being less attractive than alternative options. Therefore, leachate management is probably best suited to lower strength leachates.

This project suggests that the use of SRWC for leachate management, reuse, and recovery is a cost effective alternative to conventional methods of on-site leachate treatment or tankering to sewage treatment works. By only applying leachate during the growing season, full advantage can be taken of the plant uptake capacity of the system. This can be achieved by providing winter storage in lagoons or within the landfill or using an alternative method of treatment during winter. While the generation of a commodity crop is likely to be a secondary consideration in these systems, chemical analysis did not identify any enrichment of undesirable contaminants in the wood produced at the leachate loading rates used, suggesting that its utilization as a biofuel is not compromised.

A project funded by the European Commission between 1999 and 2003 at field sites in Northern Ireland, France, Sweden and Greece also reported improved willow yields after application of sewage effluent, and high nutrient removals from effluent, equivalent to the efficacy of conventional sewage tertiary treatment systems (S. Larsson pers. comm. 2002). Groundwater quality was not affected by effluent application compared to the controls, which were not irrigated or fertilized. Heavy metals in the soil were low, and any changes observed were more or less independent of applied wastewater rates. Heavy metal contents of plant tissues were not correlated with soil or wastewater concentrations, suggesting that wastewater application does not enrich the heavy metal content of the wood. Balancing the heavy metal application rate with the metals removed by the crop is recommended, however, to avoid the accumulation of metals in the soil. The copper (Cu), zinc (Zn), lead (Pb), and cadmium (Cd) exports in harvested stems ranged from 0.03 to 0.11, 0.19 to 0.92, 0.007 to 0.71, and 0.004 to 0.06 kg ha^{-1} yr^{-1}, respectively.

Pilot-scale trials at Kågeröd, Sweden, concluded that wastewater irrigation greatly increased the growth of willows and that the removal rate of N and P in the willow-soil system was higher than conventional nitrification/denitrification and phosphate chemical precipitation treatment processes (Hasselgren 1998). Wastewater application up to three times the evapotranspiration rate from the plantation did not deteriorate the quality of the superficial groundwater. The pilot trials were so successful that the municipality scaled up the trials to a full-scale application. A 13-ha plantation of willow SRWC is currently being used to treat 40 000 m^3 yr^{-1} secondary effluent, which is 12% of the wastewater produced by the 1500 occupants of the town. The system may eventually accept the entire 135 000 m^3 yr^{-1} flow from the sewage treatment works.

United States

The United States has nearly 200 demonstration or commercial phytoremediation sites (Glass 1999). During the past 15 years, field studies and commercial applications have identified the potential of SRWCs for dendroremediating nutrient, metal, and hydrocarbon contamination (Table 2). An early application of hybrid poplars was for trichloroethylene (TCE) dendroremediation at Oconee, IL, in 1989. One of the earliest studies of the potential of *Eucalyptus* dendroremediation addressed stormwater cleanup in Tampa, FL, in 1993. After reviewing examples in Illinois and Florida, we summarize SRWC improvement programs that contribute to enhancing dendroremediation.

A study at LaSalle, IL, provides evidence of the importance of various factors in establishing dendroremediation systems (Rockwood et al. 2004). Unrooted willow cuttings planted in late March 2002 had only 10.9% survival due to exceptionally late and severe freezes, and the only willow clone with more than 50% survival grew poorly. In comparison, rooted poplar cuttings had 47.8% survival at four months, but several poplar clones all survived and grew rapidly in the perchloroethylene (PCE)-contaminated clay soil. At 1.5 years old, the fast growing poplar clones were over 4 m tall and 8 cm in diameter. After 1.5 years in a companion study that planted 2 m long whips in 60 cm plastic pipes on a former PCB incineration disposal site, a hybrid poplar clone was more than 3 m tall and 2 cm in diameter, and willow clones were as large as 3 m tall and 1 cm in diameter.

In Florida, several species have shown potential for nutrient-, metal-, and hydrocarbon dendroremediation (R.W. Cardellino, MS thesis, Univ. of Florida, 2001; Rockwood et al. 2001; 2004). For example, baldcypress (*Taxodium distichum*) and castor bean (*Ricinus communis*) are promising candidates for Cu remediation, accumulating over 15 mg kg^{-1} in stem biomass. SRWCs such as *PD* (*Populus deltoides*) and *EG* (*Eucalyptus grandis*) have a high demand for water and nutrients.

PD and *Eucalyptus* species have wastewater-dendroremediation potential (Pisano and Rockwood 1997; Rockwood et al. 1996). Three of 10 *PD* clones in a sewage effluent sprayfield at Tallahassee, FL, (Table 2) were particularly productive and recommended to optimize biomass production and nutrient uptake. While three *Eucalyptus* species tolerated flooding in a stormwater dry retention treatment pond at Tampa, FL, (Table 2), *EG* was the most vigorous after 3.75 years, reaching an average height of 9.4 m, with biomass allocations to stemwood, stembark, branches, and foliage of 44%, 10%, 30%, and 15%, respectively. Relative concentrations of N, P, and K in *EG* plant tissues were typically foliage $>$ stembark $>$ branches $\gg$ stemwood. Although water use could be as high as 1600 mm annually, the exact amount of water and nutrients taken up by *EG* depends on climate, tree vigor, and the timing and extent of stormwater applications. One *EG* clone appeared superior for stormwater dendroremediation in central Florida.

After three years, *EG* biomass yields were more than twice those of *PD* in response to application of sewage effluent, compost, and/or mulch in a study at Orlando (Rockwood et al. 2004). This superior productivity of *EG* has obvious potential environmental values, as well as economic implications. *EG* plantations can increase water loading and reduce nutrient leaching, even when compost amendments are applied. The trees can effectively increase the amount of effluent that can be applied. *EG* may reduce N and P leaching by up to 75% when water only is applied and 85% when mulch is added for weed control. Even with incorporation of up to 12 Mg ha^{-1} of compost, *EG* may still reduce N leaching by some 50%.

Table 2. Representative field studies and commercial applications involving SRWCs at contaminated sites in the United States.

Study location	Estab. date	Contaminant	Species
Oconee, IL	1989	TCE	Hybrid poplar
Tallahassee, FL	4/92	Nutrients	*PD*
Cranbury, NJ	1992	Nitrogen	Hybrid poplar
Tampa, FL	5/93	Nutrients	EA, *EC*, *EG*
Lafayette, LA	1995	TCE	Hybrid poplar
Wayne, MI	1997	TCE	Hybrid poplar
Finley, OH	1997	TCE	Hybrid poplar, willow
Orlando, FL	4/98	Nutrients	*PD*
Argonne, IL	1999	TCE, Tritium	Hybrid poplar, willow
Quincy, FL	4/00; 3/01	As	*PD*
Archer, FL	4/00; 3/01	As	*PD*
St. Augustine, FL	8/00	Toluene	*PD*, *EA*
Orlando, FL	4/02	TCE, PCE	*PD*, Poplar
LaSalle, IL	3/02-4/03	TCE, PCE	Poplar, willow
Gulfport, MS	3/03		*PD*
Edgewood, MD		Chlorinated Solvents	Hybrid poplar
Ft. Worth, TX		Chlorinated Solvents	*PD*
New Gretna, NJ		Chlorinated Solvents	Hybrid poplar
Ogden, UT		Petroleum Hydrocarbons	Hybrid poplar
Hill AFB, UT	2001	TCE	Fruit trees
Aiken, SC		TCE	PD, Loblolly pine
Anderson, SC		Heavy Metals	Hybrid poplar
Beaverton, OR		Metals, Nitrates	Cottonwood
Texas City, TX		PAHs	Mulberry
Amana. IA		Nitrates	Hybrid Poplar
Palo Alto, CA	1997–98	As, VOCs, Metals	Eucalyptus, Tamarisk
Aurora, IL	1999	TCE	Hybrid poplar
Woodburn, OR		Nutrients	Hybrid poplar

Source: Compiled from Douchette et al. 2004; EG Gatliff pers. comm. 2003; Glass 1999; USEPA 2003; Wilde et al. 2004).

An economic analysis (Rockwood et al. 2004) indicated the importance of input costs, harvest prices, progeny, harvest timing (rotation age), and the decision to harvest and replant, or harvest subsequent coppice yields. *EG* progeny 3309 had significantly higher yields. At stumpage prices of \$10 green Mg^{-1} for mulchwood and using a 4% real discount rate, *EG* 3309 'without coppice' was most profitable with a land expectation value (LEV) of \$7300 ha^{-1}, equal annual equivalent (EAE) of \$292 ha^{-1} yr^{-1}, and internal rate of return (IRR), assuming sunk land costs, of 29%. The optimal rotation age was 33 months. Overall returns for the average *EG* progeny were considerably less, but still demonstrated LEVs in excess of \$2471 ha^{-1} (with coppicing). By any reasonable standard, *EG* production for mulchwood appears to be profitable as compared to alternative agricultural uses of the land. Deciding between coppicing and harvesting/replanting depends on relative seedling and coppice yields, and seedling establishment costs and output prices. Mulchwood is presently the most likely commercial use of *EG*, although increasing demand is expected for energywood by electric utilities (Segrest et al. 2001).

However, the social value of *EG* for wastewater dendroremediation includes both returns from biomass harvest as well as environmental values associated with reduction of nutrient leaching. The value of this service is difficult to quantify but might be estimated using known costs for alternative methods of wastewater remediation. Such additional benefits make the overall value of *EG* on these sites extraordinarily appealing, especially if private landowners are able to capture those environmental benefits via some form of subsidy program. Optimal production decisions for *EG* crops would be influenced by the incorporation of environmental benefits, as, e.g., the decision to harvest *EG* at 33 months to provide maximum benefits from biomass (e.g., mulchwood) harvesting might be lengthened (or shortened) depending upon relationships between the age of *EG* plantings and nutrient uptake potential.

Dendroremediation may effectively address contamination associated with the widespread manufacturing and use of wood treated with preservatives such as chromated copper arsenate (CCA). Growth and arsenic (As) uptake by *PD* at a CCA contaminated site in Archer, FL, varied with season, plant tissue, and clone (R.W. Cardellino, MS thesis, Univ. of Florida, 2001; Rockwood et al. 2004). Within tree variation for As concentration was highest in the leaves, next highest in branch bark, followed by stem bark, then branch wood, and lowest in stem wood. Stem position was significant for leaf concentration as leaves from the lowest part of the crown had more than twice the concentration of leaves from the upper crown. Branch wood, stem bark, and stem wood concentrations changed little with crown or stem position. Variability over time for As concentration in leaves was relatively small but seasonal as As concentrations in middle- and upper-crown leaves were both about 3 mg kg^{-1} less in May than in October. Variation among clones appears large for the tree components that have the highest As concentrations. A threefold difference was common between the clones with the lowest and highest concentrations in leaves. Twofold and larger differences in branch bark concentrations were common. The ranges for branch wood, stem bark, and stem wood concentrations were of similar extent. Based on cumulative results, several high and low As concentrating *PD* clones are notable. Total As uptake by *PD* will be a balance between biomass production and As concentration. Annual harvesting of *PD* clones will likely result in the highest As removal. To maximize As removal, high density planting, e.g., 1×1 m, of high concentrating clone(s) should be practiced.

A dendroremediation study at St. Augustine, FL, uses highly contaminated dual phase extraction wells to pump toluene contaminated groundwater for treatment and to contain the plume. After 29 months, 0.15 m diameter plastic 'training' tubes inhibited above-ground growth of *PD* and *EA* and presumably root growth and access to groundwater (Rockwood et al. 2004). *PD* and *EA* were equally vigorous, but *EA* had higher survival (85% for *EA* vs. 77% for *PD*), and an *EA* progeny was the most productive genotype. Toluene was detected in leaf and branch samples of *EA* but not *PD*. Tree roots subsequently reached groundwater elevations, as a monitoring well, which previously had not detected toluene, but has recently seen substantial increases in contaminant concentrations, presumably due to contaminant movement. Soil hydraulic conductivities have recently been increasing and are expected to also enhance the remedial efficiency of the dual-phase-extraction system.

At a TCE and PCE contaminated study in Orlando, FL, chlorinated volatile organic compounds (CVOCs) are present in soil and shallow groundwater. The CVOC plume extends from 1.2 to 13.7 m below land surface in the shallow and deep aquifers and flows toward a lake ~152 m away. Dendroremediation was designed and installed in March 2002 to 1) enhance natural attenuation of the CVOCs, 2) promote rhizodegradation of the CVOCs, 3) remediate the fraction of the CVOCs taken up with the transpired water by phytodegradation and phytovolatilization processes (Nzengung et al. 2004), and, secondarily, to use the phreatophytes to minimize transport of contaminants to the lake. Whips of poplar Clone DN34, *PD* clones, and one local willow species (*Salix caroliniana*) were densely and deeply planted (0.6 m above water table) over the 1.2 ha plume. The *PD* whips averaged 5.5 m in height after six months (Rockwood et al. 2004). Locally adapted *PD* clones typically grew more than DN34, which is widely planted for dendroremediation in more temperate regions. DN34 had proportionately more root biomass, though rooting was limited due to a higher water table. In a drier location on site, root development extended beyond 1.8 m and was a larger allocation of DN34's biomass, compared to the trees planted in moist soil conditions.

One year after implementation, the effects of dendroremediation were already positive for uptake and plume control. No discernible spatial trends were evident, but tissue samples clearly indicated uptake of the CVOCs PCE, TCE, and cis-1,2-dichloroethene (DCE). These baseline data were not surprising since the young trees were still establishing root systems. However, some roots are beginning to impact the shallow groundwater aquifer contaminated by the CVOC plume and to uptake the CVOC contaminants. Plant tissue sampling after two growing seasons shows uptake of PCE, TCE, and DCE in root, stem, and leaf tissues.

The phreatophytes are expected to evapotranspire 76 cm yr^{-1} of groundwater due in part to the deep, cylindrical rooting zones created by filling the 1.8 m deep auger holes with composted media, which may also enhance rhizodegradation. To ensure that the rhizosphere enhancements are improving the bioremediation potential of the plant microbial system, monitoring involves the analysis of microbial membrane lipids, specifically phospholipid fatty acids (PLFA). As the trees become established and impact groundwater, PLFA analysis will be used for phylogenetic identification, monitoring the physiological status of the microbes, and detecting changes in microbial community structure.

Genotypes derived from traditional tree improvement programs can be very helpful in developing planting stock for dendroremediation systems. Such genotypes, typically genetically superior for growth, survival, and pest incidence, bring many of the characteristics needed for maximum remediation. For example, the *PD* clones in the above described Archer, St. Augustine, and Orlando studies were derived from a US Forest Service program, but these clones may be superceded by the best 100 of over 1000 new clones that surpassed standard clones in a Southeast-wide *PD* improvement program led by Mississippi State University (Land et al. 2001). If so, more productive *PD* clones may soon be available for dendroremediation in the Southeast.

Poplar breeding programs in the US include the North Central US Populus Regional Testing Program (Riemenschneider et al. 2001) and a private breeding program in the Pacific Northwest (Stanton et al. 2002). These programs are producing superior plant materials for SRWC use that are just beginning to be used regionally for dendroremediation. Most of the clones in the US that are currently used for dendromediation are older clones produced in the northeastern United States and Canada in the 1990s or clones naturalized from Europe (Dickmann et al. 2001).

Incorporating SRWC and phytoremediation into agroforestry systems

While SRWCs and dendroremediation may often be justified as independent ventures, combining the two can increase their economic feasibility, and incorporating both into agroforestry systems has promise for further enhancing annual income possibilities, a critical aspect for acceptance. The generally favorable cost of dendroremediation with SRWCs, e.g., for remediating landfill-leachate, can be augmented by nearby fuelwood or other timber markets when the planting is sufficiently large to produce merchantable biomass quantities. Similarly, dendroremediating SRWCs in an agroforestry format will have periodic value, when planted on a large scale in a marketable area, that can add to the more frequent returns from the agronomic component.

Agroforestry ecosystems have light, water, and nutrient resource pools that can be shared in horizontal, vertical, and time dimensions involving competitive, differential, and complementary aspects (Buck 1986). In an agroforestry system with dendroremediating SRWCs, resource sharing may be especially competitive horizontally and vertically due to the rapid growth of the SRWCs, necessitating careful design and proper and timely management of the system. Five agroforestry systems identified by Garrett and Buck (1997) represent opportunities for incorporating SRWC and dendroremediation. Of tree-agronomic crop systems (alleycropping and intercropping), riparian vegetative buffer strip systems, tree-animal systems (silvopasture), forest/specialty crop systems (forest farming), and windbreak systems (shelterbelts), riparian systems and windbreaks have the greatest potential (Table 3).

SRWCs in the riparian component of agroforestry systems dendroremediate as well as provide numerous other benefits. Poplar trees effectively lower subsurface NO_3-N concentrations and stabilize degraded agricultural streambanks while growing rapidly (O'Neill and Gordon 1994). SRWCs as part of a 20 m-wide multispecies riparian buffer strip in central Iowa effectively removed nutrients, pesticides, and sediments in addition to stabilizing streambanks, providing wildlife habitat, affording commercial timber and other products, and improving the aesthetics of the agroecosystem (Schultz et al. 1995).

As noted by Isebrands and Karnosky (2001), timberbelts, multirow shelterbelts of SRWCs, can also provide the environmental benefits (moderate temperature and wind, decrease noise, dust, and soil erosion, provide wildlife habitat, harvest water) of windbreaks/shelterbelts while producing the economic benefits of timber products. If the agricultural component of a timberbelt mix were suitable for livestock consumption and of necessary proportion, the mix could be classified as silvopasture.

Although substantial fuelwood is derived from converting forests to farmland in many areas, considerable amounts of fuelwood also are obtained from

Table 3. Examples of potential agroforestry systems incorporating SRWCs with crops for dendroremediation.

System	SRWC	Crop	Dendroremedation	Description
Riparian	Poplars	Agronomic	Nutrients/Pesticides/Sediment	Rows of trees along streams
Windbreak	Poplars	Grain	Sediment	Widely spaced tree rows
Windbreak	Eucalypts	Citrus	Nutrients/Pesticides	Surrounding tree rows
Intercrop	Acacia	Grasses/Forbs/Herbs	Sediments	Interplanted trees
Intercrop	Eucalypts	Native understory	Land restoration	Planted trees on mined lands, Seeded understory
Silvopasture	Pines	Animals	Nutrients	Widely spaced tree groups
Forest farming	Cottonwood	Fern	As	Interplanted trees and ferns

agroforestry practices on marginal and farmland (FAO 1996). Low intensity agroforestry systems suggest the prospects for combining SRWC and dendroremediation. Between 350 000 and 500 000 ha of woodland in Tanzania have been restored by dry-season fodder reserves that are farmer-led initiatives that retain standing vegetation during the rainy season (Kamwenda 2002). This traditional practice of carefully managed trees dominated by *Acacia* spp and understory grasses, herbs, and forbs is suitable for sustainable production of dry season fodder, food, fuelwood, and mitigation of land degradation.

Another opportunity could be mixing an SRWC planted for dendroremediation and/or timber product with a crop specially suited for dendroremediation instead of agricultural production. A mixture involving a hyperaccumulating understory plant that sequesters a contaminant from the upper soil profile and an SRWC that accumulates the same contaminant at a non-hazardous concentration but primarily from a lower soil profile could increase the efficiency and effectiveness of a dendroremediation system. For As remediation in the southeastern United States, such a potential combination may involve a high but non-toxic accumulating *PD* clone with the Chinese brake fern (*Pteris vittata*).

Prospective applications of dendroremediation in agroforestry systems

The broadest definition of dendroremediation extends possible linkages with SRWCs in agroforestry systems. Dendroremediation could include planting fast growing trees for land reclamation and restoration and on disturbed or mined lands to improve soil conditions, control invasive species, and provide a transition to natives. Contaminated sites, landfills, wastewater treatment facilities, stormwater collection areas, agricultural lands, and mined lands may provide planting opportunities. The number of As contaminated sites, as an example, is huge; Florida alone has over 3200 sites formerly used to dip cattle in As containing pesticides. Thousands of sites worldwide were As contaminated by mining, wood treatment, and pesticide uses. Approximately 400 000 contaminated sites occur in western Europe (Glass 1999). In the United States, 153 cities have already developed over 4000 ha of brownfields at 922 sites, but over 200 cities have some 10 000 ha of brownfields awaiting development. The United States has 3536 municipal solid waste landfills occupying about 300 000 ha. The United Kingdom has 28 000 ha of landfills in England and Wales. These contaminated lands typically have low value and are often close to cities.

Overall demand for dendroremediation could encourage agroforestry approaches combining SRWC and dendroremediation. An estimated \$30 million to \$49 million phytoremediation market in 1999 (33% organics, inorganics, and metals in groundwater; 17% landfill leachate; 32% organics and metals in soil; 10% inorganics, organics, and metals in wastewater; 2% radionuclides; 6% other) was projected to grow to \$235 million to \$400 million in 2005 (Glass 1999). EU nations were projected to spend some \$400 million on phytoremediation over the next 25 years (Glass 1999). The potential needs in central and eastern Europe, the former Soviet Union, and developing countries afford considerable opportunities.

Market driven commercialization of agroforestry with *EG* in Florida is currently limited to a demand for mulchwood that can be met with perhaps 8000 ha of *EG* plantations in rural areas. Should fossil-fuel-based electricity producers undertake biomass co-firing, the price for *EG* energywood will surpass that of mulchwood. Then, urban areas that are uniquely positioned to create closed-loop energywood systems (biomass systems specifically established for energy production) could supply renewable power derived from the application of urban reclaimed water and yard wastes (e.g., Rockwood et al. 2002).

A three-year project funded through the European Union Environment and Climate Programme and involving partners from the United Kingdom, Germany, Austria, Spain and Sweden studied bioremediation of industrially degraded land using biomass fuel crops (D. Riddell-Black et al. pers. comm. 2002). The study focussed on heavy-metal-contaminated sites and sought ways to maximize the quantity of metals that could be removed from contaminated soil and/or allow soil stabilization by establishing *Salix* and *Phalaris* (a fast growing grass species) on the sites. Three situations were identified where phytoremediation using biomass fuel crops was considered to be a beneficial alternative to conventional remediation techniques: 1) Where former industrial land has little potential for development and the environmental risks associated with the site are low, biomass fuel crops offer an opportunity to introduce the land into economic use; 2) Where agricultural land has become contaminated through atmospheric depositions or waste disposal, biomass fuel crops may be grown when food crops are not permitted; 3) Where agricultural land has been contaminated with cadmium from rock phosphate fertilizer additions, biomass fuel crops could offer major improvements in soil quality within 10–15 years.

Societal and political factors influencing the energy and forestry sectors will greatly influence development. Greater use of wood for energy will require removal of technical and nontechnical barriers by developing technologies that make fuelwood more cost effective and competitive and policies that expand fuelwood markets and by promoting the economic and environmental benefits of fuelwood (Trossero 2002). Poplars, willows, and eucalypts suitable for fuelwood and other timber products, unlike hyperaccumulators, may not require special treatment or disposal, and therefore could provide an income in combination with gradual site cleanup and lower risk to groundwater and human health.

Regardless of which agroforestry system is used for a particular dendroremediation application, SRWCs will reach their dendroremediation potential only when the system is properly established and maintained. Instead of establishing a system in one massive operation, we recommend a deliberate, smaller startup that tests genotypes, cultural options, etc., to ensure that the appropriate genotypes of a species are selected and matched with the necessary silvicultural options. A phased phytoremediation strategy outlined by Westphal and Isebrands (2001) for matching the proper plant materials to a contaminated site involves experimental screening studies (Phase I), verification to narrow options (Phase II), demonstration plantings on site on a small scale (Phase III), and scale up plantings with materials of proven merit (Phase IV). This conservative strategy allows assurance that the plants are capable of the remediation and gives agencies time for administrative and policy decisions (Licht and Isebrands 2003).

Conclusions

SRWC systems may meet societal needs ranging from renewable energy to dendroremediation of contaminated soil and groundwater. Locally adapted SRWCs such as eucalypts, poplars, and willows may dendroremediate while providing revenue from various timber products. Effective dendroremediation depends on tree-contaminant interactions and on silvicultural, genetic, and environmental factors influencing tree growth. Current agricultural systems that threaten biodiversity and degrade ecosystems require transformation to systems that improve habitat and provide more space for wildlife while maintaining or improving productivity (McNeely and Scherr, 2003). Agroforestry systems combining SRWC and dendroremediation, such as riparian buffers, could emerge as holistic approaches for sustainable energy and agricultural development.

Acknowledgements

This paper includes research results made possible by the Florida Institute of Phosphate Research, the Consortium of Plant Biotechnology Research, Woodard and Curran, Ecology & Environment, the Center for Natural Resources at the University of Florida, CH2M Hill, Florida Energy Office, and Tampa Electric Company. This is Florida Agricultural Experiment Station Journal Series R-09908.

References

Alker G.R., Godley A.R. and Hallett J.E. 2002. Landfill Leachate Management Using Short Rotation Coppice Final Technical Report. WRC Report No. CO 5126. 201 pp. http://www.wrcplc.co.uk/Final_Technical_Report_Leachate_&_SRC_CO5126.pdf.

Aronsson P. and Perttu K. 2001. Willow vegetation filters for wastewater treatment and soil remediation combined with biomass production. Forest Chron 77: 293–299.

Buck M.G. 1986 Concepts of resource sharing in agroforestry systems. Agroforest Syst 4: 191–203.

Dickmann D.I., Isebrands J.G., Eckenwalder J.E. and Richardson J. 2001. Poplar culture in North America. NRC Press, Ottawa, Canada. 397 pp.

Douchette W., Chard J., Chard B., Fabrizius H., Crouch C. and Gorder K. 2004. Trichloroethylene in fruits and vegetables: Preliminary field survey results. In: Magar V.S. and Kelley M.E. (eds), *In Situ* and On-Site Bioremediation-2003. Proceedings of the Seventh International In Situ and On-Site Bioremediation Symposium, June 2–5, 2003, Orlando, FL, Paper F-15, Battelle Press, Columbus, OH.

El Bassam N. 1998. Energy plant species: Their use and impact on environment and development. James & James Science Publishers, London, UK. 334 pp.

Ettala M.O. 1987. Influence of irrigation with leachate on biomass production and evapotranspiration on a sanitary landfill. Aqua Fennica 17(1): 69–86.

FAO 1996. Wood Energy News 11(2). Food and Agricultural Organization Regional Office for Asia, Bangkok, Thailand. http://www.rwedp.org/wen11-2.html.

FAO 2003. State of the World' Forests. Food and Agricultural Organization of the United Nations, Rome. http://www.fao.org/docrep/005/y7581e/Y7581e00.htm.

Garrett H.E. and Buck L. 1997 Agroforestry practice and policy in the United States of America. For Ecol Manag 91: 5–15.

Glass D.J. 1999. U.S. and International Markets for Phytoremediation, 1999–2000. http://www.channel1.com/dglassassoc/INFO/phy99exc.htm. (9 June 2003).

Gordon A.M., McBride R.A. and Fisken A.J. 1989. The effects of landfill leachate spraying on foliar nutrient concentrations and leaf transpiration in a northern hardwood forest. Canada Forestry 6: 19–28.

Gustavsson L., Börjesson P., Johansson B. and Svenningsson P. 1995. Reducing CO_2 emissions by substituting biomass for fossil fuels. Energy 20: 1097–1113.

Gustavsson L. and Börjesson P. 1998. CO_2 mitigation cost. Bioenergy system and natural gas system with decarbonisation. Energy Policy 29: 699–713.

Hall D.O., House J. and Scrase I. 1999. Introduction. In: Rosillocalle F, Bajay S and Rothman H (eds), Industrial uses of biomass energy. The example of Brazil. Taylor and Francis, London. 304 pp.

Hasselgren K. 1998. Use of Municipal Wastewater in Short Rotation Energy Forestry – Full Scale Application. pp. 835–838. In: Biomass for Energy and Industry. 10th European Conference and Technology Exhibition 8–11 June, Würzburg, Germany.

Horgan G.P. 2002. Wood energy economics. Unasylva 53(4): 23–27.

Hytönen J. 1994. Effect of fertilizer application rate on nutrient status and biomass production in short rotation plantations of willows on cut away peatland areas. Suo 45(3): 65–77.

Isebrands J.G. and Karnosky D.F. 2001. Environmental benefits of poplar culture. pp. 207–218. In: Dickmann D.I., Isebrands J.G., Eckenwalder J.E. and Richardson J. (eds), Poplar Culture in North America. Part A, Chapter 6. NRC Research Press, National Research Council of Canada, Ottawa, Canada.

Kamwenda G.L. 2002. Ngitili agrosilvipastoral systems in the United Republic of Tanzania. Unasylva 53(4): 46–50.

Kopp R.F., Smart L.B., Maynard C.A., Isebrands J.G., Tuskan G.A. and Abrahamson L.P. 2001. The development of improved willow clones for eastern North America. Forest Chron 77: 287–292.

Labrecque M., Tedodorescu T.I., Babeux P., Cogliastro A. and Daigle S. 1993. Growth patterns and biomass productivity of two *Salix* species grown under short rotation, intensive culture in southwestern Quebec. Biomass Bioenergy 4: 419–425.

Labrecque M., Tedodorescu T.I., Babeux P., Cogliastro A. and Daigle S. 1994. Impact of herbaceous competition and drainage conditions on the early productivity of willows and short rotation intensive culture. Can J For Res 24: 493–501.

Labrecque M., Tedodorescu T.I. and Daigle S. 1997. Biomass productivity and wood energy of *Salix* species after two years growth in SRIC fertilized with waste water sludge. Biomass Bioenergy 12: 409–417.

Land S.B. Jr, Stine M., Rockwood D.L., Ma X., Warwell M.V. and Alker G.R. 2001. A tree improvement program for eastern cottonwood in the southeastern United States. pp. 84–93. In: Proc. 26th South. For. Tree Improvement Conf., June 26–29, 2001, Athens, GA.

Licht L.A. 1994. *Populus* spp. (Poplar) capabilities and relationships to landfill water management. Paper 94WA86.02. pp. 19–22. In: Air and Waste Management Assn. 87th Annual Meeting. Cincinnati OH, June.

Licht L.A. and Isebrands J.G. 2003. Linking phytoremediation pollutant removal to biomass economic opportunities. Biomass Bioenergy (In press).

McNeely J.A. and Scherr S.J. 2003. Ecoagriculture: Strategies to Feed the World and Save Wild Biodiversity. Island Press, 296 pp.

Nzengung V.A., Spriggs T., Tsangaris S. and Nwokike B. 2004. Phytoremediation of a chlorinated solvent plume in Orlando, Florida. In: Magar V.S. and Kelley M.E. (eds), *In Situ* and On-Site Bioremediation-2003. Proceedings of the Seventh International In Situ and On-Site Bioremediation Symposium. June 2–5, 2003, Orlando, FL, Paper F-13, Battelle Press, Columbus, OH.

O'Neill G.J. and Gordon A.M. 1994. The nitrogen filtering capacity of Carolina poplar in an artificial riparian zone. J Environ Qual 23: 1218–1823.

Pathak P.S., Gupta S.K. and Debroy R. 1981. Production of aerial biomass in *Leucaena leucocephala*. Indian Forester 107: 416–419.

Pisano S.M. and Rockwood D.L. 1997. Stormwater phytoremediation potential of Eucalyptus. pp. 32–42. In: Proceedings 5th. Biennial Stormwater Research Conference, Nov. 5–7, 1997, Tampa, FL. Southwest Florida Water Management District, Brooksville, FL.

Rahmani M., Hodges A.W. and Kiker C.F. 1999. Optimal location for biomass conversion facilities in Florida: A Geographic Information System. pp. 91–97. In: Overend R.P. and Chornet E. (eds), Biomass: A Growth Opportunity in Green Energy and Value-Added Products, Proceedings of the Fourth Biomass Conference of the Americas, August 29-September 2, 1999, Oakland, CA, Elsevier Science, Oxford, UK.

Riemenschneider D.E., Berguson W.E., Dickmann D.I., Hall R.B., Isebrands J.G., Mohn C.A., Stanosz G.R. and Tuskan G.A. 2001 Poplar breeding and testing strategies in the north-central U.S.: Demonstration of potential yield and consideration of future research needs. Forest Chron 77: 245–253.

Rockwood D.L., Cardellino R., Alker G., Lin C., Brown N., Spriggs T., Tsangaris S., Isebrands J., Hall R., Lange R. and Nwokike B. 2004. Fast-growing trees for heavy metal and chlorinated solvent phytoremediation. In: Magar V.S. and Kelley M.E. (eds), In Situ and On-Site Bioremediation-2003. Proceedings of the Seventh International In Situ and On-Site Bioremediation Symposium, Orlando, FL, June 2–5, 2003, Paper F-12, Battelle Press, Columbus, OH.

Rockwood D.L., Carter D.R., Alker G.R. and Morse D.M. 2002. Compost utilization for forest crops in Florida. In: Proc. Recycle Organics '02, Composting in the Southeast Conf. and Exposition, October 6–9, 2002, Palm Harbor, FL. http://conference.ifas.ufl.edu/compost/abstract.pdf.

Rockwood D.L., Ma L.Q., Alker G.R., Tu C. and Cardellino R.W. 2001. Phytoremediation of contaminated sites using woody biomass. Final Report to the Florida Center for Solid and Hazardous Waste Management, June 2001. 95 pp. http://www.floridacenter.org/publications/publicationsnew.htm.

Rockwood D.L., Pisano S.M. and McConnell W.V. 1996. Superior cottonwood and Eucalyptus clones for biomass production in waste bioremediation systems. pp. 254–261. In: Proc. Bioenergy 96, 7th National Bioenergy Conference, September 15–20, 1996, Nashville, TN.

RWEDP-FAO 1999. Wood energy data and information. Regional wood energy development programme, FAO. http://www.rwedp.org/gs_ecfpp_casestudies.html (9 July 2003).

Schultz R.C., Collettii J.P., Isenhart T.M., Simpkins W.W., Mize C.W. and Thompson M.L. 1995 Design and placement of a multispecies riparian buffer strip system. Agrofor Syst 31: 117–132.

Segrest S.A. 2003 www.treepower.org/globalwarming/quickfacts.html (6 June 2003).

Segrest S.A., Rockwood D.L., Stricker J.A. and Alker G.R. 2001. Partnering to cofire woody biomass in central Florida. In: Abstracts 5th Biomass Conference of the Americas. 2 pp. http://bioproducts-bioenergy.gov/pdfs/bcota/abstracts/4/z280.pdf.

Stanton B., Eaton J., Johnson J., Rice D., Schuette B. and Moser B. 2002 Hybrid poplar in the Pacific Northwest: the effects of market-driven management. J For 100(4): 28–33.

Stricker J.A., Rahmani M., Hodges A.W., Mishoe J.W., Prine G.M., Rockwood D.L. and Vincent A. 1995. Economic development through biomass systems integration in central Florida. Proceedings of the Second Biomass Conference of the Americas: Energy, Environment, Agriculture, and Industry, 1608–1617.

Trossero M.A. 2002. Wood energy: the way ahead. Unasylva 53(4): 3–12.

Tuskan G.A. and Walsh M.E. 2001 Short rotation woody crop systems, atmospheric carbon dioxide and management: A US case study. Forest Chron 77: 259–264.

US Environmental Protection Agency 2003. Phytoremediation bibliography – online. US EPA. http://www.rtdf.org/public/phyto/phytodoc.htm.

WEC-FAO 1999. The challenge of rural energy poverty in developing countries world energy council. Food and Agriculture Organization of the United Nations, UK. 185 pp.

Westphal L.M. and Isebrands J.G. 2001. Phytoremediation of Chicago's brownfields – consideration of ecological approaches and social issues. Proceedings Brownfields 2001 Conference, Chicago, Illinois. BB-11-02. http://www.brownfields2002.org/proceedings2001/BB-11-02.pdf.

Wilde E.W., Brigmon R.L., Berry C.J. and Altman D.J., Rossabi J., Harris S.P. and Newman L.A. 2003. Drip irrigation of TCE contaminated groundwater. In: Magar V.S. and Kelley M.E. (eds), *In Situ* and On-Site Bioremediation-2003. Proceedings of the Seventh International In Situ and On-Site Bioremediation Symposium, June 2–5, 2003, Orlando, FL, Paper F 10, Battelle Press, Columbus, OH.

Wong M.H. and Lueng C.K. 1989. Landfill leachate as irrigation water for tree and vegetable crops. Waste Manage Res 7: 311–324.

Agroforestry Systems **61:** 65–78, 2004.

Windbreaks in North American agricultural systems

J.R. Brandle[1,*], L. Hodges[2] and X.H. Zhou[1]
[1]*School of Natural Resources, 3B Plant Industry Building, University of Nebraska-Lincoln, Lincoln, Nebraska 68583-0814, USA;* [2] *Department of Agronomy and Horticulture, University of Nebraska-Lincoln, Lincoln, Nebraska 68583-0724, USA;* [*]*Author for correspondence: e-mail: jbrandle1@unl.edu*

Key words: Crop production, Microclimate, Shelterbelt benefits, Shelterbelt structure, Wind protection

Abstract

Windbreaks are a major component of successful agricultural systems throughout the world. The focus of this chapter is on temperate-zone, commercial, agricultural systems in North America, where windbreaks contribute to both producer profitability and environmental quality by increasing crop production while simultaneously reducing the level of off-farm inputs. They help control erosion and blowing snow, improve animal health and survival under winter conditions, reduce energy consumption of the farmstead unit, and enhance habitat diversity, providing refuges for predatory birds and insects. On a larger landscape scale windbreaks provide habitat for various types of wildlife and have the potential to contribute significant benefits to the carbon balance equation, easing the economic burdens associated with climate change. For a windbreak to function properly, it must be designed with the needs of the landowner in mind. The ability of a windbreak to meet a specific need is determined by its structure: both external structure, width, height, shape, and orientation as well as the internal structure; the amount and arrangement of the branches, leaves, and stems of the trees or shrubs in the windbreak. In response to windbreak structure, wind flow in the vicinity of a windbreak is altered and the microclimate in sheltered areas is changed; temperatures tend to be slightly higher and evaporation is reduced. These types of changes in microclimate can be utilized to enhance agricultural sustainability and profitability. While specific mechanisms of the shelter response remain unclear and are topics for further research, the two biggest challenges we face are: developing a better understanding of why producers are reluctant to adopt windbreak technology and defining the role of woody plants in the agricultural landscape.

Introduction

Windbreaks or shelterbelts are barriers used to reduce wind speed. Usually consisting of trees and shrubs, they may be composed of perennial or annual crops, grasses, wooden fences, or other materials. Throughout history they have been used to protect homes, crops and livestock, control wind erosion and blowing snow, provide habitat for wildlife, and enhance the agricultural landscape.

Windbreaks have their origins in the mid-1400s when the Scottish Parliament urged the planting of tree belts to protect agricultural production (Droze 1977). From these beginnings, shelterbelts have been used extensively throughout the world (Caborn 1971; Grace 1977; Brandle et al. 1988; Cleugh et al. 2002;) to provide protection from the wind. As settlement in the United States moved west into the grasslands, homesteaders planted trees to protect their homes, farms, and ranches. In the 1930s, in response to the Dust Bowl conditions, the U. S. Congress authorized the Prairie States Forestry Project to plant windbreaks (Droze 1977). In northern China, extensive plantings of shelterbelts and forest blocks were initiated in the 1950s to counter eroding agricultural conditions. Today the area is extensively protected, and studies have documented a modification in the regional climate (Zhao et al. 1995). Windbreak programs also have been established in Australia (Burke 1998), Canada (Kort 1988), New Zealand (Sturrock

Table 1. Wind speed reductions in shelter at various distances windward and leeward of shelterbelts with different optical densities in Midwestern United States[a].

Type of windbreak	Optical density	Percent of open wind speed at various distances								
		Windward			Leeward					
		−25 H	−3 H	−1 H	5 H	10 H	15 H	20 H	25 H	30 H
Single row deciduous	25–30	100	97	85	50	65	80	85	95	100
Single row conifer	40–60	100	96	84	30	50	60	75	85	95
Multi-row conifer	60–80	100	91	75	25	35	65	85	90	95
Solid wall	100	100	95	70	25	70	90	95	100	100

[a]Reductions are expressed as percent of open wind speed where open wind speed is assumed to be less than 10 meters per second and distance from the windbreak is expressed in terms of windbreak height (H).

1984), Russia (Konstantinov and Struzer 1965), South America (Luis and Bloomberg 2002), and several developing countries (Nair 1993). The focus of this chapter, however, is on windbreaks in the context of commercial, mechanized agriculture in the temperate zone, especially in North America.

The goal of this review is to provide a summary of practical information for those wishing to understand how windbreaks work and how they may be integrated into sustainable agricultural production systems. It is divided into three main sections: i) how windbreaks work, ii) how organisms respond to wind protection including the benefits of wind protection, and iii) the overall role of windbreaks in the sustainable agricultural landscape. The reader is referred to recent reviews by Nuberg (1998), Brandle et al. (2000), and Cleugh et al. (2002) for more details.

How windbreaks work

Wind flow in the environment

Wind is air in motion. It is caused by the differential heating of the earth's surface resulting in differences in pressure and is influenced by Coriolis forces caused by the earth's rotation. On a global scale, atmospheric circulation drives our daily weather patterns. On a microscale, there is a very thin layer of air (several millimeters or less) next to any surface within which transfer processes are controlled by the process of diffusion across the boundary layer. Between these two scales are the surface winds. They move in both vertical and horizontal directions and are affected by the surfaces they encounter. Surface winds extend 50 to 100 meters above the earth's surface and are dominated by strong mixing or turbulence (Grace 1981). These surface winds influence wind erosion, crop growth and development, animal health, and the general farm or ranch environment. They are also the winds affected by shelterbelts.

Although surface winds can be quite variable and the flows highly turbulent, the main component of the wind moves parallel to the ground. Wind speed at the soil surface approaches zero due to the frictional drag of the surface. The amount of drag is a function of the type of surface. In the case of vegetation, the height, uniformity, and flexibility of that vegetation determines the amount of frictional drag exerted on wind flow (Lowry 1967). A rough surface (e.g., wheat stubble) has greater frictional drag, slower wind speeds, and greater turbulence near the surface than a relatively smooth surface (e.g., mown grass). A windbreak increases surface roughness and, when properly designed, provides large areas of reduced wind speed useful for agriculture.

Wind flow across a barrier

A windbreak is a barrier placed on the land surface that obstructs the wind flow and alters flow patterns both up-wind of the barrier (windward) and down-wind of the barrier (leeward). As wind approaches a windbreak, a portion of the air passes through the barrier. The remaining air flows around the ends of the barrier or is forced up and over the barrier. As the air moves around or over the barrier, the streamlines of air are compressed (van Eimern et al. 1964). This upward alteration of flow begins at some distance windward of the windbreak and creates a region of reduced wind speed on the windward side. This protected area extends for a distance of 2 H to 5 H, where H is the height of the barrier. A much larger region of reduced wind speed is created in the lee of the barrier. This region typically extends for a distance of 10 H to 30 H (Wang and Takle 1995). Some wind speed reductions

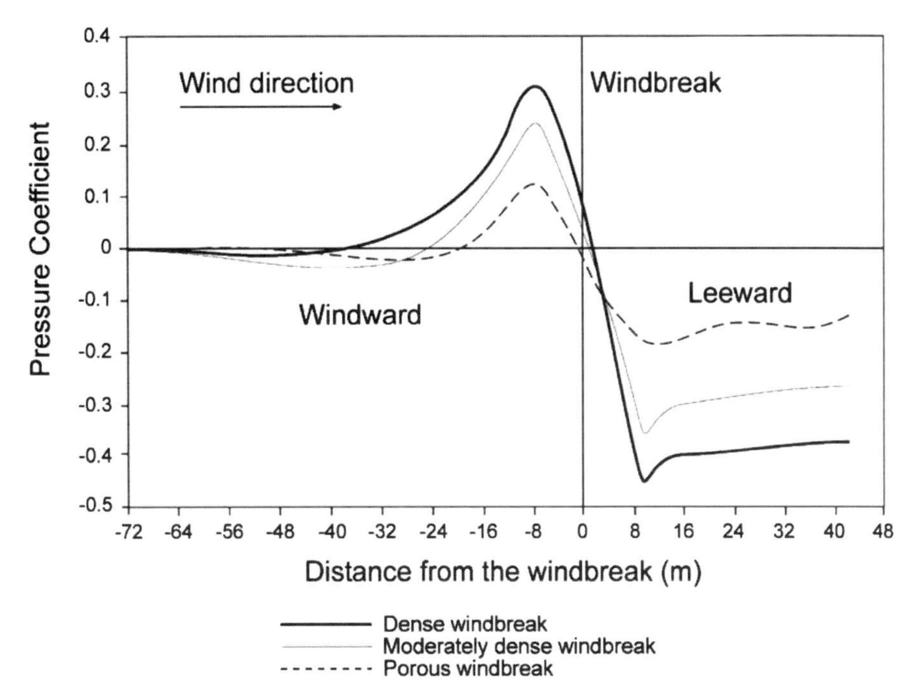

Figure 1. Changes in the pressure coefficient at ground level windward and leeward of three windbreaks with different levels of optical density. Distances from the leeward edge of the windbreak are given as positive distances and as negative distances in the windward direction.

extend as far as 60 H to the lee (Caborn 1957), but it is unlikely that small reductions at these distances have significant microclimatic or biological impacts (Table 1).

Pressure on the ground is increased as the wind approaches the barrier and reaches a maximum at the windward edge of the barrier (Figure 1). Pressure drops as the wind passes through the barrier, reaching a minimum just to the lee. Pressure gradually increases returning to the original condition at or beyond 10 H. The magnitude of the pressure difference between the windward and leeward sides of the windbreak is one factor determining the flow modification of the barrier and is a function of windbreak structure (Takle et al. 1997).

Windbreak structure

The effectiveness of a windbreak is determined partially by its external structure, which is characterized by height, length, orientation, continuity, width, and cross-sectional shape. It is determined also by its internal structure, which is a function of the amount and distribution of the solid and open portions, the vegetative surface area, and the shape of individual plant elements. (Figure 2).

External structure

Windbreak height (H) is the most important factor determining the extent of wind protection. Distance from the windbreak is usually expressed in terms of windbreak height and is normally measured from the center of the outer row of the windbreak along a line parallel to the direction of the wind. The length of the windbreak should be at least ten times the height in order to reduce the effects of wind flow around the ends of the windbreak. Together, they determine the total area protected. Windbreaks are most efficient when they are oriented perpendicular to the problem winds. As the angle of the approaching wind becomes more oblique, the size and location of the protected zone decrease (Wang and Takle 1996a). The continuity of a windbreak also influences its efficiency: a gap or opening concentrates wind flow through the opening, creating a zone leeward of the gap in which wind speeds exceed open field wind velocities. Windbreak width influences the effectiveness of a windbreak through its influence on density (Heisler and DeWalle 1988). Traditionally this meant the adding of additional tree rows; thus, as more rows were added, density increased. More recently, researchers have distinguished between optical density, the amount of solid material appearing in a two dimensional photograph, and

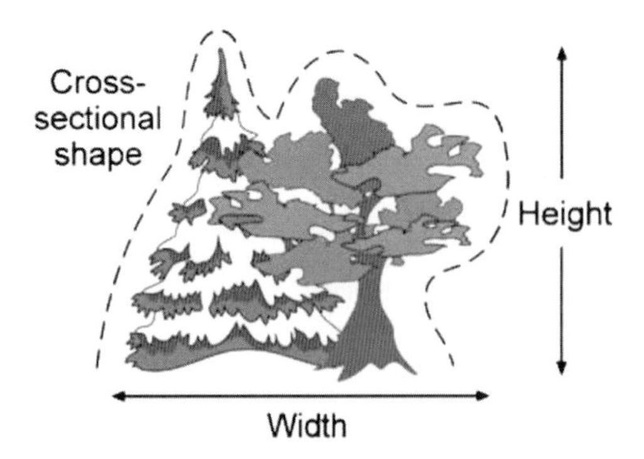

Figure 2. Illustration of the external structure components of a two row mixed windbreak. Internal structure consists of the volumes and surface areas of the individual components, leaves or needles, branches and trunk. The shape of individual elements and the arrangement of the elements within the windbreak are also components of the internal structure.

aerodynamic density which has been defined as the amount of surface area per unit volume. This change is justified in that the wind flows not in a straight line, but around or across all of the vegetative elements in the windbreak. Research using numerical simulation methods suggests that aerodynamic density is one of the critical components of internal structure (Wang and Takle 1996b). Early work (van Eimern et al. 1964) indicated that the cross-sectional shape influences the magnitude and extent of wind speed reductions in the sheltered zone. Again, more recent research using numerical simulation models suggests that the overall arrangement of the solid and open portions of the windbreak may have significant influence on wind flow patterns. These issues are discussed in the next section.

Internal structure

Historically the internal structure of a shelterbelt was described by either density (the amount of solid material), or porosity (the amount of open spaces) (Caborn 1957). Now, the focus is on defining the aerodynamic structure of a windbreak in three dimensions (Zhou et al. 2002). These descriptions of internal structure include the amount and distribution of the solid elements and open spaces, recognizing both volume and surface area, as well as the geometric shape of individual vegetative elements (Zhou et al. 2004). Using these parameters, the effect of shelterbelt structure on the flow fields surrounding the shelterbelt are being simulated with numerical modeling and verified under field conditions. Preliminary assessments (Brandle, Takle, Zhou unpublished data) indicate that optical density overestimates aerodynamic density, especially at higher densities. For most applications, the consequences of overestimation appear to be minimal.

Microclimate changes

Windbreaks reduce wind speed in the sheltered zone. As a result of wind speed reduction and changes in turbulent transfer rates, the microclimate in the sheltered zone is altered (McNaughton 1988; Cleugh 2002; Cleugh and Hughes 2002). The magnitude of microclimate changes for a given windbreak varies within the protected zone. It depends on the existing atmospheric conditions, the windbreak's structure and orientation, the time of day, and the height above the ground at which measurements are made.

Radiation

On a regional scale, shelterbelts have minimal influence on the direct distribution of incoming radiation; however, they do influence radiant flux density, or the amount of energy per unit surface area per unit time, in the area immediately adjacent to the windbreak. Solar radiant flux density is influenced by sun angle, which is a function of location, season, and time of day, and by windbreak height, density, and orientation. Likewise, at any given location, the extent of the shaded zone is dependent on time of the day, season of the year, and height of the windbreak. During portions of the day, radiation is reflected off windbreak surfaces facing the sun, increasing radiant flux density immediately adjacent to the windbreak.

Air temperature

In temperate regions, daytime temperatures within 8 H of a medium-dense barrier tend to be several degrees warmer than temperatures in the open due to the reduction in turbulent mixing. In tropical or semi-tropical regions, the magnitude of temperature effects is increased and may limit plant growth, especially in regions of limited moisture availability. In temperate regions, temperature effects appear to be greater early in the growing season. Between 8 and 24 H, daytime turbulence increases and air temperatures tend to be several degrees cooler than for unsheltered areas (McNaughton 1988). Nighttime temperatures near the ground, or within 1 m, are generally 1 °C to 2 °C warmer in the protected zone, which is up to 30 H, than in the exposed areas. In contrast, temperatures 2 m above the surface tend to be slightly cooler. On very calm nights, temperature inversions may occur and protected areas may be several degrees cooler at the surface than exposed areas (Argete and Wilson 1989).

The largest impact of increased air temperature may be an increase in the rate of accumulation of heat units. This provides several benefits to the producer. Crops grown in sheltered areas mature more quickly than unsheltered crops. For vegetable crops, this may provide a marketing advantage and result in a premium price for the product. For grain crops, the increase in the rate of development may mean that critical stages of growth occur earlier in the season when periods of water stress may be less likely. An increase in heat units at the beginning or end of the season may allow greater flexibility in selecting crop varieties.

Soil temperature

Average soil temperatures in shelter are slightly warmer than in unprotected areas (McNaughton 1988). In most cases this is due to the reduction in heat transfer away from the surface. In areas within the shadow of a windbreak, soil temperatures are lower due to shading of the surface. The magnitude of this effect is dependent on the time of day, height of the barrier, and the angle of the sun, which affects the size and duration of the shaded area. In areas receiving reflected radiation from the windbreak, soil temperatures may be higher due to the added radiation load. Again, it appears that these differences are greatest early in the season in temperate regions (Caborn 1957).

Frost

On clear, calm nights, infrared radiation emission by soil and vegetation surfaces is unimpeded. Under these conditions surfaces may cool rapidly resulting in decreased air temperature near the surface. When this temperature reaches the dew point, condensation forms on surfaces. If temperatures are below freezing, the condensation freezes resulting in a radiation frost. Radiation frosts are most likely under very calm conditions when strong temperature inversions may occur. In contrast, advection frosts are generally associated with large-scale, cold air masses. Strong winds are typically associated with the passage of the front and, while the radiative process contributes to heat loss, temperature inversions do not occur. Shelterbelts may offer some protection against advective frosts when episodes are of short duration and when windward temperatures are just below 0 °C. In sheltered areas, wind speed is reduced resulting in reduced turbulent transfer coefficients, or less mixing of the warm air near the surface with the colder air of the front, and reduced heat loss from the sheltered area (Brandle et al. 2000).

Precipitation

Rainfall over most of the sheltered zone is unaffected except in the area immediately adjacent to the windbreak. These areas may receive slightly more or less than the open field depending on wind direction and intensity of rainfall. On the leeward side there may be a small rain shadow where the amount of precipitation reaching the surface may be slightly reduced. The converse is true on the windward side, as the windbreak may function as a barrier and lead to slightly higher levels of measured precipitation at or near the base of the trees due to increased stemflow or dripping from the canopy.

In contrast, the distribution of snow is greatly influenced by the presence of a windbreak and can be manipulated by managing windbreak density (Scholten 1988; Shaw 1988). A dense windbreak (>60% density) will lead to relatively short, deep snow drifts on the leeward side, while a more porous barrier (~35% density) will provide a long, relatively shallow drift to the lee (Figure 3). In both cases, the distribution of snow and the resulting soil moisture will affect the microclimate of the site. In the case of field windbreaks, a more uniform distribution of snow may provide moisture for significant increases in crop yield. This is especially true in more northern

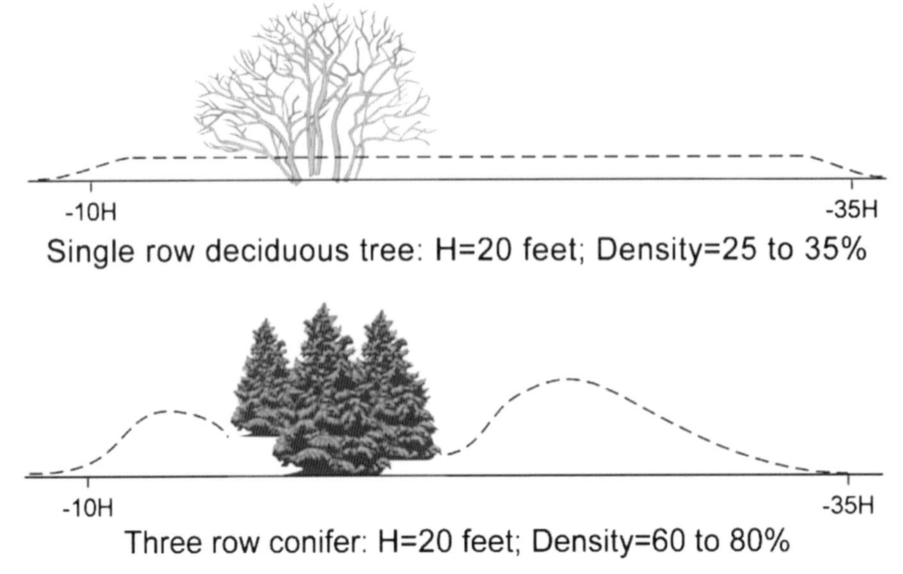

Figure 3. Snow distribution as influenced by a very porous windbreak and a very dense windbreak. The porous windbreak is used to distribute snow across a field while a dense windbreak is used to collect snow in a relatively narrow deep drift.

areas where snowfall makes up a significant portion of the annual precipitation. In addition, fall planted crops insulated by a blanket of snow are protected against desiccation by cold, dry winter winds (Brandle et al. 1984).

Humidity

Humidity, or the water vapor content of the air, is related to its role in the energy balance of the system. Decreases in turbulent mixing reduce the amount of water vapor transported away from surfaces in the sheltered area. As a result, humidity and vapor pressure gradients in shelter are generally greater both during the day and at night (McNaughton 1988). And, because water vapor is a strong absorber of infrared radiation, higher humidity levels in shelter tend to protect the crop from radiative heat losses, reducing the potential for frost.

Evaporation

Evaporation from bare soil is reduced in shelter due to wind speed reductions and the reduction in transfer of water vapor away from the surface. In most cases this is an advantage, conserving soil moisture for plant growth. Evaporation from leaf surfaces is also reduced in shelter, and, in rare cases, may contribute to a higher incidence of disease. Combined with lower nighttime temperatures in shelter, high humidity levels may cause more dew formation. In these cases, the added humidity and reduced evaporation in shelter may increase the possibility of disease. However, when situations do occur where very dense windbreaks in combination with high humidity, rainfall, or irrigation may contribute to abnormally high humidity levels in sheltered areas, reducing windbreak density will increase windflow, reducing humidity and the potential for disease (Hodges and Brandle 1996).

Response to wind protection

Response of plants to shelter

The effect of wind on plants has been reviewed extensively (Grace 1988; Coutts and Grace 1995; Miller et al. 1995). Both photosynthesis and transpiration are driven in part by environmental conditions, particularly those within the leaf and canopy boundary layers of the plant. As shelter modifies micro-environment, it impacts plant productivity.

One useful concept explaining how plants respond to shelter is that of coupling. Monteith (1981) defines coupling as the capacity of exchanging energy, momentum, or mass between two systems. Exchange processes between single leaves and the atmosphere or between plant canopies and the atmosphere are controlled by the gradients of temperature, humidity, and CO_2 that exist in the immediate environment above the leaf or canopy. When these gradients are modified by shelter, plant processes within the sheltered zone may become less strongly coupled from the atmosphere above the canopy resulting in a build up of heat, moisture, and CO_2 near the surface (Grace 1981; McNaughton 1988).

Plant temperature differences between sheltered and exposed sites are relatively small, on the order of 1 °C to 3 °C. In the sheltered zone, where the rate of heat transfer from a plant is reduced by decreased vertical temperature gradients, a slight increase in temperature can be an advantage, especially in cooler regions where even a small increase in plant temperature may have substantial positive effects on the rate of cell division and expansion and other phenological patterns (Grace 1988; van Gardingen and Grace 1991). Lower night temperatures in shelter may reduce the rate of respiration, which may result in higher rates of net photosynthesis and more growth. Indeed, there are many examples of sheltered plants being taller and having more extensive leaf areas (Rosenberg 1966; Ogbuehi and Brandle 1982). Higher soil temperatures in the sheltered zone may result in more rapid crop emergence and establishment, especially for crops with high heat-unit-accumulation requirement for germination and establishment (Drew 1982). In contrast, temperatures above the optimum for plant development may lead to periods of water stress if the plant is unable to adjust to the higher demands for moisture.

The overall influence of shelter on plant water relations is extremely complex and linked to both the temperature and wind speed conditions found in shelter. Until recently, the major effect of shelter and its influence on crop growth and yield was assumed to be due primarily to soil moisture conservation and a reduction in water stress of sheltered plants (Caborn 1957; Grace 1988). There is little question that evaporation rates are reduced in shelter (McNaughton 1988); however, the effect on plant water status is less clear. According to Grace (1988), transpiration rates may increase, decrease, or remain unaffected by shelter depending on wind speed, atmospheric resistance, and saturation vapor pressure deficit. Davis and Norman (1988) suggested that under some conditions, sheltered plants made more efficient use of available water. Monteith (1993) suggested that water use efficiency in shelter was unlikely to increase except when there was a significant decrease in saturation vapor pressure deficit. Indeed, the increase of humidity in sheltered areas would contribute to a decrease in saturation vapor pressure deficit and thus an increase in water use efficiency. However, sheltered plants tend to be taller and have larger leaf areas. Given an increase in biomass, sheltered plants have a greater demand for water and under conditions of limited soil moisture or high temperature may actually suffer greater water stress than exposed plants (Grace 1988). Overall, shelter improves water conservation and allows the crop to make better use of available water over the course of a growing season. The magnitude of this response depends on the crop, stage of development, and environmental conditions.

The complex nature of crop water relations in shelter was demonstrated again in the recently completed Australian National Windbreak Program (Cleugh et al. 2002). Results from the program indicated generally larger plants in shelter but very mixed and frequently negative results in terms of yield response of common Australian crops. Some of these results were explained by the extreme and variable climate conditions of many Australian crop production regions and some by soils with very low soil water holding capacity. Variable precipitation patterns resulted in a shortage of moisture late in the growing season. Water holding capacity of the soil was inadequate and failed to supply sufficient water to the larger plants found in shelter resulting in reduced yields. The Australian experience clearly demonstrates that we still have much to learn on how windbreaks influence plant water use (Hall et al. 2002; Nuberg and Mylius 2002; Nuberg et al. 2002; Sudmeyer and Scott 2002).

Growth and development response of plants to shelter

As a result of favorable microclimate and the resulting physiological changes, the rate of growth and development of sheltered plants may increase. Vegetative growth is generally increased in sheltered environments (Kort 1988). The increase in the rate of accumulation of heat units in shelter contributes to earlier maturity of many crops and the ability to reach the early market with many of these perishable crops can mean sizable economic returns to producers (Brandle et al. 1995).

Wind influences plant growth directly by the mechanical manipulation of plant parts (Miller et al. 1995). This movement may increase the radial enlargement of the stem, increase leaf thickness, reduce stem elongation and leaf size (Grace 1988), and affect cellular composition (Armbrust 1982). On the whole-plant level, it appears that the interaction of ethylene and auxin (Biddington 1986) as well as possible inhibition of auxin transport (Mitchell 1977) are involved. The threshold wind speed and duration for these types of direct responses appears to be very low, perhaps as low as 1 m/s for less than one minute. As a result, these types of responses may be more indicative of a no wind situation than an indicator of various wind speed differences found in sheltered and non-sheltered conditions (Miller et al. 1995).

Wind can cause direct physical damage to plants through abrasion and leaf tearing (Miller et al. 1995). As tissue surfaces rub against each other, the epicuticular waxes on the surfaces are abraded, increasing cuticular conductance and water loss (Pitcairn et al. 1986; van Gardingen and Grace 1991). Tearing is common on leaves that are larger, damaged by insects, or subjected to high wind speeds. Wind contributes to the abrasion of plant surfaces by wind blown particles (usually soil), often referred to as sandblasting. The extent of injury depends on wind speed and degree of turbulence, amount and type of abrasive material in the air stream, duration of exposure, plant species and its stage of development, and microclimatic conditions (Skidmore 1966). All three of these – abrasion, leaf tearing, and sandblasting – damage plant surfaces and can lead to uncontrolled water loss from the plant (Miller et al. 1995).

Plant lodging is another direct mechanical injury caused by wind. It takes two forms: stem lodging, where the lower internode permanently bends or breaks; and root lodging, where the soil or roots supporting the stem fail. Stem lodging is most common as crops approach maturity, while root lodging is more common on wet soils and during grain filling periods (Easson et al. 1993; Miller et al. 1995). In both cases, heavy rainfall tends to increase the potential for lodging. Medium dense shelterbelts tend to reduce crop lodging within the sheltered zone because of reduced wind speeds. As windbreak density increases, turbulence increases and the likelihood of lodging is greater.

Crop yield response to shelter

While the influences of wind and shelter on individual plant processes are only partially understood, the net effect of shelter on crop yield is generally positive (Kort 1988; Brandle et al. 1992a, 2000) although the Australian experience was less conclusive (Cleugh et al. 2002). The reasons vary with crop, windbreak design, geographic location, moisture condition, soil properties and cultural practice. In 1988, Kort summarized yield responses for a number of field crops from temperate areas around the world. Average yield increases varied from 6% to 44%. However, a close reading of the individual studies behind these averages indicates great variability in yield results. This is understandable because final crop yield is the culmination of a series of interacting factors present throughout the growth and development of the crop. The possible combinations of growth response and microclimate are unlimited, and the probability of a single combination and the corresponding crop response occurring on an annual basis is relatively small. As Sturrock (1984) explained, the relationship between shelter and crop response is complex and dynamic, subject to continual change as a result of changes in mesoclimate, windbreak efficiency, and growth and development of the protected crop. Again, the results from Australia illustrate how complex the issue remains.

Another factor, which may influence crop response to shelter, is crop cultivar. Almost without exception, crops have been selected and bred under exposed conditions. As a result, most common cultivars represent those selections best able to perform under exposed conditions. In order to take full advantage of the microclimate conditions created by windbreaks, a producer should select crop cultivars best suited to the microclimate conditions found in shelter. For example, using shorter, thicker-stemmed cultivars will reduce the potential for lodging while taking advantage of the favorable growing conditions found in sheltered fields.

Baldwin (1988) and Norton (1988) provide the most recent comprehensive reviews of horticultural crops and shelter. Fruits and vegetables, in general, are more sensitive to wind stress than many agronomic crops, showing loss of yield and quality at lower wind speeds. In horticultural crops, marketable yields, quality of the product, and earliness to market maturity are of primary importance (Hodges and Brandle 1996). For horticultural crops grown in sheltered conditions, the moderation of temperature extremes, warmer soil and air temperatures, and improved plant water status contributed to yield increases in total marketable yield and individual fruit weight. The moderated microclimate in shelter contributes to longer flowering periods and increased bee activity, and can result in improved fruit set and earlier maturity (Norton 1988). Wind-induced sandblasting and abrasion compound the direct effects of wind on the yield and quality of vegetable and specialty crops. As the amount of wind-blown soil, wind speed, or exposure time increases, crop survival, growth, yield, and quality decrease. Young plants tend to be more sensitive to damage. Concern for damage by wind-blown soil is greatest during the early spring when stand establishment coincides with seasonally high winds and large areas of exposed soil during field preparation. Another critical time is during the flowering stage when wind abrasion and abrasion by wind-blown soil may result in damage to or loss of buds and flowers. Vegetable producers need to be especially aware of the problems associated with wind erosion because the soil characteristics that favor vegetable production are typical of erosive soils.

The zone of competition

One of the most common negative comments concerning the benefits of field windbreaks is related to the impact of competition between the windbreak and the adjacent crop. There is no question that under conditions of limited moisture, competition between the windbreak and the crop has significant negative impacts on yield. Crop yields within the zone of competition may be reduced due to allelopathy, nutrient deficiency, shading, temperature, or soil moisture deficiency (Kort 1988). The degree of competition varies with crop, geographic location (Lyles et al. 1984), windbreak species, and soil or climate conditions (Sudmeyer et al. 2002).

It may be possible to reduce some forms of competition by tree-root pruning, i.e, cutting of lateral tree roots extending into the crop field. The effectiveness of the practice depends on the rooting patterns of the windbreak species, the depth of root pruning, and soil moisture levels (Rasmussen and Shapiro 1990; Hou et al. 2003; Jose et al. 2004). Under limited moisture conditions, root pruning significantly increases crop yields within the zone of competition. During wet years, the benefits are less obvious. Root pruning must be repeated every one to five years depending on tree species and local weather conditions and can have negative impacts on windbreak survival.

Wind erosion control

Of all the benefits of field windbreaks, wind erosion control is the most widely accepted. If wind speed is reduced, wind erosion and its impacts on both crop productivity and off-site costs are reduced (Huszar and Piper 1986; Pimental et al. 1995; Ribaudo 1986). Windblown soil can carry inoculum for bacterial and fungal diseases as well as provide potential entry points for pathogens. Controlling wind erosion may reduce the incidence and severity of crop diseases (Hodges and Brandle 1996). It is worth noting that while crop responses were mixed in the Australian studies, the benefits associated with wind erosion control were reaffirmed (Cleugh et al. 2002; Sudmeyer et al. 2002).

Snow management

In many northern, semiarid areas, snow is a critical source of soil moisture for crop and forage production during the subsequent growing season. Greb (1980) estimated that over one-third of the snowfall in these northern areas is blown off the field. Much of this wind-blown snow is deposited in road ditches, gullies, or behind fence-rows or other obstructions (Aase and Siddoway 1976). Even more may simply evaporate (Schmidt 1972). Many factors influence snow distribution including: i) the amount and specific gravity of the snow, ii) the topography and surface conditions, iii) wind velocity and direction, and iv) the presence and characteristics of barriers to wind flow (Scholten 1988). Field windbreaks help capture the moisture available in snow by slowing the wind and distributing the snow across the field. As a result, crop yields on fields protected by field windbreaks are increased 15 % to 20 % (Brandle et al. 1984; Kort 1988). These increases are a result of increased moisture due to snow capture and the protection of the crop from wind desiccation.

Integrated pest management and windbreaks

Both crop pests and their natural enemies are influenced by the presence of windbreaks (Dix et al. 1995; Burel 1996). This influence is reflected in the distribution of insects as a result of wind speed reductions in the sheltered zone (Heisler and Dix 1988; Pasek 1988) or as a function of the numerous microhabitats, including the diversity of the associated plant species (Corbett and Plant 1993; Corbett and Rosenheim 1996; Forman 1995). Windbreaks influence the distribution of both predator and prey. In narrow vegetative windbreaks or artificial windbreaks, insect distribution appears to be primarily a function of wind conditions. As windbreak structure becomes more complex, various microhabitats are created and insect populations increase in both number and diversity (Pasek 1988). Greater vegetative diversity of the edges provides numerous microhabitats for life-cycle activities and a variety of hosts, prey, pollen, and nectar sources.

Windbreak technology at the farm and landscape levels

In this section we identify other windbreak uses and their benefits and discuss very briefly the ecological implications of windbreak technology to support the farm operation.

Livestock windbreaks

There are many benefits of windbreaks to the successful livestock operation. As in the case of crops, the goal is to utilize the microclimate conditions created by shelter to benefit the animal production system. In the northern Great Plains of the United States, the Canadian Prairie region, and southern Australia, livestock protection is a vital part of successful operations. Livestock vary in their need for wind protection (Primault 1979). Producers in North and South Dakota, United States, report significant savings in feed costs, survival, and milk production when livestock are protected by windbreaks from winter storms (Anderson and Bird 1993). New-born lambs and freshly shorn sheep are especially sensitive to cold, wet, windy conditions (Bird 2000; Holmes and Sykes 1984) and benefit significantly from wind protection. While the literature on the effects of shelter on livestock production is not nearly as extensive as that pertaining to crop production, there does appear to be a consensus, especially among producers, that reducing wind speed in winter lowers animal stress, improves animal health, increases feed efficiency, and provides positive economic returns (Atchison and Strine 1984; Bird 2000).

As minimum daily temperatures decrease, cattle on rangeland spend less time grazing, reducing forage intake and weight gain (Adams et al. 1986). In a pair of recent studies of winter stalk grazing in east-central Nebraska (Morris et al. 1996; Jordon et al. 1997), average winter temperatures (1994-1995 and 1995-1996) were moderate and animals behaved similarly on both open and sheltered fields. However, on days with low temperatures ($\leq 20\,^{\circ}$C) and strong winds (>10 m/s), cattle sought any available shelter. In particular, it was noted that cattle on the sheltered fields were grazing in the sheltered zones, while cattle on the exposed fields were lying down in low areas to reduce stress associated with the cold, windy conditions. Even so, they concluded that shelter had little effect on weight gain from winter stalk grazing during mild winters in east-central Nebraska.

Properly designed livestock windbreaks provide additional benefits to the livestock producer (Dronen 1988; Brandle et al. 2000). On rangeland, windbreaks located across the landscape will increase the amount of forage production on the sheltered areas (Kort 1988) and provide protection for calving against early spring snow-storms. In a Kansas study, average calving success increased 2% when cows were protected by a windbreak (Quam et al. 1994). Windbreaks can be designed to harvest snow and provide water to supplement stock ponds located in remote areas (Tabler and Johnson 1971; Jairell and Schmidt 1986, 1992). Protecting confinement systems with multi-row windbreaks can control snow drifting, enabling access to feedlots and other facilities such as grain and hay storage, and reducing costs associated with snow removal (Dronen 1984).

Windbreaks and specialty crops

Incorporating various nut or fruit species, woody decorative florals or other specialty crops into windbreak plantings may provide additional income for producers. A recent study in Nebraska (Josiah et al. 2004) indicated gross returns approaching $15 per meter on the best producing species. Initial investment, labor costs and marketing expenses are high and remain the principle challenge for producers wishing to pursue these types of operations. A careful market analysis should be conducted prior to pursuing specialty crop production systems as local markets are often limited and quickly saturated leaving the producer with few options.

Farmstead windbreaks

The basic goal of a farmstead windbreak is to provide protection to the living and working area of a farm or ranch. The greatest economic benefit is derived from reducing the amount of energy needed to heat and cool the home. The amount of savings varies with climatic conditions, (particularly wind and temperature), as well as local site conditions, home construction, and the design and condition of the windbreak. Well-designed farmstead windbreaks can cut the average energy use of a typical farm or ranch home in the northern portions of the United States and Canada by 10% to 30% (DeWalle and Heisler 1988).

Farmstead windbreaks improve living and working conditions by screening undesirable sights, sounds, smells, and dust from nearby agricultural activities or roads. They reduce the effects of windchill and make outdoor activities less stressful. Properly located farmstead windbreaks can help in snow management, reducing the time and energy involved in snow removal from farm working areas and driveways. Locating the family garden within the sheltered zone improves yield and quality, and incorporating fruit and nut production into the windbreak will add additional benefits (Wight 1988).

Wildlife windbreaks

In many agricultural areas, windbreak and riparian systems offer the only woody habitat for wildlife (Johnson and Beck 1988). In Nebraska, foresters identify wildlife as a primary reason given by landowners for the establishment of windbreaks on agricultural land. Recently, Beecher et al. (2002) reemphasized the potential role of these types of habitats in the control of crop pests in agricultural regions. Because of their linear nature, windbreaks are dominated by edge species. As the width of a windbreak increases, species diversity increases as additional microhabitats are added (Forman 1995). In a Kansas study of habitat use within agricultural settings, these linear forests were favored by hunters and contributed significantly to the local economy (Cable and Cook 1990).

Windbreaks and climate change

Brandle et al. (1992b) assessed the potential of windbreaks as a means of reducing atmospheric CO_2 concentration. They calculated not only the direct sequestration of carbon in the growing trees but also estimated the indirect benefits to agricultural production systems due to crop and livestock protection and energy savings (See also Kort and Turnock 1999).

Windbreaks can play a significant role in adaptation strategies as agricultural producers strive to adapt to changing climates. Easterling et al. (1997) reported that windbreaks could help maintain maize (*Zea mays*) production in eastern Nebraska under several climate scenarios. Using a crop modeling approach, they considered temperature increases up to 5 °C, precipitation levels of 70% to 130% of normal, and wind speed changes of plus or minus 30%. In all cases, sheltered crops continued to perform better than nonsheltered crops. In all but the most extreme cases, windbreaks more than compensated for the change in climate, indicating the potential value of wind protection under these conditions.

Conclusion

In the context of agroforestry practices in temperate regions, windbreaks or shelterbelts are a major component of successful agricultural systems. By increasing crop production while reducing the level of inputs, they reduce the environmental costs associated with agriculture. They help control erosion, particularly wind erosion, and contribute to the long-term health of our agricultural systems. When various species are included in the design, they can contribute directly to the production of nuts, fruits, timber, and other wood products as well as farmstead aesthetics. When used in livestock production systems, they improve animal health, improve feed efficiency, and contribute to the economic return of producers. Designed for snow management, they can capture snow for crop or livestock production.

As part of the overall agricultural enterprise, they reduce energy consumption by the farm or ranch home and improve working conditions within the farm area. When designed for snow control, they can reduce the costs of snow removal and improve access to livestock feeding areas. Windbreaks provide habitat for wildlife and a number of benefits to landowners and producers alike. The interspersion of woody wildlife habitat in agricultural areas contributes to a healthy and diverse wildlife population to the benefit of both hunters and nonhunters.

On a larger scale, windbreaks provide societal benefits both locally and regionally. Reductions in erosion not only benefit the landowner but reduce the off-site costs of erosion as well. Windbreaks have potential to assist with adapting to future changes in climate and may, in some cases, ease the economic burdens associated with this change.

The integration of windbreaks and other agroforestry practices into sustainable agricultural systems can provide many rewards. It requires, however, careful consideration of all aspects of the agricultural system, an understanding of basic ecological principles, and a working knowledge of local conditions and markets.

Future research needs

Even with the long history of windbreak research there remain a number of specific questions, which should be addressed. For example: *i*. What are the relationships between windbreak structure and how the windbreak functions? *ii*. Are there methods available to practitioners to determine the three-dimensional structure of a windbreak so that the landowner can manage the windbreak to meet his/her specific goal? *iii.* What are the mechanisms of crop and animal response to sheltered microclimate and how can we manipulate the microclimate to take advantage of the shelter we have created?

In addition to these very detailed questions, there are two very broad issues, which must be addressed. We must begin to look at the role of woody plants, whether in windbreaks, riparian systems or other woody plantings, in the context of the overall agricultural landscape. New techniques in landscape ecology must be applied to determine the overall impact of woody plants on ecosystem health and the impacts of diverse landscapes on human health. Second, while research has identified numerous benefits, both economic and environmental, the use of many conservation buffer plantings such as windbreaks or riparian forest buffers, is not wide spread. Adoption by landowners has been limited. Understanding adoption techniques and developing new ways to secure higher levels of adoption of conservation practices involving woody plants are critical to the future success of agroforestry programs.

Acknowledgements

A contribution of the University of Nebraska Agricultural Research Division, Lincoln NE, Journal Series No. 14350. This research was supported in part by funds provided through the Hatch Act and by McIntyre Stennis Forestry Research Funds.

References

Aase J.K. and Siddoway F.H. 1976. Influence of tall wheatgrass wind barriers on soil drying. Agron J 68: 627–631.

Adams D.C., Nelsen T.C., Reynolds W.L. and Knapp B.W. 1986. Winter grazing activity and forage intake of range cows in the northern Great Plains. J Anim Sci 62: 1240–1246.

Anderson V.L. and Bird J. 1993. Effects of shelterbelt protection on performance of feedlot steers during a North Dakota winter. 1993 Beef Production Field Day, Carrington Res. Ext. Ctr., Livestock Unit. Vol 16: 19–21. North Dakota State Univ., Fargo.

Argete J.C. and Wilson J.D. 1989. The microclimate in the centre of small square sheltered plots. Agr For Meteorol 48: 185–199.

Armbrust D.V. 1982. Physiological responses to wind and sandblast damage by grain sorghum plants. Agron J 74: 133–135.

Atchison F.D. and Strine J.H. 1984. Windbreak protection for beef cattle. Kansas State University Cooperative Extension Service Publication No. L-708, Manhattan, KS.

Baldwin C.S. 1988. The influence of field windbreaks on vegetable and specialty crops. Agric Ecosyst Environ 22/23: 159–163.

Beecher N.A., Johnson R.J., Brandle J.R., Case R.M. and Young L.J. 2002. Agroecology of birds in organic and nonorganic farmland. Conserv Biol 16: 1620–1631.

Biddington N.L. 1986. The effects of mechanically-induced stress in plants – a review. Plant Growth Regulat 4: 103–123.

Bird P.R. 2000. Farm forestry in southern Australia – a focus on clearwood production of specialty timbers. Pastoral and Veterinary Institute, Dept. Natural Resources and Environment, Victoria, Australia, 264 pp.

Brandle J.R., Hintz D.L. and Sturrock J.W. 1988. Windbreak Technology. Elsevier Science Publishers, Amsterdam, 598 pp.

Brandle J.R., Hodges L. and Stuthman J. 1995. Windbreaks and specialty crops for greater profits. pp. 81–91. In: Rietveld W.J. (ed.), Agroforestry and Sustainable Systems: Symposium proceedings. USDA For. Serv. Gen. Tech. Rpt. RM-GTR-261.

Brandle J.R., Hodges L. and Wight B. 2000. Windbreak practices. pp. 79–118. In: Garrett H.E., Rietveld W.E. and Fisher R.F. (eds), North American agroforestry: An Integrated Science and Practice. Am. Soc. Agronomy, Madison, WI.

Brandle J.R., Johnson B.B. and Akeson T. 1992a. Field windbreaks: Are they economical? J Prod Agric 5: 393–398.

Brandle J.R., Johnson B.B. and Dearmont D.D. 1984. Windbreak economics: The case of winter wheat production in eastern Nebraska. J Soil Water Conserv 39: 339–343.

Brandle J.R., Wardle T.D. and Bratton G.F. 1992b. Opportunities to increase tree planting in shelterbelts and the potential impacts on carbon storage and conservation. pp. 157–176. In: Sampson N.R. and Hair F. (eds), Forests and Global Change, Vol. 1. Opportunities for Increasing Forest Cover. American Forests, Washington, DC.

Burel F. 1996. Hedgerows and their roles in agricultural landscapes. Critical Reviews in Plant Sciences 15(2): 169–190.

Burke S. 1998. Windbreaks. Inkata Press, Port Melbourne, Victoria, Australia.

Cable T.T. and Cook P.S. 1990. The use of windbreaks by hunters in Kansas. J Soil Water Conserv 45: 575–577.

Caborn J.M. 1957. Shelterbelts and Microclimate. Forestry Commission Bull. No. 29. Edinburgh University, Edinburgh, 135 pp.

Caborn J.M. 1971. The agronomic and biological significance of hedgerows. Outlook Agr 6: 279–284.

Cleugh H.A. 2002. Field measurements of windbreak effects on airflow, turbulent exchange and microclimates. Aust J Exp Ag 42: 665–677.

Cleugh H.A. and Hughes D.E. 2002. Impact of shelter on crop microclimates: A synthesis of results from wind tunnel and field experiments. Aust J Exp Ag 42: 679–701.

Cleugh H.A., Prinsley R., Bird P.R., Brooks S.J., Carberry P.S., Crawford M.C., Jackson T.T., Meinke H., Mylius S.J., Nuberg I.K., Sudmeyer R.A. and Wright A.J. 2002. The Australian National windbreaks program: Overview and summary of results. Aust J Exp Ag 42: 649–664.

Corbett A. and Plant R.E. 1993. Role of movement in the response of natural enemies to agroecosystem diversification: A theoretical evaluation. Environ Entomol 22(3): 519–531.

Corbett A. and Rosenheim J.A. 1996. Impact of a natural enemy overwintering refuge and its interaction with the surrounding landscape. Ecol Entomol 21: 155–164.

Coutts M.P. and Grace J. 1995. Wind and Trees. University Press, Cambridge, 485 pp.

Davis J.E. and Norman J.M. 1988. Effects of shelter on plant water use. Agric Ecosyst Environ 22/23: 393–402.

DeWalle D.R. and Heisler G.M. 1988. Use of windbreaks for home energy conservation. Agric Ecosyst Environ 22/23: 243–260.

Dix M.E., Johnson R.J., Harrel M.O., Case R.M., Wright R.J., Hodges L, Brandle J.R., Schoenberger M.M., Sunderman N.J., Fitzmaurice R.L., Young L.J. and Hubbard K.G. 1995. Influence of trees on abundance of natural enemies of insect pests: A review. Agroforest Syst 29: 303–311.
Drew R.L.K. 1982. The effects of irrigation and of shelter from wind on emergence of carrot and cabbage seedlings. J Hort Sci 57: 215–219.
Dronen S.I. 1984. Windbreaks in the Great Plains. North J Appl For 1: 55–59.
Dronen S.I. 1988. Layout and design criteria for livestock windbreaks. Agric Ecosyst Environ 22/23: 231–240.
Droze W.H. 1977. Trees, Prairies, and People: A History of Tree Planting in the Plains States. USDA For. Serv. and Texas Woman's University Press, Denton, TX, 313 pp.
Easson D.L., White E.M. and Pickles S.J. 1993. The effects of weather, seed rate, and cultivar on lodging and yield in winter wheat. J Agric Sci 121: 145–156.
Easterling W.E., Hays C.J., Easterling M.M., and Brandle J.R. 1997. Modeling the effect of shelterbelts on maize productivity under climate change: An application of the EPIC model. Agric Ecosyst Environ 61: 163–176.
Forman R.T.T. 1995. Land Mosaics: The Ecology of Landscapes and Regions. Cambridge University Press, Cambridge, 632 pp.
Grace J. 1977. Plant Response to Wind. Academic Press, London.
Grace J. 1981. Some effects of wind on plants. pp. 31–56. In: Grace J., Ford E.D. and Jarvis P.G. (eds), Plants and their Atmospheric Environment, Blackwell Scientific Publishers, Oxford.
Grace J. 1988. Plant response to wind. Agr Ecosyst Environ 22/23: 71–88.
Greb B.W. 1980. Snowfall and its potential management in the semi-arid central Great Plains. USDA. Res. Sci. Ed. Admin., Western Series No. 18. USDA, Agricultural Research, Oakland, CA.
Hall D.J.M., Sudmeyer R.A., McLernon C.K. and Short R.J. 2002. Characterization of a windbreak system on the south coast of Western Australia. 3. Soil water and hydrology. Aust J Exp Ag 42: 729–738.
Heisler G.M. and DeWalle D.R. 1988. Effects of windbreak structure on wind flow. Agric Ecosyst Environ 22/23: 41–69.
Heisler G.M. and Dix M.E. 1988. Effects of windbreaks on local distribution of airborne insects. pp. 5–12. In: Dix M.E. and Harrell M. (eds), Insects of Windbreaks and Related Plantings: Distribution, Importance, and Management. Conference Proceedings, December 6, 1988, Louisville, KY. USDA Forest Service, Gen Tech Rpt RM-204.
Hodges L. and Brandle J.R. 1996. Windbreaks: An important component in a plasticulture system. HortTechnology 6(3): 177–181.
Holmes C.W. and Sykes A.R. 1984. Shelter and climatic effects on livestock. pp. 19–35. In: Sturrock J.W. (ed.). Shelter Research Needs in Relation to Primary Production: The report of the National Shelter Working Party, Wellington, New Zealand. Water Soil Misc Publ No. 59.
Hou Q., Brandle J., Hubbard K., Schoeneberger M., Nieto C. and Francis C. 2003. Alteration of soil water content consequent to root-pruning at a windbreak/crop interface in Nebraska, USA. Agroforest Syst 57: 137–147.
Huszar P.C. and Piper S.L. 1986. Estimating the off-site costs of wind erosion in New Mexico. J Soil Water Conserv 41(6): 414–416.
Jairell R.L. and Schmidt R.A. 1986. Scale model tests help optimize wind protection and water improvements for livestock. pp. 159–161. In: Hintz D.L. and Brandle J.R. (eds), Proceedings International Symposium on Windbreak Technology, June 23–27, 1986. Great Plains Agricultural Council-Forestry Committee. Publ. No. 117.
Jairell R.L. and Schmidt R.A. 1992. Harvesting snow when water levels are low. pp. 121–124. In: Shafer B. (ed.) Proceedings, 60th Western Snow Conference, April 14–16, 1992, Jackson, WY.
Johnson R.J. and Beck M.M. 1988. Influences of shelterbelts on wildlife management and biology. Agric Ecosyst Environ 22/23: 301–335.
Jordan D.J., Klopfenstein T., Brandle J. and Klemesrud M. 1997. Cornstalk grazing in protected and unprotected fields. pp. 24. In: 1997 Nebraska Beef Report, University of Nebraska, Lincoln. MP 67.
Jose S., Gellespie A.R. and Pallardy S.G. 2004. Interspecific interactions in temperate agroforestry (This volume).
Josiah S.J., St-Pierre R, Brott H. and Brandle J.R. 2004. Productive conservation: Diversifying farm enterprises by producing specialty woody products in agroforestry systems. J Sustain Agr 23: 93–108.
Konstantinov A.R. and Struzer L.R. 1965. Shelterbelts and crop yields. Translated from Russian by Israel Program for Scientific Translations. U.S. Department of Commerce, Clearinghouse for Federal Scientific Technical Information, Springfield, VA.
Kort J. 1988. Benefits of windbreaks to field and forage crops. Agric Ecosyst Environ 22/23: 165–190.
Kort J. and Turnock R. 1999. Carbon reservoir and biomass in Canadian prairie shelterbelts. Agroforest Syst 44: 175–186.
Lowry W.P. 1967. Weather and Life: An Introduction to Biometeorology. Academic Press, New York.
Luis P.P. and Bloomberg M. 2002. Windbreaks in southern Patagonia, Argentina: A review of research on growth models, wind speed reduction, and effects on crops. Agroforest Syst 56: 129–144.
Lyles L., Tatarko J. and Dickerson J.D. 1984. Windbreak effects on soil water and wheat yield. Trans ASAE 20: 69–72.
McNaughton K.G. 1988. Effects of windbreaks on turbulent transport and microclimate. Agr Ecosyst Environ 22/23: 17–40.
Miller J.M., Bohm M. and Cleugh H.A. 1995. Direct mechanical effects of wind on selected crops: A review. Centre for Environmental Mechanics, Canberra, ACT. Tech. Rpt. No. 67.
Mitchell C.A. 1977. Influence of mechanical stress on auxin stimulated growth of excised pea stem sections. Physiol Plant 41: 129–134.
Monteith J.L. 1981. Coupling of plants to the atmosphere. pp. 1–29. In: Grace J., Ford E.D. and Jarvis P.G. (eds), Plants and their Atmospheric Environment. Blackwell Scientific Publishers, Oxford.
Monteith J.L. 1993. The exchange of water and carbon by crops in a Mediterranean climate. Irrig Sci 14: 85–91.
Morris C., Klopfenstein T., Brandle J., Stock R., Shain D. and Klemesrud M. 1996. Winter calf grazing and field windbreaks. 1996 Nebraska Beef Report. University of Nebraska, Lincoln, MP 66-A: 44.
Nair P.K.R. 1993. An Introduction to Agroforestry. Kluwer, Dordrecht, The Netherlands, 499 pp.
Norton R.L. 1988. Windbreaks: Benefits to orchard and vineyard crops. Agric Ecosyst Environ 22/23: 205–213.
Nuberg I.K. 1998. Effect of shelter on temperate crops: A review to define research for Australian conditions. Agroforest Syst 41: 3–34.
Nuberg I.K. and Mylius S.J. 2002. Effect of shelter on the yield and water use of wheat. Aust J Exp Ag 42: 773–780.
Nuberg I.K., Mylius S.J., Edwards J.M. and Davey C. 2002. Windbreak research in a South Australian cropping system. Aust J Exp Ag 42: 781–795.

Ogbuehi S.N. and Brandle J.R. 1982. Influence of windbreak-shelter on soybean growth, canopy structure, and light relations. Crop Sci 22: 269–273.

Pasek J.E. 1988. Influence of wind and windbreaks on local dispersal of insects. Agric Ecosyst Environ 22/23: 539–554.

Pitcairn C.E.R., Jeffree C.E. and Grace J. 1986. Influence of polishing and abrasion on the diffusive conductance of leaf surfaces of *Festuca arundinacea* Schreb. Plant Cell Environ 9: 191–196.

Pimental D., Harvey C., Resosudarmo P., Sinclair K., Kurz D., McNair M., Crist S., Shpritz L., Fitton L., Saffouri R. and Blair R. 1995. Environmental and economic costs of soil erosion and conservation benefits. Science 267: 1117–1123.

Primault B. 1979. Optimum climate for animals. pp. 182–189. In: Seemann J., Chirkov Y.I., Lomas J. and Primault B. (eds), Agrometeorology, Springer-Verlag, Berlin.

Quam V.C., Johnson L., Wight B. and Brandle J.R. 1994. Windbreaks for livestock production. University of Nebraska Cooperative Extension EC 94-1766-X, Lincoln, NE.

Rasmussen S.D. and Shapiro C.A. 1990. Effects of tree root-pruning adjacent to windbreaks on corn and soybeans. J Soil Water Conserv 45(5): 571–575.

Ribaudo M.O. 1986. Targeting soil conservation programs. Land Econ 62(4): 402–411.

Rosenberg N.J. 1966. Microclimate, air mixing, and physiological regulation of transpiration as influenced by wind shelter in an irrigated bean field. Agr Meteorol 3: 197–224.

Scholten H. 1988. Snow distribution on crop fields. Agric Ecosyst Environ 22/23: 363–380.

Schmidt R.A. Jr. 1972. Sublimination of wind-transported snow – a model. USDA For. Serv. Res. Pap. RM-90. Rocky Mtn. Forest and Range Exp. Stn.

Shaw D.L. 1988. The design and use of living snowfences in North America. Agr Ecosyst Environ 22/23: 351–362.

Skidmore E.L. 1966. Wind and sandblast injury to seedling green beans. Agron J 58: 311–315.

Sturrock J.W. 1984. The role of shelter in irrigation and water use. pp. 79–86. In: Sturrock J.W. (ed.), Shelter Research Needs in Relation to Primary Production: The Report of the National Shelter Working Party Ministry of Works and Development, Water and Soil Misc. Publ. No. 94, Wellington, New Zealand.

Sudmeyer R.A. and Scott P.R. 2002. Characterization of a windbreak system on the south coast of Western Australia 1. Microclimate and wind erosion. Aust J Exp Ag 42: 717–715.

Sudmeyer R.A., Hall D.J.M., Eastham J. and Adams M.A. 2002. The tree-crop interface: The effects of root pruning in southwestern Australia. Aust J Exp Ag 42: 763–772.

Tabler R.D. and Johnson K.L. 1971. Snow fences for watershed management. pp. 116–121. In: Snow and ice in relation to wildlife and recreation. Symposium Proceedings. Ames, IA, February 11–12, 1971.

Takle E.S., Wang H., Schmidt R.A., Brandle J.R. and Jairell R.L. 1997. Pressure perturbation around shelterbelts: Measurements and model results. In: 12th Symposium on Boundary Layers and Turbulence. pp. 563–564. American Meteorological Society, July, 1997, Vancouver, British Columbia.

van Eimern J., Karschon R., Razumava L.A. and Robertson G.W. 1964. Windbreaks and Shelterbelts. Technical Note No. 59 (WMO-No.147.TP.70), Geneva, Switzerland.

van Gardingen P. and Grace J. 1991. Plants and wind. Adv Bot Res 18: 189–253.

Wang H. and Takle E.S. 1995. Numerical simulations of shelterbelt effects on wind direction. J Appl Meteorol 34: 2206–2219.

Wang H. and Takle E.S. 1996a. On shelter efficiency of shelterbelts in oblique wind. Agr For Meteorol 81: 95–117.

Wang H. and Takle E.S. 1996b. On three-dimensionality of shelterbelt structure and its influences on shelter effects. Boundary-Layer Meteorol 79: 83–105.

Webster A.J.F. 1970. Direct effects of cold weather on the energetic efficiency of beef production in different regions of Canada. Can J Anim Sci 50: 563–573.

Wight B. 1988. Farmstead windbreaks. Agr Ecosyst Environ 22/23: 261–280.

Zhao Z., Xiao L., Zhao T. and Zhang H. 1995. Windbreaks for Agriculture. China Forestry Publishing House, Beijing (In Chinese).

Zhou X.H., Brandle J.R., Takle E.S. and Mize C.W. 2002. Estimation of the three-dimensional aerodynamic structure of a green ash shelterbelt. Agr Forest Meteorol 111: 93–108.

Zhou X.H., Brandle J.R., Mize C.W. and Takle E.S. 2004. The relationship of three-dimensional structure to shelterbelt function: A theoretical hypothesis. J Crop Production (In press).

Agroforestry Systems **61:** 79–90, 2004.

Organic farming and agroforestry: Alleycropping for mulch production for organic farms of southeastern United States

C.F. Jordan
Institute of Ecology, University of Georgia, Athens, Georgia, 30602-2202, USA; e-mail: CFJordan@UGA.EDU

Key words: Albizia julibrissin, Green manure, Hedgerows, Permaculture, Prunings, Weed suppression

Abstract

Organic farming offers an alternative that can eliminate many of the environmental problems of conventional agriculture in the industrialized world. Instead of using petroleum-derived chemicals to fertilize and protect crops, farmers manage their fields so as to take advantage of naturally-produced composts and mulches that recycle nutrients, and control pests and weeds. However, organic farming is often logistically inefficient, because these organic composts and mulches are bulky and difficult to transport. Alleycropping as practiced in the tropics may be able to make organic farming more efficient in the southeastern United States. In this form of alleycropping, trees or shrubs, often leguminous, are planted in hedgerows between open spaces ('alleys') where the crop is grown. The hedgerow species are periodically pruned (both aboveground and belowground), and prunings fall directly onto or into the soil where the crop is growing. These prunings add carbon and nutrients to the soil, and provide mulch that helps suppress weeds. Use of prunings reduces the need for composting and hauling manures and mulches, thereby increasing the efficiency by which organic material is supplied to the soil that supports the economic crop. In Georgia, dry weight annual production of prunings reached up to 18.4 Mg ha^{-1}, a quantity high enough to maintain crop production. Thus, alleycropping may be feasible for organic farmers in the southeastern United States.

Introduction

Agriculture before the industrial revolution was by definition mostly organic because agricultural chemicals, except for some fungicides, were not available. Fertilizing crops was a major problem for the pre-industrial farmers, because animal and green manures were bulky and inconvenient to collect and spread. Industrial farming made agriculture much easier, but it also created environmental problems. Modern organic farming was founded as a reaction against agricultural practices that had the potential to harm nature and human health (Harwood 1990). While organic farming is growing in importance in Europe and the United States, some certified organic farmers do not adhere to the principles of organic farming advocated by the founders of the modern organic farming movement (International Federation of Organic Agriculture Movements, 1989, cited in Lampkin 1990). One such principle is to avoid soil amendments not generated on the farm itself. Another is to control weeds with mulch instead of a harrow, since tilling the soil destroys soil organic matter. These principles are often economically and logistically difficult for the small and medium-scale farmer to put into practice (National Research Council 1989). Agroforestry, specifically alleycropping, can help these farmers make organic farming more efficient.

Alleycropping is a type of agroforestry found in both tropical and temperate zones. The tropical version frequently consists of alleys for economic crops bordered by rows of leguminous trees or shrubs that are regularly pruned (Kang et al. 1990). The prunings add soil fertility and help suppress weeds. In contrast, the trees in alleycropping systems of the temperate zone often are used for fruit, nuts, or timber (Workman and Allen 2003). Tropical style alleycropping generally is not practiced in temperate farming operations, because it is not readily adaptable to mechanization. However, alleycropping where hedgerows are used to improve soil fertility and suppress weeds may be more attractive to the small-scale organic farmer. The major problem is obtaining enough biomass in the prunings to effectively perform these functions (Seiter et al. 1999). This chapter presents evidence that in climates similar to southeastern Unites States, hedgerow production may be high enough for tropical-style alleycropping to be feasible, thereby helping farmers grow their crops in accordance with the principles set forth by the founders of modern organic agriculture.

Antecedents to modern organic agriculture

Although domestication of wild plants probably occurred independently among indigenous tribes throughout the world (Neumann 2003), the roots of agriculture as it is known today is thought to have originated 11 000 years ago in the region between the Tigris and Euphrates rivers in the ancient kingdom of Mesopotamia. Farming villages were located on upland sites with friable, fertile, easily tilled soils, and irrigation was used to increase production (Troeh et al. 2004). Throughout the millennia, agriculture spread around the world. Efforts to increase yields were carried out by wider adoption of established techniques such as the use of legumes to enhance soil fertility. However, increases in yields were slow until around the mid 1800s. The industrial revolution of that time was a turning point for agriculture, as factory-made implements aimed at saving labor were widely diffused and artificial fertilizers were introduced (Grigg 1984).

Although the term 'organic farming' is of recent origin, farming throughout the ages was in a sense organic, because synthetic fertilizers, pesticides, and herbicides were not available. Weeds were dug with hoes or mechanical cultivators. Insect pests were controlled, at least in some regions, by natural plant products. Soil fertility was also maintained through natural processes. Much of early agriculture was carried out in fertile river valleys, where periodic flooding deposited silt that fertilized the soil and suppressed weeds. Sometimes natural mulches and manures were added to the soils. For example, the Maya in early Mexico harvested waterweeds from canals to fertilize their crops in raised field beds called chinampas (Turner and Harrison 1981). In Europe, it was common for farmers to 'fallow' their fields, that is, to leave them free from cultivation for a period of time during which natural processes re-established soil fertility. But this type of farming gradually diminished with the industrial revolution.

Many historians cite the mid-seventeenth century as the beginning of the agricultural revolution (Grigg 1984). Gradual increases in crop yield were due to the step-by-step replacement of human and animal labor with tractors and a wide range of machines. The more spectacular increases following WW II were due largely to the steady rise in the use of artificial fertilizers, the adoption of new high-yielding cereal varieties, and the chemical control of pests and diseases. Implementation of these techniques in tropical regions has been called the 'Green Revolution' (Evenson and Gollin 2003).

The increasing use of synthetic chemicals in agriculture has had disadvantages. Nitrogen leached from fertilizers applied to cropland in mid western Unites States has resulted in an increase in biologically available nitrogen in the Mississippi River, which in turn has been linked to eutrophication and hypoxia (McIsaac et al. 2001). Nitrogen in the Gulf of Mexico has caused algal blooms that deplete oxygen in the waters, and resulted in decreased populations of fish and shellfish (Ferber 2001; Rabalais et al. 2002.) Phosphorus adsorbed onto soil particles that are eroded from cropland also threatens rivers, lakes, and coastal oceans with eutrophication (Bennett et al. 2001; Nair and Graetz 2004). Animal-waste lagoons and sprayfields near aquatic environments may significantly degrade water quality and endanger health (Mallin 2000). Synthetic chemicals used to control weeds also cause environmental problems. For example, the herbicide atrazine makes hermaphrodites of male frogs at concentrations commonly found in the environment (Withgott 2002). Since the 1962 publication of 'Silent Spring' (Carson 1962), it has been known that pesticide residues on food crops can cause health problems in humans and reduce populations of beneficial insects (Soule et al. 1990).

Conventional agriculture also lacks sustainability because it depends heavily on petroleum for powering farm machinery, and for hauling produce to markets that can be thousands of miles away from the farm (Pimentel and Pimentel 1996a).

Modern organic farming arose as an alternative to agriculture that depends on farm chemicals. The roots of modern organic farming can be traced in the literature to the late nineteenth/early twentieth century. In 1924, the Austrian philosopher Dr. Rudolph Steiner developed a vision of agriculture that delivered increased productivity combined with consideration for quality of soil, environmental welfare, and human health. While his concepts (Steiner 1924) regarding sustainable farming seem based mostly on intuition, revelation, or even mysticism, they are remarkably prescient in light of recent scientific studies regarding the importance of soil organic matter and the microorganisms that use that organic matter to produce a healthy soil (Pankhurst and Lynch 1994). Also in the 1920s, a movement in Germany and Switzerland was founded for agricultural reform centered on concepts of land stewardship and preservation of family farms. The key principles were self-sufficiency and economic viability, with soil fertility maintained through crop rotation and careful management and use of animal manures (Stockdale et al. 2001).

This approach to agriculture, sometimes known as organic-biological or biodynamic agriculture, remained in relative obscurity for many decades, and even today is practiced by only a small percentage of farmers in the industrialized world. The idea that soil rich in organic material is more healthful, environmentally sound, and economically sustainable because of physical and chemical properties brought about by the organic matter itself, as well as the micro-organisms in the soil that depend on this organic matter, is being taken more seriously, however, by farmers and mainstream agronomists. Because of the increasing recognition that chemical-dependent agriculture can be deleterious to the environment and to human and animal health, alternative approaches to production of food and fiber are increasingly accepted (Soule and Piper 1992). Some of these alternative approaches have been termed 'biological,' 'ecological,' 'bio-dynamic,' 'natural' (Lockeretz 1990) 'holistic resource management' and 'permaculture' (Jordan 1998). The principles that lie behind these different names are, however, essentially similar, and are most widely considered as integral to 'organic agriculture' (Lampkin 1990). Other terms such as 'agroforestry,' 'low-input agriculture,' 'sustainable agriculture,' 'alternative agriculture,' and 'agroecology' refer to techniques that sometimes include elements of organic agriculture, and that strive to be less reliant on commercial chemicals. 'Ecoagriculture' is an umbrella term for agricultural systems that enhance biodiversity, production, and ecosystem services at a landscape scale (McNeely and Scherr 2003). These systems may be organic but are not necessarily so.

Modern organic agriculture

Organic farming has been one of the fastest growing segments of U.S. agriculture during the 1990s. The United States Department of Agriculture (USDA) estimates that the value of retail sales of organic foods in 1999 was approximately $6 billion. The number of organic farmers in the United States has been increasing by about 12% per year, and in 2000 stood at about 12 200, most of them small-scale producers. Certified organic cropland more than doubled from 1992 to 1997 (McAvoy 2000). Since 1997, another 400 000 ha (about a million acres) of certified organic cropland and pasture has been added, bringing the 48-state total to 950 000 ha (Greene and Kremen 2003). The increase in production is due to increasing consumer demand. Growth in retail sales has equaled 20% or more annually since 1990. Organic products are now available in nearly 20 000 natural food stores, and are sold in 73% of all conventional grocery stores (Dimitri and Greene 2002).

While the per-hectare gross income from organic farming may sometimes be less than that from conventional farming, the resulting nonmarket values yield a higher benefit for all. A 21-year study of biodynamic, bioorganic, and conventional farming systems in Central Europe found that in the organic systems, crop yields were 20% lower, but fertilizer input was lower by 53% and pesticide input by 97%. The organic plots had enhanced soil fertility and higher biodiversity (measured by Shannon indices of soil microbial functional diversity). A higher diversity was related to lower metabolic quotient ($qCO2 =$ soil basal respiration /soil micro. biomass), i.e., greater energy efficiency (Mäder et al. 2002).

Key characteristics

Soil organic matter constitutes the key to organic farming. Protecting and maintaining soil organic matter is one of several key characteristics of organic farming, as outlined by Padel and Lampkin (1994):

- Protecting the long-term fertility of soils by fostering soil biological activity, and minimizing mechanical intervention.
- Maintaining nitrogen self-sufficiency through the use of legumes and biological nitrogen fixation, as well as effective recycling of organic materials, including crop residues and livestock wastes.

- Controlling weeds, diseases, and pests by relying primarily on crop rotations, natural predators, diversity, organic manuring, disease-resistant crop varieties, and limited (preferably minimal) thermal, biological, and chemical intervention.
- Supplementing crop nutrients, by facilitating conversion of nutrients from unavailable to available forms through the action of soil microorganisms.
- Fostering extensive management of livestock, paying full regard to their evolutionary adaptations, behavioral needs, and animal welfare issues with respect to nutrition, housing, health, breeding, and rearing.
- Reducing adverse impacts of farming systems on the wider environment and enhancing conservation of wildlife and natural habitats.

To many organic farmers, the word 'organic' connotes more than a technique for growing vegetables. An important aspect of organic-farming philosophy is the minimization of off-site inputs, both for production of crops and delivery of produce. The idea of farm sustainability implies not having to rely on sources outside the farm for inputs. Self-reliance – relying only on one's own knowledge and ability – frees the farmer from being at the mercy of suppliers. Self-reliance includes not only substituting manures for chemical fertilizers, it also includes minimizing the amount of petroleum products needed for farm operations. Pioneers of the organic movement set out to create not just an alternative mode of production (the farms) but of distribution (the 'co-op's and health-food stores) and even consumption [Pollan 2001. How organic became a marketing niche and a multibillion-dollar industry – naturally. *The N.Y. Times Magazine*, May 13, 2001, (Section 6), pp. 30–65]. Local sales provide fresher foods, and minimize the petroleum needed to get produce to market. They also foster a sense of community. In recent years, there has been an explosive growth in farmer's markets and community-supported agriculture (Halwell 2002).

Problems in organic agriculture

Composts

Farm composts are generally made by mixing a high-nitrogen component such as cow or pig manure with a low-nitrogen component such as wood chips. Manure must be gathered and transported to the mixing site. Wood chips usually have to be prepared with a mechanical chipper. Gathering, mixing, and piling of the components can be done by shovel or mechanical loader. Time required for the material to ferment often is several months, and during that interval, the compost must be periodically turned over. Spreading of the manure on the cropland is the final chore.

In some cases, green manures are used as fertilizers. These are leguminous crops such as alfalfa (*Medicago sativa*) or clover (*Trifolium* sp.), and are sometimes called cover crops. Composting is often not necessary. The green manures must, however, be cut, raked, transported to the crop, and spread. The amount of time and labor required for composting and managing is considerably more than is needed when inorganic fertilizers are used. Green manures are bulky, and contain only a small amount of nutrients. For example, the nitrogen content of Crimson clover (*Trifolium incarnatum*) is approximately 0.5 percent of wet weight. For phosphorus and potassium, the percentages are typically 0.05 and 0.5 respectively (Natural Resources Conservation Service 2003). In contrast, the percentage of essential elements in commercially available inorganic fertilizers can be as high as 20%, and the material is aggregated into small, dry particles, making application easy and quick. The difference in requirements is illustrated by Green (Green, E.V. May, 2002, pers. comm.), who obtained similar yields of sorghum (*Sorghum bicolor*) in an alleycropping system in which there were 7450 kg ha^{-1} dry weight of prunings ($\sim$20 Mg ha^{-1} wet weight), and in a conventionally cultivated system in which 448 kg ha^{-1} of inorganic fertilizer was added.

Weed control

Organic farmers often identify weeds as their key problem (Stockdale et al. 2001). Traditional methods of weed control include the preparation of a stale seedbed in which the weeds are allowed to germinate before crop seeds are planted. Once the weeds have emerged, they can be controlled by shallow cultivation, or by laying down a plastic sheet on the soil so that temperatures increase to the point where they kill the weeds. Post-emergence harrowing and inter-row hoeing can be used to control weeds in row-crops. Various types of flails are used to cut weeds, and propane torches have been adapted to burn weeds. Mulching the soil with crop residue or green manures can physically suppress weed seedling emergence. Water-permeable plastic ground covers also can be used to prevent weed growth (Lampkin 1990).

Each of these methods has serious problems. The stale-seedbed approach delays the start of growth of the economically important species. Conventional hoeing and harrowing destroy soil organic matter, the very resource that organic farmers should be trying to conserve. Flailing and mowing set the weeds back only temporarily. Mulching the soil for weed suppression has the same problems as mulching for fertility – the mulch must be collected somewhere else, transported, and distributed on the area of interest. Plastic weed control barriers are bulky, and are difficult to take up and dispose of once the growing season is over.

In contrast, weed control with herbicides is simpler, and requires less labor. Control with herbicides is especially easy when the crop species has been genetically engineered to be immune to herbicides. With crops engineered to be resistant to herbicides, the herbicide can be sprayed directly over the growing crops without fear of damaging them. But because herbicides have the potential to damage soil organisms such as algae, they are not permitted on farms that are certified 'organic.' As a result, organic farmers must expend more time and energy on controlling weeds than do conventional farmers.

Pest control

In organic farming, pest control strategies are mostly preventative rather than reactive. The primary strategy is to pursue diversity, both genetically within the crop species and among crop species (Letourneau and Altieri 1999). It is important to avoid large fields of genetically identical individuals, because once a pest species finds or adapts to the first individual, there are virtually no barriers to a population explosion of the pest. When this occurs, control can only be achieved with pesticides, an approach not permitted in organic farming. Diversity also can be achieved by interplanting, that is to plant fields with several crop species. This can also avoid exponential growth of pest species' populations.

Developing habitat for predator insects that feed upon pest species also is important (Thies and Tscharntke 1999). Fallow areas around cropped fields can serve as habitat. For example, if sunflower (*Helianthus* sp.) is planted around the perimeter of a cropped field, the blossoms will attract predator insects that will then attack pest species on the crops. This strategy is even more effective if parts of the perimeter are planted at different times, so there will always be some flowers in bloom throughout the growing season.

A diversity of species within a field can cause confusion for many insect pests (Letourneau and Altieri 1999). Diversity in structure as well as in species is important. Even weeds ('companion plants') can help reduce the impact of insect pests on crop species. In general, organic farming systems often have a greater abundance and diversity of predator arthropod communities, because of the greater diversity of plant species and structure (Basedow 1995; Berry et al. 1996).

The problems with organically acceptable methods of pest control are similar to the problems of organically acceptable methods of weed control: they are not easily adaptable to large-scale farming. Modern conventional farming is highly specialized, and requires special kinds of machinery for planting, cultivating and especially harvesting. For a farmer to invest in machinery tailored for a specific crop, he or she has to grow a lot of that crop. If the whole farm can be dedicated to one crop, the economies of scale come into effect. The organic farmer who wants to control pests by diversifying his fields is faced with a problem. It is difficult to raise a variety of crops and species while at the same time taking advantage of the economies of scale enjoyed by the conventional monocropping farmer.

The rules of organic farming

Following the rules, neglecting the spirit

Almost all farmers strive to make their means of production more efficient. Because small-scale organic farmers are at an economic disadvantage due to high labor requirements, they are especially motivated to improve their efficiency. Some organic farmers use means that, while not illegal under the USDA rules for organic farming (see details in the next section), nevertheless violate the spirit of organic farming, and the principles embedded in the 'Key Characteristics' of organic farming.

The following sections present an overview of rules that a farmer must follow to be certified 'organic.' They also explain how the rules allow a farmer to be certified, even though he or she may violate the philosophical principles of organic farming.

The rules

In order for producers to advertise and sell their grains, vegetables, fruits or meats as organic, the farm must be examined and certified by an examining agency or organization authorized to make such a certification. In the United States, the USDA has published guidelines that must be followed in order for a farm to be certified organic (National Organic Program 2003). Similarly, the European Commission has established standards for organic food and farming in Europe (Lampkin 1990). The rules list the materials and amendments that can be used on organically certified farms, as well as those that are forbidden. For example, animal manures (but not sewage) can be used as a source of nutrients, but commercially synthesized inorganic fertilizers such as ammonium nitrate cannot be. Naturally occurring insecticides such as Bt, an insecticide derived from the bacterium *Bacillus thuringiensis* (Wadman 1996) are permitted, but manufactured pesticides such as DDT are not. In order for a farm to be certified organic, the farmer must prove that he or she has used only the permissible compounds but none of the prohibited ones. Although the rules have been in effect only within the past few years, many farmers who consider their farms to be organic are already dissatisfied because the rules do not prohibit practices that violate the spirit of organic farming [Slattery P. 2003, Rebuilding the organic community. Acres: A newspaper for Eco-agriculture, published in Austin, Texas. Vol. 33 (No. 6): 14]. For example, the rules for organic certification specify that nutrients added to the soil for fertilizer must occur as part of a naturally occurring organic compound, or be released to the soil as a result of biological activity. Animal manures and green manures produced on a farm and recycled onto cropland is a traditional technique of organic agriculture, and is permitted under USDA rules. But sometimes recycling is taken to an extreme, as when manure is washed out of stables and collected in lagoons, and sprayed from there to nearby fields by means of elaborate irrigation systems. Problems occur when heavy rainfall floods the lagoons and pollutes local streams (Becker E. 2002. Feedlot perils outpace regulation. P. A10, *The N.Y. Times*, August 13, 2002).

Another example is the practice of using manure from concentrated animal feeding operations as a source of nutrients for another farm. In some places in the United States, some farms produce 'organic' cotton using chicken manure hauled in by tractor/trailers from poultry houses where tens of thousands of chickens are kept immobile in wire cages, and their droppings mixed with sawdust or other filler (Kittredge 2003).

Weed control also poses a problem for producers wishing to adhere to the spirit of organic agriculture (Coleman 1995). Since the dawn of agriculture, weeds have been controlled by various mechanical means: hoes, plows, harrows, rototillers, and shanks with sweeps, by which weeds are pulled out of the ground and their roots are chopped, or turned over the soil. The problem with these mechanical approaches is that they churn the soil, exposing horizons originally below the surface to the air (Faulkner 1947). As a result of exposure to oxygen, soil bacteria immediately begin to respire, and the energy source for this respiration is the soil organic matter. The result is the destruction of the soil organic matter, the material most essential for sustainable farming (Rovira 1994). In recent years, 'no-till cultivation' or 'conservation tillage' has been adopted by many farmers, as a means to conserve soil organic matter. In this system, the soil surface is not disturbed, or is cultivated only in narrow bands. A layer of mulch or decomposing humus lies on top of the soil, and seeds are planted by special planters that inject the seed through the mulch into the soil below. The mulch can smother many of the weeds that germinate in the mineral soil (National Research Council 1989). However, in the Southeast, where high temperatures and humidity result in rapid decomposition, mulch from a winter cover crop is usually insufficient to last a whole season on a cropped field. Within a month or six weeks after a cover crop flowers and dies, its leaf and stem litter is decomposed and has disappeared. Unless new mulch is added to the soil, weeds will invade (Carrillo D.Y. pers. comm. Sept. 2003). A recent technique to control weeds that is allowed by the USDA rules for organic farming is plastic weed barrier (Altieri 1995a). Several types are available. One type is impermeable, and an irrigation line is run under the barrier. Vegetables are planted in holes that penetrate the barrier. Another type of weed barrier is permeable, and can be used either with irrigation or rain-fed agriculture. While these barriers are relatively easy to be laid down, gathering them up after the growing season is difficult, but organic rules specify that beds must be exposed during the non-growing season (National Organic Program 2003). Disposing of the plastic is difficult and expensive, and violates the principle that farm materials should be recycled.

Another weed control technique allowable under USDA rules is to burn them with flame from a propane tank. A torch that can be hand-held or mounted behind a tractor directs the flame (Rasmussen and Ascard 1995). But using propane gas to incinerate weeds also violates the spirit of organic farming and destroys organic matter as well.

To alleviate the problem of an insufficient on-farm supply of nutrients, some organically certified farmers import organic materials such as fish meal and dried and ground algae (Gliessman 1998). These materials may be shipped long distances from the point where they are collected to the point where they are used. Annually, an estimated 50×10^{12} kcal of energy is expended to transport the 50 million tons of goods and supplies needed on U.S. farms (Pimentel and Pimentel 1996b). While importation of these compounds is permitted by organic rules, hauling them cross-country violates a principle of organic farming developed by the International Federation of Organic Agriculture Movements. That principle is 'to encourage and enhance biological cycles *within* the farming system, involving microorganisms, soil flora and fauna, plants, and animals.'

The use of plant extracts as natural insecticides also poses a dilemma for the organic farmer. Most are produced and concentrated in locations far from the organic farm. An example is azadirachtin from the seeds of the neem tree (*Azadirachta indica*) that acts against a wide range of herbivorous insect pests (Schmutterer and Ascher 1987). Other 'natural' insecticides extracted from plants such as pyrethrum (*Chrysanthemum cinerariaefolium*), rotenone (*Derris elliptica*) and quassia *(Quassia amara)*) (Lampkin 1990) are often extracted and distilled in the states or countries where these plants occur.

Another violation of the spirit of organic production is the increasing control by corporations over the whole organic food process, from production through preparation, distribution, and marketing. Some segments of the organic farming trend have already become dominated by agribusiness firms [Pollan, M. 2001. How organic became a marketing niche and a multibillion-dollar industry – naturally. *The N.Y. Times Magazine*, May 13, 2001 (Section 6) pp. 30–65]. As in conventional agriculture, efficient production of organic crops requires specialized techniques and equipment, the cost of which leads to an increase in size of farms, due to economies of scale. Smallholder farmers often cannot afford the expense and some sell their farms to corporations, while others become merely tenant farmers, a link in the agri-business chain of organic food production and distribution (National Research Council 1989).

Corporate dominance of modern organic agriculture defeats a major premise of traditional organic farming, which is that a farmer should have intimate connection to the land, and understand the characteristics and peculiarities of his or her land, as well as of the animals living on the farm. The farm should be managed 'holistically,' that is, as if it were an organism in and of itself, not merely a cog in a gigantic economic system that puts no value on land stewardship or on the community in which it exists (El Titi 1995).

Agribusiness cannot manage holistically. For example, in corporate agriculture 'precision farming' is becoming common, in which a tractor driver (not a farmer) sits in an air-conditioned cab monitoring his computer receiving signals from a satellite that pinpoints his location, and directs the driver as to how he should set his controls (Robert et al. 1998). There is no connection between the hired hand on the tractor and the earth below him. He has no understanding of what is good for the earth, and what is bad. He just does what the computer tells him. This is the antithesis of what organic farming is supposed to be about.

Following the spirit, neglecting the rules

Fruit, vegetables, grain, or meat labeled 'organic' may have been produced according to the letter of the law, but production may have violated the spirit of organic farming. Because the term 'organic' no longer certifies that food is produced in a manner consistent with the ideals of sustainability, some farmers are adopting other terms to describe the ideal that guides their work. One phrase that is used by The Land Institute in Salina Kansas is 'Farming in Nature's Image' (Soule and Piper 1992). The basic idea is that the maximum productivity (total yield, not necessarily all economic) and sustainability in a given region of the country are found in natural ecosystems, because species there, over millennia, have finely tuned their adaptive mechanisms to optimize productivity and sustainability. The goal of farmers should be to understand how natural systems function, and then use this understanding to design agricultural systems that incorporate, as much as is possible, those structures and functions that can contribute to the production of economically important crops. It is the basis of a philosophy that advocates 'Working with Nature,' that is, taking advantage of the free services of nature rather than fighting against natural processes in the attempt to impose systems dictated by strictly economic criteria (Jordan 1998).

While pure organic agriculture is fine in theory, there are many practical impediments to carrying it out. If a farmer sets a goal of maximizing on-farm recycling and minimizing importation of off-site products, he or she will have a number of problems, among them:

- Nutrient management: Fertilizing crops, when animal manures and winter cover crops are insufficient to replace nutrients lost during harvest of vegetables, grains and fruits.
- Weed control: Controlling weeds when mulch from green manures decomposes within weeks or months, leaving a bare seedbed where weeds can germinate.
- Pest control: Controlling pests, where lack of structural differentiation in crop plants results in lack of barriers to pest infiltration.

The potential of agroforestry to solve, at least partially, two key problems of organic agriculture as envisioned by its founders, are examined in the following sections. The problems are obtaining sufficient mulch to supply nutrients without subsidies from outside the farm, and controlling weeds without destroying soil organic matter.

Agroforestry

An important question that arises in the discussion on agroforestry in the context of organic farming is: What is the reason for incorporating trees into agriculture? One reason is that when trees and crops complement each other functionally and structurally, the performance of at least one component is improved. An example is coffee *Coffea* spp. that is grown under the canopy of leguminous trees (Bryan 2000). Bacteria symbiotic on the roots of the trees fix nitrogen from the atmosphere, and this nitrogen then becomes available to the coffee plants, either through leaf litter fall, or through sloughing off of roots. Trees also may bring up nutrients like potassium from soil horizons below the reach of the coffee plant roots, and make it available through litter fall to the coffee. Other benefits of trees include prevention of soil erosion and increasing the availability of phosphorus. Phosphorus is often bound in insoluble compounds, but organic acids leached from leaf litter of trees renders this phosphorus into a form available to plants (Jordan 1995). Of major importance also in this context is the improvement in quality of shade-grown, as opposed to sun-grown coffee (Muschler 2001).

Alleycropping is a type of agroforestry (Nair 1993). In the type of alleycropping that has been developed for the tropics, leguminous trees or shrubs are planted in hedgerows that border strips or 'alleys' in which an economic crop such as vegetables or grains are grown (Nair 1993). Alleycropping benefits the economic crop when the branches or roots of the hedgerow species are pruned. As the leaf litter and roots decompose, nutrients, especially nitrogen, contained in them are released to the soil and become available to the crop plants inside the alley. The system can benefit farmers where fertilizers are not available, or as in the case of organic agriculture, commercial fertilizers are not permitted.

Solving problems of organic agriculture

Applying mulch to control weeds in the soil is a common method in organic farming. The layer of organic matter inhibits the germination of weed seeds in the mineral soil below. If the seeds do germinate, the mulch prevents light from reaching the seedling, and forms a barrier that the seedling must penetrate to survive. Mulch can result from a winter cover crop that is mowed or flattened, or from mulch such as straw brought in from other parts of the farm (Gliessman 1998).

While living mulches can protect the soil (Alley et al. 1999), they are not practical when the economic crops are vegetables or row crops, because of competition. Mulch must be imported, but green manure that has been harvested usually does not last throughout the entire growing season. Once the mulch has decomposed, weeds will appear unless more mulch is supplied. However, if the crops are growing in an alleycropping system, it is relatively easy to apply new mulch. Prunings fall directly from the hedgerow into the alley. Many leguminous trees and shrubs continue putting on new growth throughout the season and can be pruned several times, thereby providing a continuous mulch cover to suppress weeds, and making an important addition to the nutrient supply for the economic crop. In a study of root dynamics of hedgerow species, Peter and Lehmann (2000) found that pruning of above-ground growth can result in the increased dieback of roots, resulting in decreased competition with crop species, and increased nutrient availability as the dead roots decompose.

Alleycropping can also help lessen insect herbivory. There are several organically acceptable approaches to insect control that are facilitated by the hedge species in an alleycropping system. One is to increase the habitat for predators and parasites on the pest species. The hedgerows in an alleycropping system can perform this function. For example, Thies and Tscharntke (1999) found that old field margin strips along rape fields were associated with increased mortality of the rape pollen beetle. These beetles were attacked by larval parasitoids *Tersilochus heterocerus* and *Phradis interstitialis* (Hymenoptera, Ichneumonidae). Such pests have a more difficult time locating, remaining on, and reproducing on their preferred hosts when these plants are spatially dispersed and masked by confusing visual and chemical stimuli presented by a system such as alleycropping (Altieri 1995b).

Problems with alleycropping

Alleycropping is sometimes criticized because land that could be producing an economic crop is used for the hedgerow species. In an alleycropping system, 20% or 25% of the land can be taken up by hedges (Matta-Machado and Jordan 1995). This could be an important consideration in areas where land is scarce. On organic farms where land is managed intensively, availability of labor to work the land is often more limiting than land availability.

A common problem with alleycropping systems is competition between the hedge species and the economic crops growing in the alley (Nair 1993). In an alleycropping experiment in the Piedmont of Georgia, corn (maize, *Zea mays*) was planted in alleys between mature hedgerows of *Albizia julibrissin*, a leguminous tree that sprouts readily when pruned. It was suspected that there was competition between the corn and the hedgerow species. Compared to controls, plots in which the roots of *Albizia julibrissin* were pruned with a trenching machine had better growth due to greater soil moisture (Reichlen, J.J., pers. comm. Sept. 2003).

Potential of alleycropping

A substantial amount of work on alleycropping has been reported from various parts of the tropics (Kang et al. 1990; Rao et al. 1998). Alleycropping in the United States is quite different from the 'alleycropping' in the tropics. Many of the temperate systems consist of trees for timber, fruit, and nuts planted in widely spaced rows, with vegetables, hay, or row crops planted in the alleys (Garrett and Harper 1999). In the southeastern United States, many orchard growers combine rows of trees with grazing or hay production. Some fruit tree growers cultivate vegetable crops between young citrus fruit seedlings for the first few years. There are also a few farmers who combine fruit and nut trees with horticultural or ornamental crops (Workman et al. 2003). An important objective of these systems is to take advantage of the open spaces between the rows of trees, while the trees are still small. However, there are few accounts of alleycropping systems where the objective was to produce mulch for weed control and soil fertility (Workman and Allen 2003).

In fact, there are few studies of temperate alleycropping systems where the objective was to quantify biomass of hedgerow prunings as nutrient inputs to the alleys. Seiter et al. (1999) experimented with alleycropping in Oregon and determined the amount of biomass input from inorganically fertilized hedgerows of alder (*Alnus rubra*) and black locust (*Robinia pseudoacacia*) to corn planted in the alleys. Mean pruning biomass ranged from 0.9 to 4.7 Mg ha^{-1} of dry matter, but this was not enough to maintain corn yield over four years. Thevathasan and Gordon (1997) examined the effects of poplar leaves on corn and barley (*Hordeum vulgare*) in southern Ontario, Canada. They found that the addition of poplar leaves through leaf fall may have increased nitrification rates and total soil organic carbon (see Thevathasan and Gordon 2004).

In the United States, alleycropping in organic farming may have the greatest applicability in the Southeast, because of environmental conditions. The high temperatures during a major part of the year as in the tropics and the abundant rainfall present optimal conditions for plant growth. That is where alleycropping can help, since the hedgerow species are continually producing organic matter throughout the long growing season.

Experimental results

In an alleycropping experiment on an Ultisol in the Georgia Piedmont, dry weight biomass input for hedgerows of *Albizia julibrissin* was more than 13 Mg ha^{-1} in 1991 (Matta-Machado and Jordan 1995). Production of sorghum was higher than in an annual legume-based cropping system. Although production of sorghum did not increase during the three-year experiment, the nutrient pool in the system (hedgerows plus soil) was still increasing. In 1993, biomass from hedgerow prunings in the same experimental plots dropped to 5.3 Mg ha^{-1}, presumably because rainfall was only 20% to 30% of normal (Rhoades et al. 1998). Soil nitrate and ammonium were 2.8 and 1.4 times higher, however, in the alleycropped plots than in the treeless controls. In 2000, another drought year, there were 7.45 Mg ha^{-1} dry weight of prunings in the same plots (Green, May, 2002, pers. comm.). In 2002, part of the plots were trenched, that is, a trench 70 cm deep and 5 cm wide was dug around along the edge of the hedgerow with a trenching machine. One third of each alley was trenched, 1/3 was trenched and the trench lined with plastic, and 1/3 was kept as control. Corn production was higher in the trenched than in the non-trenched plots, showing that competition for moisture can be important in non-irrigated alleycropping systems. Dry weight of prunings ranged from 7.1 Mg ha^{-1} on untrenched plots to 18.4 Mg ha^{-1}. In 2003, production ranged from 5.0 to 15.6 Mg ha^{-1} (Reichlen, J.J., Sept. 2003, pers. comm.).

An important problem for alleycropping in temperate regions is selection of an appropriate hedge species. Compared to tropical regions, there are very few woody leguminous species available in temperate regions. *Albizia julibrissin* is sometimes considered an invasive species, e.g, in Florida. Consequently we are now working with a shrub species native to Eastern United States, *Amorpha fruticosa*, known as false indigo. The shrub can be pruned every two to three weeks, depending on the season, after its root system has had time to become adequately established, usually a period of two to three growing seasons. With pruning of this frequency, the sprouts do not have time to become woody, and as a result, the mulch is of a consistent and desirable quality.

Applicability

Alleycropping in its present stage of development is not practical for large-scale production of commodity crops. One problem is pruning the hedges, which is still labor intensive. We prune the hedges in our alleycropping system with an articulated hedge trimmer. It requires about an hour for two people to prune the alleys in our experimental system of about 0.5 ha. At this rate, an average organic farmer could easily handle an alleycropping system of 10 ha.

Alleycropping could be useful to small farm operators, at least in southeastern United States. Since 94% of the 2 million farmers in the United States are classified as 'small farm operators' (USDA 2003), and there are 12 200 organic farms in the country (McAvoy 2000), there probably are many small-scale organic farms in the Southeast that could benefit from alleycropping.

Conclusion

Many small-scale organic farmers would prefer to adhere to the ideals of organic agriculture, as envisioned by the founders of the movement. It is difficult to do so, however, because maintaining soil fertility with organic matter is logistically difficult. Alleycropping holds the promise of increasing the efficiency of organic farming, because it simplifies the process of supplying organic matter to soils. Use of prunings from the hedgerows in the alleycropping system reduces the need for composting and hauling manures and mulches, thereby increasing the efficiency by which organic material is supplied to the soil. A major problem is generating enough biomass by the hedges. Experimental results suggest that the system is most applicable in southeastern United States, where long growing seasons and high temperatures are most favorable for biomass production of the hedgerow species.

Acknowledgements

This work was supported by an Initiative for Future Agricultural and Food Systems/Cooperative State Research, Education and Extension Service USDA grant.

References

Alley J.L., Garrett H.E., McGraw R.L., Dwyer J.P. and Blance C.A. 1999. Forage legumes as living mulches for trees in agroforestry systems. Agroforest Syst 44: 281–291.

Altieri M.A. 1995a. Organic farming. pp. 179–199. In: Agroecology: The Science of Sustainable Agriculture. Altieri M.A. (ed.), Westview Press. Boulder, Colorado, 433 pp.

Altieri M.A. 1995b. Integrated pest management. pp. 267–281. In: Agroecology: The Science of Sustainable Agriculture. Altieri M.A. (ed.), Westview Press. Boulder, Colorado, 433 pp.

Basedow T. 1995. Insect pests, their antagonists and diversity of the arthropod fauna in fields of farms managed at different intensities over a long term – a comparative survey. Mitt. Deut. Gesell. Allgemeine Angewandte Entomol 10: 565–572.

Bennett E.M., Carpenter S.R., Caraco N.F. 2001. Human impact on erodable phosphorus and eutrophication: a global perspective. BioScience 51: 227–234.

Berry N.A., Wratten S.D., McErlich A. and Frampton C. 1996. Abundance and diversity of beneficial arthropods in conventional and organic carrot crops in New Zealand. N.Z.J. Crop Hort Sci 24: 307–313.

Bryan J.A. 2000. Nitrogen fixation of leguminous trees in traditional and modern agroforestry systems. pp. 161–182. In: The Silvicultural Basis for Agroforestry Systems. Ashton M.S. and Montagnini F. (eds), CRC Press, Boca Raton, Florida. 278 pp.

Carson R. 1962. Silent Spring. Houghton Mifflin. Boston, 355 pp.

Coleman E. 1995. The New Organic Grower. Chelsea Green Publishing Co., White River Junction, Vermont, 340 pp.

Dimitri C. and Greene C. 2002. Recent Growth Patterns in the U.S. Organic Foods Market. ERS Agriculture Information Bulletin No. AIB777. http://www.ers.usda.gov/publications/aib777/

El Titi A. 1995. Ecological Aspects of Integrated Farming. pp. 242–256. In: Glen D.M., Graves M.P. and Anderson H.M. (eds), Ecology and Integrated Farming Systems. Wiley, N.Y., 329 pp.

Evenson R.E. and Gollin D. 2003. Assessing the impact of the Green Revolution, 1960 to 2000. Science 300: 758–762.

Faulkner E.H. 1947. Plowman's Folly. University of Oklahoma Press. Reprinted by Island Press, Washington, D.C., 193 pp.

Ferber D. 2001. Keeping the stygian waters at bay. Science 291: 968–973.

Garrett H.E. and Harper L.S. 1999. The science and practice of black walnut agroforestry in Missouri, USA: a temperate zone assessment. pp. 97–110. In: Buck L.E., Lassoie J.P. and Fernandes E.C.M. (eds), Agroforestry in Sustainable Agricultural Systems. CRC Press, Boca Raton, FL., 416 pp.

Greene C. and Kremen A. 2003. U.S. Organic Farming in 2000–2001: Adoption of Certified Systems. Agriculture Information Bulletin No. AIB780. http://www.ers.usda.gov/publications/aib780/

Gliessman S.R. 1998. Agroecology: Ecological Processes in Sustainable Agriculture. Ann Arbor Press. Chelsea, Michigan, 357 pp.

Grigg D.B. 1984. The agricultural revolution in Western Europe. pp. 1–17. In: Bayliss-Smith T. and Wanmali S. (eds), Understanding Green Revolutions. Cambridge Univ. Press. Cambridge, 384 pp.

Halwell B. 2002. Home Grown: The Case for Local Food in a Global Market. Worldwatch Paper 163. The Worldwatch Institute, Washington, D.C., 83 pp.

Harwood R.R. 1990. A History of Sustainable Agriculture. pp. 3–19. In: Edwards C.A., Lal R., Madden P., Miller R.H. and House G. (eds), Sustainable Agricultural Systems. St. Lucie Press, Delray Beach, Florida, 696 pp.

Jordan C.F. 1995. Conservation: Replacing Quantity with Quality as a Goal for Management. Wiley, N.Y., 340 pp.

Jordan C.F. 1998. Working with Nature: Resource Management for Sustainability. Harwood Academic Publishers. Amsterdam, 171 pp.

Kang B.T., Reynolds L. and Atta-Krah A.N. 1990. Alley Farming. Adv Agron 43: 315–359.

Kittredge J. 2003. Advocates say organic program yielding to industry pressure. Organic Farming Research Foundation Information Bulletin 12: 4–5.

Lampkin N. 1990. Organic Farming. Farming Press Books and Videos. Ipswich, U.K., 715 pp.

Letourneau D.K. and Altieri M.A. 1999. Environmental management to enhance biological control in agroecosystems. pp. 319–354. In: Bellows T.S. and Fisher T.W. (eds.) Handbook of Biological Control. Academic Press, San Diego, 1046 pp.

Lockeretz W. 1990. Major issues confronting sustainable agriculture. p. 423–438. In: Francis C.A., Flora C.B. and King L.D. (eds.), Sustainable Agriculture in Temperate Zones. John Wiley & Sons, N.Y., 487 pp.

Mäder P.A., Fliessbach A., Dubois D. Gunst L., Fried P. and Niggli U. 2002. Soil fertility and biodiversity in organic farming. Science 296: 1694–1697.

Mallin M.A. 2000. Impacts of industrial animal production on rivers and estuaries. Am Sci 88: 26–37.

Matta-Machado R.P. and Jordan C.F. 1995. Nutrient dynamics during the first three years of an alleycropping system in southern USA. Agroforest Syst 30: 351–362.

McAvoy S. 2000. Glickman announces national standards for organic food. USDA release No. 0425.00. http://www.usda.gov/news/releases/2000/12/0425.htm

McNeely J.A. and Scherr, S.J. 2003. Ecoagriculture: Strategies to Feed the World and Save Biodiversity. Island Press. Washington, 321 pp.

McIsaac G.F., David M.B., Gertner G.Z., Goolsby D.A. 2001. Nitrate flux in the Mississippi River. Nature 414: 166–167.

Muschler R.G. 2001. Shade improves coffee quality in a sub-optimal coffee-zone of Costa Rica. Agroforest Syst 51: 131–139.

Nair P.K.R. 1993. An Introduction to Agroforestry. Kluwer Academic Publishers. Dordrecht, 499 pp.

Nair V.D. and Graetz D.A. 2004. Agroforestry as an approach to minimizing nutrient loss from heavily fertilized soils: The Florida experience (This volume).

National Organic Program 2003. http://www.ams.usda.gov/nop/indexIE.htm. Accessed November 5, 2003.

National Research Council, 1989. Economic Evaluation of Alternative Farming Systems. pp. 195–241. In: Alternative Agriculture: Committee on the Role of Alternative Farming Methods in Modern Production Agriculture. National Academy Press, Washington, D.C., 448 pp.

Natural Resource Conservation Service, 2003. Plant Nutrient Content Database. http://www.nrcs.usda.gov/technical/ECS/nutrient/tbb2.html. Accessed August 10, 2003.

Neumann K. 2003. New Guinea: A cradle of agriculture. Science 301: 180–181.

Padel S. and Lampkin N.H. 1994. The Economics of Organic Farming: An International Perspective. CAB International, Wallingford, Oxon, U.K., 468 pp.

Pankhurst C.E. and Lynch J.M. 1994. The role of the soil biota in sustainable agriculture. pp. 3–9. In: Pankhurst C.E., Doube B.M., Gupta V.V.S.R. and Grace P.R. (eds), Soil Biota: Management in Sustainable Farming Systems, CSIRO. East Melbourne, Victoria, Australia, 262 pp.

Peter I. and Lehmann J. 2000. Pruning effects on root distribution and nutrient dynamics in a acacia hedgerow planting in northern Kenya. Agroforest Syst 50: 59–75.

Pimentel D. and Pimentel M. 1996a. Energy and Society. pp. 1–8. In: Pimentel D. and Pimentel M. (eds), Food, Energy, and Society. Univ. Press of Colorado, Niwot, Colorado, 165 pp.

Pimentel D. and Pimentel M. 1996b. Transport of Agricultural Supplies and Foods. pp. 199–201. In: Pimentel D. and Pimentel M. (eds), Food, Energy, and Society. Univ. Press of Colorado, Niwot, Colorado, 165 pp.

Rabalais N.N., Turner R.E., Scavia D. 2002. Beyond Science into Policy: Gulf of Mexico Hypoxia and the Mississippi River. BioScience 52: 129–142.

Rao M.R., Nair P.K.R. and Ong C.K. 1998. Biophysical interactions in tropical agroforestry systems. Agroforest Syst 38: 3–50.

Rasmussen J. and Ascard, J. 1995. Weed control in organic farming systems. pp. 49–67. In: Glen D.M., Graves M.P. and Anderson H.M. (eds), Ecology and Integrated Farming Systems. Wiley, N.Y., 329 pp.

Rhoades C.C., Nissen T.M. and Kettler J.S. 1998. Soil nitrogen dynamics in alley cropping and no-till systems on ultisols of the Georgia Piedmont, USA. Agroforest Syst 39: 31–44.

Robert P.C., Rust R.H. and Larson W.E. 1998. Proceedings of the Fourth International Conference on Precision Agriculture, 19–22 July, St. Paul, Minnesota. American Society of Agronomy, Madison, Wisconsin, 1938 pp.

Rovira A.D. 1994. The effect of farming practices on the soil biota. p. 81–87. In: Pankhurst C.E., Doube B.M., Gupta V.V.S.R. and Grace P.R. (eds), Soil Biota: Management in Sustainable Farming Systems, CSIRO. East Melbourne, Victoria, Australia, 262 pp.

Schmutterer, R. and Ascher K.R.S. (eds), 1987. Natural pesticides from the Neem tree and other tropical plants. Proceedings of the 3rd International Neem Conference, Nairobi, July 1986. Gesellschaft für technische Zusammenarbeit; Eschborn, Germany, 696 pp.

Seiter S., William R.D. and Hibbs D.E. 1999. Crop yield and tree-leaf production in three planting patterns of temperate-zone alley cropping in Oregon, USA. Agroforest Syst 46: 273–288.

Soule J., Carré D., and Jackson W. 1990. Ecological impact of modern agriculture. pp. 164–188. In: Carroll C.R., Vandermeer J.H. and Rosset P. (eds), Agroecology. McGraw-Hill, N.Y., 641 pp.

Soule J.D. and Piper J.K. 1992. Farming in Nature's Image: An Ecological Approach to Agriculture. Island Press, Covelo California, 286 pp.

Steiner R. 1924. Agriculture: A Course of Eight Lectures. Rudolph Steiner Press, Biodynamic Agriculture Association, London, 310 pp.

Stockdale E.A., Lampkin N.H., Hovi M., Keatinge R., Lennartsson E.K.M, Macdonald D.W., Padel S., Tattersall F.H., Wolfe M.S. and Watson, C.A. 2001. Agronomic and environmental implications of organic farming systems. Adv Agron 70: 261–327.

Thevathasan N.V. and Gordon A.M. 1997. Poplar leaf biomass distribution and nitrogen dynamics in a poplar-barley intercropped system in southern Ontario, Canada. Agroforest Syst 37: 79–90.

Thevathasan N.V. and Gordon A. M. 2004. Ecology of tree intercropping systems in the North temperate region: Experience from southern Ontario, Canada. (This volume).

Thies C. and Tscharntke T. 1999. Landscape structure and biological control in agroecosystems. Science 285: 893–895.

Troeh F.R., Hobbs J.A. and Donahue R.L. 2004. Soil and Water Conservation for Productivity and Environmental Protection. Prentice Hall, New Jersey, USA, 718 pp.

Turner B.L. and Harrison P.D. 1981. Prehistoric raised-field agriculture in the Maya lowlands. Science 213: 399–405.

USDA. 2003. Farmers Market Facts. http://www.ams.usda.gov/farmersmarkets/facts.htm. Accessed September 6, 2003.

Vaughan D. and Ord B.G. 1985. Soil organic matter – a perspective on its nature, extraction, turnover and role in soil fertility. pp. 1–35. In: Vaughan D. and Malcolm R.E. (eds), Soil Organic Matter and Biological Activity. Developments in Plant and Soil Sciences Vol 16. Nijhoff/Junk, Dordrecht, The Netherlands, 469 pp.

Wadman M. 1996. Row over plan to treat plants as 'pesticides.' Nature 382: 485.

Withgott J. 2002. Ubiquitous herbicide emasculates frogs. Science 296: 447–448.

Workman S.W. and Allen S.C. 2003. The practice and potential of agroforestry in the Southeastern United States. http://cstaf.ifas.ufl.edu/whitepaper.htm. Accessed November 5, 2003.

Workman S.W., Bannister M.E. and Nair P.K.R. 2003. Agroforestry potential in the southeastern United States: Perceptions of landowners and extension professionals. Agroforest Syst 59: 73–83.

Agroforestry Systems **61:** 91–106, 2004.

Mechanized land preparation in forest-based fallow systems: The experience from Eastern Amazonia

M. Denich[1,*], K. Vielhauer[1], M.S. de A. Kato[2], A. Block[3], O.R. Kato[2], T.D. de Abreu Sá[2], W. Lücke[3] and P.L.G. Vlek[1]
[1]*Center for Development Research (ZEF), University of Bonn, Walter-Flex-Str. 3, 53113 Bonn, Germany;* [2]*Empresa Brasileira de Pesquisa Agropecuaria (EMBRAPA) Amazônia Oriental, Belém, Brazil;* [3]*Institute for Agricultural Engineering, University of Göttingen, Germany;* **Author for correspondence: e-mail: m.denich@uni-bonn.de*

Key words: Secondary vegetation, Slash-and-burn agriculture, Bush chopper, Biomass, Nutrient balance, Mulch, Enrichment planting, Financial analysis, Land productivity, Labor productivity

Abstract

The slash-and-burn practice of land preparation that farmers use traditionally in forest-based fallow systems in the humid tropics causes land degradation and human health hazards. As an alternative to slash-and-burn, a mechanized, fire-free method of land preparation was evaluated on smallholdings in the eastern Amazon region. The use of machinery for harvesting fallow vegetation and chopping it for mulch eliminates the need for hard labor and fire for land clearing and increases labor productivity. Four different tractor-propelled choppers with power demand of 50 kW to 122 kW were tested. Their chopping capacity varied between 4.5 Mg and 20 Mg of fresh biomass per hour. The mechanized chop-and-mulch technology can be used in fallow vegetation that is up to 12 years old, which in the study region corresponded to 20 Mg to 150 Mg fresh biomass per hectare. Two additional choppers – a stationary silage chopper and a high-powered crawler tractor with a chopping device – were also tested but both were not suitable for smallholder fallow systems. In the context of the mulch technology, new low-input crop varieties were screened and their response to fertilizer was studied. The mulch technology facilitates extended cropping, to plant crops off-season, and modify crop rotation. Degraded fallow vegetation can be improved by enrichment planting using fast-growing leguminous tree species. Financial analysis of different scenarios revealed that farm income and labor productivity from chop-and-mulch systems can be up to two times greater than from the traditional slash-and-burn system.

Introduction

Forest-based fallow systems involve temporal integration of woody components into the land-use cycle. Such agroforestry land-use systems consist of a cropping period and a fallow period. The fallow vegetation serves a number of beneficial functions in low-input agriculture in terms of: (i) nutrient accumulation (Nye and Greenland 1960), (ii) suppression of weeds (Rouw 1995; Gallagher et al. 1999), (iii) nutrient retrieval and recovery from deeper soil layers (Sommer 2000), (iv) erosion control (Hoang Fagerström et al. 2002; McDonald et al. 2002), (v) supply of fuel wood (Sanchez 1995), and (vi) maintenance of biodiversity in the agricultural landscape (Baar 1997).

Resource-poor, smallholder farmers have traditionally been practicing the forest-based fallow systems. In developing regions, however, where agro-industrial and traditional land-use systems lie side by side or close to urban centers, fallow systems are not popular as 'modern' land-use because of their high manual labor requirement, low productivity, and the use of the almost archaic slash-and-burn practice. The burning of fallow biomass has its advantages: it is a cheap and easy practice for land clearing, the ashes reduce soil acidity and supply nutrients to crops, and the heat of the fire eliminates pests and diseases in the field. But this method has also adverse effects on the agroecosystem in terms of nutrient losses during burning through volatilization, loss of organic matter as carbon dioxide, health problems to the local population due to smoke (causing an estimated $1 million to $11 million in health costs per year in Amazonia: Diaz et al. 2002), and accidental fires in the agricultural landscape.

Fallow systems require a minimum time- and space-frame. The sustainability of fallow systems is guaranteed only if fallow periods are maintained long enough for the system to recuperate its productivity. The length of the fallow period is a function of land availability. Although longer fallow periods might result in higher crop yields, the greater land area required for longer fallow periods is often not proportionately balanced by higher yields (Kato et al. 1999; Mertz 2002). Shortening the fallow periods cannot be explained only by the wish to achieve a higher land-use factor (fallow period in relation to duration of land-use cycle) or by demographic pressure. Labor considerations and location of a field within the farm property are also important: the younger the vegetation, the easier it is to clear manually, and the closer the fields to a farmhouse or access road, the more often they are cropped.

Continuation of the traditional slash-and-burn practice under shortened fallow cycles provokes a downward spiral of decreasing productivity. In the long run, the system degrades in terms of crop production, vigor of fallow vegetation, and soil properties. Farmers finally will have to abandon their land and occupy new, more productive land or adapt their agronomic activities to cope with the escalating degradation process by, for example, progressively increasing the inputs such as labor and fertilizer, and/or expanding the cropped area. In such cases, crop yields may largely remain stable, and the system degradation becomes visible in the form of lower fallow biomass and nutrient accumulation. The deterioration of soil properties in tropical agriculture is caused mainly by an accelerated loss of soil organic matter (Tiessen et al. 1994; Shang and Tiessen 2000; McDonald et al. 2002), which leads to a decline in nutrient availability and cation exchange capacity as well as biological activity, aggregate stability, and soil aeration. Hence, to maintain soil productivity, the organic matter that is lost during the cropping period has to be replenished during the subsequent fallow period.

Efforts to modernize traditional fallow systems concentrated on replenishing soil fertility, suppressing weeds, and improving biomass and nutrient accumulation in a relatively shorter fallow period as well as on increasing the economic value of the fallow vegetation. These objectives are achieved by management practices such as ‘enriched fallows’ and ‘improved fallows’ (Nair 1993; Sanchez 1999). In enriched fallows, selected tree species are planted into natural fallows to improve biomass and nutrient accumulation (Denich et al. in press) or/and to deliver economically valuable by-products such as timber or fruits during the fallow period. Improved fallows are mostly sole stands of fast-growing nitrogen-fixing leguminous trees planted with the primary purpose of replenishing soil fertility. Both these systems are rotational fallows, and the transition from fallow to cropping is usually through fire-free land preparation, where the best stems and branches are removed and the leaves and remaining woody parts spread as mulch over the field; mulching is sometimes also combined with burning (Kanmegne and Degrande 2002). Improved fallows may perform as well as natural fallows (Tian et al. 2001), they increase crop yields and returns to labor and land (Degrande 2001), and reduce weed pressure compared to natural fallows (Akobundu and Ekeleme 2002). Improved fallows under adverse environmental conditions such as in heavily degraded lands, however, might not always be profitable (David and Raussen 2003).

Numerous studies have shown that improved fallows are agronomically and ecologically beneficial. Nevertheless, adoption of improved fallows has been low, which is mainly attributed to the discrepancies between the biophysical information on the performance of improved fallows and the socioeconomic and political realities of the farmers (Kaya et al. 2000; Mercer 2004). The time-consuming and labor-intensive management (Kanmegne and Degrande 2002) or seasonal labor demands that do not fit into the cropping calendar of the farmers (Opio 2001) may also be restricting their adoption. These problems can be overcome only by thorough assessment of improved fallows taking the socioeconomic and biophysical conditions into account (Franzel and Scherr 2002).

It was in the above background that a research and developmental project was undertaken through a joint

Table 1. Aboveground biomass (dry matter) of fallow vegetation of different ages in the northeast of Pará state, Brazil.

Component	1-year fallow	4- to 5-year fallow	7-year fallow	10-year fallow
Wood (Mg ha^{-1})	1–3	9–25	29–61	58–68
Leaves (Mg ha^{-1})	<1–2	3–5	4–6	6–9
Litter, standing dead (Mg ha^{-1})	3–6	6–8	8–11	12–17
Herbs, grasses (Mg ha^{-1})	<1–4	<1–1	<1	<1–1
Total (Mg ha^{-1})	8–12	19–38	42–77	78–94

Source: Denich (1989); Nunez J.B.H. 1995 Fitomassa e estoque de bioelementos das diversas fases da vegetação secundária, provenientes de diferentes sistemas de uso da terra no nordeste paraense, Brasil. Master of Science thesis, Federal University of Pará, Belém, Brazil, 184 pp.

German–Brazilian project aimed at improving the use of natural fallow vegetation in the east of Belém (Pará state) in the eastern Amazon region. This chapter is based primarily on the results of that project and is presented in the expectation that the results will be applicable elsewhere in the tropical humid lowlands with comparable ecological, socioeconomic, and land-use conditions.

Climatically, the region belongs to the humid tropics. The land topography is mostly flat or gently sloping; the predominant soils are Ultisols with sandy topsoil. The region covers 23 000 km^2, which is about 2% of the state's area; however, it accounts for 20% of the value of total agricultural production in Pará state. Ninety-eight percent of the farms in the region are smallholdings (<100 ha) with an average area of 14 ha, of which a third is cropped and two thirds is under fallow or forest. Family members mainly provide the labor for agricultural operations. Most smallholders participate in the local and regional markets and many in the national market. Fertilizers and farm machinery are applied by 30% of the farmers (IBGE 1998). In this project, the fallow vegetation was transformed into mulch for *in situ* use through a mechanized cut-and-chop process. This way, the burning that causes nutrient losses during land preparation is avoided and organic matter preserved. As all field studies were conducted on farmers' land, the gap between on-station research and farm reality was bridged, which is an important aspect in successful technology transfer (Muschler and Bonnemann 1997). This chapter synthesizes the work done on chop-and-mulch technology and its implications for management of subsequent crops, including economic feasibility.

Characterization of fallow vegetation in northeastern Pará

The smallholder land-use system in northeastern Pará involves a one- to two-year cropping period with maize (*Zea mays*), upland rice (*Oryza sativa*), cowpea (*Vigna unguiculata*), and cassava (*Manihot esculenta*) in rotation with a three- to seven-year fallow period. The fallow vegetation following annual crops attains a height of 1 m to 5 m and is relatively homogeneous. It differs structurally from vegetation that develops subsequent to semi-perennial crops such as passion fruit (*Passiflora edulis*) and black pepper (*Piper nigrum*) and on fields plowed before cropping which is heterogeneous and consists of a mosaic of shrub islands and grassy patches. Long cropping periods with repeated weeding as well as tillage deteriorate the regeneration potential of the fallow vegetation, as the vast majority of woody species regenerate by re-sprouting from their roots (Denich 1989; Clausing 1997; Jacobi 1997). Hence, on farmlands having rotational fallows, woody fallow vegetation develops only when the plant species that are competitive re-sprout vegetatively. Although species numbers are lower in fallow vegetation than in primary forests, vegetation surveys in 92 young- to medium-aged fallow areas (one- to 10-year-old) revealed a total of 673 plant species (Baar 1997), which included 316 species of trees and shrubs. Most of these species, however, are relatively rare: In four-year-old fallow vegetation, for example, only 20 species represent 80% of the tree and shrub individuals and biomass. Leguminous species provide 25% of the biomass (Denich 1989). Biomass accumulation plays a crucial role in fallows in which the vegetation is used as a source of mulch. Suitable mulch layers are achieved with the application of 20 Mg ha^{-1} to 30 Mg ha^{-1} of plant biomass. In first-cycle fallows of central Amazonia, these biomass quantities are pro-

Table 2. Average stem diameter, height and biomass of woody species, and number of stems per hectare in fallow vegetation of the northeast of Pará state, Brazil.

Fallow age (years)	Stem diameter[a] (cm)	Height (m)	Biomass, dry wt. (g tree^{-1})	Number of stems (stems ha^{-1})
3	0.7–1.0	1.5–1.9	97–163	107 000–140 000
4–5	0.9–1.4	1.8–2.4	135–366	67 000–97 000

Source: Denich (1989); B. Schuster (unpublished data);
[a]Measured at 0.3 m height.

duced within one year (Gehring 2003), in the degraded fallow vegetation of northeastern Pará, only within four to five years (Table 1). The latter contains 77% to 84% woody biomass and the remaining is leaf biomass (Denich 1989).

In the fire-free land-clearing practice, fallow vegetation is transformed into mulch through chopping. The efficiency of the chopping process depends on plant morphological parameters such as stem diameter, height, and biomass of the predominant trees and shrubs. The lower the values of diameter, height, and biomass, the lower are the energy demands for mulch production. In the study region, the respective values are in the lower ranges (Table 2), so that manual, semi-mechanized or fully mechanized mulch production was feasible even though individual plants with stem diameters up to 8 cm and height of more than 4 m occurred in three- to five-year-old fallow vegetation (Denich 1989; B. Schuster 2001)[1]. Another relevant property of woody plants is the wood density. Withelm (1993)[2] studied 125 woody species in fallow vegetation in northeastern Pará and found their density values to range from 0.24 to 0.99 (at 0% moisture content) with an average of 0.69 (SD = 0.18). The weighted average for young fallow vegetation was 0.68, considering the share of different individual species. Most of the density values were medium-ranked and did not seem to be a serious constraint for mulch production. On the whole, the woody fallow plants are morphologically suitable for chopping. However, the high number of stems per hectare (Table 2) might pose a problem for mulch production by fully mechanized chopping. Furthermore, the heterogeneous distribution of the plant biomass in the field at a very small-scale makes mechanized chopping a challenging operation, since very low biomass quantities alternate extreme biomass peaks irregularly (Figure 1). In our study, the fallow area was divided into strips of 2 m width and each further subdivided into 0.1 m sections lengthwise. The width is equivalent to the working width of a chopping device and the 0.1-m section is the distance moved by a chopping device in 0.1 s while chopping the vegetation. The observed biomass values ranged from 0 to 10.8 kg per 0.2 m^2, with an average of 0.33 kg per 0.2 m^2 (SD = 0.59 kg per 0.2 m^2; based on data of B. Schuster (2001)[1]. Another relevant property of woody plants is the wood density. Withelm (1993)[2] Fallows with planted trees are presumably much more homogeneous.

In addition to biomass, the nutrient stocks of the fallow vegetation are of interest. In contrast to the considerable nutrient losses through volatilization in slash-and-burn practice, all nutrients stored in the plant material (Table 3) are available to the crops in the mulch system. Nutrients in the latter system, however, are not released immediately, but after decomposition of the plant material in the course of the cropping period. How fast this takes place depends on the quality of the plant material with fresh leaf material being more readily decomposable than litter and wood. The leaf material contains one third of the aboveground N, P, K, Mg and S stocks, whereas its share in micronutrients and Ca is only one fifth of the aboveground stocks (Denich 1989; Sommer 2000).

The fallow vegetation is not only a source of organic material and nutrients, but also provides a variety of by-products to farmers. Withelm (1993)[2] reported 58% of the 144 species observed in fallows as useful for construction wood and Hohnwald (2002) reported 16% of the 192 species studied as useful as fodder plants for cattle. Furthermore, honeybees were reported to be visiting flowers of 59 species and collecting nectar and pollen in young fallow vegetation (G. Venturieri 1993. Annual Report of the project 'Secondary forests and fallow vegetation in the agricultural landscape of the eastern Amazon region – Function and management' submitted to the German Federal Ministry of Science and Technology). According to Hedden-Dunkhorst et al. (2003), two thirds of the small-scale farmers in northeastern Pará extract products from fallow vegetation, of which firewood (2 Mg ha^{-1} to 8 Mg ha^{-1} from five- to 10-year-old fal-

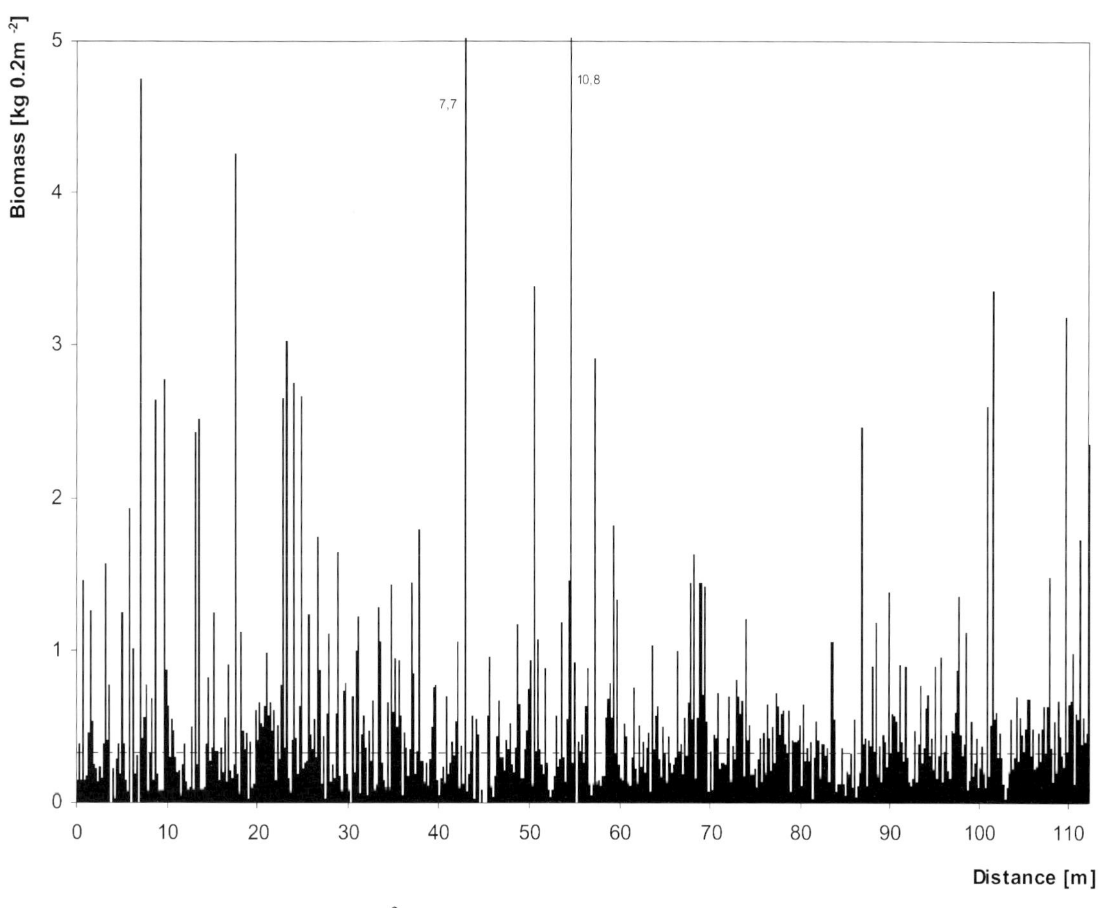

Figure 1. Spatial variability of biomass (kg 0.2 m^{-2}) in a 3-year-old fallow vegetation, in strips of 2 m width subdivided into sectors of 0.1 m (= 0.2 m^2). The horizontal line represents the mean. The strip length of 112 m was obtained by consecutively combining seven 16-m strips. See text for further explanation.

Table 3. Macro- and micro-nutrient stocks in 4- to 5-year-old fallow vegetation in the northeast of Pará state, Brazil.

Component	N	P	K	Ca	Mg	S	Mn	Zn	Cu
	(kg ha^{-1})								
Leaves	56–83	2.2–3.0	19–36	27–34	10–15	14	0.3–0.7	0.1	0.1
Wood	39–102	1.9–5.1	32–65	43–92	11–18	16	0.4–1.2	0.2–0.4	0.1–0.4
Litter	62–106	1.6–2.4	8–11	39–102	6–13	10	0.6–1.5	0.1–0.3	0.1–0.2

Source: Denich (1989); S stocks after Sommer (2000).

low) and wood for charcoal production are by far the most important in young fallow vegetation. Charcoal and firewood are mostly used for home consumption (65% and 93% respectively). Other products harvested from young fallows are materials for small implements or tools, and honey. The latter, however, as well as fruits, fencing material, construction material for houses, and wood for heavy construction are chiefly obtained from older fallows. The products derived from young fallow vegetation contribute, on average, only to a small or negligible extent to the livelihood income of the smallholders.

Slash-and-burn vs. chop-and-mulch technology

In traditional rotational fallow systems, slash-and-burn is the common practice applied by farmers world-

Table 4. Nutrient balances of fallow–crop rotation cycles with land preparation by slash-and-burn and chop-and-mulch practices.

Land preparation (Nutrient cycling processes)	N	P	K	Ca	Mg	S
	(kg ha^{-1})					
Slash-and-burn						
Atmospheric deposition	26[a]	4	12	30	15	22
Fertilization	70	48	66	31	–	–
Burning losses	−246	−8	−58	−151	−29	−35
Leaching losses	−16	−1	−11	−48	−9	−5
Harvest, firewood	−127	−22	−78	−16	−14	−7
Balance	−293	21	−69	−154	−37	−25
Chop-and-mulch						
Atmospheric deposition	26[a]	4	12	30	15	22
Fertilization	70	48	66	31	–	–
Leaching losses	−10	−1	−3	−25	−6	−13
Harvest	−112	−22	−83	−14	−12	−7
Balance	−26	29	−8	22	−3	2
Gains through chop-and-mulch	267	8	61	176	34	27

Notes:
1. Length of fallow period was 3.5 years and cropping period 2 years.
2. Negative values indicate losses.
3. [a]biological nitrogen fixation included.

Source: Modified after Sommer (2000).

wide for land preparation for cropping but it is inefficient as it causes considerable nutrient losses. In the process of burning of seven-year-old slashed fallow vegetation in our study area, 96% of nitrogen, 76% of sulfur, 47% of the phosphorus, 48% of the potassium, 35% of the calcium, and 40% of the magnesium were released to the atmosphere (Mackensen et al. 1996). These losses, together with the removal of nutrients from the field with the harvested products, are responsible for the negative nutrient balance for the slash-and-burn system (Table 4). Current rates of fertilizer use cannot compensate for the losses except for phosphorus. Leaching losses are of minor importance, especially as leached nutrients are returned to the soil surface by the nutrient pump function of the deep-rooting trees and shrubs (Sommer 2000; Sommer et al. 2000). Based on these results, it can be concluded that continuous nutrient mining through burning and harvesting is taking place in the agricultural land of northeastern Pará leading to degradation of soils and putting the traditional land-use system as a whole under pressure. This pressure becomes obvious from a rapid decline in the vigor of the fallow vegetation (Denich et al. in press).

The negative nutrient balance of the slash-and-burn system can be improved by replacing the slash-and-burn with fire-free land preparation. The unburned fallow vegetation could be a source of biomass for the maintenance of soil organic matter and plant nutrients to improve the physical, chemical and biological soil properties, and thus counter soil degradation processes. Both fire-free land preparation and management of soil organic matter can be combined in chop-and-mulch approach. Additionally, the mulch layer protects the soil from erosion, conserves soil water during dry spells, and reduces germination of weeds (Thurston 1997).

In our studies, the favorable effects of mulch could be proven to some extent as follows: (i) in mulched fields, the fertilizing effect was observed only in a prolonged cropping period after the mulch had decomposed and the nutrients released (fertilization was essential during the earlier part of the cropping period; Kato et al. 1999), and (ii) farmers reported that in mulched fields weeding activities could be reduced. Yet, even after two land-use cycles, the soil carbon content was not significantly different in mulched fields compared to traditionally burnt or fallowed areas (Table 5).

After gaining experience with the mulch system, farmers pointed out that on mulched fields usually all plant biomass was converted into mulch and no stems

Table 5. Carbon content in the 0–10 cm soil horizon in slashed-and-burnt as well as mulched fields with different cropping patterns in 2001 in the northeast of Pará state, Brazil.

Initial fallow age/ experiment established in	Land preparation	Cropping pattern	N	Carbon content (%) (±SD)	*P* value
4 years/1994	Fallow		8	1.29 (0.248)	
	Burned	R-C-M-F0.5-R-C-M-F3.5	16	1.31 (0.159)	0.852
	Mulch	R-C-M-F0.5-R-C-M-F3.5	16	1.26 (0.270)	
10 years/1994	Fallow		8	1.28 (0.338)	
	Burned	R-C-M-F0.5-R-C-M-F3.5	16	1.34 (0.315)	0.663
	Mulch	R-C-M-F0.5-R-C-M-F3.5	16	1.40 (0.274)	
5 years/1995	Fallow		8	1.29 (0.248)	
	Burned	R-C-M-F3.5	8	1.30 (0.145)	0.179
	Mulch	R-C-M-F3.5	8	1.46 (0.187)	

R = rice, C = cowpea; M = manioc (cassava); F+ number = years of fallow; N = number of repetitions; SD = standard deviation; *P* = probability of 'F' test significance.

were available for use as firewood or for producing charcoal. It seems questionable whether suitable stems can be removed out of natural fallow vegetation before it is transformed into mulch. In the traditional system, unburned stems are collected for firewood or charcoal production.

In the study region, farmers plant maize, rice and cowpea in rows using a planting tool locally known as 'tico-tico' or 'matraca.' To plant in mulched fields, the mulch layer has to be penetrated by the tip of the 'tico-tico' so that the seeds can be released into the upper part of the mineral soil. This operation is more time-consuming in mulched fields than in burnt ones. Field tests revealed that planting maize in mulched fields with a 'tico-tico' takes two-times the time and planting cassava cuttings with a hoe takes even four times, as compared to those in non-mulched (recently burned) fields.

Mechanized land preparation in fallow systems

The adoption of the mulch technology by farmers requires suitable techniques for the transformation of the woody fallow vegetation into mulch. In practice, the woody plant material in fallowed fields cannot be chopped manually, as it is a high labor-demanding and strenuous operation, which is not accepted by farmers. Mechanization of the chopping operation, therefore, is crucial for the introduction of the mulch technology.

As no adequate chopping device existed in the market, which *a priori* met the requirements for maintaining vigorous woody fallow vegetation, we set out to develop a bush chopper that fulfills the following four criteria. First, the fallow vegetation should be cut without damaging the root systems of the trees and shrubs as these were proved to be critically important for the regeneration of the woody species. Second, the fallow vegetation should be cut at ground level as this facilitates weeding unobstructed by stumps, while re-sprouting of the tree and shrub species is assured. Third, to make the whole operation more efficient, cutting the vegetation, chopping the plant material, and spreading the chips over the field should be carried out in a single operation. Fourth, the device has to be a simple and of robust construction to guarantee durability and to facilitate maintenance and repair in the already partly mechanized target region.

The first version of a tractor-propelled bush chopper that meets the above criteria and can be fixed to the front power lift of a conventional wheel tractor was Tritucap I (= prototype; Table 6). As the tractor moves forward, the vegetation, including trees and shrubs, is cut by two rotating circular saws, and subsequently chopped by two vertical steel helices (helical knives) sitting on the saw-blades. The two cut-and-chop units (rotors) are driven by the tractor's front power take-off (PTO) and rotate with a maximum of 1000 rpm. The chopped material is thrown out toward the back and under the tractor's front wheels. Chopping one hectare of degraded fallow vegetation containing an average fresh biomass of 13 Mg takes approximately one hour, whereas one hectare with a five-year-old fallow and an average of 45 Mg ha^{-1} standing fresh biomass takes

Table 6. Overview of chopping equipment tested in different-aged fallow vegetation in the context of mulch technology in the northeast of Pará state, Brazil.

Chopper type	Manufacturer	Tractor power (kW)	Working width (m)	Av. chopping capacity (Mg biomass h^{-1})	Man power (Mh Mg^{-1})	Average fuel consumption (l Mg^{-1})	Maximum stem diameter[b] (cm)	Fallow age (years)/ Maximum biomass (Mg ha^{-1})
Silage chopper	Nogueira	70[a]	n.a.	0.25	10	6	3–4	1–4 / 50
Rotary chopper	Super Tatu	50	1.8	4.5	0.22	1.3	2	1–2 / 20
Tritucap I	Inst. of Agric. Eng., Univ. of Göttingen	70	2.0	10	0.10	1.6	10	1–4 / 50
Tritucap II		122	2.4	15	0.07	1.3	10	1–4 / 50
FM 600	AHWI	122	2.3	20	0.05	1.2	30	5–12 / 150
RT 350	AHWI	220	2.3	20–25	0.05	2.0	100	25 / 300

*[a]During the tests a tractor with 70 kW was used. The silage chopper can also be driven by a lower-powered tractor (≈30 kW). Mh = man hours;
[b]Close to soil surface; n.a. = not applicable; biomass = fresh weight; AHWI = AHWI Maschinenbau GmbH, Herdwangen, Germany.

Figure 2. Newly developed bush chopper Tritucap II.

Figure 3. Tritucap II at work in 4-year-old fallow vegetation.

eight hours. As the speed of the operation is between 1 km h^{-1} and 3 km h^{-1}, the tractor has to be equipped with an inching motion facility.

Tritucap II (Figure 2 and 3, Table 6) was an improvement over Tritucarp I. Two important modifications were necessary for solving the problems identified during the field tests: (i) the ordinary saw-blades quickly became blunt due to the sandy topsoil, so they were replaced by saw-blades with detachable teeth made of extra-tempered steel and (ii) the toothed-belt drive was replaced by a cardan shaft-drive, thus improving the power transmission from the PTO to the rotors. Some additional modifications were also made such that all together, durability and operational efficiency of the chopper was increased. The Tritucap II is mounted on the rear power lift of a tractor with a reverse drive facility. That way the driver sits closer to the chopper and gets a better view of the chopping process.

Testing different chopper types reduces the risk of failures and makes it possible to offer alternative chopper technology as the circumstances may require. Therefore, parallel to the two Tritucap choppers, a forest mulcher AHWI FM 600, which is already in the international market, was tested in-depth (Figure 4, Table 6). The FM 600 is a PTO-driven device mounted on the rear power lift of a conventional wheel tractor with a reverse drive facility because of its heavy weight (2790 kg). The horizontal rotor (diameter 600 mm) is tipped with fixed carbide hammers, which are individually replaceable. While the tractor is driving backwards, the vegetation is first roughly pre-chopped with 1000 or 1600 rpm. If desired, the pre-chopped plant material covering the soil surface

Figure 4. Forest mulcher AHWI FM 600.

may be more finely chopped in a second go. In this case, the tractor drives forward pulling the chopper. The rotor picks up the plant material from the ground and chops it again. Initially, we assumed that the chopping principle of the FM 600 would be unsuitable, as we expected heavy disturbance of the topsoil through the horizontally rotating rotor possibly destroying the upper root system thus impairing the regeneration of the fallow vegetation. This concern, however, turned out to be unfounded, if the tractor was operated by a well-trained and experienced driver.

On farmlands in northeastern Pará, degraded low-standing fallow vegetation is very common. For such stands, the Tritucap and FM 600 as well as high-powered tractors would be oversized and a rotary cutter widely used in the region is recommended instead. Rotary cutters (Table 6) are devices pulled by a wheel tractor and glide on vats over the soil surface. It is powered by the rear PTO. The power demand is considerably lower than that of the Tritucap and FM 600. A rotating blade enclosed in a metal housing cuts the vegetation much the same way a lawnmower operates, but it is designed for rougher applications. The chopped material is spread across the entire width of the machine. In the study region, rotary cutters are usually used to mow grass, cut weeds or clear brush on pastures. Occasionally, rotary choppers are used for fire-free land preparation before cropping, mostly combined with plowing.

Two additional chopping devices were tested. In the context of mulch systems, however, both were found to be applicable in exceptional cases only. First, we tested a stationary silage chopper, which also is coupled to the PTO of a tractor (Table 6). The fallow vegetation has to be cut manually and brought to the chopper, where the stems and branches are fed into an integrated hopper for chopping. Rotating cutter blades or hammers chop the material and the chips are blown out through a pipe on a heap beside the machine. From there, the mulch has to be spread manually over the field. As the silage chopper can only chop woody material up to a diameter of approximately 4 cm, thick stems have to be split lengthwise before chopping. This chopping operation requires high hard labor inputs and cannot be recommended for agricultural land preparation. It might be useful in horticulture and in cases where fine-chopped and homogeneous mulch is desired. During our field tests, the latter seemed to be attractive to the farmers but perhaps more for aesthetic reasons. The second machine tested was a forest mulcher AHWI RT 350, which is a crawler tractor with an integrated FM 600 rotor (Table 6). The RT 350 operates like a wheel tractor in combination with a FM 600, but its engine power makes it far oversized for most of the fallow stands on smallholdings. We tried it in a 15- to 25-year-old secondary forest and found that trees up to 40 cm diameter could be easily chopped. Such an application, however, is beyond the vision of our mulch approach. It is worth mentioning that we got excellent results in chopping of a 15-year-old oil palm stand, which needed only 1.5 minutes to two minutes per palm. Thus, the RT might be an option for the fire-free land preparation before replanting of oil palm and logged timber plantations.

The choice of chopping device depends on the total biomass per hectare to be chopped as well as on the stem and branch diameters of the predominant woody plants. Studies showed that suitable chopping equipment is available for all kinds of smallholding mulch systems (Table 6). By far the most frequently applicable choppers are the Tritucap and FM 600 and our field tests revealed that these machines ideally complement one another. Tritucap works more efficiently and economically than FM 600 for fallow vegetation up to 4 years old, whereas for vegetation older than four years, the stronger FM 600 is the appropriate chopper. In contrast to Tritucap, which cuts and chops in one step, FM 600 cuts and chops in more time-consuming two steps, first it fells and roughly pre-chops and then finely chops the mulch – thereby making it possible to work in older fallow vegetation where the trees and shrubs are thicker and taller.

To obtain a suitable mulch layer, uniform distribution of the biomass in the field and the quality (chip size) of the mulch itself are important. As shown in Figure 1, the spatial biomass distribution is highly variable in natural fallow vegetation with frequent extreme values. The dense spots cause a considerable mechanical load (extreme torque peaks) for any kind of chopping device. In the case of 4- to 8-year-old fallow vegetation, the heterogeneous biomass distribution leads to mulch layers that cover 60% to 100% of the soil surface in 1- to 6-cm thickness (C.M.P. Bervald, pers. comm.). The size of the woody chips reflects the quality of the chopping operation and influences the quality and properties of the mulch and the subsequent decomposition. All the tested chopping devices delivered sufficiently splintered and frayed chips of a satisfactory size. The average chip size increased in the order Tritucap II < Tritucap I < FM 600. Depending on the fallow vegetation and the respective hardness and tenacity of the woody species, 50% to 90% of the chips were smaller than 4 cm^3. The share of only roughly chopped wood ranges from 10% to 30% of the chopped biomass (C.M.P. Bervald, pers. comm.). The FM 600 sometimes encountered problems in chopping of fibrous vine species and the very fibrous banana-like tree *Phenakospermum guyannense*, which were coiled up by the rotor and blocked it. These problems were not observed with the Tritucap.

Implications of fire-free land preparation and mulching

The fire-free land preparation and mulch technology have some agronomic requirements and constraints, but they also offer different options to manage the system more flexibly and make it attractive to farmers. For example, fertilizer application is necessary for ensuring the success of the technology; similarly, extension of the cropping period, and choice of crop varieties suited for mulch conditions are important considerations. The chop-and-mulch technology allows, however, off-season planting and rearrangement of the crop rotation. These constraints, requirements, and opportunities were investigated in a series of experiments on farmers' fields in the eastern Amazonia region. While valuable lessons have been learned from these efforts, several trials are in progress and many results, especially of long-term nature, are inconclusive. The salient aspects of the major studies are summarized in the following section; detailed results of each will be published in international scientific media in due course.

Fertilization

Fertilizer as a substitute for the ashes of the slash-and-burn system was proved to be indispensable to achieve economically acceptable yields in the proposed mulch system. A rice crop without fertilizer produces considerably lower yields in mulched fields than in burnt fields, whereas cassava produced satisfactory yields without fertilization; on the other hand, cowpea practically failed without fertilization in mulched and burnt fields (Kato et al. 1999).

Extending the cropping period

The mulch technology allows extending the cropping period beyond the common one to two years, i.e., two cropping periods instead of one before the field enters into fallow, almost doubling the land-use factor from 0.27 to 0.43. In the second cropping period, the yields of the mulched fields were higher than in the first period (except cassava) and higher than those of the burned fields both with and without fertilizer application. The increased yields in the mulched fields could be explained as a consequence of the release of nutrients from the slowly decomposing mulch only in the second cropping period (Kato et al. 1999).

Selection of crop varieties suitable for mulch conditions

Screening experiments conducted on farmers' fields with 11 maize, eight rice, 21 cowpea and five cassava varieties under mulch conditions without the use of fertilizer revealed scope for overcoming the limitations to plant growth in mulched fields through the use of selected rice and cassava varieties. Mulched fields cropped with modern rice varieties or with selected cassava varieties produced yields higher than the local average yields under slash-and-burn conditions (Kato et al. 1999).

Enrichment planting

Enrichment planting – a fallow management technique in which tree or shrub species are introduced into the fallowed area to improve the ecological or/and economic value of fallow vegetation – was tested in two field experiments on smallholders' land using the leguminous tree species *Acacia angustissima*, *Acacia auriculiformis*, *Acacia mangium*, *Clitoria racemosa*, *Inga edulis*, and *Sclerolobium paniculatum*. The species differed in their origin (autochthonous, exotic), growth form, rooting depth, and leaf litter degradability (Brienza 1999). At the age of 21 months, the enriched fallows produced 13% to 132% higher aboveground biomass than natural fallow without enrichment planting. The planted trees significantly suppressed the natural vegetation depending on the density of the trees planted. Enrichment plantings seem to be feasible only with *A. auriculiformis* and *A. mangium*. Also, higher benefit-cost-ratio for biomass accumulation was found at the lower planting density that produced greater stem diameters. For a mulch system with cut-and-chop land preparation, higher planting densities that produce lower stem diameters might be preferable to facilitate mechanized chopping. This may play only a minor role, however, with the availability of suitable chopping devices. Enrichment plantings allow shortening of the fallow periods. The resulting increased land-use factor is highly desirable from the farmers' point of view. For example, an extended cropping period together with a shortened fallow period may increase the land-use factor to 0.6. There is concern, however, that very short fallow periods may restrict suppression of weeds, although this has not so far been confirmed by field observations.

The study also revealed that after 21 months of fallow, only 31 woody species were recorded in plots enriched with 10 000 trees ha^{-1} compared with 38 species with an enrichment of 2500 trees ha^{-1} and 39 species in plots without enrichment planting. Thus, the higher the planting density of the introduced leguminous trees, the greater is the reduction in the diversity of naturally occurring woody species. Moreover, enrichment planting reduced the growth of herbaceous weeds and grasses in the first year of fallow itself compared to two years required for natural fallow to have a similar effect so that even very short fallow periods seem to be possible without diminishing the weed suppression function of the fallow.

Change of cropping calendar

Traditionally, land preparation in the fallow system depends on a sufficiently long dry season to allow the slashed vegetation to dry before it can be burnt satisfactorily. The chop-and-mulch system, however, makes planting possible even in the rainy season (January to May). The mulch conserves water and permits with low risk to extend cropping into the drier part of the year, thus enabling the farmer to change the cropping calendar and plant off-season. Fire-free land preparation and off-season planting would allow farmers to organize their agronomic activities in a way that the work load is spread out evenly over the whole year. Moreover, farmers can avoid the peak produce supply period when prices are usually low and capture rising off-season market prices of annual crops.

Rearrangement of crop rotation

In mulch systems, nutrient release from the decomposing mulch material occurs relatively slowly. Furthermore, microorganisms decomposing the organic materials may even temporarily immobilize nutrients. Slow nutrient release and immobilization result in nutrient shortage at the very beginning and an increased supply during the later stages of the cropping period. Thus, we hypothesized that inversion of crop rotation by planting a less nutrient-demanding crop such as cassava as a pioneer crop and more nutrient demanding crops such as maize or rice later might adequately mitigate temporary nutrient constraints in the mulch system. This way, we attempted to synchronize the nutrient demand of the crops with the nutrient release from the mulch, making best use of the nutrient pool in the mulch layer. On-farm (farmer-managed) studies to compare the traditional cropping sequence of maize – cassava with the inverted cassava – maize sequence have been carried out. On the basis of maize and cassava yields, we tentatively conclude that the traditional sequence slightly outperforms the inverted. To come to a final conclusion, further studies on the rearrangement of the crop rotation are necessary, particularly on the inclusion of modern low-input crop varieties combined with fertilizer.

Financial analysis of chop-and-mulch approach

Farmers adopt new land-use systems and agricultural practices only if they are economically attractive. Financial analysis was conducted to assess returns to land and labor of selected chop-and-mulch scenarios compared with the traditional slash-and-burn system. The four scenarios compared were: (i) slash-and-burn with low fertilizer input (S+B, land-use factor R = 0.27), (ii) chop-and-mulch with high fertilizer input (C+M, R = 0.27), (iii) chop-and-mulch with high fertilizer input and extended cropping period (C+Mi, R = 0.43), and (iv) chop-and-mulch with high fertilizer input, extended cropping period and half fallow period (C+Me, R = 0.60). Although additional scenarios were evaluated, only those that were considered sustainable and

Table 7. Receipts and expenditures of four cropping scenarios in the northeast of Pará state, Brazil.

	S+B	C+M	C+Mi	C+Me
Cycle length (years)	5.5	5.5	7	5
Cropping period, *fallow period* [years]	1.5 + *4*	1.5 + *4*	1.5 + 1.5 + *4*	1.5 + 1.5 + *2*
Receipts	(R$ ha^{-1})			
Rice grains	897	1446	3317	3317
Cowpea grains	1570	1540	3520	3520
Cassava meal	3260	5760	10960	10960
Total receipts	5727	8746	17797	17797
Expenditures				
Tractor and chopper rent		−760	−760	−760
Fertilizers	−423	−803	−1606	−1606
Other inputs (seeds, plant protection, transport)	−300	−300	−600	−600
Tree seedlings (2500 trees ha^{-1})				−417
Hired labor at 10 R$ per man day		−730	−3358	−3391
Financing of costs additional to S+B		−393	−1176	−1271
Total expenditures	−723	−2986	−7500	−8045
Balance of receipts and expenditures	5004	5760	10297	9752

S+B = slash-and-burn; C+M = chop-and-mulch with fertilization; C+Mi = chop-and-mulch with fertilization and extended cropping period; C+Me = chop-and-mulch with fertilization, extended cropping period and shortened fallow period due to enrichment planting.
1 R$ = 0.31 US$.

feasible for the farmers are reported here. All the tested scenarios included rice, cowpea, cassava and a fallow period. Low fertilizer input means that only cowpea was fertilized, and high fertilizer input means additional fertilizer to rice. Halving the fallow period was possible by fallow enrichment with fast-growing leguminous trees.

The spatial reference for the analysis was one hectare and, where necessary, cross-system comparability was obtained by standardizing the temporal reference to one year by dividing all systems by the respective length of the land-use cycle. All prices used for inputs, labor and products are farm-gate prices in August 2002 obtained through inquiries in northeastern Pará. Product prices were derived from the statistics of DIEESE (2002). The chopping unit used for the calculations was a 120 kW wheel tractor with driver and the forest mulcher FM 600. To obtain expenditures for the chopping unit, a renting price of 760 R$ ha^{-1} (236 US$ ha^{-1}) was calculated based on a profitability calculation for a private contractor of farm services, which is not shown here. Being a mere cash-flow analysis, noncash elements such as environmental services of the mulch system, e.g., the plant nutrients preserved in the system, are not included. This also holds for the unpaid family labor, the remuneration of which is represented by the net benefit that each system offers. Expenditures of variable costs and receipts are balanced, resulting in the monetary return of each cropping system (Table 7), which is then applied to calculate land and labor productivity. Land productivity increased in the order S+B < C+M < C+Mi < C+Me whilst labor productivity was highest for C+Mi (Table 8), the latter being the greater incentive to the farmer. C+M alone was still clearly above S+B, which makes the chop-and-mulch a viable system even when applied with very little additional input.

Since experimental yields are not 100% transferable to farmers' practice, a sensitivity analysis was done to know how far yields could drop in all systems before chop-and-mulch became less viable than slash-and-burn (Figure 5). C+Mi and C+Me became less profitable in terms of both land and labor productivity than the traditional S+B system only if the yields drop to 30% of the experimental yields, whilst C+M less profitable at 50% of the yields. The average yields obtained in northeastern Pará are around 60% to 70% of the experimental yields. We conclude that (i) chop-and-mulch with fertilization of rice and cowpea and an extended cropping period (C+Mi) and (ii) chop-and-mulch with fertilization of rice and cowpea, an

Table 8. Contribution margins to land (land productivity) and labor (labor productivity) of four cropping scenarios in the northeast of Pará state, Brazil.

	S+B	C+M	C+Mi	C+Me
Land (ha)	1	1	1	1
Land productivity (R$ ha^{-1})	910	1047	1471	1951
NBI [R$ ha^{-1}] of land productivity relative to S+B		+138	+561	+1041
NBI [%] of land productivity relative to S+B	100	115	162	214
Family labor in man days (MD)	203.6	203.6	203.6	203.6
Family labor productivity (R$ per MD)	24.6	28.3	50.6	47.9
NBI (R$ per MD) of family labor productivity relative to S+B		+4	+26	+23
NBI (%) of family labor productivity relative to S+B	100	115	206	195

NBI = net benefit increase; S+B = slash-and-burn; C+M = chop-and-mulch with fertilization; C+Mi = chop-and-mulch with fertilization and extended cropping period; C+Me = chop-and-mulch with fertilization, extended cropping period and shortened fallow period due to enrichment planting.
1 R$ = 0.31 US$.

extended cropping period and a halved fallow period (C+Me) are the most appropriate intensified and sustainable land use systems to increase income to farmers. So far, economy of scale that greater numbers of chopping units would reduce costs of chopping below 760 R$ ha^{-1} was not considered. A cost estimate by the chopper manufacturer for rental turned out to be 480 R$ ha^{-1} (149 US$ ha^{-1}), when a minimum of 30 chopping units in a given region (AHWI do Brazil, pers comm) were introduced. But even in the unlikely case of increased chopper costs, these could rise up to factor of five before land and labor productivity of the two recommended systems would drop below the ones of slash-and-burn.

Concluding remarks

- Mechanized land clearing of fallow vegetation with tractor-driven bush choppers in connection with the application of mulch technology appears to be agronomically and environmentally superior to the traditional slash-and burn practice of land preparation in forest-based land use systems in the northeastern Amazon region. This technology may be relevant to similar other humid tropical conditions where slash-and-burn is practiced.
- As the chop-and-mulch land preparation is not a completely new land-use system, but merely a replacement of the burn practice, this approach is not likely to lead to adoption problems, because farmers are not confronted with many changes. We observed that the farmers in the study region were keen to apply mechanized chopping. Indeed, it seems that the introduction of machinery is seen as a modernization of the traditional land-use, which makes it attractive, especially as it avoids the backbreaking labor and increases labor productivity.
- Environmental benefits such as soil conservation, reduction in the losses of nutrients and organic matter as well as the absence of accidental fires in the agricultural landscape are advantages of the mulch system that are important not only to individual farmers but to the society as a whole.
- The chop-and-mulch technology is versatile in the sense that the farmer can go back to the traditional fallow system any time he chooses. This option makes it easier for the farmer to overcome the psychological barriers or to encourage the farmer to take a calculated risk, which enhances the readiness for adoption.
- Before moving to technology diffusion, large-scale testing has to be carried out. Currently, the political circumstances in Brazil are most favorable for the introduction of environmentally friendly technologies in agriculture, especially those supporting smallholders. Chop-and-mulch would certainly be among those practices with a strong and direct impact, since burning can be stopped immediately.
- Avoiding fire in the land-use system allows farmers to establish perennial crops and plantations without the risk of loosing the crops through accidental fires. The latter represent an incalculable risk to farmers with respect to the establishment of permanent crops, woodlots for timber production or agroforestry systems combining crops and woody components.

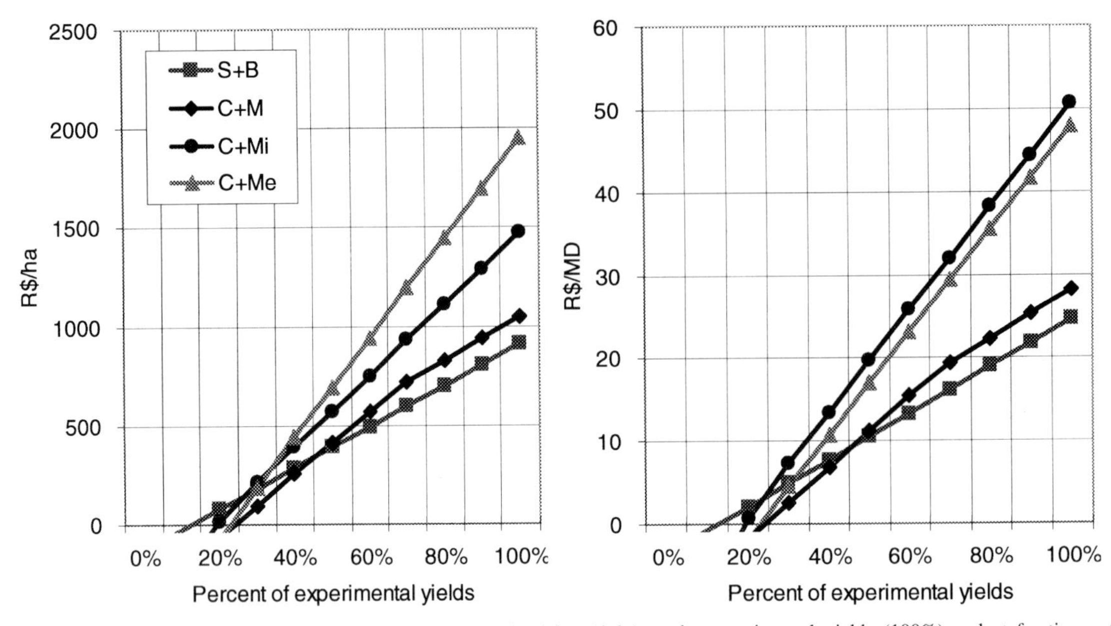

Figure 5. Sensitivity of land productivity (left) and labor productivity (right) at the experimental yields (100%) and at fractions of the experimentally achieved yields (<100%). S+B = slash-and-burn; C+M = chop-and-mulch with fertilizer; C+Mi = chop-and-mulch with fertilizer and extended cropping period; C+Me = chop-and-mulch with fertilizer, extended cropping period and shortened fallow period due to enrichment planting.

- Agroecologically, zero-tillage is mandatory in fallow systems for achieving one of the primary goals, which is to maintain the re-sprouting capacity of the root system of the fallow vegetation. However, in mulch systems, this requires further development of planting technologies suitable for woody mulch layers, especially when considering animal or tractor-driven planting machines.
- Biomass transfer (ex-situ mulch) may become increasingly important with respect to perennial crops and permanent cropping. Tritucap after a slight modification can readily serve such a system. A pipe has to be added at the outlet of the chopper that directs the chopped material onto a trailer to transport it. Mulch material could be obtained from areas that are specially kept for that purpose or from material from firebreaks that can be cut using the chopper.
- The Tritucap can be applied in alley cropping systems to cut and chop the hedgerows and spread the mulch over the alley.
- Researchers and farmers are increasingly under pressure to search for fire-free land preparation techniques as legal and political initiatives, national and international bodies are aiming at prohibiting the use of fire in land management. Chopping and mulching might play a crucial role in facilitating environmental compliance.

End Notes

1. The structure of secondary vegetation in the context of shifting cultivation in Eastern Amazonia, Brazil. Originally in German. Unpublished diploma thesis, University of Göttingen, Germany, 136 pp.
2. Withelm (1993). The utilization of secondary vegetation by the population of the Igarapé-Açu region (Pará state, Brazil). Originally in German. Unpublished diploma thesis, University of Hamburg, Germany, 134 pp.

Acknowledgements

The project is a cooperation of the Center for Development Research (ZEF, University of Bonn, Germany), the Institute for Agricultural Engineering (University of Goettingen, Germany) and EMBRAPA Amazônia Oriental, Belém-PA, Brazil, under the Governmental Agreement on Cooperation in the field of Scientific Research and Technological Development between Germany and Brazil. The project is financed by the bilateral research program 'Studies on Human Impact on Forests and floodplains in the Tropics – SHIFT' (No. 01LT0012).

References

Akobundu I.O. and Ekeleme F. 2002. Weed seedbank characteristics of arable fields under different fallow management systems in the humid tropical zone of southeastern Nigeria. Agroforest Syst 54 161–170.

Baar R. 1997. Vegetationskundliche und -ökologische Untersuchungen der Buschbrache in der Feldumlagewirtschaft im östlichen Amazonasgebiet. Göttinger Beiträge zur Land- und Forstwirtschaft in den Tropen und Subtropen 121, 202 pp.

Brienza Jr., S. 1999. Biomass dynamics of fallow vegetation enriched with leguminous trees in the eastern Amazon of Brazil. Göttinger Beiträge zur Land- und Forstwirtschaft in den Tropen und Subtropen 134, 133 pp.

Clausing G. 1997. Early regeneration and recolonization of cultivated areas in the shifting cultivation system employed in the eastern Amazon region, Brazil. Nat Resour Dev 45/46: 76–102.

David S. and Raussen T. 2003. The agronomic and economic potential of tree fallows on scoured terrace benches in the humid highlands of southwestern Uganda. Agric Ecosyst Environ 95: 359–369.

Degrande A. 2001. Farmer assessment and economic evaluation of shrub fallows in the humid lowlands of Cameroon. Agroforest Syst 53: 11–19.

Denich M. 1989. Untersuchungen zur Bedeutung junger Sekundärvegetation für die Nutzungssystemproduktivität im östlichen Amazonasgebiet, Brasilien. Göttinger Beiträge zur Land- und Forstwirtschaft in den Tropen und Subtropen 46, 265 pp.

Denich M., Vlek P.L.G., Abreu Sá T.D. de Vielhauer K. and Lücke W. in press. A concept for the development of fire-free fallow management in the Eastern Amazon, Brazil. Agric Ecosyst Environ (In press).

Diaz M. del C.V., Nepstad D., Mendonça M.J.C., Mota R.S., Alencar A., Gomes J.C. and Ortiz R.A. 2002. O Preço oculto do Fogo na Amazônia: Custos Econômicos Associados ao Uso de Fogo. Report of IPAM/IPEA/WHRC, Belém – PA, Brazil, 43 p. http://www.ipam.org.br/publica/publica-papers.php (November 2003).

DIEESE 2002. National Staple Food Basket in 16 capitals of Brazil. DIEESE – Inter Trade Union Department of Statistics and Socio-Economic Studies, http://www.dieese.org.br/ (November 2003).

Franzel, S.and Scherr, S.J. (eds.) 2002. Trees on the Farm. Assessing the Adoption Potential of Agroforestry Practices in Africa. CAB International, Wallingford, UK, 197 pp.

Gallagher R.S., Fernandes E.C.M.and McCallie E.L. 1999. Weed management through short-term improved fallows in tropical agroecosystems. Agroforest Syst 47: 197–221.

Gehring C. 2003. The Role of Biological Nitrogen Fixation in Secondary and Primary Forests of Central Amazonia. Ecology and Development Series 9, Cuvillier, Göttingen, Germany, 170 pp.

Hedden-Dunkhorst B., Denich M., Vielhauer K., Mendoza-Escalante A., Börner J., Hurtienne T., Sousa Filho F.R. de, Abreu Sá T.D. de and Assis Costa F. de 2003. Forest-based fallow systems: a safety net for smallholders in the Eastern Amazon? Paper presented at the international conference on 'Rural livelihoods, Forests and Biodiversity' held in Bonn, Germany on 19–23 May 2003. Center for International Forestry Research (CIFOR), Bogor, Indonesia. http://www.zef.de/research_activities/shift/publications.htm (Accessed November 2003).

Hoang Fagerström M.H., Nilsson S.I., van Noordwijk M., Thai Phien O. M., Hansson A. and Svensson C. 2002. Does *Tephrosia candida* as fallow species, hedgerow or mulch improve nutrient cycling and prevent nutrient losses by erosion on slopes in northern Vietnam? Agric Ecosyst Environ 90: 291–304.

Hohnwald S. 2002. A grass-capoeira pasture fits better than a grass-legume pasture in the agricultural system of smallholdings in the humid Brazilian tropics. Cuvillier, Göttingen, Germany, 136 pp.

IBGE 1998. Censo Agropecuário 1995–1996. Número 5, Pará. Rio de Janeiro, Brazil, 217 pp.

Jacobi I. 1997. Der Beitrag von Keimlingen zur Regeneration der Brachevegetation im östlichen Amazonasgebiet. Ph.D. thesis, University of Hamburg, Germany, 148 pp.

Kanmegne J. and Degrande A. 2002. From alley cropping to rotational fallow: farmers' involvement in the development of fallow management techniques in the humid forest zone of Cameroon. Agroforest Syst 54: 115–120.

Kato O.R. 1998. Fire-free land preparation as an alternative to slash-and-burn agriculture in the Bragantina region, Eastern Amazon: Crop performance and nitrogen dynamics. Cuvillier, Göttingen, Germany, 132 pp.

Kato M.S.A., Kato O.R., Denich M. and Vlek P.L.G. 1999. Fire-free alternatives to slash-and burn for shifting cultivation in the Eastern Amazon region. The role of fertilizers. Field Crops Res 62: 225–237.

Kaya B., Hildebrand P.E. and Nair P.K.R. 2000. Modeling changes in farming systems with the adoption of improved fallows in southern Mali. Agr Syst 66: 51–68.

Mackensen J., Hölscher D., Klinge R. and Fölster H. 1996. Nutrient transfer to the atmosphere by burning of debris in eastern Amazonia. Forest Ecol Manag 86: 121–128.

McDonald M.A., Healey J.R. and Stevens P.A. 2002. The effects of secondary forest clearance and subsequent land-use on erosion losses and soil properties in the Blue Mountains of Jamaica. Agric Ecosyst Environ 92: 1–19.

Mercer E. 2004. Adoption of agroforestry innovations in the tropics: a review. Agroforest Syst (This volume).

Mertz O. 2002. The relationship between length of fallow and crop yields in shifting cultivation: a rethinking. Agroforest Syst 55: 149–159.

Muschler R.G. and Bonnemann A. 1997. Potentials and limitations of agroforestry for changing land-use in the tropics: experiences from Central America. Forest Ecol Manag 91: 61–73.

Nair P.K.R. 1993. An Introduction to Agroforestry. Kluwer, Dordrecht, The Netherlands, 499 pp.

Nye P.H. and Greenland D.J. 1960. The soil under shifting cultivation. Technical Communication No. 51, Commonwealth Bureau of Soils, Harpenden, UK, 156 pp.

Opio C. 2001. Biological and social feasibility of *Sesbania* fallow practice in smallholder agricultural farms in developing countries: a Zambian case study. Environ Manag 27: 59–74.

Rouw A. de 1995. The fallow period as a weed-break in shifting cultivation (tropical wet forests). Agric Ecosyst Environ 54: 31–43.

Sanchez P. 1995. Science in agroforestry. Agroforest Syst 30: 5–55.

Sanchez P. 1999. Improved fallows come of age in the tropics. Agroforest Syst 47: 3–12.

Shang C. and Tiessen H. 2000. Carbon turnover and carbon-13 natural abundance in organo-mineral fractions of a tropical dry forest soil under cultivation. Soil Sci Soc Am J 64: 2149–2155.

Sommer R. 2000. Water and nutrient balance in deep soils under shifting cultivation with and without burning in the Eastern Amazon. Cuvillier, Göttingen, Germany, 240 pp.

Sommer R., Denich M. and Vlek P.L.G. 2000. Carbon storage and root penetration in deep soils under small-farmer land-use systems in the eastern Amazon region, Brazil. Plant Soil 219: 231–241.

Thurston H.D. 1997. Slash/Mulch Systems – Sustainable Methods for Tropical Agriculture. Westview Press, Boulder, CO, 196 pp.
Tian G., Salako F.K., Ishida F. and Zhang J. 2001. Biological restoration of a degraded Alfisol in the humid tropics using planted woody fallow: synthesis of 8-year-results. pp. 333–337. In: Stott R.H., Mohtar R.H. and Steinhardt G.C. (eds), Sustaining the Global Farm. Purdue University, West Lafayette, USA.
Tiessen H., Cuevas E. and Chacon P. 1994. The role of soil organic matter in sustaining soil fertility. Nature 371: 783–785.

Agroforestry Systems **61:** 107–122, 2004.

Medicinal and aromatic plants in agroforestry systems

M.R. Rao[1,*], M.C. Palada[2] and B.N. Becker[3]
[1]*Former ICRAF (World Agroforestry Centre) staff, 11, ICRISAT Colony (Phase-I), Brig. Syeed Road, Secunderabad–500 009, Andhra Pradesh, India;* [2]*Agricultural Experiment Station, University of the Virgin Islands, RR2, Box 10000, Kingshill, St. Croix, Virgin Islands 00850, USA;* [3]*School of Forest Resources and Conservation, Institute of Food and Agricultural Sciences, University of Florida, Gainesville, FL 32611, USA;*
Author for correspondence: e-mail: mekarao@sol.net.in

Key words: Herbal medicine, Homegardens, Intercropping, Multistrata systems, Nontimber forest products, Phytomedicine

Abstract

A large number of people in developing countries have traditionally depended on products derived from plants, especially from forests, for curing human and livestock ailments. Additionally, several aromatic plants are popular for domestic and commercial uses. Collectively they are called medicinal and aromatic plants (MAPs). About 12.5% of the 422 000 plant species documented worldwide are reported to have medicinal values; but only a few hundred are known to be in cultivation. With dwindling supplies from natural sources and increasing global demand, the MAPs will need to be cultivated to ensure their regular supply as well as conservation. Since many of the MAPs are grown under forest cover and are shade tolerant, agroforestry offers a convenient strategy for promoting their cultivation and conservation. Several approaches are feasible: integrating shade tolerant MAPs as lower strata species in multistrata systems; cultivating short cycle MAPs as intercrops in existing stands of plantation tree-crops and new forest plantations; growing medicinal trees as shade providers, boundary markers, and on soil conservation structures; interplanting MAPs with food crops; involving them in social forestry programs; and so on. The growing demand for MAPs makes them remunerative alternative crops to the traditional ones for smallholders in the tropics. Being underexploited species with promising potential, the MAPs require research attention on a wide array of topics ranging from propagation methods to harvesting and processing techniques, and germplasm collection and genetic improvement to quality control and market trends. Joint forest management with farmers and contract farming with drug companies with buyback arrangement will promote cultivation of medicinal plants.

Introduction

Medicinal and aromatic plants (MAPs) play an important role in the healthcare of people around the world, especially in developing countries. Until the advent of modern medicine, man depended on plants for treating human and livestock diseases. Human societies throughout the world have accumulated a vast body of indigenous knowledge over centuries on medicinal uses of plants, and for related uses including as poison for fish and hunting, purifying water, and for controlling pests and diseases of crops and livestock. About 80% of the population of most developing countries still use traditional medicines derived from plants for treating human diseases (de Silva 1997). China, Cuba, India, Sri Lanka, Thailand, and a few other countries have endorsed the official use of traditional systems of medicine in their healthcare programs. For example, the Indian systems of medicine 'Ayurveda,' 'Sidha' and 'Unani' entirely, and homeopathy to some extent, depend on plant materials or their derivatives for treating human ailments (Prajapati et al. 2003). People in villages and remote areas primarily depend on traditional medicines as the modern system is out of reach and expensive. Many among the educated in Asian and African countries use traditional medicines for reasons of firm belief that they are more effective than modern medicines for certain chronic diseases, they do not have side effects of some of the modern medicines, and/or for economic reasons. Thus, in many societies, traditional and modern systems of medicines are used independently.

About 12.5% of the 422 000 plant species documented worldwide are reported to have medicinal value; the proportion of medicinal plants to the total documented species in different countries varies from 4.4% to 20% (Schippmann et al. 2002). About 25% of drugs in modern pharmacopoeia are derived from plants (phytomedicines) and many others are synthetic analogues built on prototype compounds isolated from plants. Up to 60% of the drugs prescribed in Eastern Europe consist of unmodified or slightly altered higher plant products (Lancet 1994). These drugs carry important therapeutic properties including contraceptives, steroids and muscle relaxants for anesthesia and abdominal surgery (all made from the wild yam, *Dioscorea villosa*); quinine and artemisinin against malaria; digitalis derivatives for heart failure; and the anti-cancer drugs vinblastin, etoposide and taxol. These compounds cannot be synthesized cost-effectively, which means that their production requires reliable supplies of plant material (van Seters 1997).

The global importance of MAP materials is evident from a huge volume of trade at national and international levels. During the 1990s, the reported annual international importation of MAPs for pharmaceutical use amounted on average to 350 000 Mg valued at over USD 1 billion (Table 1). A few countries dominate the international trade with over 80% of the global import and export allotted to 12 countries each. Whereas Japan and Korea are the main consumers of medicinal plants, China and India are the world's leading producing nations. Hong Kong, United States and Germany stand out as important trade centers. It is estimated that the total number of MAPs in international trade is around 2500 species worldwide (Schippmann et al. 2002).

Forests: the primary source of MAPs

Forests are the primary source of medicinal plants, and MAPs are one of the many valuable categories of nontimber forest products (NTFPs) that include food and beverages, fodder, perfumes, cosmetics, fibre, gums, resins, and ornamentals and materials for dyeing and tanning, plant protection, utensils and handicrafts (FAO 2003). The knowledge on NTFPs – their collection, processing, use and exchange – forms part of the social and natural capital of many rural communities (Vantomme et al. 2002). Forests provide a wealth of highly prized MAPs, whether in the alpine and sub-alpine Northwest Himalayas (Gupta 1986; Joshi and Rawat 1997), Afro-montane areas (ICRAF 1992; Cunningham 1997), humid tropics (On et al. 2001; Parelta et al. 2003) or temperate regions (Foster 1993; Hill and Buck 2000). The native woodlands and long-term fallows also are an important source of MAPs. In India, medicinal plants are harvested regularly in appreciable quantities and exported with government approval (Gupta 1988) or usufruct rights are extended to the so-called tribals and other indigenous communities at the forest margins. Collection and selling of secondary resources such as edible herbs and medicinal plants from natural sources is an economic activity that fetches greater returns than wage rates in many places, as in the savanna region in South Africa (Shackleton and Shackleton 2000).

In Nepal, over 15 000 Mg of medicinal herbs representing some 100 species are harvested every year from the wild for commercial and industrial purposes (Bhattarai 1997). In Sri Lanka, the number of species used in traditional medicine is estimated to be between 550 and 700 (de Alwis 1997). Of the 1000 commonly used medicinal plants in China, 80% in terms of number of species and 60% in terms of total quantity come from wild sources (He and Sheng 1997). In India, 1100 species are recognized as providing raw materials for Ayurvedic and Unani formulations (Gupta 1986). In the Near East region, the use of MAPs goes back to thousands of years and forms an important part of various cultures (Heyhood 1999). In sub-Saharan Africa, rural families are increasingly turning to medicinal plants and other NTFPs to cope with HIV/AIDS (Barany et al. 2001). In Amazonia, at least 1300 plant species are being used as medicines, poison or narcotics (Schultes 1979). Amazonian Shaman, the traditional healer, is also a skilled botanist and has a great talent on locating the requisite plant from the green vastness that makes up their natural pharmacy. In Latin America and Africa, this knowledge has largely remained undocumented and is handed down orally from parents to children.

The increasing demands for medicinal plants by people in developing countries have been met by indiscriminate harvesting of spontaneous flora including those in forests (de Silva 1997). Consequently, many

Table 1. Leading countries of import and export of medicinal and aromatic plant materials based on annual average during 1991–1998.

Country of import	Volume (Mg)	Value (1000 USD)	Country of export	Volume (Mg)	Value (1000 USD)
Hong Kong	73 650	314 000	China	139 750	298 650
Japan	56 750	146 650	USA	11 950	114 450
USA	56 000	133 350	Germany	15 050	72 400
Germany	45 850	113 900	Singapore	11 250	59 850
Singapore	6 550	55 500	India	36 750	57 400
Rep. Korea	31 400	52 550	Chile	11 850	29 100
France	20 800	50 400	Egypt	11 350	13 700
China	12 450	41 750	Mexico	10 600	10 050
Italy	11 450	42 250	Bulgaria	10 150	14 850
Pakistan	11 350	11 850	Albania	7 350	14 050
Spain	8 600	27 450	Morocco	7 250	13 200
UK	7 600	25 500	Pakistan	8 100	5 300
Total	342 550	1 015 200	Total	281 550	643 200

Source: Lange (2002).

species have become extinct and some are endangered. It is therefore, imperative that systematic cultivation of medicinal plants be developed and steps introduced to conserve biodiversity and protect threatened species. This article takes a critical look at the current sources of supply of MAPs, reviews literature on the prospects for their cultivation, conservation and sustainable use in agroforestry systems and identifies future research needs.

The need for conservation and cultivation of MAPs

Forest degradation throughout the tropical world has diminished the availability of widely used medicinal plant species. Five of the top 12 medicinal trees in the eastern Amazon region of Brazil have begun to be harvested for timber decreasing the availability of their barks and oils for medicinal purposes (Shanley and Luz 2003). Many valuable species are becoming endangered due to over-exploitation from wild. Some examples include '*yohimbe*' (*Pausinystalia johimbe*, bark is used to treat male impotence) in Central Africa (Sunderland et al. 1999), goldenseal (*Hydrastis canadensis*) collected from hardwood forests in eastern North America (Hill and Buck 2000) and African cherry (*Prunus africana*, bark used to treat prostatitis) in Cameroon and Madagascar (Cunningham et al. 2002). A survey in the Shinyanga region in northwestern Tanzania indicated that the top 10 priority medicinal species that people traditionally have been harvesting from natural woodlands have become scarce and are near extinction (Dery et al. 1999). Over-exploitation of these species is leading to unsustainable depletion of natural resources and narrowing of their genetic base. In India, 16 medicinal plants including the highly valued *Atropa acuminata*, *Dioscorea deltoidea* and *Rauvolfia serpentina* are listed as endangered species in the northwestern Himalayas (Gupta 1986). These are just a few examples and many such cases abound throughout the world.

To meet the requirements of expanding regional and international markets and healthcare needs of growing populations, increasing volumes of medicinal plants are harvested from forests and other natural sources. Loss of forestland for agriculture and plantations, overgrazing, irregular exploitation of herbs in forest and other natural woodlands by private individuals and commercial enterprises are further contributing to the depletion of the supply of MAPs from forests. During the past 15 years, there has been a substantial loss of habitats, notably tropical forests (which are disappearing at a rate of about 1% per year; FAO 2003), wetlands and other types of biome as a result of human action. The main threats to the resource base of MAPs are those that affect any kind of biodiversity that is used by humans such as habitat loss, habitat fragmentation, unsustainable use, largely unmonitored trade, over-harvesting or destructive harvesting techniques, alien and invasive species and global change. Over-exploitation and consequent depletion of medicinal plants not only affects their supply and loss of genetic diversity within these species but seriously affect the livelihoods of indigenous people living on forest margins. The most vulnerable species are popular, slow reproducing species with specific habitat requirements and a limited distribution.

Conservation of plant genetic resources has received worldwide attention in the past several years (McNeely 2004). In the case of medicinal plants, conservation policies and practices should meet future supply needs and have provisions for species conservation. On an international and/or national level, protective measures range from species conservation programs, shifting of processing from consumer to source countries and resource management to trade restrictions or even trade bans. Specific steps include basic studies to identify medicinal plants and traditional knowledge of their uses; sustainable utilization through cultivation, controlled wild harvest and reduction of waste; conservation both *in situ* – in natural habitats, *ex situ* – in botanic gardens and seed banks and through alternative technologies such as micropropagation and cryopreservation for those species that cannot be conserved *in situ* and that cannot be stored in seed banks; and increased public support for conservation of medicinal plants, built through communication and cooperation. Establishing high altitude nurseries and drug farms in the plants' natural habitat and education of herb collectors will be of further help (Joshi and Rawat 1997). Limiting harvest to a sustainable level is important but it is complicated by the conflict of interests between use and protection. It requires effective management systems and sound scientific information. In the past few years, field-based methods have been developed for sustainable harvest assessment and monitoring of NTFPs, resulting in the publication of research guidelines and predictive models (FAO 1995). Although in theory sustainable use of bark, roots, or whole plants as herbal medicines is possible, the inputs required in terms of money and manpower for intensive management of slow growing species in multiple species systems is unlikely to be found in many countries (Cunningham 1997). This emphasizes the need for cultivation to provide alternative supply sources for popular and high conservation priority species outside of core conservation areas.

Although many species are employed in traditional medicine, the number of MAP species currently in formal cultivation for commercial production does not exceed a few hundreds worldwide. For example, the species cultivated are only 100 to 250 in China (He and Sheng 1997), 38 in India (Prajapati et al. 2003) and about 40 in Hungary (Bernáth 1999). A survey carried out among companies involved in trade and production of herbal remedies and other botanical products revealed that although companies reported 60% to 90% of material was from cultivation, the number of species cultivated relative to the volume is small (Laird and Pierce 2002). Of the 1543 species traded in Germany, only 50 to 100 species are exclusively obtained from cultivation (Schippmann et al. 2002). Given the demand for a continuous supply of medicinal plants and accelerated depletion of forest resources, increasing the cultivation of MAP species appears to be an important strategy for meeting the growing demand and relieving harvest pressure on wild populations (Lambert et al. 1997). The high price paid for some of the species make them potential new crops (e.g., *Warburgia salutaris*, *Garcinia kola*, *G. afzelii* and *G. apunctata* in Africa) that can be grown in sole as well as agroforestry systems (Cunningham 1997).

From the marketing perspective, domestication and cultivation of MAPs offer a number of advantages over wild harvest for production of plant-based medicines. These include: 1) availability of authentic and botanically reliable products; 2) guaranteed steady source of raw material; 3) possibility for good rapport between growers and wholesalers (or agents of pharmaceutical companies) on volumes and prices over time; 4) controlled post-harvest handling and therefore rigorous quality control; 5) possibilities for adjustments of product standards to regulations and consumer preferences; and 6) possibilities for implementing product certification (Laird and Pierce 2002). Although the production of traditionally used medicinal plants has been suggested for over a decade, there has been little response in many countries. The lack of understanding with respect to cultivation, economics of cultivation and market opportunities are considered the main limiting factors in commercialization.

MAPs in agroforestry systems

Medicinal trees in traditional agroforestry systems

Many plants in traditional agricultural systems in the tropics have medicinal value. These can be found

Table 2. Examples of tree species reported to have medicinal values grown in traditional agroforestry systems in the tropics.

Tree species (Common name)	Agroforestry system	Medicinal use	Reference
Acacia nilotica (*babul*)	Field bunds, scattered trees in croplands, woodlots and grazing lands in East Africa and Indian sub-continent	Gum used for treating diarrhea, dysentery, diabetes, sore throat, bark used to arrest external bleeding	Pushpangadan and Nayar (1994)
Azadirachta indica (neem)	Woodlots, scattered trees, shelterbelts in Africa and India	Digestive disorders, malaria, fever, hemorrhoids, hepatitis, measles, syphilis, boils, burns, snakebite, rheumatism	Singh et al. (1996)
Erythrina spp.	Shade tree in coffee, live fence	Different *Erythrina* species have different uses.	Russo (1993)
Parkia biglobosa (*neré* or locust bean tree)	Parklands in West Africa	Piles, malaria, stomach disorders, jaundice	Teklehaimanot (2004)
Prosopis cineraria (*khejri*)	Scattered trees in croplands in arid to semiarid areas mostly in northwestern India	Flowers for blood purification and curing skin diseases. Bark against summer boils, leprosy, dysentery, bronchitis, asthma, leucoderma and piles	Pushpangadan and Nayar (1994)
Tamarindus indica (tamarind)	Field bunds and scattered trees in croplands in semiarid India and Africa	Fruit pulp is used in Indian medicine as refrigerant, carminative and laxative. It is also recommended in febrile diseases and bilious disorders	Singh et al. (1996)

(either planted or carefully tended natural regenerations) in homegardens, as scattered trees in croplands and grazing lands, and on field bunds (Table 2). Many *Acacia* species found in Africa such as *Acacia nilotica*, *A. seyal*, *A. senegal* and *A. polyacantha*, as well as several species in African croplands (e.g. *Faidherbia albida*, *Vitellaria paradoxa*, *Adansonia digitata*, *Markhamia lutea* and *Melia volkensii*) have medicinal value (ICRAF 1992). Similarly, '*arjun*' (*Terminalia arjuna*) in India, chinaberry (*Melia azederach*) in Asia and *Erythrina* species in Latin and Central America combine many uses including medicinal. Holy Basil or '*tulsi*' (*Ocimum sanctum*), drumstick (*Moringa oleifera*) and curry leaf (*Murraya koenigii*) are backyard plants in many Indian households and they are routinely used for common ailments or in food preparations. A number of plants used as live fences around home compounds such as henna (*Lawsonia inermis*) in India (Singh et a. 1996), *Ipomoea carnea* ssp. *fistulosa* in Bolivia and Asian countries (Frey 1995) and *Euphorbia tirucalli* around crop fields in Africa (ICRAF 1992) have medicinal values. Although the medicinal value of these plants is 'exploited' locally, they are seldom used for commercial purposes (except in the case of commercially exploited neem). In fact, many of these species are valued for poles, fuel wood, fodder, fruit, shade, and/or boundary demarcation and their medicinal value is secondary.

Forests and forest plantations

MAPs growing in forests require (or tolerate) partial shade, moist soils high in organic matter, high relative humidity, and mild temperatures. Cultivation of such MAPs can be taken up in thinned forests and cleared forest patches, and as intercrops in new forest plantations (Table 3). In China, cultivation of medicinal plants has been an age-old practice under the name of 'silvo-medicinal' systems. In northeast China, ginseng (*Panax ginseng*) and other medicinal plants are grown in pine (*Pinus* spp.) and spruce (*Picea* spp.) forests; in central China, many medicinal plants are planted with *Paulownia tomentosa* and in southern China medicinal herbs are often planted in bamboo (*Bambusa* spp.) and Chinese fir (*Cunninghamia lanceolata*) forests (Zou and Sanford 1990). In Yunnan province, China, traditional 'Dai and Jinuo' agroforestry systems involve the medicinal crop *Amomum villosum* in the forest areas cleared of undergrowth (Saint-Pierre 1991). The forest is thinned to give 30% to 40% shade and seedlings or cuttings are planted, which produce an average dried fruit yield of 375 kg ha^{-1} per annum (Zhou 1993). Gupta (1986) listed a number of indigenous understory herbs and shrubs that can be produced as part of forest farming or in new forest plantations to improve economic returns from forests in India (Table 3). Indigenous people living in the Himalayan forest margins in Uttaranchal, India, are known to have conserved and cultivated several medicinal species for centuries (Kumar et al. 2002).

A farmers' cooperative in the northern lowlands of Costa Rica has successfully demonstrated cultivation of the medicinal herb 'raicilla' (*Cephaelis ipecacuanha*) in natural forests for export to the Netherlands and Germany (Hager and Otterstedt 1996). American ginseng (*Panax quinquefolium*), a medicinal herb exported to China from the United States and Canada is grown as an understorey in red maple (*Acer rubrum*) forests (Nadeau et al. 1999) or deciduous hardwoods such as black walnut (*Juglans nigra*) and sugar maple (*Acer saccharum*), instead of growing under artificial shade with considerable expense (Hill and Buck 2000). Indeed, cultivation of ginseng and several other medicinal plants in the forests is a common and growing form of forest-farming practice of agroforestry in North America (Table 3). Light demanding understory species (e.g., *Echinacea* sp.) may be intercropped initially to provide early returns from plantations and after canopy closure, shade-tolerant species such as ginseng and goldenseal can be intercropped (Teel and Buck 2002). Studies in New Zealand have indicated that the American ginseng can be successfully grown under *Pinus radiata* with best growth under a tree stand of 130 stems ha^{-1} (Follett 1997). In addition to providing shade, the trees may also benefit the understory component from hydraulically lifted water. Fungal diseases are a major concern in forest farming but the application of fungicide can be detrimental to the forests' health, therefore proper spacing and mixed cropping is recommended (Cech 2002). Mechanical cultivation may not be feasible under forested conditions so labor availability needs to be considered as a constraining factor (Hill and Buck 2000).

As in the taungya system, newly established forest plantations can be intercropped with MAPs similar to food crops until the trees cover the ground. The participation of the local people with right to share benefits of the plantations, especially ownership to crops, has helped governments to establish and protect large-scale tree plantations without conflict with the local people in many Asian countries (Nair 1993). The same approach can be employed for the cultivation of MAPs in the new plantations. In the rehabilitation of degraded forestlands, participatory planning and implementation with local communities and economic benefits from an early stage onwards will ensure commitment of the people (Rao et al. 1999).

The intensity of shade experienced by the understory MAPs growing in forests and tree plantations affects their growth and chemical composition. Growth, and bark and quinine yields of *Cinchona ledgeriana* grown on the Darjeeling hills, India, increased when it was associated with shade of five species compared with that of a nonshaded stand (Nandi and Chatterjee 1991). The best yields were obtained when *C. ledgeriana* was planted under the shade of *Crotalaria anagyroides* or *Tephrosia candida* initially and *Alnus nepalensis* in the later stages. *Alnus* is planted at 3.6 × 3.6 m and is progressively thinned to 14.4 × 14.4 m as *C. ledgeriana* grows. The other factors that affect MAPs yields are their growth cycle and nutrient inputs. The optimum rotation for *C. ledgeriana* in the Himalayan region is 16 years (Nandi and Chatterjee 1991). *Dioscorea deltoidea* grown in deodar (*Cedrus deodara*), fir (*Abies* spp.) and spruce plantations attains exploitable tuber size in about 10 years (Gupta 1988).

Homegardens

Homegardens are complex agroforestry systems involving many plant species characterized by different morphology, stature, biological function and utility, practiced mostly in the humid and subhumid tropics (Kumar and Nair 2004). Food, fruit and timber species may dominate the homegardens and occupy the middle and upper strata, but medicinal plants, spices and vegetables occupy the lower stratum. Three categories of medicinal plants could be noted in homegardens: species used exclusively for medicine, horticultural or timber species with complementary medicinal value, and 'weedy' medicinal species. While the first two categories are deliberately planted the latter group is part of spontaneous growth. The species composition, plant density and level of management vary considerably depending on the soil, climate and market opportunity and cultural background of the people. Homegardens of individual holdings generally cover small parcels of land and are established around homesteads. Although these systems in the past were mostly seen to meet the home needs of small-scale farmers in the forest margins, increased urbanization, transport and market opportunities in recent times are helping to produce cash value crops. Multistrata systems involve fewer species (three to 10 species) than in homegardens in definite planting arrangement and can be designed for home consumption as well as commercial production.

Table 3. Examples of commercially valuable medicinal plants under cultivation or that can be produced as understory component(s) in forests and tree plantations.

Latin name	Common name	Plant type	Parts used	Medicinal use	Location
Aconitum heterophyllum	*Atis*	Tall herb	Rhizomes	Hysteria, throat diseases, astringent	Alpine and sub-alpine Himalayas
Amomum subulatum	Large cardamom	Perennial herb	Seeds	Stimulant, indigestion, vomiting, rectal diseases	Sub-Himalayan range, Nepal, Bhutan
Amomum villosum	*Saren*	Perennial herb	Seeds	Gastric and digestive disorders	China
Caulophyllum thalictroides	Blue cohosh	Perennial herb	Roots	Gynecological problems, bronchitis	North America
Cimicifuga racemosa	Black cohosh	Perennial herb	Roots	Menses related problems	North America
Chlorophytum borivilianum	*Safed musli*	Annual herb	Tubers	Male impotency, general weakness	India
Costus speciosus	Crepe ginger	Cane	Leaves, stem, rhizomes	Purgative, depurative and as a tonic	India
Dioscorea deltoidea	Himalayan yam	Vine	Tubers	Source of saponins and steroids	India, Pakistan
Echinacea purpurea	Coneflower	Perennial herb	Roots, rhizomes	Enhancing immune system	North America
Hydrastis canadensis	Goldenseal	Perennial herb	Rhizomes	Tonic	North America
Panax ginseng	Ginseng	Herb	Roots	Tonic	China, Korea, Japan
Panax quinquefolium	American ginseng	Perennial herb	Roots	Tonic	North America
Cephaelis ipecacuanha	Raicilla, Ipecac	Shrub	Roots	Whooping cough, bronchial asthma, amoebic dysentery	Brazil, India, Bangladesh, Indonesia
Rauvolfia serpentina	Rauvolfia	Shrub	Roots	Hypertension and certain forms of insanity	Sub-montane zone, India
Serenoa repens	Saw palmetto	Shrubby- palm	Fruits	Swelling of prostrate gland	Southeastern USA

Source: Gupta (1986); Saint-Pierre (1991); Rao et al. (1999); Garrett and McGraw (2000); Hill and Buck (2000); Teel and Buck (2002).

Medicinal plants are an invariable component of homegardens, whether they are in the Peruvian Amazon (Lamont et al. 1999), on the slopes of the Mt. Kilimanjaro in Tanzania (O'Kting'ati et al. 1984), or in the humid and semiarid Cuba (Wezel and Bender 2003). The species composition differs depending on cultural background, distance from markets and influence of tourism. Medicinal plants accounted for about 27% of total plant species in the homegardens in Amazon (Padoc and de Jong 1991), 56% in northern Catalonia (Iberian Peninsula) (Agelet et al. 2000) and 45% in the floodplains of the river Jamuna in Bangladesh (Yoshino and Ando 1999). In the Soqotra Island, Yemen, endemic medicinal plants such as *Aloe perryi*, *Jatropha unicostata* and *Commiphora ornifolia* are cultivated in homegardens (Ceccolini 2002). On St. Croix and St. Thomas, U.S. Virgin Islands, the medicinal trees neem, moringa, and noni (*Morinda citrifolia*) have become popular in homegardens (Palada and Williams 2000).

The slash-and-burn system, which is so widely practiced in the humid and sub-humid tropics, has reduced soil fertility, crop yields, biodiversity and mature forest vegetation. Agroforests that mimic local ecosystem processes can be developed to provide livelihoods for farmers while protecting and preserving forest reserves and biodiversity. Based on the analysis of plant species in different strata and analog hypothesis, an agroecological system can be designed that meets the farmers' needs of food, timber, medicinal plants and other NTFPs, efficiently uses the on-site resources and protects the natural resource base of the region (Clerck and Negreros-Castillo 2000). Integration of medicinal plants in agroforests and multistrata systems provides an answer to the plantation commodity conundrum. These complex agroforestry systems can be utilized to grow MAP species for home use and markets. MAPs produced in agroforests may be targeted to niche markets to secure higher premium on the premise of better quality similar to those harvested from wild. The forest-type environment of these systems facilitates the integration of species that generally grow in the forest (for example see Table 4) and thereby helps conserve the endangered species and produce them for markets. Homegardens and multistrata systems are recognized as the most productive, remunerative, environmentally sound and ecologically sustainable alternative land use systems to slash-and-burn systems and 'alang alang' (*Imperata cylindrica*)-infested degraded lands in humid tropics (Tomich et al. 1998). High value MAPs may be integrated in the newly establishing homegardens and multistrata systems.

Riparian buffer zones

An agroforestry system that has received considerable attention in North America is riparian buffers zones (Schultz et al. 2004). Riparian buffer zones can improve water quality and protect streams and rivers from degradation by nutrient loading and chemical pollutants from agriculture and urban areas, from erosion by attenuating peak flows and provide habitat for wildlife. NTFP production can help defray the cost of buffer zone installation and maintenance. Slippery elm (*Ulmus rubra*), harvested for its aromatic and medicinal inner bark, is commonly found in riparian areas in North America (Teel and Buck 2002). Riparian buffer zones are an ideal location for the production of this species, which suffers from commercial over-exploitation and the Dutch elm disease.

Intercropping of MAPs

Two types of intercropping systems can be distinguished involving MAPs: (1) medicinal plants as upperstory trees and (2) MAPs as intercrops in other tree crops.

Medicinal plants as overstory trees

Coffee (*Coffea arabica*), cacao (*Theobroma cacao*) and tea (*Camellia sinensis*) are traditionally grown under shade offered by multipurpose trees that produce timber, fruit, flowers, nuts, palms etc. Medicinal tree species that grow tall and develop open crown at the top can also be used for this purpose, for example *yongchak* (*Parkia roxburghii*) in India, the protein rich seeds of which are used to treat stomach disorders (Balasubramanian 1986) and *Ginkgo biloba* in China, the nuts of which are used in Chinese medicine and fetch high value (Shen 1998). In Ivory Coast, 19 of the 41 tree species planted as shade trees in coffee and cacao provide pharmaceutical products for traditional medicine (Herzog 1994). New plantations of coffee, tea, and cacao offer scope for cultivation of forest medicinal trees that are under demand. However, research needs to identify the medicinal trees that can be grown in association with these plantation crops and develop management practices for them. Tall and perennial medicinal trees that need to be planted at wider spacing such as *Prunus africana*, *Eucalyptus globulus* (for oil), sandalwood (*Santalum album), ashok (Saraca indica*), bael (*Aegle marmelos*), custard apple (*Anona squamosa*), *amla* (*Emblica officinalis*), drumstick or moringa (*Moringa oleifera*) and soapnut tree (*Sapindus mukorossi*) can be intercropped with annual crops in the early years until the tree canopy covers the ground. Some of the medicinal trees may allow intercropping for many years or on a permanent basis depending on the spacing and nature of the trees. The intercrops give some income to farmers during the period when the main trees have not started production.

Medicinal plants as intercrops

Many tropical MAPs are well adapted to partial shading, moist soil, high relative humidity and mild temperatures (Vyas and Nein 1999), allowing them to be intercropped with timber and fuel wood plantations, fruit trees and plantation crops. Some well known medicinal plants that have been successfully intercropped with fuel wood trees (e.g., *Acacia auriculiformis*, *Albizia lebbeck*, *Eucalyptus tereticornis*, *Gmelina arborea*, and *Leucaeana leucocephala*) in India, include *safed musli* (*Chlorophytum borivilianum*), rauvolfia (*Rauvolfia serpentina*), turmeric (*Curcuma longa*), wild turmeric (*C. aromatica*), *Curculigo orchioides*, and ginger (*Zingiber officinale*) (Chadhar and Sharma 1996; Mishra and Pandey 1998; Prajapati et al. 2003). Only 10 out of 64 herbaceous medicinal plants tried in intercropping with two-year old poplar (*Populus deltoides*) spaced 5 m apart gave poor performance (Kumar and Gupta 1991), indicating that many medicinal plants can be grown in agroforestry systems. The trees may benefit from the inputs and management given to the intercrops. Short stature and short cycle MAPs and culinary herbs are particularly suited for short-term intercropping during the juvenile phase of trees. Wherever markets are established, MAPs are remunerative alternative intercrops to the

Table 4. Important medicinal trees currently exploited from Peruvian Amazon forest that can be grown in multistrata or homegarden agroforestry systems.

Name	Plant type	Part used	Medicinal use
Cedrella odorata	Tree	Bark	Snake bite
Croton lechleri	Tree	Latex	Swellings
Euterpe precatoria	Palm	Roots	Kindness, diabetes
Ficus insidpida	Tree	Latex	Anemia
Maytenus macrocarpa	Tree	Bark	Arthritis, diarrhea, stomach disorders
Tabebuia serratifolia	Tree	Bark	Vaginal diseases, hepatitis, diabetes, arthritis
Uncaria tomentosa	Shrub	Bark	Infections, cancer

Adapted from: Clavo et al. (2003).

traditionally grown annual crops (Maheswari et al. 1985; Zou and Sanford 1990). The number of years MAPs can be intercropped with a given tree species depends on the size and intensity of its canopy shade, tree spacing and management, especially pruning of branches and nature of the MAPs. Shade-tolerant and rhizomatic MAPs can be grown on a longer-term basis in widely spaced plantations.

Intercropping of medicinal plants in coconut (*Cocos nucifera*) and arecanut (*Areca catechu*) stands is an age-old practice in India and other parts of south- and southeast Asia. These palms allow 30% to 50% of incident light to the underneath, which is ideal for some MAPs, including cardamom (*Elettaria cardamomum*). *Kacholam* or galang (*Kaempferia galanga*) – a medicinal herb – is traditionally intercropped in mature coconut gardens in Kerala, India. *Kacholam* intercropped in a 30 year-old coconut plantation produced 6.1 Mg ha^{-1} of rhizomes compared with 4.8 Mg ha^{-1} as a sole crop (Maheswarappa et al. 1998). Twelve-year old coconut trees did not adversely affect the growth and yields of a number of medicinal species grown as intercrops compared to the yields in the open (Nair et al. 1989). In Karnataka and Kerala states, India, arecanut palm is commonly intercropped with ginger, turmeric, black pepper (*Piper nigrum*) and cardamom (Korikanthimath and Hegde 1994). Some of these intercrops may cause small reduction in arecanut yields but the combined returns from both the components are greater than from arecanut alone. Another plantation crop intercropped with MAPs is rubber (*Hevea brasiliensis*), for example with *Dioscorea floribunda* in the state of Assam in India (Singh et al. 1998) and with *Amomum villosum* in Yunnan province of China (Zhou 1993). In Sikkim, India, large cardamom (*Amomum subulatum*) is grown under 30 different shade tree species (Patiram et al. 1996). In Fujian Province, China, *Cunninghamia lanceolata* – an important timber tree – is intercropped with a variety of cereals, cash and medicinal and oil-producing crops (Chandler 1994). Many of the medicinal herbs commonly grown in thinned forests can also be grown intercropped with trees (Zhou 1993).

In the Caribbean islands, there has been increased interest on alternative crops that have better economic potential than traditional crops. For example, in the U.S. Virgin Islands, a number of farmers are now opting for specialty crops such as the West Indian hot peppers (*Capsicum chinense*), thyme (*Thymus vulgaris*) and chives (*Allium schoenosprasum*) instead of vegetables (Crossman et al. 1999). The prospects of growing indigenous MAPs such as 'japana' (*Eupatorium triplinerve*), worrywine (*Stachytarpheta jamaicensis*), inflammation bush (*Verbersina alata*) and lemongrass (*Cymbopogon citratus*) in association with the medicinal trees noni (*Morinda citrifolia*) and moringa have been explored at the University of the Virgin Islands, St. Croix, (Palada and Williams 2000). These local herbs are commonly used as bush teas and very popular in the Caribbean. Medicinal plants and herbs in intercropping produced similar yields to those in sole cropping at the first harvest, but they tended to be lower than in sole cropping at subsequent harvests (Palada and Williams 2000).

Aromatic plants as intercrops

Studies have been conducted in India since mid-1980s on the feasibility and economic aspects of intercropping aromatic plants with timber trees. Experiments at different sites in India demonstrated that aromatic *Mentha* spp. (*M. arvensis*, *M. piperita*, *M. citrata* var. *citrata*, *M. spicata*, *M. cardiaca* and *M. gracilis*) and *Cymbopogon* spp. [lemon-

Table 5. Yields of aromatic plants in intercropping with trees relative to those in sole cropping at a few locations in India and Nepal.

Location	Tree species and spacing	Aromatic plant	Intercrop yield as % of sole crop		Comments
			Herbage	Oil	
Pantnagar, India	*Populus deltoides*	*Cymbopogon flexuosus*	82	82	Results averaged over five years
	5 × 4 m	*C. martinii*	96	86	
	Eucalyptus	*C. flexuosus*	96	90	
	hybrid 2.5 × 2.5 m	*C. martinii*	100	86	
Pantnagar, India	*Populus deltoides*	*Mentha arvensis*	90	90	Results averaged over three years
	5 × 4 m	*M. piperita*	85	88	
		M. citrata	90	90	
		M. spicata	87	87	
		C. flexuosus	99	97	
		C. martinii	99	98	
		C. winterianus	98	99	
Hyderabad, India	*Eucalyptus citriodora*	*C. flexuosus*	87	87	Results averaged over two years
	3 × 2 m	*C. martinii*	87	87	
		C. winterianus	112	91	
		Pelargonium sp.	91	90	
Nepal	*Dalbergia sissoo* 10 × 2 to 8 m	*C. winterianus*	–	100	Results average over two years

Source: 1. Singh et al. (1989); 2. Singh et al. (1990); 3. Singh et al. (1998); 4. Amatya (1996).

grass (*C. flexuosus*); Java citronella (*C. winterianus*); and palmarosa (*C. martinii*)] can be grown intercropped with *Populus deltoides* or *Eucalyptus* spp. for three to five years after planting the trees (Table 5). Some of the above aromatic species and *Pelargonium* sp. (geranium) can also be intercropped with another essential-oil-yielding tree *Eucalyptus citriodora* (lemon- or citron-scented gum) (Singh et al. 1998). Tree growth was generally little affected or improved probably due to the inputs given to the intercrops. Herb growth and oil yields of these aromatic plants were not affected during the first two years of intercropping, but they decreased slightly from the third year compared with sole crops (Singh et al. 1990). Essential oil profiles of citronella and palmarosa grown in open and partial shade were not significantly different, but the oil of menthol mint (*M. arvensis*) produced in the open was richer in menthone while that produced under shade was richer in menthol (Singh et al. 2002). This implies the need to consider the effect of intercropping on oil quality of some aromatic plants. These intercropping systems gave two to four times greater economic returns than sole cropping of the components. These studies indicate that aromatic plants can be profitably grown in association with linearly growing fuel- and timber trees such as *L. leucocephala*, *Casuarina* spp., and *Grevillea robusta*.

MAPs in other agroforestry systems

Aromatic grasses such as vetiver (*Vetiveria zizanioides*), lemongrass and citronella (*Cymbopogon nardus*) can be grown on field bunds and soil conservation bunds in croplands. Vetiver has extensively been tested and is being promoted for planting in contour strips or as a live hedge barrier and to stabilize terrace risers on sloping lands in a number of countries, for example, in India, Fiji, Haiti and Indonesia (NRC 1993). In the hill subsistence farming systems, maintaining woody perennials in contour strips across slopes and around fields is a common practice. The trees and shrubs produce fodder and firewood and reduce soil erosion hazards allowing crop production on steep slopes where it would otherwise be difficult. Many locally available medicinal plants can be incorporated into these systems (Fonzen and Oberholzer 1984). The harvest of grass strips on sloping lands for economic purpose should be planned carefully to avoid soil erosion. Indiscriminate digging of vetiver for distilling oil from its roots has often resulted in worst erosion problems in Haiti (M. Bannister pers. comm., October 2003).

Hardy medicinal and fruit trees can be grown in community lands in villages and degraded lands. In the state of Madhya Pradesh, India, *amla* – a rich source of vitamin C, and custard apple, for both fruit and medicine, have been taken up as part of social forestry (Chadhar 2001). *Amla* and bael withstand saline and alkali conditions and can be grown on salt-affected soils (Khanna 1994). Similarly, in West Bengal, India, *Costus speciosus*, a diosgenin-producing plant, is recommended as a major understorey crop in social forestry programs (Konar and Kushari 1989). Other medicinal trees that can be grown on wasted lands include soapnut tree and wood apple. A study of the medicinal species in the Shinyanga region of northwestern Tanzania indicated that the top 10 priority species can be integrated into land use systems such as *ngitili* (grazing lands planted with economically valuable trees and shrubs), scattered trees in croplands, on field bunds or in homesteads (Table 6). It is recognized in India that many medicinal trees can be integrated into croplands by planting them on field bunds or as scattered trees (Pushpangadan and Nayar 1994).

Future trends and research needs

The demand for MAPs will continue to increase in developing and industrialized countries because of population growth and awareness about the benefits of natural products to provide economical, safe and effective alternatives to expensive industrially synthesized drugs. The best recourse to meet the expanding trade in medicinal plants is cultivation of priority MAPs for conservation as well as continuous supply of raw material to the industry and to provide high returns to farmers. Successful cultivation of MAPs requires information on the species adaptation to outside their natural habitats, cultivation practices and management, but there is, however, limited information on these aspects of medicinal plants in many countries. As domestication and cultivation cannot be taken up for all potential MAPs, research as a first step should establish priority species based on participatory surveys and market analysis involving rural people, traditional healers, material collectors and drug industry. Research should focus on species for which there are ready markets and natural sources are fast depleting. Work should follow for priority species on ecology and natural distribution, reproductive biology, propagation techniques, germplasm collection, evaluation and genetic improvement aimed at improving both yield and quality of the product. Simultaneously, information should be produced on the soil and environmental conditions to which they are adapted, systems in which they can be grown, nutrient management, harvest techniques and processing. Attention should also be given to the potential danger of pests and diseases cropping up with intensive cultivation of medicinal plants. The on-going efforts on domestication and cultivation of *Prunus africana* in agroforestry involving smallholders in Cameroon and East Africa should serve as a case study to illustrate the methodology and economic potential of some species (Cunningham et al. 2002; Simons and Leakey 2004).

An important area that needs investigation is how production of MAPs in nontraditional systems affects content and quality of the principal compounds for which they are grown. Research tends to focus on increasing biomass yields rather than on quality of the compounds for which the plants are valued. The production of secondary compounds is not necessarily linear to increases in biomass of all species. Edaphic factors, timing of inputs and intensity and duration of shade effects are important determinants of plant productivity of secondary products. For example, the diosgenin concentration in *Costus speciosus* rhizomes was increased by treatment with mango leaf leachate, unaffected by the *Shorea robusta* and teak (*Tectona grandis*) leaf leachates and decreased by the *Eucalyptus globulus* leaf leachate (Konar and Kushari 1989). As quality determines the acceptability and price of product, especially for export market, research should examine the appropriate practices that achieve good quality.

Research should simultaneously focus on processing methods at the farm-level and exploring market opportunities. Area expansion without knowledge of market potential can lead to over production and disappointment to growers. The recent spurt in the cultivation of *Chlorophytum borivilianum* in India is an example where many farmers have taken up cultivation to sell tubers as seed material at a high premium, rather than laboriously processing them. Consequently, not only the prices of seed (fresh tubers) and dried materials have dropped considerably in the past three years but also growers have begun to perceive marketing this high-input-requiring crop at a remunerative price as a major concern (Oudhia 2003).

Table 6. The top 10 priority medicinal trees currently harvested from wild in Shiyanga region of Tanzania that can be grown as scattered trees in grazing lands (*'Ngitili'*), croplands or in homesteads.

Latin name	Local name (English name)	Parts used	Medicinal use
Albizia anthelmintica	*Mgada/mkutani* (Worm-cure albizia)	Bark, roots, leaves	Abdominal problems, convulsions and infertility
Cassia abbreviata	*Mlundalunda* (Long-pod cassia)	Bark, roots	Abdominal problems, pain relief and urinary problems
Combretum zeyheri	*Msana* (Large-fruited combretum)	Bark, roots, leaves	Pneumonia, peptic ulcer, coughs and soar throat
Entada abyssinica	*Ngeng'wambula/Mfutwambula* (Tree entada)[a]	Bark, roots, leaves	Abdominal problems, cough, asthma and hernia
Entandrophragma bussei	*Mondo* (Wooden banana)	Bark, roots, leaves	Abdominal problems, diarrhea and anemia
Securidad longipedunculata	*Nengonego* (African violet tree)	Bark, roots, leaves	Convulsions, abdominal problems, gonorrhea, syphilis, and asthma
Terminalia sericea	*Mzima* (Silver terminalia)	Bark, roots, leaves	Fever, anemia and abdominal problems
Turraea fisceri	*Ningiwe* (honeysuckle tree)	Bark, roots, leaves	Abdominal problems, hypertension, and dysentery
Zanha africana	*Ng'watya/mkalya* (Velvet-fruited zanha)	Bark, roots, leaves	Convulsions, abdominal problems and psychosis
Zanthoxylum chalybeum	*Mlungulungu/Nungubalagiti* (Knobwood)	Bark, roots, leaves	Jaundice, abdominal problems and pain relief

[a]This can also be grown in row intercropping with annual crops.
Source: Dery et al. (1999).

As part of promoting cultivation of MAPs, government departments such as forestry may enter into partnership with local farmers to cultivate MAPs in association with forest plantations on profit-sharing basis. This is similar to the taungya system except that instead of food crops farmers are encouraged to plant selected MAPs under the guidance of the forest department. This type of joint forest management with 25% to 50% share of profit to farmers was found successful in the Gujarat state of India (Singh 1997). Partnerships between drug industry and farmers in the form of contract farming with buyback arrangement of the product will go a long way in promoting cultivation of MAPs. The drug companies embarked on exploiting medicinal plants should be made development partners in the production areas. For instance, the drug company Merck Sharpe & Dome has paid USD 1 million for research in Costa Rica and has agreed to contribute 25% of profits made from Costa Rican plants to rainforest conservation in Costa Rica (Sittenfeld and Gamez 1993). Shaman Pharmaceutical Inc., a pioneer in natural product research since 1989, considers indigenous people as partners and only collects plant samples on the indication of a shaman. This approach appears to be more effective than random collection methods and has already resulted in the discovery of three novel drugs (van Seters 1997). The national governments should ensure that the local people do derive benefits for their contributions to the new innovations. Many developing countries, however, may not be having appropriate national bodies to safeguard the intellectual property rights (IPR) over indigenous knowledge of local people, protect from potential biopiracy of genetic material, or to undertake research, promotion and marketing of MAPs.

The pharmaceutical industry rediscovered tropical rainforests as an unmatched source of chemicals with potential for modern drug development (Pistorius and van Wijk 1993). Thousands of plant extracts taken from all continents are being screened for activity against HIV and cancer in the laboratories of the U.S. National Cancer Institute. For example, an alkaloid casuarine from the bark of *Casuarina equisetifolia* was found to have potential against HIV and cancer (Nash et al. 1994). While efforts to find new chemicals continue, investigations should also be encouraged on the traditionally used plant medicines to dispel myths and strengthen the case for valid ones. It is likely that not all the indigenous healthcare practices currently in use have scientific bases. A study conducted among tribals in Kaimur Bhabhua district in Bihar state, India, indicated that their use of *Acacia nilotica* against bleeding gum and soar throat, neem as antibiotic and *Terminalia arjuna* for curing heart diseases had scientific merit, but not the use of *Madhuca longifolia* and *Terminalia belirica* (Pandey and Varma 2002).

Policies that deprive ownership to medicinal trees as in the case of some parkland trees in West Africa or imposing license fees even with good intentions of protecting the trees would be counterproductive to planting and management of medicinal trees. For example, farmers in the southern highlands of Tanzania have stopped planting the traditionally grown, native multipurpose tree *Hagenia abyssinica* (hardwood, fuelwood, fodder, green manure, flowers used for deworming humans and livestock, seed as a condiment or spice), since the introduction of a royalty fee as license for its management in the mid-1980s; instead, exotic species are planted (East and Thurow 1999). Farmers need to be encouraged by making available planting material and appropriate technical know-how.

Conclusion

Traditional systems of medicine in most developing countries depend primarily on the use of plant products either directly or indirectly. Besides serving the healthcare needs of a large number of people, medicinal plants are the exclusive source of some drugs even for modern medical treatment. The use of plant products as nutrition supplements and in the cosmetic and perfume industry has increased the value of medicinal and aromatic plants in recent years. The over dependence on forests, natural woodlands and long-term fallows for extraction of MAPs is threatening the survival of many valuable plant species. It is imperative therefore that such endangered species are cultivated outside their natural habitats to ensure their regular supply for human needs as well as to preserve the genetic diversity. Cultivation is an important strategy for conservation and sustainable maintenance of natural stocks, but, few MAPs are actually cultivated. Lack of basic knowledge on biology, ecology, propagation methods and cultural practices for the concerned species is an important constraint.

Agroforestry offers a convenient way of producing many MAPs without displacing the traditional crops. Research is needed in each country, however, on germplasm improvement for priority species, appropriate systems in which they can be grown, input management, and value-adding processes. Existing government policies may not be conducive to promotion of MAPs in many places. The potential of MAPs can be realized when policy constraints are removed and efforts are made simultaneously to commercialize the products and explore markets for less known species. Although research has indicated that agroforestry can be a viable approach to production of MAPs, few commercially important species are actually produced outside homegardens and forest plantations.

References

Agelet A., Bonet M.A. and Valles J. 2000. Homegardens and their role as a main source of medicinal plants in mountain regions of Catalonia (Iberian Peninsula). Econ Bot 54: 295–309.

Amatya S.M. 1996. Non-timber forest products and their production opportunities in Nepal. In: Leakey R.R.B., Temu A.B., Melnyk M. and Vantomme P. (eds), Domestication and Commercialization of Non-Timber Forest Products in Agroforestry Systems. Non-Wood Forest Products No. 9. FAO, Rome, Italy, p. 284.

Balasubramanian M.A. 1986. Yongchak as shade tree. Planters' Chronicle 80(4): 134.

Barany M., Hammett A.L., Sene A. and Amichev B., 2001. Non-timber forest benefits and HIV/AIDS in sub-Saharan Africa. J Forest 99(12): 36–41.

Bernáth J. 1999. Biological and economical aspects of utilization and exploitation of wild growing medicinal plants in middle and south Europe. Acta Horticulturae 500: 31–41.

Bhattarai N.K. 1997. Biodiversity-people interface in Nepal. pp. 82–89. In: Bodeker G., Bhat K.K.S., Burleyand J. and Vantomme P. (eds), Medicinal Plants for Forest Conservation and Healthcare. Non-Wood Forest Products No. 11, FAO, Rome, Italy.

Ceccolini, L. 2002. The homegardens of Soqotra island, Yemen: an example of agroforestry approach to multiple land-use in an isolated location. Agroforest Syst 56: 107–115.

Chadhar S.K. 2001. Six years of extension and research activities in Social Forestry Division Jhabua (M.P.). Vaniki Sandesh 25(4): 6–10.

Chadhar S.K. and Sharma M.C. 1996. Survival and yield of four medicinal plant species grown under tree plantations of Bhataland. Vaniki Sandesh 20(4): 3–5.

Chandler P. 1994. Adaptive ecology of traditionally derived agroforestry in China. Hum Ecol 22: 415–442.

Chech R.A. 2002. Balancing conservation with utilization: restoring populations of commercially valuable medicinal herbs in forests and agroforests. pp. 117–123. In: Iwu M.M. and Wootten J.C. (eds), Ethnomedicine and drug discovery. Elsevier Science B.V., The Netherlands.

Clavo Z.M., Seijas S.P. and Alegre J.C. 2003. Plantas Medicinales Usadas por Mujeres Nativas y Mastizas en la Region Ucayali. IVITA-INIA-ICRAF Ediciones Talleres. Grafficos Av. Yarinacocha No. 799, Yarinacocha, Pucallpa, Peru, 127 p.p.

Clerck F.A.J. de and Negreros-Castillo P. 2000. Plant species of traditional Mayan homegardens of Mexico as analogs for multistrata agroforests. Agroforest Syst 48: 303–317.

Crossman S.M.A., Palada M.C and Davis A.M. 1999. Performance of West Indian hot pepper cultivars in the Virgin Islands. Proceedings Caribbean Food Crops Society 35: 169–176.

Cunningham A.B. 1997. An Africa-wide overview of medicinal plant harvesting, conservation and healthcare. pp. 116–129. In: Bodeker G., Bhat K.K.S., Burley J. and Vantomme P. (eds), Medicinal Plants for Forest Conservation and Healthcare. Non-Wood Forest Products No. 11, FAO, Rome.

Cunningham A.B., Ayuk E., Franzel S., Duguma B. and Asanga C. 2002. An economic evaluation of medicinal tree cultivation: *Prunus africana* in Cameroon. People and Plants Working Paper No. 10, Division of Ecological Sciences, UNESCO, Paris, France, 35 pp.

de Alwis L. 1997. A biocultural medicinal plants conservation project in Sri Lanka. pp. 103–111. In: Bodeker G., Bhat K.K.S., Burley J. and Vantomme P. (eds), Medicinal Plants for Forest Conservation and Healthcare. Non-Wood Forest Products No. 11, FAO, Rome, Italy.

Dery B.B., Otsyina R. and Ng'tigwa C. (eds.) 1999. Indigenous Knowledge of Medicinal Trees and Setting Priorities for their Domestication in Shinyanga Region, Tanzania. International Center for Research in Agroforestry, Nairobi, Kenya, 87 pp.

de Silva T. 1997. Industrial utilization of medicinal plants in developing countries. pp. 38–48. In: Bodeker G., Bhat K.K.S., Burley J. and Vantomme P. (eds), Medicinal Plants for Forest Conservation and Healthcare. Non-Wood Forest Products No. 11, FAO, Rome, Italy.

East R. and Thurow T. 1999. Challenging tradition: well-intentioned conservation laws in Tanzania are hampering age-old practices. Agroforest Today 11(1/2): 8–10.

FAO 1995. Non-Wood Forest Products for Rural Income and Sustainable Development. Non-Wood Forest Products No. 7, FAO, Rome, Italy, 127 pp.

FAO 2003. State of the World's Forests. FAO, Rome, Italy, 153 pp.

Follett J. 1997. Ginseng production in NZ forests: experiences from Tiketere. New Zealand Tree Grower 18(3): 19–21.

Fonzen P.F. and Oberholzer E. 1984. Use of multipurpose trees in hill farming systems in western Nepal. Agroforest Syst 2: 187–197.

Foster S. 1993. Herbal Renaissance. Gibbs-Smith Publ. Salt Lake City, UT, USA, 234 pp.

Frey R. 1995. *Ipomoea carnea* ssp. *fistulosa* (Martius ex Chiosy) Austin: Taxonomy, biology and ecology reviewed and inquired. Trop Ecol 36: 21–48.

Garrett H.E. and McGraw R.L. 2000. Alleycropping practices. pp. 149–188. In: Garrett H.E., Rietveld W.J. and Fisher R.F. (eds), North American Agroforestry: An Integrated Science and Practice. American Society of Agronomy, Inc., Madison, WI, USA.

Gupta M.P. 1988. Present status and future prospects for the cultivation and collection of medicinal plants of Himachal Pradesh. Indian Forester 114: 19–25.

Gupta R. 1986. Integration of medicinal plants cultivation in forest and forest plantations of northwestern Himalaya. pp. 59–67. In: Agroforestry Systems: A New Challenge. Indian Society of Tree Scientists, Solan, India.

Hager N. and Otterstedt J. 1996. Coope San Juan, a farmers' cooperative. Sustainable use of the natural forest–past and future: a minor field study. Working Paper No. 302. International Rural Development Centre, Swedish University of Agricultural Sciences, Uppsala, Sweden. 59 pp.

He S.A. and Sheng N. 1997. Utilization and conservation of medicinal plants in China. pp. 112–118. In: Bodeker G., Bhat K.K.S., Burley J. and Vantomme P. (eds), Medicinal Plants for Forest Conservation and Healthcare. Non-Wood Forest Products No. 11, FAO, Rome, Italy.

Herzog F. 1994. Multiple shade trees in coffee and cocoa plantations in Coted'Ivoire. Agroforest Syst 27: 259–267.

Heywood V.H. 1999. Plant resources and their diversity in the near east. pp. 4–18. In, Medicinal Culinary and Aromatic Plants in the Near East: Proceedings International Expert Meeting 19–21 May 1997. FAO Forestry Department and FAO Regional Office for the Near East, Cairo, Egypt.

Hill D.B. and Buck L.E. 2000. Forest farming practices. pp. 283–320. In: Garrett H.E., Rietveld W.J. and Fisher R.F. (eds), North American Agroforestry: An Integrated Science and Practice. American Society of Agronomy Inc., Madison, WI, USA.

ICRAF 1992. A Selection of Useful Trees and Shrubs in Kenya: Notes on their Identification, Propagation and Management for Use by Farming and Pastoral Communities. International Centre for Research in Agroforestry, Nairobi, Kenya, 226 pp.

Joshi D.N. and Rawat G.S. 1997. Need for conservation and propagation of alpine and sub-alpine medicinal plants of north-west Himalayas. Indian Forester 123: 811–814.

Khanna S.S. 1994. Management of sodic soils through plantations: a successful case study. J Indian Soc Soil Sci 42: 498–508.

Konar J. and Kushari D.P. 1989. Effect of leaf leachate of four species on sprouting behaviour of rhizomes, seedling growth and diosgenin content of *Costus speciosus*. Bulletin of the Terrey Botanical Club. 116: 339–343.

Korikanthimath V.S. and Hegde R. 1994. Cardamom and arecanut mixed-cropping systems. Indian Cocoa, Arecanut and Spices J 18(4): 109–112.

Kumar A., Bisht P.S. and Kumar V. 2002. Traditional medicinal plants of Uttaranchal Himalayas. Asian Agri-History 6: 167–170.

Kumar B.M. and Nair P.K.R. 2004. The enigma of tropical homegardens. (This volume).

Kumar K. and Gupta C. 1991. Intercropping of medicinal plants with poplar and their phenology. Indian Forester: 535–544.

Laird S.A. and A.R. Pierce. 2002. Promoting sustainable and ethical botanicals. Strategies to improve commercial raw material sourcing. Rainforest Alliance, New York. http://www.rainforest-alliance.org/news/archives/news/news44.html.Website accessed 30.7.2003.

Lambert J., Srivastava J. and Vietmeyer N. 1997. Medicinal Plants: Rescuing a Global Heritage. World Bank Technical Paper No. 355. World Bank, Washington, DC, 61 pp.

Lancet, The. 1994. Pharmaceuticals from plants: great potential, few funds. The Lancet 343: 1513–1515.

Lange D. 2002. The role of east and southeast Europe in the medicinal and aromatic plant's trade. Medicinal Plants Conserv 8: 14–18.

Lamont S.R., Eshbaug W.H. and Greenberg A.M. 1999. Species composition, diversity and use of homegardens among three Amazonian villages. Econ Bot 53: 312–326.

Maheswarappa H.P., Hegde M.R. and Nanjappa H.V. 1998. Kacholam (*Kaempferia galanga* L.) – a potential medicinal-cum-aromatic crop for coconut gardens. Indian Coconut J 29(5): 4–5.

Maheswari S.K., Dhantonde B.N., Yadav S. and Gangrade S.K. 1985. Intercropping of *Rauvolfia serpentina* for higher monetary returns. Indian J Agri Sci 58: 108–111.

McNeely J.A. 2004. Nature vs. nurture: Managing relationships between forests, agroforestry and wild biodiversity. (This volume).

Mishra R.K. and Pandey V.K. 1998. Intercropping of turmeric under different tree species and their planting pattern in agroforestry systems. Range Manag Agroforest 19: 199–202.

Nadeau I., Oliver A., Semard R.R., Coulobe J. and Yelle S. 1999. Growing American ginseng in maple forests as an alternative land use system in Quebec, Canada. Agroforest Syst 44: 345–353.

Nair G.S., Sudhadevi P.K. and Kurian A. 1989. Introduction of medicinal and aromatic plants as intercrops in coconut plantations. pp. 163–165. In: Raychauduri S.P. (ed.), Recent Advances in Medicinal, Aromatic and Spice Crops. Today and Tomorrow's Printers and Publishers, New Delhi, India, Vol. 1.

Nair P.K.R. 1993. An Introduction to Agroforestry. Kluwer Academic Publishers, Dordrecht, The Netherlands, 499 pp.

Nandi R.P. and Chatterjee S.K. 1991. Effect of shade of trees on growth and alkaloid formation in *Cinchona ledgeriana* grown in Himalayan hills of Darjeeling. pp. 181–184. In: Trivedi R.N., Sharma P.K.S. and Singh M.P. (eds), Environmental Assessment and Management: Social Forestry in Tribal Regions. Today and Tomorrow's Printers & Publishers, New Delhi, India.

Nash R.J., Longland A.C. and Wormald M.R. 1994. The potential of trees as sources of new drugs and pesticides. pp. 899–910. In: Singh P., Pathak P.S. and Roy M.M. (eds), Agroforestry Systems for Degraded Lands. Oxford & IBH Publishing Co. Pvt. Ltd., New Delhi, India, Vol. 2.

NRC 1993. Vetiver grass: a thin green line against erosion. National Research Council, National Academy Press. Washington DC, 171 pp.

O'Kting'ati A. Maghembe J.A., Fernandes E.C.M. and Weaver G.H. 1984. Plant species in the Kilimanjaro agroforestry system. Agroforest Syst 2: 177–186.

On T.V., Quyen D., Bich L.D., Jones B., Wunder J. and Russell-Smith J. 2001. A survey of medicinal plants in BaVi National Park, Vietnam: methodology and implications for conservations and sustainable use. Biol Conserv 97: 295–304.

Oudhia P. 2003. Problems perceived by safed musli (*Chlorophytum borovilianum*) growers of Chattisgarh (India): a study. Journal of Medicinal and Aromatic Plant Sciences 22(4a) & 23(1a): 396–399.

Padoc C. and de Jong W. 1991. The house gardens of Santa Rosa: diversity and variability in an Amazonian agricultural system. Econ Bot 45: 166–175.

Palada M.C. and Williams M.E. (eds.) 2000. Utilizing Medicinal Plants to Add Value to Caribbean Agriculture. Proceedings of the Second International Workshop on Herbal Medicine in the Caribbean. University of the Virgin Islands, St. Croix, U.S. Virgin Islands, 217 pp.

Pandey S. and Varma S.K. 2002. Medicinal importance of trees suitable for agroforestry systems. Indian J Agroforest 4: 47–51.

Patiram, Bhadauria S.B.S. and Upadhyaya R.C. 1996. Agroforestry practices in hill farming of Sikkim. Indian Forester 122: 621–630.

Pistorius R. and van Wijk J. 1993. Biodiversity prospecting: commercializing genetic resources for export. Biotechnol Dev Monit 15: 12–15.

Prajapati N.D., Purohit S.S., Sharma A.K. and Kumar T. 2003. A Handbook of Medicinal Plants. Agribios (India), 553 pp.

Pushpangadan P and Nayar T.S. 1994. Conservation of medicinal and aromatic tree species through agroforestry. pp. 265–284. In: Thampan P.K. (ed.), Trees and Tree Farming. Peekay Tree Crops Development Foundation, Cochin, India.

Rao K.S., Maikhuri R.K. and Saxena K.E. 1999. Participatory approach to rehabilitation degraded forest lands: a case study in a high altitude village of Indian Himalayas. Int Tree Crops J 10: 1–17.

Russo R.O. 1993. The use of *Erythrina* species in the Americas. In: Westely S.B. and Powell M.H. (eds), *Erythrina* in the New and the Old World. Nitrogen fixing Tree Association, Hawaii, USA, pp. 28–45.

Saint-Pierre C. 1991. Evolution of agroforestry in the Xinshuangbanna region of tropical China. Agroforest Syst 13: 159–176.

Shackleton C.M. and Shackleton S.E. 2000. Direct use values of secondary resources harvested from communal savannas in the Bushbuckridge lowveld, South Africa. J Trop For Products 6: 28–47.

Schippmann U., Leaman D.J. and Cunnningham A.B. 2002. Impact of cultivation and gathering of medicinal plants on biodiversity: global trends and issues. In: Biodiversity and the Ecosystem Approach in Agriculture, Forestry and Fisheries. Ninth Regular Session of the Commission on Genetic Resources for Food and Agriculture. FAO, Rome, Italy, pp. 1–21.

Simons A.J. and Leakey R.R.B. 2004. Tree domestication in agroforestry. Agroforest Syst (This volume).

Schultes R.E. 1979. The Amazonia as a source of new economic plants. Econ Bot 33: 259–266.

Schultz R. C., Isenhart T.M., Simpkins W.W. and Colletti J.P. 2004. Riparian forest buffers in agroecosystems: Lessons learned from the Bear Creek Watershed Project in Central Iowa, USA. Agroforest Syst (This volume).

Shanley P. and Luz L. 2003. The impacts of forest degradation on medicinal plant use and implications for health care in eastern Amazonia. BioScience 53: 573–584.

Shen L. 1998. Management considerations and economic benefits of intercropping tea with *Ginkgo biloba*. J Zhejiang Forest Sci Technol 18(2): 69–71.

Singh H.S. 1997. Appropriate silvicultural models under the joint forest management: Gujarat State. Indian Forester 123: 477–483.

Singh R.S., Bhattacharya T.K., Dutta A.K., Das P.K. and Nag D. 1998. Production potential in the intercropping sequence of medicinal yam (*Dioscorea floribunda*) with pigeonpea (*Cajanus cajan*) and rubber (*Hevea brasiliensis*). Indian J Agric Sci 68: 231–232.

Singh K., Chauhan H.S., Rajput D.K. and Singh D.V. 1989. Report of a 60 month study on litter production, changes in soil chemical properties and productivity under poplar (*P. deltoides*) and *Eucalyptus* (*E. hybrid*) intercropped with aromatic grasses. Agroforest Syst 9: 37–45.

Singh K., Rajeswara Rao B.R., Rajput D.K., Bhattacharya A.K., Chauhan H.S., Mallavarapu B.R. and Ramesh S. 2002. Composition of essential oils of menthol mint (*Mentha arvensis* var. *piperascens*), citronella (*Cymbopogon winterianus*) and palmarosa (*Cymbopogon martinii* var. *motia*) grown in the open and partial shade of poplar (*Populus deltoides*) trees. J Med Aromatic Plant Sci 24: 710–712.

Singh K., Rajeswara Rao B.R., Singh C.P., Bhattacharya A.K. and Kaul P.N. 1998. Production potential of aromatic crops in the alleys of *Eucalyptus citriodora* in semi-arid tropical climate of south India. J Med Aromatic Plant Sci 20: 749–752.

Singh K., Singh V., Hussain A. and Kothari S.K. 1990. Aromatic plants as efficient intercrops under poplars (*Populus deltoides* Bartram ex Marshall). Indian Forester 116: 189–193.

Singh U., Wadhwani A.M. and Johri B.M. 1996. Dictionary of Economic Plants in India. Indian Council of Agricultural Research, New Delhi, India, 288 pp.

Sittenfeld A. and Gamez R. 1993. Biodiversity prospecting by INBio. pp. 69–97. In: Biodiversity Prospecting. World Resources Institute, Washington, DC.

Sunderland T.C.H., Ngo-Mpeck M.L., Tchoundjeu Z. and Akoa A. 1999. The ecology and sustainability of *Pausinstalia johimbe*: an over-exploited medicinal plant of the forests of Central Africa. pp. 67–77. In: Sunderland T.C.H., Clark L.E. and Vantomme P.

(eds), Non-Wood Forest Products of Central Africa: Current Research Issues and Prospects for Conservation and Development. Yaounde, Cameroon.

Teel W.S and Buck L.E. 2002. Between wildcrafting and monocultures: agroforestry options. pp. 199–222. In: Jones E.T., McLain R.J. and Weigand J. (eds), Non-Timber Forest Products in the United States. University Press of Kansas, Lawrence, KS, USA.

Teklehaimanot Z. 2004. Improvement and exploitation of *Parkia biglobosa* and *Vitellaria paradoxa* in the parkland systems of West Africa. Agroforest Syst (This volume).

Tomich T.P., van Noordwijk M., Budidorsono S., Gillison A., Kusumanto T. Murdiyarso D., Stolle F. and Fagi A.M. 1998. Alternatives to Slash-and-Burn in Indonesia: Summary Report & Synthesis of Phase II. ASB-Indonesia Report No. 8, ICRAF Southeast Asia, Bogor, Indonesia, 139 pp.

Vantomme P., Markkula A. and Leslie R.N. (eds.) 2002. Non-Wood Forest Products in 15 Countries of Tropical Asia: An Overview. FAO, Rome, Italy. http://www.fao.org/DOCREP/005/AB598E/AB598E00.htm. Website accessed October 14, 2003.

van Seters A.P. 1997. Forest based medicines in traditional and cosmopolitan health care. pp. 10–16. In: Bodeker G., Bhat K.K.S., Burley J. and Vantomme P. (eds), Medicinal Plants for Forest Conservation and Healthcare. Non-Wood Forest Products No. 11, FAO, Rome, Italy.

Vyas S. and Nein S. 1999. Effect of shade on the growth of *Cassia ungustifolia*. Indian Forester 125: 407–410.

Walter K.S. and Gillett H.J. 1998. 1997 IUCN red list of threatened plants. Compiled by the World Conservation Monitoring Centre. IUCN, Gland, Switzerland and Cambridge, UK, 1xiv + 862 pp.

Wezel A. and Bender S. 2003. Plant species diversity of homegardens of Cuba and its significance for household food supply. Agroforest Syst 57: 39–49.

Yoshino K. and Ando K. 1999. Utilization of plant resources in homestead (*bari-bhiti*) in floodplain in Bangladesh. Japanese J Trop Agric 43: 306–318.

Zhou S.Q. 1993. Cultivation of *Amomum villosum* in tropical forests. Forest Ecol Manag 60: 157–162.

Zou X. and Sanford R.L. 1990. Agroforestry systems in China: a survey and classification. Agroforest Syst 11: 85–94.

Agroforestry Systems **61:** 123–134, 2004.

Forest gardens as an 'intermediate' land-use system in the nature–culture continuum: Characteristics and future potential

K.F. Wiersum
Forest and Nature Conservation Policy Group, Department of Environmental Sciences, Wageningen University, The Netherlands; e-mail: freerk.wiersum@wur.nl

Key words: Agroforests, Ecological sustainability, Forest management, Landscape dynamics, Social sustainability, Synergy

Abstract

Forest gardens are reconstructed natural forests, in which wild and cultivated plants coexist, such that the structural characteristics and ecological processes of natural forests are preserved, although the species composition has been adapted to suit human needs. These agroforests include a range of modified and transformed forests, and form an integral part of local land-use systems. They lie between natural forests and tree-crop plantations in terms of their structure and composition, and low intensity of forest extraction systems and the high intensity plantation systems in terms of their management intensity. Their management is characterized by combined use of silvicultural and horticultural operations, and spatial and temporal variations. These ecologically sustainable systems are often dynamic in species composition in response to changing socioeconomic conditions. Evolved over a long period of time as a result of local community's creativity, forest gardens have still received little attention in agroforestry research, just as in the case of the more intensively domesticated homegardens. The study of forest gardens offers good opportunities for obtaining a better understanding of the 'nature-analogous' agroforestry systems and for developing multifunctional agroforestry systems combining production and biodiversity values.

Introduction

When the concept of agroforestry was developed in the 1970s, it was visualized that agroforestry development might include two pathways, i.e., the incorporation of trees in agricultural cropping systems or the incorporation of crops in forest systems. To date, most agroforestry research has focused on the first development pathway. Although occasional calls for a forest vision in agroforestry have been made (Michon and De Foresta 1999), relatively little attention has yet been given to the development of multi-strata cropping systems that mimic the structure of natural forests (Lefroy et al. 1999; Muschler and Beer 2001). Two major types of such 'nature-analogous' agroforestry systems can be distinguished: homegardens and forest gardens. Homegardens are predominantly fenced-in gardens, surrounding individual houses, planted with fruits and other trees, vegetable herbs and annual crops (see Kumar and Nair 2004). Forest gardens are mixed tree plantations surrounding or at some distance from villages. They are usually less intensively tended than homegardens and they include a higher percentage of native trees (Wiersum 1982). Homegardens have received a reasonable amount of research attention (e.g., Landauer and Brazil 1990; Torquebiau 1992; Kumar and Nair 2004). Most of the research has focused, however, on the description of the location-specific structure and functions of homegardens, and little progress has been made in understanding the ecological and economic functioning of the homegardens. Therefore the question was posed recently whether tropical homegardens are an enigma that eludes science (Nair 2001; Kumar and Nair 2004). Even less research attention has been given to forest gardens. These systems have been characterized as forest-analogous agroforests (Michon and De Foresta 1997) or as 'intermediate' forest types, which straddle the divide between the natural forests and specialised tree crop plantations (Angelsen et al. 2000). In such agroforests or 'intermediate' forest types, tree composition has been adapted to local needs, but their structure still closely resembles to that of natural forests. In this paper, the terms agroforests and 'intermediate' forest types will be used as synonyms for a group of agroforestry systems, and forest gardens will be described as a specific agroforestry type within this group of agroforestry systems.

Although agroforests still have received relatively little scientific attention, this does not mean that such systems are scarce. During the 1990s, it has been recognized that a large variety of agroforests such as forest gardens exist (Wiersum 1996 and 1997a). These systems were developed by local communities as a result of purposeful activities to safeguard the forest resources on which they depend for livelihood. The indigenous use and manipulation of forest has in many instances resulted in a transformation of natural forests into forest gardens or other types of agroforests. The existence of a large variety of complex agroforests demonstrates the creative role of local communities in maintaining forest resources. At present, several factors contribute toward a greater interest in the nature and possible future scope of these agroforestry systems. These include increased interests in developing novel approaches toward the conservation of tropical rain forests and biodiversity (Sayer and Campbell 2001), as well as increased recognition for the scope of indigenous land-use systems as a viable basis for sustainable development (e.g., Lawrence 2000). The aim of this paper is to describe the characteristics of agroforests as an 'intermediate' land-use type in a dynamic nature – culture continuum as well as the features of forest gardens as a specific type of agroforests. The article focuses on the following questions:

1. What are the main characteristics of agroforests?
2. What are the specific features of forest gardens?
3. Can forest gardens be considered to be sustainable and what is their future potential?
4. What conclusions can be drawn regarding the future scope of forest gardens and what are the research implications?

Agroforests as 'intermediate' forest types

In tropical forest ecology, a differentiation is made between old-growth, or virgin forests, and secondary forests, the latter being those that develop after clearing of the original forest. In addition, a third type of forests may be distinguished, i.e., altered forests (Boerboom and Wiersum 1983), in which man has exercised an appreciable influence on forest composition, but many elements of the original forest are maintained. Such altered forests have traditionally received relatively little attention, but recently interest in these – what now are also called agroforests (Michon and De Foresta 1997) or 'intermediate' forests (Angelsen et al. 2000) – is increasing. Agroforests may be defined as a mixed tree stand in which species composition has been adapted to suit human needs, but which are still 'nature analogous' (Oldeman 1983; Lefroy et al. 1999) in the sense of preserving most of the structural characteristics and ecological processes prevailing in natural forests. They thus represent an anthropogenic forest system shaped by the interactions between ecological processes and human manipulations in the form of forest management activities (Wiersum 1997b). The management of agroforests consists of conscious efforts to both maintain forest resources and stimulate increased production of valuable forest resources. In order to ensure the use and maintenance of agroforests, a large variety of forest management practices may be carried out. These may take the form of socially oriented measures with the aim to limit forest use by non-legitimate users, or of biologically oriented measures to maintain production by either controlling overexploitation or stimulating production and regeneration (Table 1).

Forest management always starts with defining user rights. Initially such regulations consist of the establishment of communal rights to specific patches of forests and/or of private use rights to individual trees in the natural forests. If forest resources become scarcer, interest in manipulating the forests by biologically oriented management practices may develop. In such cases, a combination of social and biological oriented management practices will develop. The access and ownership regulations gradually change from communal rights towards temporary or permanent private rights. This change is complex because of the differentiation in access rights to trees and to land (Fortmann and Bruce 1988). Concomitant with this development of private land and tree rights, the intensity of biologically oriented management prac-

Table 1. Different categories of indigenous forest management practices.

1. Control of access to resources by definition of legitimate user groups

2. Maintenance of the resource through controlled utilization and protection
   - Only certain species harvested according to stand composition, e.g., size, stand structure, etc.
   - Rotational harvesting regimes
   - Using harvesting techniques that do not cause tree mortality, e.g., limited harvesting, coppicing/pollarding/lopping
   - Control of pests and diseases, e.g., sanitary pruning
   - Fire control practices

3. Stimulation of the production of required products within existing vegetation
   - Selecting coppice shoots, rejuvenation pruning, ringing trees to stimulate fruiting
   - Decreasing water/nutrient/light competition for trees by weeding and thinning non-valuable species
   - Optimisation of soil conditions (e.g., mulching) to favour desired species

4. Stimulating regeneration of valued species
   - Protection of natural regeneration
   - Stimulating root sprouting
   - Planting of cuttings
   - Transplanting of seedlings obtained from natural forests or plantations
   - Incidental or purposeful seeding

5. Cultivation of genetically improved varieties
   - Selection of (high-yielding) cultivars

Source: (Wiersum 1997a,b).

tices increases (Gilmour 1990; Wiersum 1997a). In different types of agroforests, a variety of arrangements regarding the use of trees and their products exist. Agroforests, in which the biologically oriented measures are limited to controlled use and protection, are often under communal ownership, whereas agroforests, in which production is consciously stimulated or active tree planting takes place, are often under private ownership (Wiersum 1997a).

The impact of management practices on the vegetation depends on their nature and intensity. Anderson (1990) distinguished two categories of forest management: (1) Tolerant forest management practices by which the native vegetation is largely conserved or reconstituted through successional stages, and (2) Intrusive forest management practices by which the native vegetation is replaced by (mixed) tree plantations that are manipulated by long-term human activities. The first category mostly consists of controlled utilization and protection practices and of measures to stimulate production of required products. But only limited attention is given to purposeful regeneration of valued tree species. In contrast, the second category involves a dominant role for artificial regeneration (sometimes with genetically improved varieties). In these systems trees are often grown in mixtures with agricultural or horticultural crops. Consequently, tolerant forest management results in modification of forests, while intrusive forest management results in transformed forests. Agroforests encompass both modified and transformed forests; within each of these two subcategories several more specific forest types may be distinguished (Table 2). Strictly speaking, it could be argued that the 'forests modified by gathering and protection of selected resources' as well as 'interstitial trees on croplands' as mentioned in Table 2 should not be considered as agroforests. Rather, they could be considered as representing boundary situations between natural forests and agroforests, and between agroforests and croplands, respectively. As will be argued later, the dynamics of agroforests cannot, however, be properly understood if these boundary types are not taken into consideration.

The existence of different types of agroforests situated along a gradient from natural ecosystems to tree-dominated agro-ecosystems illustrate that various (agro)forestry systems have gradually developed as a result of the interactions between local communities and forests. The agroforests represent a variety of nature – human relations; they reflect manifold forms of human creativity in dealing with forest resources (Wiersum 1997a,b). A series of gradually more intensive and varied management practices has resulted in a process of co-domestication of tree crops and

Table 2. Major types of 'intermediate' forest types.

A. (Modified) forests with prevalent tolerant forest management practices

Natural forests modified by gathering of forest products and selected protection of specific resources:
Specific areas in natural forests which are favoured and protected because of their value for extraction of useful materials.
Examples: Forest patches with individually claimed trees

Resource-enriched natural forests: Natural forests, either old-growth or fallow vegetations, whose composition has been altered by selective protection and incidental or purposeful propagule dispersion of food and/or commercial species.
Examples: Enriched natural forests
Enriched fallows

B. Transformed forests with prevalent intrusive forest management practices

Reconstructed natural forests: (Semi-)cultivated forest stands with several planted useful species, tolerated or encouraged wild species of lesser value and non-tree plants (herbs, lianas) composed of mainly wild species.
Example: Forest gardens

Mixed arboriculture: Cultivated mixed stands, almost exclusively of planted, and often domesticated, tree species.
Examples: Home gardens
Mixed smallholder plantations

Interstitial trees on croplands: Either naturally regenerated and protected trees, or planted and sometimes domesticated trees scattered over agricultural fields.
Example: Protected trees on swidden fields
Scattered fruit trees cultivation on/along crop fields

(Adapted from Wiersum 1997a).

forest ecosystems, during which a concomitant change from wild tree species to domesticated tree species and a change in structure and composition of forest vegetation took place (Den Hertog and Wiersum 2000; Paudel and Wiersum 2002). The various hypothetical phases in this process are illustrated in Figure 1. This model is an analytical rather than an explanatory one. It should not be regarded as indicating unidirectional and deterministic trends in which the various phases represent pre-ordained steps on a ladder of increasingly 'advanced' stages of general societal development (Wiersum 1997a). Nonetheless, the model may assist in clarifying the main comparative characteristics concerning the ecological conditions and intensity of management of various agroforest types.

Main characteristics of forest gardens

As indicated in Table 2, forest gardens can be considered as a specific type of agroforests. They can be defined as reconstructed natural forests in which intrusive forest management practices have resulted in the co-existence of cultivated trees and wild plant species. Such reconstructed forests may analytically be distinguished from resource-enriched natural forests and fallows in which tolerant forest management prevail, and from homegardens and mixed smallholder plantations in which management is more intensive resulting in a larger proportion of cultivated plant species. As will be elaborated below, the boundaries between these three vegetation systems are anything but discrete, and gradual transformations between enriched fallows and forest gardens, and between forest gardens and mixed smallholder plantations take place.

Forest gardens can be found in all tropical continents (Table 3). Within the forest gardens, two main structural ensembles of vegetation can be distinguished, i.e., a matrix of naturally growing species interspersed with introduced cash-crop species. The overall vegetation structure may either resemble primary forest ecosystems, with a predominance of big trees and a high species complexity, or be more close to secondary forests, with dense stands of smaller trees and rapid turnover of species (Michon and De Foresta 1997). The cash-crop species may be either native or exotic. The forest gardens are a typical example of an 'intermediate' forest type. Their 'intermediate' position may be related either to the vegetation or to the management characteristics of these systems. On the one hand they hold an intermediate position between natural forest and plantations regarding their vegetation structure and composition, and, on the other hand an intermediate position regarding management intensity between the low intensity of natural forest extraction systems and the high intensity of plantation systems with selected cultivars.

The management of forest gardens is characterised by two main features. First, they form a dynamic component of an integrated local land-use system. This has

Uncontrolled procurement of forest products	Forest/tree crop management			
	Protection of forest resources	Cultivation of wild trees	Cultivation of domesticated trees	
Casual/opportunistic gathering/collection of wild forest products	Protection of patches of forest against collection of products			Forest conservation
	Controlled gathering/ Collection of forest Products	Enhancement of (re)productive potential of valued species through ecosystem manipulation Purposeful dispersal of seeds/seedlings of wild species in natural forests Purposeful incorporation of valuable species in fallow fields	Cultivation of exotic or local cultivars in forest environment	Modification of forest
		Purposeful regeneration of wild species on cleared fields – home gardens – agricultural fields	Plantations of selected & improved cultivars – timber plantations – fruit orchards – tree crop plantations Growing of domesticated trees on croplands	Forest transformation

↓ Increasing manipulation of forest ecosystem

Initial acculturalization — Controlled regeneration — Cultivation of genetically selected tree crop → Increasing input of human energy per unit area of exploited forest

Figure 1. Stages in forest management (Wiersum 1997b).

a consequence, that their management is influenced by their relationship to the other components of the land-use system from which they form a part. Second, the management practices are often not focused on the vegetation as a holistic agroecosystem, but rather they are oriented at an amalgamation of multiple forest resources. As a consequence, the management of forest gardens is characterised by their adaptive nature.

Forest gardens as components of integrated land-use systems

Forest gardens are almost always a component of a larger land-use system consisting of a diverse range of managed fields and/or forests such as permanent or intermittent cultivated crop fields, homegardens, fallow vegetations and/or exploited forest. In many areas, forest gardens and other agroforest types co-exist, with each type occupying a specific landscape and/or tenurial niche (e.g., Posey 1985; Mary and Michon 1987; Salafsky 1995; Colfer et al. 1997; De Jong 2002). Temporary croplands often form an essential transitional stage in the transformation of natural forests into enriched fallows and, with increased age, subsequent to forest gardens. Forest gardens also may be gradually transformed into mixed smallholder plantations. Whereas in forest gardens a matrix of original vegetation still exists, in such mixed plantations planted trees, including cultivated shade trees, prevail. The boundaries between forest gardens and enriched fallow as well as between forest gardens and mixed smallholder plantations are therefore not discrete, but fuzzy.

For instance, in a case study in East Kalimantan, Indonesia (Gomez Gonzalez, I.C. 1999. Master's Thesis, Indigenous management of forest resources in East Kalimantan, Indonesia: The role of secondary ve-

Table 3. Examples of forest gardens.

Location	Type of forest garden.
Asia and Papua New Guinea	
Papua New Guinea highlands	Forest gardens dominated by coffee and *Casuarina* (Bourke 1985)
Indonesia: Sumatra and Kalimantan	Damar (*Shorea javanica*) agroforests
	Benzoin (*Styrax*) agroforests
	Rubber (*Hevea brasiliensis*) agroforests
	Fruit forest gardens
	Rattan gardens
	(Michon and De Foresta 1996, 1999)
Philippines, Luzon	Ifugao woodlots (Olofsen 1980)
Thailand: northern highlands	Fruit-based agroforests (Withrow-Robinson et al. 1998)
India, Western Ghats	Cardamom (*Elettaria cardamomum*) agroforests (Kumar et al. 1995)
Africa	
East Africa: Ethiopia, Uganda	Traditional Coffee – banana gardens (Oduol and Aluma 1990; Teketay and Tegineh 1991)
West Africa	Oil palm (*Elaeis guineensis*) groves (Zeven 1972)
Latin America	
Mexico	Maya forest gardens (Gomez-Pompa and Kaus 1990)
Amazon basin	Fruit forest gardens by Amerindians (Denevan et al. 1984; Posey 1985)
	Fruit forest gardens of non-tribal indigenous people (De Jong 1995)

getation. Wageningen Agricultural University), it was found that the local land-use system consisted of the following elements:

- Exploitation of selected forest products such as *gaharu* (the fragrant heartwood of *Aquilaria* spp.), honey, rattan (*Calamus* and *Daemonorops* spp.) and timber from the primary forest;
- Management of patches of forest gardens (*pulong*) with (native) fruit trees and some additional cacao (*Theobroma cacao*) and coffee (*Coffea arabica*);
- Management of cash-crop plantations (*kebun*) with cacao and coffee and scattered other trees and/or vegetables and spices;
- Management of enriched fallow vegetation (*jekau*);
- Cultivation of both permanent wet rice fields and temporary swidden fields.

These different land-use elements were not discrete; some overlap between the various elements did occur. Moreover, one system could be changed into another over time. For instance, if crop production on certain swidden fields decreases they may be put into fallow. To maintain crop production, enriched fallows may subsequently be transformed into swidden fields. Or, if prices for cash crops such as coffee and cacao are high, the management of these crops may be intensified with *pulong* gradually changing into *kebun*. Alternatively, if coffee or cocoa prices are low, other crops may be introduced in the *kebun* and they may regain more of a *pulong* character.

The above example illustrates that forest gardens are a dynamic component of a forested landscape composed of a mosaic of fields and forests. Similar examples of forest gardens forming a component within a forested landscape continuum have been described from other areas in Indonesia (Aumeeruddy and Sansonnens 1994; Mary and Michon 1987) as well as from the Philippines (Olofsen 1980; Fujisaka and Wollenberg 1991), Thailand (Withrow-Robinson et al. 1999) and the Amazon region (Denevan et al. 1984; Posey 1985). Chase (1989) has suggested the term of 'domiculture' for the creation of such forested landscapes with a series of localized areas (*domuses*) of interaction between people and forest vegetation with each area having a specific set of management practices.

Croplands are not only a transitional stage in the formation of forest gardens; but an essential complement to forest gardens. Whereas the croplands are used to produce essential staple food crops for the rural households, forest gardens provide supplementary products such as fruits, cash crops, (fire) wood and/or medicines. In this aspect of providing products, which supplement the staple food production, forest gardens resemble homegardens. The production systems of forest gardens and homegardens are different, however: whereas in homegardens mostly products for own consumption and enjoyment (ornamentals) are grown, in forest gardens production is much more oriented to cash crops.

As a result of their forming a part of a landscape system in which different landscape units are shaped and transformed in interaction with each other, the management decisions regarding forest gardens are taken on the basis of how best to use labor resources in fulfilling a variety of subsistence and commercial household needs in the different landscape niches. The management practices of forest gardens may either be intensified or 'extensified' in response to changes in the other land-use components, such as changes in availability of forest resources (Fujisaka and Wollenberg 1991; Gilmour 1990; Henkemans et al. 2000) or agricultural intensification (Mary and Michon 1987; Belsky 1993). Moreover, decisions to intensify or deintensify the management practices are also based on changes in socioeconomic conditions, such as increased or decreased access to markets and fluctuating market prices (Arnold and Dewees 1995; Mary and Michon 1987).

Amalgamation of multiple forest resources

The management practices for forest gardens are closely linked to their multi-resource nature. They are primarily focused on specific forest components that are considered to be valuable from a utilitarian and/or cultural point of view. As a consequence, the management of forest gardens is often not directed at a forest stand as a holistic (agro)ecosystem, but rather at selected forest components such as valuable trees or spatial units. The tree management practices are aimed to protect the trees and increase their production. The management involves both improved silvicultural and horticultural practices (e.g., removing competing vegetation, sanitary or rejuvenation pruning, and lopping). In addition, the vegetation is enriched with commercial tree species. Depending on their value and ease of management, different trees within a stand may be subject to different degrees of management. For instance, while native fruit trees may be propagated through protecting natural regeneration, exotic cash trees may be planted. In addition to such tree-focused management practices, spatially focused management practices may also take place in forest gardens. This category of practices involves measures to protect specific forest patches such as clumps of valuable trees, or to open up forest patches to establish new trees.

Forest gardens as adaptive management systems

The position of a forest garden within a dynamic landscape context as well as the nested nature of the management system with a combination of spatially oriented and tree oriented management practices allows for continuous adaptation to changing ecological conditions, household needs and marketing opportunities. This indigenously developed adaptive management approach (Berkes et al. 2000) differs from the common management approach in agroforestry systems such as alley farming and improved fallows developed by researchers. In these 'researcher-developed' systems, the attention is usually focused on spatially defined land-use units as well as specialised production processes. The management of these systems is not based on an amalgamation of resources, but rather at specific agroecosystem which are often dominated by a few trees species only. Moreover, the focus is on development of stable rather than dynamic systems and little attention is given to the role of these agroforestry systems within forested landscapes.

Are forest gardens sustainable?

In view of the dynamic nature of forest gardens, it could be questioned whether these systems can be considered to be sustainable. On the one hand, it could be argued that this dynamic nature is related to a more general process of land-use dynamics in forest frontier areas (Henkemans et al. 2000). In this view, it might be assumed that the forest gardens are intermediate stages in the pre-ordained process of land-use change from forests to plantations. During this process some temporary 'intermediate' land-use systems such as forest gardens may be developed, but in time these will be replaced by 'modern' land-use practices. On the other hand, it could be argued that the forest gardens have been neglected by modern science, and that these forest types offer good options for combining biodiversity conservation and production for human benefits. In this context, Michon and De Foresta (1997) have argued that the agroforests should be considered as representing a true domestication of forest ecosystems rather than as intermediate stages in the process of extraction of forest products to cultivation of commercial tree species in mono-species plantation systems.

The term 'sustainability' is often used ambiguously referring either to present or future conditions (Wiersum 1995). In efforts to define criteria for sustainable (agro)forestry systems (e.g., Torquebiau 1992; Higman et al. 1999), much attention is given to identify characteristics that can be used to assess whether the present (agro)forestry systems can be considered sustainable. This approach is based on the idea that sustainability should adhere to the presently accepted norms on what is at stake in land-use management. It does not give much attention, however, to the question of whether the present norms may change in the future and to the related question of how the concept of sustainability should be judged when considering social dynamics. This is rather surprising in view of the fact that the present interest in the concept of sustainability resulted from the well-known Brundtland report (WCED 1987), which focused on the concept of sustainable development. In this report sustainable development is defined as a development that meets the requirements of present generations without undermining the natural resource base for future generations to use these resources. Thus, the concept of sustainability does not only relate to the present, but also to possible future, ecological and socioeconomic conditions. When judging whether the forest gardens can be considered sustainable, one is therefore faced not only with the question of whether the ecological base of natural resource production is maintained in a socially and economically acceptable manner, but also with the question of whether the system is capable of adapting to long-term changes caused by the process of socioeconomic dynamics (Wiersum 1995).

Thus, an evaluation of the sustainability of forest gardens as a dynamic nature–human system should not be based on criteria related to natural ecosystems, but rather on integrated criteria representing their dualistic nature (Michon and De Foresta 1997). Both the features regarding ecological sustainability and adaptation to social dynamics need to be considered.

Ecological sustainability

When considering the ecological sustainability of forest gardens, two aspects merit attention, their ecological integrity and ecological functioning in respect to production. Regarding ecological integrity, forest gardens consist of transformed natural forests. Consequently, not only have the natural regeneration processes been tampered with, but also the biodiversity of these systems has been changed in comparison to natural forests. This does not mean, however, that forest gardens do not have any biodiversity value. The biodiversity may be preserved through conscious preservation of valued species by restricting open-ended exploitation. Moreover, the management may include conscious regeneration of species that are of direct value to the local communities; this may not only concern (semi)domesticated species, but also rare species (Kessy 1998). Thus, the forest gardens offer a good scope for what, in conformity with the term domiculture as referring to the creation of a series of areas of interaction in the nature-culture continuum (Chase 1989), could be called '*in domo*' conservation of biodiversity (Wiersum 2003).

Even if the species composition of forest gardens has been altered through human activities, this usually does not mean that the basic ecological production processes have been impaired. The multispecies composition of the forest gardens allows for a series of synergetic ecological processes (Anderson and Sinclair 1993; Lefroy et al. 1999; Van Noordwijk et al. 2001):

- Efficient utilization of aboveground and belowground space;
- Efficient circulation of nutrients and reduced risks for depletion of nutrients as the result of the presence of filters against such losses;
- Plant protection as a result of the presence of buffers against damaging agents such as pests and diseases;
- Protection against potentially degrading forces such as torrential rainfall, surface run-off or strong winds as a result of the presence of vegetative barriers.

As a result, the risk of disturbance of production ecological processes is small. Moreover, the multiresource character of the forest gardens also ensures reduced risks for total crop failures due to weather fluctuations.

Adaptation to social dynamics

In the past, it was often considered that the development of tree cultivation systems could be based on the principle that the cultivation of improved tree species would take place in specialized tree production systems such as mono-species plantations. Such specialization would make it possible to optimize production and thus to make the most efficient use of the improved tree species. Gradually, however, this presupposition has started to change. It became recognized that although mono-species tree plantations may provide optimal yields in case of intensive management, they may have several limitations under other conditions. First, mono-species plantations under suboptimal management may be subject to ecological hazards such as fluctuating weather conditions or attacks by pests and diseases. Such technically nonoptimal management may be caused by the fact that the tree cultivator lacks the means to obtain all the necessary external inputs needed to counteract these natural hazards. Second, in many tropical regions specialized forms of mono-species cultivation do not fit into the existing, predominantly smallholder, farming systems. Especially in the case of less optimal production conditions, these farming systems are often characterized by multifunctional production, which focus on a range of subsistence and commercial products. As illustrated above in respect to the forest gardens, the specific nature of the management of such multiresource systems allows for efficient adaptation to changing socioeconomic conditions. Under conditions of smallholder farming in forested landscapes, forest gardens are therefore more relevant than highly specialized mono-cropping systems for avoiding risks and allowing adaptation to newly arising conditions.

In addition to assessing whether the forest gardens can be considered socioeconomically adaptable to the present land-uses conditions, it is also very important to assess whether they still offer scope for development under newly evolving conditions. Several recent developments may significantly impact on creating a more positive appreciation of the forest gardens as well as other agroforest types.

- Concomitantly with the decrease of the original tropical forest cover, the area of human modified forests including both secondary forests and adapted forests is on the increase. For the past several years much attention has been given to how these anthropogenic forest types can best be managed sustainably. Most attention is still focused on secondary forests; little attention is given to different types of agroforests such as forest gardens. The potential of these forests is gradually becoming recognized, however. Recently it even has been suggested that forest gardens be classified as one of the six main secondary forest types (Chokkalingam and De Jong 2001).
- In view of the alarming rate in the loss of biodiversity, at present much attention is given to biodiversity conservation. Many efforts focus primarily on the conservation of species. It is increasingly recognized that biodiversity conservation should not only focus on species-level, but also on landscape-level (Sayer and Campbell 2001). Moreover, it should not only be based on the notion of land-use practices decreasing the biodiversity of natural ecosystems, but also on the notion of land-use practices creating new landscape elements offering scope for *in domo* conservation of biodiversity (Wiersum 2003).
- The concerns about loss of biodiversity and environmental deterioration associated with modern high-external input land-use practices have resulted in an increased appreciation for multifunctional land-use systems in which production and biodiversity values are balanced (Van Noordwijk et al. 1997). Many efforts are now undertaken to search how natural systems can be used as examples for the development of land-use systems that transcend the traditional nature-culture dichotomy and that offer scope for contributing toward ecologically balanced land-use patterns (Ewel 1999; Hobbs and Morton 1999).
- Within forest management, increased attention is given to the scope of multifunctional forestry by extending the range of forest products from timber to nontimber forest products (NTFPs) (e.g., Arnold and Ruiz-Perez 2001; Ros-Tonen 1999). However, in many old-growth natural forests the density of NTFP-producing plants is limited. As illustrated in Table 3, forest gardens have often been enriched by NTFP-species, while still maintaining many of the ecological characteristics of natural forests. Thus, they offer good scope for balancing the concerns on maintaining forest-like structure and a reasonable degree of biodiversity, and stimulating NTFP production (Michon and De Foresta 1996).
- Attention is given to the need to decentralize forest management and to stimulate greater community involvement (Arnold 2001). The development of community forestry brings with it new interests in the development of locally adapted forms of forest management that fit into the overall land-use systems of rural communities (Lawrence 2000). The forest gardens offer an excellent example of the scope for such location-specific management systems for forest resources.
- It is now recognized that the transfer-of-technology approach, which dominated agricultural and forestry development in the past, has reached the limits of its applicability, and that new approaches are needed

based on endogenous instead of exogenous innovations (e.g., Lawrence 2000). This recognition has also contributed towards increased attention to the scope for the application of principles embedded in indigenous land-use management systems such as forest gardens.

Conclusion: future scope of forest gardens and research implications

Forest gardens are typical representatives of 'intermediate' forests and are characterized by their evolution within the local communities rather than as a result of formal agroforestry research. The future for these agroforestry systems seems promising. It can be expected that consequent to the loss of primary forests, anthropogenic forest types such as agroforests will increase in importance. And, with increased efforts at conservation of the last remnants of primary forests, the forest frontier areas will gradually be stabilized. This may bring with it a gradual transfer of fallow forests into different types of forest gardens. Consequently, forest gardens might increasingly become a component of a forested landscape that is valued for contributing toward both resource production and biodiversity conservation.

Forest gardens also offer an excellent opportunity to develop new insights into the options for agroforestry development. They show several features that hitherto have received relatively little attention in agroforestry research. In the first place, the most characteristic ecological processes operating in forest gardens are not related to competition, but synergy. While considering ecological interactions in agroforestry systems, most attention until recently has been given to processes of competition and predation and relatively little attention has been given to synergetic processes such as mutualism and commensalism. With the increased attention to the question how to balance production and biodiversity conservation, it can be expected that the concept of synergy will gradually obtain an equal footing to the concept of competition in agroforestry research.

In the second place, the management of forest gardens is characterized by several aspects of complexity and multiple scales of management that have until now received still little attention in agroforestry research:

- Forest gardens form a component of an integrated land-use system rather than a specialized agroforestry system. Whereas in agroforestry research attention is mainly focused on the question how to best use the production factors at a field level, in forest gardens management decisions are taken on the basis of how best to distribute production factors over the various production units at landscape level.
- In the forest gardens the management practices are oriented at multiple resources. Different forest components are subject to different management practices and intensities, and management is characterized by a combination of tree-focused and spatially focused practices. In contrast, agroforestry research is usually oriented at agroecosystems as fixed spatial systems.
- Forest gardens are dynamic systems. Their species composition and management intensity may change in time in response to changing livelihood conditions and/or changing production and marketing situations. In contrast, agroforestry research is mainly focused on developing agroforestry systems with a more or less stable combination of selected trees and crops.

In conclusion, as a result of the trends in social appreciation regarding forested landscapes, forest gardens should not be conceived as dating from a traditional past and scientifically outmoded. Several recent developments, such as increased appreciation for multifunctional land-use systems combining production and biodiversity, as well as increased attention to the scope for development of indigenous forest management systems, cause that the scope for further development of forest gardens is bright. Moreover, the study of these agroforestry systems offers exciting opportunities for developing new key concepts for agroforestry research. Thus the observation by Nair (2001) of homegardens hitherto eluding science is equally relevant for forest gardens. Together with homegardens, forest gardens offer great scope for developing new unifying concepts for agroforestry research as well as further development of 'nature-analogous' agroforestry systems.

References

Anderson A.B. 1990. Extraction and forest management by rural inhabitants in the Amazon estuary. pp. 65–85. In: Anderson A.B. (ed.), Alternatives to Deforestation: Steps Towards Sustainable Use of the Amazon Rain Forest. Columbia University Press, New York, USA.

Anderson L.S. and Sinclair F.L. 1993. Ecological interaction in agroforestry systems. Agroforest Abstracts 6: 57–91.

Angelsen A., Asbjørnsen H., Belcher B., Michon G. and Ruiz-Pereze M. 2000. Cultivating (in) Tropical Forests? The evolution and sustainability of intermediate systems between extractivism and plantations. International Workshop Cultivating (in) Tropical Forests. Kraemmervika, Lofoten, Norway, June 28-July 1, 2000 (Proceedings in press).

Arnold J.E.M. 2001. Forests and People: 25 years of Community Forestry. FAO, Rome, Italy. 134 pp.

Arnold J.E.M. and Dewees P.A. 1995. Tree Management in Farmer Strategies. Responses to Agricultural Intensification. Oxford University Press, London, 292 pp.

Arnold J.E.M. and Ruiz Perez M. 2001. Can non-timber forest products match tropical conservation and development objectives? Ecological Economics 39: 437–447.

Aumeeruddy Y. and Sansonnens B. 1994. Shifting from the simple to complex agroforestry systems: an example for buffer zone management from Kerinci (Sumatra, Indonesia). Agroforest Syst 28: 113–141.

Belsky J.M. 1993. Household food security, farm trees, and agroforestry: a comparative study in Indonesia and the Philippines. Hum Organ 52: 130–141.

Berkes F., Colding J. and Folke C. 2000. Rediscovery of traditional ecological knowledge as adaptive management. Ecol Appl 19: 1251–1262.

Boerboom J.H.A. and Wiersum K.F. 1983. Human impact on tropical moist forest. pp. 83–106. In: Holzner W., Werger M.J.A. and Ikusima I. (eds), Man's Impact on Vegetation. W. Junk Publishers, The Hague.

Bourke R.M. 1985. Food, coffee and casuarina: an agroforestry system from the Papua New Guinea highlands. Agroforest Syst 2: 273–279.

Chase A.K. 1989. Domestication and domiculture in northern Australia: a social perspective. pp. 42–54. In: Harris D.R. and Hillman G.C. (eds), Foraging and Farming, the Evolution of Plant Extraction. Unwin Hyman, London.

Chokkalingam U. and De Jong W. 2001. Secondary forest: a working definition and typology. Int For Rev 3: 19–26.

Colfer C., Peluso N. and See Chung C. 1997. Beyond slash and burn: building on indigenous management of Borneo's tropical rain forests. New York Botanical Garden, USA, Advances in Economic Botany Vol. 11.

De Jong W. 1995. Diversity, variation, and change in ribereno agriculture and agroforestry. Ph.D. dissertation, Agricultural University, Wageningen, The Netherlands, 168 pp.

De Jong W. 2002. Forest products and local forest management in West Kalimantan, Indonesia: implications for conservation and development. Tropenbos International, Wageningen, the Netherlands. Tropenbos-Kalimantan Series No. 6, 120 pp.

Den Hertog W.H. and Wiersum K.F. 2000. Timur (*Zanthoxylum armatum*) production in Nepal, dynamics in nontimber forest resource management. Mountain Research and Development 20(2): 32–41.

Denevan W.M., Treacy J.M., Alcorn J.B., Padoch C., Denslow J. and Paitan S.F. 1984. Indigenous agroforestry in the Peruvian Amazon: Bora Indian management of swidden fallows. Interciencia 9: 346–357.

Ewel J.J., 1999. Natural systems as models for the design of sustainable systems of land use. Agroforest Syst 45: 1–21.

Fortmann L. and Bruce J.W. 1988. Whose Trees? Proprietary Dimensions of Forestry. Westview, Boulder and London, 135 pp.

Fujisaka S. and Wollenberg E. 1991. From forest to agroforests and logger to agroforester: a case study. Agroforest Syst 14: 113–129.

Gilmour D.A. 1990. Resource availability and indigenous forest management systems in Nepal. Society and Natural Resources 3: 145–158.

Gómez-Pompa A. and Kaus A. 1990. Traditional management of tropical forests in Mexico. pp. 45–64. In: Anderson A.B. (ed.), Alternatives to Deforestation: Steps towards Sustainable Use of the Amazon Rain Forest. Columbia University Press, New York, USA.

Henkemans A., Persoon G.A. and Wiersum K.F. 2000. Landscape transformations of pioneer shifting cultivators at the forest fringe. pp. 53–69. In: Wiersum K.F. (ed.), Tropical forest resource Dynamics and Conservation: from Local to Global Issues. Wageningen University, The Netherlands, Tropical Resource Management Papers No. 33.

Higman S., Bass S., Judd N., Mayers J. and Nussbaum R. 1999. The Sustainable Forestry Handbook. A Practical Guide for Tropical Forest Managers on Implementing New Standards. Earthscan Publications, London, UK, 289 pp.

Hobbs R.J. and Morton S.R. 1999. Moving from descriptive to predictive ecology. Agrofor Syst 45: 43–55.

Kessy J.F. 1998. Conservation and utilization of natural resources in the East Usambara Forest Reserves: conventional views and local perspectives. Wageningen Agricultural University, the Netherlands, Tropical Resource Management Papers No. 18, 168 pp.

Landauer K. and Brazil M. (eds) 1990. Tropical Homegardens. United Nations University Press, Tokyo, Japan, 257 pp.

Lawrence A. (ed.) 2000. Forestry, Forest Users and Research: New Ways of Learning. European Tropical Forest Research Network, Wageningen, The Netherlands, ETFRN Series No. 1, 190 pp.

Lefroy E.C., Hobbs R.J., O'Connor M.H. and Pate J.S. (eds) 1999. Agriculture as a mimic of natural ecosystems. Agroforest Syst 45: 1–436.

Kumar M.B., Kumar S. and Mathew t. 1995. Floristic attributes of small cardamom (*Elettaria cardamomum* (L.) Maton) growing areas in the Western Ghats of peninsular India. Agroforest Syst 31: 275–289.

Kumar B.M. and Nair P.K.R. 2004. The enigma of tropical homegardens (This volume).

Mary F. and Michon G. 1987. When agroforests drive back natural forests: a socio-economic analysis of a rice-agroforestry system in Sumatra. Agroforest Syst 5: 27–55.

Michon G. and De Foresta H. 1996. Agroforests as an alternative to pure plantations for the domestication and commercialization of NTFPs. pp. 160–175. In: Leakey R.R.B., Temu A.B., Melnyk M. and Vantomme P. (eds), Domestication and Commercialization of Non-Timber Forest Products in Agroforestry Systems. FAO, Rome, Italy, Non-Wood Forest Products No. 9.

Michon G. and De Foresta H. 1997. Agroforests: pre-domestication of forest trees or true domestication of forest ecosystems? Neth J Agr Sci 45: 451–462.

Michon G. and De Foresta H. 1999. Agro-forests: incorporating a forest vision in agroforestry. pp. 381–406. In: Buck L.E., Lassoie J.P. and Fernandes E.C.M. (eds), Agroforestry in Sustainable Agricultural Systems. CRC Press, Boca Raton, Florida.

Muschler R. and Beer J. (eds) 2001. Multistrata agroforestry systems with perennial crops. Agroforest Syst 53: 85–245.

Nair P.K.R. 2001. Do tropical homegardens elude science, or is it the other way around. Agroforest Syst 53: 239–245.

Oduol P.A. and Aluma J.R.W. 1990. The banana (*Musa* spp.)–*Coffee robusta* traditional agroforestry system of Uganda. Agroforest Syst 11: 213–226.

Oldeman R.A.A. 1983. The design of ecologically sound agroforests. pp. 173–207. In: Huxley P.A. (ed.), Plant Research in Agroforestry. International Council for Research in Agroforestry, Nairobi, Kenya.

Olofsen H. 1980. An ancient social forestry. Sylvatrop, Philippines Forestry Research Journal 4: 255–262.

Paudel S. and Wiersum K.F. 2002. Tenure arrangements and management intensity of Butter tree (*Diploknema butyracea*) in Makawanpur district, Nepal. Int Forestry Review 4: 223–230.

Posey D.A. 1985. Indigenous management of tropical forest ecosystems: the case of the Kayapó indians of the Brazilian Amazon. Agroforest Syst 3: 139–158.

Ros-Tonen M.A.F. (ed.) 1999. NTFP Research in the Tropenbos Programme: Results and Perspectives. Tropenbos Foundation, Wageningen, The Netherlands, 203 pp.

Salafsky N. 1995. Forest gardens in the Gunung Palung region of West Kalimantan, Indonesia: Defining a locally developed, market-oriented agroforestry system. Agroforest Syst 28: 237–268.

Sayer J.A. and Campbell B. 2001. Research to integrate productivity enhancement, environmental protection, and human development. Conserv Ecol 5: 32 [online] URL: http://www.consecol.org/vol5/iss2/.

Teketay D. and Tegineh A. 1991. Traditional tree crop based agroforestry in coffee producing areas of Harerge, Eastern Ethiopia. Agroforest Syst 16: 257–267.

Torquebiau E. 1992. Are tropical agroforestry homegardens sustainable? Agriculture, Ecosystems and Environment 41: 189–207.

Van Noordwijk M., Tomich T.P. and Verbist B. 2001. Negotiation support models for integrated natural resource management in tropical forest margins. Conserv Ecol 5(2): 21 [online] URL: http://www.consecol.org/vol5/iss2/art21.

Van Noordwijk M., Tomich T.P., De Foresta H. and Michon G. 1997. To segregate – or to integrate: The question of balance between production and biodiversity conservation in complex agroforestry systems. Agroforest Today 8(1): 6–9.

Warren D.M. 1991. Using Indigenous Knowledge in Agricultural Development. World Bank, Washington DC, USA, World Bank Discussion Paper No. 127, 46 pp.

Wiersum K.F. 1982. Tree gardening and taungya on Java: examples of agroforestry techniques in the humid tropics. Agroforest Syst 1: 53–70.

Wiersum K.F. 1995. 200 years of sustainability in forestry: lessons from history. Environ Manage 19: 321–329.

Wiersum K.F. 1996. Domestication of valuable tree species in agroforestry systems: evolutionary stages from gathering to breeding. pp. 147–158. In: Leakey R.R.B., Temu A.B., Melnyk M. and Vantomme P. (eds), Domestication and commercialization of Non-Timber Forest Products in Agroforestry Systems. FAO, Rome, Italy, Non-Wood Forest Products No. 9.

Wiersum K.F. 1997a. Indigenous exploitation and management of tropical forest resources: an evolutionary continuum in forest-people interaction. Agr Ecosyst Environ 63: 1–16.

Wiersum K.F. 1997b. From natural forest to tree crops, co-domestication of forests and tree species, an overview. Neth J Agr Sci 45: 425–438.

Wiersum K.F. 2003. Use and conservation of biodiversity in East African forested landscapes. In: Zuidema P. (ed.), Tropical Forests in Multi-functional Landscapes. Seminar Proceedings. Prins Bernard Centre, Utrecht University, The Netherlands (In press).

Withrow-Robinson B., Hibbs D.E., Gypmantasiri P. and Thomas D. 1999. A preliminary classification of fruit-based agroforestry in a highland area of northern Thailand. Agroforest Syst 42: 195–205.

WCED (World Commission on Environment and Development), 1987. Our Common Future. Oxford University Press, London, UK, 383 pp.

Zeven A.C. 1972. The partial and complete domestication of the oil palm (*Elaeis guineensis* Jacq.). Econ Bot 26: 274–279.

Agroforestry Systems **61:** 135–152, 2004.

The enigma of tropical homegardens

B.M. Kumar[1] and P.K.R. Nair[2]
[1]*College of Forestry, Kerala Agricultural University, Thrissur – 680 656, Kerala, India; e-mail: bmkumar53@yahoo.co.uk;* [2]*School of Forest Resources and Conservation, University of Florida, Gainesville, Florida 32611, USA; e-mail: pknair@ufl.edu*

Key words: Carbon sequestration, Integrated farming systems, Multispecies systems, Species diversity, Species inventory, Sustainability

Abstract

Tropical homegardens, one of the oldest forms of managed land-use systems, are considered to be an epitome of sustainability. Although these multispecies production systems have fascinated many and provided sustenance to millions, they have received relatively little scientific attention. The objective of this review is to summarize the current state of knowledge on homegardens with a view to using it as a basis for improving the homegardens as well as similar agroforestry systems. Description and inventory of local systems dominated the 'research' efforts on homegardens during the past 25 or more years. The main attributes that have been identified as contributing to the sustainability of these systems are biophysical advantages such as efficient nutrient cycling offered by multispecies composition, conservation of bio-cultural diversity, product diversification as well as nonmarket values of products and services, and social and cultural values including the opportunity for gender equality in managing the systems. With increasing emphasis on industrial models of agricultural development, fragmentation of land holdings due to demographic pressures, and, to some extent, the neglect – or, lack of appreciation – of traditional values, questions have been raised about the future of homegardens, but such concerns seem to be unfounded. Quite to the contrary, it is increasingly being recognized that understanding the scientific principles of these multispecies systems will have much to offer in the development of sustainable agroecosystems. Research on economic valuation of the tangible as well as intangible products and services, principles and mechanisms of resource sharing in mixed plant communities, and realistic valuation and appreciation of hitherto unrecognised benefits such as carbon sequestration will provide a sound basis for formulating appropriate policies for better realization and exploitation of the benefits of homegardens.

Introduction

Farming systems variously described in the English language as agroforestry homegardens, household or homestead farms, compound farms, backyard gardens, village forest gardens, dooryard gardens and house gardens abound in the tropics. Some local names such as *Talun-Kebun* and *Pekarangan* that are used for various types of homegarden systems of Java (Indonesia), *Shamba* and *Chagga* in East Africa, and *Huertos Familiares* of Central America, have also attained international popularity because of the excellent examples of the systems they represent (Nair 1993).

Although several authors have tried to describe the term 'homegarden,' none is perhaps universally accepted as 'the definition'; but it is well understood that the concept refers to 'intimate, multi-story combinations of various trees and crops, sometimes in association with domestic animals, around homesteads.' These multistrata agroforests are estimated to occupy about 20% of the arable land in Java (Jensen 1993a) and are regarded as the 'epitome of sustainability' throughout the tropics (Torquebiau 1992). Homegardening has been a way of life for centuries and is still critical to the local subsistence economy and food security in Kerala state in peninsular India that has about 5.4

million small gardens (mostly less than 0.5 ha in area) (KSLUB 1995). In Africa, Central America, and the Caribbean and the Pacific Islands also, homegardens are of vital importance (Ruthenberg 1980; Caballero 1992; Anderson 1993; Clarke and Thaman 1993; Rugalema et al. 1994a, 1994b, 1995; High and Shackleton 2000).

Presumably, homegardening is the oldest land use activity next only to shifting cultivation. It evolved through generations of gradual intensification of cropping in response to increasing human pressure and the corresponding shortage of arable lands. The two great Indian epics *Ramayana* and *Mahabharata* (based on events that have supposedly happened around 7000 B.C. and 4000 B.C. respectively) contain an illustration of *Ashok Vatika*, a form of today's homegardens (see Puri and Nair 2004). The Javanese homegardens have reportedly originated as early as the seventh millennium B.C. (Hutterer 1984), and the Kerala homegardens are thought to be at least 4000 years old. Natural history studies in southern India during the late 1800s to early 1900s suggest that people traditionally used their homesteads for a variety of needs such as food, energy, shelter, and medicines[1]. Socio-culturally, homegardening fits well with the traditional farming systems and established village lifestyles. These systems have probably evolved over centuries of cultural and biological transformations and they represent the accrued wisdom and insights of farmers who have interacted with environment, without access to exogenous inputs, capital or scientific skills.

A great deal has changed, however, in the philosophy of and approaches to tropical land use during the past few decades. When the economy was predominantly subsistence-oriented, the homegardens that provide an array of outputs (Jose and Shanmugaratnam 1993) were quite appropriate land-use systems. With the advent of market economy, and consequent emphasis on maximization of production and use of external inputs in crop production, the homegardens have lost some of their relevance. There has been a resurgence of interest, however, in traditional land use practices in the wake of mounting environmental deterioration and/or failures of single-species agricultural enterprises; consequently homegardens have received some attention although they may not address environmental deterioration at a large scale because they exist in scattered small plots.

Pioneering research on homegardens dating back to the 1940s was reported from Indonesia (Terra 1953; 1958). These were followed by Ruthenberg's studies on tropical mixed-species cropping systems in the 1970s (Ruthenberg 1980) and similar work at the Institute of Ecology, Bandung, Indonesia (e.g., Soemarwoto 1987), the Central Plantation Crops Research Institute, Kasaragod, Kerala, India (Nair 1979), and elsewhere (e.g., Michon et al. 1983). These efforts got a further boost as the ICRAF (International Centre for Research in Agroforestry) effort on global inventory of agroforestry systems got underway (Nair 1987), and several descriptions and syntheses of traditional homegarden systems were published (Nair 1989). While most of the recorded reports on homegardens provided regional perspectives, others (notably Brownrigg 1985; Fernandes and Nair 1986; Soemarwoto 1987; Gliessman 1990; Landauer and Brazil 1990; Nair 2001; Soemarwoto and Conway 1991; Torquebiau 1992) were more comprehensive, describing the structure, functioning and sustainability of different forms of homegardens in various places. And, all of them – without exception – highlighted the need for coordinated scientific research on these extremely interesting systems. The objective of this paper is to summarize and analyze the trends in homegarden research during the past 25 years with a view to providing some new insights and directions on improving the homegardens and hopefully other similar systems.

Major themes and advances in homegarden research

It is well known that traditional land-use systems are influenced to a great extent by the biophysical and socio-cultural characteristics of the locales where they are practiced. Homegardens are no exception. A typical homegarden, nevertheless, is an integral part of the farmer's farming system and an adjunct to the house, where selected trees, shrubs and herbs are grown for edible products and cash income, as well as for a variety of outputs that have both production and service values including aesthetic and ecological benefits. Although agroforestry literature lists homegardens as an agroforestry practice (Nair 1993), it could be more appropriate to say that homegardening is a generic concept – much like agroforestry itself. Both involve diverse life forms and managerial regimes, which exist in harmony with one another on the same land management unit and/or on the landscape level.

Ecologists have variously described these as 'steady-state' systems (where production equals res-

Table 1. A comparison of the ecological attributes of climax forests, homegardens and conventional agricultural systems (monocropping).

Parameter	Natural climax vegetation (humid tropics)	Homegardens	Conventional agricultural systems
Biogeochemistry	Nutrient inputs equal outputs	Inputs and outputs balance each other	Outputs far exceed inputs
Biotic stress	Low	Low	High
Canopy architecture	Multistrata	Multistrata	One- or two- layered
Disturbance regimes	Rare except natural disturbances such as tree fall, wind throw etc.	Intermediate	High
Diversity	High	Intermediate	Low
Ecological succession	Normally uninterrupted, reaches a stable end-stage, e.g. climatic climax	Consciously manipulated	Arrested, succession does not proceed beyond the early stage
Entropy	Low	Low (?)	High
Floristic spectrum	Shade tolerant and intolerant	shade tolerant and intolerant	Mostly shade intolerant
Input use	No external inputs	Low	High
Overall homeostasis	High (Odum 1969)	High	Low
Site quality	Progressive improvement (e.g. facilitation)	Progressive improvement	Steady decline
Standing biomass/net primary productivity (NPP)	Highest among the terrestrial ecosystems (mean NPP: 2000 g m^{-2} $year^{-1}$)	Comparable to the climax formations but firm estimates are lacking[a]	Low (mean NPP: 650 g m^{-2} $year^{-1}$; Leith 1975)
Sustainability	Sustainable	Sustainable	Unsustainable

[a]However, a lone report on this (Christanty et al. 1986), *cf* Torquebiau (1992), provides a value of 5.23 kJ (= 1250 cal) m^{-2} yr^{-1} for the *pekarangan* gardens in Java. Clearly, this is lower than the annual energy fixation in both cultivated lands and tropical rainforests (i.e., 11.3 and 34.6 MJ m^{-2} yr^{-1} respectively; Leith 1975).

piration and/or inputs balance outputs), with structural characteristics comparable to those of the natural mixed forest vegetation systems (Fassbender 1993; Jensen 1993a, b; Jose and Shanmugaratnam 1993). Perhaps the high species diversity and low 'export' of harvested products endow sustainability to them. Ewel (1999) considers such systems 'structurally and functionally the closest mimics of natural forests yet attained.' Quantitative data on the biogeochemical and physiological processes in tropical homegardens are inadequate, however, to substantiate such comparisons. Some of the ecological attributes of homegardens in comparison to those of agricultural and forest systems are summarized in Table 1. It needs to be noted that such comparisons between homegrdens and forests, although fascinating to the ecologists, may sometimes be unpopular with farmers. For example, farmers in Java consider it an insult to compare their *pekarangan* to a forest, because, they regard homegardens as carefully tended plots, unlike the 'wild' areas of natural forests (Soemarwoto 1987).

Scientific reports on homegardens are replete with statements such as 'the studies on these systems have been disproportionately lower than what their economic value, ecological benefits or sociocultural importance would warrant' (Nair 2001). Furthermore, most of the reported studies are of inductive nature describing the system. Very little deductive research (hypothesis formulation and testing) has been done so that our understanding of the ecological rationalities behind the harmony between the humans, the homegardens, and the environment is limited. Methodological/design problems also hinder homegarden research more than in any other related fields. Each homegarden is unique in its own way despite the larger structural and functional similarities, making it extremely challenging to use commonly accepted research designs and procedures in the study of homegardens.

Most of the reported studies on homegardens are from Asia. Notwithstanding the pre-1990 work, 38 out of 83 post-1990 papers listed in CABI abstracts have been conducted in Southeast Asia (mainly Indonesia), followed by 22 from the South Asian region. Homegardens are, admittedly, predominant in those regions; yet, reports from other homegarden-rich regions such as Central America and West Africa are far less in relation to the extent of the practice.

The unique characteristics of homegardens have attracted the attention of several types of researchers: from proponents of energy-intensive modern production systems to champions of low-input sustainable systems, and from biophysical scientists who are in endless pursuit of quantifying every aspect of biological activity to social and behavioral scientists for whom the whole is often more than the sum of its parts. These various efforts in homegarden research during the past quarter century are briefly reviewed here.

System inventory

System description has been the most dominant aspect of the homegarden research so far. Thirty-two out of 83 published works on tropical homegardens since 1990 dealt with this. These published inventories suggest that food crops, medicinal plants, ornamentals, fruit trees, multipurpose trees and fodder crops abound in the homegardens. Woody components generally include timber trees, fruit trees, fuelwood species and palms, and may include both native as well as exotics, besides wild or domesticated crop plants. Such studies also documented the local practices and species inventory, elucidated the need for conservation of biocultural diversity and documented the sociocultural determinants of biodiversity, besides the traditional uses of various plants. Although the value of such databases has been recognized, voices have been raised about the need for moving on from species inventories and descriptions to establishing underlying concepts and general principles and developing and pursuing researchable hypotheses (Nair 2001).

Species composition and diversity

While most advances in agriculture and forestry entail single-species stands, characterized – quite appropriately – as 'biological deserts' of low species diversity, tropical homegardens are glorious examples of species diversity in cultivated and managed plant communities. Several landraces and cultivars, and rare and endangered species have been preserved in the homegardens (Watson and Eyzaguirre 2002). Soemarwoto and associates reviewed the early work on species diversity and plant density in Indonesian homegardens (Soemarwoto 1987; Soemarwoto and Conway 1991); more recent reports are summarized in Table 2. Wide variations in species assemblages of different geographic/eco-climatic regions are apparent. Comparisons are, however, difficult because of the different criteria employed for preparing inventories by different workers. Many studies on floristic richness of homegardens also lack information on the degree of heterogeneity in the study area (e.g. extent and socioeconomic nature of sampling units). Besides, little attention has been given to relate floristic richness to ecosystem processes. Some authors have advocated an ecological approach to computing diversity indexes and correspondence analysis (Rico-Gray et al. 1990; Kumar et al. 1994; Drescher 1996; Vogl et al. 2002; Wezel and Bender 2003), which is still relatively new in homegarden studies. The limited studies that have been made in this area indicate that homegarden diversity indexes are comparable with that of adjacent forest formations (Gajaseni and Gajaseni 1999; Kumar et al. 1994).

The crop combinations found in the homegardens of a region are strongly influenced by the specific needs and preferences of the household and nutritional complementarity with other major food sources, besides ecological and socioeconomic factors (Asfaw and Woldu 1997; Christanty et al. 1986; Vogl et al. 2002). The homegardeners are perpetual 'experimenters.' They are constantly trying and testing new species and varieties and their management (Niñez 1987). A new species may be chosen by the garden owner because of its properties, i.e., food, wood, medicinal, religious, ornamental, and based on self-instinct or information passed on by neighbours and relatives. There is, however, no specific time of the year for planting or introducing new plant species into the gardens; it will depend on the space available and/or soil conditions (Rico-Gray et al. 1990; Yamamoto et al. 1991).

Species richness of homegardens within a region is influenced by farm size. Kumar et al. (1994) reported that floristic diversity (measured by number of species per unit land area) was greater in small gardens (<0.4 ha size) (Simpson's diversity index, $D = 0.61$) than in medium (0.4 to 2.0 ha, $D = 0.44$) and large (>2 ha, $D = 0.46$) gardens. Drescher (1996) also showed that the density-based Shannon-index decreased with increasing garden-size up to 599 m^2, after which it, however, increased up to 1499 m^2; nevertheless, the smallest gardens located in the urban areas showed the highest crop species diversity in his study. Presumably, the limited space forces people to accommodate many different species in relatively small numbers on small plots. Furthermore, the religious/cultural beliefs, customs, and taboos of the villagers influence the diversity/composition of homeg-

Table 2. Floristic elements reported from homegardens in different regions[a] of the world.

Region/location and the floristic spectrum sampled	Number of species		Source
	per garden	Total for geographical location	
South Asia			
Pitikele, Sri Lanka (edible species)	–	55	Caron (1995)
Kandy, Sri Lanka (woody species)	4–18	27	Jacob and Alles (1987)
Thiruvananthapuram, Kerala, India (all species)	–	107	John and Nair (1999)
As above	26–36	–	D. Jose (pers. comm., 1992)
Kerala, India (woody species)	3–25	127	Kumar et al. (1994)
Bangladesh (perennial species)	–	30	Leuschner and Khaleque (1987)
As above	–	92	Millat-e-Mustafa et al. (1996)
Kerala, India (all species)	–	65	Nair and Sreedharan (1986)
Southeast Asia			
West Java (all species)	–	195	Abdoellah et al. (2001)
Northeastern Thailand (all species)	15–60	230	Black et al. (1996)
Chao Phraya Basin, Thailand (all species)	26–53	–	Gajaseni and Gajaseni (1999)
West Java (all species)	–	602	Karyono (1990)
Central Sulawesi, Indonesia (all species)	28–37	149	Kehlenbeck and Maas (2004)
Cianjur, West Java (all species)	4–72	–	K Sakamoto (pers. comm., 2003)
Cilangkap, Java (all species)	42–58	–	Yamamoto et al. (1991)
South/Central America & the Caribbean			
Quintan Roo, Mexico (all species)	39	150	De Clerck and Negreros-Castillo (2000)
Cuba (all species)	–	80	Esquivel and Hammer (1992)
Central Amazon (woody species)	–	60	Guillaumet et al. (1990)
Belize (all species)	30	164	Levasseur and Olivier (2000)
Masaya, Nicaragua (all species)	–	324	Méndez et al. (2001)
Santa Rosa, Peruvian Amazon (all species)	18–74	168	Padoch and de Jong (1991)
Yucatan, Mexico (all species)	–	133–135	Rico-Gray et al. (1990)
As above	–	301	Rico-Gray et al. (1991)
Chiapas, Mexcio (all species)	30	241	Vogl et al. (2002)
Cuba (all species)	18–24	101	Wezel and Bender (2003)
Other regions			
Catalonia, Spain (all species)	–	250	Agelet et al. (2000)
Southern Ethiopia (all species)	14.4	60	Asfaw and Woldu (1997)
Bungoma, Western Kenya (all species)	–	253	Backes (2001)
Soqotra island, Yemen (all species)	3.9–8.4	–	Ceccolini (2002)
Democratic Republic of Congo (Zaire) (all species)	–	272	Mpoyi et al. (1994)
Bukoba, Tanzania (woody species)	–	53	Rugalema et al. (1994a)
Central, eastern, western and southern Ethiopia (all species)	–	162	Zemede and Ayele (1995)

[a]All except Catalonia are tropical.

ardens. For example, crops/trees/animals are retained or excluded depending on the above considerations (Millat-e-Mustafa et al. 1996). Women too play an important role in determining the species to be included in their homegardens; indeed, in many instances, women are responsible for introducing new species into their homegardens.

Structure

Presence of a large number of species on the same land management unit, often seemingly not following any specific geometry, makes it difficult to define the temporal/spatial architecture of homegardens. Although a typical homegarden may represent a clearly demarcated area (fenced-in or bounded by field risers), often it is difficult to distinguish where the homegarden ends and arable cropping starts. In addition, size and shape of gardens, and the nature of cropping are dynamic, further complicating the structural pattern. The structural entities of homegardens are, however, arranged in a complex micro-zonal pattern having well-defined vertical/horizontal stratification with each structural ensemble occupying a specific niche, such that they cannot be easily dissociated from one another (Nair and Sreedharan 1986). Understanding these interrelations will be one step forward in utilizing the advantages of homegarden agroforestry.

Many authors have reported that distinct horizontal zones occur in the homegardens, and that their location, size and plant species composition reflect deliberate management strategies (Abdoellah 1990; Padoch and de Jong 1991; Zemede and Ayele 1995; Millat-e-Mustafa et al. 1996; Agelet et al. 2000; Mendez et al. 2001). Position or distance from home, besides size, shape, crop composition, and planting pattern of the garden are the principal determinants in this respect. The number of such management zones per homegarden varies from two to six, with a mode value of three. In general, food and fruit producing species dominate near the living quarter and working areas, and small plots of annual vegetable crops separate this part of the garden from the more distant parts favored for timber species. Medicinal and ornamental species are typically cultivated in small areas or in pots surrounding the house, and vegetables in areas adjacent to the kitchen. It needs to be clarified that such vegetable and ornamental gardens, may not have agroforestry implications; but they all usually mesh well into a land-use system surrounding the home in a relatively small area of land (often less than 0.5 ha), and the whole unit is referred to as 'homegarden'. Coconut palm (*Cocos nucifera*) forms the 'nucleus' of the Kerala homegardens, around which the other components are orchestrated (Jose and Shanmugaratnam 1993). Multipurpose tree and shrub species used as live fences are usually planted on farm boundaries, regardless of holding size. Trees also may be scattered throughout the homestead or at specific points to provide or avoid shade, necessary or harmful to different plants, besides providing support for climbers (e.g., *Piper nigrum*; Mathew et al. 1996). Plants and their local uses, which are included in different zones, generally mirror the farmer's management priorities and socioeconomic needs.

The multi-tiered canopy structure is one of the most distinguishing features of homegardens, especially in humid tropical lowlands. Most workers delineate a three-to-six-strata system (e.g., Karyono 1990; Michon and Mary 1990; 1994; Gillespie et al. 1993; Jose and Shanmugaratnam 1993; Kumar et al. 1994; Millat-e-Mustafa et al. 1996; Hochegger 1998; Gajaseni and Gajaseni 1999; De Clerck and Negreros-Castillo 2000), with about three quarters to full coverage of the ground (Jensen 1993b). Predictably, the vertical stratification provides a gradient in light and relative humidity, which creates different niches for enabling various species groups to exploit them. Obviously, shade tolerant crops constitute the lower stratum, shade intolerant trees the top layer, and species with varying degrees of shade tolerance in the intermediate strata.

The Mediterranean (Catalonia, Spain) and the arid tropical (Soqotra island, Yemen) gardens do not, however, exhibit such a complex vertical structure; stratification is usually limited to a lower stratum of herbs and shrubs and a higher one of trees (Agelet et al. 2000; Ceccolini 2002). Likewise, Abdoellah et al. (2001) observed that stratification beyond 5 m was not discernible in a majority of Indonesian homegardens, suggesting profound inter-site variations. Apparently, garden age and management are important factors that influence the vegetation structure. Older gardens, regardless of size, may evolve a multistrata canopy structure. Likewise, in gardens where ground cover or litter layer is removed due to repeated hoeing or burning, ground cover development will be adversely affected (Hochegger 1998; Karyono 1990).

The question of regeneration dynamics of the woody perennials in homegardens has also been examined to a limited extent only. Kumar et al. (1994) noted that the diameter structure of trees in the homegardens exhibit a slightly skewed (+) distribution pattern, having the highest frequency in the 20 to 30 cm class, signifying adequate 'regeneration' of the homegarden tree species. Thus, homegardens, especially of the so-called forest-garden (Wiersum 2004) type, could resemble a young secondary forest both in structure and total biomass store, and may be considered as a man-made forest kept in a permanent early-successional state (Jensen 1993b).

Variations in size, form, layout, zonal pattern, species composition and management objectives abound in the homegardens (Fernandes and Nair 1986; Landauer and Brazil 1990). Accordingly, several homegarden types have been recognised. They generally reflect differences in size (area) of gardens (Jose and Shanmugaratnam 1993; Kumar et al. 1994; Millat-e-Mustafa et al. 1996) and the dominant plant species or the level of urbanization (Christanty 1990). In particular, greater urbanization increased the abundance of ornamental and aesthetic species (Drescher 1996; Karyono 1990; Rico-Gray et al. 1990). Some researchers (Michon et al. 1983; Michon and deForesta 1999; Soemarwoto 1987) employed architectural analysis to gain an understanding of the essential characteristics of the major types of homegardens. Based on zonation, diversity, total garden area and socioeconomic functions, Mendez et al. (2001) recognized six types of homegardens in Nicaragua (ornamental, handcrafting, subsistence, handcrafting and mixed production, mixed production, and minimal management) – the differences among these classes may, however, seem arbitrary. A few workers (Leiva et al 2002; Mendez et al. 2001; Quiroz et al. 2002) also employed cluster analysis of presence/absence of crop species to designate homegarden types. Despite the number of classification schemes, none has been universally accepted. The need for such a universally acceptable classification scheme is, nevertheless, questionable; the lack of such a scheme has seldom surfaced as an important prerequisite for intensifying research or development in the subject. The fact remains, however, that irrespective of the classification scheme employed, it is important to consider diversity of all different functional groups of plants in homegardens, and not to limit research to selected functional groups only.

Functional dynamics

Research attention on production potential and quality of biomass from homegardens has been modest. In one such study, Christanty et al. (1986) (*cf* Torquebiau 1992) reported a net primary production (NPP) of 5.23 kJ m^{-2} yr^{-1}. This, however, seems to be an underestimate in view of the annual energy fixation values reported by Leith (1975) for cultivated lands (11.3 MJ m^{-2} yr^{-1}; Table 1), and the aboveground energy fixation in some tropical plantations (e.g., 9.1 to 20.6 MJ m^{-2} yr^{-1} for 8.8 year-old *Ailanthus triphysa* stands – derived from the biomass and mean combustion values given by Shujauddin and Kumar 2003 and Shanavas and Kumar 2003 respectively). Reports on functional dynamics of homegardens, i.e., change in functions with time, are also not available. In most reported studies, yields of component species, availability of varied products and continuous or repeated harvests during the year are often taken as indicators of such dynamics.

Homegardens and nutritional security

Food production either directly (producing edible fruits, nuts, grain, rhizomes and tubers, leaves, flowers etc.) or indirectly (facilitating enhanced and/or sustained production) is a basic function of tropical homegardens. Consequently, food crops including herbaceous plants and vegetable and fruit yielding trees and shrubs abound in the homegardens (Caron 1995; Mendez et al. 2001; Vogl et al. 2002; Wezel and Bender 2003). Based on a comprehensive literature review, Torquebiau (1992) concluded that dietary supplies from homegardens accounted for 3% to 44% of the total calorie and 4% to 32% of the protein intake. It needs to be noted, however, that homegardens seldom meet the entire basic-staple-food needs of the family in any given area. At best, they are complementary to other plots such as rice (*Oryza sativa*) or maize (*Zea mays*) fields. Thus, homegardens are a component of the larger farming system of the household. Indeed, if the homegarden is the only land available to the household, food crops such as cassava (*Manihot esculenta*) will dominate the species composition of the garden (Wiersum 1982).

The homegardens are also significant sources of minerals and nutrients (Asfaw and Woldu 1997). In addition, the diverse products available year-round in the homegardens contribute to food security especially during 'lean' seasons (Christanty et al. 1986; Karyono 1990). Consequently, there is now a growing awareness that homegardening combined with nutrition education can be a viable strategy for improving household nutritional security for at-risk populations, particularly women and children. In experimental studies, the target families significantly increased their year-round production and consumption of vitamin-rich fruits and vegetables compared to the control group without gardens. This, in turn, alleviated deficiencies of iodine, vitamin A and iron (Molina et al. 1993) and made children of garden owners less prone to xerophthalmia (Shankar et al. 1998).

Related to food security is the issue of nutritional quality of food. As little or no chemical inputs are used in homegarden systems, the products from homegardens can be expected to be of superior quality. Studies on nutritional quality of homegarden products compared with those of products from other systems have, however, been not reported in literature. Judging from the experience of organic coffee (*Coffea* spp.) production in Central America (Muschler 2001), the products from homegardens could fetch premium prices in health-food stores in food-quality-conscious societies. This is perhaps a nonissue at least for the near future, because most of the homegarden products are used for the grower's home-consumption and marketing is not all organized in most homegarden-dominated regions. Nevertheless, the availability of 'health' foods from homegardens is certainly worth recognizing as one of the intangible benefits of the system.

Another key dimension of homegardening is the equitable distribution of the produce within the community. While a large proportion of the production is consumed domestically (Soemarwoto 1987; but see Jensen 1993b), many products such as fruits, vegetables and medicinal/ornamental plants are also generously shared within the local communities (Thaman 1990). This egalitarian distribution of the agricultural produce is significant in a social milieu and endows the homegarden system a unique social disposition.

Generation of cash income

Although interest in homegardens has been primarily focused on producing subsistence items, its role in generating additional cash income cannot be overlooked (Christanty 1990; Torquebiau 1992; Dury et al. 1996; Mendez et al. 2001). Enormous variations have been reported from different regions in the proportion of homegarden products that are used for household consumption as opposed to sale. In West Java, as much as two-thirds of the homegarden production is reported to be sold (Jensen 1993b), but only 28% of such products were sold in South African homegardens, the remainder being used for household consumption (High and Shackleton 2000). The net income generated from homegardens is also correspondingly variable. For example, in Indonesia it ranged from 6.6% to 55.7% of total income with an average of 21.1% depending on the size of the gardens, family needs and species composition (Soemarwoto 1987).

Timber and fuelwood production

While there are many reports on food and nutritional security, relatively few workers have addressed questions relating to wood production and its utilization in the homegardens. Indeed this seems to be an unrecognized value of homegardens. For example, Singh (1987) noted that 70% of the sawlogs in Bangladesh came from homesteads, and Krishnankutty (1990) found that homesteads provided 74% to 84% of wood requirements in Kerala, India. The average standing stock of commercial timber in the Kerala homegardens has been estimated to range from 6.6 to 50.8 m^3 ha^{-1} (Kumar et al. 1994); the wide range of the values is related to variations in tree stocking levels depending on local conditions.

Several workers (Krishnankutty 1990; Wickramasinghe 1996; Levasseur and Olivier 2000; Shanavas and Kumar 2003) have also reported that the traditional homegardens constitute a principal source of biofuels for the rural households. For example, 51% to 90% of the fuelwood collected in various geographical regions in South and Southeast Asia are derived from homegardens (Krishnankutty 1990; Singh 1987; Torquebiau 1992). Shanavas and Kumar (2003) noted profound variations in the combustion characteristics *vis à vis* tissue-types of woody perennials grown in the Kerala homesteads. As in most other aspects of homegardens, scientific studies have not been undertaken on the quantity and quality of phytofuels produced in homegardens and their extraction methods.

Nontimber forest products (NTFPs)

Homegardens are recognized repositories of nontimber products such as medicinal and aromatic plants, ornamentals, bamboos, gums, resins, chemical extractives and green leaf manure. Of these, medicinal-plant production has received some modest scientific attention (see for example Rao et al. 2004). Kumar et al. (1994) reported that out of the 127 trees and shrubs found in Kerala homegardens, 25 were medicinally important. Similarly, 27% of the 272 species maintained or cultivated as domestic flora in the Democratic Republic of Congo (Zaire) are reported to be for medicinal purposes (Mpoyi et al. 1994). In the Catalonian homegardens, Agelet et al. (2000) documented about 250 medicinal plants with curative, palliative, symptomatic, and/or other medicinal properties. An unfortunate aspect of the situation, however, is the gradual decline in the stock of medicinal plants as reported from a number of places (Rico-Gray et al. 1990; Caballero 1992; Agelet et al. 2000). Organized efforts are needed to conserve these valuable resources in homegardens.

Employment generation

Quite apart from providing cash income and subsistence products to the growers, homegardens have a tremendous potential for rural employment generation. Torquebiau (1992) summarized several case studies on homegarden labor demands and flexibility, and concluded that labor requirements were variable among regions depending on the size of gardens and the farming intensity. The amount of labor invested in a homegarden is also related to the size of the family, depending on homegardens and the gardener's primary occupation. It ranged from 50 min day^{-1} in a 200 m^2 garden in Lima, Peru, to 57 man days year^{-1} in Sri Lanka (Torquebiau 1992). Mendez et al. (2001) reported that homegardens used on average 32.6 h family^{-1} week^{-1} with women contributing to roughly half (48%).

Use of family labor, especially women labor in the production process not only ensures lowering of production costs (Rugalema et al. 1994a; Benjamin et al. 2001), but also satisfies a wide range of domestic needs more economically and effortlessly than through local markets (Soemarwoto and Conway 1991). Such calculations are, however, based on the assumption that family labor is a 'given,' which will otherwise remain unutilized and therefore need not be accounted for. The validity of this assumption is increasingly being disputed today. Temporal complementarity in labor allocation is yet another advantage of homegardening. For example, labor demands of homegardens seldom show sharp peaks and troughs and they are more flexible and distributed throughout the year in sharp contrast to that of seasonal agricultural operations such as wet land rice paddy cultivation (Karyono 1990).

The notion that homegardening is practiced only by subsistence farmers in the areas of acute land hunger has also been questioned (Yamada and Gholz 2002). Although most labor requirements especially in the smaller homegardens may be satisfied from within the family itself, homegardens have the potential to create jobs in the rural areas, and thus reverse the negative effects of migration to urban areas. This is because the great diversity of products from homegarden provides opportunities for development of small-scale rural industries and creates off-farm employment and marketing opportunities (Nair and Sreedharan 1986; Torquebiau 1992).

Animal production

Most homegardens support a variety of animals – cows, buffaloes, bullocks, goats, sheep, and birds such as chicken and ducks (Brownrigg 1985). Culture, religion, and economic and ecological factors are the main determinants in the choice of animal species (Soemarwoto 1987). Milk and egg production, both for domestic consumption and for sale, constitute the dominant objectives of integrating animals into these production systems. In addition, the large ruminants may provide draught power for land preparation, soil conservation practices and haulage. Animals are also important for the maintenance of soil fertility and sustainability of the production system.

Sustainability of homegarden systems

Homegardens are perceived to be highly sustainable in both biophysical and socioeconomic terms. Because of the lack of widely acceptable sustainability criteria, however, these perceptions are not quantifiable. The following section evaluates the situation based on generally 'accepted' notions of sustainability.

Biophysical aspects

Nutrient cycling

It is generally regarded that the homegardens possess a closed nutrient cycling, much similar to the tropical forests (Nair et al. 1999; Soemarwoto and Conway 1991). The dynamics of litter production and decomposition and subsequent bioelement release, which endow sustainability to these forests (Heal et al. 1997; Lavelle et al. 1993), are, therefore, relevant to homegardens. Biological N_2 fixation and mycorrhizal associations are important; yet, little quantitative information seems to exist in this respect. In particular, behavior of N_2-fixing trees that are abundant in these systems, information on the quantities of N_2 fixed, and its further utilization by associated crops are not well documented. Nutrients in tree biomass are returned to soil either naturally through litterfall and root turnover, or deliberately through pruning. Nutrients circulated internally in a Javanese homegarden were estimated as 223 kg N, 38 kg P, 373 kg K, 135 kg Ca and 50 kg Mg ha^{-1} yr^{-1} (Jensen 1993a). Despite their importance in maintaining organic matter and nutrient flows, studies on litter and fine root dynamics in the homegarden system are few and uncoordinated. Moreover, sometimes homegarden litter is removed for fuel and/or as organic manure (to be used elsewhere), on which again, little quantitative data exist.

Among the available reports on homegarden litterfall, Gajaseni and Gajaseni (1999) stated that the average litter in a standing crop was even higher

(1728 g m^{-2}) than that of a typical tropical forest (500 g m^{-2}; Jordan 1985). Soemarwoto and Conway (1991) also observed a heavy litter layer in Javanese homegardens, and Jensen (1993a) estimated litterfall and pruning as 10.0 and 7.5 Mg ha^{-1} yr^{-1} of dry matter respectively. Total litter production in the Mayan homegardens, however, was reported to be lower than that of other systems compared (1 to 4 Mg ha^{-1} yr^{-1}; Benjamin et al. 2001). Such variations are not unusual considering that stocking levels, species attributes and management, which are variable, determine the quantity and periodicity of litterfall (George and Kumar 1998).

Although hundreds, if not thousands of litter decomposition studies have been reported, only very few of them pertain to litter decay dynamics of tropical homegardens (Table 3). And, without exception all these investigations focused on characterizing the decay rates of monospecific leaf litter. This possibly underestimates the homegarden litter decay rates, because, unlike monospecific stands, homegardens provide litter of a mixture of species, and the resultant litter heterogeneity will alter the decomposition dynamics.

The limited number of studies comparing decay of monospecific litter in the subcanopy of homegardens with that of adjacent open area suggests that species- or site-related variations in decay rates are probable (Table 3). Nevertheless, five out of the seven species on which quantitative information exist decayed faster in the subcanopy. Organic matter decomposition is generally dominated by drilosphere systems formed by associations between earthworms and soil bacteria, where soil-moisture regimes are favorable; and termitosphere systems with increasing dryness (Lavelle et al. 1993). Owing to greater organic matter flux and/or favorable soil moisture relations (Table 3), the activity of drilosphere system is expected to be greater in the homegardens. Consistent with this, higher earthworm activity and fungal/bacterial/actinomycetes counts compared to open areas were observed in the Kerala homegardens (S.R. Issac, pers. comm. 2001). Apparently, the organic matter turnover processes bring about an overall improvement in soil physicochemical properties of homegardens (Table 3). With the revival of interest in organic farming, research on addition and/or decomposition of fresh agricultural wastes and mixtures of green manure/litter is likely to become important. Coincidentally, resource quality parameters and indices that govern decomposition and nutrient release have received considerable scientific attention lately (Nair et al. 1999; Palm et al. 2001). In order for the tropical homegarden systems to continue to provide social, economic and environmental benefits for successive generations, a clear understanding of the various aspects of nutrient cycling including the effects of litter/green manure addition on soil organic matter and nutrient dynamics is essential.

Low soil-erosion potential

Homegardens are not only similar to natural forests in respect of their high species-diversity and nutrient cycling processes but also are characterized by low 'export' of nutrients. The multi-tiered homegarden canopy and root architecture, besides the litter layer, act as multi-layer defense mechanisms against the impact of the falling rain drops, resulting in low rates of soil erosion (0.05 to 0.06 t ha^{-1} $year^{-1}$ as against tolerable limits set by the U.S. Soil Conservation Service, 2.2 to 11.2 t ha^{-1} $year^{-1}$; Torquebiau, 1992). Secondly, there is no complete harvesting in the homegardens (Gajaseni and Gajaseni 1999). Overall, nutrient exports from such a system are modest. Nevertheless, the argument of low nutrient export may not be valid, if whole tree harvesting and litter removal are resorted to, leading to substantial site nutrient loss (Kumar et al. 1998). Consistent with this, some authors (Fassbender 1993; D. McGrath, Ph.D. dissertation, University of Florida, 1998) based on studies on other multistrata systems have suggested that nutrient deficiencies may eventually limit system productivity. But the credibility of such reports is questioned by the history of long-term survival of homegardens in a variety of locations without any apparent indications of nutrient deficiency (Nair 2001).

Belowground architecture

Root systems of different components in the homegardens are expected to overlap considerably and the resultant higher root-length density may reduce nutrient leaching and facilitate recycling of subsoil nutrients. In certain cases, the proximity of trees to one another determines the magnitude of subsoil-nutrient recovery. For example, Kumar and Divakara (2001) found that in bamboo-based multi-strata systems in Kerala, India, ^{32}P uptake from the subsoil was greater when the bamboo clumps (*Bambusa arundinacea*) and dicot trees (*Tectona grandis* and *Vateria indica*) were close to one another. By extension, in homegardens where tree components are closely integrated, there is a substantial potential for 'capturing' the lower leaching nutrients. On-site nutrient conservation also may be accomplished through the interlocking root

Table 3. Comparison of litter dynamics and soil properties of some tropical homegardens and adjacent open areas[a]

Attributes	Reference	Comments
Litter decay and half-lives	Gopikumar et al. (2001)[b]	*Acacia mangium* leaf litter showed lower decay rates 'inside'[c] homegardens—decay rate coefficients (k) 19 to 28% lower and half-lives of litter decay 22 to 40% higher compared to 'outside'[c] homegardens.
	Issac and Nair (2002); SR Issac (pers. comm., 2001)[b]	Leaf litter of *Ailanthus triphysa, Ancardium occidentale, Artocarpus heterophyllus, Mangifera indica* and *Swietenia macrophylla* decayed faster 'inside' homegardens—k values 7 to 14% more and half-lives 2 to 29% less than 'outside.' *Artocarpus hirsutus*, however, had similar decay rates both inside and outside. Variations reiterate 'litter chemistry control' of decay dynamics.
Soil water holding capacity	J. John (pers. comm., 1997)	14 to 18% more inside homegardens
Porosity		14 to 23% more inside
Bulk density		12 to 20% less inside
Particle density		2 to 5% less inside
Soil pH	Gajaseni and Gajaseni (1999)	10 to 14% decrease in the homegardens with 'near-neutral' soils and 10 to 12% increase for more acid soils in the Chao Phrao basin, Thailand ('outside' soil pH 5.8) – 'moderation effect' on soil reaction possible?
Soil organic C (SOC)	as above	Substantial improvement: 33 to 600% more SOC 'inside' Thai gardens
	J. John (pers. comm., 1997)	48 to 62% more for southern Kerala gardens
C:N ratio	Gajaseni and Gajaseni (1999)	58 to 205% more except in homegardens on inherently N-rich sites, where a 4 to 24% decline was noted
Total N	as above	21 to 249% more inside gardens depending on site characteristics and soil depth
Available N	J. John (pers. comm., 1997)	21 to 41% more inside
Total soil P	Gajaseni and Gajaseni (1999)	27 to 311% more inside
Available P (Bray)	as above	4 to 363% more available P in Thai homegardens, depending on site characteristics and soil depth
	J. John (pers. comm., 1997)	25 to 30% more for Kerala homegardens
Exchangeable K	as above	10 to 19% more inside
Soil microbiological properties	SR Issac (pers. comm., 2001)	Earthworm, fungal, bacterial and actinomycetes counts were consistently higher inside homegardens (~4, 36, 35 and 20% respectively).
	Wichern et al. (2003)	Microbial biomass C, ergosterol and basal respiration rates were generally lower (51 to 72, 53 to 79 and 10 to 49% respectively) in 'old' South African homegardens (prolonged cropping), compared to 'new' ones (cropping period: ~1 year).

[a]Although edaphic attributes of homegardens were reported by some authors; few (as shown above) have actually compared this with adjacent open areas.
[b]Calculated from litter mass disappearance data using the exponential decay model (Olson 1963).
[c]'inside' gardens indicates directly beneath tree crowns and 'outside' means adjoining open areas, usually outside the 'fenced-in' area.

systems (root grafts and/or mycorrhizal connections), which essentially act as multipliers of the 'root systems' reach.' In addition, horizontal transfer/sharing of nutrient ions between the rhizospheres of the neighboring plants is probable. That is, the tree roots may release, leach out, and/or exude mineral and organic materials into the rhizosphere of neighboring plants, provided they interact with one another (Kumar et al. 1999).

Socioeconomic aspects

Over long periods in the history of land use in highly populated humid tropical lowlands, homegardens have apparently remained as engines of economic and social development. Although productivity compared to intensive monocultures was modest, diversified production and income generation in perpetuity were intrinsic features of tropical homegardens. Planting and maintaining of homegardens also reflect the culture and status of the household, especially the women, in the local society. In many places, women play a vital role in the design and management of these land use systems. The fact that homegardens are predominant in the traditionally matrilineal societies of south and Southeast Asia (e.g., Kerala, Central Java and West Sumatra) further underscores this point (Menon et al. 2002; Soemarwoto and Conway 1991). Many activities such as vegetable growing, harvesting the products (fruits, nuts, vegetables and medicinal plants), fuel collection and animal rearing, especially in the smaller gardens are exclusively the domain of females, with or without assistance from the male members of the family. The possibility of gender equality for participating in garden management and sharing of benefits is perhaps one of the major stimuli for continued household security enjoyed by homegardeners for generations. Some other aspects of socioeconomic sustainability such as household nutritional security and generation of income have already been discussed. The whole aspect of sustainability of homegardens seems to be fertile grounds for social science research. In spite of the authors' lack of expertise in these areas, based on a few studies of this nature that we have come across (e.g., Arnold 1987; Soemarwoto and Conway 1991; Jose and Shanmugaratnam 1993), some of the economic, social and/or cultural foundations of homegardening, in comparison with other farming-system components under similar situations are listed below:

- Low capital requirements and labor costs – suitable for resource poor and small-holder farming situations
- Better utilization of resources, greater efficiency of labor, even distribution of labor inputs and more efficient management
- Diversified range of products from a given area and increased value of outputs
- Increased self sufficiency and reduced risk to income from climatic, biological or market impacts on particular crops/products
- Higher income with increased stability, greater equity and improved standards of living
- Better use of underutilized land, labor or capital, besides creating capital stocks to meet intermittent costs or unforeseen contingencies
- Enhanced food/nutritional security and ability to meet the food, fuel, fodder, and timber requirements of the society
- Increased fulfillment of social and cultural needs through sharing or exchange of produces and recreational opportunities
- Better preservation of indigenous knowledge

Methods for valuation of such intrinsic and non-tangible social and cultural benefits of agroforestry systems are practically nonexistent. Some of the recent efforts in this direction (for example, see Alavalapati et al. 2004) clearly show the opportunities and challenges involved.

The future of homegardens

Are homegardens becoming extinct?

Change is said to be the only constant thing. Land-use patterns have changed enormously the world-over during the past half a century. Homegardens are no exception: they too have changed, and will continue to do so. The concern that is raised, however, is, are the homegardens on the way out? Are they becoming an anachronism? Will – or how will – they survive the test of time? These concerns stem not from any new discoveries questioning the ecological foundations of such systems, but from the fact that recent trends in agrarian structure and the high market-orientation, due to government policies and demographic pressures, do exert considerable pressures on homegardens. Commercialization has caused a decline of the structure and functions of the Indonesian *pekarangan* (Abdoellah 1990; Abdoellah et al. 2001; Soemarwoto and Conway 1991) and *kebun* (Whitten et al. 1996) systems. A large proportion of Kerala homegardens have been converted into small-scale plantations of coconut and rubber (*Hevea brasiliensis*) or to cropping systems consisting of less number of crops (Ashokan and Kumar 1997). In Catalonia (Spain) during the last few decades, migration of the work force to urban areas, mechanization of agriculture and the consequent specialization of agricultural tasks have lead to an impoverishment of the traditional homegarden system (Agelet et al. 2000).

Species losses from homegardens are said to occur at an unprecedented rate. Although precise data are not available, it seems reasonable to surmise that floristic diversity of homegardens in most parts of the world has declined during the past four to five decades. For instance, many local varieties of mango (*Mangifera indica*), jackfruit (*Artocapus heterophyllus*) and other traditional fruit/vegetable crops, which were once abundant in Kerala homegardens, have now become extinct (authors' personal observation). A similar situation has been reported from West Java, Indonesia, where 27 varieties of mango were reportedly lost during a 60-year period (Soemarwoto 1987). Agelet et al.

(2000) also noted the loss of many autochtonous varieties of crop plants and medicinal plants from the Catalonian gardens.

Fragmentation of land holdings due to population growth has triggered species losses in the *chagga* homegardens of Tanzania (Rugalema et al. 1994b). Also, the specific inheritance system in an Islamic society, where a property is divided equally among the heirs, caused fragmentation of the Indonesian *pekarangans* (Arifin et al. 1997). Changes in homegarden structure and ecological processes that are caused by such changes in size of gardens can safely be added to the already long list of topics that have not been, but need to be, studied. It is not clear, however, if and to what extent fragmentation of holdings *per se* has caused a decline in the average size of homegardens. Although fragmentation will clearly result in an increase in the number of holdings and therefore the number of homegardens, it is doubtful if the average size has shrunk proportionately; evidence is lacking.

'Acculturation,' the process by which a culture is transformed due to the massive adoption of cultural traits from another society, has serious consequences on the species grown in homegardens (and by subsistence farmers in general) and the extent to which they are used (Caballero 1992). For instance, modification of the traditional patterns of nutrition among the Indian groups of Mexico involved the progressive disuse of traditional foods because they were considered to be of lesser nutritional value or lower social prestige than the industrialized products or other foods offered through the national market (Vogl et al. 2002). The modern-day preference to maize over traditional millets that used to be dietary staples in African societies is well known. Likewise, spread of modern medicine and healthcare facilities lead to a decline of the traditional medicines. Yet another threat to the traditional homegarden is the large-scale influx (both deliberate and unintentional) of aggressive exotics (e.g., Australian and Meso-American species such as *Acacia* spp., *Eucalyptus* spp., *Mimosa invisa*, *Mikania micrantha* and so on, especially into South Asia) that can potentially out-compete the native flora. The cumulative effect of such social and cultural changes would be a reduction in the homegarden floristic diversity.

A further consequence of commercialization is a change in the proportion of the produce consumed. As more homegarden products are sold (to earn cash to buy other preferred items) and less consumed, there is a danger that the dietary role of the homegardens in providing vitamins and minerals may be reduced if not lost (Soemarwoto and Conway 1991). Commercialization could also result in a reduction of equitability, with less sharing of the produce even among close relations, and the traditional 'rights' of the poor may disappear. Thus, homegardening may, arguably, face the threat of self-dissolution in arbitrariness and irrelevance if the traditional homegardeners are all transformed into 'modern' urbanites. In some societies such as those in the Caribbean islands and to some extent in Kerala, where the work force migrate to nearby industrial and economically prosperous lands (North America for the Caribbean and the oil-rich Middle-East for Keralites) and remittances from abroad is a major component of the local economies, homegardening and such traditional lifestyles are getting relegated to pastime of older generations who return to land for retirement. Long-term dynamics of homegarden transformations could be fascinating areas of study to biophysical and social scientists alike.

Scientific evidence is lacking, unfortunately, to support or oppose the claims about the future of homegardens. With increasing awareness and interest the world over – especially in the industrialized societies – in movements such as organic farming, 'back to nature,' and green consumerism, we have no reason, however, to abandon and denigrate everything that is traditional. Chemical agriculture has its place, but societies that have embraced it at the total exclusion of traditional ways of life are paying the price – and that too heavily. These experiences, coupled with the slow rate at which economic and social transformations happen in such societies force us to believe that concerns about a bleak future of homegardens are unfounded and will hopefully be a passing cloud. Having stated that, however, the main thread of argument that runs throughout this chapter needs to be re-emphasized: the need for better understanding of the scientific bases of these fascinating systems and developing valuation methods that can realistically portray their economic and social values; in short, the need for intensified research.

Future research

Research is needed on all fronts. The knowledge gap that exists on various biophysical issues has already been explained. Understanding the nutrient, energy and water balance in the homegardens is crucial to providing a scientific foundation for better design and management of the system to permit efficient use of resources, to avoid loss of energy and to increase production (Benjamin et al. 2001). Aspects such as the 'how' and 'why' of interference and coexistence of different species; reasons for low herbivory/pathogenicity pressure; competitive interactions as influenced by soil resource availability and soil organic matter relations of homegardens are practically unknown. Furthermore, no effort has been made to examine the impact of harvesting pressures on resource regeneration in the homegardens. With the increasing emphasis the world over on carbon sequestration in terrestrial ecosystems, homegardens that are mostly steady-state systems as explained earlier (section 2) could be of crucial importance in the carbon-credit discussion (see Montagnini and Nair 2004). Available reports indicate that total biomass of Javanese homegardens accounted for 10 to 126 Mg C ha^{-1} (Jensen 1993a; Delaney and Roshetko 1999). Roshetko et al. (2002) estimated the aboveground, time-averaged C stock (= half the C stock at maximum rotation length) of Indonesian homegardens as 30 to 123 Mg C ha^{-1}. With an average of 35.3 Mg C ha^{-1} at 13 years of age, this corresponds to the C stocks for similar-aged secondary forests. With solid quantitative foundations and research back-up for assessing C sequestration potential of homegardens and with enabling policies put in place for rewarding landowners for carbon credits (see Alavalapati et al. 2004), the homegardens of future are bound to have a much better value and prestige than today.

Although a wide spectrum of goods and services are obtained from these traditional land use systems, there is yet incomplete understanding about their values. Conventional methods (discount rates, net present values, shadow prices, etc.) are inadequate for realistic valuation of non-conventional products and services provided by homegardens. Equally unsatisfactory is the situation regarding sustainability assessment.

Clearly, managerial information is lacking in respect of coverage, productivity and temporal changes in woody/other plant use. More focus should, therefore, be given to compiling the vital statistics of systems at local, regional, national and global levels (Nair 2001). In spite of the small holding sizes of homegardens, there seems to be scope for GIS applications for preparing databases at different scales (local, regional, national and global) and different intensities in order to provide a sound hierarchical framework for extrapolating from field sites to recommendation domains.

With relatively more of managerial inputs and judicious selection of species, homegardens may help to retain some of the advantages of both traditional land use and commercial farming practices. A schematic illustration of some of the opportunities is presented in Figure 1. We hope that this could provide a basic framework for future research for improvement of homegardens.

Conclusions

Homegardening is a time-tested example of sustainable, multispecies, agroforestry land-use, practiced as a subset of the farming system, predominantly in lowland humid tropics. The high structural and floristic diversity of tropical homegardens is a reflection of the unique biophysical environment and socio-cultural factors under which they exist. As assemblages and repositories of a vast number of plants in small parcels of land around the home in direct and constant interaction with its owners, the homegardens fulfil specific economic, social, and cultural needs of the individual owners and provide biological conservation, carbon sequestration, and such other intangible yet valuable benefits to the society. With their ecological similarities to natural forest ecosystems, homegardens act as insurance against pests and disease outbreaks (although specific data are lacking to support this conjecture), and by providing a variety of goods and services for which people may otherwise depend on or destroy forests, they act as a buffer against the strong pressures on natural forests.

Yet, tropical homegardens are an enigma. Their virtues are recognized more by intuition than in measured quantities. Because of that, the policy makers seldom recognize their importance although the practitioners adore them. Fundamental to solving this puzzle and realizing the potential of homegardens as a powerful land use model is a much-needed change in the single-commodity outlook of agronomists, foresters, and other such land-use-discipline specialists. Indeed, this is perhaps asking for too much, because it questions the whole foundations of educational and research set-up. We are neither calling for nor are not unaware of the difficulties in accomplishing such a radical change. It is realistic, however, to call for extending ecological and economic studies to understand, value, and assess such time-tested systems as homegardens, so that these systems are recognized deservedly in policy discussions at regional and national levels, and the experiences from these systems could be exploited for the design of other sustainable land-

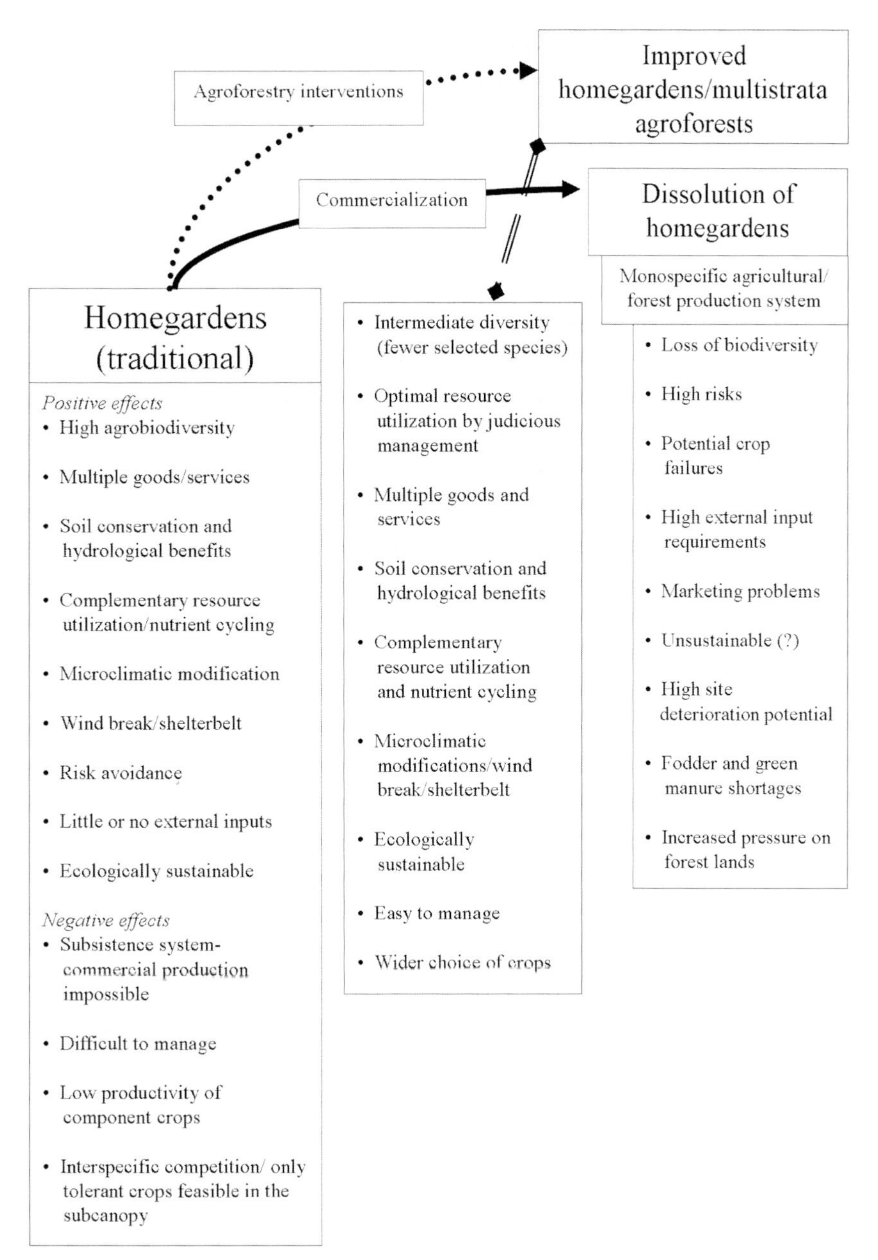

Figure 1. Predicted changes in homegarden composition and relative merits of conventional homegardens, monospecific agricultural/forest production systems and improved homegardens.

use systems. Given the 'track record' of agroforestry research and accomplishments during the past quarter century, we certainly hope that homegarden research will attract its due share of attention in years to come. Because of the small plot sizes of homegardens, 'scaling up' of their virtues may not be feasible in terms of extending areas under the practice, but the principles of their functioning could be used as foundations for the design of improved agroforestry practices.

End Note

1. Mateer S., 1883. 'Native Life in Travancore.' W.H. Allen & Co, London, 450 p.; Logan W., 1906, 'Malabar Manual' (2 vols). Asian Educational Services, 2nd ed, Madras, India 772 p.; Nagam A., 1906, 'Travancore State Manual Vol III.' Travancore Government Press, Trivandrum, Kerala, 700 pp.

References

Abdoellah O.S. 1990. Homegardens in West Java and their future development. pp. 69–79. In: Landauer K. and Brazil M. (eds), Tropical Homegardens, United Nations University Press, Tokyo.

Abdoellah O.S., Takeuchi K., Perikesit G.B. and Hadikusumah H.Y. 2001. Structure and function of homegarden revisited. pp. 167–185. In: Proc. First Seminar toward Harmonisation between Development and Environmental Conservation in Biological Production. JSPS-DGHE Core University Program in Applied Biosciences. The University of Tokyo, Tokyo.

Agelet A., Angels B.M., Valles J. 2000. Homegardens and their role as a main source of medicinal plants in mountain regions of Catalonia (Iberian peninsula). Econ Bot 54: 295–309.

Alavalapati J.R.R., Shrestha R.K., Stainback G.A. and Matta J.R. 2004. Agroforestry development: An environmental economic perspective (This volume).

Anderson E.N .1993. Gardens in tropical America and tropical Asia. Biótica (nueva época) 1: 81–102.

Arifin H.S., Sakamoto K. and Chiba K. 1997. Effects of the fragmentation and change of the social and economical aspects on the vegetation structure in the rural homegardens of West Java. J Jap Inst. Landscape Architect 60: 489–494.

Arnold J.E.M. 1987. Economic considerations in agroforestry. pp. 173–190. In: Steppler H.A. and Nair P.K.R. (eds), Agroforestry: A Decade of Development, ICRAF, Nairobi.

Asfaw Z. and Woldu Z. 1997. Crop associations of homegardens in Welayta and Gurage in southern Ethiopia. Sinet (an Ethiopian J Sci) 20: 73–90.

Ashokan P.K. and Kumar B.M. 1997. Cropping systems and their water use. pp. 200–206. In: Thampi K.B., Nayar N.M. and Nair C.S. (eds), The Natural Resources of Kerala. World Wide Fund for Nature – India, Kerala State Office, Thiruvananthapuram 14, Kerala, India.

Backes M.M. 2001. The role of indigenous trees for the conservation of biocultural diversity in traditional Agroforestry land use systems: The Bungoma case study. Agroforest. Syst 52: 119–132.

Benjamin T.J., Montañez P.L., Jiménez J.J.M. and Gillespie A.R. 2001. Carbon, water and nutrient flux in Maya homegardens in the Yucatan peninsula of México. Agroforest Syst 53: 103–111.

Black G.M., Somnasang P., Thamathawan S. and Newman J.M. 1996. Cultivating continuity and creating change: women's homegarden practices in northeastern Thailand. Multi-cultural considerations from cropping to consumption. Agriculture Human Values 13: 3–11.

Brownrigg L. 1985. Home Gardening in International Development: What the Literature Shows. The League for International Food Education, Washington, DC, 330 pp.

Caballero K. 1992. Mayan homegardens: past, present and future. Etnóecolgica 1: 35–54.

Caron C.M. 1995. The role of nontimber tree products in household food procurement strategies: profile of a Sri Lankan village. Agroforest Syst 32: 99–117.

Ceccolini L. 2002.The homegardens of Soqotra island, Yemen: an example of agroforestry approach to multiple land use in an isolated location. Agroforest Syst 56: 107–115.

Christanty L. 1990. Homegardens in tropical Asia with special reference to Indonesia. pp. 9–20. In: Landauer K. and Brazil M. (eds), Tropical Homegardens. United Nations University Press, Tokyo.

Christanty L., Abdoellah O.L., Marten G.G. and Iskandar J. 1986. Traditional agroforestry in West Java: the *Pekaranagan* (homegarden) and *Kebun-Talun* (annual-perennial rotation) cropping systems. pp. 132–158. In: Marten G.G. (ed.), Traditional Agriculture in South East Asia. Westview Press, Boulder, CO.

Clarke W.C. and Thaman R.R. (eds) 1993. Agroforestry in the Pacific Islands. United Nations University, Tokyo, 224 pp.

De Clerck F.A.J. and Negreros-Castillo P. 2000. Plant species of traditional Mayan homegardens of Mexico as analogs for multistrata agroforests. Agroforest Syst 48: 303–317.

Delaney M .and Roshetko J.M. 1999. Field test of carbon monitoring methods for homegardens in Indonesia. Field Tests of Carbon Monitoring Methods in Forestry Projects, pp. 45–51. Winrock International, Arlington, VA.

Drescher A.W. 1996. Management strategies in African homegardens and the need for new extension approaches. pp. 231–245. In: Heidhues F. and Fadani A. (eds), Food Security and Innovations – Successes and Lessons Learned, Peter Lang, Frankfurt.

Dury S., Vilcosqui L. and Mary F. 1996. Durian trees (*Durio zibethinus* Murr.) in Javanese homegardens: their importance in informal financial systems. Agroforest Syst 33: 215–230.

Esquivel M. and Hammer K. 1992. The Cuban homegarden 'conuco': a perspective environment for evolution and in situ conservation of plant genetic resources. Genet Resour Crop Evol 39: 9–22.

Ewel J.J. 1999. Natural systems as models for the design of sustainable systems of land use. Agroforest Syst 45: 1–21.

Fassbender H. 1993 Modelos Edafológicos de Sistemas Agroforestales, 2nd Ed., CATIE/GTZ, Turrialba, 530 pp.

Fernandes E.C.M. and Nair P.K.R. 1986. An evaluation of the structure and function of tropical homegardens. Agr Syst 21: 279–310.

Gajaseni J. and Gajaseni N. 1999. Ecological rationalities of the traditional homegarden system in the Chao Phraya Basin, Thailand. Agroforest Syst 46: 3–23.

George S.J. and Kumar B.M. 1998. Litter dynamics and soil fertility improvement in silvopastoral systems of the humid tropics of Southern India. Intl Tree Crops J 9: 267–282.

Gillespie A.R., Knudson D.M. and Geilfus F. 1993. The structure of four homegardens in the Peten, Guatemala. Agroforest Syst 24: 157–170.

Gliessman S.R. (ed.) 1990. Integrating trees into agriculture: the homegarden agroecosystem as an example of agroforestry in the tropics. Agroecology: Researching the Ecological Basis for Sustainable Agriculture, pp. 160–168. Ecological Studies 78, University of California, Santa Cruz, CA.

Gopikumar K., Hegde R. and Babu L.C. 2001. Decomposition and nutrient release pattern of leaf litters of mangium (*Acacia mangium* Willd.) Indian J Agroforest 3: 11–22.

Guillaumet J.L., Grenand P., Bahri S., Grenand F., Lourd M,. Santos A.A. dos and Gely A. 1990. The homegarden orchards of the central Amazon: an example of space utilization. Turrialba 40: 63–81.

Heal O.W., Anderson J.M. and Swift M.J. 1997. Plant litter quality and decomposition: an historical overview. pp. 3–30. In: Cadish G. and Giller K.E. (eds), Driven by Nature: Plant Quality and Decomposition, CAB International, Wallingford.

High C. and Shackleton C.M. 2000. The comparative value of wild and domestic plants in homegardens of a South African rural village. Agroforest Syst 48: 141–156.

Hochegger K. 1998. Farming like the forest: traditional homegarden systems in Sri Lanka. Tropical Agroecology, Margraf Verlag, Weikersheim, 203 pp.

Hutterer K.L. 1984. Ecology and evolution of agriculture in Southeast Asia. In: Rambo A.T. and Sajise P.E. (eds), An Introduction to Human Ecology Research on Agricultural Systems in Southeast Asia. University of Philippines, Los Banos.

Issac S.R. and Nair M.A. 2002. Litter decay: weight loss and N dynamics of jack leaf litter on open and shaded sites. Indian J Agrofor 4: 35–39.

Jacob V.J. and Alles W.S. 1987. The Kandyan gardens of Sri Lanka. Agroforest Syst 5: 123–137.

Jensen M. 1993a. Productivity and nutrient cycling of a Javanese homegarden. Agroforest Syst 24: 187–201.

Jensen M. 1993b. Soil conditions, vegetation structure and biomass of a Javanese homegarden. Agroforest Syst 24: 171–186.

John J. and Nair M.A. 1999. Crop-tree inventory of the homegardens in Southern Kerala. J. Trop Agric 37: 110–114.

Jordan C.F. 1985. Nutrient Cycling in Tropical Forest Ecosystems. John Wiley and Sons, Chichester, 189 pp.

Jose D. and Shanmugaratnam N. 1993. Traditional homegardens of Kerala: a sustainable human ecosystem. Agroforest Syst 24: 203–213.

Karyono 1990. Homegardens in Java; their structure and function. pp. 138–146. In: Landauer K. and Brazil M. (eds), Tropical Homegardens, United Nations University Press, Tokyo.

Kehlenbeck K. and Mass B.L. 2004. Crop diversity and classification of homegardens in Central Sulawesi, Indonesia. Agroforest Syst (in press).

Kerala State Land Use Board (KSLUB) 1995. Land Resources of Kerala State. Thiruvananthapuram, Kerala, 209 pp.

Krishnankutty C.N. 1990. Demand and supply of wood in Kerala and their future trends. Research Report 67. Kerala Forest Research Institute, Peechi, Kerala, 66 pp.

Kumar B.M. and Divakara B.N. 2001. Proximity, clump size and root distribution pattern in bamboo: A case study of *Bambusa arundinacea* (Retz.) Willd., Poaceae, in the Ultisols of Kerala, India. J Bamboo and Rattan 1: 43–58.

Kumar B.M., George S.J., Jamaludheen V. and Suresh T.K. 1998. Comparison of biomass production, tree allometry and nutrient use efficiency of multipurpose trees grown in wood lot and silvopastoral experiments in Kerala, India. Forest Ecol Manag 112: 145–163.

Kumar B.M, George S.J. and Chinnamani S. 1994. Diversity, structure and standing stock of wood in the homegardens of Kerala in peninsular India. Agroforest Syst 25: 243–262.

Kumar S.S., Kumar B.M., Wahid P.A., Kamalam N.V. and Fisher R.F. 1999. Root competition for phosphorus between coconut, multipurpose trees and kacholam (*Kaempferia galanga*) in Kerala, India. Agroforest Syst 46: 131–146.

Landauer K. and Brazil M. (eds) 1990. Tropical Homegardens. United Nations University Press, Tokyo, 257 pp.

Lavelle P., Blanchart E., Martin A., Martin S., Spain A., Toutain F., Barois I. and Schaefer R. 1993. A hierarchical model for decomposition in terrestrial ecosystems: application to soils of the humid tropics. Biotropica 25: 130–150.

Leith H. 1975. Primary production of major vegetation units of the world. pp. 203–215. In: Leith H. and Whittaker R.H. (eds), Primary Productivity of the Biosphere, Springer-Verlag, New York.

Leiva J.M., Azurdia C., Ovando W., López E. and Ayala H. 2002. Contributions of homegardens to in situ conservation in traditional farming systems – Guatemalan component. pp. 56–72. In: Watson J.W. and Eyzaguirre P.B. (eds), Homegardens and In Situ Conservation of Plant Genetic Resources in Farming Systems. Proc Second International Homegarden Workshop, 17–19 July 2001. Witzenhausen, Germany, International Plant Genetic Resources Institute, Rome.

Leuschner W.A. and Khaleque K. 1987. Homestead Agroforestry in Bangladesh. Agroforest Syst 5: 139–151.

Levasseur V. and Olivier A. 2000. The farming system and traditional agroforestry systems in the Maya community of San Jose, Belize. Agroforest Syst 49: 275–288.

Mathew T., Kumar B.M., Suresh Babu K.V. and Umamaheswaran K. 1996. Evaluation of some live standards for black pepper. J Plant Crops 24: 86–91.

Mendez V.E., Lok R., Somarriba E. 2001. Interdisciplinary analysis of homegardens in Nicaragua: micro-zonation, plant use and socioeconomic importance. Agroforest Syst 51: 85–96.

Menon T.M., Tyagi D. and Kulirani B.F. (eds) 2002. People of India: Kerala: Volume XXVII. Affiliated East-West Press for Anthropological Survey of India, New Delhi, 1704 pp.

Michon G. and deForesta H. 1999. Agro-Forests: Incorporating a forest vision in agroforestry. pp. 381–406. In: Buck L.E., Lassoie J.P. and Fernandes E.C.M. (eds), Agroforestry in Sustainable Agricultural Systems, CRC Press, Boca Raton, FL.

Michon G. and Mary F. 1990. Transforming traditional homegardens and related systems in West java (Bogor) and West Sumatra (Maninjau). pp. 169–185. In: Landauer K. and Brazil M. (eds), Tropical Homegardens. United Nations University Press, Tokyo.

Michon G. and Mary F. 1994. Conversion of traditional village gardens and new economic strategies of rural households in the area of Bogor, Indonesia. Agroforest Syst 25: 31–58.

Michon G., Bompard J., Hecketsweiler P. and Ducatillion C. 1983. Tropical forest architectural analysis applied to agroforests in the humid tropics, the example of traditional village agroforestry in West Java. Agroforest Syst 1: 117–129.

Millat-e-Mustafa M.D., Hall J.B., Teklehaimanot Z. 1996. Structure and floristics of Bangladesh homegardens. Agroforest Syst 33: 263–280.

Molina M.R., Noguera A., Dary O., Chew F., Valverde C. 1993. Principal micronutrient deficiencies in Central America. Food Nutr Agric 7: 26–33.

Montagnini F. and Nair P.K.R. 2004. Carbon sequestration: An underexploited environmental benefit of agroforestry systems (This volume).

Mpoyi K., Lukebakio N., Kapende K. and Paulus J. 1994. Inventaire de la flore domestique des parcelles d'habitation. Cas de Kinshasa (Zaire). Revue de Medecine et Pharmacopee Africaine 8: 55–66.

Muschler R.G. 2001. Shade improves coffee quality in a suboptimal coffee-zone of Costa Rica. Agroforest Syst 51: 131–139.

Nair M.A. and Sreedharan C. 1986. Agroforestry farming systems the homesteads of Kerala, southern India. Agroforest Syst 4: 339–363.

Nair P.K.R. 1979. Intensive Multiple Cropping with Coconuts in India: Principles, Programmes and Prospects. Verlag Paul Parey, Berlin, 147 pp.

Nair P.K.R. 1987. Agroforestry systems inventory. Agroforest Syst 5: 301–317.

Nair P.K.R. (ed.) 1989. Agroforestry Systems in the Tropics. Kluwer, Dordrecht, 664 pp.

Nair P.K.R. 1993. An Introduction to Agroforestry. Kluwer, Dordrecht, 499 pp.

Nair P.K.R. 2001. Do tropical homegardens elude science, or is it the other way around? Agroforest Syst 53: 239–245.

Nair P.K.R., Buresh R.J., Mugendi D.N. and Latt C.R. 1999. Nutrient cycling in tropical agroforesty systems: Myths and science. pp. 1–31. In: Buck L.E., Lassoie J.P. and Fernandes E.C.M. (eds), Agroforestry in Sustainable Agricultural Systems. CRC Press, Boca Raton, FL.

Niñez V.K. 1987. Household gardens: theoretical and policy considerations. Agr Syst 23: 167–186.

Odum E.P. 1969. The strategy of ecosystem development. Science 164: 24–270.

Olson J.S. 1963. Energy storage and balance of production and decomposers in ecological systems. Ecology 44: 322–331.

Padoch C. and de Jong W. 1991. The housegardens of Santa Rosa: diversity and variability in an Amazonian agricultural system. Econ Bot 45: 166–175.

Palm C.A., Gachengo C.N., Delve R.J., Cadisch G. and Giller K.E. 2001. Organic inputs for soil fertility management in tropical agroecosystems: application of an organic resource database. Agric Ecosyst Environ 83: 27–42.

Puri S. and Nair P.K.R. 2004. Agroforestry research for development in India: 25 years of experiences of a national program (This volume).

Quiroz C., Gutiérrez M., Rodriguez D., Perz D., Ynfante J., Gamex J., Perz de Fernandex T., Marques A. and Pacheco W. 2002. Homegardens and in situ conservation of Agrobiodiversity – Venezuelan component. pp. 73–82. In: Watson J.W. and Eyzaguirre P.B. (eds), Homegardens and in situ Conservation of Plant Genetic Resources in Farming Systems. Proc Second International Homegarden Workshop, 17–19 July 2001 Witzenhausen, Germany. International Plant Genetic Resources Institute, Rome.

Rao M.R., Palada M.C. and Becker B.N. 2004. Medicinal and aromatic plants in agroforestry systems (This volume).

Rico-Gray V., Chemas A. and Mandujano S. 1991. Use of tropical deciduous forest species by the Yucatan Maya. Agroforest Syst 14: 149–161.

Rico-Gray V., Garcia F.J.G., Chemas A., Puch A. and Sima P. 1990. Species composition, similarity, and structure of Mayan homegardens in Tixpeual and Tixcacaltuyub, Yucatan, Mexico. Econ Bot 44: 470–487.

Roshetko M., Delaney M., Hairiah K. and Purnomosidhi P. 2002. Carbon stocks in Indonesian homegarden systems: Can smallholder systems be targeted for increased carbon storage? Amer J Altern Agric 17: 125–137.

Rugalema G.H., O'Kting' Ati A., Johnsen F.H. 1994a. The homegarden agroforestry system of Bukoba district, North-Western Tanzania. Farming system analysis. Agroforest Syst 26: 53–64.

Rugalema G.H., Johnsen F.H., Rugambisa J. 1994b. The homegarden agroforestry system of Bukoba district, North-Western Tanzania. 2. Constraints to farm productivity. Agroforest Syst 26: 205–214.

Rugalema G.H., Johnsen F.H., O'Kting' Ati A., Minjas A. 1995. The homegarden agroforestry system of Bukoba district, northwestern Tanzania. An economic appraisal of possible solutions to falling productivity. Agroforest Syst 28: 227–236.

Ruthenberg H. 1980. Farming Systems in the Tropics (3rd ed.) Clarendon Press, Oxford, 424 pp.

Shanavas A. and Kumar B.M. 2003 Fuelwood characteristics of tree species in the homegardens of Kerala, India. Agroforest Syst 58: 11–24.

Shankar A.V., Gittelsohn J., Pradhan E.K., Dhungel C. and West K.P. Jr. 1998. Homegardening and access to animals in households with xerophthalmic children in rural Nepal. Food Nut Bull 19: 34–41.

Shujauddin N. and Kumar B.M. 2003. *Ailanthus triphysa* at different densities and fertilizer regimes in Kerala, India: Biomass productivity, nutrient export and nutrient use efficiency. Forest Ecol Manag 180: 135–151.

Singh G.B. 1987. Agroforestry in the Indian subcontinent: past, present and future. pp. 117–138. In: Steppler H.A. and Nair P.K.R. (eds), Agroforestry: A Decade of Development, International Centre for Research in Agroforestry, Nairobi.

Soemarwoto O. 1987. Homegardens: A traditional agroforestry system with a promising future. pp. 157–170. In: Steppler H.A. and Nair P.K.R. (eds), Agroforestry: A Decade of Development, ICRAF, Nairobi.

Soemarwoto O. and Conway G.R. 1991. The Javanese homegarden. J Farming Syst Res Extn 2: 95–118.

Terra G.J.A. 1953. Mixed-garden horticulture in Java. Malayan J Trop Geogr 1: 33–43.

Terra G.J.A. 1958. Farm systems in southeast Asia. Neth J Agric Sci 6: 157–182.

Thaman R.R. 1990. Mixed homegardening in the Pacific Islands: present status and future prospects. pp. 41–65. In: Landauer K. and Brazil M. (eds), Tropical Homegardens, United Nations University Press, Tokyo.

Torquebiau E. 1992. Are tropical agroforestry homegardens sustainable? Agric Ecosyst Environ 41: 189–207.

Vitousek P.M. and Reiners W.A. 1975. Ecosystems succession and nutrient retention: a hypothesis. BioScience 25: 376–381.

Vogl C.R., Vogl-Lukraser B. and Caballero J. 2002. Homegardens of Maya Migrants in the district of Palenque, Chiapas, Mexcio: Implications for Sustainable Rural Development. pp. 1–12. In: Stepp J.R., Wyndham F.S. and Zarger R.K. (eds), Ethnobiology and Biocultural Diversity, University of Georgia Press, Athens, GA.

Watson J.W. and Eyzaguirre P.B. (eds) 2002. Homegardens and *in situ* Conservation of Plant Genetic Resources in Farming Systems. Proc Second International Homegarden Workshop, 17–19 July 2001. Witzenhausen, Germany; International Plant Genetic Resources Institute, Rome, 184 pp.

Wezel A. and Bender S. 2003. Plant species diversity of homegardens of Cuba and its significance for household food supply. Agroforest Syst 57: 37–47.

Whitten T., Soeriaatmadja R.E. and Afiff S.A. 1996. The Ecology of Java and Bali. The ecology of Indonesia series 2, Perplus Edition, Jakarta, 969 pp.

Wichern F., Richter C. and Joergensen R.G. 2003. Soil fertility breakdown in a subtropical South African vertisol site used as a homegarden. Biol Fertil Soils 37: 288–294.

Wickramasinghe A. 1996. The non forest wood fuel resources of Sri Lanka. Wood Energy News 11: 14–18.

Wiersum K.F. 1982. Tree gardening and taungya on Java. Agroforest Syst 1: 53–70.

Wiersum K.F. 2004. Forest Gardens as an 'Intermediate' Land-use System in the Nature-Culture Continuum: Characteristics and Future Potential (This volume).

Yamada M. and Gholz H.L. 2002. An evaluation of agroforestry systems as a rural development option for the Brazilian Amazon. Agroforest Syst 55: 81–87.

Yamamoto Y., Kubota N., Ogo T. and Priyono 1991. Changes in the structure of homegardens under different climatic conditions in Java Island. Japanese J Trop Agric 35: 104–117.

Zemede A. and Ayele N. 1995. Homegardens in Ethiopia: characteristics and plant diversity. Sinet (an Ethiopian J Sci) 18: 235–266.

Biological and Ecological issues

Agroforestry Systems **61:** 155–165, 2004.

Nature vs. nurture: managing relationships between forests, agroforestry and wild biodiversity

J.A. McNeely
The World Conservation Union (IUCN), 1196 Gland, Switzerland; e-mail: jam@iucn.org

Key words: Coffee, Conservation, History, Invasive species

Abstract

Many agroforestry systems are found in places that otherwise would be appropriate for natural forests, and often have replaced them. Humans have had a profound influence on forests virtually everywhere they both are found. Thus 'natural' defined as 'without human influence' is a hypothetical construct, though one that has assumed mythological value among many conservationists. Biodiversity is a forest value that does not carry a market price. It is the foundation, however, upon which productive systems depend. The relationship between agroforestry and the wild biodiversity contained in more natural forests is a complicated one, depending on the composition of the agroforestry system itself and the way it is managed. Complex forest gardens are more supportive of biodiversity than monocrop systems, shade coffee more than sun coffee, and systems using native plants tend to be more biologically diverse. Nonnative plants, especially potentially invasive alien species, threaten biodiversity and need to be avoided. The relationship between forests, agroforestry and wild biodiversity can be made most productive through applying adaptive management approaches that incorporate ongoing research and monitoring in order to feed information back into the management system. Maintaining diversity in approaches to management of agroforestry systems will provide humanity with the widest range of options for adapting to changing conditions. Clear government policy frameworks are needed that support alliances among the many interest groups involved in forest biodiversity.

Introduction

This paper addresses the changing perceptions of relationships between people and nature and how this affects forest management. It begins by reviewing some basic concepts, outlining the history of human impacts on forest ecosystems, and discussing the relationship between biodiversity conservation and agroforestry. This leads to the identification of several management issues and a call for closer collaboration between agroforestry and conservation interests.

In western culture, 'nature' is often considered to be that which operates independently of people (Hoerr 1993), and a major focus of development has been to bring nature under greater human control. In fact, 'progress' is often measured by technological innovations that have enabled humans to gain a greater share of the planet's productivity. Conservation, on the other hand, has been based on the idea of sequestering the largest possible tracts of nature in a state of imagined innocence as national parks and other kinds of protected areas. Forests that are 'pristine' or 'virgin' or 'primary' are thus given particularly high value for conservation, and considered likely to have high biological diversity (Gomez-Pompa and Kaus 1992).

Despite the dominance of this view of nature, work in ecology (Sprugel 1991), paleontology (Martin and Klein, 1984), forestry (Poffenberger 1990), history (Boyden 1992; Ponting 1992; Flannery 2001), archaeology (Audric 1972; Raven-Hart 1981), anthropology (Denevan 1992a; Roosevelt 1994), and ethics (Taylor 1986) is calling into question the separation of people from nature, supporting instead the age-old view of many cultures that people are part of nature and that the biodiversity – that is, the variety of genes, species, and ecosystems – found in today's forests result from a combination of cyclical ecological and climatic processes and past human action. Evidence is building to support the view that very few of today's forests anywhere in the world can be considered 'pristine,' 'virgin,' or even 'primary,' and that conserving biological diversity requires a far more subtle appreciation of both human and natural influences. McNeely (1994) drew from recent studies to suggest four basic conclusions about the history of forests and biodiversity:

i. Humans have been a dominant force in the evolution of today's forests.

ii. As humans develop more sophisticated technology, their impact on forests increases until forests are degraded to the long-term detriment of the overexploiting society.

iii. Overexploitation is usually followed by a culture change that may reduce human pressure, after which some forests may return to a highly productive and diverse, albeit altered, condition, and others may be permanently altered to much less productive and diverse conditions.

iv. The best approach to conserving forests and their biodiversity is through a variety of management approaches ranging from strict protection through intensive use, with a careful consideration of the distribution of costs and benefits of each.

Different systems of management may enhance or reduce forest diversity. Completely excluding human intervention from species-rich communities found in hilly, high rainfall areas of the tropics, for example, may reduce genetic and species diversity by changing the mix of successional stages, or it may help to conserve species that are confined to old-growth stages of succession. Further, the sheer number of species is not necessarily a useful measure. Australia, California, Hawaii and New Zealand, for example, have more species now than ever before. Many of these are nonnative and maintained by human action, but some are 'invasive alien species' that threaten the native species (McNeely et al. 2001). The notions of 'natural' vegetation or ecosystem processes, therefore, are still useful goals in forest management, but they should be revised to recognize that a range of ecosystems can legitimately be considered 'natural' (Sprugel 1991), and nearly all of them have been significantly influenced by people. Managing agroforestry systems to address biodiversity concerns will both enhance productivity and contribute to conservation objectives.

The role of myth in conserving biodiversity

The idea that nature exists as something separate from people has become part of the mythology of industrial society. The Oxford Dictionary (Pollard 1994) defines 'myth' as a story arising from an unknown source and containing ideas or beliefs that purport to explain natural events without a basis in fact. It appears that this mythic image of nature is essential to the psychological well being of modern humanity (Campbell 1985; Jung 1964), whose industrial approach to conquering nature can be so destructive. Nonindustrial societies have different myths, often treating what industrial society calls 'nature' as a set of very real threats to human existence. Campbell (1985) suggests that societies that cherish and keep their myths alive 'will be nourished from the soundest, richest strata of the human spirit,' and that myths often contain many elements that ring true to the people who believe them.

The myth of the virgin forest and nature untouched by humans is above all a myth of urban-dwelling people who are well separated from the reality of the forest. Those who actually live in the forest are faced with rich diversity that is mostly hidden far above or in the heavy cover of the ground vegetation where various dangers lurk (Campbell 1985). The forest village is a place of safety and stability that is separated from the forest, though fields often are hacked from the forest to bring human order into the chaos of nature. Agroforests or agroforestry systems may bring domestication to the disorder of wild nature. This is supported by various kinds of mythology that help to explain observed phenomena, such as how people first learned to domesticate plants and animals (see, for example, Suzuki and Knudtson 1992). Rites and rituals are carried out by forest-dwelling peoples to help reinforce the myths, while the myths provide the mental support for rites and rituals, helping to ensure that children are made well aware of the social and natural environment into which they must fit in order to become a competent member of society (Campbell 1985). The corresponding rituals of modern urban society in relation to the myth of nature may be watching nature programs on television, giving money to organizations that claim to conserve nature, and visiting protected areas, many of which assume almost a sacred character (Putney et al. 2003).

The western mythical vision of an untouched wilderness has permeated global policies and politics in resource management (Gomez-Pompa and Kaus 1992). But this view of forests is based on an outmoded ecological perspective, and on misunderstanding of the historical relationship between people and forests, and the role people have played in maintaining biodiversity in forested habitats. A brief review of certain episodes in the history of people, forests, and biodiversity will show how humans have affected the birth, growth, death, and renewal cycle of forests in a variety of ways, with various outcomes in different parts of the world (Holling 1978). A more sophisticated understanding of the relationship between forests and people will not replace the myth of the pristine forest but may lead to better forest conservation and better agroforestry.

Connections: the history of people, forests, and biodiversity

The conclusion that the world has few, if any forests which have not been significantly influenced by people is supported by evidence from many parts of the world. Cycles of human activity that have affected biodiversity are evident in tropical Asia, the Western Hemisphere, and Europe and the Mediterranean.

Tropical Asia

Tropical Asia was one of the heartlands of shifting cultivation (Solheim 1972), a cyclical land use that has had a profound influence on habitats throughout the region over the past 10 000 years. Spencer (1966), in a detailed study of the impact of shifting cultivation in Asia, concluded that 'most of the mature forests of the Orient today are not virgin forests in the proper sense, but merely old forests that have reached a fairly stable equilibrium of ecological succession after some earlier clearing by human or natural means. In some areas it is possible that old forests are not secondary forests or even tertiary forests, but forests of some number well above three.'

Traditional systems of shifting cultivation often include many agroforestry elements. The older fields contain a high proportion of fruit trees, which are attractive to primates, squirrels, hornbills, and a variety of other animals. In addition, large mammals flourish, with elephants, wild cattle, deer, and wild pigs all feeding in the abandoned fields; tigers, leopards, and other predators are in turn attracted by the herbivores. Wharton (1968) has provided convincing evidence that the distribution of the major large mammals of Southeast Asia is highly dependent on shifting cultivation. Mature tropical forests conceal most of their edible products high in the canopy beyond the reach of the terrestrial herbivores, while forest clearings bring the forest's productivity down to where it can be reached by hungry browsers. The earlier successional stages are also faster growing, and therefore more productive than the later stages of the cycle as the forest becomes more mature.

The conclusion that shifting cultivation has benefited both people and forest depends, however, on the practice being carried out in a sustainable manner, which today is rare. Shifting cultivation can become maladaptive in at least three ways: by an increase in human population that causes old plots to be recultivated too soon; by inept agricultural practices such as cultivating the land for so long that productivity declines and persistent weeds such as *Imperata cylindrica* become established; and by attempting to cultivate forests that are too dry, so recovery is slow and the danger of cataclysmic fire is great (Geertz 1963). Sometimes the three factors work together to destroy wide areas of tropical forest.

Most shifting cultivation has taken place in the hills, where vegetation dries out more quickly and updrafts help fan the flames among the cut vegetation. The lowlands, many of which were seasonally flooded or otherwise difficult to burn, remained relatively intact during the early years of agriculture and were used mostly for hunting, fishing, and gathering of tubers and other plants. With the development of irrigation and agricultural surpluses, all that changed, and new civilizations flourished in lowlands where wet rice could be grown, often leading to substantial forest clearance.

Sumatra, Indonesia, for example, was the centre of the rice-growing Sriwijaya civilization, which spread its influence from what is now Palembang throughout Southeast Asia. Following the collapse of the Sriwijaya civilization in the 14th century, forests quickly reclaimed much of the landscape that had been transformed by Sriwijaya (Schnitger 1964); parts of these ancient farmlands are now so important for biodiversity that they are included in Indonesia's protected area system. Some of Indonesia's most remote protected areas are proving to contain important Sriwijanan archaeological sites, as in Kalimantan's Kayan Mentarang Nature Reserve, an indication of substantial historical human activity in forests noted today for their high biodiversity.

In tropical Asia, precolonial forest management was primarily in the hands of the people who lived in the forests. The colonial era brought forests into the global market system, leading to the nationalization of many forests, the import of forest management technology from Europe, and the loss of traditional resource management practices that maintained biodiversity in forests, though not explicitly designed to do so (Poffenberger 1990). The post-colonial period perhaps has been worse with few forest departments able to control exploitation from concession holders or to work constructively with forest-dwelling people. Poffenberger (1990) points out that conflicts between state land management policies and locally operating forest-use systems is a major cause of forestland mismanagement throughout southeast Asia. With both colonial and post-colonial regimes, forest departments in Myanmar (Burma), India, Indonesia, and Sri Lanka, were designed to generate revenue for the state rather than provide direct benefits to rural communities that were practicing various forms of agroforestry. He concludes that radical changes in tenure rights and lack of clarity over ownership of tree and forest products are key factors in understanding the speed with which Asian forests have been depleted and why so many species are threatened today. Agroforestry that returns control of at least some forests to local people can help restore sustainable-use management practices that can complement biodiversity conservation in protected areas.

The western hemisphere

The vast boreal forest-covered wilderness of North America is often considered to be natural. But people have occupied this forest from its very beginnings, as the great ice sheets withdrew northwards at the end of the Pleistocene. New studies have established that Native Americans in northern Alberta regularly and systematically burned habitats to influence the local distribution and relative abundance of plant and animal resources (Hoffecker et al. 1993). Trees played a crucial role in the initial occupation of the Western Hemisphere. It is now believed that the critical environmental variable that enabled the first humans to move from Asia into North America was the reappearance of trees in Alaskan river valleys, which provided essential fuel sources as glaciers withdrew at the end of the Pleistocene around 11 000–12 000 years ago. As Lewis and Ferguson (1988) show, this pyrotechnology is similar to that reported for hunter-gatherers elsewhere in the world, creating an overall fire mosaic that characterizes the northern boreal forests. Cross-cultural comparisons of these practices with those in other parts of North America, as well as in several parts of Australia, illustrate functionally parallel strategies in the ways that hunter-gatherers employed habitat fires, specifically in the maintenance of 'fire-yards' and 'fire corridors' in widely separated and different kinds of biological zones (Flannery 1994; 2001).

As they moved further south, the immigrants from Asia continued to modify the American forests. In reviewing the evidence, Denevan (1992b) concluded that pre-Columbian human settlement had modified forest extent and composition, expanded grasslands, and rearranged the local landscape through countless artificial earthworks. Agricultural fields, towns, and trails were common, having local impacts on soil, microclimate, hydrology, and wildlife. These early immigrants had a significant impact on biodiversity as well, with some 34 genera of large mammals becoming extinct around the time of first human occupation of the continent (Martin and Klein 1984).

Agroforestry may well have been one of the earliest vehicles of plant domestication in Central America. For example, Native American populations are known to have cultivated a large number of plants and domesticated them for their starch-rich belowground parts such as rhizomes and tubers. Suggestions that tropical forests, the likely source of many of these crops, were an early and influential centre of plant husbandry have long been controversial because the organic remains of roots and tubers are poorly preserved in the archaeological sediments from the humid tropics. But Piperno et al. (2000) reported the occurrence of starch grains identifiable as manioc (*Manihot esculenta*), yams (*Dioscorea* spp.) and arrowroot (*Maranta arundinacea*) on assemblages of plant milling-stones from preceramic horizons at the Aguadulce shelter, Panama, dated between 7000 and 5000 years before present. The artefacts also contain maize (*Zea mays*) starch, indicating that early horticultural systems in this region were mixtures of root and seed crops, as well as tree crops, though the evidence for the latter is circumstantial. The data provide the earliest direct evidence for root crop cultivation in the Americas, and an ancient and independent emergence of plant domestication in the lowland Neotropical forest. This serves as another indication that the composition of presumed 'virgin' or 'untouched' tropical forests has been importantly influenced by people.

By A.D. 800, the Maya had modified 75% of the Yucatan forest, and following the collapse of the classical Mayan civilization shortly thereafter, forest recovery in the central lowlands was nearly complete by the time the Spaniards arrived 700 years later (Whitmore et al. 1990). The Aztecs followed a similar cycle. Examining the association between erosion and pre-Columbian population in Central Mexico, Cook (1949) concluded: 'An important cycle of erosion and deposition accompanied intensive land use by huge primitive populations in central Mexico, and had gone far toward the devastation of the country before the white man arrived.' O'Hara et al. (1993) found three major episodes of erosion in central Mexico during the past 4000 years, correlated to the first arrival of corn (maize)-farming of steep slopes, and a period of heavy population density, all in pre-colonial times. The current composition of the vegetation in Central America thus is the legacy of past civilizations, the heritage of cultivated fields and managed agroforests abandoned hundreds of years ago (Gomez-Pompa and Kaus 1992) as the pre-European population estimated at 22.8 million people crashed (Denevan 1992). Turner and Butzer (1992) estimate that 76% of the human population was eliminated between 1492 and 1650, primarily because of disease, leaving wide areas to revert to tropical forests which today are often considered 'pristine' or 'natural.'

Further south, by about 2000 years ago, fairly complex societies had been established in Amazonia, with large populations, public works, differentiated settlements, elaborate ceremonial art, long-distance trade, and other signs of complex chiefdoms. By the time of first European contact, these chieftains had large domains, maintained by organized large-scale warfare and diplomacy, elite ranking based on descent from deified human ancestors, and long-distance interregional trade and tribute systems (Roosevelt 1994). Archaeological sites indicate that substantial human populations inhabited Amazonia before Europeans arrived, with some shell middens so extensive that they have been used as commercial lime mines, providing fertilizer and road-building material for more than 200 years. Some Bolivian prehistoric archaeological midden sites have been an important soil resource for commercial agriculture in the Brazilian Amazon since the mid-19th century. Recent findings from Brazil's Upper Xingu, the largest area of Amazonian forest still being managed by indigenous peoples, indicate that the region supported a dense human population that substantially altered both forests and wetlands (Heckenberger et al. 2003). This also supports Denevan's (1992) estimate that South America had a pre-colonial population of 24.3 million.

As Roosevelt (1994 p. 9) concludes, 'Historical studies in the floodplains show that the European conquest affected the transition from ancient chiefdoms to indigenous village societies or peasant communities through a long process involving military defeat, decimation, forced migration, enslavement, misogynization, and acculturation. In response to the European invasion, the areas that held the majority of the massive late prehistoric occupations of Amazonia were rapidly emptied of Indians, who now persist in a mosaic with non-Indian communities, mostly at the margins of the basin.' Thus the current Amazonian Indians have a way of life that is a complex indigenous adaptation to conquest and nationalization as well as to Amazonian environments, and the 'virgin' Amazonian forests are at least partly an inadvertent legacy of the European invasion of South America.

Europe and the Mediterranean

The case for Europe is perhaps even more dramatic. The early Holocene of the Mediterranean area was a mixed evergreen and deciduous forest of oaks, beech, pines, and cedars. The forest was eaten away by waves of different civilizations that used the forest and forest lands to further their development objectives, expanding and contracting as the wisdom of their policies was tested. The process of forest clearance was already well underway at the time of Homer in the 9th century B.C., who likened the noise of a battle to 'the din of wood cutters in the glades of a mountain.' Early in the 4th century B.C., Plato, referring to the disappearance of forests in Attica, wrote: 'What now remains compared with what then existed, is like the skeleton of a sick man, all the fat and soft earth having been wasted away, and only the bare framework of the land being left' (quoted in Ponting 1992 p. 76). Civilizations from Bronze Age Crete and Knossos, Mycenaean Greece, Cyprus, Greece, and Rome rose and fell with the forests that supported them (Perlin 1989). Subsequent overgrazing by sheep, cattle, and goats prevented the forests from ever becoming re-established, though agroforestry is now beginning a comeback.

The olive (*Olea europaea*) can be considered the 'flagship species' of the Mediterranean. Developed from a straggly wild relative along the coasts of Syria and Anatolia in the 6th century B.C., it soon became a crop of outstanding economic importance. But it also contributed to significant deforestation, land degradation, and loss of biodiversity, at least of vertebrates. As the richer valley lands were cleared of forests to plant crops, the poorer soils of the hillsides were being planted with olives. The development of Crete between 2500 and 1500 B.C. was supported by the export of timber and olive oil to Egypt, as forest trees were felled and olive trees were planted. But as a result of deforestation, soil accumulated over millennia was being washed from the hillsides in just a few centuries, and the natural wealth of the country was eroded with the soil. The decline of the Cretan forests was mirrored by the same transformation, following in the wake of the axe, the plough, and the olive in their westward progress through all the civilized states of the Mediterranean (Darlington 1969). As a result, much of the evergreen forest in the Mediterranean region was transformed into the brushwood known as *maquis*, which today is maintained by fire. The loss of native forests through conversion to agriculture also had significant impact on biodiversity, with some 90% of the endemic mammalian genera of the Mediterranean becoming extinct after the development of agriculture (Sondaar 1977). While little of the remaining forest can be considered 'natural' in any sense, agroforestry may be used as a tool to reclaim land devoted to low-yield annual crops and contribute to ecological diversity in closed forests. It may help to bring back at least some of the biodiversity that formerly characterized the Mediterranean region – an area that even now is considered a 'biodiversity hotspot' (Myers et al. 2000).

The interface between biodiversity conservation and agroforestry

Conserving biodiversity where people live is a major challenge, especially in the tropical countries. But a growing body of research shows that agroforestry can make a significant contribution to conserving biodiversity in a wide range of settings.

Complex agroforests

Marjokopri and Ruokolainen (2003) compared agroforests (*Tembawang*) in West Kalimantan, Indonesia, to patches of primary forest, which may have been old-growth secondary forests, in terms of successional stage, mode of dispersal, and characteristics of human use of nonplanted tree species. Data were collected for 144 tree species in six agroforests and two natural forest patches. The investigators found that the older agroforests had nearly the same proportions of species of different successional stages and modes of dispersal as natural forests, and emphasized the potential of agroforests in conserving tree species. Nonplanted tree species of the agroforests and natural forests have similar human uses, indicating that the management of these agroforests does not significantly discriminate between species with certain uses.

However, regional differentiation between agroforests and natural forest patches was significant. Late successional species and animal-dispersed species of the agroforests were more widespread geographically than species of the same ecological characteristics found in natural forests. The investigators concluded that although the agroforests are similar to natural forests in terms of numbers of species with different ecological characteristics, the composition of nonplanted tree species in the agroforests is not consistent with natural forests, but over-represents species that are easily dispersed and/or established.

Swidden succession

Ferguson et al. (2003) compared post-agricultural succession along the range of farming activities practiced in Guatemala's northern lowlands: agroforestry, swidden, ranching, and input-intensive monocultures. Using several characteristics of the vegetation, they found that succession was dramatically faster on agroforestry and swidden sites than on pastures or input-intensive monocultures. Overall recruitment was faster for swiddens than for agroforests, but other response variables did not differ significantly between the two treatments.

The investigators' results suggest that the conservation strategy of discouraging swidden agriculture in favor of sedentary, input-intensive agriculture to relieve pressure on old-growth forest may be counterproductive over the long term. Agroforestry may be a more appropriate option for maintaining biodiversity in production landscapes (McNeely and Scherr 2003).

*Shaded coffee (*Coffea *spp.)*

Petit and Petit (2003) examined bird communities associated with 11 natural and human-modified habitats in Panama and assessed the importance of those habitats for species of different vulnerability to disturbance.

Calculating habitat importance scores using both relative habitat preferences and vulnerability indices for all species present, they found that species of moderate and high vulnerability were those categorized principally as forest specialists or forest generalists. As expected, even species-rich nonforest habitats provided little conservation value for the most vulnerable species. Shaded coffee plantations were modified habitats with relatively high conservation value however, as were gallery forest corridors.

Sugarcane (*Saccharum officinarum*) fields and Caribbean pine (*Pinus caribaea*) plantations, important land uses in much of Central America, offered virtually no conservation value for birds. However, agroforestry, especially systems that use native species, can provide substantial biodiversity benefits.

Benzoin gardens in Sumatra

Styrax paralleloneurum is a forest canopy tree species from Sumatra that produces benzoin, an aromatic resin, and is cultivated in a traditional agroforestry management system. Garcia-Fernandez et al. (2003) sought to determine the impact of benzoin garden management on forest structure, species composition, and diversity. They chose 45 gardens for study in two northern Sumatra villages, where data on management practices and ecological structure were gathered. Ecological information was also collected from abandoned benzoin gardens and what they considered primary forest areas for purposes of comparison. Although benzoin management requires that competing vegetation be thinned, these activities are not intensive, allowing species that coppice to remain in the garden and thereby reducing the effects of competitive exclusion mechanisms on species composition.

The investigators found that tree species diversity in abandoned gardens was similar to that in primary forest, but endemic species and species characteristic of mature habitats were less common. They concluded that traditional benzoin garden management represents only a low-intensity disturbance and maintains an ecological structure that allows effective accumulation of forest species over the long term.

Community-based enterprise

An essential element of successful agroforestry is the economic viability of the various enterprises that are involved. If local people can benefit financially from enterprises that depend on the biodiversity of the forest within which they live, then they might reasonably be expected to support the conservation and sustainable use of the forest ecosystem. This attractive idea has been extensively tested across 39 project sites in Asia and the Pacific, involving activities such as ecotourism, distilling essential oils from wild plant roots, producing jams and jellies from forest fruits, collecting other forest products, and sustainably harvesting timber (Salafsky et al. 2001).

The conclusion from this study was that a community-based enterprise strategy can indeed lead to conservation, but only under specified conditions that are critically dependent on external factors such as market access. Further, any such enterprise can be sustainable only if it is designed to be adaptable to changing conditions. Because many agroforestry areas are subject to political or economic turmoil, fires, droughts, and other external factors, this adaptability of the enterprise is essential to long-term sustainability of the forest. The complexity of factors affecting agroforests also calls for multiple levels for protecting biodiversity, with actions at local, national, and international levels providing the redundancy needed for ensuring that all genes, species, and ecosystems are conserved.

Management systems

Adaptive management is the most effective approach to agroforestry, involving careful planning on the basis of available information, implementation, associated research, monitoring systematic of results, and feeding the results of the monitoring back to improved management of the agroforestry enterprise. This approach helps ensure that agroforestry can respond effectively to changing conditions through constant adaptation.

Converting the potential benefits of forest biodiversity conservation into real and perceived goods and services for society at large, particularly for local people, requires a systems approach to management that includes agroforestry as an essential element that provides economic benefits to local people. The approach is comprised of six key components.

i. Integrated protected areas encompassing various levels of management and administration, including the national, provincial, and local governments, nongovernmental organizations, local communities and indigenous peoples, the private sector, and other stakeholders (McNeely 1999).

ii. Civil society engagement in economic development that includes managing production forests, agroforests, and protected areas, especially for tourism and the sustainable use of certain natural resources (Szaro and Johnston 1996).

iii. Bioregional or ecoregional resource management frameworks, that include farms, agroforests, protected areas, harvested forests, human settlements, and infrastructures as part of a diverse landscape (Miller 1996).

iv. Multi-stakeholder cooperation in agroforestry between private landowners, indigenous peoples, other local communities, industry and resource users.

v. Economic mechanisms to support agroforestry including financial incentives, tax arrangements and land exchanges to promote biodiversity conservation.

vi. Institutional capacities which encourage local stakeholders, universities, research institutions, and public agencies to harmonize their efforts in agroforestry and biodiversity conservation.

Combining these elements can lead to a balanced program for sustainable agroforest management and biodiversity conservation. Enforced governmental regulation should be designed to complement the rights, responsibilities and management activities that are assumed by civil society. Agroforesters who live within or close to protected areas and other forested regions should form alliances with other stakeholders that enable each to assume appropriate roles according to clear government policies and laws.

Management issues

Agroforestry, biodiversity, and invasive alien species

As the global movement of people and products has expanded, so has the movement of plant and animal species. When a tree species is introduced into a new habitat, for example the introduction of the oil palm (*Elaeis guineensis*) from Africa into Indonesia, eucalypts (*Eucalyptus* spp.) from Australia into California, or rubber (*Hevea brasiliensis*) from Brazil into Malaysia, the alien species typically requires human intervention to survive and reproduce. Many of the most popular species of trees used for agroforestry are alien or nonnative species that prosper in their new environments, partly because they no longer face the same competitors, predators and pests that they did at home. Such alien species have been economically important and have enhanced the production of various forest commodities in many parts of the world.

But in some cases, introduced species are a significant problem, becoming established in the wild and spreading at the expense of native species and affecting entire ecosystems. Notorious examples of these 'invasive alien species' that have negative effects on native biodiversity include various species of Northern Hemisphere pines (*Pinus* spp.) and Australian acacias (*Acacia* spp.) in southern Africa, and *Melaleuca* from South America invading Florida's Everglades National Park. These and many other woody plants were introduced intentionally but had unintended consequences. Of the 2000 or so species used in agroforestry, perhaps as many as 10% are invasive (Richardson 1999). While only about 1% are highly so, this includes some popular species such as *Casuarina glauca*, *Leucaena leucocephala*, and *Pinus radiata.* Great care is required to ensure that such species serve the economic purposes for which they were introduced, and do not escape to cause unanticipated negative impacts on native ecosystems and their biodiversity. One management option would be to plant only sterile forms, so reproduction and spread would be impossible.

Worse perhaps, are the invasive alien species that are introduced unintentionally, such as disease organisms that can devastate an entire tree species. The Dutch elm disease (*Ophiostoma ulmi* and *O. novaulmi*) and the American chestnut blight (*Cryphonectria parasitica*) in North America are notorious examples. Pests, such as gypsy moths (*Lymantria dispar*) or longhorned beetles (*Anoplophora glabripennis*), can have profound economic impacts on native forests or plantations. The economic impact of such pests amounts to several hundred billion dollars per year (Perrings et al. 2000). Much of this economic toll is felt in forested ecosystems, even within well-protected national parks. The 1951 International Plant Protection Convention was established to address some of these issues, and new international programs have been developed to respond to current serious problems. A global strategy has been developed (McNeely et al. 2001) and best practices for prevention and management have been designed (Wittenberg and Cock 2001). But growth in global trade means that the threat of devastating invasive species of insects and pathogens will also grow and could fundamentally alter natural forests and virtually wipe out plantations whose lack of species diversity makes them especially vulnerable. Both conservation of biological diversity and sustainable agroforestry need to be managed to ensure that the issue of invasive alien species is well recognized and addressed.

Shade or sun?

Siebert (2002) investigated biophysical, soil and biodiversity effects associated with growing methods of coffee and cacao (*Theobroma cacao*) in Sulawesi, Indonesia, where neither species is native. Canopy height, tree, epiphyte, liana and bird species diversity, vegetation structural complexity, percent ground cover by leaf litter, and soil calcium, nitrate nitrogen, and organic matter levels in the O horizons were all significantly greater in shaded than in sun-grown farms. In contrast, air and soil temperatures, weed diversity, and percent ground cover by weeds were significantly greater in sun compared to shade farms. At the landscape level, conversion of shade-grown crops to sun conditions isolates protected areas and remnant old-growth forest fragments. Local cultivators realize the agronomic and socioeconomic risks associated with sun-grown perennial monocultures and some are increasing the density and diversity of fruit tree cultivation in an effort to provide shade and organic matter, and increase and diversify crop yields. A side effect is better conditions for supporting wild biodiversity.

While coffee grown in monoculture plantations with full exposure to sun have higher yields, coffee grown in the shade is far more beneficial for sustainable agriculture and conserving biodiversity (Siebert 2002). Shade coffee, especially in diverse systems, can support more than twice as many species of birds as sun coffee. The more diverse systems with multiple species of trees providing shade also help support beneficial insects, orchids, mammals, and other species. They also protect fragile tropical soils from erosion, providing nutrients, and suppressing weeds, thereby reducing or eliminating the need for chemical herbicides and fertilizers and thus reducing farming costs. Farmers also are able to harvest various species of fruits, firewood, lumber, and medicines from the shade trees. In agroforestry as in other uses of the land, how much biodiversity is conserved depends on the management practices applied.

Knowledge limitation

To develop effective means of managing biodiversity in forests and in agroforestry systems will require a robust and dynamic knowledge base from which to develop strategies and make decisions. Our present state of knowledge for this purpose is limited, though CIFOR and the World Agroforestry Centre are addressing many of the highest-priority constraints. Priorities for research should be organized around key questions that are consistent with elements of adaptive management systems.

i. Landscape Management: What can landscape scale studies teach us about biodiversity management options, and is it worth continuing investment in them? Or is it better to let market forces, local people, and government policies operate without a landscape-scale vision?

ii. Landscape ecology: How can diversified landscapes, with numerous different approaches to earning a living from the land, be made conducive to maintaining biodiversity while also producing a wide range of products?

iii. Socioeconomics: How can livelihood security be delivered at the community level while still managing forests in the national interest? How can ecosystem service agroforest enterprises be established? What sorts of tenure systems might need to be involved? What new policies are required? How should the benefits be distributed?

iv. Fundamental research: How can the diversity of soil microorganisms be assessed, and how important is their diversity in terms of ecosystem functions in agroforestry systems?

v. The ecology of agroforestry: What are the effects of forest regeneration when crops are grown under trees? How is recruitment affected, and what are the implications of this for biodiversity?

vi. Sustainable use: How can we unravel the package of issues around sustainable use, including what are the benchmarks that need to be established for assessing and monitoring sustainable use? What are the synergies, complementarities, and conflicts among the provision of the various goods and services that are potentially available from agroforests? How can the output of goods and services at the agroforest level be optimized? What are appropriate silviculture practices for multiple use that can be applied by communities?

vii. Native species: How can we make greater use of native species in agroforestry and other rural development efforts, thereby enabling more of the native biodiversity to survive?

viii. Invasive alien species: What are the impacts on biodiversity and ecosystem functions of invasive alien species in agroforestry systems?

Conclusions

Changing circumstances are bringing about new perceptions and new demands on agroforestry. Deepened understanding of the roles of trees and forests in carbon, other nutrient, climatic and water cycles will stimulate new approaches to forest management, agroforestry management and biodiversity conservation. A challenge inherent in a multiple-use approach is that outputs of forests that can be allocated by markets are relatively easy to quantify and exploit, while those that cannot be given a market value, such as biodiversity, tend to be undervalued and are therefore likely to be degraded over time.

Utilitarian values as expressed through agroforestry are often in conflict with strongly held romantic and symbolic values that have been referred to as 'myths.' To many urban people today, clearing rare old-growth forests for their commodity values and subsequent conversion to agroforestry is as sensible as melting down the Eiffel Tower and selling the iron to make more automobiles. Any money yielded by such an action would be inconsequential relative to the social value of the national symbol. As nonproduct benefits like biodiversity become more important to urban citizens, our social and political systems inevitably will become more prominent in forest management. The issue of whether these measures are symptoms of significant improvements or of management systems becoming 'over-connected' and more brittle, and therefore approaching senility, remains to be seen.

Foresters increasingly are seeking diverse combinations of compatible forest uses. They are finding, for example, that conserving biodiversity and storing carbon for reducing atmospheric carbon dioxide are highly compatible forest services, and that such uses can also allow the production of nontimber forest products, the conservation of soil and water, and provision of recreation and tourism. These uses are incompatible with clear-felling, but may be compatible with well-managed agroforestry. The trend away from single-product forestry is continuing, delivering more diversity and benefits for people living in and around the forests.

It appears that the best way to maintain biodiversity in forest ecosystems as the 21st century begins is through a combination of strategically selected, strictly protected areas; multiple-use agroforestry areas intensively managed by local people; natural forests extensively managed by forestry professionals for sustainable production of logs and other commodities, and forest plantations intensively managed for other wood products needed by society. This diversity of approaches and uses will provide humanity with the widest range of options – the greatest diversity of opportunities – for adapting to the cyclical changes that are certain to continue.

References

Alavalapati J.R.R. Shrestha R.K., Stainback G.A. and Matta J.R. 2004. Agroforestry development: An environmental economic perspective (This volume).

Audric J. 1972. Ankor and the Khmer Empire. Robert Hale, London, 207 pp.

Behan R.W. 1975. Forestry and the end of innocence. American Forester 81: 16–19.

Boyden S. 1992. Biohistory: The Interplay Between Human Society and the Biosphere. UNESCO and Parthenon Publishing Group, Paris, 265 pp.

Campbell J. 1985. Myths to Live By. Grenada Publishing, London, 276 pp.

Conklin H. 1954. An ethnoecological approach to shifting cultivation. Trans New York Acad Science 17: 133–142.

Cook S.F. 1949. Soil Erosion and Population in Central Mexico. University of California Press, Berkeley.

Darlington C.D. 1969. The Evolution of Man and Society. Simon and Shuster, New York, 753 pp.

Denevan W. M. 1992a. The Native Population of the Americas in 1492 (second edition). University of Wisconsin Press, Madison, 353 pp.

Denevan W. M. 1992b. The pristine myth: The landscape of the Americas in 1492. Annals of the Association of American Geographers 82(3): 369–385.

Ferguson B., Vandemeer J., Morales H. and Griffith D. 2003. Post-agricultural successional in El Peten, Guatemala. Conserv Biol 17: 818–828.

Flannery T. 1994. The Future Eaters: An Ecological History Of The Australasian Lands And People. Reed Books, Port Melbourne, 432 pp.

Flannery T. 2001. The Eternal Frontier: An Ecological History of North America and its Peoples. The Text Publishing Company, Melbourne, 368 pp.

Garcia-Fernandez C., Casado M. and Ruiz Perez M. 2003. Benzoin gardens in north Sumatra, Indonesia: effects of management on tree diversity. Conserv Biol 17: 829–836.

Gardner B.L. 2002. American Agriculture in the 20th Century: How it Flourished and What it Cost. Harvard University Press, Cambridge MA, 480 pp.

Geertz C. 1963. Agricultural Involution. University of California Press, Berkeley CA, 176 pp.

Gomez-Pompa A. and Kaus A. 1992. Taming the wilderness myth. BioScience 42: 271–279.

Harmon D. and Putney A.D. (eds) 2003. The Full Value of Parks: From Economics to the Intangible. Rowman and Littlefield, Lanham, Md, 360 pp.

Heckenberger M. J., Kuikuro A., Kuikuro, U.T., Russell J.C., Schmidt M., Fausto C., Franchett B. 2003. Amazonia 1492: Pristine forest or cultural parkland? Science 301: 1710–1714.

Hoerr W. 1993. The concept of naturalness in environmental discourse. Natural Areas Journal 13(1): 29–32.

Hoffecker J.F., Powers W.R. and Goebel T. 1993. The colonization of Beringia and the peopling of the New World. Science 259: 46–53.

Holling C.S. (ed.) 1978. Adaptive Environmental Assessment and Management. John Wiley, New York, 576 pp.

Jung C.G. 1964. Man and His Symbols. Doubleday, New York. 432 pp.
Koch N.E. and Kennedy J.J. 1991. Multiple-use forestry for social values. Ambio 20: 330–333.
Kunstadter P. 1970. Subsistence agricultural economics of Lua and Karen hill farmers of Mae Sariang District, Northern Thailand. In: Balankura B. (ed.), International Seminar on Shifting Cultivation and Economic Development in Northern Thailand. Land Development Department, Bangkok, Thailand, 432 pp.
Lewis H.T. and Ferguson T.A. 1988. Yards, corridors and mosaics: How to burn a boreal forest. Hum Ecol 16: 57–78.
Mancall P.C. 1991. Valley of Opportunity: Economic Culture along the Upper Susquehanna, 1700–1800. Cornell University Press, Ithaca, NY, 253 pp.
Marjokopri A. and Ruokolainen K. 2003. The role of traditional gardens in the conservation of tree species in West Kalimantan, Indonesia. Biodivers Conserv 12: 799–822.
Martin P.S. and Klein R.G. (eds) 1984. Quaternary Extinctions: A Prehistoric Revolution. University of Arizona Press, Phoenix, 892 pp.
McNeely J.A. 1987. How dams and wildlife can coexist: Natural habitats, agriculture, and major water resource development projects in Tropical Asia. Conserv Biol 1: 228–238.
McNeely J.A. 1994. Lessons from the past: Forests and biodiversity. Biodivers Conser 3: 3–20.
McNeely J.A. 1999. Mobilizing Broader Support for Asia's Biodiversity: How Civil Society Can Contribute to Protected Area Management. Asian Development Bank, Manila, 248 pp.
McNeely J.A., Mooney H.A., Neville L., Schei P. and Wagge J. (eds). 2001. A Global Strategy on Invasive Alien Species. IUCN, Gland, Switzerland, 50 pp.
McNeely J.A. and Scherr S. 2003. Ecoagriculture: Strategies to Feed the World and Save Wild Biodiversity. Island Press, Washington D.C., 323 pp.
Miller K.R. 1996. Balancing the Scales: Guidelines for Increasing Biodiversity's Chances Through Bioregional Management. World Resources Institute, Washington D.C., 73 pp.
Myers N., Mittermeier R., Mittermeier C., Fonseca G. and Kent J. 2000. Biodiversity hotspots for conservation priorities. Nature 403: 842–843.
Nef J.U. 1977. An early energy crisis and its consequences. Sci Am 237(5): 140–151.
O'Hara S.L., Street-Perott F.A. and Bert T.P. 1993. Accelerated soil erosion around a Mexican highland lake caused by pre-Hispanic agriculture. Nature 362: 48–51.
Perlin J. 1989. A Forest Journey: The Role of Wood in the Development of Civilization. W.W. Norton, New York, 445 pp.
Perrings C., Williamson M. and Dalmazzonej S. (eds) 2002. The Economics of Biological Invasions. Edward Elgar Publishing, Cheltenham, UK, 249 pp.
Petit L. and Petit D. 2003. Evaluating the importance of human-modified lands for Neotropical bird conservation. Conserv Biol 17(3): 687–694.
Piperno D.R., Ranere A., Holst I. and Hansell P. 2000. Starch grains reveal early root crop horticulture in the Panamanian tropical forest. Nature 407: 894–897.
Poffenberger M. 1990. Keepers of the Forest: Land Management Alternatives in Southeast Asia. Kumarian Press, West Hartford, 256 pp.
Pollard E. (ed.) 1994. Oxford Paperback Dictionary (Fourth Edition). Oxford University Press, London, 938 pp.
Ponting C. 1992. A Green History of the World: The Environment and the Collapse of Great Civilizations. St. Martin's Press, New York, 432 pp.
Posey D.A. 1982. The Keepers of the Forest. Garden 6: 18–24.
Pyne S.J. 1982. Fire in America: A Cultural History of Wild Land and Rural Fire. Princeton University Press, Princeton, N.J., 680 pp.
Rappaport R.A. 1971. The flow of energy in an agricultural society. Sci Am 225(3): 116–132.
Raven-Hart R. 1981. Ceylon: History in Stone. Lakehouse Investments, Ltd., Colombo, Sri Lanka, 343 pp.
Richardson D. M. 1999. Commercial forestry and agroforestry as sources of invasive alien trees and shrubs. pp. 237–257. In: Sandlund, O.T., Schei P.J. and Viken A. (eds), Invasive Species and Biodiversity Management. Kluwer, Dordrecht, The Netherlands.
Richter D.D. and Markewitz D. 2001. Understanding Soil Change: Soil Sustainability over Millennia, Centuries, and Decades. Cambridge University Press, Cambridge, UK, 272 pp.
Roosevelt H.C. (ed.) 1994. Amazonian Indians: From Prehistory to the Present. The University of Arizona Press, Tucson, 420 pp.
Salafsky N., Cauley H., Balachander G., Cordes B., Parks J., Margolvis C., Bhatt S., Encarnacion C., Russell D. and Margolis R. 2001. A systematic test of an enterprise strategy for community-based biodiversity conservation. Conserv Biol 15: 1585–1595.
Schnitger F.M. 1964. Forgotten Kingdoms in Sumatra. E.J. Brill, Leiden, 228 pp.
Siebert S.F. 2002. From shade- to sun-grown perennial crops in Sulawesi, Indonesia: Implications for biodiversity conservation and soil fertility. Biodivers Conserv 11: 1889–1902.
Solheim W.G. 1972. An earlier agricultural revolution. Sci Am 266(4): 34–41.
Sondaar P.Y. 1977. Insularity and its affect on mammal evolution. pp. 671–707. In: Hecht M.K., Goody R.C. and Hecht B.M. (eds), Major Patterns in Vertebrate Evolution. Plenum, New York.
Spencer J.E. 1966. Shifting Cultivation in Southeast Asia. University of California Press, Berkeley CA, 247 pp.
Sprugel D.G. 1991. Disturbance, Equilibrium and Environmental Variability: What is 'natural' vegetation in a changing environment? Biol Conserv 58: 1–18.
Suzuki D. and Knudtson P. 1992. Wisdom of the Elders. Bantam Books, 320 pp.
Szaro R. and Johnston D.W. 1996. Biodiversity in Managed Landscapes: Theory and Practice. Oxford University Press, Oxford, UK, 808 pp.
Taylor P.W. 1986. Respect for Nature: A Theory of Environmental Ethics. Princeton University Press, Princeton, N.J., 329 pp.
Turner B.L. and Butzer K.W. 1992. The Colombian encounter and land-use change. Environment. October: 16–44.
Wharton C.H. 1968. Man, Fire, and Wild Cattle in Southeast Asia. Proc. Ann. Tall Timbers Fire Ecol. Conf. 8: 107–167.
Whitmore T.M., Turner B.L., Johnston D.L., Kats R.W. and Gottscheng T.R. 1990. Long-term population change. pp. 25–39. In: Turner B.L. (ed.), The Earth as Transformed by Human Action. Cambridge University Press, Cambridge, U.K.
Wittenberg R. and Cock M. (eds) 2001. Invasive Alien Species: A Tool Kit of Best Prevention and Management Practices. CAB International, Wallingford, UK, 228 pp.
Wunder S. 2000. The Economics of Deforestation: The Example of Ecuador. MacMillan Press, New York, 256 pp.

Agroforestry Systems **61:** 167–181, 2004.

Tree domestication in tropical agroforestry

A.J. Simons[1,*] and R.R.B. Leakey[2]

[1]*World Agroforestry Centre, United Nations Avenue, P. O. Box 30677-00100, Nairobi, Kenya;* [2]*Agroforestry and Novel Crops Unit, School of Tropical Biology, James Cook University, Cairns, Australia QLD 4870;* **Author for correspondence: e-mail: t.simons@cgiar.org*

Key words: Agroforestry tree products (AFTPs), *Dacryodes edulis*, Genetic improvement, Participatory appraisal, *Prunus africana*, Tree breeding

Abstract

We execute tree 'domestication' as a farmer-driven and market-led process, which matches the intraspecific diversity of locally important trees to the needs of subsistence farmers, product markets, and agricultural environments. We propose that the products of such domesticated trees are called Agroforestry Tree Products (AFTPs) to distinguish them from the extractive tree resources commonly referred to as non-timber forest products (NTFPs). The steps of such a domestication process are: selection of priority species based on their expected products or services; definition of an appropriate domestication strategy considering the farmer-, market-, and landscape needs; sourcing, documentation and deployment of germplasm (seed, seedlings or clonal material); and tree improvement research (tree breeding or cultivar selection pathways). The research phase may involve research institutions on their own or in participatory mode with the stakeholders such as farmers or communities. Working directly with the end-users is advantageous towards economic, social and environmental goals, especially in developing countries. Two case studies (*Prunus africana* and *Dacryodes edulis*) are presented to highlight the approaches used for medicinal and fruit-producing species. Issues for future development include the expansion of the program to a wider range of species and their products and the strengthening of the links between product commercialization and domestication. It is important to involve the food industry in this process, while protecting the intellectual property rights of farmers to their germplasm.

Introduction

Tree domestication is an umbrella term that is often applied erroneously to a subset of activities such as provenance testing or to a narrow application such as industrial forestry. Simply put, tree domestication refers to how humans select, manage and propagate trees where the humans involved may be scientists, civic authorities, commercial companies, forest dwellers or farmers. In the tropics, the trees involved in tree domestication occur in natural forest, secondary forest, communal fallow lands, plantations and farms. These trees in turn provide both products (timber, fruit, fodder, etc.) and services (shade, soil improvement, erosion control, etc.). Of course, human history is interwoven with forests and trees before agriculture, urbanisation and commerce began, but until relatively recently the interaction was about extraction of tree products from natural forests.

While much is written about natural forest management, commercial tree improvement and forest analogue systems, relatively less is published on the largest group of people who are the rural population in developing countries and their interactions with the large group of tree species in agricultural landscapes. In Asia, Latin America, Africa, and Oceania this is where the greatest potential exists for tree domestication to contribute to sustainable development. Agriculture and forestry are no longer thought of as mutually exclusive activities, yet national and international statistics are only kept on the differentiated land cover of these systems and the data on the extent of integrated agroforestry systems are not available.

Pantropically, there has been deliberate selection and management of trees by humans in forests to provide what are now collectively termed non-timber forest products (NTFPs). In several cases, the extraction, and often overextraction of NTFPs has led to market expansion and supply shortages, which in turn has led to cultivation of trees for the same products. The ambiguity arises, however, in that such products are no longer from the forest and some of the species also provide valuable timber (e.g. the medicinal species *Prunus africana*). The authors therefore suggest that a new term is required to define tree products that are sourced from trees cultivated outside of forests. Most logically these should be referred to as Agroforestry Tree Products (AFTPs). This paper describes the beginnings, current status and future directions of domestication of trees for the agroforestry systems of the tropics with a focus on those providing AFTPs.

Origins and concepts of domestication

The word *domestication* has had several definitions and interpretations since its first appearance in the English language in 1639 (OED 1989). When applied to animals it refers quite narrowly to taming wild subjects and bringing them into the homestead. With respect to plants, there is a spectrum of meaning from nurturing wild plants through to plant breeding through to genetic modification *in vitro*. Most commonly the word is used with reference to annual food crop plants that have undergone selection, breeding and adaptation in agricultural systems. Archaeologists concur that annual crop domestication began with wheat 10 000 years ago in Eurasia at a time of rising human populations and over-exploitation of local resources (Simmonds 1979). Cereal domestication surely ranks as one of the greatest technological advances in human history since not only does wheat (*Triticum aestivum*) still provide 20% of food calories consumed globally but domestication of other cereal crops (e.g. rice [*Oryza sativa*], barley [*Hordeum vulgare*], maize [*Zea mays*]) has seen human populations increase 1000-fold since domestication started (Diamond 2002).

Tree domestication is a far more recent phenomenon than annual-crop domestication. One of the earliest records of tree domestication is that of manipulating pollination in *Ficus* trees 2800 years ago by the prophet Amos (Dafni 1992). More important though than the date of onset of domestication in trees is the scale of activity. In terms of conventional improvement, tree species are far more neglected today than agricultural crops, with the exception of temperate fruit trees (Janick and Moore 1996). In commercial forestry, fewer than 40 taxa have genetic improvement programs underway and most of these are less than 60 years old (Barnes and Simons 1994). Attempts at improving the N_2-fixing agroforestry species (e.g., *Leucaena*) started even later, in the 1980s, coinciding with concerns about soil fertility management, a tropical fuelwood crisis, and renewed interest in social forestry. Domestication of other agroforestry trees has received substantial recent interest following a number of articles and conferences, most significantly the 1992 IUFRO Conference in Edinburgh, UK (Leakey and Newton 1994). Much of the progress in tree domestication has been informed by the well-described homegarden systems of the Amazon, Southeast Asia and Africa (Kumar and Nair 2004). Domestication of trees for the provision of AFTPs has, however, been more frequently equated to conventional timber-tree improvement and horticultural improvement than to homegarden domestication.

Three striking differences between conventional timber-tree improvement and agroforestry-tree improvement exist. These are the number of taxa involved, the industrial rather than subsistence use and the number of stakeholders involved. Commercial plantations typically handle one or a few species and one company may control all the operations from planning, germplasm sourcing, tree improvement, nursery management, planting, tree husbandry to harvesting. These operations are all carried out at a scale to maximize profit. In contrast, agroforestry is concerned with thousands of tree species and millions of subsistence farmer clients influenced by a mixture of government, private sector, community and international partners, each engaged in different and largely uncoordinated activities. In most cases, agroforestry tree improvement has been concerned with on-farm use of firewood, fodder, fruit, live fence, medicinal and fallow trees. The next large change in agroforestry worldwide, which has already started (Franzel et al. 2004), will probably come from a greater focus on cultivating trees for cash, and most likely for fruit, timber and medicines. Thus it is inappropriate to simply equate agroforestry-tree domestication with industrial-tree improvement since aspects of species prioritization, indigenous knowledge, farming systems improvement, adoption and marketing are as important as selection and multiplication. For these reasons, the World Agroforestry Centre introduced a wider concept of tree domestication:

> Domesticating agroforestry trees involves bringing species into wider cultivation through a *farmer-driven* and *market-led* process. This is a science-based and iterative procedure involving the *identification, production, management* and *adoption* of high quality germplasm. High quality germplasm in agroforestry incorporates dimensions of productivity, fitness of purpose, viability and diversity. Strategies for individual species vary according to their functional use, biology, management alternatives and target environments. Domestication can occur at any point along the continuum from the wild to the genetically transformed state. The intensity of domestication activities warranted for a single species will be dictated by a combination of biological, scientific, policy, economic and social factors. In tandem with species strategies are approaches to *domesticate* landscapes by investigating and modifying the uses, values, interspecific diversity, ecological functions, numbers and niches of both planted and naturally regenerated trees. (Simons 2003a).

This rather wordy concept can be simplified to: tree domestication in agroforestry is farmer-driven and about matching the intra-specific diversity of many locally important tree species to the needs of subsistence farmers, the markets for a wide range of products and the diversity of agricultural environment.

Traditionally utilized extractive resources, such as fruits, medicines, and fibres from forests have been collectively described as Non-Timber Forest Products (NTFPs) or Non-Wood Forest Products (NWFPs). Much research has been done on these products in the hope of finding better ways of managing and conserving natural forests while also benefiting the people living in or near the forests. Now there is confusion in the statistics and literature as people describe and discuss these same products as new crops from farmland (Belcher 2003). It is therefore proposed that the products of domesticated agroforestry trees, including timber, should be called Agroforestry Tree Products (AFTPs) to distinguish between the wild and the domesticated products.

Objectives of tree domestication

Trees occur naturally in forests and rangelands and can be grown in commercial plantations and on farms. Within the tropics, natural forests cover 35% of the land; commercial tree plantations account for 1% of the land cover, tree crop plantations [such as cacao (*Theobroma cacao*), rubber (*Hevea brasiliensis*), tea (*Camelia sinensis*), coffee (*Coffea* spp.), citrus (*Citrus* spp.), mango (*Mangifera indica*), and oil palm (*Elaeis guineensis*)) account for a further 1.5%; and agricultural land accounts for 40% of the land area (FAO 2002). With current tropical deforestation rates at around 1%, principally for expansion of agricultural areas, and even with the most optimistic increases in timber-plantation estates (FAO 2003), the largest scope for future tree planting in the tropics will be on agricultural land (Simons et al. 2000a).

In agroforestry, the objective of domestication is to enhance the performance of trees in terms of improved tree products, such as timber, fruits, and medicines, and/or improved environmental services, such as the amelioration of soil fertility. In the former case, improvements will usually be for yield and/or quality with a specific market opportunity as the driver of genetic selection. The demands for these outputs of tree domestication in agroforestry trees are sure to attract increasing interest and resources over the coming decades, and to have an increasing market orientation.

The scale and direction of tree domestication in agroforestry is dependent upon the varying objectives of different stakeholders. As already mentioned, in most situations in the past, the objective has been for subsistence use and less frequently for income generation. However, to date, implementation of tree 'needs assessment' and the consideration of actual or potential contributions of trees to household budgets has been inadequate (Njenga and Wesseler 1999), and in many locations the relative importance of trees, let alone opportunities for tree domestication, has not been established. Nevertheless, exercises in tree-species prioritization have been carried out quite extensively, as described below.

Selection of tree species

Whilst species of only three genera (*Acacia, Pinus* and *Eucalyptus*) account for more than 50% of all tropical tree plantations (FAO 2003), more than 3000 tree species have been documented in agroforestry systems (Burley and von Carlowitz 1984). Until the

1990s, research priorities amongst this vast array of agroforestry trees were determined arbitrarily, often based on individual interests of researchers. Recognition of the importance of understanding of user needs and preferences, technological opportunities and systematic methods for ranking species emerged with publication of guidelines for tree species priority-setting procedures (Franzel et al. 1996). Regional surveys of agroforestry-tree species prioritization have subsequently been completed for the Sahel, southern Africa and West Africa (Jaenicke et al. 1995; Sigaud et al. 1998; Maghembe et al. 2000), as well as for individual countries including Bangladesh, Brazil, Ghana, India, Indonesia, Peru, Philippines, and Sri Lanka (e.g., Sotelo and Weber 1997; Lovett and Haq 2000). Species priorities depend on the objectives of domestication, and will differ if it is for income generation, satisfying farm household needs, germplasm conservation through use, forest conservation through enrichment planting or farm diversification.

Trees found on farms may originate from forest remnants, natural regeneration or deliberate planting. Much of the deliberate planting of indigenous trees on farm [such as the Indonesian damar (*Shorea javanica*), cinnamon (*Cinnamomum* spp.) and rubber agroforests – Michon and de Foresta 1996] arose because of spontaneous farmer initiatives, while government and donor projects tended to promote exotic species (see Shanks and Carter 1994). Recent studies of the frequency and abundance of trees on farms in Cameroon, Kenya, Nigeria and Uganda show the balance between indigenous and exotic tree species in these new plantings (Kindt 2002; Schreckenberg et al. 2002b).

Whilst indigenous taxa may account for the majority of species on farm, introduced exotic taxa account for the many of the trees on farm, especially in Africa. Exotics can have some advantages such as superior growth, although they also have several risks including weediness and aggressive use of natural resources. The existence of exotics clearly demonstrates deliberate planting. Certain exotics have been planted for decades or centuries such that local communities consider the naturalized populations as indigenous species. Examples include *Grevillea robusta* in Kenya or *Gliricidia sepium* in Sri Lanka (Harwood 1992; Stewart et al. 1996). Whether exotic, naturalized or indigenous it is clear that many agricultural landscapes may be tree rich, but species poor (frequently within-species diversity, especially in exotic species, is dangerously poor – Lengkeek et al. in press). For this reason improvement of the landscape by examining and ameliorating tree species diversity within and between functional uses (e.g. boundary, fodder, firewood, and fruit) can be as important as improvement of a single species. Thus the concept of domesticating the landscape becomes relevant. Four points of intervention are relevant here: replacement, substitution, expansion and better management of trees.

In conventional forestry, a classic step in tree improvement is derivation of a shortlist of fast-growing and productive species, which is often arrived at from species elimination trials. Early work in agroforestry included such trials, either as unreplicated arboreta or as replicated trial series (Stewart et al. 1992). Sadly, much of this research was undocumented and/or discontinued, and this has resulted in agroforestry tree plantings relying on fewer commonly known species. Another common shortcoming of this work, even when it was written up, is that seed sources that were used to represent species are not reported. This means that many of the results of species trials are confounded by the seed source used and species ranking may have been different had other or mixed sources been used. The solutions to this shortcoming include: (1) clear documentation of germplasm used; (2) use of multiple provenances per species, where possible; (3) or, when (2) is not possible, inclusion of more than one provenance to consider a provenance mixture in species elimination and species proving trials.

Tree domestication strategies

The determinants of a sound domestication strategy for an individual species can be grouped under 14 headings. These are:

- Reasons for domestication (home use, market, conservation of the species, agroecosystem diversification, improved livelihood strategies)
- Tree uses required (products [AFTPs] and services)
- History and scale of cultivation (as native and exotic)
- Natural distribution, intraspecific variation and ecogeographic survey information
- Species biology (reproductive botany, ecology, invasiveness)
- Scale and profile of target groups and recommendation domains (biophysical, market, cultural)
- Collection, procurement or production of germplasm and knowledge (including ownership, attribution, benefit sharing, access and use)

- Propagule types (including symbionts) envisaged
- Nursery production and multiplication
- Tree productivity (biomass, timing, economics, risks)
- Evaluation – scientific and farmer participatory
- Pests and diseases
- Genetic gain and selection opportunities, methods and intensities
- Dissemination, scaling up, adoption and diffusion

The problem with most agroforestry tree species is that information is incomplete, which has led to suboptimal tree domestication strategies. While tree domestication work has increased within agroforestry, the documentation of the logic and the approach has been generally scant. Even when results are shared or published, it is typically the outcomes that are reported and not the processes. A few case studies of tree domestication strategies are available (e.g., *Prunus africana*; Simons et al. 2000b) as well as a generic tree domestigram. What is most needed and lacking, however, is a generalized domestication decision-framework, which uses the elements of the domestigram. Case studies of various tree categories (product type, mode of propagation, generation interval) are currently being used to construct such a framework.

Beyond the individual tree species domestication work to meet the needs of the farmer, the market (with and without processing to enhance shelf life and market value), there are several elements to consider for landscape-level domestication:

- Likelihood of interaction of planted trees with natural populations and consequences of introducing external populations;
- Primary and multiple uses of the species since individual species differ in the number of total uses and their primary and secondary uses to different clients, and lack of attention to this has led to the miracle species concepts associated with species such as *Leucaena leucocephala*;
- Combined value of all species (economic, social, biological) in the landscape;
- Diversity (within and between species) in each functional use group since some groups, e.g. fodder, may be dominated by a single species and substitution of species in one use group may be more useful than replacement of existing species with better material of the same species;
- Number of trees per unit area and per farm as informed by considerations of species richness and abundance;
- Niche integration on farm and within the landscape as it will affect the viability of populations.

A wider consideration of other elements of agrobiodiversity, such as soil biota, is also pertinent and may need attention.

Germplasm sourcing, documentation, and deployment

One of the most fundamental elements of a tree domestication program is the sourcing and deployment of germplasm. This is especially true in agroforestry since little or no formal selection and breeding are generally carried out. The introductions made on farm (native and exotic) are in many cases a once-off exercise with subsequent generations derived by the farmer from the original parent stock. While subsequent introductions by farmers may broaden the genetic material, considerations of the diversity in founder stock are rarely considered. It is routinely reported from various authors that seed availability is a significant bottleneck to tree planting. Perhaps these statements could be better phrased as ‘seed supply and demand in agroforestry are poorly understood and poorly matched.’ In essence, the key missing information is the quantification of the various flows (Figure 1) of each species between sources, suppliers, and users of germplasm.

The collection of germplasm from wild stands and the distribution of seed from National Tree Seed Centres and scientists (FAO 2001; Harwood 1997; Figure 1) have been best described. One feature evident from the literature and agroforestry practice is that most people differentiate between seeds and seedlings. To generalize, seeds are seen as the domain of centralized seed centres, and seedlings are seen as the mandate of decentralized small-scale nursery operators.

Herein lies the heart of the ‘disconnect’ in agroforestry. The Tree Seed Suppliers Directory (Kindt et al. 2003) seeks to contribute to understanding of the availability from a wider set of international seed dealers, but the companion volumes of small-scale seed producers and nursery operators are missing. For most species, little information is available and the fundamental gap in completing the summary in Figure 1 is the lack of quantitative data to place against each flow, as well as details on the number of actors and volumes of germplasm – whether for a single species, a species group or trees in general. Against this background, a new approach in which farmers or village communit-

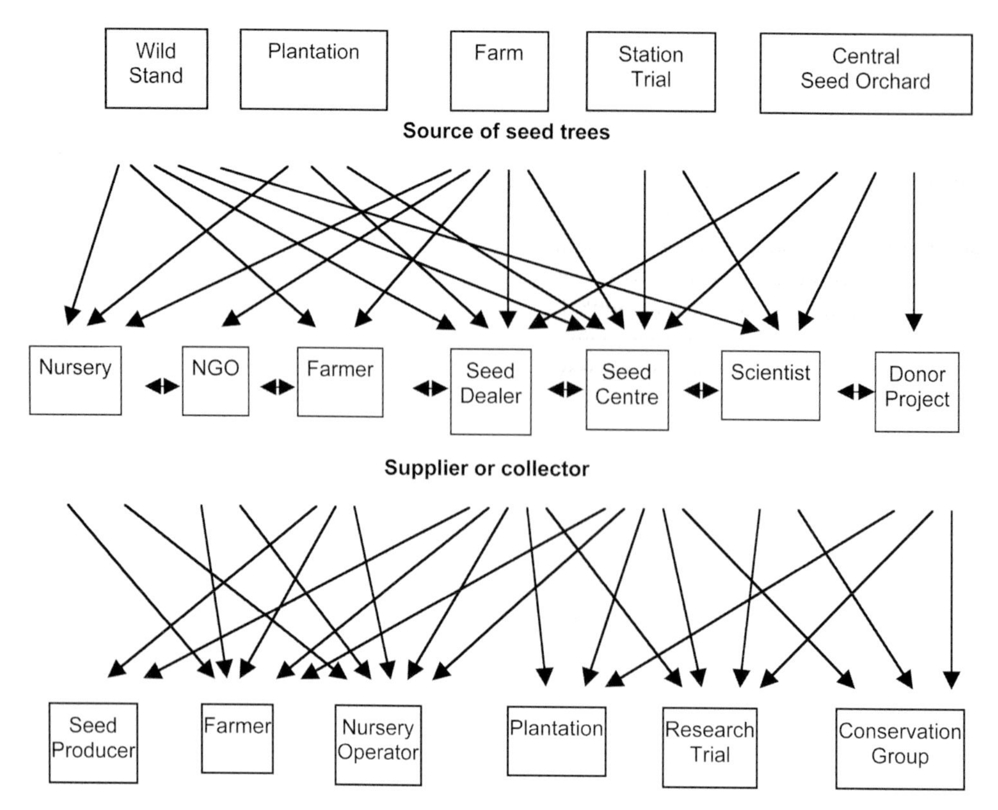

Figure 1. Diagram showing flows between sources, suppliers and users of agroforestry-tree germplasm.

ies develop cultivars from the superior trees in their own land is being developed by World Agroforestry Centre within its Participatory Domestication model (Tchoundjeu et al. 1998; Leakey et al. 2003). This therefore involves a much more localized/internalised selection of germplasm.

As with all other plant species, trees became subject to new international regulations with the coming into force of the Convention on Biological Diversity in December 1993. Most importantly national legislation stipulates how access to and benefit sharing from plant genetic resources should be handled. In the main, trees have not undergone the same level of scrutiny from advocacy groups or government watchdogs as commercial crop or medicinal plants. This is partly explained by the lack of commercial interest from seed multinationals, and also by the lack of harmonization of national laws on genetic resources of agricultural species and those on forest genetic resources. To be able to regulate germplasm it needs to be well described. Descriptors provided for germplasm of various annual crops allow plant breeders and others to value and morphologically characterize the materials developed. Such descriptors for agroforestry trees are nonexistent. Interestingly, molecular characterization has been more frequent for agroforestry species and in particular for the leguminous taxa, including those within the genera: *Acacia*, *Calliandra*, *Gliricidia*, *Inga*, *Leucaena* and *Sesbania*, although this approach is gathering momentum for some timber (e.g., *Swietenia macrophylla* – Novick et al. 2003) and indigenous fruit trees (*Irvingia gabonensis* – Lowe et al. 1998; 2000).

Two contrasting foci have emerged in the agroforestry tree seed industry. First there has been the need to satisfy immediate seed requirements for donor, non-governmental-organization- (NGO) and government-led projects. This relies on sizeable procurement activities and in a few cases large-scale seed production, and the seed is typically given away to farmers and communities for free. Two negative consequences of this approach have been that farmers have undervalued the seed as it has been given out free, and small-scale seed producers have not been able to expand, as they cannot compete against a free good. However, a case could be made that if tree cultivation is considered desirable for society then tree establishment should be subsidized. The second force has been initiatives by the Danish Forest Seed Centre and others to diagnose and establish sustainable seed supply systems. These initiatives are not well suited to project formats where indicators of success can be reported after two to three years. However, if more work is not put into sustainable seed systems, the situation of start-stop tree-seed supply will continue due to its association with projects. One way to increase the functionality of seed systems may be with improved material. Often users of tree germplasm do not differentiate within a species and labelling improved provenances, varieties, families or clones would assist.

Whether seed will come from project start up activities or sustainable supply systems there is a need for accurate demand forecasting of what type, where, when and how much seed will be needed. A 'tree seed forecaster' has been developed by Simons that, for a given species, examines geographic scope for cultivation, potential target population of farmers, testing and adoption behavior of early and late adopters, average farm size, number of trees per farm, reproductive generation interval and scope for on-farm germplasm production and domestication.

One peculiarity in agroforestry is that germplasm is typically viewed as relating to seed. Yet few farmers plant seed since few agroforestry species are direct sown. The few species that are direct sown are those that tend to provide services rather than products, such as live fences, nutrient replenishment, shade, windbreaks, and erosion control. Thus, to a farmer germplasm is more about seedlings rather than seed, since the majority of agroforestry trees are planted from nursery-raised seedlings originating from either seed or vegetative propagules. Germplasm supply is, therefore, as much about seedling supply as seed supply. Whilst commercial forestry operations have undertaken substantial research on nurseries and propagation, far less work has been undertaken for agroforestry trees. In agroforestry there has mainly been a focus on the quantity rather than quality of seedlings (Wightman 1999), although it is now recognized that quality encompasses both physiological and genetic components. Both Jaenicke (1999) and Wightman (1999) describe good practices for research and community nurseries, while Boehringer et al. (2003) describe the importance of the social and organizational aspects of nurseries in addition to the technical ones.

Tree genetic resources used in agriculture should ideally receive commensurate funding for conservation in tune with annual crop plants. Sadly, at the international level, trees are given a back seat compared to herbaceous and graminaceous crops, as evidenced by their exclusion at the 1996 Leipzig, Germany, Conference on Global Plan of Action for Genetic Resources for Food and Agriculture. Furthermore the recently negotiated FAO International Treaty on Plant Genetic Resources for Food and Agriculture (www.fao.org) lists only two genera for a multilateral conservation system (*Artocarpus* and *Prosopis*).

Tree improvement research

Tree improvement research has generally been based on two methods of propagation: sexual or vegetative. Improved tropical fruit trees are typically vegetatively propagated using buds, grafts, marcots (air-layers) or cuttings. Here elite clones of species such as mango, avocado (*Persea americana*), tamarind (*Tamarindus indica*) and citrus are selected at high intensities for mass propagation as cultivars. Improved timber trees, such as eucalypts (*Eucalyptus* spp.) and poplars (*Populus* spp.), may be propagated by cuttings, or from seed (e.g., *Swietenia macrophylla*, *Cordia alliodora*), in which case cloning may have been used in a seed orchard phase. Nearly all other agroforestry trees are propagated by seed, either direct sown or as nursery grown seedlings. Given this orientation, tree improvement research has focused both on the propagation method and on the genetic gains from selection amongst clones or sexual progeny.

The most advanced tree improvement research that has been undertaken on any agroforestry tree is that carried out on *Leucaena* species by Brewbaker and colleagues at the University of Hawaii (Brewbaker and Sorensson 1994) during the 1970s to late 1990s. *Leucaena leucocephala* is atypical with respect to its self-mating system, thus making it easy to fix elite traits. The backcrossing and interspecific hybridization carried out by Brewbaker and colleagues saw rapid progress in traits such as growth, vigor, palatability to livestock, cold tolerance and insect resistance. Interestingly, this is the agroforestry species that is most widely planted throughout the tropics, and this is presumed to be due both to availability of seed and of improved varieties.

With the exception of *Leucaena*, there has been little formal breeding of agroforestry trees. To date, tree improvement in agroforestry has largely come about through species and provenance screening trials for sexually propagated species, and through clonal selection for vegetatively propagated species. It is worth contrasting this situation with the tree improvement program of a single developed country such as Sweden, which has more than 1500-plus trees and 3500 clones in seed orchards in 22 separate breeding populations for a single species, *Picea abies* (Norway spruce) (FAO Reforgen Database: www.fao.org).

In the absence of formal tree improvement in agroforestry, informal selections by farmers, mostly with trees on their farms, dominate rural areas. Researchers are beginning to understand better the techniques used and traits sought by farmers (e.g., Lovett and Haq 2000; Schreckenberg et al. 2002b). It is worth noting, however, that the poor results from some farmer selections are due largely to a narrow founder populations and inbreeding through propagation from a limited number of parents. This observation is borne out by the superiority of many wild populations that were recollected and compared with exotic landraces (e.g., *Gliricidia*; Simons and Stewart 1994). With fruit trees and some other trees, which are clonally propagated the issue does not arise although the optimal number of clones to use is a consideration.

Participatory domestication

In recent years international aid to developing countries has developed a strong focus on poverty reduction. In parallel with this, the World Agroforestry Centre (formerly ICRAF) initiated its tree domestication program in the mid-1990s with a new focus on products with market potential from mainly indigenous species (Simons1996). With this came a shift from on-station formal tree improvement towards more active involvement of subsistence farmers in the selection of priority species for domestication and the implementation of the tree improvement process. In many ecoregions in which ICRAF is active, farmers selected indigenous fruit trees as their top five priorities. Consequently, over the last decade a strategy for the domestication of indigenous trees producing high-value products of traditional and cultural significance has been developed (Tchoundjeu et al. 1998; Leakey et al. 2003). This approach to improving the trees planted by farmers has a number of advantages:

- It has a clear poverty reduction focus, which has been endorsed by a review on behalf of UK Department for International Development (DFID) (Poulton and Poole 2001). The income derived from tree products (AFTPs) is often of great importance to women and children, for example, to meet the demand for school fees and new uniform.
- It has immediate impact by going straight into implementation at the village level, so avoiding delays arising from constraints to the transfer of technology from the field station to the field that can be due to technical, financial, dissemination and political difficulties.
- The approach being developed is focused on simple, low-cost, appropriate technology yielding rapid improvements in planting stock quality based on selection and multiplication of superior trees in ways which create new plants, which also produce fruits within a few years and at heights which are easily harvested.
- It builds on traditional and cultural uses of tree products (AFTPs) of domestic and local commercial importance, and meets local demand for traditional products.
- It promotes food and nutritional security in ways that local people understand, including promoting the immune system, which is especially important in populations suffering from AIDS.
- It can promote local level processing and entrepreneurism, hence employment and off-farm economic development. These benefits can stimulate a self-help approach to development and allows poor people the opportunity to empower themselves.
- It can be adapted to different labour demands, market opportunities, land tenure systems, and is appropriate to a wide range of environments.
- It builds on the rights conferred on indigenous knowledge and the use of indigenous species by the Convention on Biological Diversity and is a model for ‘best practice,’ in contrast to biopiracy.
- It builds on the commonly adopted farmer-to-farmer exchange of indigenous fruit tree germplasm as practiced in west and central Africa, for example *Dacryodes edulis*, which although native to southeast Nigeria and southwest Cameroon, is now found across much of the humid tropics of central Africa (Cameroon, Congo, Central African Republic, Zaire, Gabon, etc.).
- It builds on the practice of subsistence farmers to plant, select and improve indigenous fruits (Leakey et al. 2004), such as marula (*Sclerocarya birrea*) in South Africa, where the yields of cultivated trees are increased up to 12-fold and average fruit size is 29 g, while trees in natural woodland are 21 g (Shackleton et al. 2002).

- The domestication of new local cash crops provides the incentive for farmers to diversify their income and the sustainability of their farming systems (Leakey 2001a,b).

Against these advantages there are possibly disadvantages, such as reduced genetic diversity in the wild population as it is replaced by a domesticated population. However, the implementation of some *in situ* or *ex situ* conservation (McNeely 2004) of wild germplasm, together with the deliberate selection of relatively large numbers of unrelated cultivars can minimize these risks. Indeed, the current strategy, whereby each village develops its own set of cultivars, should ensure that at a national/regional scale the level of intra-specific diversify remains acceptable for the foreseeable future.

Concerns about genetic diversity and how to help farmers to maximize their gains through plus-tree selection for multiple traits have raised questions about how farmers perceive variation. Consequently, quantitative studies of tree-to-tree variation in fruit and nut traits have been implemented at the village level, and it is clear that there is very considerable intra-specific variation in each trait and that many of these traits are unrelated (e.g., *Irvingia gabonensis* - Atangana et al. 2001, 2002; Leakey et al. in press). Thus, the current approach to selection of trees meeting various market-oriented ideotypes ensures that cultivars are almost certainly highly variable in many other desirable traits, such as resistance to pests and diseases. It is also interesting that in terms of the level of variability in the measured traits, the current semi-domesticated on-farm populations are more variable than the wild populations (Leakey et al. 2004) suggesting farmer selections may have multiple population origins.

To maximize the economic, social and environmental benefits from participatory domestication it is crucial to develop post-harvest techniques for the extension of shelf life of the raw products, and processing technologies to add value to them. Without this parallel preparation for increased commercialisation, domestication will not provide all the above-listed benefits. The combination, however, has potential applications that extend beyond subsistence agriculture to agricultural diversification of farming systems. In tropical North Queensland, Australia, for example, this is linked, at least in part, to the development of an Australian 'bush tucker' industry supplying restaurants and supermarkets worldwide.

In his review of how agroforestry fits the Millennium Development Goals, Garrity (2004) concludes that agroforestry needs 'a research and development strategy to reduce dependency on primary agricultural commodities, and to establish production of added-value products based on raw agricultural materials (with traditional and cultural values and locally recognized importance and markets), with links to growing and emerging markets.' In agreement with much of the above on agroforestry tree domestication, Garrity affirms that agricultural R&D institutions must develop new skills in the domestication of indigenous species and the processing/storage of their products, in market analysis and market linkages. This would help focus development on poverty in ways that are of interest and importance to subsistence farmers.

Case studies

Two case studies are presented for agroforestry species that are used primarily for timber, fruit and medicinal products.

Prunus africana

Prunus africana (Rosaceae) – 'pygeum' – is an afromontane forest species found only above 1000 m altitude and confined to isolated populations forming a wide but disjunct distribution (Cunningham and Mbenkum 1993). The genus *Prunus* contains more than 200 species (e.g., peach [*Prunus persica*] and plum [*Prunus domestica*]) of which many have undergone intensive domestication through selection and breeding. *Prunus africana* is however the only species native to Africa and is an important medicinal tree (Hall et al. 2000). Currently, the commercial harvesting of the bark (approximately 4000 Mg of bark per year) for commercial and domestic medicinal use (treatment of Benign Prostatic Hyperplasia [BPH]) has led to extinction of some populations and the listing of the species on the CITES Appendix II. This is a list of plants requiring protection through extraction permits and monitoring of international trade, which currently has an over-the-counter value for *P. africana* products of $220 million per year (Cunningham et al. 1997).

The combination of the intense conservation interest, the considerable commercial and human health importance of the medicinal product, and the potential of the species to be cultivated by small-scale farmers has resulted in the development of a domestication strategy (Leakey 1997; Simons et al. 1998) aimed at the restoration of the resource through diverse agroforestry plantings. This strategy is expected to enhance farmer livelihoods, meet future needs of the industry and, to some extent, have positive ecological and conservation benefits.

The objectives of domesticating *P. africana* are:

1) To conserve wild populations through reducing pressure on the natural resource base by encouraging cultivation of trees by small-scale farmers. Cultivated material also serves a useful *circa situ* conservation function if attention is paid to genetic diversity issues.
2) To locate unique and diverse natural populations that warrant specific *in situ* conservation measures.
3) To identify productive genetic diversity to demonstrate the growth potential of the species.
4) To quantify the genetic control of traits of economic interest (timber volume, chemical profile, bark yield) and determine appropriate selection methods.
5) To undertake participatory domestication with small-scale farmers to determine their preferences and perspectives on cultivating and improving the species.
6) To establish seed production stands and develop appropriate management techniques to be able to deliver sufficient propagules to farmers.
7) To improve propagation methods of the species to encourage wider adoption.
8) To undertake marketing studies in order to better monitor and predict demand and supply, and evaluate prospects for green-labelling and the establishment of premiums for small-scale producers.
9) To use *P. africana* as a case study for the domestication of other medicinal and high-value trees.

The first step before developing a domestication strategy for any species is to collate all available information on the species including: botanic descriptions, geographic distribution, ecology, forest inventories, farmer surveys, harvesting techniques, trade figures and conservation status. For *P. africana*, the key information gaps identified were details on market intelligence, growth data, reproductive ecology, pests and diseases, genetic variation and propagation methods. These knowledge gaps have been the subject of recent studies.

Market intelligence projections suggested that, given the increase in the aging male population in Europe and America and the increase in consumer confidence in herbal remedies, demand could rise two- to three-fold (to 8000 to 12 000 Mg per annum). It is clear that natural forests will not be able to meet such demands and thus that cultivation is required. This raises issues about how to achieve this cultivated resource, which are reflected by the two unlikely extremes of a single plantation of 8000 hectares (at 4 × 4 m spacing) benefiting a very limited number of producers, or many smallholders each producing a few trees through agroforestry (one farmer growing 5 million trees or a million farmers each producing five trees).

Studies of the few existing plantations of *P. africana* in Cameroon and Kenya found that trees over 12 years of age produced acceptable yields and concentrations of active constituents, although yields of bark-extract increased as trees aged further (Kimani 2002[1]; Cunningham et al. 2002). The highest bark-extract yield (1.2%) was found in 55-year-old trees, although trees of similar diameter class can have differences in mean extract yield from 0.8% to 1.33%. In addition, the major sterol component, B-sitosterol, varied significantly between provenances (101 μg/g to 150 μg/g), while individual tree yields varied from 50 μg/g to 191 μg/g). It is not known to what extent this variation is under genetic control. However, molecular studies of neutral genetic variation have revealed that differences between populations across the range account for 58% to 73% of total variation (Muchugi 2001[2]) reflecting the disjunct nature of afromontane populations. Elite trees of different provenances are now being mixed in a seed orchard, with trees from Kibale, Rwenzori and Bwindi populations in Rwanda. This approach is comparable to industrial forestry with combining populations in Breeding Seedling Orchards (Barnes and Simons 1994).

Studies of the storage of *P. africana* seed have found that it is recalcitrant to intermediate in its behaviour. Greatest germination (40% to 70%) was found when purple coloured seeds were collected and depulped. Extracted seeds could be stored at 4 °C for up to one year at 15% moisture content, although this decreased viability by 50%. Surveys of seed price in Kenya and Cameroon show it to be $8 to $25 per kg, which at 3000 to 5000 seeds per kg, makes it relatively expensive. The infrequent nature of fruiting, high price or seed and recalcitrant seed behaviour indicate that sourcing sufficient seedlings on a regular basis may be problematic.

Vegetative-propagation studies for *P. africana* in Kenya and Cameroon have found that juvenile tissues root well (75% to 90%) as leafy cuttings (Nzilani 1999[3]; Tchoundjeu et al. 2002), opening the way for clonal approaches to producing medicinal extracts with high yield, quality and uniformity. On-station trials have been established in Muguga and Kakamega (Kenya), Kabale (Uganda) and Buea (Cameroon) to examine growth and survival, as well as family and provenance variation. Species prioritization studies with farmers in Cameroon, Uganda and Kenya have confirmed the use and popularity of the species. On-farm planting with *Prunus* has now taken place in Cameroon, Kenya, Madagascar and Uganda, all with unimproved material. By the mid 1990s these had occurred on several thousand small-scale farms in Cameroon (Cunningham et al. 2002), although to date these immature plantings have not taken pressure off the trees in natural forests.

It is foreseen that the on-farm cultivation of *P. africana* would result in the following outcomes: Trees grown on field boundaries have more spreading crowns than in closely spaced plantations, but are a good source of bark and timber.

- Before felling for timber, nondestructive bark panelling can be undertaken at 15, 23 and 31 years – providing 15, 25 and 60 kg of bark per tree, respectively.
- A tree could conceivably be felled for timber at 40 years of age, when it would yield approximately 200 kg of bark.
- This equates to a production of 7.5 kg of bark per tree per year over a 40-year period.
- For a 3000 Mg market, 400 000 such trees will be needed.
- For a 10 000 Mg market, 1.3 million trees will be needed.

Dacryodes edulis

The domestication of *Dacryodes edulis* – Safou or African plum – has been the developing model for Participatory Domestication approaches in Cameroon and Nigeria described above (see papers in Schreckenberg et al. 2002a), since its identification as the top priority species for agroforestry in western and central Africa. It is widely grown in mixed farming systems across many countries of Central Africa (28% to 57% of all fruit trees), especially as the shade tree for cacao or coffee in the forest-savanna transition zone, as well as a middle-stratum in cacao farms under secondary humid forest and as a common constituent of homegardens. Mean numbers of trees per farm range from 20 to 200 in some villages in Cameroon (Schreckenberg et al. 2002b), with tree density being greatest in farms under 2 ha. In 1997, the trade of Safou in Cameroon alone was worth $7.5 million, excluding domestic consumption. Of this trade, $2.5 million is in exports (Awono et al. 2002). The fruits are typically roasted and eaten as a nutritious staple; it is rich in fat (64%), protein (24%) and carbohydrate (9%); and there is the potential for vegetable oil extraction on a commercial scale.

There is considerable farmer-to-farmer exchange of germplasm of this species, which almost certainly explains its common occurrence in farmland even outside its natural range, regardless of social features such as the wealth of the farmers, prevailing land tenure regime, labour availability, and level of education. The multi-institutional and multi-disciplinary domestication program is built on earlier tree improvement studies examining reproductive biology, provenance variation, etc. initiated by the Institut de la Recherche Agricole pour le Dévéloppement (IRAD) (Kengue 1998; Kengue and Singa 1998). Now it is being focussed at the village level on the development of cultivars from trees with superior fruit size, color and taste. Vegetative propagation techniques, especially marcotting, have been used to capture the phenotype of selected trees and so to produce cultivars. To increase the multiplication rate, initial problems with the rooting of leafy stem cuttings have been overcome using simple, low-tech propagators established in a central nursery and replicated in all pilot villages (Leakey et al. 1990). With the help of NGOs, the villagers have been trained in nursery and vegetative propagation skills and become the nursery managers and the owners of the germplasm developed from their local population.

Some 10-fold variation in mean fruit mass between different trees, and additional variation in other commercially important traits, indicates considerable opportunity to increase the size and quality of fruits for market. To this is being added a better understanding of consumer preference and assessment of the genetic variation in sensory traits (taste, oiliness, acidity, smell, etc.), and studies to relate these to other visual characteristics are underway. Selection criteria based on the identification of market-oriented ideotypes have been identified (Leakey et al. 2002; Anegbeh et al. in press) so that the improved prices of desirable fruit types currently only recognized in retail urban markets can be acquired by the producer dealing with traders at the farm gate. Currently the diversity in fruit size and quality from the virtually wild trees on most farms, each phenotypically different from other trees, precludes wholesale buyers from recognizing superior fruits. Hopefully, however, when a vehicle can be loaded with the uniform fruits of recognized cultivars, farmers will be financially rewarded for their efforts. In Cameroon, women have indicated that the coincidence of schooling expenditures with income flows from the sale of fruits is one of the attributes of *D. edulis* that they appreciate (Schreckenberg et al. 2002b).

Currently, Safou fruits are perishable and difficult to store for more than a few days. However, studies are in progress to improve shelf life by various forms of processing and value-adding. Once this is achieved, it is anticipated that the demand will increase and the fruits, or fruit products, will become available outside the current four-to-five-month production season. In parallel with these developments are studies to understand more about the potential for the production of different oils for the food industry, involving the relationships between oil quality and other chemical and morphological traits. Safou fruits are already traded on a small scale into Europe and America; it is hoped that the combination of targeted genetic selection and processing will open up market opportunities, which can further benefit subsistence farmers. Much work remains to be done to ensure that this is achieved, with particular recognition that the intellectual property of the farmers developing the cultivars needs to be assured.

Recommendations: future developments

From the issues discussed in this paper, there is evidently much work to be done to build on the work of the last decade. There are very large numbers of species in all ecoregions, which potentially could be domesticated to produce marketable AFTPs and environmental services. The extension of the above mentioned strategies, principles and techniques to even a small proportion of these species represents an enormous challenge to countries and to development agencies with limited resources. Probably the biggest constraint to achieving this on the scale required is the serious lack of people in NGOs and National Agricultural Research Systems (NARS) with adequate knowledge of vegetative propagation techniques for trees. There is also a need for better methods of propagating mature trees (Leakey in press). Resolving the technical and implementation issues surrounding the expansion of this 'grass-roots' revolution will be enormous, with strategies on issues such as how to avoid a loss of intra-specific genetic diversity needing to be resolved, especially when recognising that crop domestication is an on-going iterative process.

The problems of widespread implementation of the domestication of agroforestry trees will be exacerbated by the fact that the philosophy to diversify farming systems and economies in the ways described here runs counter to the philosophy promulgated by many agencies that biotechnology is the solution to the issues of poverty, malnutrition and rural development. Nevertheless, there is a growing body of evidence that supports the agroforestry solution to many of these problems, and the evident congruence between agroforestry outputs and Millennium Development Goals offers hope for the future (Garrity 2004). Agroforestry centred on indigenous trees is very compatible with the aims of the Ecoagriculture initiative ratified at the World Summit on Sustainable Development in Johannesburg in 2002 (McNeely and Scherr 2003). In this regard, tree domestication provides a powerful incentive for subsistence farmers to diversify their farms with indigenous trees that provide economic returns and environmental services, including biodiversity conservation (see also McNeely 2004).

In addition, the future success of AFTP domestication and commercialization will depend on the benefits remaining with the small-scale farmers and their local industries and markets. (Clement et al. 2004) This will inevitably depend on finding ways to satisfactorily protect the intellectual property rights (IPR) of the farmers and communities investing in the process. Currently it is unclear how individual farmers and communities in economically and socially disadvantaged countries can avail themselves of the 'rights' conferred by the Convention on Biological Diversity. In this connection, the attempts of the African Union to develop a model law for the protection of the rights of poor farmers is encouraging.

The 'chicken-and-egg' linkages between domestication and commercialization will be difficult to resolve, making it crucial that agroforesters find ways to work closely with commercial and industrial partners so that the products match the needs of both the farmers, the processing industries and the consumers. It may not be a beneficial situation for everyone and clear understanding of the winners and losers of greater cultivation and commercialisation will be needed.

Finally, however, the strength of the approach lies in its ability to empower the millions of farmers desperate to improve their lot in the world. What they need most is information about what is possible and the simple skills needed to do the job. The public in the developed world can help by buying the newly improved products, and by demanding that policy makers recognise the importance of simultaneously resolving the poverty and environmental crises facing the developing world.

End Notes

1. Kimani P.G. 2002. Population variation in yield and composition in bark chemical extracts of Prunus africana, and its potential for domestication. M. Phil Thesis, Moi University, Kenya, 73 pp.
2. Muchugi A.M. 2001. Genetic variation in the threatened medicinal tree *Prunus africana* in Kenya and Cameroon: implications for the genetic management of the species. M. Sc. thesis, Kenyatta University, 90 pp.
3. Nzilani N.J. 1999. The status of *Prunus africana* in Kakamega Forest and the prospects for its vegetative propagation. M. Phil. thesis, Moi University, Kenya, 78 pp.

References

Anegbeh P.O., Ukafor V., Usoro C., Tchoundjeu Z., Leakey R.R.B. and Schreckenberg K. (in press). Domestication of *Dacryodes edulis*: 1. Phenotypic variation of fruit traits from 100 trees in southeast Nigeria. New Forests.

Atangana A.R., Tchoundjeu Z., Fondoun J.-M., Asaah, E., Ndoumbe M. and Leakey R.R.B. 2001. Domestication of *Irvingia gabonensis*: 1. Phenotypic variation in fruit and kernels in two populations from Cameroon. Agroforest Syst 53: 55–64.

Atangana A.R., Ukafor V., Anegbeh P.O., Asaah E., Tchoundjeu Z., Usoro C., Fondoun J.-M., Ndoumbe M. and Leakey R.R.B. 2002. Domestication of *Irvingia gabonensis*: 2. The selection of multiple traits for potential cultivars from Cameroon and Nigeria. Agroforest Syst 55: 221–229.

Awono A., Ndoye O., Schreckenberg K., Tabuna H., Isseri F. and Temple L. 2002. Production and marketing of Safou (*Dacryodes edulis*) in Cameroon and internationally: Market development issues. Forest, Trees and Livelihoods 12: 125–147.

Barnes R.D. and Simons A.J. 1994. Selection and breeding to conserve and utilize tropical tree germplasm. pp. 84–90. In: Leakey R.R.B. and Newton A.C. (eds), Tropical Trees: The Potential for Domestication and the Rebuilding of Forest Resources. London, HMSO.

Belcher B.M. 2003. What isn't an NTFP? Int For Rev 5: 161–168.

Boehringer A., Ayuk E.T., Katanga R. and Ruvuga S. 2003. Farmer nurseries as a catalyst for developing sustainable land use systems in southern Africa. Agri Syst 77: 187–201.

Brewbaker J.L. and Sorensson C.T. 1994. Domestication of lesser-known species of the genus *Leucaena*. pp. 195–204. In: Leakey R.R.B. and Newton A.C. (eds), Tropical Trees: the Potential for Domestication. Rebuilding Forest Resources. London, HMSO.

Burley J. and von Carlowitz P. 1984. Multipurpose Tree Germplasm. International Centre for Research in Agroforestry, Nairobi, Kenya, 298 pp.

Clement C.R, Weber J.C., van Leeuwen J., Domian C.A, Cole D.M., Arévalo Lopez L.A.A and Argüello H. 2004. Why extensive research and development did not promote use of peach palm fruit in Latin America (This volume).

Cunningham A.B. and Mbenkum F.T. 1993. Sustainability of harvesting *Prunus africana* bark in Cameroon: a medicinal plant in international trade. People and Plants Working Paper 2. Paris: UNESCO, 28 pp.

Cunningham M., Cunningham A.B. and Schippmann U. 1997. Trade in *Prunus africana* and the implementation of CITES. Bundesamt für Naturschutz, Bonn, Germany.

Cunningham A.B., Ayuk E., Franzel S., Duguma B. and Asanga C. 2002. An economic evaluation of medicinal tree cultivation: *Prunus africana* in Cameroon. People and Plants Working Paper 10. Paris: UNESCO, 35 pp.

Dafni A. 1992. Pollination Ecology: A Practical Approach. Oxford University Press, Oxford, 250 pp.

Diamond J. 2002. Evolution, consequences and future of plant and animal domestication. Nature 418: 700–707.

FAO. 2001 Report on the 12th Session of the Panel of Experts on Forest Genetic Resources. FAO, Rome, Italy, 108 pp.

FAO. 2002. FAOSTAT Database www.fao.org, Rome, Italy.

FAO. 2003. State of the World's Forests. Food and Agriculture Organization, Rome, Italy, 151 pp.

Franzel S., Jaenicke H. and Janssen W. 1996. Choosing the right trees: setting priorities for multipurpose tree improvement. ISNAR Research Report 8, The Hague: International Service for National Agricultural Research, 87 pp.

Franzel S., Denning G.L., Lilisøe J.-P. and Mercado A.R Jr. 2004. Scaling up the impact of agroforestry: Lessons from three sites in Africa and Asia (This volume).

Garrity D. 2004. World agroforestry and the achievement of the Millenium Development Goals. (This volume).

Hall J.B., O'Brien E.M. and Sinclair F.L. 2000 *Prunus africana*: a monograph. University of Wales, Bangor, UK, 104 pp.

Harwood C.E. (ed.) 1992. *Grevillea robusta* in Agroforestry and Forestry. Proceedings of an International Workshop. International Centre for Research in Agroforestry, Nairobi, Kenya, 190 pp.

Harwood C.E. 1997. Domestication of Australian tree species for agroforestry. pp. 64–72. In: Roshetko J.M. and Evans D.O. (eds), Domestication of Agroforestry Trees in Southeast Asia. Winrock International, Arkansas, USA.

Jaenicke H. 1999. Good Tree Nursery Practices – Practical Guidelines for Research Nurseries. International Centre for Research in Agroforestry, Nairobi, Kenya, 95 pp.

Jaenicke H., Franzel S. and Boland D. 1995. Towards a method to set priorities amongst multipurpose trees for improvement activities: a case study from west Africa. J Trop For Sci 7: 490–506.

Janick J. and Moore J.N. 1996. Fruit Breeding: Volume 1. Tree and Tropical Fruits. Wiley, New York, 632 pp.

Kengue J. 1998. Point sur la biologie de la reproduction du safoutier (*Dacryodes edulis* (G. Don) H.J. Lam). pp. 97–111. In: Kapseu C. and Kayem G.J. (eds), Proceedings of the 2nd International Workshop on African Pear Improvement and Other New Sources of Vegetable Oils. ENSAI, Presses Universitaires de Yaoundé, Cameroon.

Kengue J. and Singa E.M. 1998. Preliminary characterization of some African Plum accessions in Barombi-Kang germplasm collection. pp. 113–122. In: Kapseu C. and Kayem G.J. (eds), Proceedings of the 2nd International Workshop on African Pear Improvement and Other New Sources of Vegetable Oils. ENSAI, Presses Universitaires de Yaoundé, Cameroon.

Kindt R. 2002. Methodology for tree species diversification planning for African ecosystems. Ph.D. Thesis, University of Ghent, 327 pp.

Kindt R., Mutua A., Muasya S. and Kimotho J. 2003. Tree Seed Suppliers Directory. World Agroforestry Centre, Nairobi, Kenya, 426 pp.

Kumar B.M. and Nair P.K.R. 2004. The enigma of tropical homegardens (This volume).

Leakey R.R.B. 1997. Domestication potential of *Prunus africana* ('Pygeum') in sub-Saharan Africa. pp. 99–106. In: Kinyua A.M., Kofi-Tsekpo W.M. and Dangana L.B. (eds), Conservation and Utilization of Medicinal Plants and Wild Relatives of Food Crops. UNESCO, Nairobi, Kenya.

Leakey R.R.B. 2001a. Win-win land use strategies for Africa: 1. Building on experience with agroforests in Asia and Latin America. Int For Rev 3: 1–10.

Leakey R.R.B. 2001b. Win-win land use strategies for Africa: 2. Capturing economic and environmental benefits with multistrata agroforests. Int For Rev 3: 11–18.

Leakey R.R.B. In press. Physiology of vegetative propagation in trees. In: Burley J., Evans J. and Youngquist J.A. (eds), Encyclopaedia of Forest Sciences. Academic Press, London, UK.

Leakey R.R.B. and Newton A. 1994. Domestication of Tropical Trees for Timber and Non-Timber Forest Products. MAB Digest No. 17, UNESCO, Paris, 94 pp.

Leakey R.R.B., Atangana A.R., Kengni E., Waruhiu A.N., Usoro C., Anegbeh P.O. and Tchoundjeu Z. 2002. Domestication of *Dacryodes edulis* in West and Central Africa: characterization of genetic variation. Forests, Trees and Livelihoods 12: 57–71.

Leakey R.R.B., Greenwell P., Hall M.N., Atangana A.R., Usoro C., Anegbeh P.O., Fondoun J.-M. and Tchoundjeu Z. In press Domestication of *Irvingia gabonensis*: 4. Tree-to-tree variation in food-thickening properties and in fat and protein contents of Dika Nut. Food Chemistry.

Leakey R.R.B., Mésen F., Tchoundjeu Z., Longman K.A., Dick J.M.P., Newton A.C., Matin A., Grace J., Munro R.C. and Muthoka P.N. 1990. Low-technology techniques for the vegetative propagation of tropical trees. Commonw Forest Rev 69: 247–257.

Leakey R.R.B., Schreckenberg K. and Tchoundjeu Z. In press 2003. Contributing to poverty alleviation: The participatory domestication of West African indigenous fruits. Int For Rev. 5: 338–347.

Leakey R.R.B., Tchoundjeu Z., Smith R.I., Munro R.C., Fondoun J.-M., Kengue J., Anegbeh P.O., Atangana A.R., Waruhiu A.N., Asaah E., Usoro C. and Ukafor V. 2004. Evidence that subsistence farmers have domesticated indigenous fruits (*Dacryodes edulis* and *Irvingia gabonensis*) in Cameroon and Nigeria. Agroforest Syst. 60: 101–111.

Lengkeek A.G., Jaenicke H. and Dawson I.K. (In press). Genetic bottlenecks in agroforestry systems: results of tree nursery surveys in East Africa. Agroforest Syst.

Lovett P.N. and Haq N. 2000. Evidence for anthropic selection of the sheanut tree (*Vitellaria paradoxa*). Agroforest Syst 48: 273–288.

Lowe A.J., Russell J.R., Powell W. and Dawson I.K. 1998. Identification and characterization of nuclear, cleaved amplified polymorphic sequences (CAPS) loci in *Irvingia gabonensis* and *I. wombolu*, indigenous fruit trees of west and central Africa. Mol Ecol 7: 1771–1788.

Lowe A.J., Gillies A.C.M., Wilson J. and Dawson I.K. 2000. Conservation genetics of bush mango from central/west Africa: implications from random amplified polymorphic DNA analysis. Mol Ecol 9: 831–841.

Maghembe J., Simons A.J., Kwesiga F. and Rarieya M. 1998. Selecting Indigenous Trees for Domestication in Southern Africa. ICRAF, Nairobi, Kenya, 94 pp.

McNeely J.A. 2004. Nature vs. nurture: Managing relationships between forests, agroforestry and wild biodiversity (This volume).

McNeely J.A. and Scherr S.J. 2003. Ecoagriculture: Strategies to Feed the World and Save Wild Biodiversity. Island Press, Washington DC, 323 pp.

Michon G. and de Foresta H. 1996. Agroforests as an alternative to pure plantations for the domestication and commercialisation of NTFPs. pp. 160–175. In: Domestication and Commercialisation of Non-Timber Forest Products in Agroforestry Systems, Leakey R.R.B., Temu A.B., Melnyk M. and Vantomme P. (eds), Non-Wood Forest Products 9, FAO, Rome, Italy.

Njenga A. and Wesseler J. 1999. Participatory farming systems analysis of tree based farming systems in Southern Malawi. GTZ–ITFSP Project, Nairobi, Kenya, 34 pp.

Novick R.R., Dick C.W., Lemes M.R., Navarro C., Caccone A. and Bermingham E. 2003. Genetic structure of Mesoamerican populations of Big-leaf mahogany (*Swietenia macrophylla*) inferred from microsatellite analysis. Molecular Ecology (in press).

OED. 1989. Oxford English Dictionary. Simpson J.A. and Weiner E.S.C. (eds.) 2nd Edition. Clarendon Press, Oxford, UK.

Poulton C. and Poole N. 2001. Poverty and fruit tree research. DFID Issues and Options Paper, Forestry Research Programme. Imperial College at Wye, Kent, England, 70 pp. www.nrinternational.co.uk/forms2/frpzf0141b.pdf.

Schreckenberg K., Leakey R.R.B. and Kengue J. 2002a. A tree with a future: *Dacryodes edulis* (Safou, the African plum). Forests, Trees and Livelihoods [special issue] 12: 1–152.

Schreckenberg K., Degrande A., Mbosso C., Boli Baboulé C., Boyd C., Enyong L., Kanmegne J. and Ngong C. 2002b. The social and economic importance of *Dacryodes edulis* (G. Don) H.J. Lam in southern Cameroon. Forests, Trees and Livelihoods 12: 15–40.

Shackleton C.M., Botha J. and Emanuel P.L. 2003. Productivity and abundance of *Sclerocarya birrea* subsp. *caffra* in and around rural settlements and protected areas of the Bushbuckridge lowveld, South Africa. Forests, Trees and Livelihoods 13: (in press).

Shanks E. and Carter J. 1994. The organization of small-scale nurseries. Overseas Development Institute, London, UK, 144 pp.

Sigaud P., Hald S., Dawson I. and Ouedraogo A. 1998. FAO/IPGRI/ICRAF workshop on the conservation, management, sustainable utilization and enhancement of forest genetic resources in dry-zone sub-saharan Africa. Forest Genetic Resources No. 26, FAO, Rome, pp. 9–12.

Simmonds N.W. 1979. Principles of Crop Improvement. Longman, London, 408 pp.

Simons A.J. 1996. ICRAF's strategy for domestication of indigenous tree species. pp. 8–22. In: Domestication and Commercialization of Non-Timber Forest Products in Agroforestry Systems. FAO Special Publication, Forest Division, FAO, Rome.

Simons A.J. 2003a. Concepts and principles of tree domestication. In: Simons A.J. and Beniest J. (eds), Tree Domestication in Agroforestry. World Agroforestry Centre, Nairobi, Kenya, 244 pp.

Simons A.J. 2003b. Auxiliary plants. pp. 164–171. In: Schmelzer G.H. and Omino E.A. (eds), Plant Resources of Tropical Africa. Proceedings of the First PROTA International Workshop, 23–25 September 2002, Nairobi, Kenya. Wageningen, Netherlands.

Simons A.J., Dawson I., Tchoundjeu Z. and Duguma B. 1998. Passing problems: Prostate and *Prunus*. Journal of American Botanic Council 43: 49–53.

Simons A.J. and Stewart J.L. 1994. *Gliricidia sepium*. pp. 30–48. In: Gutteridge R.C. and Shelton H.M. (eds), Fodder Tree Legumes in Tropical Agriculture. CAB International, Wallingford, UK.

Simons A.J., Jaenicke H., Tchoundjeu Z., Dawson I., Kindt R., Oginosako Z., Lengkeek A. and Degrande A. 2000a. The future of trees is on farms: Tree domestication in Africa. In: Forests and Society: The Role of Research. XXI IUFRO World Congress, Kuala Lumpur, Malaysia, pp. 752–760.

Simons A.J., Tchoundjeu Z., Munjuga M., Were J., Dawson I., Ruigu S., Lengkeek A. and Jaenicke H. 2000b. Domestication Strategy. pp. 39–43. In: Hall J.B., O'Brien E.M. and Sinclair F.L. (eds), *Prunus africana*: A Monograph. UNCW, Bangor, Wales.

Sotelo-Montes C. and Weber J. 1997. Prioritization of tree species for agroforestry systems in the lowland amazon forests of Peru. Agroforesteria en las Amecicas, 4(14):12–17.

Stewart J.L., Dunsdon A.J., Hellín J.J. and Hughes C.E. 1992. Wood Biomass Estimation of Central American Dry-Zone Species. Oxford Forestry Institute, Oxford, UK, 83 pp.

Stewart J.L., Allison G.E. and Simons A.J. 1996. *Gliricidia sepium*: Genetic Resources for Farmers. Oxford Forestry Institute, Oxford, UK, 125 pp.

Tchoundjeu Z., Avana M.L., Leakey R.R.B., Simons A.J., Asaah E., Duguma B. and Bell J.M. 2002. Vegetative propagation of *Prunus africana*: effects of rooting medium, auxin concentrations and leaf area. Agroforest Syst 54: 183–192.

Tchoundjeu Z., Duguma B., Fondoun J.-M. and Kengue J. 1998. Strategy for the domestication of indigenous fruit trees of West Africa: case of *Irvingia gabonensis* in southern Cameroon, Cameroon J Biology Biochem Sci 4: 21–28.

Wightman K. 1999. Good Tree Nursery Practices – Practical Guidelines for Community Nurseries. International Centre for Research in Agroforestry, Nairobi, Kenya, 95 pp.

World Conservation Monitoring Centre. 1996. Original Forest Map, Current Forest Map. Cambridge, UK: WCMC.

Agroforestry Systems **61:** 183–194, 2004.

Managing biological and genetic diversity in tropical agroforestry

K. Atta-Krah[1,*], R. Kindt[2], J.N. Skilton[1] and W. Amaral[3]
[1]*International Plant Genetic Resources Institute (IPGRI), sub-Saharan Africa Regional Office, P.O.Box 30677, Nairobi, Kenya;* [2]*World Agroforestry Centre, P.O.Box 30677, Nairobi, Kenya;* [3]*IPGRI Headquarters, Via dei Tre Denari 472/a 00057 Maccarese, Rome, Italy;* [*]*Author for correspondence: e-mail: k.atta-krah@cgiar.org*

Key words: Agroecosystems, Genetic diversity, Genetic management, Conservation, Tree genetic resources

Abstract

The issues of biological and genetic diversity management in agroforestry are extremely complex. This paper focuses on genetic diversity management and its implications for sustainable agroforestry systems in the tropics, and presents an analysis of the role and importance of inter- and intra-specific diversity in agroforestry. Diversity within and between tree species in traditional agroforestry systems and modern agroforestry technologies in the tropics is assessed, with a view to understanding the functional elements within them and assessing the role and place of diversity. The assessment shows that although the practice of agroforestry has been a diversity management and conservation system, research in agroforestry over time has de-emphasized the diversity element; nevertheless farmers do value diversity and do manage agroforestry from that perspective. Based on a profiling of various traditional agroforestry systems and research-developed technologies, a strong case is made for increased species- and genetic diversity, at both inter- and intra-specific levels. The review and analysis point to the need for increased awareness, training/education, partnerships and collaborative efforts in support of genetic diversity in agroforestry systems; of special importance is increased cross-disciplinary research.

All the flowers of all the tomorrows are in the seeds of today

– A Chinese proverb

Introduction

It has been well recognized that agroforestry can serve to bridge the conflict and the divide that often exists between the need for conservation of biodiversity and provision of needs of human society (McNeely and Scherr 2003). Generally, human demands for forest products or conversion of forests into agriculture or human settlement take priority over forest conservation. When tropical forests are modified due to commercial logging, illegal felling, shifting cultivation and conversion to pasture and agriculture, ecosystem structure is simplified, and biodiversity is destroyed. More attention should be given to the need to conserve biodiversity, through promotion of sustainable management and use of our genetic resources. Human activities that result in simplification and fragmentation of biodiversity need to be better managed, taking the above into consideration. But first, let us examine the basic concepts of biological and genetic diversity.

Biological and genetic diversity

Biodiversity consists of a hierarchy of definitions from the molecular level through taxa to the landscape level. The United Nations Convention on Biological Diversity (CBD) defines biodiversity as 'the variability among living organisms from all sources including *inter alia*, terrestrial, marine and other aquatic ecosystems and ecological complexes of which they are a part; this includes diversity within species, between species and of ecosystems' (CBD 1992). The utility of the biodiversity concept, while imperfect given its all-inclusive nature (Lautenschlager 1997), gives us a framework on which to focus attention on ecological variability across a hierarchy of scales and levels of organization (Noss 1990). This definition of biodiversity recognizes three major levels: ecosystem diversity, species diversity and genetic diversity. In general, an ecosystem is 'a relatively homogeneous area of organisms interacting with their environment' that may occur in one of three forms: patches, corridors, or area of matrix (Forman 1995). Ecosystem diversity refers to the variation among ecosystems within a landscape, including the variety of ecological processes or habitats. Because ecosystems are not discrete biological entities like species and genes (Hunter et al. 1988), defining what constitutes an ecosystem may be problematic.

Species are the diversity units that ecologists are best able to count and so are frequently used as a practical measure of diversity. In addition, biodiversity can also be defined in terms of assemblage of species in space, such as in beta-diversity (species composition along an environmental gradient) and in gamma-diversity (species composition within similar habitats at different sites) (Whittaker 1965).

Genetic diversity is a fundamental component of biodiversity, forming the basis of species and ecosystem diversity. It represents all of the genetically determined differences that occur between individuals of a species in the expression of a particular trait or set of traits. There are three fundamental levels of genetic diversity: genetic variation within individuals (heterozygosity), genetic variation among individuals within a population, and genetic differences among populations. An understanding of the extent and distribution of genetic variation within and among plant populations is essential for determining appropriate genetic management strategies for utilization and conservation purposes. Genetic variation plays a critical role in the ability of populations to respond to changing environments. Therefore, it is usually a good strategy to ensure in modified or man-made ecosystems a broad genetic base to mitigate the effects of uncertainty, such as that associated with climate change or changing environments (Bawa and Dayanandan 1998). To buffer against risks, the intra- and inter-specific diversity of associated species such as N_2-fixing microsymbionts, mycorrhizal species, and pollination and seed dispersal vectors, also need to be conserved (Kindt 2002). In addition, the extent and causes of genetic variation provide the basis of any evolutionary studies (Weir 1996).

Biodiversity is not an issue only within forests, parks and other unmanaged or natural ecosystems but also within agriculture. Agricultural biodiversity includes all the components of biological diversity relevant to food and agriculture such as crops, trees, fish and livestock, and all interacting species of pollinators, symbionts, pests, parasites, predators and competitors (see also Qualset et al. 1995). Soil organisms, for instance, contribute a wide range of essential services to the sustainable function of agroecosystems through their role in nutrient cycling, soil carbon sequestration and greenhouse gas emission, their effects on soil physical structure and water regimes, and influence on plant life (e.g. N_2-fixation and interactions in the soil of pests, predators and other organisms) (Swift and Anderson 1999). Pollinators are essential for seed and fruit production and their number and diversity can profoundly affect crop production levels.

The importance of the different components of agricultural biodiversity, and the contribution they make to sustainable production, livelihoods and ecosystem health are now becoming generally recognized. Thus, farmers use diversity within and between crops, livestock, and productive tree species for risk avoidance, increased food security and income generation, as well as to optimize land use and help adapt to changing conditions (Brush 1995). Agroforestry contributes to agricultural biodiversity as it incorporates additional species (mainly trees and shrubs) into agriculture. This is based on the understanding that the diversity existing in the agricultural system prior to the introduction of the tree component is not sacrificed or compromised in any way. In this paper the basic need and importance of diversity in agricultural systems is accepted as a core part of the diversity within agroforestry. In addressing the issue of biological and genetic diversity in agroforestry, the paper places emphasis on the tree/shrub species that is introduced into the system, where the agroforestry practitioner can directly influence the overall diversity of the system.

The importance and nature of genetic diversity

Genetic diversity ensures the long-term survival of species and is therefore important for the sustainability of entire ecosystems (SGRP 2000). Genetic diversity generally arises in populations through mutations that can alter the DNA sequence and is maintained through a complex combination of factors. The important issue in the case of agroforestry systems is that these factors tend to operate differently in small and large populations. This is because most tree species in agroforestry systems generally occur at very low densities on farm. Recent studies of tree densities and germplasm sources in agroforestry systems in western and central Kenya, central Uganda and in Cameroon showed that 75% of all tree species observed on farms were represented at a density of one or less individual per hectare (Kindt 2002). For this reason, population genetics should be of particular interest to agroforesters. The challenge is to determine the 'genetic diversity window' into which tree species must fall to ensure a healthy balance of similar and dissimilar genes to avoid both inbreeding and out breeding depression.

Three biological reasons explain why genetic diversity is important in agroforestry systems. The first and perhaps the most powerful reason is to guard against the instability that can result from its absence. Genetic diversity enables evolution and adaptation of species to take place within changing environments, both in natural ecosystems and on farms. The fundamental theorem of natural selection (Fischer 1930) states that the rate of evolutionary change in a population is proportional to the amount of genetic diversity available. It is the variation between individuals of the same species that ensures that the species as a whole can adapt and change in response to natural and artificial selection pressures (Hawkes et al. 2000). Second, heterozygosity, or high genetic variation within an individual species, is positively related to fitness, i.e., the relative contribution of an individual's genotype to the next generation in context of the population gene pool (Hansson and Westerberg 2002). Trees often carry a heavy genetic load of deleterious recessive alleles (Boshier 2000). Avoidance of inbreeding is therefore particularly relevant for these organisms. Third, the 'blueprint' of life representing all the information for all the biological processes on this planet is locked in all the genetic diversity available (Meffe and Carroll 1994). Loss of this diversity would mean the loss of the potential for any improvement to meet changing human needs and end-use requirements. The additive and interactive effects of inter- and intraspecific genetic diversity determine both the resilience of agroecosystems and the evolutionary potential of species (Sauchanka 1997; SGRP 2000). This is becoming more important as we live in an increasingly changing environment with agricultural developments, global warming, pollution and desertification (CBD 2003).

An intense academic debate is going on currently about the relationship between complexity (and thus diversity) and stability of ecosystems (Loreau et al. 2001; Naeem 2002). The prevailing hypothesis is that diversity begets stability. This is based on the 'insurance hypothesis'; the impeccable logical notion that having a variety of species insures an ecosystem against a range of environmental upsets. Lack of convincing empirical evidence of the expected positive diversity–stability relationship has, however, led to controversy among ecologists (Kindt 2002). On the other hand, the niche-complementarity hypothesis based on the assumption that a community of species whose niches complement one another, shows that a multispecies system is much more efficient in using resources than an equivalent set of monocultures. Research on models with multiple species has shown that the effect of adding a species to a monoculture system, on the stability and the productivity of an ecosystem will be larger than adding the same species to a multispecies system (Tilman et al. 1997; Norberg et al. 2001). The effects of diversification will thus be largest when monoculture systems are targeted.

Traditionally, morphological and agronomic traits have been used to characterise patterns of diversity in plants. It is now known that these represent only a small proportion of the genome. Such traits are influenced by environmental factors, thus limiting their use for description of genetic relationships and variability. Molecular approaches such as the use of isozymes, and other genetic markers, which may be independent of environment and production responses, are likely to provide a more powerful method to gauge species relationships and origins (Dawson and Chamberlain 1996). McQueen (1996) confirmed that studies of molecular data, polyploidy and hybridization research, rather than morphological work, were needed to understand the complex patterns of variation in the species *Calliandra calothyrsus*. The same could be said for all other agroforestry species.

Management of genetic diversity

In the tropics, conservationists have focused their attention on the protection of natural forests and woodlands and, until recently (Schellas and Greenberg 1996), not given much attention to the widely dispersed forest patches throughout human-occupied landscapes. These patches are often critical components of a farmer's environment, being a source of products and environmental services of importance to their livelihood and welfare. As biogeographical islands, their role in maintaining biological diversity is crucial. Issues of scale are central to this role and thus in a landscape mosaic, forest patches and areas of agroforestry are potentially complementary, especially when considering the need for ecological equilibrium and population size vis à vis genetic diversity (Gajaseni et al. 1996).

Management of genetic resources aims at optimizing the use and benefits of the resources, while at the same time ensuring their continued availability for present and future generations. In agroforestry, management of diversity must be looked at in the context of people's livelihoods. Farmers plant trees in pursuit of their livelihood goals of income generation, risk management, household food security and optimal use of available land, labour and capital (Arnold and Dewees 1997). It is worth noting that, generally, diversity matches farmers' specific needs in specific situations; it is dynamic as it varies from household to household and from place to place and changes over time, and serves several purposes (Almekinders and de Boef 2000). For long-lived organisms such as trees, a better understanding of key factors determining the long-term conservation of these resources is vital and requires an interdisciplinary approach, integrating ecological, genetic and socioeconomic information (Bawa 1997). Equally important is an understanding of the social, economic and political factors that support the decision making processes of farmers, communities and other stakeholders to adopt (or not), certain land-use systems or management practices. Most of these factors are related to short-term gains that often influence long-term conservation and use options of genetic resources, especially of trees.

Conservation and management strategies of genetic resources may be grouped primarily into two domains: *in situ* and *ex situ* conservation. *In situ* conservation is defined as 'the conservation of ecosystems and natural habitats and the maintenance and recovery of viable populations of species in their natural surroundings and, in the case of domesticated or cultivated plant species, in the surroundings where they have developed their distinctive properties.' *Ex situ* conservation, on the other hand, is defined as 'the conservation of plant genetic resources for food and agriculture outside their natural habitat.' In *ex situ* conservation the genetic materials therefore need to be physically collected from areas of their origin, diversity or availability, and conserved through various mechanisms and approaches available (McNeely 2004).

Hughes (1998) discusses the merits of *in situ* (maintenance of natural population), *ex situ* (e.g., germplasm banks and botanic gardens) and *circa situm* (maintenance while in agricultural use, e.g., as hedgerow) conservation. Ruredzo and Hanson (1988) have applied in vitro techniques for conservation and multiplication of germplasm, and elimination of disease in the case of *Leucaena leucocephala*, *Erythrina brucei* and *Sesbania sesban.* Each of the different conservation approaches has its own advantages and disadvantages and should not be viewed as options but rather should be practiced as complementary approaches to conservation (Maxted et al. 1997; Hawkes et al. 2000). The complementary conservation strategy is defined as a: 'combination of different conservation actions, which together leads to an optimum sustainable use of genetic diversity existing in a target gene pool, in the present and future' (IPGRI 1993). Complementary conservation strategy should not be construed simply as a complementarity between *ex situ* and *in situ* conservation. It should be recognized that there is a continuum across the two approaches, extending from the conservation of wild plants in protected natural areas at one end of the spectrum to DNA libraries at the other end. The choice of the methods used will be dependant on many factors relating to the species biology, conservation and use objectives, human and financial resources, and legal issues. Further, since the ultimate goal of any conservation and management strategy of genetic resources is to make them available for use, the different conservation approaches should also examine complementarity between conservation aspects with the use of plant genetic resources.

Biodiversity issues in agroforestry systems and technologies

In this section, we attempt to chronicle and review various agroforestry systems and technologies, with a view to understanding the functional elements within them and assessing the role and place of diversity. Agroforestry practice in tropics and sub-tropics is probably as old as agriculture itself, and it is considered as a way of life of traditional farmers, although research on it started only about 25 years ago. In reviewing agroforestry systems, a distinction is made between the age-old traditional agroforestry systems and agroforestry technologies that have been developed through formal scientific research (Nair 1985). In addition to the above, a third category could be added – those born out of research, but later modified, adapted and further developed by farmers, as a result of their experiences and needs.

Table 1 provides an assessment of biodiversity and tree function considerations within traditional agroforestry systems. Many of these traditional systems are thought to maintain valued biological interactions and biodiversity at higher levels than some of the 'new' agroforestry technologies (Leakey 1998). In general, agroecosystems that are more diverse, more permanent, isolated, and managed with low input technology (e.g., agroforestry systems, traditional polycultures) take fuller advantage of ecological processes associated with higher biodiversity than highly simplified, input-driven and disturbed systems, i.e., modern row crops and vegetable monocultures and fruit orchards (Altieri 1995).

A study conducted by Backes (2001) on the contribution of agroforestry land use to the *in-situ* conservation of indigenous trees within a typical East African smallholder farming system in western Kenya shows how species diversity is ultimately linked to the loss of habitat diversity and landscape diversity.

Biological research in most research-developed agroforestry technologies has tended to focus primarily, on selection of particular tree species, and their management within the farming system, but genetic diversity has received only limited attention in such research (Table 2). It is observed that most agroforestry technologies are developed using only few selected tree species – often in mono-tree species systems, usually with preferred characteristics such as high-yielding, fast-growing, nitrogen-fixing ability and arboreal structure. The alley farming technology, for instance, became almost synonymous with either *Leucaena leucocephala* or with *Gliricidia sepium*. Improved fallows commonly consisted of mono-species fallows, with *Sesbania sesban*, *Crotalaria* spp. or *Tephrosia* spp. (Kwesiga et al. 1999). Such an approach results in low species diversity on farms, and makes the entire system vulnerable to attack by insects or diseases. This is especially so, if one considers that many of the species that are being promoted belong to the same botanical family of Fabaceae or Leguminosae (Rao et al. 2000).

Agroforestry is a diversity-enhancing land-use system, especially in the context of interspecies diversity, as it brings together crops, shrubs, trees and in some cases, livestock on the same piece of land. The question is how much research do we really see involving such levels of species mixtures and implications of genetic diversity, either between species or within particular species? Some research has been done using multiple tree species in agroforestry systems; however such research is only recent and embryonic.

Genetic diversity considerations in species and provenance screening for agroforestry

In the early days of agroforestry research, several studies were conducted involving a range of potential tree species for the different technologies (Heineman et al. 1997; Karachi et al. 1997), involving species screening and provenance testing (Duguma and Mollet 1997). This approach is believed to have led to selection of a few 'silver bullet' species used widely in agroforestry technologies. It is known, however, that in some cases, farmers have gone ahead of research recommendations and used multiple tree species in the technologies. For example, research into the local knowledge on interspecific diversity in western Kenya showed that farmers were maintaining and even increasing diversity in such technologies, where they could see complementarities among species (Kindt 2002).

The second type of tree screening research has been provenance testing; it involved the screening of populations within particular species. For example, in an effort to extend the geographic domain of the species capirona (*Calycophyllum spruceanum*), a priority agroforestry species in the Peruvian Amazon basin, Weber et al. (2001) carried out provenance trials to select those adapted to certain soil types and rainfall regimes. The study showed that most genetic variation was within rather than between populations, but that there were significant differences among populations, and some were more diverse than others (Weber et al. 2001).

Gliricidia sepium is another agroforestry species in which a lot of provenance screening has been done. Concerned about the low genetic diversity of the germplasm of this Central American species, which is in use in West African region as well, a research project launched in 1983, introduced new accessions of the species into the region (Sumberg 1985; Atta-Krah 1988). Through this project, different accessions and genes of *Gliricidia* were introduced into the environment. It is believed that due to its out-crossing

Table 1. Biodiversity dimensions in traditional agroforestry systems in the tropics.

Agroforestry system	Tree(s) function(s)	Biodiversity issues
Shifting cultivation or slash-and-burn	Nutrient replenishment, weed suppression, and control of pests and diseases	Long fallow periods of 15 to 20 years preserve wild species diversity (Ruthenberg 1980). Short fallow periods of 5 to 10 years due to increased human and livestock pressure on land use diminish species and functional diversity. Fallows consist of multiple species; and biological diversity, in both inter- and intra species, is intense.
Homegardens and compound farms	Sustainable production of diverse products for subsistence and markets to a limited extent, and nutrient cycling	These man-made complex systems represent high inter- and intra-species diversity involving a number of fruit, fodder and timber trees and shrubs, food crops, medicinal and other plants of economic value (Kumar and Nair 2004).
Forest gardens/agroforests	These systems practiced mostly in humid tropics resemble forests structurally and functionally.	They maintain high species diversity similar to natural forests but dominated by a few carefully managed economically valuable tree species (Kaya et al. 2002; Wiersum 2004).
Parkland systems	Trees in these traditional agrosilvopastoral systems practiced in the arid to semiarid regions of Africa provide economic products and ecological services.	Parks range from monospecific to multispecific with up to 20 tree species (Depommier et al. 1991). A variety of crops grown in association with naturally propagated trees ensure wide species diversity (Teklehaimanot 2004).
Trees on farmlands (boundary plantings, scattered trees)	Trees are primarily meant for products (e.g., poles, timber, fodder etc.) but other functions include boundary marking and shelter.	Trees are planted on farms in different niches (Nair 1993; Tejwani 1994). Diversity is more at the landscape level rather than at field level in terms of both inter- and intra-species. Recent plantings with fast growing exotic species has the tendency to reduce tree species diversity.

Table 2. Biodiversity dimensions in research-developed agroforestry technologies in the tropics.

Agroforestry technology	Tree(s) function(s)	Biodiversity issues
Alley cropping/ hedgerow intercropping/alley farming	Basic function of hedgerows is nutrient replenishment; they assume the additional functions of soil conservation on sloping lands and fodder production when livestock are integrated	Experimented and promoted mostly using mono-tree and -crop species (Kang et. al. 1985; Atta-Krah 1990; Kang et al. 1990); diversity is limited to intraspecies. Emphasis on a few tree species has raised concerns on pests and diseases (Rao et al. 2000). Tree diversity can be increased through multispecies hedgerows, and crop diversity increased by adopting intercropping in the alleys to increase efficiency of nutrient cycling.
Improved fallows or planted fallows	Soil fertility replenishment	Mostly based on mono-tree species (Buresh and Cooper 1999). Multispecies fallows combining coppicing and non-coppicing species (Chirwa et al. 2003) or species differing in leaflitter characteristics (Mafongoya et al. 1998) are likely to enhance fallow function as well as reduce risk from pests (Desaeger and Rao 2001).
Fodder banks	Year-round production of high quality fodder for stall-fed livestock	Fodder banks could be sole stands of either leguminous trees or shrubs or high yielding fodder grasses; however, systems combining trees and fodder grasses (Cooper et al. 1996), or different tree species can be cultivated to increase productivity, quality of pure grass fodders and sustainability production.
Rotational woodlots	Products (e.g., pulpwood, poles, construction material, etc.) and site enrichment and C-sequestration	Rotational woodlots have been planted using sole stands of fast growing species for short-cycle harvest (Nyadzi et al. 2003). Mixing of N_2-fixing species with non- N_2-fixing species will improve nutrient cycling and site enrichment compared with non- N_2-fixing species alone (Khanna 1998).

nature, these materials may have crossed out with other lines already existing in the region, and have therefore broadened the genetic base of the *Gliricidia* in the region.

Toky (2000) examined intraspecific variation in three important agroforestry tree species in semiarid India, viz. *Prosopis cineraria*, *Albizia lebbeck*, and *Acacia nilotica*. Based on the evaluation of eight provenances of *P. cineraria*, 12 of *A. lebbeck*, and 21 of *A. nilotica*, he reported large provenance variations in growth, fodder quality and nitrogen fixing ability (See also Puri and Nair 2004). The author opined that the wide range of intraspecific diversity has helped these species to survive much greater biotic and abiotic pressures over the past centuries. Genetic diversity within these species was also said to be of great value for rural development.

Strengthening species and genetic diversity in agroforestry research

There is a need to strengthen the inter- and intraspecific diversity dimensions in agroforestry technologies. For example, research in a system such as alley farming could involve the use of multiple tree species in the same field. Alternate hedgerows could consist of different tree species, or different provenances of the same species, or mixed species within the same hedgerow. Tree species with complementary characteristics and use that will add value to the overall system could also be explored. For example, a species with high nitrogen fixation and fast foliage decomposition rate could be combined with one that has foliage with slower decomposition rate, and better mulching characteristics (Mafongoya et al. 1998). The system as a whole therefore benefits from both the nitrogen fixation and the improved soil-water retention through improved mulching. Fast growing species could also be combined with slower growing species in order to reduce the frequency and intensity of pruning of the hedgerows. Such combinations could also stagger the pruning schedules so that farmers would not be under pressure to prune their entire field at the same time, thereby introducing an element of flexibility into the management of trees in the system.

The issues raised above for alley farming are relevant to any other agroforestry technologies. There is no reason for instance, why improved fallow technologies need to be established using only one tree species. Multiple species fallows will provide an additional level of stability and resilience and minimize the chances of pest and disease outbreaks. Recent history has examples of single species promoted on a wide scale that have then crashed as a result of insect or pest outbreaks. The damage wrought by the *Leucaena* psyllid (*Heteropsylla cubana*) is a good example of the perils of relying too much on one species or one cultivar (Rao et al. 2000). Future plantings should contain not only a range of species, but probably also a range of varieties within species. This should safeguard against major disasters due to diseases or insects (Bray 1998). A diversification strategy could investigate differences between genera or families, as introducing new species of the same genus or family may not decrease risk very much.

Even more efforts are needed in research for understanding the intraspecific diversity issues in agroforestry. Various trials have been undertaken on agroforestry species, with the aim of assessing intraspecific diversity, identifying superior lines and ultimately, broadening the genetic base for priority species; but many of these agronomic trials do not include the molecular characterization and analysis that is required for a proper assessment of genetic variation. Furthermore, no mechanisms appear to be in place to move such research beyond provenance screening on stations, to get new provenances and species into farmers' fields. Consequently, there are very few examples of successful introduction of new provenances and genetic lines released for wider scale use, as happens in the case of agricultural crops (see also Simons and Leakey 2004). Part of the reason for this may be the nature of trees and their long cycle growth. Major complexities also come in with regard to maintaining particular provenances as pure lines, due to out-crossing characteristics, and the difficulty of maintaining discrete seed production fields for the selected provenances.

In line with the need for increased genetic diversity research in agroforestry, the World Agroforestry Centre (ICRAF) has recently established a molecular characterization laboratory as part of its program on Tree Domestication. This development is laudable and has contributed in a great way to strengthen genetic diversity analysis in agroforestry species. Various studies have been undertaken or are underway for priority species for domestication in Africa and Latin America, including *Calycophyllum spruceanum, Irvingia gabonensis, Gliricidia sepium, Leucaena* species, *Prunus africana, Sesbania sesban, Uapaca kirkiana, Vitex keniensis* and *Warburgia ugandensis* (Dawson and Powell 1999; Russell et al.

1999; Lowe et al. 2000; Simons and Leakey 2004). The study on *Calycophyllum spruceanum* was for example the first systematic molecular study of genetic variation in any tree species of the Peruvian Amazon (Russell et al. 1999). These studies have focused on the partitioning of genetic diversity within and among populations. Many of these studies have confirmed the general pattern within trees that most genetic variation occurs within rather than among populations, although significant differentiation may still occur among populations.

The objective of agroforestry domestication should not be just to select super species or provenances but should include promoting genetic diversity and match intra-specific diversity to the needs of farmers, markets and diversity of environments (Simons and Leakey 2004). Different farmers could be using different provenances. Therefore on landscape basis, one could experience great diversity among a particular species that may be in use. As a result of out-crossing, it is expected that there will be gene-mixing in seed production, with a continuous creation of new hybrids and complexes that would be adding to broadening of the genetic base. Sometimes mixtures of populations will have lower fitness as genetic diversity related to local adaptation gets disrupted in mixtures (Young and Boyle 2000).

Seed systems also influence the maintenance of genetic diversity or the promotion of genetic erosion in agroforestry. The basic question is 'from where do farmers get their seeds for agroforestry practice?' In most cases farmers obtain seeds from other farmers who may also have harvested the original seeds from a single tree on their farm. It is possible that a particular species could spread very fast, even though it may have a very narrow genetic base. A number of authors have indicated that farmers and nursery managers frequently collect germplasm from a relatively narrow range of maternal parents (mother trees) during propagation (Weber et al. 1997; Holding and Omondi 1998). The long-term viability of on-farm tree stands depends upon a wide genetic base providing the capacity to adapt to environmental fluctuations or changing farmer requirements, such as a change in species use, planting niche or pest outbreak. Moreover, many tree species are out-breeding. They therefore require a wide genetic base to withstand potential inbreeding depression, which may result from an increase in homozygosity and subsequent expression of unfavourable recessive alleles during generations of farmer propagation (Boshier 2000; Simons 1996).

Seeds of some agroforestry trees are also obtained from established institutions such as tree seed centres in various countries. Often these institutions are acting more as seed stores, with responsibilities for seed collection, processing, storage and distribution. Most tree seed centres in Africa and other developing countries in the tropics have no facilities or capacity for genetic analysis and therefore no consideration is given to avoiding genetic erosion or promoting genetic diversity.

Do farmers value diversity in their agroforestry systems?

The fact that farmers actively plant and/or protect trees on their farms can be seen as an indicator of the fact that they appreciate trees in their farming systems. There is increasing evidence that as natural forests recede or get degraded, farmers in many situations have historically taken up planting and managing of trees on their lands to provide the needed outputs. In a study conducted in the Middle Hills of Nepal, Gilmour (1995) reported a fourfold increase in the density of trees on farms in crop-growing areas of Nepal. Similar trends have been observed in Kenya (Pretty et al. 1995) and other regions of the tropics (Murray and Bannister 2004).

Farmers generally do care for diversity in their farming systems. This is particularly so in the context of subsistence and smallholder farmers in the tropics. Farmers plant trees in pursuit of their livelihood goals of income generation, risk management, household food security and optimal use of available land, labour and capital (Arnold and Dewees 1997). The diversity of plant and animal species maintained in traditional farming systems over many centuries, and the knowledge associated with managing these resources, constitute key assets of the rural poor in Africa (Brush 1999). In marginal and difficult farming conditions, these assets and practices are especially important. Diversity management can constitute a central part of the livelihood management strategies of farmers (particularly pastoralists) and communities in different production systems throughout Africa (Rege et al. 2003).

It needs to be stressed, however, that most farmers will not maintain or promote diversity just for the sake of it. Nor will they readily keep diversity solely on grounds of long-term 'stability' of the system or the need to conserve particular species or genetic types for future generations. Diversity must be useful to farmers for them to decide to keep and conserve it. Species that may be of neither immediate nor perceived future benefit are unlikely to be conserved by farmers in their field. One possibility for conserving tree species diversity may be to reward farmers for the biodiversity conservation service that they provide to the global community. This may be especially relevant for species that cannot be conserved *ex situ* – for instance species with recalcitrant seed that cannot maintain viability for long-term periods, or species that require conservation of co-evolved organisms that cannot be conserved *ex situ*. The question always remains as to who pays for the cost of such conservation efforts?

Institutional and policy aspects of genetic diversity in agroforestry

National gene banks and tree-seed centers are often the institutions with lead responsibility for agrobiodiversity conservation, while the agriculture research institutions are responsible for technology development and crop improvement research. One of the biggest weaknesses in agricultural research and development in most tropical countries is the inadequate interaction between the national institutions responsible for agriculture research and conservation and management of genetic resources. Without such a partnership, germplasm may be collected and stored in gene banks, but may not be put to productive and sustainable use for the enhancement of agriculture and food security.

Agricultural scientists are also not adequately familiar with the issues of biodiversity and agrobiodiversity; fewer still may have skills for genetic diversity analysis. Genetic diversity issues and skills need to be built into different aspects and levels of agricultural training. It is therefore important that universities and other institutions of training are fully involved in this work. More graduates need to be trained in plant genetic resources and in genetic diversity, particularly of tree crops. Curriculum development should also aim at incorporation of genetic diversity elements into standard agriculture and natural resources training.

Past agroforestry research was dominated by scientists dealing with crops, forestry, soils or databases and it suffered from too little insight from social scientists, anthropologists and economists. As a result, we ended up knowing more about the biology of agroforestry (crops, trees, water and soils), and less or little about the driving forces of what people preferred, whether in existing farming systems or in terms of potential new crop plants. Agroforestry is not alone in having this problem – it applies to the conservation field as well. Gradually, interdisciplinary team research – incorporating local farmers and innovative methods – is improving the recognition and use of local knowledge within agroforestry (http://www.rbgkew.org.uk/peopleplants/handbook/handbook5/interviews.htm).

Policy studies in agricultural research should include genetic resources policies and how they impact on biodiversity conservation and use. Agricultural scientists must be made aware of the policy and legislation aspects of genetic resources, and be encouraged to be involved in the development of policies that will promote the proper use, management and legal aspects of genetic resources. Essential issues in this regard include those of access of genetic resources, benefit sharing, intellectual property rights, farmers' rights, and documentation and capacity development. Agriculture and natural resources scientists also need to be fully informed on global policy conventions and instruments on genetic resources, and be able to offer technical advice to governments in this area.

Future research directions

Traditional agroforestry systems are rich in biological and genetic diversity. Modern agroforestry technologies developed through research have not adequately addressed the genetic diversity issue. Increased efforts should be made to promote diversification of species in agroforestry technology development. Methods should be explored of incorporating the element of risk in promoting species in agroforestry technologies, whereby risks and benefits of monoculture technologies should be contrasted with those of mixed species technologies.

There should be increased efforts at studying the intraspecific diversity in agroforestry species, and for expansion of diversity in this domain. The use of molecular techniques for characterization and management of diversity in agroforestry needs to be encouraged. In most species and provenance screening trials conducted for agroforestry, the emphasis appears to have been on identifying the most suitable species or provenances, according to a particular defined set of criteria, rather than to explore the existing diversity and finding mechanisms for the use of the diversity. There is need to review agroforestry research objectives, and integrate genetic diversity concerns, to strengthen sustainability and stability of production systems.

The knowledge base in genetic resources management and conservation and in genetic diversity analysis needs to be broadened. This could be done through training and capacity development, and through the incorporation of genetic diversity modules into educational program in universities and colleges, even in primary and secondary schools. Agricultural extension officers should also be trained in the importance of genetic diversity, and how it can be managed through on-farm seed systems, tree domestication and other agroforestry approaches.

It follows, therefore, that a major requirement to obtain the full benefit of agroforestry is the incorporation of diversity and conservation elements into it. This will require improved levels of knowledge of the biodiversity in production systems, of its functions and benefits, of the consequences of changes in different elements, and of the ways in which agroforestry supports conservation of different components of biodiversity (e.g., crops, livestock, trees, soil micro-organisms, pollinators, etc.). Research scientists in agroforestry, agriculture, environmental sciences, etc. are challenged and encouraged to be fully involved in genetic diversity analysis and its links with agriculture.

training

References

Almekinders C. and de Boef W.D. 2000. Encouraging diversity: a synthesis of crop conservation and development. pp. 323–325. In: Encouraging Diversity: the Conservation and Development of Plant Genetic Resources. Intermediate Technology Publications, London, UK.

Altieri M.A. 1995. Agroecology: The Science of Sustainable Agriculture. Westview Press, Boulder, CO, USA, 433 pp.

Arnold J.E.M. and Dewees P.A. (eds) 1997. Farms, Trees and Farmers: Responses to Agricultural Intensification. Earthscan, London, UK, 292 pp.

Atta-Krah A.N. 1988. Genetic improvement of nitrogen-fixing trees for agroforestry purposes: the example of *Gliricidia sepium* in West Africa. pp. 132–147. In: Gibson G.I., Griffin A.R. and Matheson A.C. (eds), Breeding Tropical Trees: Population Structure and Genetic Improvement Strategies in Clonal and Seedling Forestry. Oxford Forestry Institute, Oxford, UK and Winrock International, Arlington, VA, USA.

Atta-Krah A.N. 1990. Alley farming with *Leucaena*: effect of short grazed fallows on soil fertility and crop yields. Exp Agr 26: 1–10.

Backes M.M. 2001. The role of indigenous trees for the conservation of biocultural diversity in traditional agroforestry land use systems: the Bungoma case study. *In situ* conservation of indigenous tree species. Agroforest Syst 52: 119–132.

Bawa K.S. 1997. *In situ* conservation of tropical forests. Focus paper for IPGRI. FORGEN news; Research Update on IPGRI's Forest Genetic Resources Projects. 1997 International Plant Genetic Resources Institute (IPGRI), Rome, Italy, 23 pp.

Bawa K.S. and Dayanandan S. 1998. Global climate change and tropical forest genetic resources. Climate Change 39: 473–485.

Boshier D.H. 2000. Mating systems. pp. 63–79. In: Young A., Boshier D. and Boyle T. (eds), Forest Conservation Genetics. Collingwood: CSIRO Publishing and CAB International, Wallingford, UK.

Bray R.A. 1998. Diversity within tropical tree and shrub legumes. In: Gutteridge R.C. and Shelton H.M. (eds), Forage Tree Legumes in Tropical Agriculture. http://www.fao.org/ag/AGP/AGPC/doc/Publicat/Gutt-shel/x5556e00.htm.

Brush S. 1995. *In situ* conservation of landraces in centres of crop diversity. Crop Sci 35: 346–354.

Brush S.B. 1999. Genes in the Field: Conserving Plant Diversity on Farms. Boca Raton, FL, Lewis, 15–16.

Buresh R.J. and Cooper P.J.M. (eds) 1999. The Science and Practice of Short-Term Improved Fallows. Special Issue of Agroforest Syst, Vol. 47: 1–365.

Chirwa T.S., Mafongoya P.L. and Chintu R. 2003. Mixed planted-fallows using coppicing and non-coppicing tree species for degraded Acrisols in eastern Zambia. Agroforest Syst 59: 243–251.

Convention on Biological Diversity (CBD) 1992. Article 2. Use of terms. http://www.biodiv.org/convention/articles.asp?lg=0&a=cbd-02.

Convention on Biological Diversity (CBD) 2003. Convention on Biological Diversity http://www.biodiv.org.

Cooper P.J.M., Leakey R.R.B., Rao M.R. and Reynolds L. 1996. Agroforestry and the mitigation of land degradation in the humid and sub-humid tropics of Africa. Exp Agr 32: 235–290.

Dawson I.K. and Chamberlain J.R. 1996. Molecular analysis of genetic variation. pp. 77–91. In: Stewart J.L., Allison G.E. and Simons A.J. (eds), *Gliricidia sepium* – Genetic Resources for Farmers. Tropical Forestry Paper 33. Oxford Forestry Institute, Oxford, UK.

Dawson I.K. and Powell W. 1999. Genetic variation in the Afromontane tree *Prunus africana*, an endangered medicinal species. Mol Ecol 8: 151–156.

Depommier D., Janoder E. and Oliver R. 1991. *Faidherbia albida* parks and their influence on soils and crops at Watinoma, Burkina Faso. pp. 111–116. In: Vandenbeldt R.J. (ed.), *Faidherbia albida* in the West African Semi-Arid Tropics: Proceedings of a Workshop, 23–26 April 1991, Niamey, Niger, ICRISAT, Patancheru-502324, A.P., India.

Desaeger J. and Rao M.R. 2001. The potential of mixed covers of *Sesbania*, *Tephrosia* and *Crotalaria* to minimise nematode problems on subsequent crops. Field Crop Res 70: 111–125.

Duguma B. and Mollet M. 1997. Provenance evaluation of *Calliandra calothyrsus* Meissner in the humid lowlands of Cameroon. Agroforest Syst 37: 45–57.

Fischer R.A. 1930. The genetical theory of natural selection. Clarendon Press, Oxford. In: Meffe G.K. and Carroll C.R. (eds), Principles of Conservation Biology. Sinauer Associates, Inc. Publishers, Sunderland, MA. 272 pp.

Forman R.T.T. 1995. Land Mosaics: the Ecology of Landscapes and Regions. Cambridge University Press, 632 pp.

Gajaseni J., Matta-Machado R. and Jordan C.F. 1996. Diversified agroforestry systems: buffers for biodiversity reserves, and landbridges for fragmented habitats in the tropics. pp. 506–513. In: Szaro R.C. and Johnston D.W. (eds), Biodiversity in Managed Landscapes: Theory and Practice. Oxford University Press, Oxford, UK.

Gilmour D.A. 1995. Rearranging trees in the landscape in the middle hills of Nepal. pp. 21–42. In: Arnold J.E.M. and Dewees P.A. (eds), Tree Management in Farmer Strategies. Oxford University Press, Oxford, UK.

Hansson B. and Westerberg L. 2002. On the correlation between heterozygosity and fitness in natural populations. Mol Ecol 11: 2467–2474.

Hawkes J.G., Maxted N. and Ford-Lloyd B.V. 2000. The *Ex Situ* Conservation of Plant Genetic Resources. Kluwer Academic publishers, The Netherlands, 250 pp.

Heineman A.M., Otino H.J.O., Mengich E.K. and Amadalo B.A. 1997. Growth and yield of eight agroforestry tree species in line plantings in western Kenya and their effect on maize yields and soil properties. Forest Ecol Manag 91: 103–135.

Holding C. and Omondi W. (eds) 1998. Evolution of Provision of Tree Seed in Extension Programmes: Case Studies from Kenya and Uganda. SIDA Technical Report No. 19. Regional Land Management Unit, Swedish International Development Co-operation Agency, Nairobi, Kenya. http://www.rbgkew.org.uk/peopleplants/handbook/ handbook5/interviews.htm.

Hughes C.E. 1998. *Leucaena*. A Genetic Resources Handbook. Tropical Forestry Papers No. 37. Oxford Forestry Institute, Oxford, UK, 274 pp.

Hunter M.L. Jr., Jacobson G.L. Jr. and Webb T. 1988. Paleoecology and the coarse-filter approach to maintaining biological diversity. Conserv Biol 2: 375–383.

IPGRI. 1993. Diversity for Development – The Strategy of the International Plant Genetic Resources Institute. IPGRI, Rome, 62 pp.

Kang B.T., Grimme H. and Lawson T.L. 1985. Alley cropping sequentially cropped maize and cowpea with *Leucaena* on a sandy soil in southern Nigeria. Plant Soil 85: 267–277.

Kang B.T., Reynolds L. and Atta-Krah A.N. 1990 Alley farming. Adv Agron 43: 315–359.

Karachi M., Shirma D. and Lema N. 1997. Evaluation of 15 leguminous trees and shrubs for forage and wood production in Tanzania. Agroforest Syst 37: 253–263.

Kaya M., Kammesheidt L. and Weidelt H.J. 2002. The forest garden system of Saparua Island, central Maluku, Indonesia, and its role in maintaining tree species diversity. Agroforest Syst 54: 225–234.

Khanna P.K. 1998. Nutrient cycling under mixed-species tree systems in southeast Asia. Agroforest Syst 38: 99–120.

Kindt R. 2002. Methodology for Tree Diversification Planning for African Ecosystems. Ph.D. thesis, University of Ghent, 332 + xi pp.

Kumar B.M. and Nair P.K.R. 2004. The enigma of tropical homegardens. Agroforest Syst (This volume).

Kwesiga F.R., Franzel S., Place F., Phiri D. and Simwana C.P. 1999. *Sesbania sesban* improved fallows in eastern Zambia: their inception, development and farmer enthusiasm. Agroforest Syst 47: 49–66.

Lautenschlager R.A. 1997. Biodiversity is dead. Wildlife Society Bulletin 25: 679–685.

Leakey R.R.B. 1998. Agroforestry for biodiversity in farming systems. pp. 127–145. In: Collins W. and Qualset C. (eds), The Importance of Biodiversity in Agroecosystems, Lewis Publishers, New York.

Loreau M., Naeem S., Inchausti P., Bengtsson J., Grime J.P., Hector A., Hooper D.U., Huston M.A., Raffaelli D., Schmid B., Tilman D. and Wardle D.A. 2001. Biodiversity and ecosystem functioning: current knowledge and future challenges. Science 294: 804–807.

Lowe A.J., Gillies A.C.M., Wilson J. and Dawson I.K. 2000. Conservation genetics of bush mango from central/west Africa: implications from random amplified polymorphic DNA analysis. Mol Ecol 9: 831–741.

Mafongoya P. L., Giller K.E. and Palm C.A. 1998. Decomposition and nitrogen release patterns of tree prunings and litter. Agroforest Syst 38: 77–97.

Maxted N., Hawkes J.G. and. Ford-Lloyd B.V. 1997. Plant Genetic Conservation: The *In Situ* Approach. Chapman and Hall, 446 pp.

McNeely J.A. 2004. Nature vs. Nurture: Managing Relationships Between Forests, Agroforestry and Wild Biodiversity. Agroforest Syst (This volume).

McNeely J.A. and Scherr S.J. 2003. Ecoagriculture: Strategies to Feed the World and Save Wild Biodiversity. Island Press, Washington D.C., 323 pp.

McQueen D. 1996. *Calliandra* taxonomy and distribution, with particular reference to the series Racemosae. pp. 1–17. In: Evans D.O. (ed.), International Workshop on the Genus *Calliandra*. Forest, Farm and Community Tree Research Reports Special Issue, 1996. Winrock International.

Meffe G.K and Carroll C.R. 1994. Principles of Conservation Biology. Sinauer Associates, Inc. Publishers, Sunderland, Massachusetts, 600 pp.

Murray G. and Bannister M. 2004. Peasants, agroforesters, and anthropologists: A 20-year venture in income-generating trees and hedgerows in Haiti (This volume).

Naeem S. 2002. Biodiversity equals instability? Nature 416: 23–24.

Nair P.K.R. 1985. Classification of agroforestry systems. Agroforest Syst 3: 97–128.

Nair P.K.R. 1993. An Introduction to Agroforestry. Kluwer Academic Publishers, Dordrecht, The Netherlands, 499 pp.

Norberg J., Swaney D.P., Dushoff J., Lin J., Casagrandi R. and Levin S.A. 2001. Phenotypic diversity and ecosystem functioning in changing environments: a theoretical framework. Proceedings of the National Academy of Sciences 98 (20): 11376–11381.

Noss R.F. 1990. Indicators for monitoring biodiversity: a hierarchical approach. Conserv Biol 4: 355–364.

Nyadzi G.I., Otsyina R., Banzi F.M., Bakengesa S.S., Gama B.M., Mbwambo L. and Asenga D. 2003. Rotational woodlot technology in northwestern Tanzania: Tree species and crop performance. Agroforest Syst 59: 253–263.

Pretty J.N., Thompson J. and Kiara J.K. 1995. Agricultural regeneration in Kenya: the catchment approach to soil and water conservation. Ambio 24: 7–15.

Puri S. and Nair P.K.R. 2004. Agroforestry research for development in India: 25 years of experiences of a national program. Agroforest Syst (This volume).

Qualset C.O., McGuire P.E. and Warburton M.L. 1995. Agrobiodiversity: key to agricultural productivity. Calif Agr 49: 45–49.

Rao M.R., Singh M. and Day R. 2000. Insect pest problems in tropical agroforestry systems: Contributory factors and management strategies. Agroforest Syst 50: 243–277.

Rege J.E.O., Hodgkin T. and Atta-Krah A.N. 2003. Managing agricultural biodiversity in sub-Saharan Africa to increase food production, conserve natural resources and improve human livelihoods. pp. 75–88. In: Building Sustainable Livelihoods through Integrated Agricultural Research for Development – Programme Proposal and Reference Materials. Forum for Agricultural Research in Africa (FARA), Accra, Ghana.

Ruredzo T.J. and Hanson J. 1988. The use of in vitro culture in germplasm management. ILCA Germplasm Newsletter No. 18: 4–9.

Russell J.R., Weber J.C., Booth A., Powell W., Sotelo-Montes C. and Dawson I.K. 1999. Genetic variation of *Calycophyllum spruceanum* in the Peruvian Amazon Basin, revealed by amplified fragment length polymorphism (AFLP) analysis. Mol Ecol 8: 199–204.

Ruthenberg H. 1980. Farming Systems in the Tropics. 3rd edition. Clarendon Press, Oxford, 424 pp.

Sauchanka U.K. 1997. The Genosphere; the Genetic System of the Biosphere. The Parthenon Publishing Group, London, UK, 134 pp.

Schellas J. and Greenberg R. 1996. Forest Patches in Tropical Landscapes. Island Press, Washington, D.C., 426 pp.

SGRP. 2000. Why Genetic Resources Matter. Genetic Resources Management in Ecosystems. Report of a workshop organized by the Centre for International Forestry Research (CIFOR) for the CGIAR System-wide Genetic Resources Programme (SGRP), held in Bogor, Indonesia, 27–29 June 2000. CIFOR, Bogor, Indonesia, 13 pp.

Simons A.J. 1996. Delivery of improvement of agroforestry trees. pp. 391–400. In: Dieters M.J., Matheson A.C., Nikles D.G., Harwood C.E. and Walker S.M. (eds), Tree Improvement for Sustainable Forestry. IUFRO Conference, 27 October–1 November 1996, Caloundra, Australia.

Simons A.J. and Leakey R.R.B. 2004. Tree domestication in tropical agroforestry (This volume).

Sumberg J.E. 1985. Collection and initial evaluation of *Gliricidia sepium* from Costa Rica Agroforest Syst 3: 357–361.

Swift M.J. and Anderson J.M. 1999. Biodiversity and ecosystem function in agricultural systems. pp. 15–41. In: Schulze E.D. and Mooney H.A. (eds), Biodiversity and Ecosystem Function.Berlin: Springer-Verlag.

Tejwani K.G. 1994. Agroforestry in India. Oxford & IBH Publish Co. Pvt. Ltd., New Delhi, 233 pp.

Teklehaimanot Z. 2004. Exploiting the potential of indigenous agroforestry trees: *Parkia biglobosa* and *Vitellaria paradoxa* in sub-Saharan Africa. Agroforest Syst (This volume).

Tilman D., Lehman C.L. and Thomson K.T. 1997. Plant diversity and ecosystem productivity: theoretical considerations. Proceedings of the National Academy of Sciences 94: 1857–1861.

Toky O.P. 2000. Genetic diversity in three important agroforestry tree species of dry zones: a case study. Indian J Agroforest 2: 43–47.

Weber J.C., Labarta Chávarri R.L., Sotelo-Montes C., Brodie A.W., Cromwell E., Schreckenberg K. and Simons A.J. 1997. Farmers' use and management of tree germplasm: case studies from the Peruvian Amazon Basin. pp. 57–63. In: Simons A.J., Kindt R. and Place F. (eds), Policy Aspects of Tree Germplasm Demand and Supply. Proceedings of an International Workshop, held in Nairobi, Kenya, October 1997, ICRAF, Nairobi, Kenya.

Weber J.C., Labarta R., Sotelo C., Brodie A., Cromwell E., Schreckenberg K. and Simons A. 2001. Farmer's use and management of tree germplasm: case studies from the Peruvian Amazon. Agroforest Today 12(1): 17–22.

Weir B.S. 1996. Genetic Data Analysis II: Methods for Discrete Population Genetic Data. Sinauer Associates Inc., Sunderland, 445 pp.

Whittaker R.H. 1965. Dominance and diversity in land plant communities. Science 147: 250–260.

Wiersum K.F. 2004. Forest gardens as an 'intermediate' land-use system in the nature – culture continuum: characteristics and future potential. Agroforest Syst (This volume).

Young A.G. and Boyle T.J. 2000. Forest fragmentation. pp. 123–134. In: Young A., Boshier D. and Boyle T. (eds), Forest Conservation Genetics. Collingwood: CSIRO Publishing and CAB International, Wallingford, UK.

Agroforestry Systems **61:** 195–206, 2004.

Why extensive research and development did not promote use of peach palm fruit in Latin America

C.R. Clement[1,*], J.C. Weber[2], J. van Leeuwen[1], C. Astorga Domian[3], D.M. Cole[4], L.A. Arévalo Lopez[5] and H. Argüello[6]
*[1]Instituto Nacional de Pesquisas da Amazônia (INPA), Cx. Postal 478, 69011-970 Manaus, AM, Brasil (bolsista do CNPq); [2]Formerly with World Agroforestry Centre-ICRAF, current address 2224 NW 11th Street, Corvallis, OR, 97330 USA; [3]Centro Agronómico Tropical de Investigación y Enseñanza (CATIE), 7170, Turrialba, Costa Rica; [4]School of Forest Resources and Conservation; The University of Florida; Gainesville, FL 32611 USA; [5]World Agroforestry Centre-ICRAF; Carretera Federico Basadre, km 4.200; Pucallpa, Perú; [6]Facultad de Agronomia; Universidad Nacional de Colombia; Apartado 14490; Bogotá, Colombia; *Author for correspondence: e-mail: cclement@inpa.gov.br*

Key words: Bactris gasipaes, Fruit crops, Heart-of-palm, Market analysis, Production-to-Consumption chain, Smallholder crops

Abstract

Peach palm (*Bactris gasipaes*) was domesticated as a fruit crop by the first Amazonians in traditional Neotropical agroforestry systems, but research and development (R&D) to date has not transformed its fruit into a modern success story. The fruit is really a tree 'potato,' competing with traditional starches rather than with succulent fruits. R&D efforts have focused more on production than on product transformation, commercialization and the consumer, thus failing to fill gaps in the production-to-consumption chain. Consumer demands are only now getting more consideration, and clear identification of the smallholder farmer as the R&D client is not yet generalized. Too many, often large germplasm collections have biased R&D programs away from smallholder farmers and did not pursue the quality and uniformity that consumers want. The general lessons learnt from 25 years of R&D efforts on peach palm that should guide the development of other indigenous agroforestry fruit tree species are: 1) identify market demands, whether subsistence or market-oriented; 2) identify clients and consumers, and their perceptions of the product; 3) work on food and nutritional security aspects of the species and let entrepreneurs be attracted, rather than *vice versa*; 4) take up species improvement in a moderately sized effort, using a participatory approach tightly focused on clients' demands; and 5) reappraise the priorities from time to time.

Introduction

Eighty years after the first major article highlighting the potential of peach palm (*Bactris gasipaes*) as a fruit crop (Popenoe and Jimenez 1921), 25 years after the U.S. National Academy of Sciences (NAS 1975) again highlighted the potential of peach palm for a hungry world, this crop is no more important as a fruit today than it was in 1921 or 1975. It may be less important in the future, though an intensive research and development (R&D) effort occurred in Brazil, Colombia, Costa Rica and Peru in the past 25 years. These years of R&D did, however, create a new crop: heart-of-palm, grown in high density, high input monocultures (Mora Urpí et al. 1997; Mora Urpí and Gainza Echeverria 1999), and widely traded in American and European gourmet food markets.

Peach palm was fully domesticated in Amazonia in agroforestry systems and it became a pre-Colombian staple with important nutritional qualities (energy from starch and oil, beta-carotene). Native Americans consumed it whole after cooking, or fermented into a thick beverage, or dried and made it into flour. These products were generally further processed into numerous dishes and drinks (including alcoholic), especially in northwestern South America and southern Central America. This pre-Columbian importance stimulated interest in the crop all through the 20th century.

In modern Brazil, Colombia, Costa Rica and Peru, 40% to 50% of annual fruit production fails to be commercialized in markets and is either fed to farm animals or wasted. Although there is apparently demand for peach palm flour, entrepreneurs have failed to create and maintain even a niche market for this product, and R&D efforts have failed to generate solutions to help change the economics of this niche. The R&D also failed to establish the use of whole fruit or flour in animal feeds and to stimulate a niche market even in areas where commercial animal feeds are expensive. These observations led us to conclude that peach palm today is underutilized.

This chapter attempts to identify the constraints in transforming a starchy fruit into a modern crop attractive to both urban and rural consumers. We will also identify some of the errors and successes of the R&D efforts, which can serve as lessons for the development of other indigenous agroforestry trees.

The crop: peach palm, pupunha, chontaduro, pejibaye, pijuayo

Peach palm is perhaps one of the best-known underutilized crops. Mora Urpí et al. (1997) provided an extensive review on the subject; the following paragraphs provide only updates or minimum information as background for the reader.

Botany

Peach palm is a multi-stemmed palm that may attain up to 20 m height. Stem diameter varies from 15 cm to 30 cm and internode length from 2 cm to 30 cm. The internodes are armed with numerous black, brittle spines, although spineless mutants occur and have been the subject of selection in several areas. The monopodial stem develops suckers at its base and is topped by a crown of 15 to 25 pinnate fronds, with the leaflets originating at different angles. The heart-of-palm is a gourmet vegetable composed of the tender unexpanded leaves in the palm's crown. The inflorescences appear from the axils of the senescent fronds. After pollination the bunch may contain between 50 and 1000 fruits, and weigh between 1 kg and 25 kg. Numerous factors may cause fruit drop, such as poor pollination, poor nutrition, drought, crowding, insects, and diseases. The fruits that ripen have a fibrous red, orange or yellow epicarp, a moist starchy/oily mesocarp, and a single endocarp with a fibrous/oily white endosperm. Individual fruits commonly weigh between 30 g and 70 g (range of 10 g to 250 g), and seeds weigh between 1 g and 4 g.

Nutritional characteristics of fruit

The fruit is energy-rich, due both to starches and oils (Table 1). The relative proportions of starch versus oil vary inversely along the domestication continuum, with fruits of wild types being rich in oils and the most derived domesticates rich in starches (Table 2). Protein quality is not exceptionally high, but the oil is rich in oleic acid (Yuyama et al. 2003). Peach palm's beta-carotene is almost completely bio-available (Yuyama et al. 1999), as are its energy and protein (although the latter is of only moderate quality (Yuyama et al. 2003)). The fruit contains two anti-nutritional factors, a trypsin inhibitor and calcium oxalate crystals, which can be denatured or dissolved by boiling, respectively. The chemical composition and requirement for cooking make the name peach palm a misnomer, as the fruit is starchy and oily rather than succulent. It is only a fruit in the botanical sense. It can neither be eaten fresh, nor readily transformed into a juice (even though a beverage called *chicha* is made).

Nutritionally, peach palm fruits closely resemble edible starchy roots, such as sweet potato (*Ipomoea batatas*) and cassava (*Manihot esculenta*), the two major lowland Neotropical starches (Table 2). Of course, the edible roots, unlike peach palm fruits, have more flexible harvest periods and are easier to store in the field. A comparison with maize (*Zea mays*) is also somewhat misleading, even though their chemical compositions on a dry weight basis are quite similar. Maize can be stored in the field for weeks, which confers flexibility to its harvest time, and if protected against water and rodents it can be stored for long periods, even under very primitive conditions.

Advantageous traits

Peach palm has numerous traits that make it an ideal species for agroforestry as well as monoculture systems. Among these are its rapid juvenile growth (which helps to get a quick crop canopy started), clumping habit with basal off-shoots (so that crown size and fruiting-stem numbers can be managed via

Table 1. Mean chemical composition of peach palm fruit mesocarp[a] and of heart-of-palm[b], with percentage of daily dietary requirements in a 10.47 MJ (= 2500 kcal) diet[c]

	Fruit mesocarp[a]		Heart-of-palm[b]	
Chemical component	Unit 100 g^{-1}	Daily Value	Unit 100 g^{-1}	Daily Value
Calories (kcal)	273.5	10.9	47.6	1.9
Proteins (g)	3.3	4.4	1.5	2.0
Fats (g)	6.0	8.6	1.3	1.9
- saturated (g)	2.2	9.7	0.73	3.2
- monounsaturated (g)	3.3	14.7	0.35	1.3
- poliunsaturated (g)	0.5	2.2	0.22	1.0
Carbohydrates (g)	34.9	8.0	5.2	1.2
Fibers (g)	2.0	10.1	0.9	4.5
Vitamin A (β carotene – mg)	1.1	147.5	n.a.	n.a.
Vitamin C (mg)	18.7	30.0	3.2	5.1
Thiamin (vit. B1 – mg)	0.045	4.5	n.a.	n.a.
Riboflavin (vit. B2 – mg)	0.135	9.1	n.a.	n.a.
Niacin (mg)	0.81	4.6	n.a.	n.a.

[a]Mesocarp of three fruits.
[b]A 9-cm long by 2-cm diameter section.
[c]Dr Lúcia K.O. Yuyama, INPA, pers. comm., 2002.
n.a. = not available

spacing and pruning), competitive ability after the juvenile phase (both in terms of other crops and weeds), tolerance to acidic soils (reducing requirements for lime), moderate light interception and less shade to the lower components when spaced appropriately (making it an appropriate upperstory component in mixed systems), relatively free from pests (if grown at a low density in mixed cropping systems), relatively low maintenance costs (due to competitive ability), responsive to fertilizer application (although nutrient imbalances are a problem in chemical fertilization systems) and abundant production of fruits from an early age.

Disadvantageous traits

The fruits must be cooked to eliminate the anti-nutritional factors mentioned above. They require careful and rapid handling after harvest to get them from the farm to the consumer, as they are extremely perishable (three to seven days, depending upon maturity and handling). Considerable variability exists in fruit quality, which may be interesting to biologists and plant breeders but is frustrating for consumers, who complain that selection of high quality fruit is difficult. In Costa Rica, consumers consider the *rayado* fruit as synonymous with high quality (*rayas* are thin lengthwise cracks in the pericarp), but this is not the case in neighboring Panama and Colombia. Mesocarp water content complicates processing, and the separation of the oil from the starch requires solvents, rather than simple pressing as in the case of oil palm (*Elaeis guineensis*).

Rapid growth continues during the reproductive phase, so clumps must be managed carefully to keep harvest costs reasonable. Harvesting is a costly operation in terms of number of persons required to harvest a bunch without damaging it, especially from tall trees. This can be a problematic operation in complex agroforestry systems due to the presence of species with different architectures. Most peach palms are well armed with spines, which protect them against some pests (and thieves), but pose a problem to harvesters and plantation workers in general. The palms are nutrient loving but soil-fertility management is a problematic issue, as balanced nutrient requirements have not yet been established.

Susceptibility to pests and diseases may become a problem as plantations are intensified. Insect pests have seriously reduced yields along the Pacific coast of Colombia, where *Palmelampius heinrichi* (CIPAV/HV 2001) sometimes caused up to 100% loss of fruit yield. Physical methods, such as blue bags used in banana (*Musa* spp.) or biological control, can reduce losses, but technical assistance does not reach all affected farmers and cost of this technology may be prohibitive for some farmers.

Table 2. Comparison of mean chemical composition of peach palm (Amazonian mean and three landraces), cassava, maize, sweet potato and a set of 21 succulent Amazonian fruits

	Water	Protein	Oil (g per 100 g)[a]	Carbo.	Fiber	Energy (MJ)
Peach palm[b]	45.0	3.5	27.0	19.8	3.8	1.47
Juruá landrace[c]	54.4	3.1	13.8	19.6	8.4	1.04
Solimões landrace[c]	42.7	4.1	12.0	31.2	9.3	1.20
Putumayo landrace[c]	52.6	1.9	3.5	38.0	3.2	0.85
Cassava[d]	65.2	1.0	0.4	32.8	1.0	0.55
Maize (fresh)[d]	63.5	4.1	1.3	30.3	1.0	0.54
Sweet potato[d]	67.2	0.9	0.2	29.6	1.1	0.53
Succulent fruits[c]	82.8	0.9	0.8	11.9	2.9	0.26

[a]Fresh weights; the difference between the sum of the means and 100 is due to ash content.
[b]Mora Urpí et al. (1997);
[c]Clement (in press) and references therein;
[d]Wu Leung & Flores (1961).

Pre-Columbian versus modern cultural significance

Patiño (1963; 1992) reviewed the ethno-history of peach palm, pointing out both its great cultural significance in the pre-Columbian Neotropics and its importance in indigenous subsistence economies. The contrast between pre-Columbian and modern importance is dramatic, and it was partially the great pre-Columbian importance that motivated the NAS (1975) analysis of peach palm's potential. This review pointed out that pre-Colombians used all parts of the plant. Today peach palm essentially has two uses: the fruit, which has moderate demand for home consumption after cooking, and the heart-of-palm, which is of economic importance.

As a major pre-Columbian subsistence product, there were numerous myths of indigenous origin about peach palm but today they remain in books. Similarly, the harvest season was traditionally celebrated with great fanfare and consumption of *chicha*, followed nine months later by a major peak in births. Today, the harvest may or may not be an economic opportunity, but subsistence consumption is small where market integration is higher.

Potential uses

During the century immediately following European conquest (Patiño 1963), the fruit was used principally as a cooked starchy staple, or fermented for storage (*masato*) and then diluted to make *chicha*. Flour made from the mesocarp was reported in some areas. The wood was also important for hunting, fishing and agricultural implements, and was even used for fortification of villages. Hence, processing was an essential part of the use of peach palm in indigenous subsistence systems. Modern use of peach palm has not kept up this practice.

In the 1980s, the potential of peach palm fruit for animal feed was considered to be significant (Clement and Mora Urpí 1987), but continued R&D and subsidies to maize, sorghum (*Sorghum bicolor*) and soybean (*Glycine max*) in the first world countries have changed the economics of this potential use (Clement 2000). Given the water content of fruits, there may be potential for silage making on a small scale, but readily available technologies for farmers have not yet been developed.

Peach palm was also thought to have potential for oil (Clement and Mora Urpí 1987), particularly when germplasm with mesocarp oil content similar to that of oil palm was identified. Reevaluation in light of the R&D that has occurred with oil palm and the need for a 30-year R&D program for peach palm to equal the yields of oil palm (during which period there may be further developments in oil palm) indicated that peach palm's potential as an oil crop is less interesting than originally thought (Clement 2000).

However, Clement and Mora Urpí (1987) correctly assessed the commercial potential of peach palm to supply heart-of-palm of good quality. A recent review of R&D further confirms this (Mora Urpí and Gainza Echeverria 1999). Clement (2000) pointed out that production of seeds for heart-of-palm could change the economics of fruit production for flour or animal feed in many production areas.

Although no entrepreneurs have stepped forward in Brazil, Colombia, Costa Rica, or Peru, we think that there are at least three niche market opportunities that could be developed for the available fruit: (1) flour for baking cakes, pastries etc., (2) drinks made from mildly or moderately fermented pulp, and (3) animal feed in special situations. Processing needs to be developed if current excess peach palm production is to be marketed.

The demand for flour is evident every year after the peach palm fair in Costa Rica and the television advertising of the harvest season in Brazil. In Tucurrique, Costa Rica, flour was marketed from late 1980s to mid-1990s, but has since vanished from the market due to an unfavorable benefit/cost ratio. A similar demand for flour over a shorter period occurred in Manaus, Brazil. The processing segment of the production-to-commercialization (p-c) chain needs urgent attention to determine why these efforts failed and what is necessary to make a successful enterprise with peach palm flour. One indication of demand is that the largest women farmers' association (AMUCAU) near Pucallpa, Peru, is asking for training and financing of members to produce an infant formula and other value-added products from peach palm.

An European specialty beer company Mongozo (http://www.mongozo.org) is 'modernizing' indigenous African and American *chichas* to sell in the high-end exotic beer market. This appears to be a promising line of investigation, since peach palm's English name, and other European equivalents, are derived from the aroma given off by the preparations *masato* and *chicha*. The first attempt (by an American micro-brewery) yielded a beer with an orange tint and a pumpkin-like flavor (Jeff Moats, Amazon Origins, pers. comm. 2000).

In regions where commercial animal feed is difficult to obtain, there may still be potential for using peach palm as the energy source in silage or dry feed. The cost of drying is a major limiting factor. Mixing fresh fruit with grains and other feed components before milling facilitates drying and improves the benefit/cost ratio for preparing dry animal feeds, which are easier to store (Argüello and Afanador 2002), but a full economic analysis is not yet ready.

Peach palm in four modern Latin American economies

In this section, the status of peach palm in four Latin American countries (Brazil, Colombia, Costa Rica, and Peru) is described as they account for a major share of modern production. Consumers in these countries perceive peach palm somewhat differently due to national traditions and the type of fruit available to them. Although peach palm is a well-known traditional fruit, large amounts of fruit do not reach the fresh fruit market.

Brazil

Peach palm is grown throughout Amazonia, almost exclusively as a smallholder fruit crop in homegardens and swiddens, with a few small orchards. Heart-of-palm plantations have generally failed in Amazonia, principally because of lack of local demand, the high cost of shipping to Brazil's major market in São Paulo, and poor business planning and execution.

In the 1980s, informal observations revealed that consumers in Manaus (Amazonas state) would pay more for moderately large fruits (40 g to 60 g), containing more starch than oil. A closer examination of the preferences of these consumers indicated that most would like to buy moderately large red fruits with more oil than previously thought (Clement and Santos 2002). However, most consumers actually buy fruits that are different from what they claim they want, primarily because of price, but also because of the difficulty in selecting fruits visually for specific quality characteristics. While 'feel' of the mesocarp can help to select fruits with more oil and better texture, most vendors frown upon giving potential customers a fruit to crack open and test. Hence, demand appears to be limited by the variability of fruit characteristics.

Clement and van Leeuwen (in press) estimated that the value of fruit used for subsistence and sold in local markets in an agricultural community in Manacapuru, near Manaus, is worth 2% of a family's income (Brazil's mean annual income in 2001 was $2170). An Amazonas state extension service report (cited by Clement and van Leeuwen in press) claimed that in 2000 several agricultural communities in Coari (Amazonas) may have been producing 3% of the state's annual production of 13 600 Mg, and earning as much as $990 per family from 1.5 ha. This represents 45% of mean annual income, which may be an exaggeration by the extension service. This strong contrast has to do with fruit quality, as consumers in the Manaus market perceive the Coari fruit to be of higher quality than the Manacapuru fruit. While the Coari story is certainly good news, the same extension service estimated that 50% of the state's annual yield is wasted.

In Manaus, a fad for regional foods has seen the establishment of a number of breakfast restaurants that serve indigenous foods, based principally on regional starches and fruits. Nonetheless, peach palm is perhaps the least represented fruit/starch and the majority of the restaurants make no effort to serve peach palm out of season, even though cooked fruit can be frozen with little quality loss, and flour can be preserved dry and in the dark for long periods. Kerr et al. (1997) published a peach palm cookbook, which was a commercial failure, but offered numerous options for these restaurants. Although the Instituto Nacional de Pesquisas da Amazônia (INPA) has been studying the health benefits of peach palm for some years, especially for children, no food industry has stepped forward to take advantage of processing for the school lunch programs and regional markets.

Colombia

As in Brazil, peach palm is grown in homegardens and swiddens throughout Amazonia, and also in small-scale orchards and agroforestry plantations along the Pacific coastal lowlands. Colombia has about 7000 ha of peach palm planted for fruit production, where yields are about 7 Mg ha^{-1}. The area planted for subsistence use is certainly underestimated and difficult to measure, as the palms are generally found as scattered individuals within highly diverse agroforestry systems. There are about 1000 ha for heart-of-palm production.

Peach palm is part of Colombia's folklore (Patiño 1992). People attribute aphrodisiac powers to the fruit. In general, red fruits are preferred to yellow, and oily fruits to starchy. Given its caloric content, a handful of fruits make an adequate lunch at a very reasonable price, which is attracting attention in many urban areas. This expanding interest has caught the attention of major supermarket chains, which further increases urban demand.

Considering that 60% of the fruit does not meet consumer demands, farmers commercialize only 40% of production and earn \$660 ha^{-1} yr^{-1} profit when located near good markets (Argüello and Afanador 2002). In 2002, the federally mandated minimum salary in Colombia was \$115 per month, which means that selling even this 40% is worthwhile. Although Colombian R&D developed processing technologies similar to those used in Costa Rica, the market for these products is a tiny fraction of the fresh fruit market.

In Colombia, there is market at present for about 20 000 Mg of freshly cooked fruit per year and demand is increasing for the better quality fruit. Cali is the city with the greatest consumption, while Bogotá has only a small market. Nonetheless, Bogotá today has twenty distribution centers compared to only five centers five years ago.

Costa Rica

Costa Rica has about 1500 ha planted for fruit throughout the Atlantic and Pacific lowlands, of which about 500 ha is in Tucurrique, an Atlantic region famous for its peach palm. Most of the Costa Rican area is planted by smallholders in low diversity agroforestry or monoculture orchards, and these 1500 ha yield approximately 10 500 Mg of fruit per year (7 Mg ha^{-1}), of which about 40% are *rayados*. Farm inputs and labor costs have risen strongly, especially as the Costa Rican Colon has lost value against the US dollar during the decade, pushing up costs and cutting profit margins. As a result, farmers' interest in peach palm is stagnated. Heart-of-palm plantations occupy 15 000 to 20 000 ha and Costa Rica is the second largest world exporter of canned hearts, after Ecuador.

Costa Rican consumers prefer red fruits with *rayas*, as these are reputed to have good quality. In general, these are the fruit brought to market, while other types are brought less often or used on farm, principally to feed swine. However, Costa Rican demand for peach palm appears to be stagnant and prices are falling in real terms, both for the consumer (which theoretically should stimulate demand) and for the farmer (which should stimulate production of material that consumers are willing to pay for). During the past two decades, production has expanded to all corners of the country, with the result that fresh fruit from different regions are available year round. A major consequence of this is that prices are nearly constant year round in the important San José central market.

Costa Rica was the first country to introduce processed peach palm fruit to the market. Processing was simple, involving little more than traditional preparation, and the range of products was reasonable: whole or halved fruit, both peeled and with peel. The products were generally sold in glass jars, so that the consumer could see the product and determine if the fruits were *rayado*, and were accompanied by a recipe booklet to guide the new consumer and re-orient the jaded. However, these products have lost market share in the last decade as farmers have made fresh fruit available year round. Consequently, freshly cooked fruits are also available daily, and these are perceived to have better texture and flavor than bottled fruits that may have sat on the shelf for several months. Worse, the bottled fruit cost more than the freshly cooked fruit (a 700 g (net weight) jar costs about $2.25 while freshly cooked fruit sell for $1.25 per kg).

In Costa Rica, there is an annual food fair in Tucurrique that highlights peach palm recipes, complete with television coverage. A major disappointment for interested consumers is the difficulty in finding peach palm flour in the local market. As in Brazil, the national research institute INCIENSA (Instituto Costarricense para la Investigación y Enseñanza de Nutrición y Salud) is involved in studying health benefits of peach palm but no food industry has started processing the fruit for the school lunch programs and regional markets.

Peru

Peach palm is grown throughout Amazonia where farmers use more than 150 indigenous tree species and consider peach palm as a top priority species for agroforestry systems (Sotelo Montes and Weber 1997). It is produced in small agroforestry plots, generally established in association with annual crops and later intercropped with other fruits (Arévalo et al. 1993). In communities around Iquitos and Yurimaguas, approximately 50% of the peach palm fruit produced by farmers is sold in the markets, either directly by farmers themselves or through intermediaries, which can generate on-farm income equal to or greater than the sale of traditional crops (Labarta and Weber 1998). Consumers consider red and orange fruits to have more oil, and better texture and flavor, while yellow fruits are considered second rate and used to make flour or fed to animals.

During the 1990s, farmers were attracted to heart-of-palm production through the enthusiasm of the Ministry of Agriculture, USAID (U.S. Agency for International Development) and other agencies, even though marketing studies questioned whether Peru could compete with Costa Rica and Brazil (Vockins 1999). Saturation of the international heart-of-palm market lowered the price offered by the processing plants, however, making the necessary intensive management and inputs much too expensive for smallholders. Consequently, many farmers changed the management of their plantations for fruit production.

At the same time, new markets were opened for fruit in areas that did not produce the fruit, so that the expanded production found demand in markets outside the production areas. Near Yurimaguas, some members of small communities are earning nearly $2000 per season selling their best quality fruits into these new markets, and this is a major increase in income (Peru's mean annual income in 2002 was $900). The response of the Peruvian R&D community has, however, been timid, principally because of economic problems in Peru and the government's lack of priority for supporting R&D on underutilized crops, even with new market demand. ICRAF (International Centre for Research in Agroforestry) and INIA (Instituto Nacional de Investigación Agraria), however, have joined together in recent years to change this scenario.

In Peruvian Amazonia, a peach palm food fair was organized in the year 2000, complete with a peach palm queen, contests and prizes for the best new dishes and the best fruit bunch, in terms of phytosanitary quality, size and fruit appearance. Although this event attracted a lot of attention, it did not attract the attention of local agro-industrialists, despite the presence of numerous local food industries in Iquitos and other Peruvian Amazonian cities, and the fair was discontinued in 2002. Part of the problem is certainly peach palm's seasonality, although flour production (a necessary starting point for more elaborate industrialization) could solve this. Hence, it is not yet clear why peach palm attracted significant popular attention but no agroindustrial attention in Peru.

The market and the R&D efforts

Modern development policy is aimed at integrating producers into markets, so as to enhance their livelihoods and reduce poverty. As market integration increases, the impact of R&D and political decisions also increases. Misconceptions about which markets to explore and how to go about this are major weaknesses of most peach palm R&D programs.

Subsistence versus commercial fruit markets in tropical America

A continuum exists between purely subsistence 'markets,' such as those of isolated Amerindians along the Javarí River in the state of Amazonas, Brazil, and purely commercial markets of national and international consumers, for which production generally is in high input monocultures. In each country and region within the country, different sections of this continuum are most representative of the fruit market into which peach palm is sold. In most of Amazonia, traditional communities tend more toward subsistence, especially as the distance to urban markets increases, while settler communities are more market-oriented; nevertheless, the lack of infrastructure, energy, agroindustry, investment etc. often restrict possibilities. Many farmers of Costa Rica and the Pacific-coast of Colombian are strongly market-oriented, while Peruvian farmers vary as much as Brazilian farmers.

As market integration intensifies, the swidden/fallow fields often become low-diversity agroforestry fields and finally monoculture orchards, especially when local extension agencies are involved, and farmers' perceptions of costs, benefits and opportunities also tend to change. Their choices of which crops to produce and how much effort each is worth are dictated by personal preferences, circumstances and perceptions of the market, which in turn affect crop diversity within and between fields (de Jong 1996). Dealing with the varying needs of farmers along this continuum is the challenge of R&D institutions in Latin America.

Who are the clients?

In Brazil, the modern phase of peach palm research started in the mid-1970s at INPA, targeting entrepreneurial farmers, who supposedly required technological packages adapted to local edaphic and climatic conditions. Hence, a conventional crop development program was thought to be appropriate, with large germplasm collections and agronomic research designed to produce peach palm efficiently in monoculture or low diversity agroforestry plantations. The imagined entrepreneurs never appeared on the scene, while, in the meantime, insufficient attention was paid to smallholders and their ways of farming. As elsewhere in tropical America, smallholders produce the bulk of peach palm fruit that reaches market. This model serves to explain the situation in Colombia and Costa Rica, and part of what occurred in Peru also.

In contrast, the more recent ICRAF/INIA program in Peru recognized that their clients are smallholders, who have traditionally used the fruits produced on their farms and, where possible, have also sold part of the production into local markets. Surveys in Peruvian Amazonia indicated that approximately 50% of the fruit production was used on-farm - roughly half of this for human consumption (Labarta and Weber 1998). Not surprisingly, on-farm use of the fruit is higher in areas where markets for the fruit are less well developed (e.g., Pucallpa) than in areas where markets are well established (e.g., Iquitos and Yurimaguas). Having correctly identified its clients, the ICRAF/INIA program seems to have a promising future.

The production-to-commercialization chain

In industrialized countries and increasingly in developing countries, such as those in tropical America, investments in R&D covering production-to-commercialization (p-c) chains for crops currently dominating the international markets have given these crops a distinct advantage over the lesser-researched crops. Such economic investments have turned out to be very efficient even for resource-poor nations. This explains the current expansion of soybean into the southern fringes of Brazilian Amazonia, for example, as the mere existence of such a p-c chain gives Brazilian soybean an advantage over crops without such a system.

Production-to-commercialization chains are also tools for identifying priorities for agricultural R&D. Commonly, p-c chains for minor crops such as peach palm are poorly known or unknown. Clement and Leeuwen (in press) outlined a simple p-c chain in Manacapuru, near Manaus, identifying what the major components of the system are and how value is added as the product moves from the farm to the city. Concurrent work is examining the preferences and perceptions of the Manaus consumer. Further work will be necessary to examine the costs and benefits of each step in the p-c chain in order to identify priority areas for the attention of R&D institutions to enhance the overall functioning of the p-c chain. This analysis will also allow the R&D institutions to evaluate benefit/cost ratios of their interventions.

R&D institutions have rarely used the p-c chain adequately while working with underutilized crops, and when considered, they viewed it starting with production and ending with the consumer. In reality, this is looking at the system the wrong way, as the consumer really drives the system. As a consequence of this incorrect interpretation, much emphasis has been placed on the biology, genetics and even agronomy of peach palm (Mora Urpí et al. 1997) and appropriate processing technologies and consumer demands have been poorly investigated. If processed products are ever to be marketed successfully, even in specialized markets, the processing and consumer sections of the p-c chain must be better understood, and they should have benefit/cost ratios that attract entrepreneurs. Generating this type of information on a piece-meal basis is inadequate to address the challenge.

What do the consumers say?

As with the consumers of any product, peach palm consumers would prefer to have high and uniform quality fruit at a low price. It seems that those who like peach palm would continue to buy it and those who do not like will not buy it even as prices fall, because of lack of uniformity (and its corollary - difficulty in selecting from what is available), which is the major problem with peach palm. This consideration seems to be common to all Latin American countries, with Costa Rica being a partial exception.

A relatively high price is not necessarily a problem, however. Consumers are willing to pay more, as demonstrated by the high prices of fruits from Fonte Boa, Amazonas, Brazil (Clement and Santos 2002). But consumers will pay more only if they get the expected quality fruit. The real problem in many localities where peach palm is grown and sold is the customer's inability to buy peach palm fruit of guaranteed quality. The R&D institutions have not met this challenge even after 25 years of work on this crop.

Excessive emphasis on germplasm banks

A breeding program for an underutilized crop should necessarily be of modest size and tightly focused on its clients' demands. Unfortunately, the first step of the genetic improvement programs in these four countries was the creation of large germplasm collections (see Mora Urpí et al. 1997 for details). Worse, germplasm characterization and evaluation were never completed due to budget and personnel constraints. Such collections would be useful for programs designed around controlled crossing, which cannot be justified for a minor crop (Clement 2001). As resources dwindled due to lack of practical results, the germplasm banks became white elephants consuming the meager available resources.

Clement (2001) estimated the costs for collecting, establishing, characterizing and maintaining the 450 accession INPA germplasm bank over a 20-year period, as well as those necessary for a single conventional controlled-pollination breeding program for heart-of-palm over a 10-year period. The benefit/cost ratio of a germplasm bank for the heart-of-palm program worked out to be 7.6, on the assumptions of adequate genetic diversity at the beginning of the program and that it would increase yields by 25% over two five-year cycles. In addition to the large collection at INPA, Brazil has many small germplasm collections and programs that are trying to make headway in heart-of-palm breeding. Clement (2001) concluded that benefit/cost ratio of these small collections and programs was probably insufficient to justify their maintenance compared to the overall benefit/cost ratio of maintaining one collection for the whole peach palm improvement program in the country. The benefit/cost ratio for a fruits program would be negative because of numerous mistakes outlined in this chapter.

The same logic can be applied to the germplasm collections in other countries in the region. In Costa Rica, the large University collection can probably be justified by heart-of-palm results, but not on fruit results. The CATIE (Centro Agronómico Tropical de Investigación e Enseñanza, Costa Rica) collection cannot be justified on either grounds, and the institution has rightly de-emphasized peach palm. In Colombia, the database of one of the collections was lost a decade ago and that collection essentially ceased to exist, while two other small collections in Amazonia have not generated practical results, and therefore are difficult to justify. The same holds for Ecuador and Peru. Nonetheless, once a germplasm collection exists, it becomes a sacred cow that cannot be eliminated and consumes institutional resources.

In contrast, ICRAF and INIA followed a different approach in their program that was started in Peru in 1997 (Weber et al. 2001). Researchers asked farmers in different communities to select their best palms based on fruit characteristics, and then collected seed from these palms to establish on-farm progeny trials in the Yurimaguas and Pucallpa regions. To date, the trials include progenies from 400 selected mother palms (two plants per progeny in each replication). The trials will be rogued after five years, leaving one plant per progeny for production of selected seed (to maintain variability, no between family selection is foreseen).

The farmers are the owners of their replications; they plan to sell the selected seed, use the seed to establish more palms on their farms, or use the fruit to prepare value-added products on farm. This program is relatively less expensive and is expected to have a high benefit/cost ratio.

Entrepreneurs and market creation

As economists say, markets do not appear from nowhere - they need to be created by people. Establishment of markets costs money, which is a risky enterprise for poorly known species and for products having major competition. In the case of underutilized and new crops, there is a need for crop champions or entrepreneurs who could pressure the R&D institutions to fill gaps in the p-c chain and create new markets for the products. The crop champions are expected to transform ideas into profitable enterprises. To date, peach palm researchers have tried to fill this role, and failed.

Conclusions

The past 25 years have seen major advances in knowledge about peach palm, thanks to the R&D efforts, but not a proportional increase in the use of fruits of the palm. The lack of success is due to several reasons, including overly broad objectives (a focus on production of fresh fruits would have been better than trying to accommodate multiple uses, such as beverage, flour and animal feed), failure to understand consumers' demands (uniform varieties with desired fruit characters), and incorrect identification of the research client (the smallholder is the appropriate client, not the agricultural entrepreneur). What lessons can we learn from Latin American experiences with peach palm that might be useful for R&D on other underutilized agroforestry species?

Let markets determine demand

CATIE de-emphasized its considerable peach-palm-germplasm-R&D effort in the 1990s because markets were stagnant or nonexistent in the countries where they work in Central America and part of the Caribbean. With stagnant markets, CATIE wisely decided that duplicating the R&D effort of the University of Costa Rica was inappropriate, as one institution was more than sufficient to meet demand. CATIE maintained its concentration on peach palm in agroforestry systems, however, because of the prevailing demand for this type of information. The lesson: do not take up a research program for a product that markets do not want. The implication is that some kind of market research must precede any decision to invest in R&D. This kind of cold business-logic needs to be central to R&D decision making and planning in Latin America, and possibly elsewhere in the tropical world. A corollary follows.

Clear identification of clients and their perceptions

For peach palm, the smallholder is, at least for the time being, the main research client. This means that his/her way of farming and its consequences for peach palm production have to be understood. Important items are: risk avoidance, family food security, limited access to capital, cheap family labor, and limited access to external labor. This leads to the numerous production options observed in tropical America: intercropping of peach palm with many species (often peach palm is not the most important), combination of production goals (fruit and heart-of-palm), moderately tall stems with moderate number of suckers, low number of palms per production unit, and limited use of external inputs. Unfortunately, the smallholder is seldom seen as the main research client by national R&D institutions in tropical America. Hence, most research tends to be of moderate relevance to these clients. The lesson: know the clients' requirements first. A possible exception is to be proactive if a client and an opportunity can be clearly identified, and work in participation with the client to guarantee this.

Perceptions of farmers' welfare

International development agencies and many national agencies (e.g., of the Brazilian government) often consider producing for export markets as a more prosperous and dynamic source of growth (i.e., the *potential* for increased income) than producing for subsistence or domestic markets. In contrast, an emphasis on food security implies a decreased dependence on the vicissitudes of large-scale markets in order to feed one's family and community. The unstable outcome of a 20-year legacy of 'structural adjustment' policies reveals a much greater risk to food security via market integration than maintaining the production of diverse foodstuffs in subsistence systems (Hammond et al. 1995). A more holistic form of market integration would allow for the maintenance of semi-subsistence systems, instead of forcing their abandonment, inadvertently or otherwise.

In Peru, for example, many farmers think that there is too much risk in heart-of-palm production, even though this is encouraged by international and national development agencies, and they believe that they can still rely on the fruit to improve their food security. The fact that many farmers in Peru have transformed their heart-of-palm plantations into fruit production supports this contention. Instead of investing in projects for heart-of-palm production in small farming communities, it would seem reasonable to invest in projects to process the fruits for a value-added product, such as infant formula and school lunches, that can be produced and processed in farming communities, and sold by farmers' associations in the local and national markets. In other words, disconnecting fruit production from international markets while focusing on local and national markets for food security may be a route to avoid starch competition and start attracting entrepreneurs. The lesson is that for peach palm and other similar crops, R&D institutions should work locally on food security and let entrepreneurs be attracted, rather than aim for entrepreneurs who may not exist.

Size the program to the species' importance

An analysis of demand for subsistence needs and market allows definition of the importance of the species in question. The strategy for a breeding program should be developed accordingly. In such a program, the accent will be on mass selection and comparison of land races/provenances and open-pollinated progenies (see Simons and Leakey, this volume, for a dicotyledonous view). In the first phase of such a program, there is little or no need for large germplasm collections. Consequently, these should not be established, even if funds are easily obtained. The plant improvement program needs to have a strong participatory component (at least for on-farm trials), and the first generation of improved seed should be available as soon as possible, in order to get feedback from farmers and consumers. The lesson is that a breeding program for an underutilized crop should necessarily be of modest size and tightly focused on its clients' demands.

History need not be a guide to the future

Just because the native Americans considered peach palm a staple does not mean that modern Euro-American-centric markets will agree. Modern markets, even in the interior of the third world, may have different demands now than in the past. The decline in the usage of native American foods and vegetable oils in Amazonia is a clear example of this, and peach palm is only one case. The lesson is that researchers should keep reappraising priorities from time to time, in keeping with the changing socioeconomic conditions, and should not be the sole arbiters of program objectives.

References

Arévalo L.A., Szott L.T. and Perez J.M. 1993. El Pijuayo como componente de un sistema agroforestal. pp. 267–285. In: Mora Urpí J., Szott L.T., Murillo M. and Patiño V.M. (eds), IV Congresso Internacional sobre Biologia, Agronomia e Industrialización del Pijuayo. Editorial Univ. Costa Rica, San José, Costa Rica.

Argüello H. and Afanador G. 2002. El Chontaduro – Un Negocio Potencial para la Industria de Concentrados y la Producción Animal. Universidad Nacional de Colombia, Bogotá, Colombia, 36 pp.

CIPAV/HV 2001. Control Integrado del Barrenador del Chontaduro *Bactris gasipaes* H.B.K. Fundación CIPAV / Fundación Herencia Verde, Cali, Colombia, 20 pp.

Clement C.R. 2000. Pupunha (*Bactris gasipaes* Kunth, Palmae). Série Frutas Nativas 8, Fundep, Jaboticabal, São Paulo, Brasil, 48 pp.

Clement C.R. 2001. Melhoramento de espécies nativas. pp. 423–441. In: Nass L.L., Valois A.C.C., Melo I.S. and Valadares-Inglis M.C. (eds), Recursos Genéticos & Melhoramento – Plantas. Fundação de Apoio à Pesquisa Agropecuária de Mato Grosso – Fundação MT, Rondonópolis, Mato Grosso, Brasil.

Clement C.R. In Press. Fruit trees and the transition to food production in Amazonia. In: Balée W. and Erickson C. (eds), Time and Complexity in the Neotropical Lowlands: Studies in Historical Ecology. Columbia University Press, New York.

Clement C.R. and Mora Urpí J. 1987. The pejibaye (*Bactris gasipaes* H.B.K., Arecaceae): multi-use potential for the lowland humid tropics. Econ Bot 41: 302–311.

Clement C.R. and Santos L.A. 2002. Pupunha no mercado de Manaus: Preferências de consumidores e suas implicações. Revista Brasileira de Fruticultura 24: 778–779.

Clement C.R. and Leeuwen J. van. In Press. Underutilization of pupunha (*Bactris gasipaes* Kunth, Palmae) in Central Amazonia: history, production-to-consumption system, implications for development and conservation. In: Alexiades M. and Shanley P. (eds), From Forest to City: Tracking Non-timber Forest Products along the Production to Consumption Chain. Latin American volume. CIFOR, Bogor, Indonesia.

Hammond D.S., Dolman P.M. and Watkinson A.R. 1995. Modern ticuna swidden-fallow management in the Columbian Amazon: ecologically integrating market strategies and subsistence-driven economies? Hum Ecol 23: 335–356.

de Jong W. 1996. Swidden-fallow agroforestry in Amazonia: diversity at close distance. Agroforest Syst 34: 277–290.

Kerr L.S., Clement R.N.S., Clement C.R. and Kerr W.E. 1997. Cozinhando com a Pupunha. INPA, Manaus, Amazonas, Brasil, 95 pp.

Labarta R.A. and Weber J.C. 1998. Valorización económica de bienes tangibles de cinco especies arbóreas agroforestales en la

Cuenca Amazónica Peruana. Revista Forestal Centroamericana 23: 12–21.

Mora Urpí J., Weber J.C. and Clement C.R. 1997. Peach palm. *Bactris gasipaes* Kunth. Promoting the conservation and use of underutilized and neglected crops. 20. Institute of Plant Genetics and Crop Plant Research – (IPK), Gatersleben and International Plant Genetic Resources Institute – (IPGRI), Rome, 83 pp.

Mora Urpí J. and Gainza Echeverria J. (eds) 1999. Palmito de Pejibaye (*Bactris gasipaes* Kunth): Su Cultivo e Industrializacíon. Editorial Universidad de Costa Rica, San José, Costa Rica, 260 pp.

NAS 1975. Underexploited Tropical Plants with Promising Economic Value. National Academy of Sciences, Washington, DC, 188 pp.

Patiño V.M. 1963. Plantas Cultivadas y Animales Domésticos en América Equinoccial. Tomo I. Frutales. Imprenta Departamental, Cali, Colombia, 547 pp.

Patiño V.M. 1992. An ethnobotanical sketch of the palm *Bactris (Guilielma) gasipaes*. Principes 36: 143–147.

Popenoe W. and Jimenez O. 1921. The pejibaye, a neglected food-plant of tropical America. J Hered 12: 151–166.

Sotelo Montes C. and Weber J.C. 1997. Priorización de especies arbóreas para sistemas agroforestales en la selva baja del Perú Agroforestería en las Américas 4(14): 12–17.

Vockins K. 1999. Analysis of the U.S. Food Market-Opportunities for Heart of Palm. KV Marketing Incorporated, International Centre for Research in Agroforestry, Nairobi, Kenya, 109 pp.

Weber J.C., Sotelo Montes C., Vidaurre H., Dawson I.K. and Simons A.J. 2001. Participatory domestication of agroforestry trees: an example from the Peruvian Amazon. Development in Practice 11: 425–433.

Wu Leung W.-T. and Flores M. 1961. Tabla de Composición de Alimentos para Uso en América Latina. INCAP-ICNND, Ciudad de Guatemala, Guatemala, 157 pp.

Yuyama L.K.O., Yonekura L., Aguiar J.P.L. and Sousa R.F.S. 1999. Biodisponibilidade de vitamina A da pupunha (*Bactris gasipaes* Kunth) em ratos. Acta Amazonica 29: 497–500.

Yuyama L.K.O., Aguiar J.P.L., Yuyama K., Clement C.R., Macedo S.H.M., Fávaro D.I.T., Afonso C., Vasconcellos M.B.A., Vannucchi H., Pimentel S.A. and Badolato E.S.G. 2003. Chemical composition of the fruit mesocarp of three peach palm (*Bactris gasipaes*) populations grown in central Amazonia, Brazil. Int J Food Sci Nutr 54: 49–56.

Agroforestry Systems **61:** 207–220, 2004.

Exploiting the potential of indigenous agroforestry trees: *Parkia biglobosa* and *Vitellaria paradoxa* in sub-Saharan Africa

Z. Teklehaimanot
School of Agricultural and Forest Sciences, University of Wales Bangor, LL57 2UW, UK; e-mail: z.teklehaimanot@bangor.ac.uk

Key words: Genetic diversity, Karité, Néré, Reproductive biology, Shea butter, Vegetative propagation

Abstract

Parkia biglobosa (néré) and *Vitellaria paradoxa* (karité) are indigenous tree species that are economically and socially important for local people in sub-Saharan Africa. Farmers deliberately maintain these trees on farms mainly for their fruits and nuts. The kernels of karité yield shea butter, which is rich in fatty acids; it is used locally for food and internationally in chocolate, pharmaceutical and cosmetic products. Néré seeds are ground into a pungent nutritious spice or condiment, *soumbala* or *dawadawa*, which is added to soups and stews throughout the savanna regions of sub-Saharan Africa. The tree is also important in improving soil fertility and in traditional medicine. Despite their important uses, the populations of both species are in decline and they remain semi- or undomesticated. Recent research has shown that both species are genetically diverse, which indicates their potential for domestication through selection and breeding, and that crop production under the trees could be improved by crown pruning. Research also has helped develop vegetative propagation methods that allow multiplication of superior trees and on-farm domestication of the trees. Domestication should enhance their role in improving rural livelihoods. New knowledge on reproductive biology should be used to increase fruit production in both species. Prevailing social conditions in relation to tree tenure, marketing and processing are the major constraints to successful domestication and proper management of these valuable trees. Changes in tree tenure policy, development of local and national markets, access to market information for producers, establishment of a system for standardizing product quality at national level, and development of improved village processing technologies are needed.

Introduction

Several indigenous tree species are protected and managed by farmers on farms as part of a traditional approach to land use in the tropics. The trees serve as sources of wood, food, fodder and medicine for the farmers. They also provide ecological services including soil fertility and microclimate amelioration. In addition to direct domestic use of the tree products, they are a source of cash for many poor people. Yet, many of these trees remain semi- or undomesticated. Due to growing human- and livestock-population pressures, fallowing of farm fields – a practice on which the regeneration of these trees relied – is no longer being followed. As a result, existing trees are aging (Okullo et al. 2003). Domestication of these trees is, therefore, required in order to rejuvenate traditional agroforestry systems and enhance the potential of these trees for increasing the income and food security of poor people. On the other hand, the growing population pressure in some areas has resulted in increased establishment of trees on farms where farmers have secured land tenure or tree tenure (Leakey et al. 1999). Until recently, most of the trees planted in agroforestry systems in the tropics have, however, been fast growing exotics. This emphasis on planted exotics stems largely from the practice of forestry in which they were thought to be easier to manage, faster growing, higher yielding and with more saleable products than indigenous species (Wood and Burley 1991). There is, however, a growing international concern about the impacts of introducing species into areas outside their native range. The concern includes reduction of local biodiversity and introduction of pests and weeds. As a result, the focus has more recently shifted to indigenous, often undomesticated, agroforestry tree species, that have been traditionally and intentionally maintained on farmland to provide both products and ecological services. Interest in such species has developed particularly rapidly in dryland Africa, with increasing awareness of the value of adaptation to the capriciousness of the prevailing climates, reflected in innate resilience often lacking in exotic alternatives. The focus of tree planting in agroforestry has also so far been on trees that improve soil fertility and yield wood for fuel or construction. The income from the sale of wood to farmers is insignificant due to wood having a high weight-to-value ratio and high costs of transportation. On the other hand, agroforestry parkland trees, such as *Parkia biglobosa* (néré) and *Vitellaria paradoxa* (karité), which yield nonwood products, provide better returns to farmers because these products have a low weight-to-value ratio and are easy to transport.

Agroforestry parklands, where naturally regenerated scattered individual mature trees occur in cultivated fields, are one of the most widespread traditional land use systems in Africa (Nair 1993). When woodlands are converted to croplands, farmers cut most of the trees and leave behind some trees, those that they value most to fulfill their basic needs. This was how most of the traditional agroforestry parklands we see today have been established. Most of the trees in parkland systems yield marketable nonwood products both for local and international markets. Parkland trees also provide ecological services such as soil fertility and microclimate amelioration (Bayala et al. 2002; Boffa 1999; Tomlinson et al. 1995). Most of these trees remain semi- or undomesticated and have not been objects of scientific study until fairly recently. Due to increasing human and animal populations, which are causing environmental pressures on the land and the consequent degradation of natural resources, there is now increasing interest in domesticating these trees within their natural habitat (Leakey et al. 1999).

In the savanna regions of West Africa (14 countries: Benin, Burkina Faso, Cameroon, Chad, Gambia, Ghana, Guinea, Cotê d'Ivoire, Mali, Niger, Nigeria, Senegal, Sierra Leone and Togo), parts of central Africa (Republic of Congo and Central African Republic) and parts of East Africa (Ethiopia, Sudan and Uganda), *P. biglobosa* and *V. paradoxa* constitute the dominant indigenous tree species of agroforestry parklands (Ruyssen 1957; Hopkins and White 1984). These two tree species have drawn the attention of many scientists and development agencies and thus have been a focus of research and development in the last decade because, in countries where they occur, local people derive many benefits from these trees mainly from their nonwood products. For some of the countries where these trees occur, they are the major source of foreign exchange earnings. However, both tree species remain semi- or undomesticated.

It is well known that nonwood products supplied from wild-tree sources can never be uniform and reliable as production can vary considerably from year to year and from location to location (Maranz et al. 2004). Thus, domestication of trees such as karité and néré that provide nonwood products can be a means of ensuring quality and supply of raw material under control and relieving pressure on wild tree resources. Moreover, planting them in agroforestry associations may be a viable option under the prevailing socioeconomic conditions of the countries concerned. Domesticating such plants in agroforestry systems also provides continuous tree cover and contributes to both the productivity and sustainability of farming systems by maintaining soil fertility and creating a more favorable microclimate for associated crops and livestock. This paper presents information on how the potential of these two tree species is being exploited in the parkland system of sub-Saharan Africa and on the improvements being made to increase their potential so that the benefits derived by the local people and the role of these trees in food security, poverty alleviation and ecological services are enhanced.

Tree products and their economic value

Néré

Parkia biglobosa (Leguminoseae; Mimosoideae), known as néré in Francophone Africa, and locust bean tree in Anglophone Africa, is a multipurpose tree indigenous to sub-Saharan Africa (Hopkins 1983). The height of the tree ranges from 10 m to 20 m with a very wide spreading umbrella- shaped crown extending outwards from the bole by as much as 10 m (Ouedraogo 1995). The natural range of néré extends over 19 African countries from Senegal in the west to Uganda in the east (Figure 1; Hall et al. 1997). Néré is unusual as a savanna tree, known much more widely as a component of anthropic landscapes than of natural landscapes. The species is particularly characteristic of agroforestry parklands. Among the species

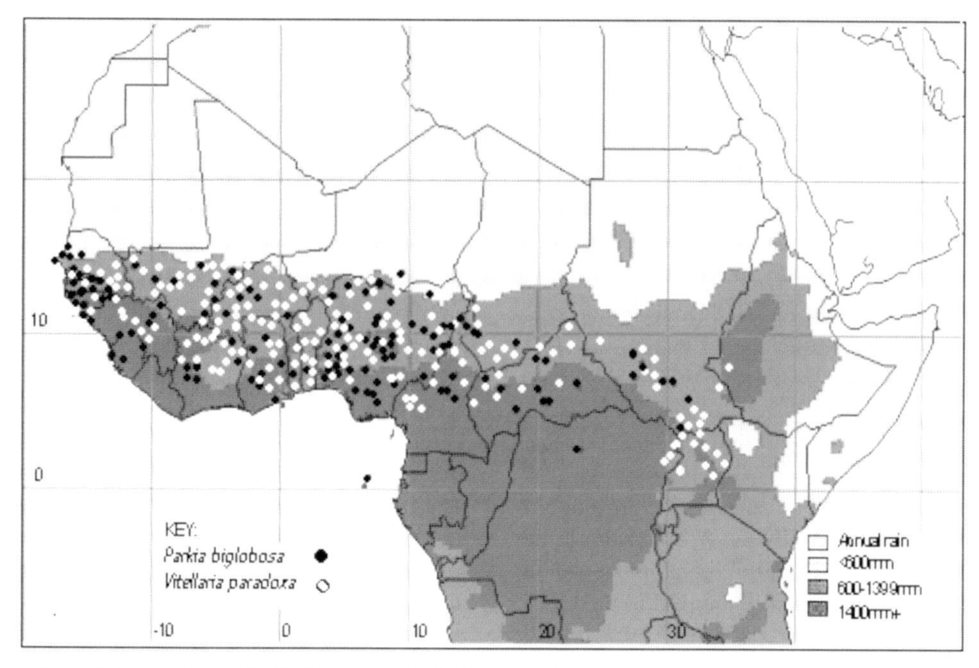

Figure 1. Distribution of *Parkia biglobosa* and *Vitellaria paradoxa* (Sources: Hall et al. 1996; Hall et al. 1997).

closely associated with néré are *Annona senegalensis*, *Daniella oliveri*, *Piliostigma thonningii*, and *Sarcocephalus latifolius* (Hall et al. 1997). Néré is retained by farmers in cultivated fields, both for its valuable nonwood products and because it is thought to improve soil. Farmers in Burkina Faso ranked it as the second most preferred woody species after *V. paradoxa* (Teklehaimanot 1997).

Néré is highly valued for its brown seeds that are ground into a pungent nutritious spice or condiment, *soumbala* or *dawadawa*, which is added to soups and stews throughout the savanna regions of sub-Saharan Africa (Campbell-Platt 1980). This is the prime use of néré. The seed is cooked to remove the testa, recooked to soften the cotyledons and then fermented to produce the condiment. The seeds of néré may be sold for commercial processing or bought by women traders who cook and ferment them for sale in the markets. The fresh *soumbala* or *dawadawa* is usually sold wrapped in leaves of plantain, banana (*Musa* sp.), or cacao (*Theobroma cacao*) leaves, or occasionally in plastic bags. The dried form is moulded into flat rounds or ball shapes (Teklehaimanot 1997).

According to surveys carried out between 1993 and 1994 in Nigeria and Burkina Faso, the value of the seeds increases considerably once processed (Teklehaimanot 1997). In Nigeria, 100 kg of dried seed was sold on average for approximately 2000 Naira (US$1 = 21 Naira in 1994). On average the cost of 50 g of processed néré varied between 1.5 and 2 Naira; thus there was a doubling in price compared with the unprocessed seeds. It was the same in Burkina Faso where the price of dried seeds varied between 100 and 300 fCFA kg^{-1}, while the price of fermented seeds was between 300 and 650 fCFA kg^{-1} (US$1 = 250 fCFA in 1994).

This *soumbala* seasoning, sold as balls of brown paste, can easily be tracked down by its remarkable smell that is a very distinctive feature of Sahelian markets. Néré seeds provide an important source of protein and the fermented seeds have often been referred to as a cheese substitute (Booth and Wickens 1988). The pulp also has considerable nutritional value, and is rich in sugars (Table 1; Ouedraogo 1995). The floury pulp is used to make drinks and couscous, particularly in the dry season. The flowers also are consumed. The nectar is leached from the flowers or the whole flowers may be added to soup. The green pods are also eaten as a vegetable, but usually only in times of famine. Human medicine is one of the chief uses of néré and nearly all parts of the plant are used as a cure against a large range of human ailments, especially piles, malaria and stomach disorders. The bark is the part most commonly used in the treatment of digestive ailments, followed by roots, leaves, *soumbala*, exocarp, and pulp. The seeds appear to be important

Table 1. Nutritional composition of the products of néré (*Parkia biglobosa*).

Product of néré	Proteins	Lipids	Carbohydrates	Vitamine C	Phosphorous
	g per 100 g	g per 100 g	g per 100 g	mg per 100 g	mg per 100 g
Dried seeds	34.6	21.8	32	6	503
Fermented seeds	38.4	32.8	16.4	0	477
Fruit pulp	3.4	0.7	80.7	164	3.2

(Source: Ouedraogo 1995).

in the treatment of blood disorders, whereas extracts from the leaves are used in alleviating skin disorders. The pulp, the seeds and the leaves are most often used in the treatment of jaundice and similar other ailments. The bark, the roots and the fruit appear to have similar properties, as they are the most often used parts in the treatment of wounds and gangrenes. In general, the roots are only used when a strong dose of medicine is required. *Soumbala* is most often used in the treatment of nervous system disorders (Hall et al. 1997; Teklehaimanot 1997).

Néré is also important both in agriculture and raising livestock. The exocarp, the leaves, and the seed coats are all used for fertilizer. A large proportion of farmers interviewed in Nigeria and Burkina Faso thought that néré was important in agroforestry systems (Teklehaimanot 1997). A whole range of crops are grown in association with néré, but farmers report success only with certain plants. Vegetables were reported to associate particularly well, especially large leafed species, which benefit from the shade provided by the tree. Root crops such as yam (*Dioscorea rotundata*), sweet potato (*Ipomoea batatus*), cocoyam (*Xanthosoma sagittifolia*), and cassava (*Manihot esculenta*) are reported to associate well with the tree, but not groundnuts (*Arachis hypogea*), maize (*Zea mays*), sorghum (*Sorghum bicolor*) or millet (*Pennisetum americanum*). However, in northern Nigeria a dwarf sorghum variety was reported to be successful. Ninety-two percent of farmers interviewed believe that the tree is important in improving soil fertility (Teklehaimanot 1997). Many of them thought that this was brought about by increased litter fall from the tree, a fact that has also been confirmed by scientific investigations (Tomlinson et al. 1995). Research results showed significant soil amelioration for total nitrogen and available potassium close to the tree. A study of root distribution in Burkina Faso by Tomlinson et al. (1998) suggested that one mechanism for increased soil fertility beneath néré is the widespread lateral root system, which redistributes nutrients from an extensive radius to a concentrated area beneath the crown. The root system of néré is also aided in the efficient uptake of nutrients by endomycorrhizae, which have been found to infect about 85% of the roots (Tomlinson et al. 1995). Néré is a legume; but contradictory reports are available on its nodulation. Tomlinson et al. (1995) reported that root nodules were absent on the roots of trees examined in the field, both in Burkina Faso and Nigeria, or in the seedlings in a laboratory study in which néré seedlings treated with a laboratory grown miscellany rhizobial culture were grown in the soil collected from beneath the trees in the field. The tree also reduces wind erosion, with 90% of farmers interviewed in northern Nigeria considering this important. Because of the spreading nature of the crown, néré is regarded as an important shade tree, and nearly half of the farmers interviewed liked to rest in the shade of their néré trees (Teklehaimanot 1997).

The tree also has high fodder value. The most commonly used part of the tree is fruit pulp, which is fed to pigs, goats, sheep, cattle, and ducks. The next in importance are shaft, leaves, whole pods, and flowers. Occasionally the parasitic 'mistletoe' (*Tapinanthus* spp.) that grows on the tree also is fed to animals. The pulp is used against cockroaches and the parasitic plant striga (*Striga hermonthica*). In poultry farming, the tree plays an important role in egg production. The seeds are fed to hens during incubation of the eggs and the tree leaves are placed beneath the eggs. The slippery nature of the leaves allows the eggs to move beneath the bird so that they can be warmed evenly. The leaves are also used against poultry lice. The pulp is most frequently used for treating trypanosomes and ulcers, particularly mouth ulcers of ruminant livestock. The flowers and seeds are used to treat Newcastle disease (*Avulavirus* spp.) (Teklehaimanot 1997).

Néré is also often traditionally used in ceremonies connected with virility, pregnancy, birth, circumcision, marriage, death, funeral, fetishism, and spiritual protection. *Soumbala* is often given as a present to maintain good relationships. It is a mark of respect between people, and strangers often give it to their host. *Soumbala* given by a woman to a male relative or a daughter in her house is a password for a gift of cloth. Its offer is to show that the receiver is dear to the giver (Teklehaimanot 1997).

It is difficult to estimate the annual production of néré fruits as they are collected and used by individuals. Production also varies from year to year. Villagers state that production is higher in fields than in fallows, the same as karité, as a result of care and protection provided by farmers, which characterizes it as an anthropic species. On a per-tree basis, fruit production ranges between 15 to 130 kg year^{-1}. On average a household requires 27 kg to 60 kg of dried seed per year – equivalent to the harvest from four to nine trees (Hall et al. 1997; Teklehaimanot 1997).

Néré fruit is a major commodity of local and regional trade in sub-Saharan Africa. In Burkina Faso, néré fruits are highly commercialized, with over 50% of respondents in a nation wide survey participating in trade (Teklehaimanot 1997). In general, women are wholly responsible for the sale of fermented seeds of néré, but both men and women sell the dried seeds. The revenue earned per annum from the sale of néré products in Burkina Faso is approximately 27 300 fCFA per household. This constitutes 28.8% of the income per household. In Cote d'Ivoire, Bernard et al. (1995) found that 4% of total village revenue came from néré fruit, compared with 2% from karité. In terms of revenue from woody species in Burkina Faso, néré fruits provide 73% of the total income per household, compared with 14% from karité (Teklehaimanot 1997). On the contrary, in Benin, néré is the third most economically important species, making up 15% of the revenue from nontimber forest products, after karité (50%) and African oil palm (*Elaeis guineensis*) (26%) (Schreckenberg 1996). This difference in revenues between these two countries probably reflects greater abundance of néré in Burkina Faso. A total of 1638 kg of fermented seed (*soumbala*) is sold every market day in Burkina Faso (Guinko and Pasgo 1992). Here, each person spends three to 17 fCFA every day on purchasing *soumbala*. In Nigeria, based on a nationwide survey, the average contribution to the household budget from the sale of néré products was estimated to be 14.5% (Teklehaimanot 1997). A rough estimate based on the number of fruit producing trees owned by farmers, the quantity of pods produced and the market price of the seeds suggests that on average a farmer could earn up to 60 000 Naira per annum from his néré trees (Teklehaimanot 1997). A néré-based stock cube has also been developed by an industry in Nigeria (Cadbury) and is now being sold under the name of 'Dadwa' (Hall et al. 1997).

Karité

Vitellaria paradoxa (formerly *Butyrospermum paradoxum* Gaertn. f.), known as karité in Francophone Africa and shea nut in Anglophone Africa, belongs to the family Sapotaceae (Henry et al. 1983). The tree grows from 10 to 20 m in height, with a parasol or pyramidal shape of crown. The cylindrical trunk measures 3 to 4 m in height and 30 cm to 80 cm in diameter (Ruyssen 1957). The geographical distribution of karité ranges within the 600 mm to 1500 mm rainfall isohyets from Senegal in the west to Uganda in the east (Figure 1; Bonkoungou 1987; Hall et al. 1996). Karité forms the dominant tree species of agroforestry parklands in Sudanian savannas where it is associated with other species such as *Acacia senegal*, *Annona senegalensis*, *Parkia biglobosa* and *Terminalia avicennioides* (Boffa 1999; Hall et al. 1996). Farmers maintain karité on farms primarily for its kernel, which is rich in fatty acids (oleic, stearic, linoleic, and palmitic acids) (Maranz et al. 2004). Karité is the second most important oil crop in Africa, after oil palm, but as it grows in areas unsuitable for oil palm; it assumes primary importance in West Africa, in regions where annual precipitation is less than 1000 mm (Hall et al. 1996).

Chemical analysis of karité nuts indicates fat content to range from 20% to 50% (Table 2; Maranz et al. 2004). Fatty acid profiles of karaté nuts across Africa confirm the existence of a very high oleic acid containing karité population in Uganda (50% to >60% oleate), which produces the most liquid karité fat on the continent. The karité population of east Africa belongs to *Vitellaria paradoxa* subspecies *nilotica*. In contrast, karaté nuts in the region extending from eastern Senegal to Nigeria (usually considered to be *Vitellaria paradoxa* subspecies *paradoxa*) produce the well-known solid karité butter with balanced levels of oleate and stearate (Maranz et al. 2004).

Karité butter samples on fractionation by column chromatography were found to contain minor components of commercial importance (esters of sterols and triterpene alcohols). Esters of triterpene alcohols identified by nuclear magnetic resonance included α-amyrin, β-amyrin, lupeol, parkeol, butyro-

Table 2. Fatty acid profiles of karité (*Vitellaria paradoxa*) nuts in different regions of Africa.

	Percentage of total fatty acids					
	Palmitic	Stearic	Oleic	Linoleic	Arachidic	
Regional populations	16:0[a]	18:0	18:1	18:3	20:0	No of samples
N. Uganda	4.7	31.2	56.5	5.8	1.0	35
N. and W. Cameroon	6.0	33.7	52.0	8.0	0.2	11
W. Senegal and Gambia	3.8	35.3	50.0	8.5	1.3	3
E. Senegal through W. Nigeria	4.0	42.8	45.0	6.4	1.4	298
Mossi Plateau, Burkina Faso	3.7	45.1	43.3	6.2	1.4	55

[a]Fatty acid carbon chain length: number of double bonds (Source: Maranz et al. 2004).

spermol, germanicol. Sterols including α-spinasterol, Δ^7-stigmastenolo, 24-methyllanost-9 (11) enol and Δ^7-avenasterol were also identified, as was Karitene, a polyisoprenic compound (Wiesman et al. 2003).

Analysis of karité fruit pulp has confirmed that it is generally very nutritious. The tree plays a key dietary role during the planting time of crops, when fruits ripen (Wiesman et al. 2003). The sweet pulp is widely consumed fresh and a market for the fruit exists in sub-Saharan African cities and along roadsides (Lamien et al. 1996). The pulp contains large quantities of protein and minerals, although there is considerable variation from tree to tree in the nutritional value of the fruit. Protein content ranges from 2.5% to over 10%, phosphorus from 0.17 to 2.26 mg g^{-1} and potassium from 4.55 to 37.46 mg g^{-1} (Table 3; Wiesman et al. 2003).

Karité butter is used as a source of fat in many African kitchens and it is also used in the pharmacology, chocolate and cosmetic industries and in local traditional medicines. Karité butter is a primary ingredient in many of the most expensive facial creams on the international market. In addition, karité butter has therapeutic properties in the treatment of general skin and hair conditions (Tran 1984) and is marketed in a variety of preparations by cosmetic industries both in sub-Saharan Africa and in the north. In Burkina Faso, karité butter soap is supplied in hotels. In addition, the tree improves microclimate and soil fertility for associated staple crops in agroforestry parklands (Bayala et al. 2002; Boffa 1999), the branches are used as firewood and for saw logs (Hall et al. 1996). In general, in all countries where karité occurs, local people derive diverse benefits from it, and for some of the countries where it occurs, it is the major source of foreign exchange earnings and it plays a significant role in food security, poverty alleviation and ecological services (Boffa 1999). In Burkina Faso, karité nuts alone represent the country's third largest export (World Bank 1989). Karité was designated as a resource development priority in Burkina Faso, Mali and Chad (FAO 1990).

Based on a number of national and international sources, Becker and Statz (2003) estimated the total karité nut production (i.e., collected nuts) in the whole of Africa in the year 2000 to be 650 Gg. Up to 150 Gg of nuts are exported annually from the region to European and Japanese food processing industries, primarily for use as a replacement for cocoa butter in confectionery and as a source of fat for pastry products. At present, Europe is the only continent where karité nuts are imported on a regular basis. Interviews with European importers carried out in 2001 by Becker and Statz (2003) confirm that imports of karité to the European Union amount to about 50 Gg per year. While 95% of the imports are directed towards the chocolate industry, only 5% are used for the higher value segment of cosmetics and pharmaceuticals.

Prices of karité are closely related to cocoa prices: high price levels for cocoa generally trigger greater demand and higher prices for karité. While the price of cocoa butter in January 2003 was about \$3500 Mg^{-1} (the highest in the last 15 years), the price of karité was about US\$ 2680 Mg^{-1}, which is about 23% cheaper than cocoa butter (Becker and Statz 2003). In contrast to the West African karité subspecies, the subspecies *nilotica* of East Africa yields a liquid fat. This oil is sold in a completely different market segment than the West African solid karité butter. While exports of the latter are primarily used by the European food industry, the East African karité oil with its lower melting point and softer texture is much wanted by the cosmetic industry as it needs no fractionation to get the olein fraction (Becker and Statz 2003).

Table 3. Nutritional composition of karité (*Vitellaria paradoxa*) fruit pulp.

	N%	Protein%	P	Zn	Fe	Mn	Mg	Cu	K	Ca
UGANDA – dried fruit pulp All figures in mg/g unless labelled otherwise										
mean	0.66	4.13	0.74	0.05	0.24	0.01	0.81	0.00	13.20	1.84
st. dev.	0.18	1.11	0.26	0.16	0.39	0.01	0.23	0.00	3.89	0.65
Max	1.10	6.88	1.31	1.01	1.80	0.04	1.43	0.00	26.25	3.93
Min	0.40	2.50	0.30	0.01	0.02	0.00	0.29	0.00	5.40	0.75
# of trees	39	39	40	40	40	40	40	40	40	40
MALI and BURKINA FASO – dried pulp										
Mean	1.04	6.52	0.71	0.05	0.14	0.01	1.42	0.00	17.90	5.63
st. dev.	0.32	1.99	0.36	0.03	0.10	0.00	0.34	0.00	7.49	1.68
Max	1.70	10.63	2.26	0.16	0.41	0.02	2.38	0.00	37.46	9.94
Min	0.60	3.75	0.17	0.01	0.03	0.00	0.78	0.00	4.55	3.36
# of trees	36	36	36	36	36	36	36	36	36	36

(Source: Wiesman et al. 2003).

Karité is reported to be less productive in fallows and natural forests; however, it bears fruit abundantly in cultivated fields of the Sahelian parklands and for this reason it is characterized as an anthropic species that benefits from the close care given to it by local communities (Lovett and Haq 2000a). Women are primarily engaged in the collection of fruits, processing of kernels into butter and marketing of these products at local markets (de Saint Sauveur 2003). They sometimes have to pay their husbands to have access to trees to collect the fruits. Most women interviewed in Burkina Faso, Mali and Nigeria reported that production is mainly for their own consumption (the ratio of own consumption to marketing being about 60:40) (Becker and Statz 2003). They sell the butter almost exclusively at the local markets to consumers and/or middlemen. Thus, while women's role in karité marketing is confined to the early stages of the value-addition chain, men control all further marketing activities up the chain (Becker and Statz 2003). Price of karité butter in 2000 varied from 400 to 500 fCFA kg^{-1} (640 fCFA = 1US$ in 2000). Amounts sold by the individual producers are low, sometimes as low as 30 kg per year per individual, which fetches an income of 12 000 fCFA at the lowest price.

Tree domestication

Background

Farmers have been managing karité and néré with the aim of diversifying production and maintaining site stability in terms of both soil fertility and microclimate amelioration of parklands. During the past few decades, however, the degradation of parklands in general and those of karité and néré in particular has been reported widely. Evidence of degradation has been shown in terms of reduced tree densities and population structure. For instance, Gijsbers et al. (1994) found that the average tree density decreased by about 45% between 1957 and 1988 in Burkina Faso. In Senegal, total parkland tree density declined from 10.7 trees ha^{-1} in 1965 to 8.3 trees ha^{-1} in 1985 (Lericollais 1989). According to Chevalier (1946), the Sudanian savanna zone had the greatest density of karité in the 1940s, with a population of 230 trees ha^{-1}, but this has now been reduced to as few as 11 trees ha^{-1} (Nikiema et al. 2003). Kelly et al. (2004) found distinct differences in the population structure of karité in Mali between cultivated fields and forests. In crop fields, most individuals had a girth ranging between 65 cm and 105 cm whereas in forest most of the individuals had lower girth. This may be explained by greater competition for growth resources between trees in natural forests due to greater tree density than in cropped fields. According to Okullo et al. (2003), parklands of karité in Uganda are also aging and are characterized by the predominance of old trees and lack of regeneration. This is due to a complex combination of natural, technological, and socioeconomic factors including drought, increasing population pressure (which results in shortened fallows, on which regeneration of karité and néré rely), mechanization, and conflicting land-use activities such as uncontrolled bush burning, extensive cattle grazing, and deforestation to give way to monocrop arable farming. As a result, the gene pool of karité and néré populations is under threat (Oni 1997).

The main constraints to domestication of karité and néré are their slow growth, long juvenile phase and large yield-variability. A high degree of variation in fruit and nut production, nut and fruit size, pulp sweetness, oil content, and quality has been documented for karité and néré (Adu-Ampomah et al. 1995; Hall et al. 1997; Maranz et al. 2004). No major farmers' initiatives have been noted for planting indigenous parkland trees in general and karité and néré in particular; some recent efforts have, however, been initiated to improve these trees in an effort to mitigate rural poverty. Two major approaches are involved: tree improvement and management. To reduce the time of delivery of improved materials of karité and néré for domestication, parallel investigations into the genetic variation, reproductive biology, and vegetative propagation of the species are also being undertaken.

Genetic variation

Sustainable in-situ conservation, improvement and management of karité and néré as a future rural resource require information on their current state of population genetics. Results of an isoenzyme analysis carried out in 1994 based on samples collected from across 11 countries of West Africa (90% of the species range) has showed very high genetic diversity in néré both at inter- and intra-population levels (Teklehaimanot 1997). Genetic diversity in a species is usually expressed by 'Genetic Diversity Index' (H_e), which is the expected heterozygosity determined from its allele frequencies within a population. Although the genetic diversity index of néré ($H_e = 0.397$) is lower than that of *Faidherbia albida*, $H_e = 0.450$ (Joly 1991), néré is among the most diverse species compared with most tropical and temperate species studied so far: $H_e = 0.141$ for *Acacia crassicarpa* and 0.146 for *Acacia auriculiformis* (Moran et al. 1989); H_e=0.182 for eucalypts (Moran and Bell 1983) and $H_e = 0.270$ for conifers (Mitton 1983). The differentiation between the populations, $F_{ST} = 0.129$ was also considerable (Teklehaimanot 1997) (F_{ST} is Wright's F-statistics used as a measure of genetic diversity among populations, based on comparison of the observed and predicted levels of heterozygosity). This indicates that a large part of the genetic diversity is due to differences between populations. A similar result was found in *Faidherbia albida*, which has a F_{ST} =0.123 (Joly 1991). This diversity is of importance to the improvement of the species and should be conserved both *in-situ* and *ex-situ* for future use through selection or breeding.

Studies carried out in 2002 across the species range indicated that human activities have had impacts on the genetics of karité. It appears that populations in crop fields have the highest mean number of alleles as well as the expected heterozygosity when compared with populations in fallows and forests (Kelly and Bouvet 2003). Even though the tree populations in the three land-use types are genetically differentiated, most of the genetic variation in karité is within rather than between populations. Shannon's diversity index, which is a measure of genetic diversity among populations based on allele frequencies, for the total population across the natural range was 0.526 ± 0.142 (Bouvet et al. 2004). Lower values were observed for the populations located at the west and east limits of the species range, i.e, Senegal and Uganda, but the differences were not significant. The expected heterozygosity followed a similar pattern, with a value for the total population being 0.35 ± 0.12. Nested analysis of variance with random amplified polymorphic DNA (RAPD) showed that 82% of the variation was explained by variation amongst individuals within populations. According to Hamrick et al. (1991), long-lived, out-crossed, insect pollinated and widely spread species tend to have high within-population variability. These characteristics of karité may explain its high genetic diversity within populations. The status of karité as an anthropic species may also explain its high genetic variability within population. For many generations, rural populations have played a role in the gene flow of karité by transporting fruits from village to village. Although the seeds of karité are recalcitrant and do not maintain their viability during periods of long transportation, models of population genetics have shown that a small number of seeds could be sufficient to reduce genetic differentiation (Bekessy et al. 2002). Bouvet et al. (2004) indicate that selection of quality germplasm must take into account within population variation for improvement of karité.

Reproductive biology

An understanding of the reproductive biology of a tree species is needed for developing successful crossing techniques for the genetic improvement in overall fruit yield. The success of both *in-situ* and *ex-situ* conservation of karité and néré as tree crops also largely depends on increasing the level of available information about the reproductive process.

In Nigeria (Oni 1997) and Burkina Faso (Ouedraogo 1995), many pollinating agents were attracted to néré, but they were mainly pollen thieves and nectar-lapping visitors. Bats were observed to be the main pollinators. The nocturnal shedding of pollen limits the role of birds for pollination in néré (Oni 1997). Although néré produces many inflorescences, only a small proportion ends up in fruit formation (Oni 1997). Where investment on fruit production is high, as in the case of néré, the effects of high abortion rate become significant at management level. Flowering in néré is such that groups of capitula units within an inflorescence mature in sequential order, with the ones at the most distal end maturing first. This acropetal arrangement favors selective abortion (Stephenson 1981). Stephenson (1981) indicated that the number of fruits formed is usually determined by resources rather than by the number of female flowers. He further indicated that incompatible pollen might result in flower abscission. Resource allocation in néré suggests that more of the capitula located at a higher crown level develop into fruits compared with those at the lower crown level (Oni 1997). It is not clear, however, whether the foraging behavior of bats or resource allocation within the tree is responsible for the pattern of fruit formation. As bats still probably play a significant part in néré pollination and their foraging behavior is more concentrated at the upper crown of the tree, selection of trees of 'tall' (>20 m) to – 'medium' (10–20 m) height may favor improved fruiting in néré. It is generally accepted that large trees have a large root volume and higher nutrient levels in the trunk and are therefore expected to be more productive than smaller trees (Stephenson 1981).

In a study of factors affecting fruit production in néré conducted by Oni (1997), the open- pollination treatment and the control (unrestricted visitation by the pollinator) had greater numbers of pods than the other treatments (capitula made inaccessible to visitors and assisted pollination), suggesting that possibly bats still play a significant role in fruit production of néré. He also found that néré produces a large number of capitula relative to the number of fruit set. Stephenson (1981) reported that the ultimate number of flowers that set fruit largely depends on the resources of the maternal parent, while the proportion of flowers that set fruit decreases as the number of pollinated flowers increases. According to Stephenson (1981), when buds or newly opened flowers are artificially thinned in many orchard species, the percentage of pollinated flowers that became fruits increased in proportion to the intensity of thinning. Thus, management practices that deliberately favor thinning of extra capitula after pollination in the initial sets of capitula may increase fruit set. This is a potential area of future research in enhancing fruiting efficiency in the species.

According to Oni (1997) néré invests heavily in pulp production of the fruit at the expense of the seed, which accounts for less than 20% of the overall pod weight. This makes the fruit attractive to frugivores at the expense of the seeds required for natural regeneration and food. The major disadvantage of this is the payload that the seeds have to equip themselves to produce competitive seedlings. Seed size has been shown to correlate with the success of seedling establishment and subsequent survival in a competitive environment. Selection of naturally big seeds through collection and screening of provenances should be the first steps in developing bigger seeded néré at the expense of the pulp (Oni 1997).

Studies in Uganda have indicated bees as a major pollinator of karité (Okullo et al. 2003). Fruit set was highest (32%) in manually cross-pollinated flowers, followed by 26% in an open pollinated treatment. The lowest fruit set was found in flowers emasculated and bagged without pollination to test for apoximis or agamospermy, i.e., nonsexual reproduction, indicating that there could be no possibility of autonomous autogamy, apoximis or agamospermy in *Vitellaria paradoxa* subsp. *nilotica*. The low level of fruit-set under natural conditions compared with controlled pollination indicates that fruit set is limited by pollination and other factors such as inbreeding, insect oviposition (larva seen in some aborted fruits), fruit abortion and fruit fall caused by weather factors. It is important to note that in a mixed mating system (where some fruits are produced from self-fertilization and others from out crossing), any force that influences the degree of selectivity of fruit or seed abortion can modify the mating system of the plant. For example, if selfed seeds are more likely to be aborted than out-crossed seeds under limiting resources within a tree, then resource limitation can increase out crossing in that tree. The presence of a strong, copious, sweet scented odor during peak flowering suggests an important role for floral rewards in karité pollination. In addition, abundant, small, sticky and readily available pollen also serve as part of the reward system (Okullo et al. 2003).

The activity of bees is highly affected by climatic conditions such as temperature. Increased temperature as the day progresses in the dry season depresses bees' activity and smoke from fires, which has a tranquillizing effect on bees, depresses their activity. The presence in the landscape of tree species such as *Mangifera indica*, *Vitex doniana* and *Combretum* spp., the flowering of which coincide with that of karité, may also divert bees from karité. In unprotected fascicles, soot and ash from burning grass sometimes clog stigmatic surfaces and could prevent the germination of pollen tubes on these stigmas. Therefore, reduced bee activity and savanna burning may be contributing to low fruit set in karité. Overall, the reproductive success of karité is very limited indeed. Fruit set is remarkably low, albeit supplemental pollination resulting in some significant increases in reproductive output. The self-incompatibility system, as well as the paucity of insect visitors, environmental hazards such as fires, contribute to limited fruit set (Okullo et al. 2003).

The high fruit-set under hand pollination and opening of inflorescence during peak flowering time resulting in higher fruit-set provide some evidence for pollination limitation as the cause for low fruit production in karité. This can also serve as a mechanism for selective abortion of inferior progeny without reducing total fruit production, which requires an in-depth investigation.

Vegetative propagation

Mature trees of néré and karité are not being replaced by natural regeneration and their natural populations are declining because local people continuously keep harvesting the fruits (Etejere et al. 1982). Therefore, there is a danger of continuing depletion and erosion of the natural gene pool of these species. Many scientists believe that vegetative propagation offers the opportunity to rapidly overcome such problems (Leakey et al. 1982). The productivity of néré and karité could be improved by using genetically improved seeds, which can be produced from an established seed orchard. However, until the seed orchard starts producing seed, and that too, in adequate quantities, vegetative propagation is the only means of meeting demand for improved plant material. Furthermore, in the cases of provenances for which there are problems of seed collection, viability, and germination, vegetative propagation is a valuable means of propagation. Plants produced by vegetative means can be used for phenotypic selection followed by clonal testing to evaluate the genotypes for the production of superior planting material. It may be possible through vegetative propagation and clonal selection to achieve recovering néré and karité populations and promote them as horticultural cultivars in the farmlands of arid and semiarid Africa.

Methods for propagating néré and karité from juvenile and adult stem cuttings have been developed (Lovett et al. 1996: Lovett and Haq 2000b; Teklehaimanot et al. 1996). Stem cuttings of karité and néré successfully root under both mist and nonmist conditions. Studies have shown that néré can be clonally multiplied and differentiated by tissue culture, which eventually should be the primary means for clonal propagation of superior individuals and to increase their numbers from intra or interspecific hybrids (Teklehaimanot et al. 2000). The results from air layering experiments indicate that it is possible to root air layers of néré (Teklehaimanot et al. 2000). A grafting experiment of karité conducted in Burkina Faso and Mali was also successful (Sanou et al. 2004). Five methods of grafting (side cleft, top cleft, tongue, chip budding and side veneer) were investigated and the results showed decreasing success of survival of grafts 16 weeks after grafting as follows: side cleft (86%), tongue (81%), top cleft (78%), chip budding (38%) and side veneer (21%). The season of grafting had significant effect on the survival and growth of scions, with dry season giving the best success. Two years after grafting, two-side veneer grafts produced fruits. These methods can help shorten the juvenile phase of karité and néré and allow multiplication of superior quality trees, in order to develop cultivars with greater potential for fruit production.

Improving the performance of associated crops

Yields of sorghum and millet under néré and karité trees were found to be reduced by 30% to 60% compared to the yields in the open field (Boffa et al. 1995; Kater et al. 1992; Kessler 1992). The yield reduction is attributable to reduced light under trees. Photosynthetically active radiation (PAR) under unpruned néré and karité trees was reduced to 49% and 34% of the incident PAR, respectively, which was low enough to cause substantial reduction in crop yields. The high rates of transpiration of unpruned trees may also contribute to reduction in crop performance. Maximum sap velocities of 82.4 cm h^{-1} and 87.5 cm h^{-1} recorded for unpruned trees for karité and néré, respectively, indicate that both species are capable of transpiring large quantities of water. In Burkina Faso, crown pruning of trees significantly increased crop production (Bayala et al. 2002; 2003). The highest millet yields (507 ± 49 kg ha^{-1} $year^{-1}$) were obtained under totally pruned néré trees and the same was the case with sorghum under totally pruned karité trees (1092 ± 290 kg ha^{-1} $year^{-1}$). However, pruning severely affected fruit production of both the tree species. There was only a partial recovery in fruit production of totally pruned trees even after three years of pruning: the pruned trees had only 15% and 29% of fruiting in unpruned karité and néré trees, respectively. The results also showed that crown pruning had different influences on root length density of karité and néré, and this may also contribute to improvement in crop performance (Bayala et al. 2004). Reduced fine root length density of trees in the upper soil layer due to pruning and dominance of crop roots has also been reported (60 to 75% of their density). This effect can partly explain the increase in crop production under totally pruned trees compared with unpruned trees. Application of pruned leaves of karité as mulch had also a positive effect on millet performance possibly as a consequence of changes in soil physical and chemical properties (Bayala et al. 2003). The conclusion that can drawn from the above results is that in the short term crop production under néré and karité could be improved by crown pruning, but long term effects may depend on the ability of the trees to maintain the amelioration of soil fertility and how quickly the trees recover from pruning.

Conclusions

The review provides ample evidence for the value and economic benefits of néré and karité to the local people in sub-Saharan Africa. The overall returns from néré and karité are so high that domestication of these species could enhance the economic returns of subsistence farmers, who may currently be earning less than $1 per day. The domestication of karité and néré in agroforestry could also enhance the productivity and sustainability of farming systems by maintaining soil fertility and creating a more favorable microclimate for associated crops and livestock, in addition to supplying food, fodder, fuelwood, building material and other NWFPs that provide income and employment in rural areas. There are several semi-or undomesticated indigenous tree species with similar potential to néré and karité, which could be exploited in a similar manner in order to enhance their role in food security, poverty alleviation and ecological services. However, there are several socioeconomic conditions of sub-Saharan Africa that need to be addressed, to achieve successful tree domestication and their management in agroforestry parkland systems.

- It is increasingly recognized that the centralized and repressive manner in which national forest regulations in the sub-Saharan Africa have been enforced locally has kept indigenous parkland management at suboptimal levels (Bagnoud et al. 1995). Most farmers do not own parkland trees and this is one of the critical factors affecting farmers' willingness to domesticate and manage parkland trees including néré and karité. Thus, a change in land use and forest policies in sub-Saharan Africa is urgently needed.
- The intensity of management of parkland trees particularly karité and néré has declined recently due to a lack of production incentives related to markets and technology. Low prices offered by local and international industries may discourage collection and processing of néré and karité products beyond home consumption. This is a disincentive to farmers. According to Becker and Statz (2003), karité changed hands at 16 instances before reaching the final consumer in Europe. Thus, there is a need to develop local and national markets breaking this long chain of middlemen and improve income to farmers. Authorities should ensure producers' access to market information and a system for standardizing product quality at national level should also be established.
- The preparation of fermented seeds of néré is a time-consuming business, which takes three to four days and requires large quantities of firewood for cooking the seeds. Stock cubes made from néré seeds would provide a labor-saving alternative and have a longer shelf life. But in a survey carried out in 1994 in Nigeria, the traditionally prepared seeds were preferred to the commercial stock cubes by all those interviewed (Teklehaimanot 1997). Methods of improving the efficiency of traditional production need to be tested and developed.
- Traditional karité butter-extraction techniques are time consuming and physically strenuous; they require large amounts of firewood and water, and tend to have low extraction efficiency, creating a significant drain on these scarce resources. Becker and Statz (2003) estimated that for the transformation of each kg of dried nuts 2.6 kg of firewood is needed as well as 3.6 l of water. It is estimated that 80% of butter made in Mali and Burkina Faso is made traditionally, though semiindustrialized processes are

developing alongside traditional methods. For example, improved plate grinders and oil presses have been developed in Uganda for the liquid karité butter. Although they have proved successful in the amount of butter extracted, there have been problems because of their relative complexity and level of inputs required for operation and maintenance (Masters 2003). Improved village level technologies at various processing stages addressing these production constraints and quality requirements remain to be tested and developed, particularly for the West African solid karité butter.

Acknowledgements

Most of the information provided in the review has been generated through two major research projects carried out in sub-Saharan Africa: 'Germplasm conservation and improvement of *Parkia biglobosa* (1993–1997)' and 'Improved management of agroforestry parkland systems in sub-Saharan Africa (1998–2003),' both of which were co-ordinated by the author of this review. The research projects were funded by the European Union. The contributions of the following partners involved in the two projects are gratefully acknowledged: University of Wales Bangor, United Kingdom, Centre National de Semences Forestiere, Burkina Faso, Forestry Research Institute of Nigeria, University of Ibadan, Nigeria, Institut de l'Environnement et de Recherches Agricoles, Burkina Faso, Institut d'Economie Rural, Mali, Ouagadougou University, Burkina Faso, Makerere University, Uganda, COVOL, Uganda, Ben Gurion University of the Negev, Israel, Instituto Sperimentale per la Elaiotecnica, Pescara, Italy, Aarhus Oliefabrik A/S, Denmark, PROPAGE, France, Wageningen University, the Netherlands, University Pierre et Marie Curie, France, University of Freiburg, Germany, CIRAD, France, and ICRAF.

References

Adu-Ampomah Y., Amponsah J.D. and Yidana J.A. 1995. Collecting germplasm of sheanuts (*Vitellaria paradoxa*) in Ghana. Plant Genetic Resources Newsletter 102: 37–38.

Bagnoud N., Schmithusen F. and Sorg J.P. 1995. Les parcs a karite et néré au Sud-Mali. Analyse du bilan economique des arbres associes aux cultures. Bois et Forets des Tropiques 244: 9–23.

Bayala J., Teklehaimanot Z. and Ouedraogo S.J. 2002. Millet production under pruned tree crowns in a parkland system in Burkina Faso. Agroforest Syst 54: 203–214.

Bayala J., Teklehaimanot Z. and Ouedraogo S.J. 2003. Effects of pruning on production of *Vitellaria paradoxa* and *Parkia biglobosa* and yield of associated crops in parklands. pp. 85–100. In: Teklehaimanot Z. (ed.), Improved Management of Agroforestry Parkland Systems in Sub-Saharan Africa. EU/INCO Project Contract IC18-CT98-0261, Final Report, University of Wales Bangor, UK.

Bayala J., Teklehaimanot Z. and Ouedraogo S.J. 2004. Fine root distribution of pruned trees and associated crops in a parkland system in Burkina Faso. Agroforest Syst 60: 13–26.

Becker M. and Statz J. 2003. Marketing of parkland products. pp. 142–151. In: Teklehaimanot Z. (ed.), Improved Management of Agroforestry Parkland Systems in Sub-Saharan Africa. EU/INCO Project Contract IC18-CT98-0261, Final Report, University of Wales Bangor, UK.

Bekessy S.A. Allnutt T.R., Premoli A.C. Lara A., Ennos R.A., Burgman M.A., Cortes M. and Newton A.C. 2002. Genetic variation and endemic Monkey Puzzle tree, detected using RAPDs Heredity 88: 243–249.

Bernard C., Oualbadet M., Nklo O. and Peltier R. 1995. Parcs agroforestiers dans un terroir Soudanien: cas du village Dolekaha au nord de la Cote d'Ivoire. Bois et Forets des Tropiques: 244: 25–42.

Boffa J.M. 1995. Productivity and Management of Agroforestry Parklands in the Sudan Zone of Burkina Faso, West Africa. Ph.D. Thesis, Purdue University, West Lafyette, Indiana, USA, 101 pp.

Boffa J.M. 1999. Agroforestry Parklands in Sub-Saharan Africa. FAO Conservation Guide No 34, Rome, 230 pp.

Bonkoungou E.A. 1987. Monographie du Karité *Butyrospermun paradoxum* (Gaertn. F.) Hepper, Espèce Agroforestière à Usages Multiples. Institut de Recherches en Biologie et Ecologie Tropicale. Ouagadougou, Burkina Faso, 69 pp.

Booth F.E.M. and Wickens G.E. 1988. Non-timber uses of selected arid zone trees and shrubs in Africa. FAO Conservation Guide No 19. FAO, Rome, 176 pp.

Bouvet J.M., Fontaine C., Sanou H. and Cardi C. 2004. An analysis of the pattern of genetic variation in *Vitellaria paradoxa* using RAPD markers. Agroforest Syst 60: 61–69.

Campbell-Platt G. 1980. African locust bean (Parkia species) and its West African fermented food product, dawadawa. Ecol Food Nutr 9: 123–132.

Chevalier A. 1946. L'arbre à beurre d'Afrique et l'avenir de sa culture. Oléagineux 1: 7–11.

de Saint Sauveur A. 2003. Indigenous management techniques of farmed parklands. pp. 29–42. In: Teklehaimanot Z. (ed.), Improved Management of Agroforestry Parkland Systems in Sub-Saharan Africa. EU/INCO Project Contract IC18-CT98-0261, Final Report, University of Wales Bangor, UK.

Etejere E.O., Fawole M.O. and Sani A. 1982. Studies on the seed germination of *Parkia clapppertoniana.* Turrialba 32: 181–185.

FAO 1990. Forest Genetic Resource Priorities (by Region, Species, and Operation). 10. Africa. Report of the Seventh Session of the FAO Panel of Experts on Forest Gene Resources, Rome, Italy, 4–6 December, 1989. FAO, Rome, 79 pp.

Gijsbers H.J.M., Kessler J.J. and Knevel M.K. 1994. Dynamics and natural regeneration of woody species in farmed parklands in the Sahel region (Province of Passore, Burkina Faso). Forest Ecol and Manag 64: 1–12.

Guinko S. and Pasgo L.J. 1992. Harvesting and marketing of edible products from local woody species in Zitenga, Burkina Faso. Unasylva 43: 16–19.

Hall J.B., Aebischer D.P., Tomlinson H.F., Osei-Amaning E. and Hindle J.R. 1996. *Vitellaria paradoxa*: A Monograph. School of

Agricultural and Forest Sciences, University of Wales Bangor, UK, 105 pp.
Hall J.B., Tomlinson H.F., Oni P.I., Buchy M. and Aebischer D.P. 1997. *Parkia biglobosa*: A Monograph. School of Agricultural and Forest Sciences, University of Wales Bangor, UK, 107 pp.
Hamrick J.L., Godt M.J.W., Murawski D.A. and Loveless M.D. 1991. Correlations between species traits and allozymes diversity: Implications for conservation biology. pp. 75–86. In: Falk D. and Holsinger K. (eds), Genetics and Conservation of Rare Plants. Oxford University Press, New York.
Henry A.N., Chithra V. and Nair C. 1983. *Vitellaria paradoxa* vs *Butyrospermum* (Sapotaceae). Taxon 32: 286.
Hopkins H.C. 1983. The taxonomy, reproductive biology and economic potential of *Parkia* (Leguminosae: Mimosoideae) in Africa and Madagascar. Botanical J Linnean Society 87: 135–167.
Hopkins H.C. and White F. 1984. The ecology and chorology of *Parkia* in Africa. Bulletin Jardin Botanique Naturelle Belgique 54: 235–266.
Joly H.I. 1991. Population genetics of *Acacia albida (= Faidherbia albida)*. Bulletin of the International Group of Study of Mimosoidae 19: 86–95.
Kater L.J., Kanté S. and Budelman A. 1992. Karité (*Vitellaria paradoxa*) and néré (*Parkia biglobosa*) associated with crops in South-Mali. Agroforest Syst 18: 89–105.
Kelly B. and Bouvet J.M. 2003. Impact of farmers' practices on the genetic diversity of *Vitellaria paradoxa*. pp. 51–65. In: Teklehaimanot Z. (ed.), Improved Management of Agroforestry Parkland Systems in Sub-Saharan Africa. EU/INCO Project Contract IC18-CT98-0261, Final Report, University of Wales Bangor, UK.
Kelly B., Bouvet J.M. and Picard N. 2004. Size class distribution and spatial pattern of *Vitellaria paradoxa* in relation to farmers' practices in Mali. Agroforest Syst 60: 3–11.
Kessler J.J. 1992. The influence of karité (*Vitellaria paradoxa*) and néré (*Parkia biglobosa*) trees on sorghum production in Burkina Faso. Agroforest Syst 17: 97–118.
Lamien N., Sidibe A. and Bayala J. 1996. Use and commercialisation of non-timber forest products in Western Burkina Faso. pp. 51–64. In: Leakey R.R.B., Temu A.B., Melnyk M. and Vantomme P. (eds), Domestication and Commercialisation of Non-Timber Forest Products in Agroforestry Systems. Non-wood Forest Products No. 9. FAO, Rome.
Leakey R.R.B., Last F.T. and Longman K.A. 1982. Domestication of tropical trees: an approach securing future productivity and diversity in managed ecosystems. Commonw Forest Rev 61: 33–42.
Leakey R.R.B., Wilson J. and Deans J.D. 1999. Domestication of trees for agroforestry in drylands. Ann Arid Zone 38: 195–220.
Lericollais A. 1989. La mort des arbres a Sob, en pays Sereer (Senegal). pp. 187–197. In: Tropiques, Lieux et Liens. ORSTOM, Paris, France.
Lovett P.N., Azad A.K., Paudyal K. and Haq N. 1996. Genetic diversity and development of propagation techniques for tropical fruit trees. pp. 286–287. In: Smartt J. and Haq N. (eds), Domestication, Production and Utilisation of New Crops. ICUC, UK.
Lovett P.N. and Haq N. 2000a. Diversity of the Sheanut tree (*Vitellaria paradoxa* C.F.Gaertn.) in Ghana. Genetics and Crop Evolution 47: 293–304.
Lovett P.N. and Haq N. 2000b. Evidence for anthropic domestication of the sheanut tree (*Vitellaria paradoxa*). Agroforest Syst 48: 273–288.
Maranz S., Wiesman Z., Bianchi G. and Bisgaard J. 2004. Germplasm resources of *Vitellaria paradoxa* based on variations in fat composition across the species distribution range. Agroforest Syst 60: 71–76.
Masters E. 2003. Improved Methods of Storage, Processing and Post-Processing Village Technologies of *Vitellaria paradoxa* butter. pp. 152–155. In: Teklehaimanot Z. (ed.), Improved Management of Agroforestry Parkland Systems in Sub-Saharan Africa. EU/INCO Project Contract IC18-CT98-0261, Final Report, University of Wales Bangor, UK.
Mitton J.B. 1983. Conifers. pp. 431–443. In: Tanksley S.D. and Orton T.J. (eds), Isozymes in Plant Genetics and Breeding. Part B. Elsevier, The Netherlands.
Moran G.F. and Bell J.C. 1983. *Eucalyptus*. pp. 423–430. In: Tanksley S.D. and Orton T.J. (eds), Isozymes in Plant Genetics and Breeding. Part B. Elsevier, The Netherlands.
Moran G.F., Muona O. and Bell J.C. 1989. Breeding systems and genetic diversity in *Acacia auriculiformis* and *Acacia crassicarpa*. Biotropica 21: 250–256.
Nair P.K.R. 1993. An Introduction to Agroforestry. Kluwer Academic Publishers, The Netherlands, 499 pp.
Nikiema A., van der Maesen L.J.G. and Hall J.B. 2003. The impact of parkland management practices on plant resources diversity. pp. 43–50. In: Teklehaimanot Z. (ed.), Improved Management of Agroforestry Parkland Systems in Sub-Saharan Africa. EU/INCO Project Contract IC18-CT98-0261, Final Report, University of Wales Bangor, UK.
Okullo J.B., Hall J.B. and Masters E. 2003. Reproductive biology and breeding systems of *Vitellaria paradoxa*. pp. 66–84. In: Teklehaimanot Z. (ed.), Improved Management of Agroforestry Parkland Systems in Sub-Saharan Africa. EU/INCO Project Contract IC18-CT98-0261, Final Report, University of Wales Bangor, UK.
Oni P. 1997. *Parkia biglobosa* (Jacq.) Benth. in Nigeria: A Resource Assessment. Ph.D. Thesis, University of Wales Bangor, UK, 220 pp.
Ouedraogo A.S. 1995. *Parkia biglobosa* (Leguminosae) en Afrique de l'Ouest: Biosystematique et Amélioration. Ph.D. Thesis. Institute for Forestry and Nature Research, Wageningen, 205 pp.
Ruyssen B. 1957. Le karité au Soudan. L'Agronomie Tropicale 12 (2): 143–226.
Sanou H., Kambou S., Teklehaimanot Z., Dembélé M., Yossi H., Sina S., Djingdia L. and Bouvet J.M. 2004. Vegetative propagation of *Vitellaria paradoxa* by grafting. Agroforest Syst 60: 93–99.
Schreckenberg K. 1996. Forests, Fields and Markets: A Study of Indigenous Tree Products in the Woody Savannas of Bassila Region. Ph.D. Thesis, University of London, London, UK, 326 pp.
Stephenson A.G. 1981. Flowering and fruit abortion: proximate causes and ultimate functions. Annu Rev Ecol Syst 12: 253–257.
Teklehaimanot Z., Tomlinson H., Lemma T. and Reeves K. 1996. Vegetative propagation of *Parkia biglobosa* (Jacq.) Benth., an undomesticated fruit tree from West Africa. J Hortic Sci 71: 205–215.
Teklehaimanot Z. 1997. Germplasm Conservation and Improvement of *Parkia biglobosa* (Jacq.) Benth. for Multipurpose Use. EU/INCO Project Contract TS3*-CT92-0072, Final Report, University of Wales Bangor, UK, 166 pp.
Teklehaimanot Z., Tomlinson H., Ng'andwe M. and Nikiema A. 2000. Field and *in vitro* methods of propagation of the African locust bean tree (*Parkia biglobosa* (Jacq) (Benth.). J Hortic Sci Biotec 75: 42–49.

Tomlinson H., Teklehaimanot Z., Traore A. and Olapade E.O. 1995. Soil amelioration and root symbiosis of *Parkia biglobosa* (Jacq.) Benth. in West Africa. Agroforest Syst 30: 145–159.

Tomlinson H., Traore A. and Teklehaimanot Z. 1998. An investigation of the root distribution of *Parkia biglobosa* in Burkina Faso, West Africa, using a logarithmic spiral trench. Forest Ecol Manag 107: 173–182.

Tran T. 1984. Shea butter-more than just a cosmetic ingredient. Drug and Cosmetic Industry 54: 54–61, 114–116.

Wiesman Z., Maranz S., Bianchi G. and Bisgaard J. 2003. Chemical analysis of fruits of *Vitellaria paradoxa*. pp. 131–139. In: Teklehaimanot Z. (ed.), Improved Management of Agroforestry Parkland Systems in Sub-Saharan Africa. EU/INCO Project Contract IC18-CT98-0261, Final Report, University of Wales Bangor, UK.

Wood P.J. and Burley J. 1991. A Tree for All Reasons. International Centre for Research in Agroforestry, Nairobi, Kenya, 158 pp.

World Bank 1989. Sub-Saharan Africa: From Crises to Sustainable Growth. A Long-Term Perspective Study. Washington, DC, World Bank, 300 pp.

Agroforestry Systems **61:** 221–236, 2004.

Ecological interactions, management lessons and design tools in tropical agroforestry systems

L. García-Barrios[1,*] and C.K. Ong[2]

[1]*El Colegio de la Frontera Sur, México. Carretera Panamericana y Perisur (s/n); San Cristóbal de las Casas, Chiapas, Cp. 29290 México;* [2]*World Agroforestry Centre, P.O. Box 30677, Nairobi, Kenya;* [*]*Author for correspondence: e-mail: lgarcia@sclc.ecosur.mx*

Key words: Growth resources, Indices, Predictive understanding, Roots, Simulation models, Tree–crop interactions

Abstract

During the 1980s, land- and labor-intensive simultaneous agroforestry systems (SAFS) were promoted in the tropics, based on the optimism on tree-crop niche differentiation and its potential for designing tree-crop mixtures using high tree-densities. In the 1990s it became clearer that although trees would yield crucial products and facilitate simultaneous growing of crops, they would also exert strong competitive effects on crops. In the meanwhile, a number of instruments for measuring the use of growth resources, exploratory and predictive models, and production assessment tools were developed to aid in understanding the opportunities and biophysical limits of SAFS. Following a review of the basic concepts of interspecific competition and facilitation between plants in general, this chapter synthesizes positive and negative effects of trees on crops, and discusses how these effects interact under different environmental resource conditions and how this imposes tradeoffs, biophysical limitations and management requirements in SAFS. The scope and limits of some of the research methods and tools, such as analytical and simulation models, that are available for assessing and predicting to a certain extent the productive outcome of SAFS are also discussed. The review brings out clearly the need for looking beyond yield performance in order to secure long-term management of farms and landscapes, by considering the environmental impacts and functions of SAFS.

Introduction

Traditional low-input agricultural systems involving trees have been designed and managed for centuries by poor peasants around the world, and are still conspicuous in the tropics. During the past century, land-use intensification, agroecosystem simplification and other social changes have undermined the functionality of many of these low-input systems, and confronted peasant agriculture with enormous sustainability challenges (Nair 1998; García-Barrios and García-Barrios 1992; García-Barrios 2003). In the past two decades, great expectations have been set on the promotion of traditional and novel agroforestry practices as a means for slowing down or reversing such trends, once it became clear that high-input strategies promoted by development agencies had failed to be adopted and/or to deliver benefits to smallholders (Sanchez 1995). Where there is still scope for fallow agriculture in the tropics, sequential agroforestry systems such as enriched fallows have been proposed; where land-use intensification and fragmentation is more severe, the bet has been on simultaneous agroforestry systems (SAFS) such as alleycropping, alley farming, parkland systems and trees on field boundaries.

During the 1980s, alleycropping was promoted throughout the tropics as a sustainable option for low-input agriculture. By the 1990s it was recognized that the density, management intensity and environmental scope of new tree-based SAFS had been pushed too far. It became clear that introducing trees in croplands was in some cases like walking on a razor-edge because trees provide peasants with both crucial products and strongly facilitate crops, but can also exert very strong competitive effects. Among other things, this motivated researchers to unravel the science of agroforestry (Sanchez 1995). The aim was a predictive understanding (at the ecological, physiological and agronomic level) on how such interactions and their productive outcome respond to species selection, tree management practices and resource limited environments, in order to develop realistic agroforestry solutions. In the process, a number of useful concepts, predictive models and production assessment tools were enhanced or developed. At the same time it became clearer that socioeconomic motivations and restrictions strongly influence how farmers perceive and manage tree–crop interactions, and that recommendations based on productivity at the plot level alone are insufficient for the long-term management and sustainability of farms and landscapes. This chapter addresses the issues previously mentioned and reviews current advances in tree–crop interactions and their implications, focusing on annual crops and trees in simultaneous agroforestry systems in the tropics. Two companion chapters in this volume, by Jose et al. (2004) and Thevathasan and Gordon (2004), discuss interspecific interactions in temperate agroforestry systems.

Positive and negative tree effects on crops in SAFS

Competition occurs when two overlapping plants reduce one or more growth resources to the point where the growth, survival or reproductive performance of at least one of them is negatively affected (Harper 1990). Overlap increases as a consequence of growth and/or increased density. Plants differ greatly in size, life form, phenology and capacity to capture and use efficiently above- and below-ground resources (Goldberg 1990); therefore, intra- and interspecific competition can differ strongly. When acting together in plant mixtures, these two interactions can produce a range of situations. At one extreme, species coexist and overyield (i.e. more land would be required to obtain the mixture yields by sowing each species separately as sole crops). At the other, a single dominant species strongly reduces the performance of others. When above- and/or belowground resources are partitioned between trees and crops such that relative interspecific competition is lower than relative intraspecific competition, we refer to the case as weak interspecific competition, complementarity or niche differentiation (Vandermeer 1989).

Until recently, ecologists accepted niche differentiation theory (MacArthur and Levins 1967) as the most important explanation of plant coexistence, and agroecologists consequently predicted that there must be considerable scope for obtaining or enhancing mixed crop over-yielding by finding the proper species combinations, spatio-temporal designs and simple management tasks which produce minimum interspecific competition (e.g., Vandermeer 1989). Ecologists have realized that weak interspecific competition is only one of many conditions, which can explain species coexistence in a natural plant community (García-Barrios 2003). Agroecologists and agroforesters have also found that complementarity is less conspicuous and requires more plant management than originally thought (Ong and Leakey 1999).

A plant facilitates another plant when it modifies the biophysical environment in such a way that it creates one or more potential conditions for a better performance of the latter (Hunter and Aarssen 1988). The facilitator can a) produce a net resource increase in the system, b) enhance the potential capture and/or use efficiency of one or more growth resources, c) produce hormone-type stimulatory or inhibitory allelopathic compounds and/or d) deter the presence of competitors, predators and pathogens and e) attract pollinators or dispersal agents (Holmgren et al. 1997). It is important to stress that a) such potential benefits can have little actual facilitative effect when other non-facilitated resources or conditions limit growth (Kho 2000), and b) that the facilitator's competitive and other negative effects can override its positive effects.

The interplay of positive and negative plant interactions and its outcome can change along resource gradients, both in natural communities (e.g. Pugnaire and Luque 2001), intercrops (e.g., Rao and Willey 1980) and SAFS (e.g., Rao et al. 1998). Predicting the consequence of a change in environment is elusive, as results are highly variable and trends contradictory. A clearer and more comprehensive picture is beginning to emerge as researchers consider to what degree each resource is limiting to growth in an environment and how the species involved are suited to compete for and/or to facilitate the use of such resource(s) (Kho 2000; Ong et al. 2004).

Tree–crop interactions in SAFS share the basic features indicated in the previous section. Yet, trees have noteworthy characteristics that make them have very strong positive and negative effects on agroecosystem structure and function (Ong 1995). Trees have more biomass, are taller, and have more extended and deeper roots than crops. Trees are perennial and therefore: a) they eventually explore a relatively large space and can substantially modify their biophysical environment to their benefit, b) they are better adapted to resist cyclic environmental harshness and to use and recycle resources when it is more efficient to do so, and c) they commonly have a canopy and a root system in place when crops starts growing (Ong et al. 1996).

Mature trees in SAFS can have the following positive effects on crop fields:

1. They can add considerable amounts of organic matter to the soil and slow down its decomposition rate, improving soil fertility and physical structure. Trees' roots can stabilize loose soil surfaces, which together with tree litter cover, reduces erosion (Young 1997). They recover leached nutrients from deep soil or bedrock layers inaccessible to crops (Rao et al. 1998). Trees can make more phosphorus available via mycorrizha and fix important amounts of nitrogen, which can be transferred to the crops via shoot and root prunings (Giller 2001).
2. Although trees can increase the potential soil water-holding capacity, they have variable and conflicting effects on the actual water volume available in the tree–crop–soil system: Rainfall intercepted by the canopy that evaporates without reaching the soil can be as much as 50% when tree density is high (Ong et al. 1996). Yet, reduced soil evaporation by tree shade can help offset such losses as long as rainfall is lower than 700 mm per annum (Ong and Swallow 2004). Trees can eventually increase infiltration but this strongly depends on slope and soil characteristics (Ong and Swallow 2004). In general, tree presence produces a net increase in total water used by the system (Ong and Swallow 2004).
3. Tree shade can reduce leaf temperature and evaporative demand experienced by crops, increasing the latter's water productivity (i.e., g of dry mass per g of water transpired). The net shade effect is more positive (or less negative) when the annual crop is a C3 plant which is normally light saturated in the open, so partial shade may have little effect on assimilation or even be beneficial (Ong 1996). This improvement in conversion efficiency is relatively modest when compared to the effect described in the previous point (Ong and Swallow 2004).
4. Tree hedgerows provide protection against wind and runoff (Rao et al. 1998.).
5. Trees can reduce weed populations and change weed floristic composition towards less aggressive, slow growing species (Leibman and Gallandt 1997).
6. Little is known about the complex and sometimes conflicting effects of trees on pest and disease control, but Schroth et al. (2000) and Rao et al. (2000) provide comprehensive reviews on the topic. Increased pest and disease incidence has often been observed directly at the tree–crop interface. Trees increase air humidity, which favors microorganisms, provide shelter for herbivores (insects, birds, and small mammals), which damage the crops, and reduce pest and disease tolerance of competition-stressed crops. Trees themselves can be more susceptible to pest and disease attack when sown at densities and spatial arrangements uncommon to their natural environments. Yet, tree hedges have the potential to slow down windborne pests and diseases, to act as repellants, and to attract natural enemies; recent evidence is provided by Girma et al. (2000).

On the other hand, with increase in density of trees, their size, and/or ability to capture resources in SAFS, they can exert strong competition for light, water and nutrients, and reduce annual crop yields beyond the interests of farmers if improperly selected and managed (García-Barrios 2003). Nevertheless, weak competition is possible under certain circumstances such as the following:

1. Tree roots can potentially reach below the crop root zone, and thus they can use water accumulated deeper in the ground when the crop is growing; after the crop is harvested, they can use whatever residual available water is found in the crop root zone; and they can use any additional rain which falls outside the crop growing season (Ong et al. 1996). It is important to stress that trees used in SAFS do not always have deep pivotal roots, and that mixed and superficial tree root architectures are common (van Noordwijk et al. 1996). Moreover, if water recharge below the root zone is infrequent and/or nutrients are superficial, most trees will tend to develop or redirect their roots to the upper soil layers (Rao et al. 2004) and only a few species will develop roots that can reach relatively deep water tables. Consequently, there seems to be less scope for vertical root complementarity than originally thought (Sanchez 1995; Ong and Swallow 2004).
2. Some deciduous tree species used in SAFS in semi-arid regions such as *Faidherbia albida* exhibit reverse phenology: they produce their leaves and demand water only during the dry season, while their litter provides nutrients and their trunk and bare branches cast a light shade over crops during the rainy season (Rao et al. 1998).

3. Hundreds of different nitrogen fixing leguminous trees are used in SAFS (Giller 2001). When their leaves are not used as mulch, they do not facilitate crops, but their alternative N source can significantly reduce competition for this resource.
4. Radiation can be shared between closed canopy tree stands with low tree-LAI (leaf area index) and understory crops which tolerate or even require light shade, yet, most C4 and C3 annual crops are fast growing species which do not easily adapt to this condition (Black and Ong 2000).

In most SAFS there are problematic tradeoffs between positive and negative tree effects. Trees can modify the soil and microclimate environment much more than crops; they can have strong facilitative effects; and they can produce important environmental services and in specific circumstances compete weakly for some resources. Yet, tree characteristics often confer them a clear competitive advantage and they can strongly out-compete crops or reduce crop yield beyond acceptable levels to the farmer. Design and management practices such as optimum density and spatio-temporal arrangement, fertilization, weeding, and proper shoot and root pruning can help reduce competition and/or enhance facilitation. Yet, developing novel and successful SAFS designs and management practices has proved more difficult than expected. This has led to a number of partial failures as well as to useful lessons (Ong 1994; Sanchez 1995; Rao et al 1998).

Management difficulties arise in SAFS for a number of reasons, including the following:
1. Many SAFS practices are labor intensive and require relatively high crop-yield-advantage over sole crops to be attractive to farmers. This economic limitation is aggravated when novel SAFS are designed – and tree species for them are selected – mainly for mulch production and soil protection, with little attention to more immediate and direct tree products such as fodder, lumber and wood (Ong and Leakey 1999).
2. Practices that reduce competition for a given resource can inadvertently increase competition for other resources and/or decrease facilitation. For example, in tropical alleycropping systems: a) Intensive shoot pruning and accumulation of N-rich mulch in the alley reduce light competition and facilitate N allocation to the crop; yet, they both promote lateral-superficial tree roots which strongly compete with the crop for water and nutrients, sometimes overriding the positive effects (Hairiah et al. 1992; Fernández et al. 1993). b) Tree distances can be increased and optimized (Vandermeer 1998), and roots pruned (Ong and Swallow 2004), but in some cases with a negative effect on the formation of the root safety net (van Noordwijk et al. 1996).
3. The consequences of expanding successful SAFS to other environments have proved difficult to predict until recently, and the environmental scope of promising SAFS has been more limited than expected. The way in which tree-crop interactions respond to changes in environmental resource balance partially explains these situations. For example, alleycropping was extended to arid and semiarid regions, generally with very poor results (Rao et al. 1992). It was later realized that tree competition for water strongly depressed crop yields and was not compensated by the amounts and quality of mulch (Rao et al. 1992). Similar disappointments were obtained in very acid and unfertile soils (e.g., Ultisols and Oxisols), where P and other nutrients rather than nitrogen were the most limiting resources (Sanchez 1995).

In order to optimize tradeoffs between positive and negative tree effects, it is necessary to select in each environment the proper annual crops, tree species and densities, and soil amendments. It is also important to correctly choose the timing and intensity of root- and shoot pruning. Proper design and management is critical to achieve the desired balance between tree growth, annual crop production, tree crop production, weed suppression, soil improvement and other environmental services. Among other things, this implies finding tree/crop combinations that promote resource facilitation, resource sharing, and minimum resource competition and are compatible with farmers' goals and conditions. These combinations are not always readily obvious or easily found. A number of analytical and simulation tools have been developed in the past decade which help to understand some of these interactions and their productive outcome. We will now discuss some examples that represent the different approaches and their scopes.

Tree–crop interactions and yield assessment

Methods that do not consider growth resources explicitly

Quantifying the net effect of interactions

The combined effects of intra and interspecific interactions on species and community yields in mixtures can be quantitatively assessed without separating positive and negative effects explicitly, and without considering resource distribution and use. An extensive literature has been developed on this matter during the past 40 years (Vandermeer 1989). A number of competition- and productivity indices have been developed and their scope, limitations, and pitfalls discussed (Connolly et al. 2001; Williams and McCarthy 2001). Most productivity indices have been developed for two species intercrops. They assume that the farmer is interested in finding whether combinations of both crops in the same field can outperform the corresponding sole crops. We will give a very brief account of how crop and tree yields relate separately to their net intra and interspecific interactions by developing a graphical analysis of a hypothetical and generic SAFS. We will then discuss the simultaneous assessment of crop and tree yields.

Consider the crop and tree sole stands in Figure 1(a) and (b). Both are sown at their optimum sole crop densities (N.), i.e., the number of individuals per unit area which produces the maximum possible yield (Y.) in that surface. These parameters result from the reasonably hyperbolic relation between sole crop yields and sole crop plant densities (Willey and Heath 1969) depicted in Figure 1d and are a direct consequence of net intra-specific interactions. In our example, the sole crop stand parameters are Nc = 6 and Yc = 6, while the sole tree stand parameters are Nt = 2 and Yt = 18. Intra-row plant distances are the same for both crops. In the tree–crop mixture depicted in Figure 1c, every sixth crop row has been substituted with a tree row. As a consequence, the crop density in mixture (nc) is 5 and the tree density in mixture (nt) is 1. In other words, nc = 5/6*Nc and nt = 1/2*Nt. For simplicity, we will assume that both tree competition and facilitation on the crop can be present, and that no tree pruning is practiced so that facilitation results from other mechanisms. Crop plants nearer to the tree are generally smaller as they experience more competition (see Figure 1c), but here we consider the average plant performance within the unit area.

Different crop yield outcomes in mixture (yc) are possible; they are shown in Figure 2a. An interesting starting point is case 3, where yc = 5/6*Yc, i.e., mixture crop yield is reduced in the same proportion as crop density. The average crop individual has the same weight in sole crop and intercrop, which means that the average net effect of a tree on crop plants matches the net intra-specific effect of the average crop individual in the sole stand. Considering the asymmetry in tree and crop sizes, this seems highly unlikely, unless it results from very weak tree competition due to high resource complementarity, or from facilitation almost compensating competition. In case 4, either tree competition is zero (perfect complementarity) or facilitation exactly compensates competition, because in this case, yc equals the yield expected for five plants per unit area in the sole crop (here we make the assumption that given a near-optimum crop density, particular plant arrangement has little effect). In case 5, the crop stand experiences enough net facilitation to match the maximum sole crop yield, and in case 6 to surpass it. In case 1 and 2 the average crop plant experiences significant net competition and crop yields are strongly reduced.

A similar analysis is possible from Figure 2(b), for the tree's performance in mixture; the roles are simply inverted. Cases 1 and 2 are highly unlikely, unless strong crop allelopathy effects on young trees are present. The most probable outcome is somewhere between case 3 and 4. Cases 5 and 6 are not realistic, given asymmetry between tree and crop.

Of course, the crop and tree yield outcomes in our example will be coupled and depend on one another such that not all outcome combinations make sense. Some possible combinations are depicted graphically in Figure 3. In more general terms, each specific spatio-temporal mixture design for a given pair of plant species in a given environment produces a pair of coupled yields. The set of coupled yield outcomes of all possible designs (which include all sole crop yields as well) is called a yield set (Vandermeer 1989); the subset of points which constitute the exterior envelope of this set is called the production possibility frontier (PPF; Ranganathan and De Wit 1996); (Figure 4 presents a possible yield set and PPF for our hypothetical SAFS). A PPF can be analyzed to compare sole and mixed crop outcomes and to find the optimum mixed crop designs according to different biological and economical performance criteria. An example of such criteria is the land equivalent ratio (LER), defined as (yc/Yc) + (yt/Yt). A LER $>$ 1 means that there mixture is advantageous because more land would be required to obtain yc and yt by sowing each species separately as sole crops. (For further details on LER and other mixture performance indices, see Vandermeer 1989). A few scores of yield pairs can be obtained through field experiments, while more thorough yield set and PPF constructions can be aided by experimentally fitted models. Simple analytical models have been developed for this purpose, which only consider global population densities and render final

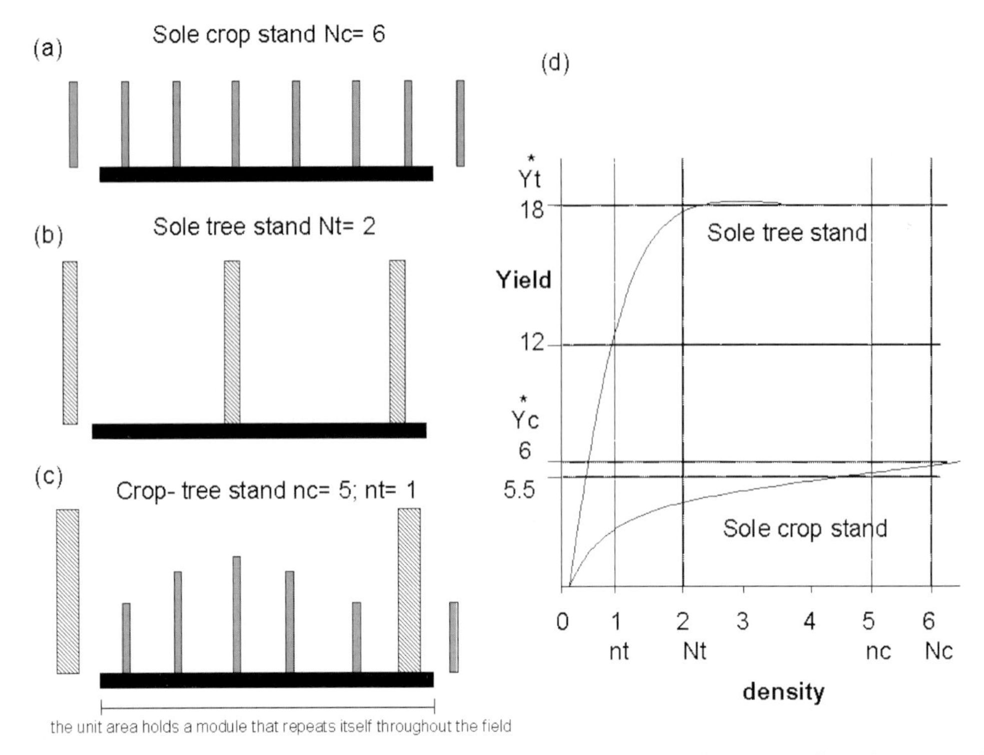

Figure 1. Schematic representation of the yield vs. density relation of a tree-crop mixture, and its corresponding sole crops stands. Bar areas represent the aboveground biomass of the individual plants. (a) sole crop stand; (b) sole tree stand; (c) tree-crop mixture; (d) sole stand yield vs. density relations.

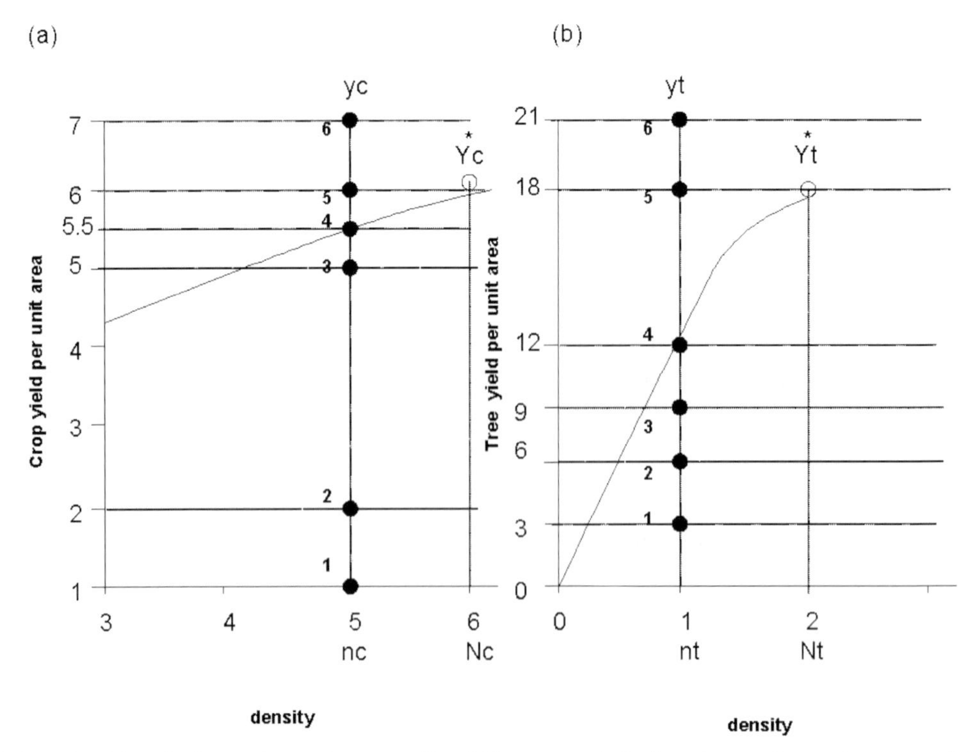

Figure 2. Sole stand yield vs. density relations and some possible yield outcomes (1–6) after mixture with the other species. (a) Crop yields; (b) Tree yields. See text for further explanation of mixture outcomes.

biomass or yield (e.g., Ranganathan and De Wit 1996). Spatially explicit, individual-based plant growth models have also been developed. They require empirical growth and competition parameters and render yields at any growth stage for any spatio-temporal design (e.g. García-Barrios et al. 2001). Both models need to be re-parameterized in different environments.

Quantifying the effects of positive and negative interactions separately

The idea to separate and quantify positive and negative tree effects on crop yield in SAFS was formalized by Ong (1995) in the simple equation:

$$I = F + C \quad (1)$$

where I is the overall interaction, i.e., the percentage net increase in production of one component attributable to the presence of the other component, F is the fertility effect, i.e., the percentage production increase attributable to favorable effects of the other component on soil fertility and microclimate, and C is the competition effect, i.e., the percentage production decrease attributable to competition with the other component for light, water and nutrients. The I value is based on total area, including that occupied by trees; therefore, a positive I value means net increase in total crop yield, irrespective of crop population in mixture relative to that of sole crop (Rao et al. 1998). In alleycropping systems, the measurement of F and C has been accomplished with four treatments: Co = sole crop; Cm = sole crop + mulch from pruned trees; Ho = crop + tree with mulch removed; Hm = crop + tree with its mulch. The equation then becomes

$$I = (Cm - Co) - (Hm - Cm) \quad (2)$$

The competition term can also be calculated as (Ho − Co) or, more conveniently, as the average of (Hm − Cm) and (Ho − Co). This equation motivated the analysis of numerous previous alleycropping experiments and the establishment of new ones (Ong 1996). Unfortunately, few proved to have these four treatments and/or the appropriate experimental designs and field display. Nevertheless, successful positive and negative separation in a few experiments made evident, among others, the following important facts (Sanchez 1995; Ong 1996; Rao et al. 1998; Ong et al. 2004): 1) Strong tree competition is more conspicuous than originally thought; 2) high tree growth rates and tree biomass, intuitively associated with the potential for a significant fertility effect are also strongly related with tree competitiveness, so tradeoffs between both interactions are high; 3) positive and negative component effects are very site specific and change with the environment .

After a modification by Ong (1996), the equation evolved to (Rao et al. 1998):

$$I = F + C + M + P + L + A \quad (3)$$

where F refers to effects on chemical, physical and biological soil fertility, C to competition for light, water and nutrients, M to effects on microclimate, P to effects on pests, diseases and weeds, L to soil conservation and A to allelopathy effects. The benefit of Equation (3) is that it encapsulates a comprehensive overview of the possible effects involved. However, as emphasized by the authors, many of these effects are interdependent and cannot be experimentally estimated independently of one another. Limitations have been identified for this approach (Ong et al. 2004): 1) due to interdependence, the individual terms most likely will give a sum that exceeds I, such that the relative importance of each term cannot be established. 2) It cannot predict delayed effects and long-term trends. 3) It is not meant to predict the consequences of moving from one environment to another, as it does not explicitly consider growth resource capture and use, and their interaction (e.g. water x P interaction in P-fixing soils). Cannell et al. (1996) attempted to clarify the resource base of Ong's equation but did not make plant interaction with resources sufficiently explicit. Mechanistic research may be necessary to understand SAFS functioning and performance over time and/or in different environments.

Methods that explicitly consider growth resources

Positive and negative interactions between plant species largely depend on how the latter affect each other's ability to capture and use growth resources. The principles of light, water and nutrient capture and use efficiency – first applied to sole crop growth analysis and modeling – have been extended to intercropping and agroforestry research in the past decade. The field has seen enormous advances in the gathering of relevant experimental data, theory development, modeling capability, and construction of very sensitive instruments for measuring direct above- and below-ground resource flow and capture (Black and Ong 2000). They have also made evident the challenges for studying these processes in multispecies systems, and the limitations of some basic idealized and simplified assumptions about the relation between plant growth and resource capture and use efficiency.

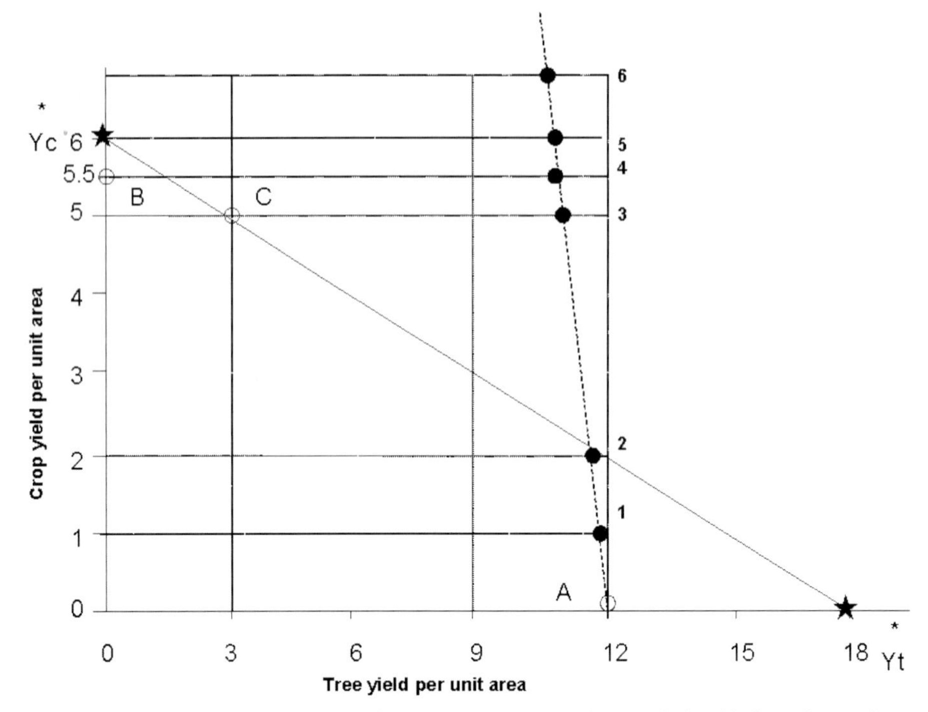

Figure 3. Coupled tree and crop yields in mixture. Points 1–6 represent the six crop yield cases depicted in figure 2(a), and reasonable associated tree yields. Three additional outcomes are included: in A, tree excludes crop; In B crop excludes young tree through strong allelopathy. In C allelopathy effect on tree is milder. The slope of the dashed line indicates the intensity of the slightly negative crop effect on tree biomass. Stars are maximum sole crop yields, and the line that crosses them represents all outcomes with a land equivalent ratio (LER) = 1. Above this line, LER > 1.

A central tenet of mechanistic plant-plant interaction analysis has been that, according to the Law of the Minimum (Blackman 1905), as long as a resource is the most limited, growth depends linearly on the capture of this resource. If eventually another resource becomes the most limiting, the former's concentration ceases to have an effect and the latter dictates growth. As a consequence, biomass production (W) should be easily modeled as the product of the capture of the most limiting resource, and the efficiency with which the captured resource is converted into biomass (Monteith et al. 1994; Ong et al. 1996). The conversion efficiency of the limiting resource is considered to be conservative for a given species for a given environment, and the growth response of the plant is attributed to the increased capture of the resource. Research data (e.g., Demetriades-Shah et al. 1992; Black and Ong 2000) and theoretical developments (Kho 2000; Ong et al. 2004) have shown that: 1) Usually several resources are limiting, in which case the relation between biomass production and the capture of one single resource should be viewed as a correlation, not a causal relation that can be readily modeled. 2) The variation in conversion efficiencies between environments is larger than that between species; therefore such efficiencies are more determined by the environment than by species. 3) Plants alter the availability of resources simultaneously and thus conversion efficiencies by changing resource limitation. Moreover, 'the most limiting' resource for a species can differ in sole system and mixture. 4) Efficiencies should be studied and modeled in relation to the availabilities of the other resources, and treated as variables in process-based models. 5) Dynamic simulation models for different resources should be linked.

In recent years, different resource-based modeling approaches have been developed and/or used to explore or predict how tree-environment-crop interactions (and the productive performance of SAFS) change when environmental resource availability is modified. Some relevant examples are cited below:

The mulch–shade model

Tree canopies in alleycropping provide N-rich mulch but reduce radiation available for crops. The net tree effect should then be a function of at least three factors: 1) the tree's mulch:shade ratio (MSR), 2) the

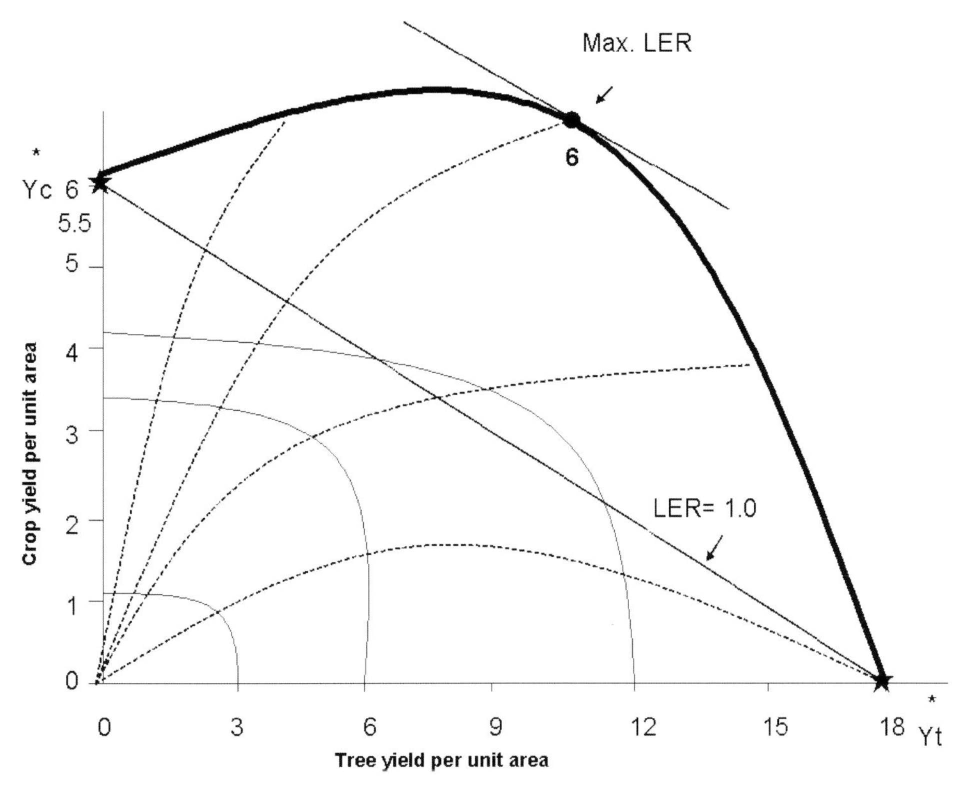

Figure 4. A yield set and Production Possibilities Frontier for the hypothetical tree-crop mixture presented in figures 1 (c). Thin solid lines result from fixing a global density and modifying the crop-tree proportion; thick dashed line from fixing a tree-crop proportion and modifying global density. The thick solid line is the production possibility frontier, and the area under it is the yield set. The thin dashed line represents all outcomes for which LER = 1.0, and the black dot on the PPF is the tree crop mixture which renders the highest attainable LER value.

importance in the particular environment of nitrogen supplied by mulch, expressed as a N_{soil}: N_{mulch} ratio (NNR), and 3) tree row distance as a surrogate of tree leaf area index. Van Noordwijk (1996) developed a static-analytical alleycropping model, which links these species attributes and management factors, and produces explicit algebraic solutions. The simple MSR offers a basis for comparing and selecting tree species. The equation variables and parameters require a lengthy explanation; here we are interested only in discussing the type of results it delivers. The model predicts that at low soil fertility, where the soil fertility improvement due to mulch can be pronounced, there is more chance that an agroforestry system improves crop yields than at higher fertility where the negative effects of shading will dominate. Moreover, it defines combinations of MSR and NNR for which some alleycropping systems should be expected to work (Figure 5a). Although not validated with empirical data, it has been parameterized for typical alleycropping tree species in order to estimate their optimum hedgerow density at different N_{soil} values (Figure 5b). The mulch/shade model provides useful insights, but does not incorporate the interactions between resource dynamics, and crop and tree growth. Incorporating these elements extends the model beyond what can be solved analytically and into the realm of dynamic simulation models.

Integrated water, nutrient and light simulation models in agroforestry systems

In the late 1990s complex agroforestry simulation models and user platforms were constructed for example, WaNuLCAS (van Noordwijk and Lusiana 1999) and HYPAR (Mobbs et al. 2001), and they are still undergoing development, parameterization, and validation. These two models are process-based and have a useful level of spatial structure. Both require weather databases and a considerable number of soil- and plant parameters as inputs. WaNuLCAS is quite elaborate in its soil sub-models and HYPAR in its canopy sub-models. They are too complex to allow a useful description here, but excellent specifications and user manuals are available. Once proper inputs and

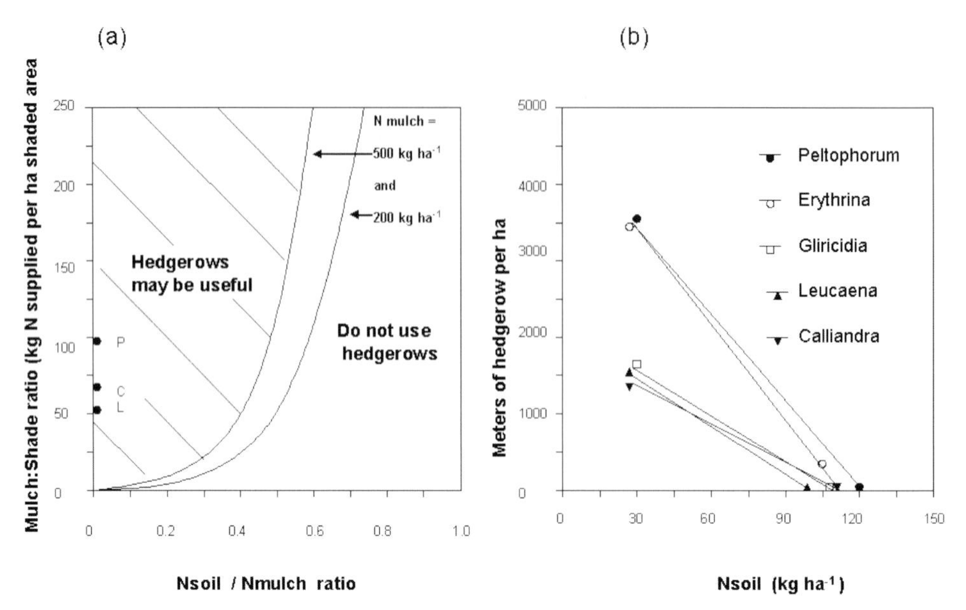

Figure 5. Some of the results obtained by van Noordwijk, (1996) with the mulch and shade model. (a) shows combinations of MSR and NNR for which some hedgerow systems should be expected to work, given two different values of N mulch. P, C and L stand for estimated MSR values for *Peltophorum dasyrrachis, Calliandra calothyrsus* and *Leucaena leucocephala*, respectively. (b) shows predicted optimum tree density in an alley cropping system for five tree species (*Peltophorum dasyrrachis, Erytrina poeppigiana, Gliricidia sepium, Leucaena leucocephala* and *Calliandra calothyrsus*) as a function of Nsoil. Tree density is expressed in linear meters of hedgerow per hectare. Species parameters were estimated empirically and are reported in van Noordwijk, (1996), table 3.2.

parameters are in place, these models can be powerful tools for systematically exploring the possible consequences of diverse management practices and of one or more resource gradients, before engaging in costly and time consuming experiments and productive projects. Van Noordwijk and Lusiana (1999) used WaNuLCAS to model grain- and wood production, water use, water-use efficiency and N limitation for a wide range of annual average rainfall conditions (240 mm to 2400 mm) in parkland systems with differently shaped trees; an example of their results is presented in Figure 6a and 6b. As yet, no experimental data set exists on the same agroforestry system at the same soil but widely differing rainfall conditions, so they used data and parameters from different sources and from theoretical assumptions. Model results generally agreed with conclusions derived from experimental evidence (Breman and Kessler 1997). Similar conclusions were reached with the HYPAR model about the effect of climatic gradients on tree-crop interactions and productive outcomes (Cannell et al. 1998).

Complex process-based models are potentially powerful tools but have their own limitations and pitfalls; they are expected to confront problems of parameter estimation, validation and input gathering similar to – or more severely than – those found in sole crop physiological models (Aggarwal 1995; Hakanson 1995). These models can be used with caution in order to gain insight about management practices and about the consequences of theoretical assumptions; their outputs should be constantly confronted with empirical results and farmer's experience.

A general tree–environment–crop interaction equation for predictive understanding of SAFS

Kho (2000) has developed a simple but powerful analytical model in which the overall tree effect on crop production is explained as a balance of (positive and negative) relative net tree effects on resource availability to the crop. We will describe here some of its basic features, consequences and applications. When trees are introduced in a crop field, they simultaneously change the availability of several resources in the environment of the crop, some for the better and some for the worse. Notwithstanding the Law of the Minimum stated earlier, it has become clear that not one but many resources can limit growth simultaneously and that the degree to which a resource affects production at a given level is dependant on the availability of the other growth resources (Kho 2000). Consequently, the relation between a given resource and production

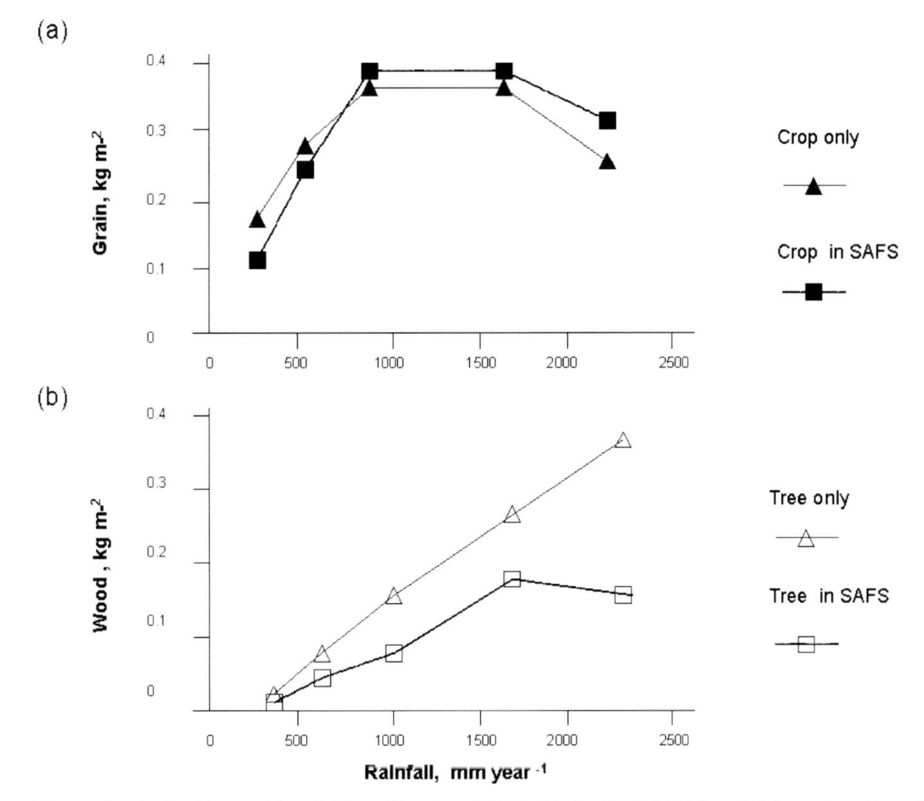

Figure 6. Some of the results obtained by van Noordwijk and Lusiana (1999) with the WaNuLCAS model for grain and wood production, for a range of annual rainfall conditions in an hypothetical agroforestry system where trees provide N-rich mulch. For comparison, the sole crop and sole tree stand are grown also. (a) shows a gradual shift from water to N as the major factor limiting crop production as rainfall increases. At low rainfall levels, tree competition for water dominates over positive effects of N supplied by tree mulch. (b) Because of crop competition, tree yield is lower in the SAFS stand than in sole tree stand.

is not linear but asymptotical when other resources are held constant. As a particular resource becomes less available in relation to others, its influence on production becomes greater and in this last sense it becomes more limiting. The limitation of a resource Ai (i.e., Li) can be formally defined as the ratio between the slope of the production response curve (at a given resource level) and the average use efficiency of that level of resource by the crop; it is therefore a relative non-dimensional term. Li can also be defined (more intuitively) as the relative change in production in response to a relative change in resource availability (Kho 2000), which corresponds to the definition of elasticity in economic theory. Kho 2000 has fruitfully applied Euler's law from economics to demonstrate that the sum of limitations (or elasticities) of all growth resources (Σ Li) should reasonably be equal to 1.0 if the so called constant return to scale assumption holds. De Wit (1992) shows agricultural data supporting this latter assumption of proportional relation of output to input.

As a consequence of the previous arguments, when trees interact with a crop, the expected relative crop yield change (dW/W) should be equal to a weighted sum of the relative changes induced by the tree on each resource Ai; the weights should be the Li's which result from the particular balance of resources in the specific environment. This leads to the equation:

$$dW/W = \sum_{i=1}^{n} (\text{dAi/Ai}) * \text{Li}_i \qquad (4)$$

In short, each resource contributes to the relative change in production proportionally to its degree of limitation and proportionally to its relative change in availability. From Equation (4), Kho (2000) derives Equation (5), which predicts the relative change in crop production (I, as in Ong 1995), as a function of tree effects on resource availability and of resource balance in the particular environment considered:

$$I = \sum_{i=1}^{n} L_i \times T_i \qquad (5)$$

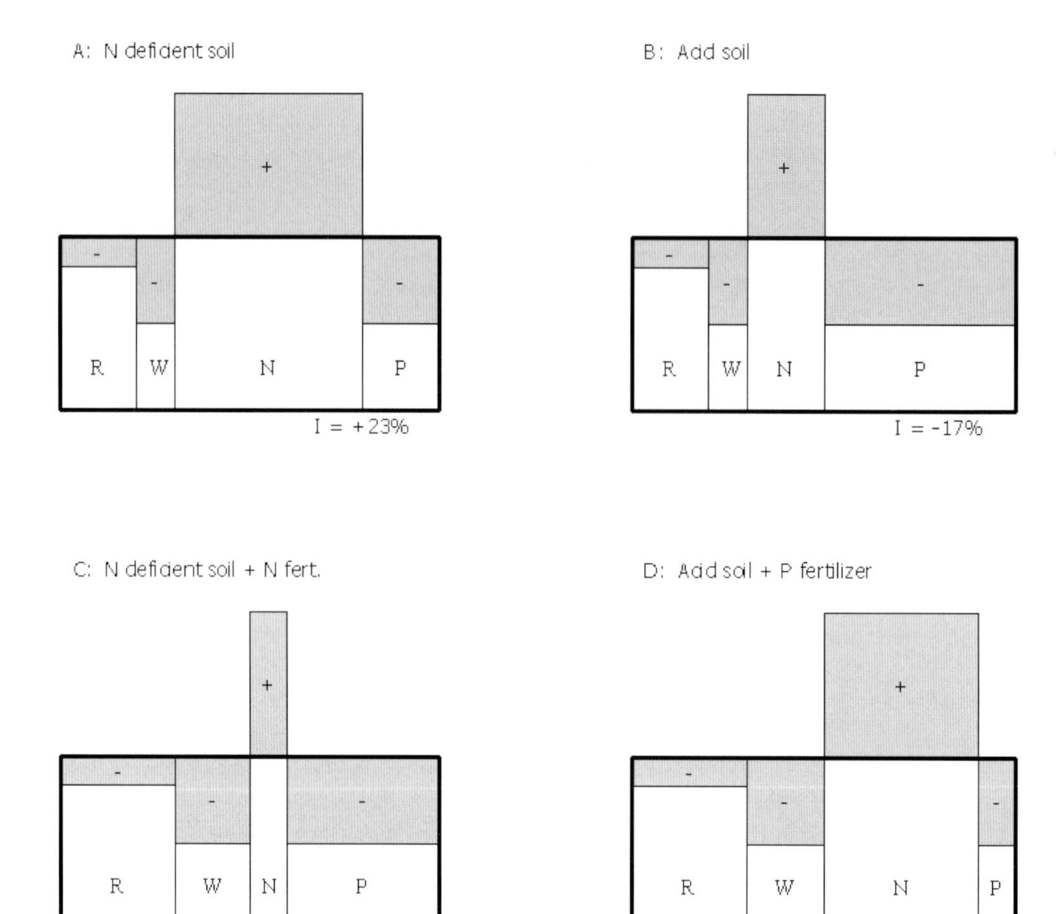

Figure 7. Trees influence crop production through altering (the balance of) resource availabilities to the crop. Each rectangle represents a different resource: water (W), nitrogen (N), phosphorus (P) and radiation (R). The height of each shaded area relative to the height of the rectangle represents the relative change in availability of the resource (Ti). The width of each shaded area relative to the total width represents the limitation of the resource in the tree-crop interface (Li). The sum of positive and negative shaded surfaces relative to the total surface of the rectangle represents the overall tree effect I expressed as fraction of sole crop production. The figures show possible tree effect balances of an alley cropping technology in a humid climate. A) on nitrogen deficient soils, B) on acid (phosphorus deficient) soils, C) on nitrogen deficient soils with nitrogen fertilizer (applied to the alley crop and the sole crop), and D) on acid soils with phosphorus fertilizer (applied to the alley crop and the sole crop). The relative net tree effects on availability of each resource (Ti) are equal in A–D; only the environments (i.e. resource limitations Li) change, and explain the different overall effects (I). Source: Ong, et al., 2004.

where

$$T_i = \frac{\Delta A_i}{A_i} = \frac{A_{i;multi} - A_{i;mono}}{A_{i;mono}} \qquad (6)$$

Ti is the relative net change in availability of resource i because of the tree, $A_{i;multi}$ is the availability of resource i to the crop in the SAFS and $A_{i;mono}$ in the sole crop. Robust estimations of L and T values for a particular SAFS and environment can be obtained experimentally and/or derived from the literature following relatively simple methods which are described and applied by Kho (2000) and Kho et al. (2001).

The relation between resource balance and limitation combined with Equation (5) leads to two rules that can be viewed as counterparts of classic crop production principles (Kho 2000): 1) The greater the availability of a resource in the environment, the smaller is its share in the overall tree-crop interaction. 2) The greater the availability of other limiting resources in the environment, the greater is the share of a resource in the overall tree-crop interaction. These rules are helpful for predicting the performance of a SAFS technology when it is extended to another environment and for developing a SAFS technology. For example, Kho (2000) showed that for the alleycropping technology the net effect of the alleys on the availability (to the crop) of the resources light, water, and phosphorus is most likely negative, and that for nitrogen it is most likely positive. Consequently, in a (sub-) humid climate on nitrogen- deficient soils, the overall alleycropping effect is most likely positive, because of the high limitation for the positive nitrogen effect and the low limitations for the negative net effects on other resources (Figure 7A). In the same climate, but on acid soils, phosphorus is relatively less available, which will increase the share of the negative phosphorus effect (rule 1) and will decrease the share of the positive nitrogen effect (rule 2), resulting in a negative overall effect (Figure 7b). Through management practices, the share of positive net effects can be increased and the share of negative net effects decreased. Compare, for example, Figure 7b and 7d for the effect of phosphorus fertilizer applied to both the alleycropping system and the sole crop system. External inputs of organic or inorganic nitrogen are most likely inappropriate, because these will reduce the share of the positive nitrogen effect (rule 1) and increase the share of the negative effects (rule 2; cf. Figure 7a and 7c).

Beyond yield performance

During the past two decades, agroforestry research in the tropics has focused, naturally, on the prospects for higher productivity in low-input agriculture, which are based on plot-level research just as in agronomic research. In recent years, there is a growing realization that recommendations based on productivity at the plot level alone are insufficient for the long-term management of farms and landscapes. Sustainability of land use practices, a major incentive for preferring agroforestry to high input agriculture, depends (among other things) on ensuring that the flow of energy through the agroecosystems is in close balance to those of the natural ecosystems (Lefroy and Stirzaker 1999). Furthermore, extrapolation of results from plot to landscape level is often flawed because they fail to account for the lateral movement of water and soil, which are greatly influenced by filters in the landscape (van Noordwijk and Ong 1999).

Young (1998) argues that much of the future increase in food and wood production in the humid tropics will have to be achieved from existing land and water resources. Therefore, future research agenda should aim to improve the efficiency with which land and water are currently used. One promising option for improvements of this kind is by using agroforestry, with the ultimate aim of achieving sustainability of production and resource use. The principal agroforestry systems suited to humid tropical environments are multistrata systems, perennial crop combinations, managed tree fallows, contour hedgerows and reclamation agroforestry (ICRAF 1996; Young 1997; Tomich et al. 1998). These mixtures of trees and crops have the potential to improve land management via their ability to reduce soil erosion and improve soil conditions for plant growth. There is some evidence for the utility of agroforestry systems for soil conservation, soil organic matter maintenance, nutrient retrieval, and nutrient recycling (Buresh and Tian 1998; Young 1997). In farming for annual crops, contour hedgerows may provide a viable alternative to conventional conservation measures (Kiepe and Rao 1994). However, despite the demonstration that such systems can dramatically reduce soil losses and improve soil physical properties, the beneficial effects on crop yield are often unpredictable and insufficient to attract widespread adoption (Alegre and Rao 1996).

Agroforestry is now receiving increasing attention by researchers, landowners and policy makers in Australia as a potential solution to the salinity problems caused by the rising water-table (Lefroy and Stirzaker 1999). Replacement of the native vegetation, mainly trees and shrubs, have resulted in a steady increase in the water table for much of the semiarid wheat (*Triticum* sp.) belt of Eastern and Western Australia, because the vegetation has been replaced by winter-growing crops such as wheat (*Triticum aestivum*), barley (*Hordeum vulgare*), canola (*Brassica napus*), and lupin (*Lupinus* spp.), which cannot fully utilize the annual rainfall. The salinity problem is more complex than that experienced in the tropics because of the nature of the duplex soils there, which slow lateral movement of water across the landscape and there are few profitable tree species to replace the existing annual crops. Widely spaced tree rows (10 mm to 300 m and also called 'alleycropping') of fast-growing native (e.g., *Eucalyptus* spp.) and exotic origin (e.g., tree lucerne – *Chamaecytisus palmensis*) were originally seen as the practical alternative for solving this huge landscape problem. As with the tropical alleycropping experience, there is a strong tradeoff between environmental function and crop performance in the Australian environment. Therefore, there is now a serious emphasis on identifying trees that can provide direct value to farmers (e.g., oil malle, *Eucalyptus polybractea*), as suggested by Ong and Leakey (1999) for sub-Saharan Africa, in addition to the hydrological function.

Conclusions and future research needs

During the past decade it was recognized that introducing trees in croplands to promote sustainable low-input SAFS in the tropics was a challenging task. Some reasons for this are: a) trees provide both crucial products for smallholders and strongly facilitate crops, but can also exert stronger competitive effects than previously expected; b) practices aimed at increasing trees' beneficial effects can sometimes also enhance trees' competitiveness; c) the interplay between positive and negative effects of trees change – sometimes significantly – from one environment to another. This makes it difficult to predict the consequences of extending successful agroforestry practices to new environments having different resource levels and resource balances.

This motivated scientists to develop predictive understanding of how such interactions and their production outcome respond to species selection, tree management and resource availability in order to develop realistic SAFS solutions. Among other research tools and strategies, a number of analytical and simulation tools were developed and/or adapted during the period in order to assess tree and/or crop yields. A first group of static- or dynamic models analyze the yield outcomes of positive and negative tree effects without considering growth resources explicitly. A second group does consider these and can therefore predict to a certain extent the effect of the growth environment on tree-crop interactions. Static and dynamic models that do not explicitly consider resources are comparatively easy to parameterize and to use in determining tree-crop interaction effects (I) and construction of production possibility frontiers, and can shed light on specific questions, but need to be calibrated for each particular environment. Spatially explicit, process-based, simulation models (which balance resource use explicitly) are comprehensive and quite powerful research and design tools for SAFS, but are difficult and costly to parameterize. Non-modelers might make a more efficient and better-targeted use of them by acquiring insights provided by simple but powerful analytical models such as the tree-crop-environment equation described previously. Model outputs should be constantly confronted with empirical results and farmers' experience, needs and constraints. It is also important to bear in mind that recommendations based on productivity at the plot level alone are insufficient for the long-term management and sustainability of farms and landscapes.

The partial failure of labor-intensive agroforestry practices using very competitive fast-growing trees at relatively high densities is re-directing SAFS research. Where light-demanding annual crops are important to farmers and soil resources are limited and/or fragile, major efforts should be directed to low tree-density SAFS. Tree species selected for this purpose should provide both environmental services and immediate products and be amenable to pruning during low labor-demand periods. In more general terms, a flexible approach to SAFS design need to be developed with tree density and management as a function of a) environmental and economical constraints and b) relative value of crop and the tree products and services. In the tropics, where drylands and/or unfertile soils are predominant, competition for water and nutrients is critical. Belowground tree-crop interactions and the effects of root pruning strategies are receiving more attention and their research especially to develop economically usable root pruning practices should be pursued further. Tree effects on weeds, herbivores and pathogens, which reduce crop yields, need more attention and their effects should be incorporated into quantitative assessment analysis. SAFS (and all agroforestry systems) are attracting increasing attention as one of the potential mitigations for large-scale soil problems and negative effects of global warming. It is therefore necessary to consider the environmental impacts and functions of SAFS at a broader scale. Similarly, the long-term effects of SAFS on soil and water resources should receive attention particularly in different soil and environmental conditions.

Modeling efforts should continue. Linking plant growth processes and modeling approaches, designing inexpensive resource-flow measuring instruments and developing experimental designs and robust parameterization techniques seem to be the major challenges. Relatively simple equations and concepts can be used to unravel positive and negative aspects of tree–crop interactions and how they influence overall production in simple agroforestry systems. However, designing more complex multispecies SAFS with appropriate management practices presents additional challenges both for satisfying immediate household needs and for replication at different scales while achieving sustainability in the face of environmental and social change. Managing such systems has to be based more on monitoring, diagnosis, remediation, mitigation and adaptation, rather than on a blueprint predictability of the behavior of the agroecosystem. The potential of process-based integrated resource use models in designing and evaluation of complex systems needs to be tested. Research methods are needed to derive useful and flexible rules from adaptive management.

References

Aggarwal P. 1995. Uncertainties in crop soil and weather inputs used in growth models: implications for simulated outputs and their applications. Agr Syst 48: 361–384.

Alegre J.C. and Rao M.R. 1996. Soil and water conservation by contour hedging in the humid tropics of Peru. Agric Ecosyst Environ 57: 17–25.

Black C.R. and Ong C.K. 2000. Utilization of light and water in tropical agriculture. Agr Forest Meteorology 104: 25–47.

Blackman F.F. 1905. Optima and limiting factors. Ann Bot 19: 281–295.

Breman H. and Kessler J.J. 1997. The potential benefits of agroforestry in the Sahel and other semi-arid regions. Eur J Agron 7: 25–33.

Buresh R.J. and Tian G. 1998. Soil improvement by trees in sub-Saharan Africa. Agroforest Syst 38: 51–76.

Cannell M.G.R., van Noordwijk M. and Ong C.K. 1996. The central agroforestry hypothesis: the tree must acquire resources that the crop would not otherwise acquire. Agroforest Syst 34: 27–31.

Cannell M.G., Mobbs D., and Lawson G. 1998. Complementarity of light and water use in tropical agroforests. II. Modelled theoretical tree production and potential crop yield in arid to humid climates. Forest Ecol Manag 102: 275–282.

Connolly J., Goma H. and Rahim K. 2001. The information content of indicators in intercropping research. Agric Ecosyst Environ 87: 191–207.

Demetriades-Shah T.H., Fuchs M., Kanemasu E.T. and Flitcroft I.D. 1992. Further discussion on the relationship between cumulated intercepted solar radiation and crop growth. Agr Forest Meteorol 68: 231–242.

De Wit C.T. 1992. Resource use efficiency in agriculture. Agr Syst 40: 125–151.

Fernández E.C.M., Davey C.B. and Nelson L.A. 1993. Alleycropping on acid soils in the upper amazon: mulch fertilizer and hedgerow pruning effects. pp. 77–96. In: Technologies for Sustainable Agriculture in the Tropics. ASA Special Publication 56, Madison WI, USA.

García-Barrios L. and García-Barrios R. 1992. La modernización de la pobreza: dinámicas de cambio técnico entre los campesinos temporaleros de México. Revista de Estudios Sociológicos (El Colegio de México) 10(29): 263–288.

García-Barrios L. 2003. Plant-plant interactions in tropical agriculture. pp. 1–58. In: Vandermeer J. (ed.), Tropical Agroecosystems. CRC Press, NewYork, USA.

García-Barrios L., Franco M., Mayer-Foulkes D., Urquijo-Vasquez G. and Franco-Perez J. 2001. Development and validation of a spatially explicit individual based mixed crop growth model. B Math Biol 63: 507–526.

Giller K. 2001. Nitrogen Fixation in Tropical Cropping Systems. (2nd edition). CAB International, Wallingford, UK, 222–250 pp.

Girma H., Rao M.R. and Sithanantham S. 2000. Insect pests and beneficial arthropod populations under different hedgerow intercropping systems in semiarid Kenya. Agroforest Syst 50: 279–292.

Goldberg D. 1990. Components of Resource Competition in Plant Communities. pp. 27–49. In: Grace J.B. and Tilman D. (eds), Perspectives on Plant Competition. Academic Press, New York, USA.

Hairiah K., van Noorwijk M., Santoso B. and Syekhfani M.S. 1992. Biomass production and root distribution of eight trees and their potential for hedgerow intercropping on an Ultisol in southern Sumatra. Agrivita 15: 54–68.

Hakanson L. 1995. Optimal size of predictive models. Ecol Model 78: 195–204.

Harper J.L. 1990. Population Biology of Plants. Academic Press, London, UK, 892 pp.

Holmgren M., Scheffer M. and Huston M. 1997. The interplay of facilitation and competition in plant communities. Ecology 78: 1966–1975.

Hunter A.F. and Aarssen L.W. 1988. Plants helping plants. BioScience 38: 34–40.

ICRAF 1996. Annual Report 1995. International Centre for Research in Agroforestry, Nairobi, Kenya. 288 pp.

Jose S., Gillespie A.R. and Pallardy S.G. 2004. Interspecific interactions in temperate agroforestry. (This volume).

Kho R.M. 2000. A general tree-environment-crop interaction equation for predictive understanding in agroforestry systems. Agric Ecosyst Environ 80: 87–100.

Kho R.M., Yacouba B., Yayé M., Katoré B., Moussa A., Iktam A. and Mayaki A. 2001. Separating the effects of trees on crops: The case of *Faidherbia albida* and millet in Niger. Agroforest Syst 52: 219–238.

Kiepe P. and Rao M.R. 1994. Management of agroforestry for the conservation and utilization of land and water resources. Outlook Agr 23(1): 17–25.

Lefroy E.C. and Stirzaker R.J. 1999. Agroforestry for water management in southern Australia. Agroforest Syst 45: 277–302.

Liebman M. and Gallandt E. R. 1977. Many little hammers: ecology management of crop-weed interactions. pp. 291–343. In: Jackson L.E. (ed.), Ecology in Agriculture, Academic Press. New York.

Liebman M. and Stavers C.P. 2001. Crop diversification for weed management. pp. 322–374. In: Liebman M., Mohler C.L. and Stavers C.P. (eds), Ecological Management of Agriculture Weeds. Cambridge University Press, UK.

MacArthur R.H. and Levins R. 1967. The limiting similarity convergence and divergence of coexisting species. Am Nat 101: 377–378.

Mobbs D.C., Lawson J. and Brown T.A. 2001. HYPAR version 4.1 Model for Agroforestry Systems. Centre for Ecology and Hydrology, Edinburgh, UK. 119 pp.

Monteith J.L., Scott R.K. and Unsworth M.H. (eds) 1994. Resource Capture by Crops. Nottingham University Press, Loughborough, UK. 496 pp.

Nair P.K. 1998. Directions in tropical agroforestry research: past, present and future. Agroforest Syst 38: 223–245.

Ong C.K. 1994. Alley cropping ecological pie in the sky. Agroforest Today 6 (3): 8–10.

Ong C.K. 1995. The 'dark side' of intercropping: manipulation of soil resources. pp. 45–65. In: Sinoquet H. and Cruz P. (eds), Ecophysiology of Tropical Intercropping. Institute National de la Recherche Agronomique, Paris, France.

Ong C.K., Black C.R., Marshall F.M. and Corlett J.E. 1996. Principles of resource capture and utilization of light and water. pp. 73–158. In: Ong C.K. and Huxley P. (eds), Tree–Crop Interactions: A Physiological Approach. CAB international, Wallingford, UK.

Ong C.K., Kho R.M. and Radersma R. 2004. Multi-species agroecosystems as models for agriculture: concepts and rules. In: van Noordwijk M., Cadisch G. and Ong C.K. (eds), Belowground

Interactions in Multiple Agroecosystems. CAB International, Wallingford, UK (in press).
Ong C.K. and Leakey R.R. 1999. Why tree–crop interactions in agroforestry appear at odds with tree-grass interactions in tropical savannahs? Agroforest Syst 45: 109–129.
Ong C.K. and Swallow B.M. 2004. Water productivity in forestry and agroforestry. In: van Noordwijk M., Cadisch G. and Ong C.K. (eds), Belowground Interactions in Multiple Agroecosystems. CAB International, Wallingford, UK (in press).
Pugnaire F. and Luque M. 2001. Changes in plant interactions along a gradient of environmental stress. Oikos 93: 42–49.
Ranganathan R. and De Wit C.T. 1996. Mixed Cropping of annuals and woody perennials: an analytical approach to productivity and management. pp. 25–50. In: Ong C.K. and Huxley P. (eds), Tree–crop Interactions: A Physiological Approach. CAB International, Wallingford, UK.
Rao M.R., Nair P.K. and Ong C.K. 1998. Biophysical interactions in tropical agroforestry systems. Agroforest Syst 38: 3–50.
Rao M.R., Schroth G., Williams S., Namirembe S., Schaller M., Wilson J. and Vandermeer J. 2004. Managing belowground interactions in agroecosystems. In: van Noordwijk M., Cadisch G. and Ong C.K. (eds), Belowground Interactions in Multiple Agroecosystems. CAB International, Wallingford, UK (in press).
Rao M.R. and Willey R.W. 1980. Evaluation of yield stability in intercropping: studies on sorghum/pigeonpea. Exp Agr 16: 105–116.
Rao M.R., Ong C.K., Pathak P. and Sharma M.M. 1992. Productivity of annual cropping and agroforestry systems. Agroforest Syst 15: 51–64.
Rao M.R., Singh M.P. and Day R. 2000. Insect pest problems in tropical agroforestry systems: Contributory factors and strategies for management. Agroforest Syst 50: 243–277.
Sanchez P.A. 1995. Science in agroforestry. Agroforest Syst 30: 5–55.
Schroth G., Krauss U., Gasparotto L., Duarte-Aguilar J.A. and Vohland K. 2000. Pest and diseases in agroforestry systems in the humid tropics. Agroforest Syst 50: 199–241.
Thevathasan N.V. and Gordon A. M. 2004. Ecology of tree intercropping systems in the North temperate region: Experience from southern Ontario, Canada. (this volume).
Tomich T.P., van Noordwijk M., Budidorsono S., Gillison A., Kusumanto T. Murdiyarso D., Stolle F. and Fagi A.M. 1998. Alternatives to Slash-and-Burn in Indonesia: Summary Report & Synthesis of Phase II. ASB-Indonesia Report No. 8, ICRAF Southeast Asia, Bogor, Indonesia, 139 pp.
van Noordwijk M. and Ong C.K. 1999. Can the ecosystem mimic hypotheses be applied to farms in African savannahs? Agroforest Syst 45: 131–158.
van Noordwijk M., Cadisch G. and Ong C.K. 2004. Challenges for the next decade of research on belowground interactions in tropical agro-ecosystems: from plot to landscape scale. In: van Noordwijk M., Cadisch G. and Ong C.K. (eds), Belowground Interactions in Multiple Agroecosystems. CAB International, Wallingford, UK (in press).
van Noordwijk M. and Lusiana B. 1999. WaNulCAS a model of water nutrient and light capture in agroforestry systems. Agroforest Syst 43: 217–242.
van Noordwijk M. 1996. Mulch and shade model for optimum alley cropping design depending on soil fertility. pp. 51–72. In: Ong C.K. and Huxley P. (eds), Tree–Crop Interactions: A Physiological Approach. CAB international, Wallingford, UK.
Vandermeer J. 1998. Maximizing crop yields in alley crops. Agroforest Syst 40: 199–206.
Vandermeer J. 1989. The Ecology of Intercropping. Cambridge University Press Cambridge, UK, 248 pp.
Willey R.W. and Heath S.B. 1969. The quantitative relation between plant population and crop yield. Adv Agron 21: 281–321.
Williams A. and McCarthy B. 2001. A new index of interspecific competition for replacement and additive designs. Ecol Res 16: 29–40.
Young A. 1997. Agroforestry for Soil Management 2nd edition, CAB International, Wallingford, UK, 320 pp.
Young A. 1998. Land Resources: Now and For the Future. Cambridge University Press, UK, 319 pp.

Agroforestry Systems **61:** 237–255, 2004.

Interspecific interactions in temperate agroforestry

S. Jose[1,*], A.R. Gillespie[2] and S.G. Pallardy[3]
[1]*School of Forest Resources and Conservation, University of Florida, 5988 Hwy 90, Bldg. 4900, Milton, FL 32583, USA;* [2]*Department of Forestry and Natural Resources, Purdue University, 195 Marstellar Street, West Lafayette, IN 47907, USA;* [3]*Department of Forestry, 203 ABNR Bldg., University of Missouri, Columbia, MO 65211, USA;* **Author for correspondence: e-mail: sjose@ufl.edu*

Key words: Aboveground interaction, Allelopathy, Belowground competition, Biophysical interactions, Facilitation, Tree–crop interactions

Abstract

The ecological principles that define the competitive and complementary interactions among trees, crops, and fauna in agroforestry systems have received considerable research attention during the recent past. These principles have not yet, however, been adequately integrated and synthesized into an operational approach. This paper reviews the ecological and ecophysiological bases for interspecific interactions based on data from site-specific research and demonstration trials from temperate agroforestry systems, primarily from temperate North America. The review shows that information on ecological interactions in several temperate agroforestry systems is inadequate. It is recommended that the future research should focus on exploring new species and systems that have received little attention in the past. Priority research areas should include cultural practices and system designs to minimize interspecific competition and maximize environmental benefits such as improved water quality. Potential for genetic modification of components to increase productivity and reduce competition also needs to be explored. Process-oriented models may be used increasingly to predict resource-allocation patterns and possible benefits for a suite of site and species combinations.

Introduction

The search for highly productive, yet sustainable and environmentally responsible agricultural systems has led to a renewed interest in agroforestry practices in temperate regions of the world (Matson et al. 1997; Moffat 1997; Noble and Dirzo 1997). Increasing environmental, ecological, and economic concerns have been providing momentum for this shift in paradigm away from 'chemical agriculture.' As in the tropical regions, the major objectives for establishing agroforestry systems in temperate regions are also somewhat similar (production of tree or wood products, agronomic crops or forage, livestock, and improvement of crop quality and quantity), but at a scale and magnitude corresponding to the prevailing social and economic conditions. Furthermore, added emphasis is placed on environmental benefits such as water quality in the temperate regions. Management practices to attain these objectives have included enhancement of microclimatic conditions; improved utilization and recycling of soil nutrients; control of subsurface water levels; improved soil and water quality; provision of favorable habitats for plant, insect or animal species; soil stabilization; and protection from wind and snow, among others (Williams et al. 1997; Garrett and McGraw 2000).

As an association of woody and herbaceous plant communities, often with domesticated or semiwild animals, agroforestry systems are deliberately designed to optimize the use of spatial, temporal, and physical resources, by maximizing positive interactions (facilitation) and minimizing negative ones (competition) among the components (Jose et al. 2000a). An understanding of the biophysical processes and mechanisms involved in the allocation of these resources is essential for the development of ecologically sound agroforestry systems that are economically viable and socially acceptable (Ong et al. 1996; Rao et al. 1998). These complex biophysical interactions, which have received much recent attention in temperate agroforestry, will be the focus of this paper; companion chapters on this topic from simultaneous agroforestry systems in the tropics by Garcia-Barrios and Ong (2004) and north temperate agroforestry systems by Thevathasan and Gordon (2004) complement this chapter.

Interaction: the ecological basics

The interactions that occur among individual organisms can be grouped into four broad categories: (a) eating or being eaten, (b) competing for resources, (c) cooperating, and (d) no direct interaction (Perry 1994). Terminologies such as predation, parasitism, competition, amensalism, mutualism, commensalism, and neutralism have also been used extensively to define the relationship between two species. All of these are applicable in an agroforestry context as well (Table 1). When interaction is positive or synergistic, complementarity between system components can be expected; when interaction is negative or antagonistic, it becomes competitive. Interactions can also be classified as occurring in above- and belowground compartments, as commonly referred to in the agroforestry literature (Singh et al. 1989; Ong et al. 1991).

Similar to natural systems, productivity of an agroforestry system is ultimately the net result of positive and negative interactions among the components. Interactions occur as component species exploit growth resources above- and belowground (Ong et al. 1996). The likelihood and intensity of interspecific (as well as intraspecific) interactions decline with decreases in organism density until a maximum yield is reached. Beyond any maximum-yield density, interactions among plants occur when two or more organisms attempt to capture resources from the same location or at the same time (spatial or temporal). The Armstrong and McGehee model illustrates this mechanism of coexistence and competition (Armstrong and McGehee 1980). In this model, a single resource R, is consumed by two species, A and B. Species A has a linear growth response, whereas species B has a nonlinear response (Figure 1). For each species, the two species equilibrium with only that species and the resource present is the point at which the growth rate is equal to zero (R_A

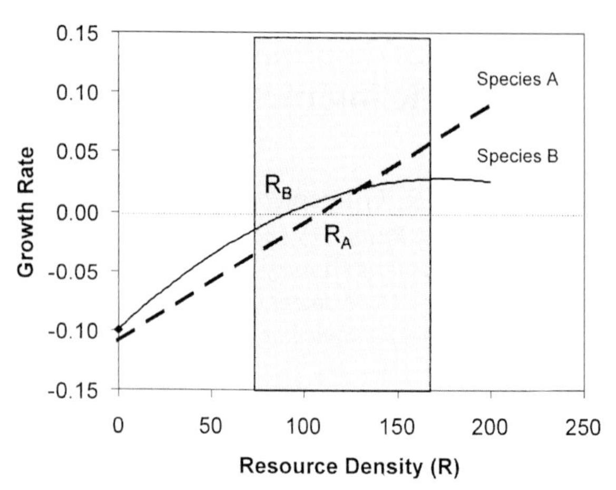

Figure 1. Population growth rates of species A and B as functions of the resource density, R, according to the Armstrong and McGhee (1980) model. R_A, and R_B are the equilibrium densities of species A and B, respectively, when grown alone. Shaded region shows the range of resource densities over which the two species can coexist Source: Adapted from Murdoch et al. (2003).

is the equilibrium density with species A alone and R_B is the equilibrium density with species B alone; see Figure 1). If the resource pool is low, the species with the lowest R value (i.e., the species that reduces the equilibrium density of the resource to the lowest level) will win in competition. In the example shown in Figure 1, species B will be the winner. If the resource pool is sufficiently large, however, species A and B will be able to coexist over a range of resource densities (shaded region in Figure 1). This rather simplistic explanation of what spurs interactions, while true in agroforestry systems, is somewhat better suited for systems where component species have very similar root distributions, such as when herbaceous weeds invade an agronomic monoculture system. This is because all organisms in a single-species system can be expected to have fairly similar physiological needs and to respond to fulfill those needs in a similar manner. Species in agroforestry systems can have different physiological needs for particular resources in certain amounts, at certain times, and possess different structural or biological means to obtain them. This idea can be easily considered in terms of niche separation. The utilization of the environment by any species includes three main components: space, resources, and time. Any species utilizing the same exact combination of these resources as another will be in direct competition, eventually resulting in competitive exclusion. However, if one species differs in utilization of even one of the components, for example light saturation of

Table 1. Interactions between two species as commonly depicted in ecological literature.

Type of interaction	Effect of the interaction[1]		Nature of the interaction	Agroforestry example
	Species 1	Species 2		
Amensalism	–	0	One species is inhibited and the other unaffected	Allelopathy; black walnut inhibiting growth of tomato
Commensalism	+	0	One species is benefited and the other one is unaffected	Improved fallows
Competition	–	–	Both species are negatively affected as a result of each other's use of growth resources	Poorly managed alley cropping or silvopasture
Mutualism (or synergism)	+	+	Both species are positively affected	Mycorrhizae, *Rhizobium* in leguminous plants
Neutralism	0	0	Neither species affects the other	Scattered trees; winter wheat in a temperate deciduous tree alley cropping
Predation and Parasitism	+	–	One species benefits at the expense of another. Predators tend to take large bites and parasites small bites. Parasites frequently invade the body of their hosts and eat from within; predators eat from outside	Pests and diseases

[1]0 = no effect; + = positive; – = negative.
Source: Modified from Perry, 1994.

C3 vs. C4 plants, an equilibrium species mixture may be established.

The *resource-ratio hypothesis*, proposed by Tilman (1982, 1985, 1990), has been used to explain coexistence and competitive exclusion in an agroforestry context as well. According to this hypothesis, coexistence occurs where resource requirements differ among the species. Greater capture of a limiting resource would be accompanied by an increased ability to utilize nonlimiting resources, which, by definition, are available but underutilized. Thus, based on the differences in physical or phenological characteristics of component species, the interactions between tree and crop species may lead to an increased capture of a limiting growth resource (Ong et al. 1996; Ashton 2000; Garcia-Barrios and Ong 2004). The system as a whole could then accrue greater total biomass than the cumulative production of those species if they were grown separately on an equivalent land area (Cannell et al. 1996). Production possibility curves have been widely used in agroforestry literature to explain this concept. For example, Figure 2a provides a graphical representation of complementarity in resource use. The line from A1 to B1 represents the total yield of A and B as the proportion of each species in sole systems on a given area varies. The curve described by A+B represents over-yielding (compared to either of the sole crop yields) of one possible mixture of species A and B.

Figure 2a, however, ignores the fact that agroforestry systems are dynamic and temporal changes can be expected over the life of the system. A hypothetical scenario of how interactions can change over time is depicted in Figure 2b. Although complementarity in sharing resources can be expected, very often there is an overlap between the needs and physical structures of two or more agroforestry component species. When this occurs, species will compete for resources. In agroforestry systems, this is more the rule than the exception. Complementarity may result in overyielding early on in agroforestry systems, but competition intensifies with time and results in under-yielding in several temperate agroforestry systems (Figure 2b). Thus, agroforestry systems may experience a complex series of inter- and intra-specific interactions guided by modification and utilization of light, water, and nutrients. The following sections discuss important above- and belowground interactions occurring between component species in temperate agroforestry systems.

Aboveground interactions

Microclimatic modification

The presence of trees modifies site microclimate in terms of temperature, water vapor content or par-

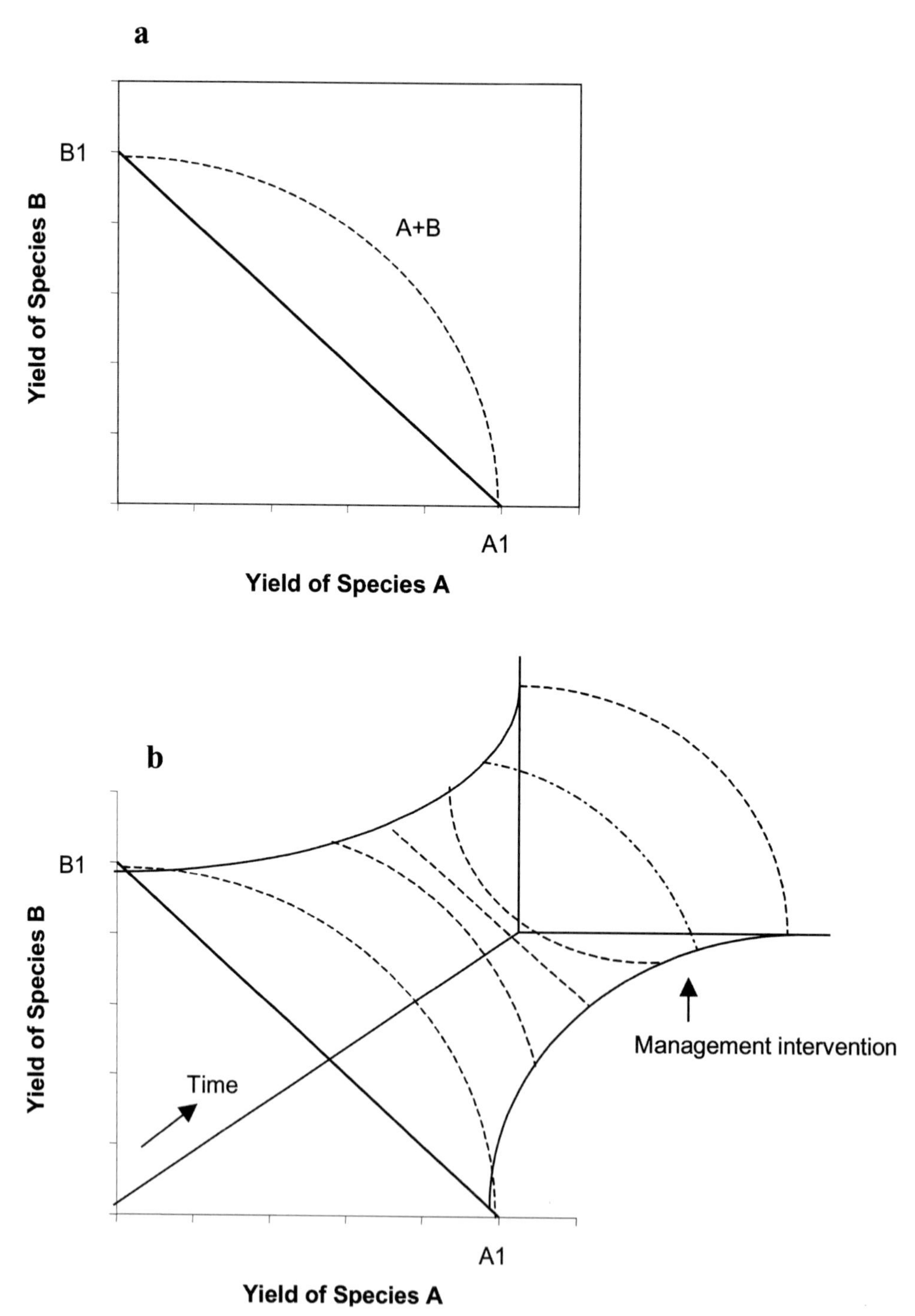

Figure 2. The production possibility curves for two species, A and B. (a) Points A1 and B1 represent the maximum production potential if species A and B were grown in monocultures. Line A1 to B1 represents the proportional yield of species A and B when grown in mixtures. The curve described by A+B represents overyielding (compared to either monoculture yields) of one possible mixture of species A and B. (b) A hypothetical temporal production possibility surface for species A and B (modified from Wojtkowski 1998). As time progresses overyielding gives away to underyielding, but a timely management intervention (e.g., root pruning of trees) alleviates competitive interactions, thereby resulting in overyielding again.

tial pressure, and wind speed, among other factors. Serving as windbreaks, trees slow the movement of air and thus in general reduce evaporative stress. Wind speed was substantially reduced under a radiata pine (*Pinus radiata*) silvopastoral system in New Zealand due to increased tree stocking (Hawke and Wedderburn 1994). Windbreaks are known to improve the distribution and utilization of irrigation water, reduce evapotranspiration, and improve crop water use efficiency (Davis and Norman 1988). Planting windbreaks or shelterbelts in crop fields provides wind protection as well as changes in microclimate; improved crop quality and yield within the sheltered area (10 to 15H, where H is the height of the windbreak) have also been reported (Kort 1988; Brandle et al. 2000; Brandle et al. 2004).

Temperature reductions can help reduce heat stress of crops and/or animals in agroforestry systems. Crops such as cotton (*Gossypium hirsutum*) and soybean (*Glycine max*) have been observed to have higher rates of field emergence when grown at moderate temperatures. For example, Ramsey and Jose (unpublished), in their study of a pecan (*Carya illinoensis*)–cotton alleycropping system in the southern United States, observed earlier germination and higher survival rates of cotton under pecan canopy cover due to cooler and moister soil conditions than in sole system. Similarly, a study in Nebraska, Midwestern United States, showed earlier germination, accelerated growth and increased yields of tomato (*Lycopersicon esculentum*) and snap bean (*Phaseolus vulgaris*) under simulated narrow alleys compared to wider alleys (Bagley 1964; Garrett and McGraw 2000). In addition, studies on *Paulownia* – wheat (*Triticum aestivum*) intercropping in temperate China have shown increased wheat quality due to enhanced microclimatic conditions (Wang and Shogren 1992).

Heat stress has been identified as a major constraint to cattle production in the tropical and temperate regions of the world (Payne 1990; Mitlohner et al. 2001). At high temperatures, evaporative cooling is the principal mechanism for heat dissipation in cattle. It is influenced by humidity and wind speed and by physiological factors such as respiration rate and density and activity of sweat glands. The failure of homeostasis at high temperatures may lead to reduced productivity or even death (Blackshaw and Blackshaw 1994). Providing shade, however, can reduce the energy expended for thermoregulation, which in turn can lead to higher feed conversion and weight gain. In a recent study in Texas, southern United States, Mitlohner et al. (2001) found that cattle provided with shade reached their target body weight 20 days earlier than those not afforded shade. These authors concluded that cattle without shade had a physiological (i.e., higher respiration rate) and behavioral (i.e., less active) stress response to heat that negatively affected productivity.

The location of shade for livestock in the landscape may have very negative impacts on other communities. For example, trees used for shade along streams in riparian pastures concentrate livestock along the bank and could create major negative impacts on the banks and the stream.

Shading: Is it competition or facilitation?

When plant growth is not limited by water or nutrients, production is limited by the amount of radiant energy that the foliage can intercept (Monteith et al. 1991; Monteith 1994). A number of studies have examined the mechanism of competition for light between perennial woody species and annual crops (Monteith et al. 1991; Knowles et. al. 1999; Gillespie et al. 2000). Biomass growth is dependent upon the fraction of incident photosynthetically active radiation (PAR, 400 to 700 nm wavelength) that each species intercepts, and the efficiency with which the intercepted radiation is converted by photosynthesis (Ong et al. 1996). These factors, in turn, are influenced by time of day, aspect, temperature, CO_2 level, species combination, photosynthetic pathway (C3 vs. C4), canopy structure, plant age and height, leaf area and angle, and transmission and reflectance traits of the canopy, (Brenner 1996; Kozlowski and Pallardy 1997).

Shading by associated tree species has been shown to be a factor in reducing yield in temperate agroforestry systems. For example, low PAR levels resulting from overhead shading significantly reduced yield of winter wheat near tree rows in a *Paulownia* – winter wheat temperate alleycropping system in China (Chirko et al. 1996). Nissen et al. (1999) also reported that both shading and belowground competition decreased the yield of cabbage (*Brassica oleracea.*) in a *Eucalyptus* alleycropping system in the Philippines. Maize (*Zea mays*) and soybean yields were reduced to 75% and 79% of the sole crop yield, respectively, when grown in alleycropping configurations involving poplar (*Populus deltoides* x *nigra* DN-177) and silver maple (*Acer saccharinum.*) in Ontario, Canada (Simpson and Gordon, Personal Communication). Similar results have also been reported for temperate silvopastoral systems. For example, Lin et al. (1999)

observed significant decreases in mean dry weight of warm season grasses as the amount of light declined.

A few studies have investigated the physiological basis of observed yield reduction in response to shading in agroforestry systems (Jose and Gillespie 1995; Gillespie et al. 2000; Miller and Pallardy 2001). Shading is known to change quality of light reaching the understory canopy: overhead canopies absorb mostly the red and blue portions of the solar spectrum so that diffuse radiation will be richer in orange, yellow, and green wavelengths (Krueger 1981). The phytochrome system of plants interacts with red and infrared wavelengths to influence the amount of growth regulating hormones and thereby growth (Baraldi et al. 1995). Lack of adequate red light is known to influence tillering in grasses (Davis and Simmons 1994a), stem production in clover (*Trifolium* spp.) (Robin et al. 1994), flowering (Davis and Simmons 1994b), and other basic plant growth processes (Sharrow 1999).

As stated earlier, the response to shading may also depend on the differences in carbon fixation pathways of the associated crop species. It is well known that photosynthetic rate (P_{net}) of C3 plants increases sharply as PAR increases from deep shade up to approximately 25% to 50% of full sunlight, then peaks and remains constant as light is increased further (Figure 3). In contrast, the photosynthetic apparatus of C4 plants does not become light-saturated and continues to increase P_{net} up to full sunlight. This difference is related to the respective abilities of C3 and C4 species to use CO_2(Kozlowski and Pallardy 1997; Lambers et al. 1998). Theoretically, C3 plants planted under shade should be able to perform better than C4 plants and could be better suited for agroforestry practices. Wanvestraut et al. (2004) showed no effect, however, of moderate shading on cotton (C3 species) growth and yield in temperate alleycropping with pecan.

Contrary to an expected yield decrease in maize (a C4 species) in response to shading, Gillespie et al. (2000) reported no effect in two alleycropping systems in the Midwestern United States. This happened despite a strong positive correlation between PAR and P_{net}. These researchers found that, in general, the eastern-most row of maize in a black walnut (*Juglans nigra*) alleycropping system received 11% lower PAR than the middle row (Figure 4). Shading was greater in a red oak (*Quercus rubra*) alleycropping system due to higher canopy leaf area, where a 41% reduction in PAR was observed for the eastern row. Similarly, western rows were receiving 17% and 41% lower PAR in comparison to the middle rows in the black walnut and

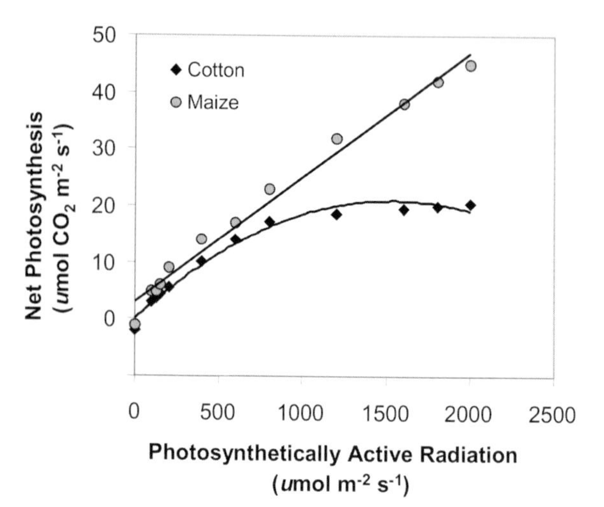

Figure 3. Net photosynthesis as a function of photon flux density in the PAR range in maize (C4 plant) and cotton (C3 plant) (Jose, unpublished data).

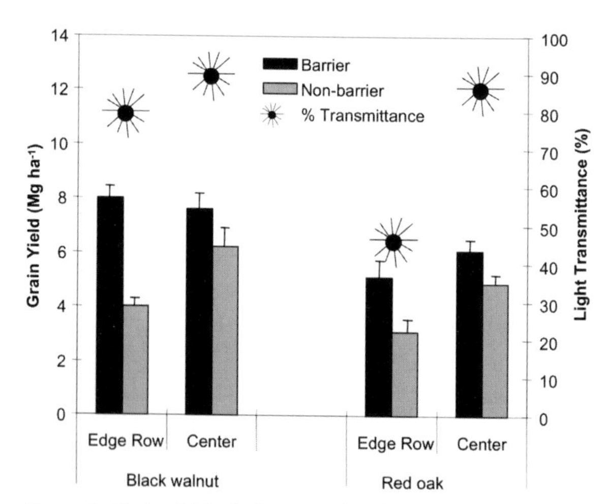

Figure 4. Grain yield of alleycropped maize at the edge (average of eastern and western rows closest to tree row) and alley center in two alley cropping systems involving black walnut and red oak in southern Indiana, United States. Percent light transmittance (as a fraction of full sunlight) reaching the top of edge and center row plants is also shown (Based on data given in Gillespie et al. 2000).

red oak systems, respectively. Irrespective of the shading, no apparent yield reduction was observed when belowground competition for nutrients and water was eliminated through trenching and polyethylene barriers. Leihner et al. (1996) also reported similar findings in maize and concluded that shading played only a minor role in competition at the tree-crop interface.

Positive effects (facilitation) of moderate shading on crop growth have been reported in some cases. Lin et al. (1999) found that two native warm-season legumes, *Desmodium canescens* and *D. paniculatum*,

Table 2. Crude protein and total crude protein of selected grasses and legumes when grown under three levels of shade during 1994 and 1995 at New Franklin, Missouri, USA.

Species	Crude protein (%)			Total crude protein (g pot^{-1})		
	Full Sun	50% Shade	80% Shade	Full Sun	50% Shade	80% Shade
Introduced cool-season grasses						
Kentucky bluegrass	20.3 b	20.7 b	22.7 a	2.45 A	2.58 A	1.57 B
Orchardgrass 'Benchmark'	12.6 c	15.7 b	19.6 a	1.80 A	1.84 A	1.19 B
Orchardgrass 'Justus'	19.8 a	16.7 a	18.5 a	1.60 A	1.92 A	1.79 A
Ryegrass 'Manhattan II'	15.3 b	16.0 b	18.5 a	1.74 A	2.06 A	1.62 A
Smooth bromegrass	16.7 c	18.1 b	20.2 a	1.64 A	2.25 A	1.94 AB
Tall Fescue 'KY31'	14.0 b	15.0 b	18.1 a	1.83 B	2.43 A	1.43 C
Tall Fescue 'Martin'	14.3 b	15.5 b	18.5 a	1.75 A	1.84 A	1.12 B
Timothy	15.4 c	17.6 b	20.4 a	1.60 A	1.59 A	1.12 A
Introduced cool-season legumes						
Alfalfa 'Cody'	19.4 a	19.9 a	19.4 a	1.49 A	1.48 A	1.00 A
White clover	20.1 a	20.6 a	19.9 a	2.49 A	2.03 A	1.23 B
Introduced warm-season legumes						
Striate lespedeza 'Kobe'	13.2 a	13.0 a	12.5 a	3.34 A	2.65 B	1.56 C
Native warm-season legumes						
Slender lespedeza	11.0 a	10.5 a	10.8 a	2.04 A	2.04 A	1.04 A
Desmodium paniculatum	11.6 b	11.7 b	12.9 a	2.57 B	3.53 A	3.38 A
Desmodium canescens	13.0 a	13.2 a	12.8 a	2.19 B	2.98 A	2.88 A
Hog peanut	9.1 ab	8.7 b	9.7 a	0.80 B	2.51 A	2.97 A

Means followed by the same letter within a row do not differ significantly from each other (Tukey's studentized range test, $\alpha = 0.05$).
Adapted from Lin et al. (1999).

exhibited shade tolerance and had significantly higher dry weight at 50% and 80% shade than in full sunlight. These authors also reported that total crude protein content of some of the forage species was greater under 50% and 80% shade than in full sun (Table 2). Burner (2003) reported that orchardgrass (*Dactylis glomerata*) yield across six harvests did not differ among loblolly pine (*Pinus taeda*) and shortleaf pine (*Pinus echinata*) silvopastures compared to yield in open pastures. In addition, orchardgrass persistence was greater in the loblolly pine system (72% stand occupancy) than the open (44% stand occupancy). In another study of a loblolly pine-mixed grass/forb silvopasture, Burner and Brauer (2003) showed that herbage yield was unaffected at alley widths of 4.9 m and above. Light transmittance was as high as 90% at this spacing. Alley widths below 4.9 m had a profound influence on light transmittance. For example, light transmittance was as low as 43% at a spacing of 2.4 m. Herbage yield increased as light transmittance increased from 43% to 92% (Figure 5a). These authors also observed a general increase in herbage quality (crude protein and digestibility) under silvopasture compared to open grown pasture (Figures 5b and 5c).

Insect density and diversity

Plant–insect interactions are another important factor in the design of agroforestry systems, as variations in tree–crop combinations, and spatial arrangements have been shown to have an effect on insect population density (Vandermeer 1989). According to Stamps and Linit (1998), agroforestry is a potentially useful technology for reducing pest problems because tree-crop combinations provide greater niche diversity and complexity than polycultural systems of annual crops. This effect may be explained in one or more of the following ways: (1) wide spacing of host plants in the intercropping scheme may make the plants more difficult for herbivores to find; (2) one plant species may serve as a trap-crop to prevent herbivores from finding the other crop; (3) one plant species may serve as a repellent to the pest; (4) one plant species may serve to disrupt the ability of the pest to efficiently attack

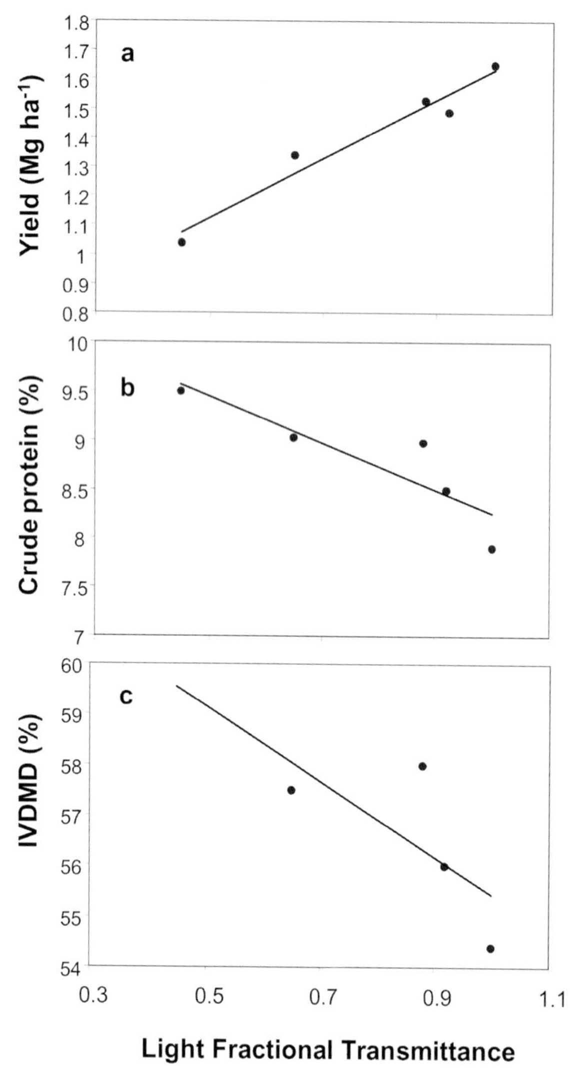

Figure 5. Forage yield, crude protein content, and in vitro dry matter digestibility (IVDMD) as a function of light transmittance in a mixed grass-forb (Bermuda grass and tall fescue being dominant grass) – loblolly pine silvopasture near Booneville, Arkansas, USA. (Based on data given in Burner and Brauer 2003).

its intended host; and (5) the intercropping situation may attract more predators and parasites than monocultures, thus reducing pest density through predation and parasitism (Root 1973; Vandermeer 1989).

The theory of "three trophic level terrestrial interactions" (Price et al. 1980), which states that the addition of a predator to a system of herbivores will lower the density of the herbivores, explains the basis of the predator-prey interactions and related biological pest control observed in agroforestry settings. Studies with pecan, for example, have looked at the influence of ground covers on arthropod densities in tree-crop systems (Bugg et al. 1991; Smith et al. 1996). Bugg et al. (1991) observed that cover crops (e.g. annual legumes and grasses) sustained lady beetles (Coleoptera: Coccinellidae) and other arthropods that may be useful in the biological control of pests in pecan. Stamps et al. (2002) observed similar results in a black walnut-based alleycropping system in Missouri in Central United States. Alleycropped forages (*Medicago sativa* and *Bromis inermis*) supported a more diverse and even arthropod fauna than adjacent monocropped forages (Table 3). In another alleycropping trial with peas (*Pisum sativum*) and four tree species (*Juglans*, *Platanus* , *Fraxinus* and *Prunus* sp.), Peng et al. (1993) found an increase in insect diversity and improved natural-enemy abundance compared to monocultured peas. The greater diversity of birds that are found in agroforestry systems compared to monoculture agronomic systems (Gillespie et al. 1995; Schultz et al. 2000) could also provide the beneficial service of pest reduction to adjacent crops. Although beyond the scope of this discussion, the competitive activity of belowground pests is another important consideration in agroforestry (Ong et al. 1991).

Belowground interactions

Competition for water and/or nutrients

Belowground competition is most likely to occur when two or more species have developed a specialized root system that directs them to explore the same soil strata for resources (van Noordwijk et al. 1996). This interaction can be problematic even in mixed-species systems. Researchers in the temperate zone, humid tropics, and semiarid tropics have reported observing the greatest concentration of tree-root density within the top 30 cm of soil, the region predominantly explored by crop root systems (Itimu 1997; Lehmannn et al. 1998; Imo and Timmer 2000; Jose et al. 2000b). For example, the root systems of all component species in two temperate alleycropping systems, consisting of maize and black walnut or red oak, were found to most heavily occupy the top 30 cm of soil and to decrease in density with depth. Consequently, maize yields were reduced by 35% and 33% for black walnut and red oak systems, respectively (Jose et al. 2000b).

Although it is difficult to separate the belowground competition for water from that for nutrients, it is widely recognized that crop production in agroforestry systems in semiarid regions is likely to be limited by

Table 3. Shannon diversity indices and evenness for arthropods found in alley cropped and monoculture alfalfa over two years of sampling in Missouri, USA.

Year	Alfalfa	Herbivores		Predators		Parasitic Hymenoptera	
		Shannon Index	Evenness	Shannon Index	Evenness	Shannon Index	Evenness
1997	Alley cropped	1.04 (0.33)	0.56 (0.17)	**0.42 (0.25)**[a]	**0.31 (0.18)**	**0.15 (0.15)**	**0.12 (0.16)**
	Monocropped	1.16 (0.36)	0.58 (0.13)	**0.17 (0.12)**	**0.13 (0.08)**	**0.04 (0.05)**	**0.05 (0.06)**
1998	Alley cropped	1.30 (0.31)	0.59 (0.15)	**0.29 (0.16)**	**0.23 (0.17)**	0.03 (0.04)	0.02 (0.05)
	Monocropped	1.23 (0.37)	0.54 (0.16)	**0.16 (0.09)**	**0.11 (0.10)**	0.03 (0.04)	0.02 (0.03)
Total	Alley cropped	1.17 (0.35)	0.57 (0.16)	**0.36 (0.22)**	**0.27 (0.18)**	**0.09 (0.12)**	**0.07 (0.13)**
	Monocropped	1.19 (0.36)	0.56 (0.14)	**0.17 (0.11)**	**0.12 (0.09)**	**0.04 (0.04)**	**0.03 (0.05)**

[a]Bold face indicates significant differences between treatments (i.e., alleycropped vs.monocropped) within a given year or for the total study period. Values in parentheses indicate SE.
Source: Adapted from Stamps et al. (2002).

competition for water. When considering alleycropping systems in the temperate regions, it is not clear as to which competitive vector will limit productivity, although competition for water has been reported to limit system productivity. In Indiana, United States, Gillespie et al. (2000) observed reductions in alley-grown maize yield, between rows of black walnut, when compared to alley-grown maize where a polyethylene root barrier separated interspecific belowground interactions. In the barrier-separated treatments, maize yield from plants grown adjacent to the tree-row were comparable to the yield of maize grown in the alley center, despite shading from the tree-row. Jose et al. (2000a), in a companion study to Gillespie et al. (2000), quantified competition for water in that system and concluded that severe interspecific competition for water was occurring.

In another temperate-zone alleycropping system with silver maple and maize in northeastern Missouri, United States, Miller and Pallardy (2001) also found competition for water to reduce crop yields by 22% to 27%. The presence of a root barrier increased soil water availability in the crop alley, a result reported from other temperate trials as well (Jose et al. 2000a; Hou et al. 2003). These authors further observed significantly higher predawn leaf water potential for maize grown at the tree–crop interface with the root barrier compared to plants without the root barrier (Figure 6). In a similar trial using belowground root barriers in Ohio, United States, Ssekabembe et al. (1994) showed that soil water was higher within the alley in the barrier treatment than the nonbarrier treatment where black locust (*Robinia pseudoacacia*) hedgerows depleted soil water.

Competition for water was also the primary determinant of cotton productivity in a pecan-cotton alleycropping system in the southern United States (Wanvestraut et al. 2004). Cotton lint yield was 26% lower in the nonbarrier treatment compared to one in which a trenched barrier was present. Plants in the barrier treatment outperformed those in the nonbarrier treatment beginning early in the growing season and were 26% taller than non-barrier plants by the end of the growing season. A difference in leaf area development was noticeable by week 5 and led to a 47.3% difference by the end of the growing season. Jose et al. (2000a) also reported a 21% increase in maize leaf area in the barrier treatment of the previously mentioned black walnut – maize alleycropping system. Water stress has been found to cause large reductions in plant height (NeSmith and Ritchie 1992a; Robertson 1994), and leaf area (Gavloski et al. 1992; NeSmith and Ritchie 1992b) in crop plants. Since leaf elongation is the first process inhibited by water stress (Hsiao 1973; Taiz and Zeiger 1991), it is plausible that plants without barrier in these examples were experiencing severe water stress. It should be noted, however, that the competitive interactions involving water become more intense as soil water gets depleted (Miller and Pallardy 2001). As a result, the degree to which competition for water limits productivity will ultimately depend on the precipitation pattern of a given area, especially in the absence of irrigation.

Research reports on competition for nutrients in temperate agroforestry are rather limited (Jose et al. 2000b; Miller and Pallardy 2001; Allen 2003). Many of the reported agroforestry systems were fertilized at the conventional agronomic level as needed for the crop component. In their silver maple–maize alleycropping trial, Miller and Pallardy (2001) applied the standard rate of N fertilization (179 kg N ha^{-1}) and a supplemental rate (standard + 89.6 kg ha^{-1})

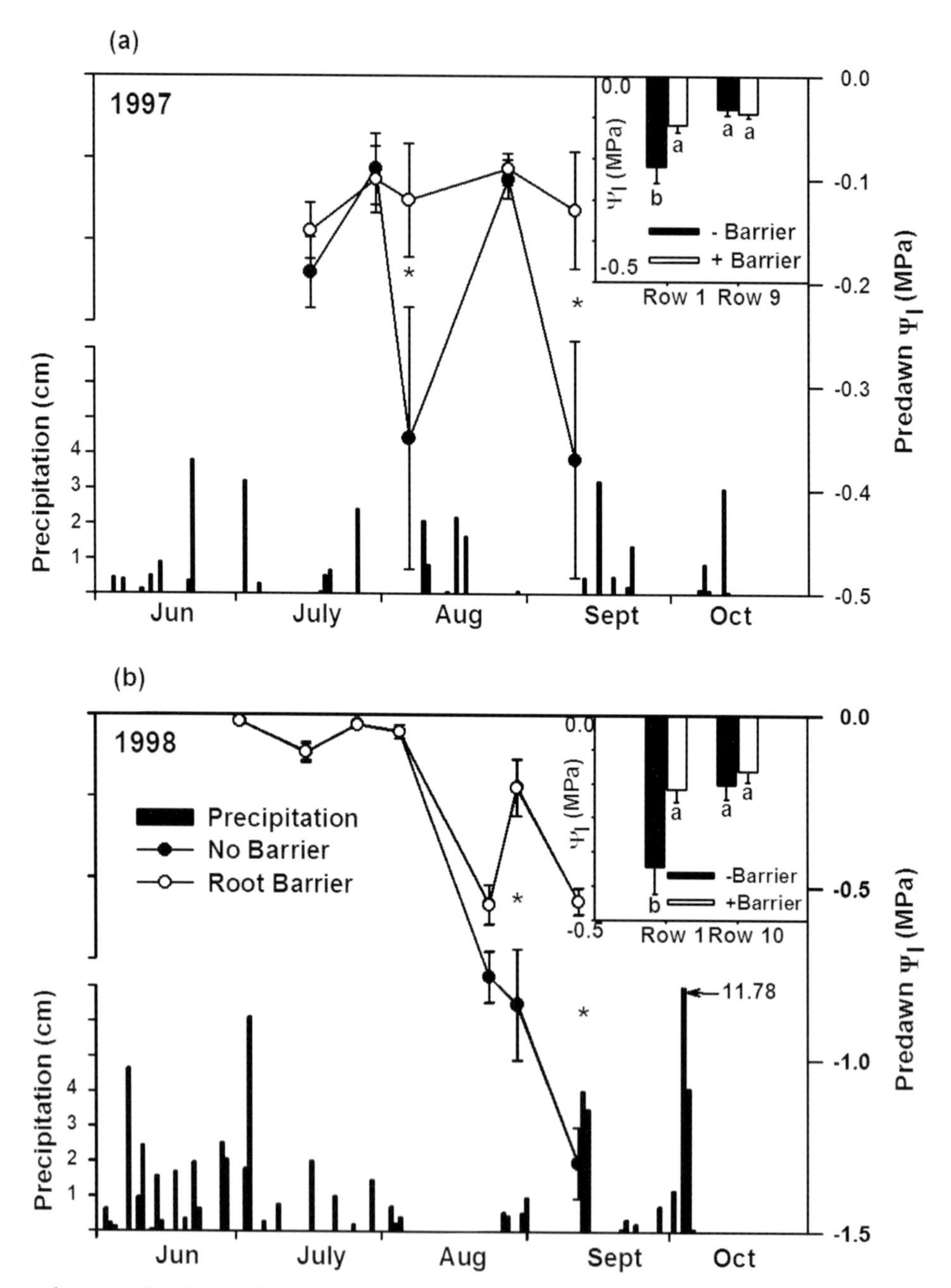

Figure 6. Seasonal patterns of predawn leaf water potential (Ψ_l) for the first row of maize adjacent to the tree row (line graphs) and seasonal means for adjacent and center rows of maize (insets) during 1997 (a) and 1998 (b) in a maize-silver maple alley cropping practice in north-central Missouri, United States. Significant treatment and row differences ($P < 0.05$) indicated by * or different letters; error bars indicate ±1 SE (Reprinted from Miller and Pallardy) 2001).

and concluded that soil nitrogen status had little effect on maize physiology and yield. Jose et al. (2000b) reported that competition for fertilizer nitrogen was minimal in their black walnut–maize alleycropping system, since nutrient acquisition was not simultaneous among the system's components. Water availability, however, was a factor in nutrient competition, as competition for water by tree roots was responsible for reduction in biomass in intercropped maize, resulting in decreased efficiency of fertilizer use (Jose et al. 2000b). Similarly, in a poplar (*Populus* sp.) – barley (*Hordeum vulgare*) system in southern Ontario, associated trees and crops utilized different sets of soil nutrient resource horizons (Williams et al. 1997; Thevathasan and Gordon 2004). Significant competition and resultant yield decrease eventually can be expected in the absence of fertilization in most temperate agroforestry systems.

Allelopathy

Allelochemicals are present in many types of plants and are released into the rhizosphere by a variety of mechanisms, including decomposition of residues, volatilization, and root exudation. Allelochemicals are known to affect germination, growth, development, distribution, and reproduction of a number of plant species (Inderjit and Malik 2002). The degree to which these allelopathic chemicals inhibit growth depends upon their production rates and residence times as well as the combinations in which they are released into the ecosystem. While there are many unanswered questions and concerns about allelopathy in agroforestry, there are a few examples from temperate systems where allelopathy could be reducing crop yield. Two of the most widely used examples are those of black walnut and pecan, two commonly used tree species in temperate agroforestry. It has long been recognized that the principal chemical responsible for walnut and pecan allelopathy is a phenolic compound called juglone (5-hydroxy-1,4-naphthoquinone) (Davis 1928).

Several reported studies using known concentrations of juglone (ranging from 10^{-6} to 10^{-3} M) applied either in hydroponic or soil culture have shown inhibitory effects on survival and growth of several herbaceous and woody plants (Table 4). For example, Rietveld (1983) investigated the sensitivity of 16 species of herbs, shrubs, and trees to juglone in solution culture with juglone concentrations varying from 10^{-3} M to 10^{-6} M. Although seed germination and radicle elongation were not affected in all species, shoot elongation and dry weight accumulation were affected. Many species were sensitive to juglone concentrations as low as 10^{-6} M. Seedlings of all species were severely wilted and eventually killed by 10^{-3} M juglone.

Interest in quantifying the effects of juglone on field crops has resulted in recent hydroponic experiments using corn and soybean (Jose and Gillespie 1998a), the two major field crops planted in black walnut-based alleycropping systems of North America. Three different concentrations of juglone (10^{-4}, 10^{-5}, and 10^{-6} M) along with a juglone-free control treatment were applied to maize and soybean in solution culture. Within three days, juglone induced significant inhibition of shoot and root relative growth rates. In general, soybean was more sensitive to juglone than was maize. Root relative growth rate was the most inhibited variable for both species, and reductions of 86.5% and 99% in this variable were observed in maize and soybean, respectively, at 10^{-4} M juglone concentration.

Actual juglone concentrations in the field may not be as high as 10^{-4} M (Jose and Gillespie 1998b; Thevathasan et al. 1998; Thevathasan and Gordon 2004). Also, under certain soil conditions, juglone may be oxidized to non-toxic forms. For example, Fisher (1978) found that the inhibitory activity of juglone disappeared under low soil water regimes. Although reduced growth of maize in the field under black walnut has been attributed to competition for water (Jose et al. 2000b), the results of Jose and Gillespie (1998a) indicate a possibility of juglone phytotoxicity. Similarly, soybean yield reduction under black walnut (Seifert 1991) may also be partly explained by the toxic effects of juglone. Interestingly, certain annual species may induce allelopathic effects on trees as well. For example, Smith et al. (2001) administered allelochemical-containing leachates to container-grown pecan trees, and found that tall fescue (*Festuca arundinacea*), bermudagrass (*Cynodon dactylon*) and cutleaf evening primrose (*Oenothera laciniata*) leachate decreased pecan trunk weight by 22%, root weight by 17%, and total tree dry weight by 19%, compared to the control treatment.

Reports of tropical tree species providing allelochemicals through leaf litter, and in turn, providing pest control for associated agronomic species are common in the agroforestry literature (Rizvi et al. 1999). Similar complementary examples are yet to be reported from temperate systems.

Table 4. Percent change in shoot and root dry weights of seedlings grown at different juglone concentrations in hydroponic cultures. A negative change means growth was reduced whereas a positive change means growth was enhanced compared to control seedlings.

Species	Shoot dry weight (%)			Root dry weight (%)		
	Concentration of juglone (M)					
	10^{-6}	10^{-5}	10^{-4}	10^{-6}	10^{-5}	10^{-4}
Herbs[a]						
Crimson clover (*Trifolium incanatum*)	−15	−50	−81	11	−50	−78
Crown vetch (*Coronilla varia*)	2	−83	−94	−1	−82	−97
Hairy vetch (*Vicia villosa*)	2	−29	−67	−10	−10	−57
Korean lespedeza (*Lespedeza stipulacea*)	−9	−47	−27	14	−29	−14
Sericea lespedeza (*L. cuneata*)	−30	−72	−92	−6	−63	−88
Shrubs[a]						
Ginnala maple (*Acer ginnala*)	67	64	−35	−35	−28	−83
Siberian peashrub (*Caragana arborescens*)	24	−46	−83	−14	−72	−91
Russian olive (*Elaeagnus angustifolia*)	−16	32	−92	8	99	−75
Autumn olive (*E. umbellata*)	−45	−65	−94	−18	−41	−88
Amur honeysuckle (*Lonicera maackii*)	−41	−61	−91	−55	−61	−94
Trees [b]						
White pine (*Pinus strobus*)	−7	−3	−31	−33	−29	−50
Scotch pine (*P. sylvestris*)	−38	0	−63	−50	0	−25
Japanese larch (*Larix leptolepis*)	−36	−14	−71	−20	0	−60
Norway spruce (*Picea abies*)	20	−20	−20	17	−17	−16
White oak (*Quercus alba*)	−23	−41	−53	−21	−27	−20
White ash (*Fraxinus americana*)	−20	−58	−83	−7	−31	−71
European black alder (*Alnus glutinosa*)	−33	−86	−94	−26	−87	−94
Yellow poplar (*Liriodendron tulipifera*)	19	8	−72	4	−20	−77
Row crops [c]						
Corn (*Zea mays*)	4	−29	−56	6	−39	−61
Soybean (*Glycine max*)	−11	−37	−33	−9	−48	−56

[a]Seedlings were grown for 4 to 6 weeks; source: Rietveld (1983).
[b]Seedlings were grown for 8 to 10 weeks for white pine, scotch pine, Japanese larch, and Norway spruce; source: Funk et al.(1979); the rest of the tree seedlings were grown for 4 to 6 weeks; source: Rietveld (1983).
[c] Seedlings were grown for 3 days; source: Jose (1997).

Safety net role

Intensive agricultural practices have led to inefficient use of applied nitrogen and to contamination of surface and subsurface water through nitrate leaching (Bonilla et al. 1999; Ng et al. 2000). The "safety net" hypothesis of nutrient capture assumes that the deep roots of trees are capable of retrieving nitrate-N and other nutrients that have leached below the rooting zone of associated agronomic crops, and of eventually recycling these nutrients as litterfall and root turnover in the cropping zone. Also, the longer temporal activity of tree roots, even at the same strata, help capture nutrients before and after a crop is planted and harvested. This would essentially increase the total resource-use efficiency of the system (van Noordwijk et al. 1996). The vertical distribution of a given species' root system in its native habitat may often exhibit a signature pattern. This can lead to overlapping vertical root distributions if two or more species have evolved in similar environments. The result is often less than optimum (biologically) growth for one or more of the species involved. Many plant species, however, have shown some degree of plasticity (ability to respond to changes in local nutrient supplies or impervious soil layers) in their vertical (as well as lateral) distribution. Plants also exploit plasticity to avoid competition (Ong et al. 1996; Schroth 1999).

Spatial separation of root architectures can be induced under certain circumstances leading to a more complementary situation. In the previously mentioned alleycropping system consisting of cotton and pecan, Wanvestraut et al. (2004) observed that cotton roots exhibited a degree of plasticity, being limited to shallower soil strata and avoiding a region of high pecan root density deeper in the soil. From the same study site, Allen (2003) reported a 34% reduction in nitrate nitrogen concentration in the soil solution at 0.9 m depth when pecan and cotton roots were not separated by a root barrier compared to the root barrier treatment, potentially indicating the safety net role of pecan roots.

The safety net concept can also be applied in explaining the sediment and nutrient capture functions of riparian buffers. Riparian buffers are widely recommended as a tool for removing non-point source pollutants from agricultural areas (Addy et al. 1999; Bharati et al. 2002; Lee et al. 2003; Schultz et al. 2004). In a recent study of a riparian buffer strip in central Iowa, Midwestern United States, Lee et al. (2003) compared the efficiency of a switchgrass (*Panicum virgatum*) buffer, a switchgrass/woody buffer, and a no buffer treatment located at the lower end of agronomic field plots. The switchgrass buffer removed 95% of the sediment, 80% of the total N, 62% of the nitrate-N, 78% of the total P, and 58% of the phosphate-P. The switch grass/woody buffer removed 97% of the sediment, 94% of the total N, 85% of the nitrate-N, 91% of the total-P, and 80% of the phosphate-P in the runoff. While the switchgrass buffer was effective in removing sediment and sediment-bound nutrients, addition of woody species increased the removal efficiency of soluble nutrients by over 20%. The 'nutrient-capture' functions of trees and other plants are also being exploited in phytoremediation of contaminated sites (Rockwood et al. 2004) and explored in the rehabilitation of heavily fertilized agricultural systems in North America (Nair and Graetz 2004).

Hydraulic lift

Hydraulic lift, the process in which some deep-rooted plants take in water from lower soil layers and release that water into upper, drier soil layers, has been reported to be an appreciable water source for neighboring plants (Corak et al. 1987; Caldwell and Richards 1989). Recent evidence shows that this phenomenon can help promote greater plant growth, and could have important implications for net primary productivity, as well as ecosystem nutrient cycling and water balance (Horton and Hart 1998).

In an agroforestry context, trees can benefit associated crop plants through the mechanism of hydraulic lift, thus providing more water during dry periods (Dawson 1993; van Noordwijk et al. 1996; Burgess et al. 1998; Ong et al. 1999). Although hydraulic lift has been reported in such genera as *Quercus*, and *Pinus* with potential for agroforestry (Ishikawa and Bledsoe 2000; Penuelas and Filella 2003), direct beneficial evidence is not yet available from temperate agroforestry systems.

N_2-fixation and sharing

Biological nitrogen fixation by tree, crop, or both components of agroforestry systems has received a lot of attention in the tropics (Nair et al. 1999). Despite its biological importance, such reports are rare in the temperate agroforestry literature. Although there are 650 woody species belonging to nine families that are capable of fixing atmospheric nitrogen, relatively few occur in the temperate regions of the world. Temperate-zone nitrogen fixing trees and shrubs with potential for agroforestry include *Robinia, Prosopis*, and *Alnus* (Table 5). Some evidence exists from a temperate alleycropping system where significant transfer of fixed nitrogen to associated crop species has been reported. Seiter et al. (1995) using ^{15}N injection technique demonstrated that 32% to 58% of the total nitrogen in maize was derived from nitrogen fixed by associated *Alnus rubra* in an alleycropping system, and that nitrogen transfer decreased with increasing distance from trees, in Oregon, United States.

Herbaceous legumes in alleycropping and silvopastoral systems are capable of fixing substantial amounts of atmospheric nitrogen as well (Table 5). Annual nitrogen fixation rates exceeding 350 kg N ha^{-1} have been reported in temperate pastures (Sharrow 1999). Although direct transfer of fixed nitrogen from legumes to associated tree species is possible, this can be a slow process, which requires several years for sufficient enrichment of soil nitrogen levels to benefit tree growth. One of the best-known examples from temperate silvopasture is that reported by Waring and Snowdon (1985), who observed decreased growth of *Pinus radiata* for the initial three years in a silvopasture with subclover. At the end of the seventh growing season, however, soil nitrogen was increased by 36% and tree diameters by 14% in

Table 5. Nitrogen fixation potential of selected tree and crop species suitable for temperate agroforestry systems.

Species	Typical levels of N_2-fixation (kg ha^{-1} yr^{-1})	Source
Leguminous		
Trees/shrubs		
Black locust (*Robinia pseudoacacia*)	30–35	Boring and Swank (1984)
Leucaena (*Leucaena leucocephala*)	100–500	Sanginga et al. (1995)
Mesquite (*Prosopis glandulosa*)	20–50	Sharifi et al. (1982)
Silk tree (*Albizia julibrissin*)	60–70[a]	Rhoades et al. (1997)
Herbaceous species		
Alfalfa (*Medicago sativa*)	148	Butler and Evers (2003[b])
Clover (*Trifolium* spp.)	42–400	Butler and Evers (2003)
Hairy vetch (*Vicia villosa*)	90–100	Butler and Evers (2003)
Soybeans (*Glycine max*)	60	Troeh and Thompson (1993)
Non-leguminous trees and shrubs Alder (*Alnus* spp.)	48–185	Daniere et al. (1986)
Autumn olive (*Elaeagnus umbellata*)	236	Paschke et al. (1989)
Snowbrush (*Ceanothus velutinus*)	24–101	McNabb and Cromack (1985)

[a]Estimated based on data given in Rhoades et al. (1997).
[b] Butler and Evers, Pers. Comm.

silvopasture compared to *P. radiata* monoculture. Van Sambeek et al. (1986) and Dupraz et al. (1999) have also reported similar growth increases in black walnut when interplanted with leguminous forage crops in the Midwestern United States and southern France, respectively.

Modeling interactions

Models help synthesize experimental and empirical evidence on how different components of an agroforestry system interact in space and time. They also help extrapolate research results to new suite of climate, soil, species and management, which are too numerous to be studied with field experiments. Earlier modeling efforts focused on integrating separately developed crop and tree models into agroforestry models and were not very successful (van Noordwijk and Lusiana 1999). Recent models, however, integrates the principles of above- and belowground resource capture and utilization in mixed cropping systems (van Noordwijk and Lusiana 1999). A variety of computer-based models (e.g., HyPAR; SBELTS; WaNuLCAS) with application to agroforestry have been developed in recent years (Mobbs et al. 1998; van Noordwijk and Lusiana 1999; Qi et al. 2000). Many of them fail to allow for the role of animal production and the associated interactions and management challenges. A recent model (ALWAYS; Alternative Land-use With Agroforestry Systems) offers promise in simulating the complex biophysical functioning of silvopastoral systems (Balandier et al. 2003). Although we have made considerable progress in modeling interactions, many models oversimplify the biophysical processes that are central to predicting the component yields. Considerable improvement is needed in making models a viable tool for selecting site-specific agroforestry combinations and making yield predictions. Since the topic is so advanced a detailed discussion is neither warranted nor feasible here.

Lessons learned

Although certain temperate agroforestry systems such as alleycropping, silvopasture, and windbreaks are well researched for their ecological viability and suitability, many other systems have received little attention. For example, forest farming is gaining popularity among landowners in temperate North America; but has received little or no research attention in this area. Similarly, design considerations and environmental benefits of riparian buffers have received considerable attention (e.g., Schultz et al. 2004), but the complex ecological interactions among its components still remain to be explored. In spite of these deficiencies, the following generalizations can be drawn based on the available literature on interspecific interactions in temperate agroforestry systems.

1. Aboveground competition for light is minor in several of the reported alleycropping systems. Shading, however, has been shown to decrease yield of associated forage species in silvopastoral systems. But, competition for light can be managed in design or maintenance of agroforestry systems.
2. Shading has generally been favored for higher quality forage and better cattle performance and production in a few of the available silvopastoral examples.
3. Competition for nutrients is minimal as most of the temperate agroforestry systems are managed with high input of inorganic or organic nutrient supplements.
4. Some evidence exists for the safety net role of trees from temperate alleycropping and riparian buffer strips. It is likely that this concept can be incorporated into the design of other agroforestry systems.
5. Competitive interactions involving water seem to be the most influential driving force of productivity in both alleycropping and silvopastoral trials.
6. The role of allelopathy in temperate agroforestry is still ambiguous, despite the widespread use of such potentially allelopathic species as black walnut and pecan in alleycropping and silvopastoral systems.
7. Evidence of increased insect diversity exists in certain specific alleycropping examples. Some beneficial arthropods are reported to be controlling pests in pecan-clover or black walnut-alfalfa alleycropping. Riparian forest buffers may also provide a refuge for beneficial insects and birds.
8. Although a few N_2-fixing trees and shrubs have potential to be incorporated into temperate agroforestry practices, their use is limited as revealed by the literature. N_2-fixing herbaceous plants have been widely used in silvopastoral examples and soil enrichment and tree growth enhancement have been reported.

Research needs

A thorough understanding of the interactions among trees, crops, animals, and their associated fauna is necessary to determine the sustainability and profitability of agroforestry systems. Although considerable progress has been made in our understanding of the complex interactions, it is limited to only a few practices in certain regions. For example, many of the examples used in this review come from alleycropping systems in the Midwestern United States involving black walnut. Since many other systems are in the formative stages of research and adoption, much more baseline information is needed in order to establish guidelines for tree–crop interactions of specific systems. To optimize production and sustainability of these systems, major research initiatives are needed in a number of areas. These include:

1. Testing of other tree and herbaceous species for different temperate agroforestry systems.
2. Study of cultural practices that would minimize competition and maximize niche separation (e.g., repeated trenching to train trees to root deeply over the lifespan of alleycropping systems, simple agronomic disking on a periodic basis, knifing of fertilizer, etc.).
3. Genetic modification of components to increase productivity and reduce competition. For example, (a) genetic selection for vertical root morphology in both tree and crop components, (b) selection for greater tolerance of shade in the crop component.
4. Continued use of modeling component interactions: modeling provides a means for integrating different positive and negative interacting vectors so that resource allocation and yield can be predicted.

Acknowledgements

The authors express their sincere thanks to Drs. Kyehan Lee and Craig Ramsey for their comments on an earlier version of the manuscript. This publication was partially supported by U.S. Department of Agriculture (USDA), Cooperative State Research, Education, and Extension Service, Initiative for Future Agriculture and Food Systems (USDA/CSREES/IFAFS) grant number 00-52103-9702 and USDA Southern Region Sustainable Agriculture Research and Education program, grant number LS02-136. Florida Agricultural Experiment Station Journal Series No. R-09662.

References

Addy K.L. Gold A.J., Groffman P.M. and Jacinthe P.A. 1999. Ground water nitrate removal in subsoil of forested and mowed riparian buffer zones. J Environ Qual 28: 962–970.

Allen S.C. 2003. Nitrogen dynamics in a temperate alley cropping system with pecan (*Carya illinoensis*) and cotton (*Gossypium sp.*). Ph.D. Dissertation, University of Florida, Gainesville, Florida, USA.

Armstrong R.A. and McGhee R. 1980. Competitive exclusion. Am Nat 115: 151–170.

Ashton P.S. 2000. Ecological theory of diversity and its applications to mixed species plantation systems. pp. 61–77. In: Ashton M.S. and Montagnini F. (eds), The Silvicultural Basis for Agroforestry Systems. CRC Press, Boca Raton, FL, USA.

Bagley W.T. 1964. Responses of tomatoes and beans to windbreak shelter. J Soil Water Conserv 19: 71–73.

Balandier P., Bergez, J.E. and Etienne, M. 2003. Use of the management-oriented silvopastoral model ALWAYS: Calibration and evaluation. Agroforest Syst 57: 159–171.

Baraldi R., Bertazza G., Bogino J., Luna V. and Bottini R. 1995. The effect of light quality on *Prunus cerasus* II. Changes in hormone levels in plants grown under different light conditions. Photochem Photobiol 62: 800–803.

Bharati L., Lee K.H., Isenhart T.M. and Schultz R.C. 2002. Soil-water infiltration under crops, pasture, and established riparian buffer in Midwestern USA. Agroforest Syst 56: 249–257.

Blackshaw J.K. and Blackshaw A.W. 1994. Heat-stress in cattle and the effect of shade on production and behavior. Aus J Exp Agr 34: 285–295.

Bonilla C.A., Muñoz J.F. and Vauclin M. 1999. Opus simulation of water dynamics and nitrate transport in a field plot. Ecol Model 122: 69–80.

Boring L.R. and Swank W.T. 1984. The role of black locust (*Robinia pseudoacacia*) in forest succession. J Ecol 72: 749–766.

Brandle J.R., Hodges L. and Wight B. 2000. Windbreak practices. pp. 79–118. In: Garrett H.E., Rietveld W.J. and Fisher R.F. (eds), North American Agroforestry: An Integrated Science and Practice. ASA Inc., Madison, WI, USA.

Brandle J.R. Hodges L. and Zhou, X. 2004. Windbreaks in sustainable agriculture (This volume).

Brenner A.J. 1996. Microclimatic modifications in agroforestry. pp. 159–187. In: Ong C.K. and Huxley P. (eds), Tree–Crop Interactions: A Physiological Approach. CAB International. Wallingford, UK.

Bugg R.L., Sarrantonio M., Dutcher J.D. and Phatak S.C. 1991. Understory cover crops in pecan orchards: possible management systems. Am J Alternative Agr 6: 50–62.

Burgess S.O., Adams M.A., Turner N.C. and Ong C.K. 1998. The redistribution of soil water by tree roots systems. Oecologia 115: 306–311.

Burner D.M. 2003. Effect of alley crop environment on orchardgrass and tall fescue herbage. Agron J 95: 1163–1171.

Burner D.M. and Brauer D.K. 2003. Herbage response to spacing of loblolly pine trees in a minimal management silvopasture in southeastern USA. Agroforest Syst 57: 69–77.

Caldwell M.M. and Richards J.H. 1989. Hydraulic lift: water efflux from upper roots improves effectiveness of water uptake by deep roots. Oecologia 79: 1–5.

Cannell M.G.R., van Noordwijk M. and Ong C.K. 1996. The central agroforestry hypothesis: the trees must acquire resources that the crop would not otherwise acquire. Agroforest Syst 34: 27–31.

Chirko C.P., Gold M.A., Nguyen P.V. and Jiang J.P. 1996. Influence of direction and distance from trees on wheat yield and photosynthetic photon flux density (Q_p) in a Paulownia and wheat intercropping system. Forest Ecol Manag 83: 171–180.

Corak, S.J, Blevins, D.G. and Pallardy, S. G 1987. Water transfer in an alfalfa–maize association: survival of maize during drought. Plant Physiol 84: 582–586.

Daniere C., Capellano A. and Moirud A. 1986. Dynamique de l'azote dans un peuplement natural *d'Alnus incana* L. Moench. Acta Oecologia/Oecologia Plant 7: 165–175.

Davis E.F. 1928. The toxic principle of *Juglans nigra* as identified with synthetic juglone and its toxic effects on tomato and alfalfa plants. Am J Bot 15: 620.

Davis J.E. and Norman J.M. 1988. Effects of shelter on plant water use. Agr Ecosyst Environ 22/23: 393–402.

Davis M.H. and Simmons S.R. 1994a. Tillering response of barley to shifts in light quality caused by neighboring plants. Crop Sci 34: 1604–1610.

Davis M.H. and Simmons S.R. 1994b. Far-red light reflected from neighboring vegetation promotes shoot elongation and accelerate flowering in spring barley plants. Plant Cell Environ 17: 829–836.

Dawson T.E. 1993. Hydraulic lift and water use by plants: implications for water balance, performance and plant–plant interactions. Oecologia 95: 565–574.

Dupraz C., Simorte V., Dauzat M., Bertoni G., Bernadac A. and Masson P. 1999. Growth and nitrogen status of young walnuts as affected by intercropped legumes in a Mediterranean climate. Agroforest Syst 43: 71–80.

Fisher, R.F. 1978. Juglone inhibits pine growth under certain moisture regimes. Soil Sci Soc Am J 42: 801–803.

Funk D.T., Case P.J., Rietveld W.J. and Phares R.E. 1979. Effects of juglone on the growth of coniferous seedlings. Forest Sci 25: 452–454.

García-Barrios L. and Ong C.K. 2004. Ecological interactions in simultaneous agroforestry systems in the tropics: management lessons, design tools and research needs (This volume).

Garrett H.E. and McGraw R.L. 2000. Alley cropping practices. pp. 149–188. In: Garrett H.E., Rietveld W.J. and Fisher R.F. (eds), North American Agroforestry: An Integrated Science and Practice. ASA Inc., Madison, WI, USA.

Gavloski J.E., Whitfield G.H. and Ellis C.R. 1992. Effect of restricted watering on sap flow and growth in corn (*Zea mays* L). Can J Plant Sci 72: 361–368.

Gillespie A.R., Jose S., Mengel D.B., Hoover W.L., Pope P.E., Seifert J.R., Biehle D.J., Stall T. and Benjamin T.J. 2000. Defining competition vectors in a temperate alley cropping system in the midwestern USA. 1. Production physiology. Agroforest Syst 48: 25–40.

Gillespie A.R., Miller B.K. and Johnson K.D. 1995. Effects of ground cover on tree survival and growth in filter strips of the Corn Belt region of the Midwestern U.S. Agric Ecosyst Environ 53: 263–270.

Hawke M.F. and Wedderburn M.E. 1994. Microclimate changes under *Pinus radiata* agroforestry regimes in New Zealand. Agr Forest Meteorol 71: 133–145.

Horton J.L. and Hart S.C. 1998. Hydraulic lift: a potentially important ecosystem process. Trends in Ecol Evolution 13: 232–235.

Hou Q., Brandle J., Hubbard K., Schoeneberger M., Nieto C. and Francis C. 2003. Alteration of soil water content consequent to root-pruning at a windbreak/crop interface in Nebraska, USA. Agroforest Syst 57: 137–147.

Hsiao T.C. 1973. Plant responses to water stress. Annu Rev Plant Phys 24: 519–570.

Imo M. and Timmer V.R. 2000. Vector competition analysis of a Leucaena-maize alley cropping system in western Kenya. Forest Ecol and Manag 126: 255–268.

Inderjit and Malik A.U. 2002. Chemical Ecology of Plants: Allelopathy in Aquatic and Terrestrial Ecosystems. Birkhauser-Verlag, Berlin, 272 pp.

Ishikawa C.M. and Beldsoe C.S. 2000. Seasonal and diurnal patterns of soil water potential in the rhizosphere of blue oaks: evidence for hydraulic lift. Oecologia 125: 459–465.

Itimu O.A. 1997. Distribution of *Senna spectabilis*, *Gliricida sepium* and maize (*Zea mays* L.) roots in an alley cropping trial on the Lilongwe Plain, Central Malawi. Ph.D. Thesis Wye College, University of London, Kent, UK.

Jose S. 1997. Interspecific interactions in alley cropping: the physiology and biogeochemistry. Ph.D. Dissertation, Purdue University, West Lafayette, IN, USA.

Jose S. and Gillespie A.R. 1995. Microenvironmental and physiological basis for temporal reduction in crop production in an Indiana alley cropping system. pp. 54–56. In: Ehrenreich J.H., Ehrenreich D.L. and Lee H.W. (eds), Growing A Sustainable Future. University of Idaho, Moscow, ID.

Jose S. and Gillespie A.R. 1998a. Allelopathy in black walnut (*Juglans nigra* L.) alley cropping. II. Effects of juglone on hydroponically grown corn (*Zea mays* L.) and soybean (*Glycine max* (L.) Merr.) growth and physiology. Plant Soil 203: 199–205.

Jose S. and Gillespie A.R. 1998b. Allelopathy in black walnut (*Juglans nigra* L.) alley cropping. I. Spatio-temporal variation in soil juglone in a black walnut-corn (*Zea mays* L.) alley cropping system in the mid-western USA. Plant Soil 203: 191–197.

Jose S., Gillespie A.R., Seifert J.R. and Biehle D.J. 2000a. Defining competition vectors in a temperate alley cropping system in the mid-western USA. 2. Competition for water. Agroforest Syst 48: 41–59.

Jose S., Gillespie A.R., Seifert J.R., Mengel D.B. and Pope P.E. 2000b. Defining competition vectors in a temperate alley cropping system in the mid-western USA. 3. Competition for nitrogen and litter decomposition dynamics. Agroforest Syst 48: 61–77.

Knowles R.L., Horvath G.C., Carter M.A. and Hawke M.F. 1999. Developing canopy closure model to predict overstorey/understorey relationships in *Pinus radiata* silvopastoral systems. Agroforest Syst 43: 109–119.

Kort J. 1988. Benefits of windbreaks to field and forage crops. Agric Ecosyst Environ 22/23: 165–191.

Kozlowski T.T. and Pallardy S.G. 1997. Physiology of Woody Plants. 2nd ed. Academic Press, San Diego, CA, USA, 411 pp.

Krueger W.C. 1981. How a forest affects a forage crop? Rangelands 3: 70–71.

Lambers H., Chapin III F.S. and Pons T.L. 1998. Plant Physiological Ecology. Springer-Verlag, New York, NY, USA, 540 pp.

Lee K.H., Isenhart T.M. and Schultz R.C. 2003. Sediment and nutrient removal in an established multi-species riparian buffer. J Soil Water Conserv 58: 1–8.

Leihner D.E., Schaeben R.E., Akond T.P. and Steinmiller N. 1996. Alley cropping on an Ultisol in sub-humid Benin. Part 2: Changes in crop physiology and tree crop competition. Agroforest Syst 34: 13–25.

Lehmann J., Peter I., Steglich C., Gebauer G., Huwe B. and Zech W. 1998. Belowground interactions in dryland agroforestry. Forest Ecol Manag 111: 157–169.

Lin C.H., McGraw R.L., George M.F. and Garrett H.E. 1999. Shade effects on forage crops with potential in temperate agroforestry practices. Agroforest Syst 44: 109–119.

Matson P.A., Parton W.J., Power A.G. and Swift M.J. 1997. Agricultural intensification and ecosystem properties. Science 277: 504–509.

McNabb D.H. and Cromack K. 1985. Dinitrogen fixation by a mature *Ceanothus velutinus* Dougl. stand in the Western Oregon Cascades. Can J Microbiol 29: 1014–1021.

Miller A.W. and Pallardy S.G. 2001. Resource competition across the crop-tree interface in a maize-silver maple temperate alley cropping stand in Missouri. Agroforest Syst 53: 247–259.

Mitlohner F.M., Morrow J.L., Dailey J.W., Wilson S.C., Galyean M.L., Miller M.F. and McGlone J.J. 2001. Shade and water misting effects on behavior, physiology, performance, and carcass traits of heat-stressed feedlot cattle. J Anim Sci 79: 2327–2335.

Mobbs D.C., Cannell M.G.R., Crout N.M.J., Lawson G.J., Friend A.D. and Arah J. 1998. Complementarity of light and water use in tropical agroforests. 1. Theoretical model outline, performance and sensitivity. Forest Ecol Manag 102: 259–274.

Moffat A.S. 1997. Higher yielding perennials point the way to new crops. Science 274: 1469–1470.

Monteith J.L., Ong C.K. and Corlett J.E. 1991. Microclimate interactions in agroforestry systems. Forest Ecol Manag 45: 31–44.

Monteith J.L. 1994. Validity and utility of the correlation between intercepted radiation and biomass. Agr Forest Meteorol 68: 213–220.

Murdoch W.W., Briggs C.J. and Nisbet R.M. 2003. Consumer-Resource Dynamics. Princeton University Press, Princeton, New Jersey, USA, 462 pp.

Nair P.K.R., Buresh R.J., Mugendi D.N. and Latt C.R. 1999. Nutrient cycling in tropical agroforestry systems: myths and science. pp. 1–31. In: Buck L.E., Lassoie J.P. and Fernandes E.C.M. (eds), Agroforestry in Sustainable Agricultural Systems. CRC Press, Boca Raton, FL, USA.

Nair V.D. and Graetz, D.A. 2004. Agroforestry as an Approach to Minimizing Nutrient Loss from Heavily Fertilized Soils: The Florida Experience (This volume).

NeSmith D.S. and Ritchie J.T. 1992a. Effects of soil water deficits during tassel emergence on development and yield component of maize (*Zea mays* L.). Field Crops Res 28: 251–256.

NeSmith D.S. and Ritchie J.T. 1992b. Short- and long-term responses of corn to pre-anthesis soil water deficit. Agron J 84: 107–113.

Ng H.Y.F., Drury C.F., Serem V.K., Tan C.S. and Gaynor J.D. 2000. Modeling and testing of the effect of tillage, cropping and water management practices on nitrate leaching in clay loam soil. Agr Water Manag 43: 111–131.

Nissen T.M., Midmore D.J. and Cabrera M.L. 1999. Aboveground and belowground competition between intercropped cabbage and young *Eucalyptus torelliana*. Agroforest Syst 46: 83–93.

Noble I.R. and Dirzo R. 1997. Forests as human dominated ecosystems. Science 277: 522–525.

Ong C.K., Black C.R., Marshall F.M. and Corlett J.E. 1996. Principles of resource capture and utilization of light and water. pp. 73–158. In: Ong C.K. and Huxley P. (eds), Tree–Crop Interactions: A Physiological Approach. CAB International. Wallingford, UK.

Ong C.K., Corlett J.E., Singh R.P. and Black C.R. 1991. Above and belowground interactions in agroforestry systems. Forest Ecol Manag 45: 45–57.

Ong C.K., Deans J.D., Wilson J., Mutua J., Khan A.A.H. and Lawson E.M. 1999. Exploring belowground complementarity in agroforestry using sap flow and root fractal techniques. Agroforest Syst 44: 87–103.

Paschke M.W., Dawson J.O. and David M.B. 1989. Soil nitrogen mineralization in plantations of *Juglans nigra* interplanted with actinorhizal *Elaeagnus umbellata* or *Alnus glutinosa*. Plant Soil 118: 33–42.

Payne W.J.A. 1990. An Introduction to Animal Husbandry in the Tropics, 4th ed. John Wiley, New York, NY, USA, 401 pp.

Peng R.K., Incoll L.D., Sutton S.L., Wright C. and Chadwick A. 1993. Diversity of airborne arthropods in a silvoarable agroforestry system. J Appl Ecol 30: 551–562.

Penuelas J. and Filella I. 2003. Deuterium labeling of roots provides evidence of deep water access and hydraulic lift by *Pinus nigra* in a Mediterranean forest of NE Spain. Environ Exp Bot 49: 201–208.

Perry D.A. 1994. Forest Ecosystems. Johns Hopkins University Press, Baltimore, MD, USA, 649 pp.

Price P.W., Bouton C.E., Gross P., McPherson B.A., Thompson J.N. and Weis A.E. 1980. Interactions among three trophic levels: influence of plants on interactions between insect herbivores and natural enemies. Annu Rev Ecol Syst 11: 41–65.

Qi, X., Mize C.W., Batchelor W.D., Takle E.S. and Litvina I.V. 2001. SBELTS: a model of soybean production under tree shelter. Agroforest Syst 52: 53–61.

Rao M.R., Nair P.K.R. and Ong C.K. 1998. Biophysical interactions in tropical agroforestry systems. Agroforest Syst 38: 3–50.

Rhoades C.C., Nissen T.M. and Kettler J.S. 1997. Soil nitrogen dynamics in alley cropping and no-till systems on ultisols of the Georgia Piedmont, USA. Agroforest Syst 39: 31–44.

Rietveld W.J. 1983. Allelopathic effects of juglone on germination and growth of several herbaceous and woody species. J Chem Ecol 9: 295–308.

Rizvi S.J.H., Tahir M., Rizvi V., Kohli R.K. and Ansari A. 1999. Allelopathic interactions in agroforestry systems. Crit Rev Plant Sci 18: 773–796.

Robertson M.J. 1994. Relationships between internode elongation plant height and leaf appearance in maize. Field Crops Res 38: 135–145.

Robin C., Hay M.J.M., Newton P.C.D. and Greer D.H. 1994. Effect of light quality (red:far-red ratio) at the apical bud of the main stolon on morphogenesis of *Trifolium repens* L. Ann Bot-London 74: 119–123.

Rockwood D.L., Naidu C.V., Carter D.R., Rahmani M., Spriggs T.A., Lin C., Alker G.R., Isebrands J.G. and Segrest S.A. 2004. Short-rotation woody crops and phytoremediation: Opportunities for agroforestry? (This volume).

Root R. 1973. Organization of a plant-arthropod association in simple and diverse habitats: the fauna of collards (*Brassica oleracea*). Ecol Monogr 43: 95–124.

Sanginga N., Vanlauwe B. and Danso S.K.A. 1995. Management of biological N_2 fixation in alley cropping systems: Estimation and contribution to N balance. Plant Soil 174: 119–141.

Schroth G. 1999. A review of belowground interactions in agroforestry, focusing on mechanisms and management options. Agroforest Syst 43: 5–34.

Schultz R.C., Colletti J.P., Isenhart T.M., Marquez, C.O. Simpkins, W.W. and Ball C.J. 2000. Riparian forest buffer practices. pp. 189–281. In: Garrett H.E., Rietveld W.J. and Fisher R.F. (eds), North American Agroforestry: An Integrated Science and Practice. American Society of Agronomy, Madison, WI, USA.

Schultz R.C, Isenhart T.M., Simpkins W.W. and Colletti J.P. 2004. Riparian Forest Buffers in Agroecosystems – Structure, Function and Management – Lessons Learned from the Bear Creek Watershed Project in Central Iowa, USA. (This volume).

Seifert J.R. 1991. Agroforestry: The southern Indiana experience of growing oak, walnut, corn, and soybeans. pp. 70–71. In: Williams P. (ed.), Agroforestry in North America. Ministry of Agriculture and Food, Ontario, Canada.

Seiter S., Ingham E.R., William R.D. and Hibbs D.E. 1995. Increase in soil microbial biomass and transfer of nitrogen from alder to sweet corn in an alley cropping system. pp. 56–158. In: Ehrenreich J.H., Ehrenreich D.L. and Lee H.W. (eds), Growing a sustainable future. University of Idaho, Boise, ID, USA.

Sharifi M.R., Nilsen E.T. and Rundel P.W. 1982. Biomass and net primary productivity of *Prosopis glandulosa* (Fabaceae) in the Sonoran Desert of California. Am J Bot 69: 760–767.

Sharrow S.H. 1999. Silvopastoralism: Competition and facilitation between trees, livestock, and improved grass-clover pastures on temperate rainfed lands. pp. 111–130. In: Buck L.E., Lassoie J. and Fernandez E.C.M. (eds), Agroforestry in Sustainable Agricultural Systems. CRC Press, Boca Raton, FL, USA.

Singh R.P., Ong C.K. and Saharan N. 1989. Above and below ground interactions in alley cropping in semiarid India. Agroforest Syst 9: 259–274.

Smith M.W., Arnold D.C., Eikenbary R.D., Rice N.R., Shiferaw A., Cheary B.S. and Carroll B.L. 1996. Influence of ground cover on beneficial arthropods in pecan. Biol Control 6: 164–176.

Smith M.W., Wolf M.E., Cheary B.S. and Carroll B.L. 2001. Allelopathy of bermudagrass, tall fescue, redroot pigweed, and cutleaf evening primrose on pecan. HortScience 36: 1047–1048.

Ssekabembe C.K., Henderlong P.R. and Larson M. 1994. Soil moisture relations at the tree/crop interface in black locust alleys. Agrofor Syst 25: 135–140.

Stamps W.T. and Linit M.J. 1998. Plant diversity and arthropod communities: implications for temperate agroforestry. Agroforest Syst 39: 73–89.

Stamps W.T., Woods T.W., Linit M.J. and Garrett H.E. 2002. Arthropod diversity in alley cropped black walnut (*Juglans nigra* L.) stands in eastern Missouri, USA. Agroforest Syst 56: 167–175.

Taiz L. and Zeiger E. 1991. Plant Physiology. Benjamin/Cummings Pub Co. CA, USA. 792 pp.

Thevathasan N.V and Gordon A.M. 2004. Ecology of tree intercropping systems in the North temperate region: Experience from southern Ontario, Canada. (This volume).

Thevathasan N.V., Gordon A.M. and Voroney R.P. 1998. Juglone (5-hydroxy-1,4 napthoquinone) and soil nitrogen transformation interactions under a walnut plantation in southern Ontario, Canada. Agroforest Syst 44: 151–162.

Tilman D. 1982. Resource Competition and Community Structure. Princeton University Press, Princeton, NJ, USA.

Tilman D. 1985. The resource ratio hypothesis of plant succession. Am Nat 125: 827–852.

Tilman D. 1990. Mechanisms of plant competition for nutrients: The elements of a predictive theory of plant competition. pp. 117–141. In: Grace J.B. and Tilman D. (eds), Perspectives on Plant Competition. Academic Press, San Diego, CA, USA.

Troeh F.R. and Thompson L.M. 1993. Soils and Soil fertility. 5th edition. Oxford University Press New York, NY, USA, 462 pp.

Vandermeer J. 1989. The Ecology of Intercropping. Cambridge University Press. Cambridge, UK, 249 pp.

van Noordwijk M. and Lusiana B. 1999. WaNuLCAS, a model of water, nutrient and light capture in agroforestry systems. Agroforest Syst 43: 217–242.

van Noordwijk M., Lawson G., Soumaré A., Groot J.J.R. and Hairiah K. 1996. Root distribution of trees and crops: competition and/or complementarity. pp. 319–364. In: Ong C.K. and Huxley P. (eds), Tree–Crop Interactions: A Physiological Approach. CAB International, Wallingford, UK.

Van Sambeek J.W., Ponder F. Jr. and Rietveld W.J. 1986. Legumes increase growth and alter foliar nutrient levels of black walnut saplings. Forest Ecol Manag 17: 159–167.

Wang Q. and Shogren J.F. 1992. Characteristics of the crop-paulownia system in China. Agric Ecosyst Environ 39: 145–152.

Wanvestraut R., Jose S., Nair P.K.R. and Brecke B.J. 2004. Competition for water in a pecan–cotton alley cropping system. Agroforest Syst 60: 167–179.

Waring H.D. and Snowdon P. 1985. Clover and urea as sources of nitrogen for the establishment of *Pinus radiata.* Aust Forest Res 15: 115–121.

Williams P.A., Gordon A.M., Garrett H.E. and Buck L. 1997. Agroforestry in North America and its role in farming systems. pp. 90–84. In: Gordon A.M. and Newman S.M. (eds), Temperate Agroforestry Systems. CAB International, Wallingford, UK.

Wojtkowski P. 1998. The Theory and Practice of Agroforestry Design. Science Publishers, Inc. Enfield, New Hampshire, USA, 282 pp.

Agroforestry Systems **61**: 257–268, 2004.

Ecology of tree intercropping systems in the North temperate region: Experiences from southern Ontario, Canada

N.V. Thevathasan* and A.M. Gordon
Department of Environmental Biology, University of Guelph, Guelph, Ontario, Canada N1G 2W1
**Author for correspondence: email: nthevath@uoguelph.ca*

Key words: Agroecosystems, Biodiversity, Biophysical interactions, Carbon sequestration, Intercropping systems, Sustainable agriculture

Abstract

Agroforestry practices in northern latitudes, although less diverse than those in warmer regions, have unique advantages over conventional land-use systems in the region in terms of water-quality enhancement, carbon sequestration, and biodiversity conservation. Tree intercropping, especially, is a potentially promising agroforestry option in the region. Understanding the ecological interactions between trees and crops in such intercropped systems provides the basis for designing efficient systems with potential for wider applicability. With this objective, the experience from several years of research on this aspect at the University of Guelph, in southern Ontario, Canada are presented. Yields of C3 crops intercropped with trees, as well as growth of trees, did not differ from those in corresponding sole-stand (conventional) systems of crops and trees. But, soil organic carbon content and bird and insect diversity increased in the intercropped area. The abundance and distribution of earthworms were higher closer to the tree rows indicating improved soil health. The C sequestration potential in fast-growing tree (hybrid-poplar)-based intercropping systems was four times more than that reported for conventional agricultural fields in the region. Because of reduced fertilizer use and more efficient N-cycling, the tree-intercropping systems could also lead to the reduction of nitrous oxide emissions from agricultural fields by about 0.7 kg ha^{-1} yr^{-1}. Marginal or degraded land that is suitable for agroforestry is estimated to be 57 million ha in Canada. Tree/crop intercropping is one agroforestry system that shows great potential for this region. We suggest that this land-management option can be placed above conventional agriculture in terms of long term-productivity and sustainability.

Introduction

Out of the estimated 140 million ha of marginal or degraded land that is considered to be available for agroforestry establishment in North America (Dixon et al. 1994), 50 million to 57 million are in Canada (McKeague 1975). Although many different types of agroforestry including windbreaks and shelterbelts, silvipastoral systems, integrated riparian forest systems, forest farming systems, and tree–crop intercropping systems have historically been practiced in North America (Garrett et al. 2000; Gordon et al. 1997; Gordon and Newman 1997), the vast potential for economic and environmental benefits attributed to agroforestry is yet to be realized on a large scale. Compared to the diverse and extremely complex agroforestry systems that exist in warmer parts of the world, the types of agroforestry applications could be limited in the northern latitudes. Even so, agroforestry is being practiced in many areas in Canada including southern Ontario (Table 1), and research at the University of Guelph, Ontario, Canada, during the past 15 years has shown the vast potential for agroforestry in terms of such benefits as water-quality enhancement, carbon sequestration, and biodiversity conservation. The integration of trees into the agricultural land-base via tree/crop intercropping systems has indeed shown great potential for the region (Thevathasan et al. 2004). Given the recent ratification of the Kyoto Protocol by Canada and the role of agroforestry in carbon sequest-

Table 1. Summary of research trials undertaken on different agroforestry technologies in southern Ontario, Canada.

Agroforestry technology	Reference
Windbreaks and shelterbelts	Kenney 1987; Loeffler et al. 1992
Integrated riparian forest systems	O'Neill and Gordon 1994; Oelbermann and Gordon 2000, 2001
Forest farming systems	Matthews et al. 1993; Christrup 1993[6]; Williams et al. 1997
Tree-based intercropping systems	McLean 1990[8]; Ball 1991[9]; Gordon and Williams 1991; Williams and Gordon 1992, 1994, 1995; Ntayombya 1993; Ntayombya and Gordon 1995; Kotey 1996[10]; Thevathasan and Gordon 1995, 1997; Thevathasan 1998; Price and Gordon 1999; Dyack et al. 1999; Price 1999[7]; Simpson 1999[3]; Zhang 1999[2]; Gray 2000[4]; Middleton 2001[5]
Silvipastoral systems	Bezkorowajnyj et al. 1993

ration (Montagnini and Nair 2004), agroforestry as a land-use approach is more important now than ever before. When carbon trading becomes effective, as it is poised to be soon, industries that emit greenhouse gasses (GHG) may be required to purchase C credits and agroforestry land-use practices may become an attractive land-use option.

The low adoption rate of tree/crop intercropping systems in southern Ontario is in part due to current tax policies that do not take into consideration the numerous nontangible, societal-level benefits associated with agroforestry systems (Dyack et al. 1999). It is also true that the traditional farming community in many areas of the region is not entirely aware of the intricacies of adopting and practicing intercropping (Matthews et al. 1993). Nevertheless, from a scientific perspective, a clear understanding of the ecological foundations of agroforestry is essential for promoting the practice and extending it to potential sites. With that background, this chapter synthesizes the ecological information generated from 15 years of tree/crop intercropping research at the University of Guelph and in other parts of Canada, and discusses the application potentials and management implications of these results in terms of long-term production and sustainability of the resource base in the north temperate region, with particular focus on Canada. We hope that this chapter will complement well with the two companion chapters in this volume, by Jose et al. (2004) and Garcia-Barrios and Ong (2004) that discuss interspecific interactions in temperate and tropical agroforestry systems respectively, in providing the ecological foundations for agroforestry in a wide range of conditions.

The results presented here are from a long-term tree-based intercropping research project that was initiated in 1987 on a 30-ha field at the University of Guelph Agroforestry Research Station, Ontario, Canada (43°32′28″ N latitude, 80°12′32″ W longitude). Ten tree species, namely *Acer saccharinum* (silver maple), *Corylus avellana* (hazelnut), *Fraxinus americana* (white ash), *Juglans nigra* (black walnut), *Picea abies* (Norway spruce), *Populus* sp. (poplar – hybrid), *Quercus rubra* (red oak), *Robinia pseudoacacia* (black locust), *Salix discolor* (willow) and *Thuja occidentalis* (white cedar) were planted and annually intercropped with maize (corn) (*Zea mays*), soybean (*Glycine max*), and winter wheat (*Triticum aestivum*) or barley (*Hordeum vulgare*). Tree rows were spaced at 12.5 m or 15 m apart with within-row spacing of 3 m or 6 m. The soil type is sandy loam (Typic Hapludalf). Crops were planted between the tree rows every year according to local 'standard' cultural practices.

Plant-to-plant interactions for growth resources

Interactions in agroforestry systems are defined as the effect of one component of the system on the performance of another component and/or the overall system (Nair 1993). The study of interactions requires the examination of a number of complex processes, including processes related to soil fertility, competition, microclimate, insect pests and diseases, soil conservation and allelopathy (Rao et al. 1998). Exploitation of positive interactions between the woody (tree) and nonwoody (agricultural or annual crop) components and the minimization of negative interactions is the key to the success of tree-based intercropping systems. This paper will therefore deal mainly with the identification and quantification of these interactions that could be used in designing management strategies that promote complementary interactions and reduce or eliminate the negative ones.

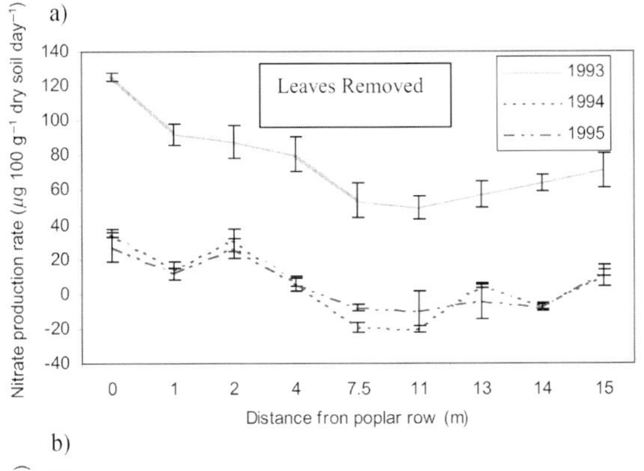

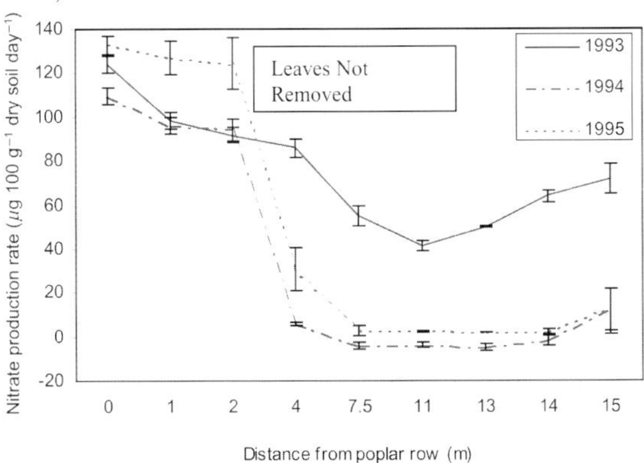

Figure 1. Effects of removing or retaining poplar-tree-leaf on soil N mineralization near the tree row (up to 2.5 m horizontal distance) in the tree-intercropping study during 1993–1995 in southern Ontario, Canada; (a) Leaves removed; (b) Leaves not removed. Source: Thevathasan (1998).

Table 2. Poplar litterfall distribution in a poplar–barley intercropping system during the 1993 and 1994 growing seasons when trees were 6 and 7 years old respectively, Guelph, Ontario, Canada.

Distance from the poplar tree row (m)	Litterfall biomass ($Mg\ ha^{-1}\ yr^{-1}$) 1993	1994
0–2.5	2.67 ± 0.04	2.76 ± 0.14
2.5–6.0	0.52 ± 0.05	0.61 ± 0.06

Source: Thevathasan and Gordon (1997).

Soil carbon and nitrogen

The effects of litterfall distribution of poplar (hybrid clone DN 177; *Populus deltoides* x *Populus nigra* 177) on soil nitrogen (N) transformations and soil organic carbon (SOC) was studied from 1993 to 1995 and then again in 2002. The associated crop was barley; the poplar trees were 6 years old in 1993 (15 yr in 2002). In field experiment 1 (Figure 1a), poplar leaves were removed after leaf senescence in 1993 and 1994; in experiment 2 leaves were not removed (Figure 1b). Poplar litterfall distribution on the ground showed a distinct pattern, with almost 80% of the leaves falling within 2.5 m from the tree-row (Table 2).

Differing rates of poplar-leaf biomass input across the alleyways created distinct regions with respect to the accumulation of soil nitrogen and carbon. Based on the abundance of these resource pools, the intercropped alley can be divided into three zones: the area close to the poplar tree row (0 m to 2.5 m on either side of the tree row), the middle of the crop alley (2.5 m to 8.0 m from the tree-row), and the area furthest away from the tree-row (8.0 m to 15.0 m)[1]. Observed mean soil nitrate production in the aforementioned zones during June to August 1993 was 73.1, 41.0 and 34.0 $\mu g\ 100\ g^{-1}$ dry soil day^{-1} respectively (Figure 1a). The higher nitrate production rates in 1993 were due to the presence of 1992 fall-shed poplar leaves. In 1995, as a result of the removal of poplar leaves from the field for two consecutive years (1993 and 1994), nitrate production rates decreased to 17.6, −2.8 and −1.7 $\mu g\ 100\ g^{-1}$ dry soil day^{-1} in the same zones, respectively (Figure 1a). In experiment 2 (June to August 1995, leaves not removed), however, mean nitrate production in the same zones was 109.4, 15.4 and 5.7 $\mu g\ 100\ g^{-1}$ dry soil day^{-1}, respectively (Figure 1b).

There is much information available on tropical hedgerow intercropping systems where nutrient release occurs through the mineralization of recently added hedgerow prunings and root decay (Rao et al. 1998). In these systems, considerable labour input is required to bring about this desirable complementary interaction. In the temperate region, and especially with tree species that have the potential to produce high leaf biomass (e.g., hybrid poplar), no special effort is made for soil incorporation of leaf biomass. Nitrogen release from annual poplar litterfall at the University of Guelph Agroforestry Research Station has been estimated to be equivalent to 7 kg N $ha^{-1}\ yr^{-1}$. This implies that rates of inorganic N fertilizer rates in the poplar-based crop alleys could be reduced by this amount, which in turn can directly reduce input costs.

Soil organic carbon did not change significantly ($P > 0.05$) in the three indicated zones from 1993 to 1995 with recorded SOC zone means of 3.25%, 2.32% and 2.50% respectively (Figure 2). This was to be expected as only 15% to 35% of added organic residue is actually incorporated into the permanent organic pool (humus) (Brady and Weil 2002).

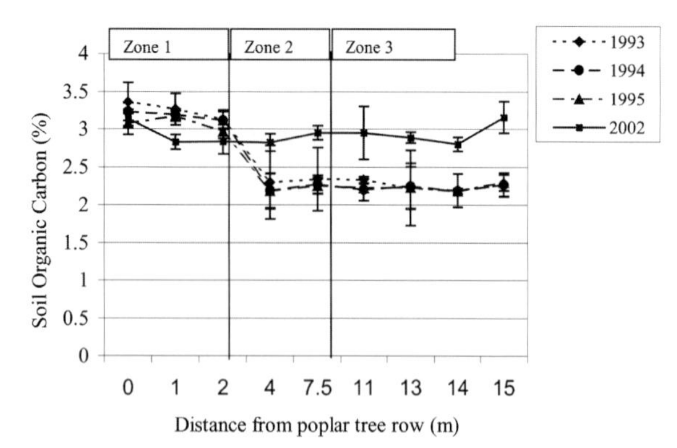

Figure 2. Soil organic carbon content at various distances from the poplar tree-row in 1993, 1994, 1995 and 2002, when the trees were 6, 7, 8 and 15 years old, in the tree-intercropping experiment in southern Ontario, Canada. Error bars that overlap indicate that associated values are not significant at $P < 0.05$. Source: Thevathasan and Gordon (1997).

The high rate of poplar leaf biomass addition (1 Mg C ha^{-1} yr^{-1}) over the total period of 8 years resulted, however, in an increase of SOC of approximately 1% close to the tree-row, and this effect extended into the alley for up to approximately 4 m. This is about a 35% relative increase (percentage difference between 3.25 and the mean of 2.32 and 2.50) in SOC close to the tree-rows over the given period of time, and reflects leaf biomass inputs in the early 1990s, when trees were small and the major portion of litterfall was distributed close to the tree row (2 to 3 m) (Table 2). By 2002, poplar trees were 14 m tall and leaf biomass was evenly distributed across the crop alley up to distances of 15 m. This resulted in a slow but inexorable increase in soil C in the middle of the crop alley, as illustrated in Figure 2.

It appears that the addition of poplar leaves significantly ($P < 0.05$) affected nitrate production rates, especially in regions close to the tree row and in the middle of the crop alley. It also appears that the major portion of nitrate was released from the labile organic pool (recently added poplar leaf biomass) rather than from the recalcitrant organic pool, since the removal of poplar leaves from the field did not significantly change the soil organic carbon pool over the three-year period (Thevathasan and Gordon 1997).

Zhang (1999)[2] has shown that trees can significantly influence nutrient additions to associated crops through throughfall (rainwater falling through tree canopies) and stemflow (rainwater falling down the branches and stems). Hybrid poplar and silver maple contributed 10.99 and 15.22 kg N ha^{-1} yr^{-1} respectively, through these combined pathways. The addition of N and other nutrients through these pathways could potentially reduce nutrient deficiencies that might arise because of tree–crop competition close to tree rows.

The build-up of SOC under tree canopies in tropical hedgerow intercropping systems and the positive influence of this on many soil physical, chemical and biological properties has been well reported (Nair 1993; Young 1997; Rao et al. 1998). Even though indirect results obtained at Guelph (e.g., increases in SOC, nitrate mineralization from fall-shed poplar leaves, and enhanced earthworm populations) indicate that the above-mentioned soil parameters are likely to have been positively influenced, no empirical data for temperate systems currently exist. Given the results obtained from tropical systems (Rao et al. 1998) and the complementary interactions that have been observed in temperate tree/crop intercropping systems (Table 1), we suggest that the adoption of tree/crop intercropping systems may ameliorate components of the degraded temperate land-base that have come about as a result of poor farming practices.

Light

The effects of shading by poplar and silver maple on the productivity of intercropped maize (C4 plant) and soybean (C3 plant) were studied during the 1997 and 1998 growing seasons, when the trees were 10 and 11 years old (Simpson 1999[3]). Generally, tree competition reduced the growth of individual plants significantly up to 2 m and often to a moderate degree up to 6 m from the tree rows in comparison with those in the control treatment (Table 3). Daily rates of C assimilation were generally lower near the trees where competition for photosyntheticallyactive radiation (PAR) was the greatest, resulting in lower crop yields. Growth characteristics (height, leaf area, weight) of individual plants were significantly correlated with available PAR ($r = 0.87$ for soybeans; $r = 0.87$ to 0.95 for maize) and net assimilation ($r = 0.73$ to 0.83 for soybeans and $r = 0.92$ to 0.96 for maize), but not significantly correlated ($r = 0.02$ to 0.16) with midday water potential. It was concluded that competition for light, and not water, within 6 m of the tree rows was the main factor that detrimentally affected maize and soybean yields (Simpson 1999[3]). Pooled observations from all the treatments of this study revealed that all ten tree species detrimentally affected the yields of C4 plants more than those of C3 plants. Based on this observation, it seems prudent to

Table 3. Growth of soybeans and maize in sole cropping and intercropping with 10-year-old poplar or maple at Guelph, Ontario, Canada.

Crop	Parameter	Control (sole cropping)		Intercropping with poplar		Intercropping with maple	
		2 m	6 m	2 m	6 m	2 m	6 m
Soybean	Height (cm)	75.6a	82.7a	45.6b	67.5a	44.4b	69.4a
	Leaf area (cm^2 $plant^{-1}$)	796.2b	1070.1a	317.4b	630.8a	247.1b	766.3a
Maize	Height (cm)	196.0b	209.3a	103.8b	177.2a	126.2b	198.8a
	Leaf area (cm^2 $plant^{-1}$)	5386.9a	5389.3a	3769.2b	5026.5a	3758.5b	5302.0a

By parameter and similar treatment, values in each row followed by the same letter are not significantly different (Tukey's HSD, $P > 0.05$). Source: Simpson (1993)[3].

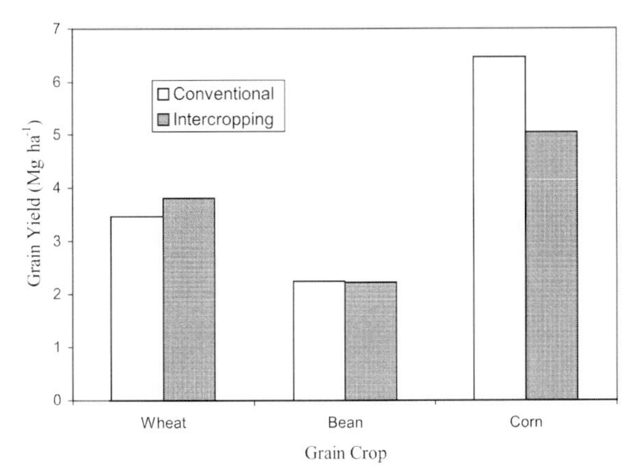

Figure 3. Overall yields of intercropped grain crops in 1999 across 10 tree species when the trees were 12 years old compared with yields obtained from an adjacent conventional sole crop field at Guelph, Ontario, Canada.

recommend alternative management strategies such as pruning and crown thinning of tree canopies to reduce shading impacts from tree rows and to ensure acceptable economic yields in the zone closest to the tree row.

We have documented (on average) a 20% to 25% yield reduction per unit land area for maize when compared to the yield recorded in conventional system from an adjacent field (control). Yields of C3 crops such as soybean and wheat were higher in intercropping or generally on par with yields in the conventional system, especially under dry weather conditions. In the former scenario, effects of microclimate modification by trees in terms of lowered crop evapotranspiration rates and reduced soil temperatures might have helped to enhance yields in the intercropping system (William and Gordon 1995; Zhang 1999[2]).

Results reported on competition for light, or shade effect, in tropical hedgerow systems have been varied. The factors governing the shade effect on crops grown in the alleys are reported to be climate, management, soils, tree, and crop species (Nair 1993; Ong and Huxley 1996). The hedgerows generally had no shading effect on crop yields when the alley spacing exceeded 4 m in humid tropics (Rao et al. 1998). Even at wider spacings, however, tree shade is reported to have negatively impacted C_4 crops such as maize more than C_3 crops such as beans (*Phaseolus* sp.) (Rao et al. 1998). A similar trend on the effect of tree shade on crop yields has been observed in the intercropping trials at the University of Guelph irrespective of row spacing (12.5 m or 15 m) (Figure 3).

Water

The site receives a total of 833 mm of precipitation per year out of which 334 mm falls during the 105-day crop-growing season (Simpson 1999)[3]. During the first three to five years after planting of trees, soil water content under the associated wheat declined rapidly to a depth of 60 cm until harvest in early August. This period also corresponds to the critical growth phase of young trees when demand for water is high. At that stage, roots of the young trees exploit a relatively small and shallow soil zone. The early demand for water by the winter wheat crop detrimentally affected the growth and development of young trees. When maize was grown in association with the trees, a temporal change in water demand was observed: water demand for maize was low early in the season and hence young trees experienced little competition for water. This positively influenced the height growth of trees in association with maize (Williams and Gordon 1995). An extensive survey of the rooting habits of hybrid poplar trees (Figure 4), indicated, however, that 15 year-old trees competed less for surface water because of penetration of roots into deeper soil horizons (Gray 2000)[4].

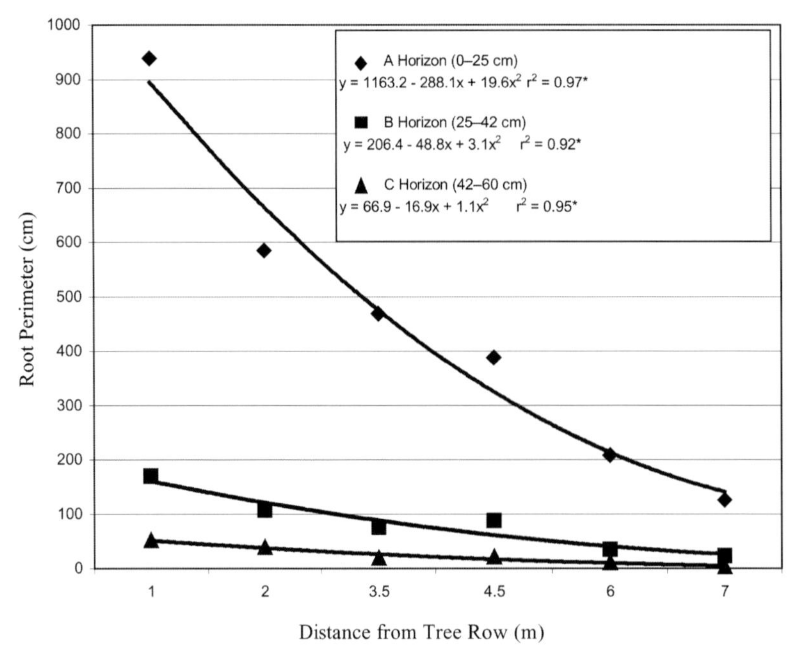

Figure 4. Mean root perimeter at various horizontal distances from a row of 12-year-old poplar trees at different soil horizons, Guelph, Ontario, Canada. Source: Gray (2000)[4].

Biological interactions

Soil fauna

Earthworm population dynamics in soil was also studied at the above research site in 1997 and 1998 (Price and Gordon 1999). Tree species played an important role in determining the spatial and temporal distribution of earthworms within the intercropping system (Table 4). Earthworm densities were greatest next to poplar and white ash tree-rows, possibly due to greater litter contributions. Although earthworm numbers decreased during the summer period, their populations in agroforestry system were still significantly greater than those from a comparable conventionally cropped field.

Earthworm activity in the soil profile has been proven to be beneficial for soil structure and stability (Lawton 1994; Edwards and Lofty 1977). A wide range of benefits is attributed to earthworms; these include decreased soil bulk density, increased decomposition of soil organic matter, and improved soil stability in cropped fields (Scheu 1995; Edwards and Bohlen 1997; Edwards and Lofty 1977). Strong correlations between higher soil organic matter levels within poplar (*Populus trichocarpa*) rows and earthworm activity have been established (Park et al. 1994). Enhanced decomposition of soil organic matter and nutrient release as influenced by earthworms have also been reported in tropical agroforestry systems (Tian 1992). The most important role played by earthworms is, however, in the comminution and mixing of organic matter within the soil. The ingestion and subsequent excretion of organic matter by earthworms throughout the soil profile, and on surface as casts, are important processes in the transformation of mineral nutrients into forms available for uptake by plant roots (Tian 1992). This process is positively correlated with soil organic matter levels and earthworms are believed to significantly contribute to soil chemical processes such as cation exchange capacity (CEC) (Brady and Weil 2001)

Birds

Williams et al. (1995) investigated the extent to which birds 'used' or visited an intercropped maize field, a conventional maize field and an old-field site. The old-field site had various tall grasses and weeds including goldenrods (*Solidago* spp.), asters (*Aster* spp.) and milkweed (*Asclepias* spp.). Only one species of bird nested in the maize field, but 10 species foraged in the intercropped plots compared to four species in the maize field and six in the old-field site (Table 5). The study revealed that intercropping provided opportunities for birds to nest and forage that were not

Table 4. Mean biomass and populations of earthworms in intercropping and sole maize measured in 1998 when the trees were 11 years old at Guelph, Ontario, Canada.

	Earthworm population (No. m^{-2})				Earthworm Biomass ($g\ m^{-2}$)			
	Poplar	Maple	Ash	Maize	Poplar	Maple	Ash	Maize
1997								
Spring	394 a(a)	257 a(b)	379 a(a)	11 a(c)	457 a(a)	440 a(a)	735 a(b)	6.07 a(c)
Summer	119 b(a)	42 b(b)	61 b(b)	4 a(c)	245 b(a)	89 b(b)	153 b(b)	4.54 a(c)
Fall	257 c(a)	196 c(a)	268 c(a)	30 b(b)	345 c(ab)	263 a(b)	437 c(a)	45.96 b(c)
1998								
Spring	90 b(a)	63 b(a)	46 b(a)	3 a(b)	181 b(a)	144 b(a)	161 b(a)	3.12 a(b)

Values followed by the same letter within a column are not significantly different (LSD, $P > 0.05$).
Values followed by the same letter across a row, within brackets, are not significantly different (LSD, $P > 0.05$).
Source: Price (1999)[7].

available in the monocropped maize field. The diversity of the breeding population in the intercropped field approached that found in the nearby old-field site, although some of the species were different. The intercropped field also provided foraging opportunities for other species whose diversity and numbers clearly demonstrated the value of the site to local and migrating bird populations. The direct impact of the increased presence of birds on alley-crop yields appeared to be neutral although there was no actual study conducted on this (cf., Figure 3). It can be speculated, however, that birds can reduce insect pest populations as they feed on them. Research in this area is lacking in the temperate region.

Insect pests

Studies on arthropod abundance and diversity in the study site showed that taxons such as the Opiliones, Dermaptera and Carabidae, which are associated with organic litter and areas that provide shelter during the day, were significantly higher in the intercropped system than in the monoculture system. Significantly higher numbers of parasitoids and detritivores were also recorded in the intercropped system compared to the monoculture system; the intercropped treatment also supported a significantly higher ratio of parasitoids to herbivores (Figure 5). We surmise that trees grown with crops such as maize may improve insect pest management options by providing habitat that will foster populations of natural enemies (Middleton 2001)[5]. More specific research on these aspects is required for many temperate regions.

Interactions affecting the environment

Carbon sequestration and greenhouse gases

The potential of agroforestry systems to sequester carbon is examined in a companion chapter in this volume (Montagnini and Nair 2004). Additionally, tree-based intercropping systems can reduce emissions of greenhouse gases (GHG) such as nitrogen oxides (NO_x), and therefore, may potentially have a significant impact on climate change mitigation. Thirteen-year-old poplar trees were sampled in this study (Table 6). During the 13-year duration of the study, the permanent tree component of the intercropping system (hybrid poplar clone DN-177) sequestered 14 Mg C ha^{-1}; C contribution to soil from leaf litter and fine root turnover was estimated at 25 Mg ha^{-1}; thus the total C sequestration was approximately 39 Mg C ha^{-1} during the 13-year period. This amounts to the immobilization of 156 Mg ha^{-1} of CO_2 in 13 years. About 70% of the C added via leaf litter and fine roots will, however, be released back into the atmosphere through microbial decomposition processes. Hence, the net sequestration potential of trees alone is 1.65 Mg C ha^{-1} yr^{-1} or approximately 7 Mg ha^{-1} yr^{-1} of CO_2. It is also interesting to note that based on the known growth rate of this particular hybrid poplar clone, it is estimated that more than 43 Mg ha^{-1} of C will be sequestered by age 40.

Impact of tree-based intercropping systems on C sequestration in woody components and in soil

The above studies illustrate the extent to which trees that are introduced into agricultural fields can contribute to enhanced C sequestration in the system.

Table 5. Number of birds observed foraging in intercropped, maize-only, and old field plots at Guelph, Ontario, Canada.

Bird species	Intercrop	Maize field	Old field
Song Sparrow (*Melospiza melodia*)	0	0	12
Eastern Kingbird (*Tyrannus tyrannus*)	0	0	6
Tree Swallow (*Tachycineta biocolor*)	12	2	2
Red-winged Blackbird (*Arelaius phoeniceus*)	2	0	17
Savannah Sparrow (*Passerculus sandwichensis*)	20	0	15
Horned Lark (*Eromphila alpestris*)	2	2	0
Eastern Meadowlark (*Sturnella neglecta*)	0	0	2
Eastern Bluebird (*Sialia sialis*)	8	0	0
American Robin (*Turdus migratorius*)	2	1	0
Kildeer (*Charadrius vociferus*)	4	1	0
American Goldfinch (*Carduelis tristis*)	12	0	0
Indigo Bunting (*Passerina cyanea*)	3	0	0
European Starling (*Sturnus vulgaris*)	16	0	0
Total Species	10	4	6

Source: Williams et al. (1995).

Table 6. Carbon content of different components of a 13-year-old poplar tree in tree–crop-intercropping system at Guelph, Ontario, Canada[a].

Tree component	Carbon content per tree (kg)
Leaves	11.7 ± 3.5
Twigs	6.3 ± 2.3
Small branches	15.0 ± 6.2
Large branches	28.0 ± 17.3
Trunk	54.3 ± 33.5
Roots	19.7 ± 1.5
Total	135.0 ± 15.73

[a]Data are reported on per tree basis, as planting densities vary considerably in temperate intercropping scenarios (average tree density used at the University of Guelph Agroforestry Research Station, Ontario, Canada is 111 trees ha^{-1}).

The possible scenarios could be summarized as follows. 1) In a maize monocropped field, annual net C input to the soil is in the range of 400 to 600 kg ha^{-1} yr^{-1} as compared to annual net C inputs as high as 2400 kg ha^{-1} C yr^{-1} in a tree-based intercropping system. 2) Tree-based intercropping systems can potentially be adopted in agricultural land classes from 1 through 4. Therefore, the land base in Canada that could be brought under tree-based intercropping is estimated at more than 45.5 million hectares; if intercropped, this would have a significant effect on C sequestration and GHG emission reduction. 3) The tree component in intercropping occupies a part of the land area and proportionately reduces the area under annual crop and consequently the need for supplemental nitrogen for crop production. 4) The decrease in nitrogen moving out of the rooting zone will lead to reduced NO_x emissions as a result of denitrification in surface water resources. 5) If the tree species are deciduous, annual litterfall will cycle some nitrogen back to the soil reserve. While this nitrogen is localized to the area close to the tree, it does constitute a quantifiable contribution to the agricultural crop. Reduced application rates of inorganic N can result in reduced environmental losses. Apart from reduction in GHG emissions, C sequestration in agricultural fields can also be augmented through this type of land-use practice, since annual leaf litter input and fine root turnover can significantly influence long-term soil organic C dynamics.

The Canadian national target of a 20% reduction in GHG emissions could be achieved if at least 200 kg C ha^{-1} is sequestered annually over the next 10 to 15 years on the aforementioned 45.5 million hectares of cropped land in Canada. Thus, the adoption of tree-based intercropping systems in geographically suitable regions of Canada could not only diversify farm income, bring about changes in biodiversity, and enhance other environmental benefits, but could also contribute towards fulfilling the national requirements of the Kyoto Protocol (Thevathasan 1998).

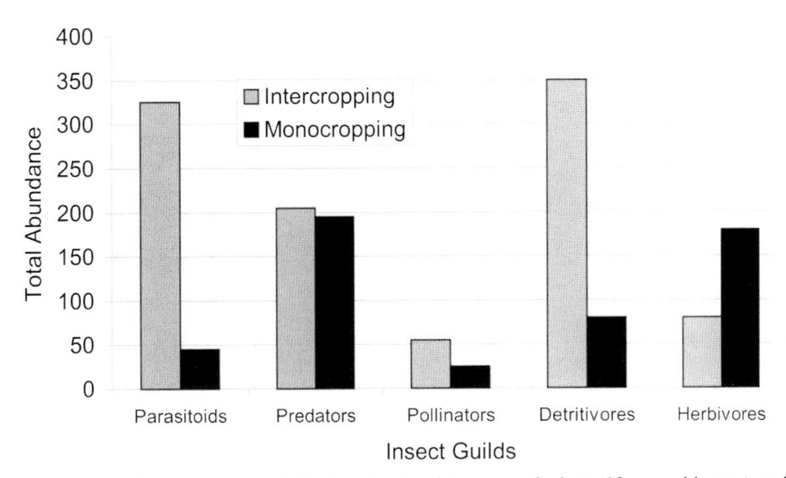

Figure 5. Total arthropod abundance in June samples (1999) in the tree-intercropped plots (12-year-old trees) and sole crop sites, Guelph, Ontario, Canada. Source: Middleton (2001)[5].

Impact of tree-based intercropping systems on reducing nitrous oxide (N_2O) emission

Tree-based intercropping systems alter the nitrogen balance compared to conventional agricultural systems in the following ways: (1) There is less N fertilizer addition as a result of reduction in the area under annual crop due to tree occupation (10%) (Table 7); (2) the decomposition of leaf litter (in both deciduous and coniferous tree species) in the litterfall zone can supply some nitrogen to the agricultural crop (Thevathasan 1998). However, it requires special agricultural management to vary the rate of nitrogen application within this zone of enhanced nitrogen cycling (e.g. Variable Rate Technology). About 1.25% of applied fertilizer N is lost as N_2O (Cole et al.1996) (Table 7). (3) In agricultural production areas with high rainfall, a portion of the applied nitrogen is lost through leaching below the rooting zone of the agricultural crops. In fact, the Intergovernmental Panel on Climate Change (IPCC) methodology estimates a 30% loss from this source.

Modeled data from our research site indicates that nitrate leaving the intercropping site can be potentially reduced by 50% compared with a conventional monocropped field (Thevathasan 1998). A portion of the nitrate leached below the agricultural rooting zone will also be denitrified and lost as N_2O. Oenema (1999), for example, estimates that 2.5% of the leached N is lost as N_2O. Based on the above assumption and from the modeled data it can be assumed that, as a result of reduced leaching, N_2O emissions in tree-based intercropping systems could be potentially reduced by 0.69 kg N_2O ha^{-1} relative to emissions from conventional agricultural fields (Table 7). In order to meet the terms of reference of the Kyoto Protocol, N_2O emissions from Ontario agricultural fields need to be reduced by approximately 2 kg N_2O ha^{-1} over the next 10 years (2008–2012 first reporting period). In this context, our research suggests that tree-based intercropping systems could potentially reduce N_2O emissions by 0.69 N_2O kg ha^{-1} yr^{-1}, 7 to 9 years after establishment of a fast-growing fibre tree species-based intercropping system in southern Ontario. As trees age and N-inputs in litterfall increase, N_2O emissions could be further reduced.

Conclusions and lessons learned

Among the various environment-friendly agricultural practices currently under consideration in southern Ontario and other parts of the temperate region, tree-based intercropping systems seem to be a viable option. The ameliorative effects of trees in terms of soil fertility, productivity and nutrient cycling can be exploited through tree-based intercropping systems on both marginal and prime agricultural lands. The success of intercropping depends mainly on the ability of the system components to maximize resource utilization while at the same time maintaining 'complementary' interactions between them.

The complexity of agroforestry systems often results in constant changes in spatial patterns, as tree growth proceeds in both horizontal and vertical dimensions. As the temporal dimensions of the system keep changing, the patterns of interaction are also likely to change. Therefore, the beneficial interactions ob-

Table 7. The potential for annual N_2O-N reduction, eight years after establishment of trees, based on a hypothetical N cycling budget developed for a fast-growing hybrid poplar-based intercropping system, Guelph, Ontario, Canada.

Causes of N_2O reduction	N fertilizer saved (kg ha^{-1})	N_2O emission reduction (N_2O-N kg ha^{-1})
10% less land area	8[a]	0.1[b]
N cycling in tree-based intercropping	7	0.09[c]
Reduction in N leaching	20	0.5[d]
Total N_2O reduction potential		0.69

(Source: Thevathasan and Gordon, personal communication, 2003).
[a]maize, bean, wheat rotation, average annual N fertilizer application = 80 kg ha^{-1}.
[b]$8 \times 0.0125 = 0.1$ (1.25% of the applied fertilizer N is lost as N_2O).
[c]$7 \times 0.0125 = 0.09$.
[d]$20 \times 0.025 = 0.5$ (2.5% of the leached N is lost as N_2O).

served in this study may also change over time as tree canopies become larger and/or when lateral tree roots further extend into the rooting zone of the cropping alley. This may result in many competitive interactions with respect to light, water and nutrients. The ameliorative effects of poplar may also get masked as a result of these possible competitive interactions. When competitive interactions are observed, however, alternative management strategies may be considered so that competitive interactions between components are delayed, reduced, or completely eliminated from the system. Agroforestry research in the tropics has shown that the proper selection of system components and adoption of proper management techniques are the two most important steps that should be considered while designing agroforestry systems to reduce or avoid competitive interactions among components (see, for example, Rao et al. 1998; Garcia-Barrios and Ong 2004).

Several alternative management strategies could be considered for poplar-based intercropping systems in southern Ontario to maintain the viability and sustainability of these systems. These strategies can also be applied to other parts of temperate regions. The effectiveness of any alternative management strategy requires, however, testing in the field on a long-term basis and to be scientifically proven before any recommendations are made. The following alternative management strategies are suggested as worthy of future research:

1. Planting of fast growing tree species such as poplar in concert with hardwood tree species in alternate rows. This will also reduce shading effects and delay crown closure. More knowledge (research) is also needed on the effects of shading on crop productivity. Closure of stomata during noon or early afternoon hours is an adaptive mechanism that plants often exhibit to overcome temporary water stress. When stomata are temporarily closed, photosynthesis does not proceed and hence shading during this time might have little effect on crop productivity. This area needs more research.
2. Pruning lateral roots of trees: Root competition can be significantly reduced if tree roots and crop roots utilize nutrients and water from different layers of the soil profile. This can be achieved by pruning tree lateral roots (if any) in the first 20 cm to 30 cm of the soil profile. Research on competition brought about by tree shading versus tree root competition is also required.
3. Continuous monitoring of root distribution of intercropped trees: This will yield a better understanding with respect to tree root pruning practices, although, as usual, there are technical research challenges associated with below-ground processes.
4. Evaluation of short-term as well as long-term costs and benefits: Various types of tree-intercropping strategies may help maintain the system economically profitable; more research is needed in this area as well.

In conclusion, on a biological level, the increased range of faunal activity is a clear indication of ecosystem 'health' within an intercropping system relative to that associated with conventional agricultural practices. From an ecological perspective, intercropping systems trap larger amounts of energy at different trophic levels, demonstrating higher energy utilization efficiency. In relation to carbon sequestration and greenhouse gas (e.g., N_2O) emission reductions, tree-based cropping systems have the potential to contribute greatly to climate change mitigation. The tangible benefits that are derived from the above described eco-biological processes, along with combined yields obtained from the trees and crops, place this land-use practice above conventional agricultural systems in terms of long term overall productivity. The economics of tree-based intercropping systems need to be examined in more detail, however. Investigation into policy measures, and/or tax incentives and cost-share programs should be initiated in order to increase adoption rates of the practice in southern Ontario in particular and other potential regions in general.

As we step into the next millennium, a change is warranted in current agricultural practices and land use options so that priority is given to the stewardship of soil and water resources. The results from our intercropping research in southern Ontario have shown much promise in this regard. The long-term viability and sustainability of tree-based intercropping systems in the temperate region needs to be examined, however, before recommendations on the use of these systems can be made to the farmers.

End Notes

1. The experiments were designed with the poplar tree row as the middle row; the two adjacent tree rows did not have any poplar trees, but did have small white ash and black walnut trees. These trees were less than 2 m high at age 6 and did not contribute any sizable quantity of leaf biomass (Thevathasan and Gordon 1997).
2. Zhang P. 1999. The impact of nutrient inputs from stemflow, throughfall, and litterfall in a tree-based temperate intercropping system, southern Ontario, Canada. M.Sc. Thesis, Dept. of Environmental Biology, University of Guelph, Guelph, Ontario, Canada, 102 pp.
3. Simpson J.A. 1999. Effects of shade on maize and soybean productivity in a tree based intercrop system. M.Sc. Thesis, Dept of Environmental Biollogy, University of Guelph, Guelph, Ontario, Canada, 106 pp.
4. Gray G.R.A. 2000. Root distribution of hybrid polar in a temperate agroforestry intercropping system. M.Sc. Thesis, Dept. of Environmental Biology, University of Guelph, Guelph, Ontario, Canada, 116 pp.
5. Middleton H. 2001. Agroforestry and its effects on ecological guilds and arthropod diversity. M.Sc.F. Thesis. Faculty of Forestry, University of Toronto. Toronto, Ontario, Canada, 108 pp.
6. Christrup J. 1993. Potentials of edible tree nuts in Ontario. M.Sc.F. Thesis, Faculty of Forestry, University of Toronto, Toronto, Ontario, Canada, 120 pp.
7. Price G.W. 1999. Spatial and temporal distribution of earthworms in a temperate intercropping system in southern Ontario. M.Sc. Thesis, Dept. of Environmental Biology, University of Guelph. Guelph, Ontario, Canada, 131 pp.
8. McLean H.D.J. 1990. The effect of maize row width and orientation on the growth of interplanted hardwood seedlings. M.Sc. Thesis, Dept. of Land Resource Science, University of Guelph, Guelph, Ontario, Canada, 90 pp.
9. Ball D.H. 1991. Agroforestry in southern Ontario: A potential diversification strategy for tobacco farmers. M.Sc. Thesis, Dept. of Agricultural Economics, University of Guelph, Guelph, Ontario, Canada, 197 pp.
10. Kotey E. 1996. Effects of tree and crop residue mulches and herbicides on weed populations in a temperate agroforestry system. M.Sc. Thesis, Dept. of Environmental Biology, University of Guelph, Guelph, Ontario, Canada, 130 pp.

Acknowledgements

The authors wish to thank Peter Williams, Jamie Simpson, Gordon Price, Ping Zhang and Refaat Abohassan for granting permission to utilize results stemming from their respective research work. We also acknowledge Gianni Picchi and Ms. Amanda Ross for their assistance in creating the Figures, Tables and Reference List. Financial assistance received from the Ontario Ministry of Food and Agriculture (OMAF) is also gratefully acknowledged.

References

Bezkorowajnyj P.G., Gordon A.M. and McBride R.A. 1993. The effect of cattle foot traffic on soil compaction in a silvo-pastoral system. Agroforest Syst 21: 1–10.

Brady N.C. and Weil R.R. 2002. The Nature and Properties of Soils, 13th edition. Prentice Hall, 960 pp.

Cole C.V., Cerri C., Minami K., Mosier A., Rosengerg N. and Sauerbeck D. 1996. Agricultural options for mitigation of greenhouse gas emissions. pp. 745–771. In: Watson R.T., Zinyowera M.C. and Moss R.H. (eds), Climate Change 1995. Impacts, Adaptations and Mitigation of Climate Change: Scientific Technical Analysis. Published for the Intergovernmental Panel on Climate Change. Cambridge University Press, Cambridge, UK.

Dixon R.K., Winjum J.K., Andrasko K.J., Lee J.J. and Schroeder P.E. 1994. Integrated land-use systems: Assessment of promising agroforest and alternative land-use practices to enhance carbon conservation and sequestration. Climate Change 27: 71–92.

Dyack B., Rollins K. and Gordon A.M. 1999. A model to calculate *ex ante* the threshold value of interaction effects necessary for proposed intercropping projects to be feasible to the landowner and desirable to society. Agroforest Syst 44: 197–214.

Edwards C.A. and Bohlen P.J. 1997. Biology and Ecology of Earthworms. Chapman and Hall, London, 426 pp.

Edwards C.A. and Lofty J.R. 1977. Biology of Earthworms. Chapman and Hall, London, 333 p.p.

García-Barrios L and Ong C.K. 2004. Ecological interactions, management lessons and design tools in tropical agroforestry systems. (This volume).

Garrett H.E., Rietveld W.J., Fisher R.F., Kral D.M. and Viney M.K. (eds) 2000. North American Agroforestry: An Integrated Science and Practices. American Society of Agronomy, Madison, WI, USA, 402 pp.

Gordon A.M. and Williams P.A. 1991. Intercropping of valuable hardwood tree species and agricultural crops in southern Ontario. Forest Chron 67: 200–208.

Gordon A.M. and Newman S.M. 1997. Temperate Agroforestry Systems. CAB International, Wallingford, UK, 269 pp.

Gordon A.M., Newman S.M. and Williams P.A. 1997. Temperate agroforestry: An overview. pp. 1–8. In: Gordon A.M. and Newman S.M. (eds), Temperate Agroforestry Systems. CAB International, Wallingford, UK.

IPCC: http://www.ipcc.ch/. Accessed October 17, 2003.

Jose S., Gillespie A.R. and Pallardy S.G. 2004. Interspecific interactions in temperate agroforestry. (This volume).

Kenney W.A. 1987. A method for estimating windbreak porosity using digitized photographic silhouettes. Agr Forestry Meteorol 39: 91–94.

Lawton J.H. 1994. What do species do in ecosystems? Oikos 71: 367–374.

Loeffler A.E., Gordon A.M. and Gillespie T.J. 1992. Optical porosity and windspeed reduction by coniferous windbreaks in southern Ontario. Agroforest Syst 17: 119–133.

Matthews S., Pease S.M., Gordon A.M. and Williams P.A. 1993. Landowner perceptions and adoption of agroforestry in southern Ontario, Canada. Agroforest Syst 21: 159–168.

McKeague J.A. 1975. Canadian Inventory: How much land do we have? Agrologist, Autumn:10–12.

Montagnini F. and Nair P. K. R. 2004. Carbon sequestration: An under-exploited environmental benefit of agroforestry systems. Agroforest Syst (this volume).

Nair P.K.R. 1993. An Introduction to Agroforestry. Kluwer Academic Publishers, The Netherlands, 499 pp.

Ntayombya P. 1993. Effects of *Robinia pseudoacacia* on productivity and nitrogen nutrition of intercropped *Hordeum vulgare* in a agrosilvicultural system: Enhancing agroforestry's role in developing low input sustainable farming systems. Ph.D. Thesis, Department of Environmental Biology, University of Guelph, Guelph, Ontario, Canada, 411 pp.

Ntayombya P. and Gordon A.M. 1995. Effects of black locust on productivity and nitrogen nutrition of intercropped barley. Agroforest Syst 29: 239–254.

Oenema O. 1999. Strategies for decreasing nitrous oxide emissions from agricultural sources. pp. 175–191. In: Desjardins R.L., Keng J.C. and Haugen-Kozyra K. (eds), Reducing Nitrous Oxide Emissions from Agroecosystems. International N_2O workshop proceedings, March 3–5, 1999. Agriculture and Agri-Food Canada, Alberta Agriculture, Canada.

Oelbermann M. and Gordon A.M. 2000. Quantity and quality of autumnal litterfall into a rehabilitated agricultural stream. J Environ Qual 29: 603–611.

Oelbermann M. and Gordon A. M. 2001. Retention of leaf litter in streams from riparian plantings in southern Ontario, Canada. Agroforest Syst 53: 1–9.

O'Neill G.J. and Gordon A.M. 1994. The nitrogen filtering capability of carolina poplar in an artificial riparian zone. J Environ Qual 23: 1218–1223.

Ong C.K. and Huxley P. 1996. Tree-crop interactions: A physiological approach. CAB International, Wallingford, UK, 386 pp.

Park J., Newman S.M. and Cousins S.H. 1994. The effects of poplar (*P. trichocarpa* x *P. deltoids*) on soil biological properties in a silvoarable system. Agroforest Syst 25: 111–118.

Price G.W. and Gordon A.M. 1999. Spatial and temporal distribution of earthworms in a temperate intercropping system in southern Ontario, Canada. Agroforest Syst 44: 141–149.

Rao M.R., Nair P.K.R. and Ong C.K. 1998. Biophysical interactions in tropical agroforestry systems. Agroforest Syst 38: 3–50.

Scheu S. 1995. Mixing of litter and soil by earthworms: effects on carbon and nitrogen dynamics – a laboratory experiment. Acta Zoologica Fennica. 196: 33–40.

Thevathasan N.V. 1998. Nitrogen dynamics and other interactions in a tree–cereal intercropping systems in southern Ontario. Ph.D. Thesis. University of Guelph, Guelph, Ontario, Canada. 230 pp.

Thevathasan N.V. and Gordon A.M. 1995. Moisture and fertility interactions in a potted poplar–barley intercropping. Agroforest Syst 29: 275–283.

Thevathasan N.V. and Gordon A.M. 1997. Poplar leaf biomass distribution and nitrogen dynamics in a poplar-barley intercropped system in southern Ontario, Canada. Agroforest Syst 37: 79–90.

Thevathasan N.V., Gordon A.M., Simpson J.A., Reynolds P.E., Price G. and Zhang P. 2004. Biophysical and ecological interactions in a temperate tree-based intercropping system. J Crop Improvement 11: (in press).

Tian G. 1992. Biological Effects of Plant Residues with Contrasting Chemical Compositions on Plant and Soil Under Humid Tropical Conditions. Kluwer Academic Publishers, Dordrecht, The Netherlands, 114 pp.

Williams P.A. and Gordon A.M. 1992. The potential of intercropping as an alternative land use system in temperate North America. Agroforest Syst 19: 253–263.

Williams P.A. and Gordon A.M. 1994. Agroforestry applications in forestry. Forest Chron 70: 143–145.

Williams, P.A. and A.M. Gordon. 1995. Microclimate and soil moisture effects of three intercrops on the tree rows of a newly-planted intercropped plantation. Agroforest Syst 29: 285–302.

Williams P.A., Koblents H. and Gordon A.M. 1995. Bird use of an intercropped maize and old fields in southern Ontario. pp. 158–162. In: Ehrenreich J.H. and Ehrenreich D.L. (eds), Proceedings of the Fourth North American Agroforestry Conference 1995, Boise, Idaho, United States.

Williams P.A., Gordon A.M., Garrett H.E., and Buck L. 1997. Agroforestry in North America and its role in farming systems. pp. 9–84. In: Gordon A.M. and Newman S.M. (eds), Temperate Agroforestry Systems. CAB International Press, Wallingford, UK.

Young A. 1997. Agroforestry for Soil Management. CAB International, Wallingford, UK, 320 pp.

Agroforestry Systems **61:** 269–279, 2004.

Agroforestry as an approach to minimizing nutrient loss from heavily fertilized soils: The Florida experience

V.D. Nair* and D.A. Graetz
Soil and Water Science Department, University of Florida, PO Box 110510, Gainesville, FL 32611-0510
Author for correspondence: e-mail: vdna@ifas.ufl.edu

Key words: Nitrogen, Nutrient leaching, Phosphorus, P-Index, Runoff, Water quality

Abstract

Nutrient buildup in the soil caused by increased animal manure and fertilizer use in agricultural and forestry practices may increase the potential for their loss from the soil, leading to groundwater contamination and non-point source pollution. Studies in the tropics have suggested that agroforestry practices can reduce such nutrient (especially nitrogen) losses because of enhanced nutrient uptake by tree and crop roots from varying soil depths, compared to more localized and shallow rooting depths of sole crop stands. In temperate systems, such benefits have been well documented for riparian forest buffer practices. Currently, other temperate agroforestry practices are also being considered for their potential to reduce runoff and leaching of chemicals and thereby improve environmental quality within the agricultural landscape. In this regard, the 'Florida P-Index,' which considers both phosphorus transport characteristics and management practices, may be a useful tool in the evaluation of nutrient management practices and environmental benefits of agroforestry. Preliminary results from an alleycropping site and a silvopastoral site on two different soil types in Florida suggest that both of these agroforestry practices will likely reduce nutrient loss compared to conventional agricultural practices. The primary aspects of P-Index include consideration of transport factors such as soil erosion, soil runoff class, leaching potential, and distance from a water body along with management factors such as soil test P, P application method, and source and rate of P application. P-Index evaluation of these studies indicates that both agroforestry sites can be on a nitrogen-based nutrient management program. The relevance of some management practices such as application of manure vs. inorganic fertilizer is also discussed in light of the P-Index and the two agroforestry practices.

Introduction

Although the agronomic benefits of fertilizer use are well known, the presence of increasing amounts of nutrients in the soil may enhance the potential for their loss, leading to groundwater contamination and non-point source pollution. With increasing levels of fertilizer use on a global scale, the gravity of the problem will inevitably increase. Reviewing the fertilizer impacts on the environment based on data from Bockman et al. (1990), Ayoub (1999) showed that in 1970, at a global level, 48% of the nutrients used by crops were derived from the soil, 13% from manure, and 39% from inorganic fertilizers. By 1990, the percentages had changed to 30% from soil, 10% from manure and 60% from inorganic fertilizer. The projection for 2020 is 21% from the soil, 9% from an organic source, and 70% from an inorganic fertilizer source. Considering that only about half of the applied fertilizer is taken up by the crop in a given season (Bockman et al. 1990), the negative impact of fertilizer use on environmental quality is likely to increase with time on a global scale. Indeed it has reached epidemic proportions in the industrialized temperate zone where fertilizer use is very high (http://www.fertilizer.org/ifa/statistics.asp). For example, fertilizer consumption in the United States has increased tremendously in recent years, from 25 Tg (million tons) in 1960 to a peak of 54 Tg in 1981. The years after 1981 showed a slight reduction in fertilizer use; however, from the mid-1990s onward, consumption has risen steadily and recently approached near-peak levels of 53 Tg in 2002 (Terry and Kirby 2003).

In addition to fertilizers, organic materials such as animal manures and sewage biosolids are being land-applied in increasing amounts (Sharpley et al. 2001). Although the manure so applied will most likely add organic matter to the soil, it might also increase phosphorus (P), potassium (K), calcium (Ca) and magnesium (Mg) in topsoil (Edmeades 2003). Based on comparison of long-term (20–120 yr) effects of organic and inorganic fertilizer applications, Edmeades (2003) concluded that organic fertilizer sources did not necessarily enhance soil quality in all cases. Water quality is also likely to be impaired by long-term application of manure because of the addition of material with a high chemical oxygen demand (COD, defined as the amount of oxygen required to degrade the organic compounds of waste water). The U.S. Environmental Protection Agency (EPA) has shown that agricultural nonpoint source pollution is a significant cause of stream and lake contamination that prevents the attainment of the water quality goals in the Clean Water Act (USEPA 2002).

Studies conducted on tropical agroforestry systems have indicated that tree and crop roots occupying different soil depths can result in enhanced levels of nutrient (specifically nitrogen, N) uptake and reduced losses from soils in agroforestry systems, compared to sole crop stands with more localized and shallow rooting depths (Buresh and Tian 1997; Nair et al. 1999). This process could be exploited to address the problem of environmental impact of high rates of fertilizer input discussed above. It is hypothesized that agroforestry systems may reduce the potential for nutrient loading of both surface and groundwaters in highly fertilized soils.

While nonavailability of P to plants is a serious problem in heavily leached tropical soils (Brady and Weil 2002), the problem of P loss from soil could be a major problem in the heavily fertilized agricultural and forestry enterprises in the industrialized world. This could be particularly serious in coarse-textured, poorly drained soils with either surface or tile drainage in which soil drainage water ultimately enters surface waters. Soils of this nature are prevalent in the southeastern and Mid-Atlantic States of the United States. Problems of P leaching in sandy soils have also been documented in the Netherlands and in the Po region of Italy (Breeuwsma and Silva 1992). Sims et al. (1998) pointed out that intensive high-input agriculture, extremely humid climate with frequent, heavy rainfall events, and wide use of irrigation and drainage would facilitate P leaching in deep sandy soils.

Alleycropping and silvopasture are two major agroforestry practices in the southeastern United States (Workman et al. 2003). Among the environmental attributes of temperate alleycropping are reduced surface runoff and erosion, improved utilization of nutrients and reduced wind erosion (Garrett et al. 2000; USDA-NRCS 1997). Silvopastoral practices that integrate tree, forage and livestock production into a farm management program designed to simultaneously provide woody-plant commodities are the most prevalent form of agroforestry found in the United States and Canada with the largest blocks of grazed forests occurring in the southern and southeastern United States (Clason and Sharrow 2000). While the effectiveness of some agroforestry practices such as riparian buffers in reducing nonpoint source pollution and thereby improving water quality is well documented (Schultz et al. 2000), little information is available on the environmental implications associated with alleycropping and silvopastoral practices, particularly in regard to their impact on surface and groundwater quality. In this scenario, this chapter discusses the potential role of alleycropping and silvopasture in addressing the problem of water quality in heavily fertilized/manured sandy soils. The limited information that is available from our ongoing work on P and N losses from these agroforestry practices on different soil types will be used as case studies in the discussion.

Nutrient management planning

The U.S. Natural Resources Conservation Service (NRCS) has the responsibility to provide planning assistance to land-use decision-makers that will protect the five major resource concerns of Soil, Water, Air, Plants, and Animals (SWAPA). In the NRCS Resource Management System (RMS), nutrient management, particularly for animal wastes, is a key part of the planning process. Application of animal wastes at rates to supply adequate N often results in the excessive application of P due to the high ratio of P:N in manure relative to plant needs (Robinson and Sharpley 1996; Heathwaite et al. 2000). Therefore, in areas where P loss potential is high, it may be necessary to apply manure nutrients based on crop requirements for P instead of N. In carrying out this responsibility, NRCS guidance on nutrient management provides three ways to determine which nutrient (N or P) will be used to develop the nutrient management plan for a given farm. These are: i) rates of N or P as recommended by state land-grant university soil test programs, ii) phosphorus application rates based on the 'threshold level' of the soil P when that level has been determined; or iii) use of a Phosphorus Index (P-Index).

Nutrient management plan based on recommended N or P rates

Nutrient management practices should be applied to all agricultural lands where plant nutrients and soil amendments are applied (NRCS 1999), and would therefore include all agroforestry systems. The plans for nutrient management 'shall specify the form, source, timing and method of application of nutrients on each field to achieve realistic production goals, while minimizing nitrogen and/or phosphorus movement to surface and/or ground waters' (NRCS 1999). In Florida, both N and P application rates are expected to be less than or to match any recommended rates of application by the Institute of Food and Agricultural Sciences, University of Florida (UF-IFAS) (Kidder et al. 2002).

Certain agricultural production systems are known for the application of large amounts of nutrients to the soil. Among these are vegetable producers (Hochmuth 2000) that generally produce a high-cash-value crop and thus tend to use significant amounts of fertilizer to ensure high yields. Animal production systems, which often produce large amounts of manure with relatively small land areas to 'dispose' of the manure, also belong to this category. In both these systems, large amounts of P have been shown to accumulate in the soil (Edmeades 2003).

Although the agronomic benefits of P fertilization are well known, it is becoming apparent that the buildup of nutrients in the soil often increases the potential for nutrient loss from the soil (Edmeades 2003). Animal manures and fertilizers, for example, pose a serious threat to surface and groundwater quality in the Suwannee River and the Lake Okeechobee basins of Florida. The Lower Suwannee River Basin of Florida (LSRB) was designated by the Florida Department of Environmental Protection as a 'Group 1' basin (Florida Department of Environmental Protection 2001), primarily because of increased nitrate concentrations in surface and groundwater in the Upper Floridan Aquifer. Group 1 basins are those watersheds designated to receive priority attention for remediation of water quality problems. The increased nitrate concentrations have been attributed to dairy wastes (Andrews 1994), poultry operations (Hatzell 1995) and fertilizer applications (Katz et al. 1999). Woodard et al. (2003) have shown that a dairy sprayfield in the LSRB receiving loading rates of dairy effluent ≥ 500 kg N ha^{-1} yr^{-1} and with a crop rotation system, corn (*Zea mays*)–perennial peanut (*Arachis glabrata* Benth.)–rye (*Secale cereale*), has the potential to negatively impact groundwater quality. Elevated NO_3-N concentrations have been measured in water from both domestic and monitoring wells in the Suwannee River Basin (Katz et al. 1999), exceeding the maximum contaminant level of 10 mg L^{-1} set by the U.S. Environmental Protection Agency. Nitrate-N concentrations above this limit can cause methemoglobinemia or 'blue baby' syndrome in infants and elderly adults (Mueller and Helsel 1996).

Di and Cameron (2002) attributed variability in nitrate leaching to soil, climate, and management factors. Of the various land-use systems they reviewed, the potential for nitrate leaching followed the order: forest $<$ cut grass $<$ grazed pastures $<$ arable cropping $<$ market gardens. The deeper and more extensive tree roots would invariably be able to remove more nitrogen from the soil compared to crops with shallower root systems. Thus, nitrogen-leaching rates from soils under agroforestry systems where trees are a major component can be hypothesized to be lower than those from other land-use systems.

Dairy farming and beef ranching in the Lake Okeechobee Basin in south Florida are major contributors of P to the lake (Fonyo and Flaig 1995). Mehlich 1-P concentrations (soil test P in Florida) in the soil surface horizons under various land uses varied with the intensity of operations: intensive area of dairy operations (430 mg kg^{-1}) $>$ pasture (26 mg kg^{-1}) $>$ beef pasture (12 mg kg^{-1}) $\geq$ forage (10 mg kg^{-1}) $>$ unimpacted or 'native' areas (2 mg kg^{-1}) (Graetz et al. 1999). Except for the highly manure-impacted intensive area soils, the Mehlich 1-P concentrations are lower than the IFAS agronomic recommended high rate of 30 mg P kg^{-1} (Kidder et al. 2002). However, low Mehlich 1-P concentrations in the surface horizon of these Spodosols is not an indication that more P can be added to the soil without detrimental effects on the quality of water leaving the farm. Evidence is available that P moves through the sandy soil profile of the Spodosols of this region and that the P could eventually reappear in surface and groundwaters (Nair and Graetz 2002).

Loss of P from agricultural soils also can have serious implications for water quality. In Florida, some areas such as the Everglades, Green Swamp, and the Okeechobee Basin, are designated as P limited by legislation and these areas are to have P-based nutrient budgets, irrespective of the nutrient source or the type of soil, http://efotg.nrcs.usda.gov/popmenu3FS.aspx?Fips=12001&MenuName= menuFL.zip (select Section IV, then tools, and finally, the Florida Phosphorus Index – verified December 3, 2003).

Phosphorus application rates based on the 'Threshold Level' of soil P

In temperate agroforestry systems such as silvopasture and alleycropping, both animal manure and inorganic fertilizers are often used irrespective of the location of such practices. Nutrient movement and losses depend on their source (organic vs. inorganic), as well as on the soil type. Although P loss potential from such systems can be high because of high rates of application of manure with high P: N ratio, the loss of P would be dependent on the P retention capacity of the soil. A soil with a higher P retention capacity would likely lose less P to the environment compared to a soil with a lower P retention capacity unless the P were surface applied under conditions with excessive amounts of surface runoff. This is the principle on which the concept of the degree of P saturation (DPS) was evolved, and the threshold DPS evaluated (Nair et al. 2004).

The concept of DPS was developed following studies in the Netherlands that showed that P concentrations in the soil solution can approach a critical concentration well before the soil is completely saturated with P (Breeuwsma and Silva 1992, Graetz and Nair 1999). DPS relates an extractable P concentration to the soil P sorption capacity (PSC). The extractable P (P_{ox}) is determined by extraction with 0.2 M ammonium oxalate buffered to pH 3.0. Phosphorus sorption capacity is determined by standard P adsorption isotherms. The PSC in the Dutch test is estimated by oxalate-extractable Al (Al_{ox}) and Fe (Fe_{ox}) because of the relatively long period of time involved in the determination of adsorption isotherms. The DPS is used as an indicator of soil P available to be released from the soil. It is now well established that the extractable P content of soils influences the amount of P in both runoff and subsurface drainage

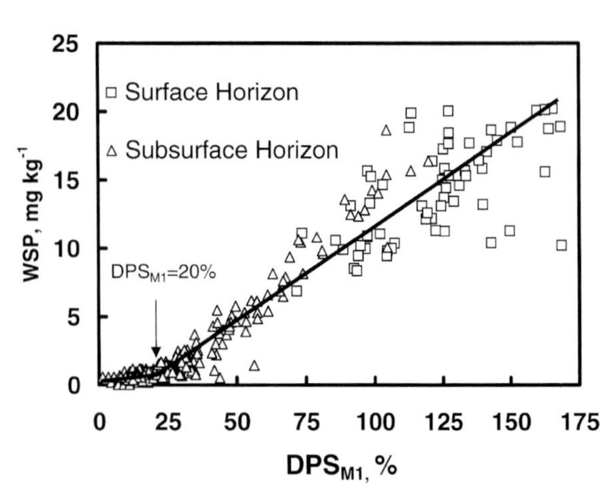

Figure 1. Change in water-soluble P (WSP) concentrations as a function of the degree of P saturation, DPS_{M1} (Source: Nair et al., 2004). The soils in this study were from sprayfields located within the Suwannee River basin of Florida.

(Pote et al. 1996; Sharpley et al. 1996; Nair et al. 1998). In the Netherlands, a groundwater P concentration of 0.1 mg ortho P L^{-1} is considered a water quality hazard. This concentration of P occurs in the soil solution at a DPS of 25%. Therefore, the Dutch consider a soil with a DPS >25% as posing a water quality threat (Breeuwsma and Silva 1992; Moore et al. 1998). For Florida conditions, where Mehlich 1-P is the agronomic soil test P (STP), Nair et al. (2004) have modified the calculation of DPS using Mehlich 1-P as the extractable P and (Mehlich 1-Fe + Mehlich 1-Al) as a measure of the P sorbing capacity of a soil and observed a DPS value above which there was a rapid increase in water soluble P (WSP) (Figure 1). This value, referred to as the 'change point,' occurred at a DPS of 20%. Taking into consideration the change point, and several other factors such as the confidence intervals, agronomic soil test values, and DPS values from other studies, they suggested that a DPS value above 30% would indicate the potential of a negative impact of P on water quality for Florida soils (Nair et al. 2004).

From an environmental standpoint, erosion of P-laden soil particles in surface runoff from agricultural lands has received most of the attention with regard to P and water quality. However, the vertical and lateral transport of P through soil profiles in which drainage water ultimately enters surface water has recently been identified as a potential source of P to surface waters in certain types of landscapes (Sims et al. 1998; Nair and Graetz 2002). These factors have been taken into consideration in the Phosphorus Index (P-Index).

The Phosphorus Index

The Florida P-Index, http://efotg.nrcs.usda.gov/popmenu3FS.aspx?Fips=12001&MenuName=menuFL.zip, was developed as a field-based index to assess site conditions and potential P loss vulnerability. The Index, developed by personnel from the University of Florida (UF-IFAS), Department of Agricultural and Food Consumer Services, USDA Agricultural Research Service, and other organizations and agencies as well as private individuals, uses several characteristics to obtain an overall rating for a given site. The Index was concurred by the USDA State Technical Committee and adopted by the State Conservationist, on November 13, 2000. The NRCS has identified the Index as the most comprehensive plan that offers the most flexibility for making management decisions for nutrient management.

The outline of the Florida P-Index (Table 1) is reproduced below; additional details are available at the Website mentioned earlier. The Index includes consideration of transport factors such as soil erosion, soil runoff class, leaching potential, and distance from a water body (Table 1, Part A), along with management factors such as soil test P (STP), P application method, and source and rate of P application (Table 1, Part B). The overall Index value is obtained by multiplying Part A and Part B. For Florida conditions, the P-Index value (unitless number) of a given site is considered LOW (<75), MEDIUM (75-150), HIGH (151-225) OR VERY HIGH (>225). If the P-Index value of a given site is rated HIGH, 'soil and water conservation and P management practices are necessary to reduce the risk of P movement and water quality degradation.' If the P-Index rating is VERY HIGH, 'all necessary soil and water conservation practices, plus a P-based management plan must be put in place to avoid the potential for water quality degradation' (see the Website). For LOW and MEDIUM conditions, manure application can be N based.

In the implementation of the Florida P-Index, the field evaluator of a site assigns an appropriate rating for each transport and management characteristic. Note that there are also weighting factors associated with some of the characteristics that were introduced based on the professional judgment of the developers of the Index. It is not our intention to discuss the Florida P-Index in detail, but to provide sufficient information to give an overview of how the Index could be used to arrive at a reasonable evaluation of the potential of a site to be an environmental concern with respect to P transport. Also, several of the individual transport and management characteristics are useful to evaluate agroforestry practices. Obviously, it is not possible to have a single P-Index for all locations, as several of the factors in the Index are dependent on the soils, hydrologic conditions, and other local situations. Within the United States, P indices have been developed in almost all the states. For information on nutrient management posters, abstracts and other related information from a given state, refer to the Soil and Water Conservation Society Website: http://www.swcs.org/default.htm (Search for Phosphorus Index; verified December 3, 2003).

Nutrient management in agroforestry systems

Although it has been suggested that agroforestry practices may reduce runoff and leaching of chemicals from farmland and thereby improve water quality, little research has been conducted to substantiate these claims, particularly in temperate alleycropping and silvopastoral systems. Plant-extractable P is normally low and immobile in the subsoil (Buresh and Tian 1997). In heavily fertilized sandy soils, however, P loss is a major problem and management practices must be adopted to control its loss from the field (Breeuwsma and Silva 1992; Sims et al. 1998). With this background, we are currently monitoring the nutrient movement (N and P) through the soil profile under two agroforestry practices in different parts of Florida: a pecan (*Carya illinoenis*)–cotton (*Gossypium hirsutum*) alley cropping experiment at Jay, in the panhandle of Florida (30°47′ N, 87°13′ W), and at a silvopasture experiment in the flatwood soils at Ona, central Florida (27°23′ N, 81°57′ W).

Nutrient movement through the soil profile

The trials at Jay are on a Red Bay sandy loam, which is a fine-loamy, kaolinitic, thermic Rhodic Kandiudult, with an average water table depth at 1.8 m. The Ap horizon is about 15 cm thick and is followed by Bt horizons with sand grains coated and bridged with clay. The soils are very deep, well drained and moderately permeable with slow to medium runoff (http://ortho.ftw.nrcs.usda.gov/osd/osd.html, verified December 03, 2003). The alleycropping system was set up in a pecan orchard that had been under nonintensive clover (*Trifolium* spp.) and ryegrass (*Lolium* spp.) production for 29 years prior to the beginning of this study (Allen, 2003). Twelve plots were

Table 1. The P-Index table.

Site and transport characteristics	Phosphorus transport rating					Value Jay	Value Ona
Soil erosion	No surface outlet 0	<5 T/A[a] 1	5–10 T/A[a] 2	10–15 T/A[a] 4	>15 T/A[a] 8	1	1
Runoff potential	Very low 0	Low 1	Medium 2	high 4	Very high 8	2	8
Leaching potential	Very low 0	Low 1	Medium 2	High 4	Very high 8	0	8
Potential to reach water body	Very low 0	Low 1	Medium 2	High 4		1	1
Sum for site and transport characteristics (A)						**4**	**18**

Phosphorus source management	Phosphorus loss rating				Value Jay	Value Ona
Fertility Index value	Soil Fertility Index × 0.025 (- - - - - - - - - - ppm P × 2 × 0.025)[b]				3.5	1.5
P application source and rate per acre[c]	$0.05 \times$ (- - - - - - - - lbs P_2O_5) for fertilizer, manure, or compost $0.015 \times$ (- - - - - - - - - - lbs P_2O_5) for biosolids $0.10 \times$ (- - - - - - - - - - lbs P_2O_5) for waste water				13.5	0.6
Application method	No surface outlet 0	Applied via irrigation 2	Solids incorporated within 5 days of application 4	Solids not incorporated within 5 days of application[d] 6	4	6
Waste water application	0.20 × - - - - - - - - - acre inches/acre/year				0	0
Sum for phosphorus source (B)					**21**	**8.1**

[a]T/A = Tons per acre per year (T/A × 2.24 = Mg ha^{-1}).
[b]From soil test (Mehlich 1) results; ppm = mg L^{-1}.
[c]Initial evaluation should be N-based rates; lbs $acre^{-1} \times 1.12 \times 10^{-3}$ = Mg ha^{-1}.
[d]Solids include fertilizers, composts, biosolids, and manure and other animal wastes.
Total P Loss Rating = A × B (Jay: 84; Ona: 144).
Source: http://efotg.nrcs.usda.gov/popmenu3FS.aspx?Fips=12001&MenuName=menuFL.zip

set up within the orchard, six of which were treated with inorganic fertilizer and the other six with organic (poultry manure) fertilizer. Poultry manure was added to plots at the same N rate as those receiving inorganic N fertilizer. Preliminary results (data not shown) indicate that soluble reactive phosphorus (SRP) increased in the surface soils in both the manure and inorganically fertilized soils after harvesting the cotton with just one fertilizer application at the beginning of the growing period. The unfertilized pecan only 'control' plots had the least concentrations of soluble reactive phosphorus (SRP) in the surface soils.

Movements of P and N were also compared in the pecan–cotton alleycropping system with inorganic fertilization under conditions where trenches were dug to 1 m depth and lined vertically with polyethylene sheets to prevent the roots of the pecan trees interfering with the cotton cropping systems (F-T). Plots were also left without trenching (F-NT) for comparison. The two treatments (F-T and F-NT) were compared to the movement of N and P in adjacent pecan tree only plots without any fertilizer or trenching treatments (control). All fertilizer applications were based on cotton crop N requirements at a rate of 90 kg N ha^{-1} (See Allen 2003 and Lee and Jose 2003 for experimental details). A 3-9-18 inorganic fertilizer blend was used for the inorganic fertilizer plots.

Table 2. Concentrations of NO_3-N and Mehlich 1-P, and the degree of P saturation (DPS) with depth for the cotton (Fertilizer - trenched, F-T), alley cropping (Fertilizer - non-trenched, F-NT) and control plots at Jay, Florida, USA.

Depth	Treatment		
cm	F-T	F-NT	Control
		NO_3-N, mg kg^{-1}	
0–5	29.1 (5.6)[1]	22.2 (2.9)	16.5 (2.4)
5–15	11.7 (2.4)	9.2 (0.8)	5.5 (0.8)
15–30	3.8 (1.3)	2.3 (0.4)	1.4 (0.6)
30–50	3.0 (1.1)	0.8 (0.4)	0.6 (0.4)
50–75	4.8 (0.8)	1.1 (0.8)	0.9 (0.4)
75-100	5.8 (1.2)	1.5 (0.6)	1.1 (0.8)
		Mehlich 1-P, mg kg^{-1}	
0–5	71.3 (2.2)	71.3 (8.0)	74.7 (24.7)
5–15	30.9 (4.3)	32.1 (2.3)	25.6 (5.4)
15–30	2.3 (1.2)	2.3 (0. 7)	3.4 (2.9)
30–50	0.7 (0.1)	0.6 (0.1)	0.9 (0.5)
50–75	0.4 (0.1)	0.4 (0.1)	0.6 (0.1)
75–100	0.3 (0.0)	0.3 (0.1)	0.5 (0.1)
		DPS,%	
0–5	32.9 (0.7)	32.1 (4.3)	26.1 (7.2)
5–15	15.2 (2.5)	15.2 (1.9)	11.2 (2.2)
15–30	2.0 (0.8)	2.2 (0.4)	2.2 (1.8)
30–50	0.8 (0.1)	0.9 (0.2)	1.0 (0.6)
50–75	0.5 (0.1)	0.5 (0.2)	0.7 (0.1)
75–100	0.4 (0.0)	0.4 (0.1)	0.6 (0.2)

[1] Numbers in parentheses are standard deviation values.

Preliminary results (Table 2) show that NO_3-N concentration in the surface soils of the control plots (pecan tree plots with no fertilizer applications) were lower compared to both the fertilized nontrenched plots (F-NT) and the fertilized trenched plots (F-T), representing alley cropping and sole crop stands, respectively. Movement of NO_3-N through the soil profile followed the same trend as for surface soils, showing that more NO_3-N was taken up by the agroforestry system where there were interactions between the rooting systems of the pecan and cotton cropping systems. The NO_3-N concentrations in the soil profiles of F-T seem to have moved more through the soil profile to the 1-meter depth sampled compared to both the F-NT plots and the control plots. Although absolute values for NO_3-N are low throughout the soil profile compared to concentrations found in other agricultural systems, we recognize that this is due to a single fertilizer application (one year of study) and project the consequences on water quality due to repeated applications of fertilizer in an agricultural farm to be problematic.

Edmeades (2003), based on evidence from long-term trials, showed that when compared to inorganically fertilized soils, manure-amended soils exhibited an excessive accumulation of nutrients such as P, Ca, and Mg in the surface soils, and of NO_3-N, Ca and Mg in subsoils. High concentrations of NO_3-N in the subsoils are evidence of its movement over a period of time. High Ca and Mg concentrations in surface soils are related to manure application, and movement of these elements in the soil profile along with P in Florida Spodosols has been reported (Nair et al. 1995). Note that while manure P is normally considered an organic fertilizer source, the percent of inorganic P in manure exceeds the organic forms (Nair et al. 1995; Sharpley 1999; Nair et al. 2003). While these long-term trials reviewed by Edmeades (2003) suggest that both N and P are more likely to accumulate in manure-amended soils than in inorganically fertilized soils, it must be cautioned that soil type would play the primary role in the effect of these nutrients on water quality impairment.

Mehlich 1-P concentrations (the soil test P in Florida) do not appear to be impacted by manure and inorganic fertilizer applications (Table 2) to the agroforestry trials. In Florida, surface Mehlich 1-P concentrations of 30 mg P kg^{-1} are considered high and 60 mg P kg^{-1} very high from an agronomic point of view (Kidder et al. 2002). In the sandy soils of Delaware state, a Mehlich 1-P concentration of 50 mg P kg^{-1} is considered excessive (Paulter and Sims 2000). The surface Mehlich 1-P concentration in this study exceeds these values even in the non-fertilized pecan tree control plots (Table 2). Movement of P through the soil profile beyond 15 cm appears to be minimal under all circumstances, likely due to high P retention capacity of these soils. A 15-cm depth in these soils signifies the end of the Ap horizon and the beginning of the Bt horizons with sand grains coated and bridged with clay that would increase the P retention capacity of the soil. The DPS in the surface soils of the F-NT and F-T plots is higher (>30%) than in the pecan tree only control plots (26%), although Mehlich 1-P is not very different for the three treatments (Table 2). The DPS values for the surface soils are very close to the recommended threshold DPS of 30%, so additional P application to these soils should be done with caution. The differences in DPS values are due to a single application of fertilizer to the agroforestry plots; continuous addition of P to this system may eventually be a cause for environmental P pollution, primarily via runoff.

Studies are also currently underway on the flatwood soils at Ona, central Florida, where in addition to runoff losses of nutrients, leaching potential is high. The soils are Spodosols, Ona fine sand (sandy, siliceous, hyperthermic Typic Alaquods), characterized by four primary horizons, a surface sandy A horizon, followed by an eluted E horizon. The Bh (or the spodic) horizon, underlying the E horizon is a dark colored horizon where C, Al, and Fe have accumulated (Soil Survey Staff 1999). The Bh horizon is followed by the Bw horizon. The water table is within a depth of 25 cm of the surface for four to 13 weeks and at depths of 25 to 100 cm for 26 to 40 weeks of the year (Soil Survey Staff 1999).

Description of the Ona study site can be found in Ezenwa et al. (2003). Ten soil profiles each were sampled in the alleys of the silvopastoral site, an adjacent grazed pasture area and a nearby native flatwoods vegetation site (Ezenwa et al. 2003). Both pastures are bahiagrass (*Paspalum notatum*) with the legumes *Desmodium heterocarpon* and *Vigna parkeri.* The silvopasture is planted to slash pine (*Pinus elliottii*) at 500 trees ha^{-1}. The open pasture has been fertilized annually with 56 kg N ha^{-1} for the past 15 years. The silvopasture has received 5 kg P ha^{-1} for the past 5 years, and the open pasture 4 kg P ha^{-1} once (2003) in the past 20 years. Grazing density for pasture is 1 cow/calf pair ha^{-1} for the past 20 years, and for range is 1 cow/calf pair 7 ha^{-1} for the past 60 years (R. Kalmbacher, pers. comm., October 2003). Preliminary data indicated that soluble reactive phosphorus (SRP) concentrations in the surface soils varied in the order: pasture > silvopasture > native flatwood vegetation (Table 3), although the pasture has received only a single inorganic P application in 20 years. Soluble reactive P concentrations also varied down a soil profile in the order: pasture > silvopasture > native flatwood vegetation. This finding suggests that the silvopastoral system is more efficient in removing P from the soil profile than a regular pasture and therefore might contribute significantly in reducing nonpoint source pollution. The deeper tree roots in the silvopastoral site could have removed more P from the lower depths. The native flatwoods had the lowest SRP concentrations, since the site had never been fertilized. The spodic horizon with its high P retention capacity was at an average depth of 30 cm for all the three sites, and would explain the general reduced SRP concentrations beyond this depth. However, because of the high water table there is the likelihood of lateral P loss before reaching the spodic horizon at these sites. The above two experiments on highly different soils types, involving both fertilizer and animal manure applications, can be taken as indicative of the likely benefits of agroforestry practices in reducing nutrient leaching through the soil profile.

Table 3. Soluble reactive phosphorus (SRP) concentrations with depth at a native flatwoods site, a silvopastoral site, and an adjacent regular pasture at Ona, Florida, USA.

Depth cm	Native flatwoods	Silvopasture SRP, mg kg^{-1}	Pasture
0–5	1.59 (1.14)[1]	4.23 (1.34)	9.10 (3.26)
5–15	1.28 (1.17)	2.09 (1.36)	2.79 (2.20)
15–30	0.12 (0.02)	0.79 (0.39)	1.52 (0.94)
30–50	0.00 (0.00)	0.17 (0.14)	1.91 (1.25)
50–75	0.00 (0.00)	0.02 (0.02)	0.42 (0.40)
75–100	0.00 (0.00)	0.04 (0.03)	0.66 (0.52)

[1]Numbers in parentheses are standard deviation values.

Application of the P-Index to the experimental sites

Values for the various components of the P-Index for the pecan-cotton alley cropping site (Jay) and for the silvopastoral site (Ona) have been assigned in the Florida P-Index (Table 1). The values were based on the P-Index described at the Website mentioned earlier. For the convenience of the field evaluator, the units in the current P-Index are non-SI units. Corresponding SI units are provided in the following discussion. For Jay, the concentration of the soil test P (Mehlich 1-P) was taken as 70 mg P kg^{-1}, and the amount of P applied, as 270 lbs P_2O_5 $acre^{-1}$ (302 kg ha^{-1}). For the Ona site, Mehlich 1-P was 30 mg P kg^{-1} and the amount of P applied was 12 lbs P_2O_5 $acre^{-1}$ (13.5 kg ha^{-1}). The total value of the Index was 84 for Jay and 144 for Ona. Both these sites fall under the MEDIUM category (75–150) of the Florida P-Index, although at the extreme ends of this category. Thus, at this time, both sites can use an N-based nutrient management program.

Evaluation of nutrient transport from agroforestry systems and possible remedial management practices

Preliminary results of the ongoing experiments in temperate agroforestry systems presented above support the hypothesis that agroforestry practices can reduce nutrient movement through soil profiles and potentially reduce nutrient loading in waterways. There is substantial evidence (Schultz et al. 2000) that forested riparian buffers in temperate regions would likely reduce nonpoint source pollution. Unfortunately, studies of this nature for other agroforestry systems in the temperate regions are rare so that a broad-based conclusion cannot be drawn at this stage.

We will now examine how some of the P-Index transport and management components (column 1 of Table 1) can be addressed though agroforestry practices. These transport and management factors are merely used as guidelines for discussion, and could therefore be applicable to any system outside Florida, both under tropical and temperate conditions. We are not applying the Florida P-Index in its entirety to other agricultural systems outside Florida. As mentioned earlier, the P-Index is site-specific and has to be developed for individual regions, although some of its components may be applicable in other places as well.

For nutrient management, the NRCS (1999) has recommended the use of windbreaks, buffers, filter strips, grassed waterways, field borders, irrigation water management, riparian herbaceous cover, and riparian forest buffer, many of which can be used either individually or in combinations in agroforestry systems to reduce nutrient movement from a given site. Several of these agroforestry practices such as riparian forest buffers (Schultz et al. 2000) and windbreaks (Brandle et al. 2000) have been reviewed in detail. Since surface runoff is also controlled by the topography of the land, terracing, where practical, may be a viable option for controlling nutrient movement from a site with a relatively steep slope.

Ancient civilizations have invariably evolved near waterways since humans need water for their daily use and waterways were used as modes of transport. This culture is still maintained in developing countries around the world. 'Potential to reach water bodies' is one characteristic that is of importance in nutrient transport (Table 1). Livestock (large and small ruminants, poultry, swine, etc.) would invariably have access to water in these life styles, and direct deposition of manure, and therefore nutrients, into the water would cause degradation of water quality. Under present conditions, it should be easier to keep livestock away from water sources via fencing or other forms of best management practices[1], including providing animals with water troughs away from other natural water sources. Additional best management practices (BMPs) could be introduced to agroforestry systems to further minimize off-site nutrient movement and subsequent degradation of water quality. Rotational grazing within a silvopastoral system, for example, could minimize localized heavy deposition of manure (Adjei et al. 1987). Devising the rate, source, and schedule of fertilizer/manure applications is yet another important consideration. Concepts and measures such as Florida P-Index are of wide applicability; but they are also very much dependant on soil type and other environmental conditions. Site specificity is already recognized as an important attribute of agroforestry in terms of a number of factors such as bioclimatic and socioeconomic conditions (Nair 1993). Site specificity attached to such measurements is perhaps only an extension of that general problem of dealing with the complex system of agroforestry.

Summary and conclusions

While the environmental benefits of agroforestry have often been postulated, evidence is lacking to support the hypotheses, particularly in the temperate climate context. Agroforestry systems are expected to minimize nutrient losses from the soil because of enhanced nutrient uptake by tree and crop roots from varying soil depths, compared to more localized and shallow rooting depths of sole crop stands. Comparison of nutrient losses from an alleycropping site and from a silvopastoral site with monoculture systems under Florida conditions show that these agroforestry practices will likely reduce nutrient loss from a field when compared to monocultural practices. It is emphasized that nutrient losses are highly site specific, and one of the most important components in the understanding of nutrient transport is the soil type. Environmental indicators such as the Florida P-Index can be useful in increasing the efficiency of nutrient management within these and other systems. At this early stage of evaluation, there is some evidence that agroforestry practices can benefit the environment through reduction of loss of nutrients from an agricultural farm. However, definite conclusions can be arrived at only after obtaining additional experimental data from these and other agroforestry practices under different soil types and environmental conditions.

End Note

1. Best management practices (BMPs) are 'those on-farm activities designed to reduce nutrient losses in drainage waters

to an environmentally acceptable level while simultaneously maintaining an economically viable farming operation for the grower' (Bottcher et al. 1995).

Acknowledgements

Results reported here are based on on-going investigations in collaboration with S. Jose and R. Kalmbacher, and with field and laboratory assistance provided by S. Allen. We thank the anonymous reviewers for their valuable comments and suggestions on the earlier version of this manuscript. This research was supported by the Florida Agricultural Experiment Station and a grant from USDA/CSREES/IFAFS, through The Center for Subtropical Agroforestry (CSTAF), University of Florida, and approved for publication as Journal Series No. R-09772.

References

Adjei M.B., Mislevy P., Quesenberry K.H. and Ocumpaugh W.R. 1987. Grazing-frequency effects on forage production, quality, persistence and crown total non-structural carbohydrate reserves of limpograss. Soil Crop Science Society of Florida Proceedings 47: 233–236.

Allen S.C. 2003. Nitrogen dynamics in a pecan (*Carya illinoenis* K. Koch)-cotton (*Gossypium hirsutum* L.) alley cropping system in the southern United States. Ph.D. Dissertation. University of Florida.

Andrews W.J. 1994 Nitrate in ground water and spring water near four dairy farms in north Florida, 1990–93: U.S. Geological Survey water-resources investigations report 94-4162, 63 pp.

Ayoub A.T. 1999. Fertilizers and the environment. Nutr Cycl Agroecosyst 55: 117–121.

Bockman O.C., Kaarstad O., Lie O.H. and Richards I. 1990. Agriculture and Fertilizers. Agricultural Group, Norsk Hydro a.s. Oslo, Norway, 245 pp.

Bottcher A.B., Tremwel T.K. and Campbell K.L. 1995 Best management practices for water quality improvement in the Lake Okeechobee Watershed. Ecol Eng 5: 341–356.

Brady N.C. and Weil R.R. 2002. The Nature and Properties of Soils. 13th edition. Prentice Hall, Upper Saddle River, New Jersey, 960 pp.

Brandle J.R., Hodges L. and Wight B. 2000. Windbreak practices. pp. 79–118. In: Garrett H.E., Rietveld W.J. and Fisher R.F. (eds), North American Agroforestry: An Integrated Science and Practice. American Society of Agronomy, Inc. Madison, Wisconsin.

Breeuwsma A. and Silva S. 1992. Phosphorus fertilization and environmental effects in the Netherlands and the Po region (Italy). Rep. 57. Agric Res Dep The Winand Staring Centre for Integrated Land, Soil and Water Research. Wageningen, The Netherlands, 38 pp.

Buresh R.J. and Tian G. 1997. Soil improvement by trees in sub-Saharan Africa. Agroforest Syst 38: 51–76.

Clason T.R. and Sharrow S.H. 2000. Silvopastoral practices. pp. 119–147. In: Garrett H.E., Rietveld W.J. and Fisher R.F. (eds), North American Agroforestry: An Integrated Science and Practice. American Society of Agronomy, Inc. Madison, Wisconsin, USA.

Di H.J. and Cameron K.C. 2002. Nutrient leaching in temperate agrocosystems: sources, factors, and mitigating strategies. Nutr Cycl Agroecosyst 64: 237–256.

Edmeades D.C. 2003. The long-term effects of manures and fertilizers on soil productivity and quality: a review. Nutr Cycl Agroecosyst 66: 165–180.

Ezenwa I.V., Kalmbacher R.S. and Mallett W.J. 2003. Projected timber yields of south Florida slash pine silvopasture in south-central Florida. Soil and Crop Science Society of Florida Proceedings 62: 47–50

Florida Department of Environmental Protection. 2001. Suwannee Basin Status Report. Division of Water Resource Management. Tallahassee, FL, 193 pp.

Fonyo C. and Flaig E. 1995. Phosphorus budgets for Lake Okeechobee tributary basins. Ecol Eng 5: 209–227.

Garrett H.E. and McGraw R.L. 2000. Alley cropping practices. pp. 149–188. In: Garrett H.E., Rietveld W.J. and Fisher R.F. (eds), North American Agroforestry: An Integrated Science and Practice. American Society of Agronomy, Inc. Madison, Wisconsin, USA.

Graetz D.A. and Nair V.D. 1999. Inorganic forms of phosphorus in water and soils. Chapter 6. pp. 171–186. In: Reddy K.R., O'Connor G.A. and Schelske C.L. (eds), Phosphorus Biogeochemistry of Sub-Tropical Ecosystems. C.R.C. Press. New York, USA.

Graetz D.A., Nair V.D., Voss R.L. and Portier K.M. 1999. Phosphorus accumulation on land utilized by dairies and beef ranches. Agric Ecosyst Environ 75: 31–40.

Hatzell H.H. 1995. Effects of waste-disposal practices on groundwater quality at five poultry (broiler) farms in north-central Florida, 1992–1993: U.S. Geological Survey water-resources investigations report 95-4064, 35 pp.

Heathwaite L., Sharpley A. and Gburek W. 2000 A conceptual approach for integrating phosphorus and nitrogen management at watershed scales. J Environ Qual 29: 158–166.

Hochmuth G.J. 2000 Management of nutrients in vegetable production systems in Florida. Soil and Crop Science Society of Florida Proceedings 59: 11–13.

Katz B.G., Hornsby H.D., Bohlke J.K. and Mokray M.F. 1999. Sources and chronology of nitrate contamination in springs, Suwannee River Basin, Florida: U.S. Geological Survey water-resources investigations report 99-4252, 54 pp.

Kidder G., Chamblis C.G. and Mylavarapu R. 2002. UF/IFAS standard fertilization recommendations for agronomic crops. SL129, Soil & Water Science, Cooperative Extension Service, IFAS, p. 9.

Lee K.H. and Jose S. 2003. Soil respiration and microbial biomass in soils under alley-cropping and monoculture cropping systems in southern U.S.A. Agroforest Syst 58: 45–54.

Moore P.A. Jr, Joern B.C. and Provin T.L. 1998. Improvements needed in environmental soil testing for phosphorus. pp. 21–29. In: Sims J.T. (ed.), Soil Testing for Phosphorus: Environmental Uses and Implications. SERA-IEG 17 USDA-CREES Regional Committee: Minimizing Agricultural Phosphorus Losses for Protection of the Water Resource.

Mueller D.K. and Helsel D.R. 1996. Nutrients in the Nation's waters-Too much of a good thing? U.S. Geological Survey Circular 1136, 26 pp.

Natural Resource Conservation Service (NRCS) Conservation Practice Standard 1999. Nutrient Management. Code 590. Technical Guide. Section IV. NRCS, FL, 8 pp.

Nair PKR 1993. An Introduction to Agroforestry. Kluwer, Dordrecht, The Netherlands, 499 pp.

Nair P.K.R., Buresh R.J., Mugendi D.N. and Latt C.R. 1999. Nutrient cycling in tropical agroforesty systems: myths and science. pp. 1–31. In: Buck L.E., Lassoie J.P. and Fernandes E.C.M. (eds), Agroforestry in Sustainable Agricultural Systems. CRC Press. LLC, Boca Raton, Florida, USA.

Nair V.D., Portier K.M., Graetz D.G. and Walker M.L. 2004. An environmental threshold for degree of phosphorus saturation in sandy soils. J. Environ Qual 33: 107–113.

Nair V.D., Graetz D.A. and Dooley D.O. 2003. Phosphorus release characteristics of manure and manure-impacted soils. J. Food, Agric Environ 1: 217–223.

Nair V.D. and Graetz D.A. 2002. Phosphorus saturation in Spodosols impacted by manure. J Environ Qual 31: 1279–1285.

Nair V.D., Graetz D.A. and Reddy K.R. 1998. Dairy manure influences on phosphorus retention capacity of Spodosols. J Environ Qual 27: 522–527.

Nair V.D., Graetz D.A. and Portier K.M. 1995. Forms of phosphorus in soil profiles from dairies of South Florida. Soil Sci Soc Am J 59: 1244–1249.

Pautler M.C. and Sims J.T. 2000. Relationships between soil test phosphorus, soluble phosphorus, and phosphorus saturation in Delaware soils. Soil Sci Soc Am J 64: 765–773.

Pote D.H., Daniel T.C., Sharpley A.N., Moore P.A. Jr, Edwards D.R. and Nichols D.J. 1996. Relating extractable soil phosphorus to phosphorus losses in runoff. Soil Sci Soc Am J 60: 855–859.

Robinson J.S. and Sharpley A.N. 1996. Reaction in soil of phosphorus released from poultry litter. Soil Sci Soc Am J 60: 1583–1588.

Schultz R.C., Colletti J.P., Isenhart T.M., Marquez W.W. and Ball C.J. 2000. Riparian forest buffer practices. pp. 189–281. In: Garrett H.E., Rietveld W.J. and Fisher R.F. (eds), North American Agroforestry: An Integrated Science and Practice. American Society of Agronomy, Inc. Madison, Wisconsin, USA.

Sharpley A.N. 1999. Global issues of phosphorus in terrestrial ecosystems. Chapter 1. pp. 15–46. In: Reddy K.R., O'Connor G.A. and Schelske C.L. (eds), Phosphorus Biogeochemistry of Sub-Tropical Ecosystems. C.R.C. Press. New York, USA.

Sharpley A.N., McDowell R.W. and Kleinman P.J.A. 2001. Phosphorus loss from land to water: integrating agriculture and environmental management. Plant Soil 237: 287–307.

Sharpley A.N., Daniel T.C., Sims J.T. and Pote D.H. 1996. Determining environmentally sound soil phosphorus levels. J Soil Water Conserv 51: 160–166.

Sims J.T., Simard R.R. and Joern B.C. 1998. Phosphorus loss in agricultural drainage: Historical perspective and current research. J Environ Qual 27: 227–293.

Soil Survey Staff 1999. Soil taxonomy: a basic system of soil classification for making and interpreting soil surveys (2nd ed.). USDA-NRCS Agric Handbook 436. U.S. Govt. Printing Office, Washington, DC, 869 pp.

Terry D.L. and Kirby B.J. 2003. Commercial Fertilizers 2002. Association of American Plant Food Control Officials and The Fertilizer Institute. Lexington, KY, USA, 41 pp.

USDA-NRCS 1997. Alley cropping, Conservation practice job sheet 311, 4 pp.

U.S. Environmental Protection Agency (USEPA) 2002. National Water Quality Inventory. EPA 841-R-02-001. USEPA Office of Water Quality, Washington, DC, 20460, 207 pp.

Woodard K.R., French E.C., Sweat L.A., Graetz D.A., Sollenberger L.E., Macoon B., Portier K.M., Rymph, S.J., Wade B.L., Prine G.M. and Van Horn H.H. 2003. Nitrogen removal and nitrate leaching for two perennial, sod-based forage systems receiving dairy effluent. J Environ Qual 32: 996–1007.

Workman S.W., Bannister M.E. and Nair P.K.R. 2003. Agroforestry potential in the southeastern United States: Perceptions of landowners and extension professionals. Agroforest Syst 59: 73–83.

Agroforestry Systems **61:** 281–295, 2004.

Carbon sequestration: An underexploited environmental benefit of agroforestry systems

F. Montagnini[1] and P. K. R. Nair[2]
[1]*Yale University, School of Forestry and Environmental Studies, 370 Prospect St., New Haven, CT 06511 USA (e-mail: florencia.montagnini@yale.edu);* [2]*University of Florida, School of Forest Resources and Conservation, 118 N-Z Hall, PO Box 110410, Gainesville, FL 32611-0410, USA (e-mail: pknair@ufl.edu)*

Key words: Carbon market, Kyoto Protocol, PES (payment for environmental services), Policy framework, Silvopasture, Soil carbon

Abstract

Agroforestry has importance as a carbon sequestration strategy because of carbon storage potential in its multiple plant species and soil as well as its applicability in agricultural lands and in reforestation. The potential seems to be substantial; but it has not been even adequately recognized, let alone exploited. Proper design and management of agroforestry practices can make them effective carbon sinks. As in other land-use systems, the extent of C sequestered will depend on the amounts of C in standing biomass, recalcitrant C remaining in the soil, and C sequestered in wood products. Average carbon storage by agroforestry practices has been estimated as 9, 21, 50, and 63 Mg C ha^{-1} in semiarid, subhumid, humid, and temperate regions. For smallholder agroforestry systems in the tropics, potential C sequestration rates range from 1.5 to 3.5 Mg C ha^{-1} yr^{-1}. Agroforestry can also have an indirect effect on C sequestration when it helps decrease pressure on natural forests, which are the largest sink of terrestrial C. Another indirect avenue of C sequestration is through the use of agroforestry technologies for soil conservation, which could enhance C storage in trees and soils. Agroforestry systems with perennial crops may be important carbon sinks, while intensively managed agroforestry systems with annual crops are more similar to conventional agriculture. In order to exploit this vastly unrealized potential of C sequestration through agroforestry in both subsistence and commercial enterprises in the tropics and the temperate region, innovative policies, based on rigorous research results, have to be put in place.

Introduction

Finding low-cost methods to sequester carbon is emerging as a major international policy goal in the context of increasing concerns about global climate change. Recognizing that the accumulation of carbon dioxide and other greenhouse gases in the upper atmosphere is the major reason for global climate change, the idea of mitigating it through forest conservation and management was discussed as early as in the 1970s. But it was in the 1990s that international action was initiated in this direction. In 1992, several countries agreed to the United Nations Framework Convention on Climate Change (UNFCCC), with the major objectives of developing national inventories of greenhouse gas emissions and sinks, and reducing the emission of greenhouse gases (FAO 2001). At the third meeting of the FCCC in 1997 in Kyoto, Japan, the participating countries, including the United States, agreed, through what would later become known as the Kyoto Protocol, to reduce greenhouse gas emissions to 5% or more below 1990 levels by 2012 (http://unfcc.int). The Protocol provides a mechanism by which a country that emits carbon in excess of agreed-upon limits can purchase carbon offsets from a country or region that manages carbon sinks. Although the United States' withdrawal from the treaty in 2001 has considerably weakened its implementation, the Kyoto Protocol represents a major international effort related to carbon sequestration. Initially there was no agreement as to whether forests could be considered as carbon sinks, but the potential role of forest conservation and management to decrease greenhouse gases in the atmosphere was soon recognized. Globally, forests contain more than half of all terrestrial carbon, and account for about 80% of carbon exchange between terrestrial ecosystems and the atmosphere. Forest ecosystems are estimated to absorb up to 3 Pg (3 billion tons) of carbon annually. In recent years, however, a significant portion of that has been returned through deforestation and forest fires. For example, tropical deforestation in the 1980s is estimated to have accounted for up to a quarter of all carbon emissions from human activities (FAO 2003).

Basically there are three categories of activities through which forest management can help reduce atmospheric carbon: *Carbon sequestration* (through afforestation, reforestation, and restoration of degraded lands, improved silvicultural techniques to increase growth rates, and implementation of agroforestry practices on agricultural lands); *Carbon conservation* (through conservation of biomass and soil carbon in existing forests, improved harvesting practices such as reduced impact logging, improved efficiency of wood processing, fire protection and more effective use of burning in both forest and agricultural systems); and *Carbon substitution* (increased conversion of forest biomass into durable wood products for use in place of energy-intensive materials, increased use of biofuels such as introduction of bioenergy plantations, and enhanced utilization of harvesting waste as feedstock such as sawdust for biofuel) (Bass et al. 2000). Of the three, carbon conservation is regarded as having the greatest potential for rapid mitigation of climate change, whereas carbon sequestration takes place over a much longer period of time. Agroforestry has been recognized to be of special importance as a carbon sequestration strategy because of its applicability in agricultural lands as well as in reforestation programs (Cairns and Meganck 1994; Ruark et al. 2003).

Some tropical countries have recently started programs of incentives to encourage tree plantation development, especially on degraded areas. For example, in Costa Rica, Payment for Environmental Services (PES) contributes, since 1996, to promoting plantations through the assignment of differential incentives for already established plantations and for new reforestation. Carbon, water, and biodiversity are the major components of the program. Funding for these incentives comes from a special tax on gasoline, and from external sources (J. J. Campos A. and R. Ortiz: pers. comm., February 1999). In 2003, agroforestry systems were added to the list of systems receiving incentives in Costa Rica. Similarly, the Dutch government has been engaged in a 25-year program to finance reforestation projects covering 2500 km^2 in South America, in order to offset carbon emissions from coal-fired stations in The Netherlands (Myers 1996).

Many observers believe that the Clean Development Mechanism (CDM) offered by the Kyoto Protocol could reduce rural poverty by extending payments to low-income farmers who provide carbon storage through land-use systems such as agroforestry (Smith and Scherr 2002). Consequent to the realization of the potential of agroforestry practices such as silvopasture and riparian buffers in providing environmental benefits including carbon sequestration, methods for their valuation and development of policies to motivate the general public to pay for such benefits are under way in industrialized nations too (Alavalapati et al. 2004). Nevertheless, the potential of agroforestry as a strategy for carbon sequestration has not yet been fully recognized, let alone exploited. A major difficulty is that empirical evidence is still lacking on most of the mechanisms that have been suggested to explain how agroforestry systems could bring about reductions in the buildup of atmospheric CO_2. In this paper we review the current status of understanding on carbon storage potential for agroforestry systems and examine how this potential could be exploited for the benefit of landowners and the society at large.

Carbon Sequestration by tree-based systems

The basic premise of carbon sequestration potential of land-use systems, including agroforestry systems, is relatively simple: it revolves around the fundamental biological/ecological processes of photosynthesis, respiration, and decomposition (Nair and Nair 2003). Essentially, carbon sequestered is the difference between carbon 'gained' by photosynthesis and carbon 'lost' or 'released' by respiration of all components of the ecosystem, and this overall gain or loss of carbon is usually represented by net ecosystem productivity. Most carbon enters the ecosystem via photosynthesis in the leaves, and carbon accumulation is most obvious when it occurs in aboveground biomass. More than half of the assimilated carbon is eventually transported below ground via root growth and turnover, root exudates (of organic substances), and litter deposition, and therefore soils contain the major stock of C in the ecosystem. Inevitably, practices that increase net primary productivity (NPP) and/or return a greater portion of plant materials to the soil have the potential to increase soil carbon stock. Since the literature on C sequestration in agroforestry systems is rather scanty compared with that of tree-plantation systems, and considering that C sequestration potential of both plantations and agroforestry systems is based mainly on the attributes of the tree component, we will first review the situation with respect to tree plantations and then attempt to relate the lessons to agroforestry systems.

Carbon sequestration by tree plantations

Conceptually trees are considered to be a terrestrial carbon sink (Houghton et al. 1998). Therefore, managed forests can, theoretically, sequester carbon both *in situ* (biomass and soil) and *ex-situ* (products). According to FAO (2000) estimates, forest plantations cover 187 million ha worldwide, a significant increase from the 1995 estimate of 124 million ha. The reported new annual planting rate is 4.5 million ha globally, with Asia and South America accounting for 89‰. The main fast-growing, short-rotation species are of the genera *Eucalyptus* and *Acacia.* Pines and other coniferous species are the main medium-rotation utility species, primarily in the temperate and boreal zones. There is strong variation in the carbon sequestration potential among different plantation species, regions and management. Variations in environmental conditions can affect carbon sequestration potential even within a relatively small geographic area. In addition, management practices such as fertilization can easily increase carbon sequestration of species such as eucalypts (Koskela et al. 2000). Various estimates are available on C sequestration rates of common plantation species of varying rotation ages (Schroeder 1992; FAO 2000; FAO 2001; FAO 2003).

Use of native species for reforestation is minimal, and exotic tree species predominate both in industrial and in rural development plantations worldwide (Evans, 1999). Plantations using indigenous species are restricted for the most part to small- and medium-sized farms where reforestation is practiced in degraded portions of the land, often using species in response to government incentives (Piotto et al. 2003). The relative efficiency of native and exotic species in terms of their carbon accumulation potential has been investigated in a few studies. In experimental plantings in Central America, for example, values of C sequestration in aboveground biomass for ten native tree species were comparable to exotic species growing under similar conditions (Table 1). Proper design and management of such agroforestry (or, farm forestry) plantations can increase biomass accumulation rates, making them effective carbon sinks (Shepherd and Montagnini, 2001). Montagnini and Porras (1998) and Shepherd and Montagnini (2001) compared three mixed plantations with monocultures of each tree included to find out the benefits or disadvantages in terms of biomass accumulation and soil fertility maintenance for mixed stands versus monocultures in Central America. Mixtures of three to four tree species had C accumulation rates similar or larger than those of the fastest-growing species included in the mixture. With relatively short or medium rotation times of 15 to 25 years and relatively high standing volumes at harvest of 250 to 300 m^3 ha^{-1}, planting of these species is attractive for smallholders of the region. Fuelwood from thinning and pruning would be an additional source of farm income and thus an incentive for tree planting. In fact, the species involved in the experiment currently account for the majority of small farm reforestation in the region, and interest has recently developed for mixed designs that include some of the fastest growing trees with good timber value (*Terminalia amazonia*, *Vochysia guatemalensis* and *Hieronyma alchorneoides*) (Montagnini et al. 1995; Montagnini and Mendelsohn, 1996).

The idea that planting trees could be an easy (and often cheap) way to absorb emissions of carbon dioxide as well as its feasibility, has, however, been challenged. Based on experiments conducted in loblolly pine (*Pinus taeda*) forests in North Carolina, Oren et al. (2001) reported that after an initial growth spurt, trees grew more slowly and did not absorb as much excess carbon from the atmosphere as expected. In two experiments with loblolly pine trees exposed to increased atmospheric CO_2, CO_2-induced biomass-carbon increment without added nutrients was undetectable at a nutritionally poor site, and the stimulation at a nutritionally moderate site was transient, stabilizing at a marginal gain after three years. However, a large synergistic gain from higher CO_2 and nutrients was detected with nutrients added, the gain being larger at the poor site than at the moderate site. Based on these findings, the authors concluded that assessment of future carbon sequestration should consider the limitations imposed by soil fertility as well as interactions with nitrogen deposition. Schlesinger and Lichter (2001) examined decomposing leaves and roots on the floor of the experimental pine forest plots

Table 1. Carbon content in above ground biomass of 10 plantation-tree species at 10 years of age at la Selva, Costa Rica.

Plantation species	Stand density (trees ha^{-1})	Carbon content (Mg ha^{-1})			
		Foliage	Branch	Stem	Total
Balizia elegans	764	0.6	2.5	17.3	20.5
Calophyllum brasiliense	551	5.8	10.8	29.7	46.3
Dipteryx panamensis	670	5.3	15.4	82	102.6
Genipa americana	278	1.1	4.8	17.5	23.3
Hieronyma alchorneoides	660	2.4	13.5	43.8	59.7
Jacaranda copaia	596	0.8	1.6	42.5	44.9
Terminalia amazonia	462	4.7	14.6	63.2	82.5
Virola koschnyi	610	2	4.1	31	37
Vochysia ferruginea	556	2.7	5.7	30.6	39.1
Vochysia guatemalensis	551	2.2	5.2	55.5	62.8

Source: F. Montagnini (unpublished data).

and found the total amount of litter increased in a carbon-dioxide-enriched atmosphere, but so did the rate at which it was broken down, resulting in the release of carbon back to the atmosphere rather than being incorporated into the soil. Although these findings certainly do not imply that tree planting is not important, they suggest that planting trees *per se* may not necessarily enhance C sequestration or serve as an adequate substitute for reducing heat-trapping greenhouse gas emissions, and the relationship between tree planting and C sequestration is not as straight forward and simple as it is often portrayed.

The role of traditional forestry practices such as natural forest management in C sequestration is also unclear. As Harmon (2001) points out, there are two contrasting views, and consequently some confusion prevails. On the one hand, young (or newly planted) forests are generally believed to be better than older ones for C sequestration because of their faster growth, higher dry-matter accumulation rates, and fewer dead trees or decomposing parts. On the other hand, replacements of older forests with younger ones are reported to result in net release of C into the atmosphere (e.g., Schulze et al. 2000) through decomposition of dead and dying material from the old forests that are replaced by new plantings. This apparent confusion disappears when we acknowledge that in the former scenario, only the living plants of the ecosystem are considered as the long-term store of carbon, whereas in the latter, a more holistic view of the whole system is considered. It is essential that such a holistic view is considered in the discussion on C sequestration potential of mixed systems such as agroforestry, in which C dynamics in pools such as detritus and soil are quite different from those in sole stands of trees or crops.

Based on the above, it seems prudent to surmise that three factors are needed to determine the amount of carbon sequestered: (1) the increased amount of carbon in standing biomass, due to land-use changes and increased productivity; (2) the amount of recalcitrant carbon remaining below ground at the end of the tree rotation; and (3) the amount of carbon sequestered in products created from the harvested wood, including their final disposition (Johnsen et al. 2001).

Carbon sequestration by agroforestry systems

Claims on C sequestration potential of agroforestry systems are based on the same premise as for tree plantations: the tree components in agroforestry systems can be significant sinks of atmospheric C due to their fast growth and high productivity. By including trees in agricultural production systems, agroforestry can, arguably, increase the amount of carbon stored in lands devoted to agriculture, while still allowing for the growing of food crops (Kursten 2000). The discussion on planted forests presented in the earlier section has shown that: (1) soil fertility may be a limiting factor in realizing carbon sequestration potential of planted forests; (2) mixed stand of plants might be more efficient than sole stands in carbon sequestration; and (3) C sequestration estimates should be based on a holistic view of the long-term carbon storage potential of all components in the system including detritus, soil, and forest products. Agroforestry systems score highly in all these points: soil fertility improvement is a distinct possibility in agroforestry systems, especially under low-fertility conditions of the tropics (Nair et al. 1999); agroforestry systems entail mixed stands of species; and, a holistic rather than compartmental consideration of systems is a key concept of agroforestry (Nair 2001). In most agroforestry systems, the tree component is managed, often intensively, for its products such as pruning of hedgerows in tropical alleycropping, and harvest of commercial, mostly nontimber, products. Such harvested materials often are returned to the soil (as in alleycropping and improved fallow systems for soil fertility improvement). In addition, the amount of biomass and therefore carbon that is harvested and 'exported' from the system is relatively low in relation to the total productivity of the tree (as in the case of shaded perennial systems). Therefore, unlike in tree plantations and other monocultural systems, agroforestry seems to have a unique advantage in terms of C sequestration. Many of these assumptions have, however, been not systematically tested and validated.

Focusing on the tree component of agroforestry, some attempts have been made to estimate the global contribution of agroforestry as a sink for C. Based on tree growth rates and wood production, and assuming ratios of tree-stem biomass to C content of 1:2 (i.e., 50% of stemwood is assumed to be C), average carbon storage by agroforestry practices has been estimated to be 9, 21, 50, and 63 Mg C ha^{-1} in semiarid, subhumid, humid, and temperate regions (Schroeder 1994). The higher levels reported for temperate ecozones reflect the longer cutting cycles in these regions, with a resulting longer-term storage. At a global scale, it has been estimated that agroforestry systems could be implemented on 585 to 1275 $\times$ 10^6 hectares of technically suitable land, and these systems could store 12 to 228 Mg C ha^{-1} under the prevalent climatic and edaphic conditions (Dixon 1995).

In addition, agroforestry systems can have an indirect effect on C sequestration if they can help to decrease pressure on natural forests, which are the largest sink of terrestrial C. Within tropical regions, it has been estimated that one hectare of sustainable agroforestry could potentially offset 5 to 20 hectares of deforestation (Dixon 1995). Based on this assumption, projects promoting agroforestry in farmland surrounding 'islands' of natural forests have been attempted in Kenya and Madagascar (Young 1997). There are some examples of cases where promotion and implementation of agroforestry systems has been successful in this regard: in Sumatra, Indonesia, farmers who integrated rice (*Oryza sativa*) production with tree crops and home gardens exerted much less pressure on adjacent forest, in comparison with farmers dedicated to rice only (ICRAF 1995). Agroforestry systems can, however, be either sinks or sources of C and other greenhouse gases. Some agroforestry systems, especially those that include trees and crops (agrisilviculture) can be C sinks and temporarily store C, while others (e.g. ruminant-based agrosilvopastoral systems) are probably sources of C and other greenhouse gases. Especially in tropical regions, agroforestry systems can be significant sources of greenhouse gases: practices such as tillage, burning, manuring, chemical fertilization, and frequent disturbance can lead to emissions of CO_2, CH_4 and N_2O from soils and vegetation to the atmosphere. Silvopastoral systems, when practiced in an unsustainable manner, can result in soil compaction and erosion with losses of C and N from soils. Ruminant-based agrosilvopastoral systems and rice paddy agrosilvicultural systems are well documented sources of CH_4 (Dixon 1995).

On the other hand, agroforestry systems, especially if well managed and if they include soil conservation practices, can contribute to increasing short-term C storage in trees and soils, as will be shown in some examples that follow. Finally, whether agroforestry systems can be a sink or a source of C depends on the land-use systems that they replace: if they replace natural primary or secondary forests, they will accumulate comparatively lower biomass and C, but if they are established on degraded or otherwise treeless lands, their C sequestration value is considerably increased.

Agroforestry and soil carbon

Measurements of carbon stocks in soil are not very accurate, due to sampling and measurement problems (Koskela et al. 2000). For the past 20 years, scientists have been attempting to calculate the global carbon stocks of tropical forests, as well as the changes in these stocks as changes in land use occur. Recently, satellite data and remote sensing have been used to characterize ground cover, providing more accurate estimates in changes of vegetation each year throughout the tropics than ever before (Loveland and Belward 1997).

According to the estimates of the Intergovernmental Panel on Climate Change (IPCC) (IPCC 2000), tropical forests are by far the largest carbon stock in vegetation, while boreal forests represent the largest

C stock in soils (www.ipcc.ch). Tropical savannas store about one third of C in vegetation as do tropical forests, but savannas also have large C stocks in soils, similar to those of temperate grasslands. Croplands worldwide have the smallest C stocks in vegetation, with intermediate values for soils. Where agroforestry lies in these ranges depends on the proportion of trees included in the production system, as well as on the soil management and conservation practices used by each agroforestry system.

Globally, 1.5 to 3 times more C occurs in soils than in vegetation (Dixon 1995). Soils are the largest pool of terrestrial carbon, estimated at 2200 Pg; tropical topsoils contain about 13% of world soil carbon (Young 1997). However small on a global sense, the potential positive role of agroforestry in increasing soil carbon cannot be disregarded, especially considering other indirect effects on carbon and soil nutrients, as for example when proper agroforestry practices can help reducing soil erosion losses.

Estimates of C sequestration in agroforestry systems

Two issues need to be addressed in the discussion on C sequestration potential of agroforestry systems, and, unfortunately both seem to be rather insurmountable at the moment. First, the area under different agroforestry systems (existing or potential) is not known, and, second, a holistic picture of the *in situ* and *ex-situ* C storage and dynamics in different agroforestry systems is not yet determined. In the following sections, we attempt to discuss these issues in the light of available information.

Tropical systems

Available estimates of C sequestration potential of agroforestry systems are mostly for tropical regions. Based on a preliminary assessment of national and global terrestrial C sinks, Dixon (1995) identified two primary beneficial attributes of agroforestry systems in terms of C sequestration: (1) direct near-term C storage (decades to centuries) in trees and soils; and (2) a potential to offset immediate greenhouse gas emissions associated with deforestation and subsequent shifting cultivation. Following that, some projections have been made on the role of agroforestry in reducing the C emission from tropical deforestation. Based on its project known as 'Alternatives to Slash and Burn', ICRAF (International Centre for Research in Agroforestry) and its collaborators around the humid tropics concluded that the greatest potential for C sequestration in the humid tropics is above ground, not in the soil: through the establishment of tree-based systems on degraded pastures, croplands, and grasslands, the time-averaged C stocks in the vegetation would increase as much as 50 Mg C ha^{-1} in 20 years, whereas the soil stocks would increase only 5 to 15 Mg C ha^{-1} (Palm et al. 2000). A projection of carbon stocks for smallholder agroforestry systems in the tropics indicated C sequestration rates ranging from 1.5 to 3.5 Mg C ha^{-1} yr^{-1} and a tripling of C stocks in a twenty-year period, to 70 Mg C ha^{-1} (Watson et al. 2000). The total carbon emission from global deforestation, estimated at 17 million ha yr^{-1}, is 1.6 Pg yr^{-1}.

Lack of reliable estimates on the extent of area under agroforestry systems in different ecological zones is a serious problem in projecting the extent to which agroforestry practices could counter carbon emission from deforestation. The difficulty is compounded by the fact that carbon sequestered in agroforestry systems varies with a number of site- and system-specific characteristics, including climate, soil type, tree-planting densities, and tree management. Nevertheless the IPCC Report (Watson et al. 2000) estimates the area currently under agroforestry worldwide as 400 million hectares with an estimated C gain of 0.72 Mg C ha yr^{-1}, with potential for sequestering 26 Tg C yr^{-1} by 2010 and 45 Tg C yr^{-1} by 2040 1 Tg = 10^{12} g or 1 million tons. That report also estimates that 630 million hectares of unproductive cropland and grasslands could be converted to agroforestry worldwide, with the potential to sequester 391 Tg C yr^{-1} by 2010 and 586 Tg C yr^{-1} by 2040. The report further argues that agroforestry can sequester carbon at time-averaged rates of 0.2 to 3.1 Mg C ha^{-1}. (Time averaged C stock is half the C stock at the maximum rotation length, and is a scale usually used to adjust C stocks of systems with varying ages and rotation lengths to a common base.) These studies recognize that agroforestry improvement practices generally have a lower carbon uptake potential than land conversion to agroforestry practices because existing agroforestry systems have much higher carbon stocks than degraded croplands and grasslands that can be converted into agroforestry.

Annual crop–tree combinations

(a) Alleycropping

Tropical alleycropping systems, the cultivation of arable crops between tree hedgerows, represent a low end with respect to potential for C storage. As trees are periodically pruned to deposit their mulch in the alleys, C is only stored in the stem left after pruning. Therefore, pruning frequency, which can be as high as once every other month during the growing season, greatly affects the C storage capacity of the system. From data from a study of two alleycropping systems from Turrialba, Costa Rica, Koskela et al. (2000) estimated the annual 'labile' C stocks (C stored in tree leaves, branches, and crops), and compared them with the 'permanent' C stocks in tree stems. In the two systems, the perennial carbon stocks were higher than the labile stocks. However, significant decreases in soil C were detected in both systems after three to five years, thus greatly decreasing the overall value of the alleycropping system as a C sink. Some results from alleycropping experiments report on increases in soil organic carbon when prunings are returned to the soil; these increases are thought to be due to higher density of roots in alleycropping systems in comparison with adjacent plots with conventional agriculture (Schroeder 1994).

(b) Other annual crop–tree combinations

Lott et al. (2000a, 2000b) examined the biomass of agroforestry systems involving the intercropping of maize (*Zea mays*) with grevillea (*Grevillea robusta*) trees, popular in agroforestry systems in the highlands of Kenya. Their study site was the semiarid savanna area of Machakos, Kenya. Plots of sole grevillea, and sole maize were compared with agroforestry plots with both maize and grevillea intercropped for two years. The expectation was that grevillea would utilize ('capture') water and nutrient sources from deeper soil layers while the maize would rely on water and nutrients in the topsoil. However, the grevillea appeared to compete with the maize for scarce water and nutrients, resulting in lower biomass for the agroforestry system as compared to the sole grevillea or sole maize crops. Sole grevillea accumulated 1.2 times the biomass of the grevillea in the agroforestry system. The sole maize plots accumulated between two to twelve times the biomass of the maize in the agroforestry plots.

Perennial crop–tree combinations

(a) Cacao under shade trees

Koskela et al. (2000) compared labile and perennial C stocks in shaded cacao (*Theobroma cacao*) systems in Turrialba, Costa Rica, based on data from a study by Beer et al. (1990) (Table 2). The shade trees were *Erythrina poeppigiana*, a nitrogen-fixing tree that was pruned periodically for mulch, and *Cordia alliodora,* a non-nitrogen- fixing, timber tree that does not need pruning. In both systems, soil C stocks increased through time: in the cacao-*Cordia* system, soil C stocks increased by 75% in 10 years, and in the cacao *Erythrina* system they increased by 65%. C sequestration in perennial plant biomass was similar for both systems: an average of 4.28 Mg C ha^{-1} yr^{-1} for the cacao-*Cordia* system, and 3.08 Mg C ha^{-1} yr^{-1} in the cacao-*Erythrina* system (Beer et al. 1990). In spite of the relatively high values for C sequestration in perennial plant tissue and soil, the C stocks in vegetation were about 50% of those found in mature forests in the region. Similarly, in cacao agroforests of southern Cameroon, soil C was higher than in secondary forest, with values reaching 62% of the carbon stock found in primary forest (Duguma et al. 2001).

(b) Multistrata agroforestry in Central Amazonia

Schroth et al. (2002) compared three multistrata agroforestry plots with five monoculture plantations and fourteen-year-old local secondary forests at the EMBRAPA (Brazilian National Agricultural System) research station, near Manaus, Brazil. Monocultures included peach palm (*Bactris gasipaes*) grown for fruit, cupuacu (*Theobroma grandiflorum*) – also for fruit, rubber (*Hevea brasiliensis* grafted with *Hevea pauciflora*), and orange (*Citrus sinensis*). The multistrata systems contained different mixtures of peach palm for fruit, peach palm for heart of palm, rubber, cupuacu, Brazil nut (*Bertolletia excelsa*), annatto (*Bixa orellana*), coconut palms (*Cocos nucifera*) and orange, as well as intercrops of papaya (*Carica papaya*), cassava (*Manihot esculenta*), maize and cowpea (*Vigna unguiculata*). In general, it was found that the trees in the multistrata systems tended to be larger in canopy volume than those in the monocultures. None of the systems reached the biomass of the secondary forest that they had replaced: the most productive plot only contained about two-thirds of the biomass of the natural regrowth. The primary forest contained more than four times the amount of total

Table 2. Carbon stocks in two cacao (*Theobroma cacao*) – shade tree systems under humid tropical conditions in Turrialba, Costa Rica.

System	Perennial C stock (Mg C ha^{-1})		Labile C stock (Mg C ha^{-1} yr^{-1})	
	Soil	Tree and cacao	Litter	Tree and cacao
Cordia alliodora + cacao				
Initial	98	–	–	–
5th year	138	18.6	2.6	3.0
10th year	171	42.8	8.8	3.0
Erythrina poeppigiana + cacao				
Initial	115	–	–	–
5th year	152	7.8	4.1	4.4
10th year	190	30.8	9.6	5.6

The perennial carbon stocks in vegetation comprise the cacao branches and stems, tree stems, an estimated 85% of roots (coarse root proportion), and tree branches in the *Cordia* + cacao system. The labile carbon stocks in vegetation comprise the leaves, an estimated 15% of roots (fine root proportion), and tree branches in the *Erythrina* + cacao system
Source: Adapted from Beer et al. (1990) by Koskela et al. (2000).

biomass than that of the most productive plot: peach palm for fruit.

Other perennial crop–tree combinations

Other studies of biomass and C sequestration by perennial crop-tree combinations also show that agroforestry systems accumulate considerably more C than the monocultures of annual crops; but lower than that accumulated by natural forests in the region. For example, results from studies in Cameroon showed that total biomass in cacao agroforests (304 Mg/ha) was much greater than in food crop fields (85 Mg ha^{-1}), and ranked third after the biomass in primary forest (541 Mg $^{-1}$ha), and long-term fallow (460 Mg/ha) (Duguma et al. 2001). In Sumatra, Indonesia, annual aboveground C stocks accumulation rates during the establishment phase after slash-and-burn land clearing were 1, close to 2, and 3.5 Mg C ha^{-1} for sun coffee (*Coffea* sp.), shade coffee and fallow regrowth, respectively. It was estimated that in Sumatra, conversion of all sun-coffee to shade coffee systems could increase average landscape level C stocks by 10 Mg C/ha during a 20-year period, or 0.5 Mg C ha^{-1} yr^{-1} (van Noordwijk et al. 2002).

As is to be expected, while comparing the annual crop-tree combinations (including alleycropping) with the perennial crop-tree combinations, agroforestry systems with perennial crops may be important carbon sinks, while intensively managed agroforestry systems with annual crops are more similar to conventional agriculture. However, even the perennial crop-tree combinations do not attain C sequestration rates comparable to the natural forests that grow in the region.

Temperate-zone systems

One of the most quoted estimates, which perhaps was also the basis for most of the subsequent calculations, is the report of Dixon et al. (1994), which states that the potential carbon storage with agroforestry in temperate areas ranges from 15 to 198 Mg C ha^{-1} with a modal value of 34 Mg C ha^{-1}. Most of the subsequent reports that are available on temperate systems are from North America. Garrett and McGraw (2000) acknowledge that reliable statistics of areas under alleycropping practice in North America are not available, but suggest that more than 45 million ha of nonfederal cropland in the United States with erodibility index (EI)[1] greater than 8 (USDA-SCS 1989) could potentially benefit from the use of alleycropping. They further argue that in the Midwest, where alleycopping adoption is greatest, an area of more than 7.5 million ha has EI greater than 10 (Noweg and Kurtz 1987), and approximately 3.6 million ha of this land that is recommended for forestry planting would

[1] Erodiblity Index (EI) is calculated based on the parameters used in the Universal Soil Loss Equation. In practical terms, soils with an EI < 5 are only slightly susceptible to erosion, and those with EI > 8 are highly susceptible. An EI between 8 and 15 indicates that crop land bare of cover could erode eight or more times the tolerance value and EI of 15 and higher suggests that erosion could occur at 15 or more times the tolerance value (see End Note 1, Garrett et al., this volume).

be ideal for alleycropping. Furthermore, across the United States, approximately 7 million ha of pastureland has high potential for conversion to cropland and could be alley-cropped. Another 16 million ha have medium potential for conversion, and a total of 25 million ha has high or medium potential for conversion, all of which could be alley-cropped. Thus, the estimated area suitable, or that could soon be available, for alleycropping in the United States is approximately 80 million ha of land. Based on the estimated average total soil C sequestration potential of 142 Tg C yr^{-1} from 154 million ha of total U.S. crop lands (Lal et al. 1999), the potential for C sequestration through alleycropping could be 73.8 Tg C yr^{-1}.

Discussing the area under silvopastoral systems, Clason and Sharrow (2000) argue that, given the widespread co-occurrence of grazing and forestry across North America, the joint production of livestock and tree products is by far the most prevalent form of agroforestry found in the United States and Canada. But, areas are generally classified as forest, rangeland, and pasture, implying a single dominant form of land; area statistics do not deal with multiple product systems such as forested rangelands, grazed forests, and pasture with trees. According to Brooks (1993), forests currently occupy approximately 314 million ha of land in the United States and 436 million ha in Canada, of which only 70% and 60%, respectively, are timber-producing areas. Considerable amounts of forage may be available for grazing under trees in mature open canopied forest stands, such as semiarid conifer forests and savannas of western United States. Based on inventories made on 11-year-old Douglas-fir (*Pseudotsuga menziesii*)/ perennial ryegrass (*Lolium perenne*)/ subclover (*Trifolium subterraneum*) silvopastoral agroforestry system in western Oregon, United States, Sharrow and Ismail (2004) reported that the agroforests (silvopastures) were more efficient in accreting C than tree plantations or pasture monocultures. They attributed this to the advantages of silvopastures in terms of higher biomass production and active nutrient cycling patterns of both forest stands and grasslands, compared to those of pastures or timber stands alone. Even close-canopied forests may produce considerable amounts of vegetation following timber harvests or fire, which could remain unexploited if not grazed by ruminant livestock. Based on these considerations, Clason and Sharrow (2000) conclude that about 70 million ha, more than a quarter of all forest land in the United States, is grazed by livestock, which is also 13% of the total grazing land in the country (USDA 1996). At the average estimated total soil C sequestration potential for U.S. grazing lands of 69.9 Tg C yr^{-1} (Follett et al. 2001), the potential for silvopastoral systems could be around 9.0 Tg C yr^{-1}. Estimates of areas under other agroforestry practices (windbreaks, riparian forest buffer, and forest farming) are even more difficult to make. The (U.S.) National Agroforestry Center estimates that protecting the 85 million ha of exposed cropland in the North Central United States by planting 5% of the field area to windbreaks would sequester over 58 Tg C (215 Tg CO_2) in 20 years, or an average of 2.9 Tg C yr^{-1}. (National Agroforestry Center, 2000, *USDA National Agroforestry Center Resources*. www.unl.edu/nac.) Similarly, planting windbreaks around the 300 000 unprotected farms in the region would result in 120 million trees (at the rate of 400 trees/home), storing 3.5 Tg C in 20 years or 0.175 Tg C yr^{-1}. Planting live snow-fences along roads would be another opportunity. All together, a conservative estimate of the C sequestration potential of a reasonable windbreak-planting program could be around 4 Tg C yr^{-1}.

Riparian buffers and short-rotation woody crops (SRWC) are other notable agroforestry opportunities for C sequestration. The U.S. Department of Agriculture (USDA) has committed to planting 3.2 million km (2 million miles) of conservation buffers (US Dept of Energy 1999). If one-fourth of these buffers were 30 m-wide forested riparian buffers, C removal would exceed 30 Tg C in 20 years or 1.5 Tg C yr^{-1}. In the Pacific Northwest, 10-year-old irrigated plantations of SRWC are estimated to remove 222.3 Mg of C ha^{-1} or an average of 22.3 Mg C yr^{-1}. Although statistics on these practices and others such as forest farming are unavailable, for discussion purposes, a modest amount of 2.0 Tg C yr^{-1} could be presented as the C sequestration potential of these agroforestry practices. Based on these assumptions, Nair and Nair (2003) estimated that the total C sequestration potential through agroforestry practices in the United States could amount to 90 Tg C yr^{-1} (Table 3). One report estimates that 57×10^6 ha of marginal or degraded land is considered suitable for establishment of trees in Canada (Thevathasan and Gordon in press); but how much of this land area will actually be converted to tree-planting and agroforestry is uncertain.

Table 3. Estimated C sequestration potential through agroforestry practices in the USA by 2025.

Agroforestry Practice	[1]Estimated area	[#]Potential C sequestration Tg C yr^{-1}, sum of above- and belowground storage
Alleycropping	80×10^6 ha	73.8
Silvopasture	70×10^6 ha	9.0
Windbreaks	85×10^6 ha[@]	4.0
Riparian Buffer	0.8×10^6 km of 30-m-wide forested riparian buffers	1.5
Short rotation woody crops (SRWC), forest farming, etc.	2.4×10^6 km conservation buffer including SRWC	2.0
Total		90.3

[1] Area that is currently under or could potentially be brought under the practice
[@] Area of exposed cropland, 5% of which to be planted to windbreaks
[#] The time frame during which these estimates will be appropriate depends on how fast these potentially feasible practices are implemented. Assuming their implementation by 2010, the estimated C sequestration benefits could be appropriate for 2025; potential benefits thereafter will depend on expansion or shrinkage of areas under the different practices.
Source: Nair and Nair (2003).

Exploiting C sequestration potential of agroforesty for the benefit of landowners: Payment for environmental services

The Kyoto Protocol triggered a strong increase in investment in plantations as carbon sinks, although the legal and policy instruments and guidelines for management are still debated (FAO 2000). A number of countries have already prepared themselves for the additional funding for the establishment of human-made forests. In an initiative claimed to be the first of its kind, in 1997 Costa Rica established tradable securities of carbon sinks that could be used to offset emissions and to utilize independent certification insurance. According to a FAO report, in 2000 greenhouse gas mitigation funding covered about 4 million hectares of forest plantations worldwide (FAO 2000). The recognition of afforestation and reforestation as the only eligible land use under the CDM of the Kyoto Protocol is expected to lead to a steep increase in forest plantation establishment in developing countries.

Can environmental services pay for agroforestry promotion? We present some examples from both the tropical region and North America to illustrate the current trend in this direction.

Tropical region

In Costa Rica, in 1997, $14 million was invested for the Payment for Environmental Services (PES), which resulted in the reforestation of 6 500 ha, the sustainable management of 10 000 ha of natural forests, and the preservation of 79 000 ha of private natural forests (R. Nasi, S. Wunder, and J. J. Campos A.: pers. comm, March 2002). Eighty percent of this funding originated nationally from a tax on fossil fuels; the other 20% came from the international sale of carbon from public protected areas and contributed with $2 million. Recently, the World Bank provided a $32.6 million loan to Costa Rica to fund the PES trough a Project called ‘Ecomarkets’. It came along with a grant from the Global Environment Facility (GEF) of approximately $8 million. These schemes only considered reforestation or forest conservation. Starting in 2003, PES in Costa Rica also includes agroforestry systems, with payments calculated based on the number of individual trees present in the farm.

Can this example from Costa Rica be extended to other tropical regions? Some examples exist, involving private enterprises and development agencies. In Guatemala, cacao and other shaded crops have been proposed for financing to the farmers through payment of environmental services such as carbon sequestration (Parrish et al. 2003). For example, TechnoServe, a non-profit organization, is undertaking a project supported by the Ford Foundation and the U.S. Agency for International Development, USAID, which investigates the use of carbon trading to promote sustainable coffee cultivation by smallholder producers in Central America. Based on data compiled from a Guatemalan coffee cooperative, they are determining the range of prices for sequestered carbon to be used in the delineation of the carbon trading and financing instrument; they are possibly extending the scheme to other agricultural commodities such as cocoa (Newmark TE 2003, *Carbon sequestration and cocoa production: financing sustainable development by trading carbon emission credits.* http://natzoo.si.edu/smbc/Research/Cacao/newmark.htm.).

Also in Guatemala, a private company, AES Thames, has already invested more than $2 million to establish agroforestry and woodlot systems to help offset CO_2 emissions from its coal-fired power plant (Dixon 1995). Through a grant to CARE, an NGO (non-governmental organization), AES is assisting 40 000 farmers to plant trees during a ten-year period. By managing the agroforestry systems for farm and community use, the project is expected to reduce deforestation in eastern Guatemala.

In Paraguay, another subsidiary of the AES corporation has committed more than $2 million to the Nature Conservancy, an NGO, for land purchase and agroforestry to help offset CO_2 emissions from a coal-fired cogeneration facility. The long-term project goals are to promote sustainable agroforestry systems, preserving existing tropical forest (biosphere reserves) and creating a sustainable watershed management system in the Paraguayan-Brazilian border (Dixon 1995). In Chiapas, southern Mexico, the Scolel Té community forestry project, with funding from the British Department for International Development (DFID) is developing model planning and administrative systems by which the farmers can gain access to carbon markets (FAO 2001; Tipper 2002). Smallholder farmers and local communities identify reforestation, agroforestry and forest restoration activities that are both financially beneficial and intended to sequester or conserve carbon. Offsets are sold through a trust fund managed by a local NGO. Carbon has been sold to various purchasers, including the International Automobile Federation. Approximately 300 farmers, having an average holding size of one hectare each, are involved. The average C sequestration potential is 26 Mg ha^{-1} at a cost of $12 per Mg. The baseline used is the mean carbon storage potential of the previous land use, assuming that the land use would have continued in the absence of project intervention. Cacho et al. (2003) have documented the details of five C sequestration projects that are currently being implemented involving smallholders in Latin America and Indonesia (Cacho OJ, Marshall GR, and Milne M, 2003, Smallholder agroforestry projects: Potential for carbon sequestration and poverty alleviation, ESA Working Paper 03-06; www.fao.org/es/esa)

A number of development agencies in Kenya are currently considering the potential of establishing a carbon credit payments scheme for small landholders in the buffer zone surrounding Mt. Kenya National Park, an International Biosphere Reserve and a World Heritage Site (KWS, 2002, *Background of Mt. Kenya National Park.* HTML: http://www.kws.org/mtkenya.htm#Background Kenya Wildlife Service. Nairobi, Kenya.). The park is located in south-central Kenya and surrounded by Nyeri, Embu, Kirinyaga and Meru Districts. These districts serve as important agricultural production centers of food and cash crops in Kenya. Based on a landscape-level study at the Manupali watershed in the Philippines, Shivley et al. in press calculated that carbon storage via land use modification costs $3.30 per ton on fallowed land and $62.50 per ton on land that otherwise supports high-value cropping. Carbon storage through agroforestry was found to be less costly than via a pure tree-based system, which is a strong argument in favor of agroforestry rather than forestry *per se* in reforestation projects. These are some examples of how systems for payment of environmental services to farmers can be carried out as mitigation projects by private companies, in alliance with NGOs or development agencies. These projects do not consist of just agroforestry but they are integrated with other forestry and conservation projects.

Agroforestry and other planted and/or managed systems may have advantages over natural forest management in terms of C sequestration in situations where land and tree tenure are contentious issues related to the question of compensating for the opportunity cost for forest conservation. For example, Tomich et al. (2002) reported that in Indonesia, C offsets through 'agroforestation' seemed more feasible than forest conservation because property rights over timber from planted trees would be easier to establish and enforce than property rights over timber from natural forests.

North America

An interesting recent development in North America and other industrialized temperate regions of the world is the increasing recognition of the value of environmental services provided by sustainable land management systems such as agroforestry. Silvopasture, for example, is becoming an increasingly popular agroforestry practice in southern United States (Clason and Sharrow 2000; Workman et al. 2003). Environmental benefits provided by this practice include water quality improvement, soil conservation, carbon sequestration, wildlife habitat protection, and aesthetics (Alavalapati and Nair 2001; Workman et al. 2003). Environmental economists argue that internalizing the environmental benefits through compensation schemes could potentially influence ranchers' and other landowners' decision to adopt silvopasture. Valuation of environmental improvement associated with forestry and agricultural practices has been studied extensively (Cooper and Kleim 1996; Bateman and Willis1999). Shrestha and Alavalapati (in press) studied the public willingness to pay (WTP) for improvements in water quality, carbon sequestration, and wildlife habitat through silvopasture in the Lake Okeechobee watershed, Florida, using the random parameter logit model. The study showed that the WTP for a moderate level of improvement in these environmental attributes amounted to $137.97 per household per year, out of which carbon sequestration alone accounted for 42.07% ($58.05). With 1.34 million households in the watershed, the total WTP for environmental improvement would be $924.4 million. With the cost of silvopastural practice as perceived by ranchers at $23.02 per ha per year, the annual opportunity cost of silvopasture adoption on 1.06 million hectares of ranchland in the watershed would be $24.41 million. Using the total WTP as a trust fund, its annual returns could be used to compensate ranchers for the environmental services provided by silvopasture. This is just an example of the scope and nature of public policy that will need to be developed for supporting agroforestry adoption for environmental services including carbon sequestration.

Another interesting development worth mentioning in this context is the development of the so-called Carbon Markets. Such markets can only develop under a 'cap and trade' system, in which the total amount of carbon emissions is limited by a mandatory cap, and carbon-emitting industries are allowed to meet their targets with some combination of carbon emission reduction technologies and the purchase of carbon offsets on open financial exchanges (V. A. Sample, JG Gray Distinguished Lecture, Univ. of Florida, 23 April 2003). The concept is based on the highly successful sulfur dioxide (SO_2) cap-and-trade mechanism aimed at combating acid rain, established in the United States by the 1990 Clean Air Act amendments. Under these amendments, the US Environmental Protection Agency (EPA) issued permits to utilities for a certain level of SO_2, and the right to trade these permits with other utilities. Ten years later, the market in SO_2 permits had grown to $3 billion, overall SO_2emissions by the utility industry as a whole were significantly *lower* than EPA targets, and the reduction had been accomplished at about one tenth of the predicted cost.

The 2001 withdrawal of the United States from participation in the Kyoto Protocol has significantly slowed progress toward developing a corresponding cap-and-trade system for carbon dioxide (CO_2) emissions in the United States and internationally. With the United States being the source of about one-quarter of the world's carbon emissions, no such system is likely to be effective in reducing global atmospheric carbon without U.S. participation. This has not stopped individual states within the United States from acting, however. More than half the states have adopted voluntary or mandatory programs for reducing carbon emissions. Other countries see opportunity in the U.S. retreat from the Kyoto Protocol, particularly those in Europe that are well along in developing their own emissions trading systems. The director of CO2e.com, a London brokerage firm, observed 'Now that the Americans are out, Europe can dominate the emissions trading market. It entitles the Europeans to write the rules for global trading.' (O. Pohl, *New York Times*, 10 April 2003).

On 16 January 2003, the Chicago Climate Exchange (CCX) opened for business, and it is poised to become the first financial market in the United States to begin trading carbon credits like any other commodity (P. Behr and E. Pianin: Firms Start Trading Program for Greenhouse Gas Emissions. *Washington Post*, 17 January 2003, p.A14. Additional information available at: *www.chicagoclimatex.com*). The goal of the CCX is to implement a voluntary, private cap-and-trade pilot program for reducing and trading greenhouse gas emissions.

What might all these mean for private forest landowners in the industrialized nations? Many questions still remain to be answered and details to be addressed. Clearly, however, major U.S. and international forest products companies have determined that generating and selling carbon credits has a potential to significantly add to the income stream from their forestry operations, and improve their financial bottom line.

Conclusions

The value of forests and trees in sequestering carbon and reducing carbon dioxide emission to the atmosphere is being recognized increasingly the world over. Forest plantations and agroforestry systems are thus recognized to have the potential to regain some of the carbon lost to the atmosphere in the clearing of primary or secondary forests. Although neither regrowth nor plantations can come close to replacing the full amount of carbon that was present in the primary forest, plantations and agroforestry systems have the added benefit of providing valuable products and food to local people. The rotation ages for plantations and trees in agroforestry systems play a large role in the amount of carbon they can sequester. In mixed plantations or agroforestry systems, since rotation lengths vary within the system according to the species, a complete clearing of the plot (which would increase the release of soil carbon) is less likely. Consideration of how the biomass from managed plots is used should also play a role in the carbon equation. If used for wood, about 25% of the carbon from that rotation can be considered sequestered for an additional span of five or more years. If used for fuelwood, most of the carbon can be considered returned to the atmosphere within the year.

As the concept of 'carbon credits' being paid by fossil fuel emitters to projects that sequester or reduce carbon output becomes more common, many nations and organizations will seek to find inventive ways to sequester carbon. The clearing of primary forest releases more carbon than natural regrowth or fast-growing plantations could recover in 25 years or more. Therefore, protection of primary forest should be top priority when looking at ways to reduce carbon emission from the tropics. The most important role that agroforestry and plantations may play is to offset destruction of primary forest by providing the necessary wood products from land that has already been cleared. If this can be done in a manner that provides competitive biomass accumulation rates to that of natural regrowth and is sustainable in terms of soil fertility, then plantations and agroforestry systems could play a substantial role in CO_2 mitigation in the tropics. The recent trends in economic valuation of the ecosystem services of forests and trees, and the development of private capital markets to actually monetize the value of those services to the benefit of forest landowners, point to exciting opportunities and new developments in this area.

In the face of carbon markets, C storage becomes an additional output that landowners might consider in their management decisions. This would change the dynamics of their agroforestry systems in terms of the rotation age of trees, crop-tree mixture, and silviculture and other management practices. With the introduction of carbon payments, agroforestry systems that are otherwise less profitable may become more attractive or vice-a-versa. This suggests that research addressing both biophysical and socioeconomic issues of carbon sequestration is needed.

References

Alavalapati J.R.R., Shrestha R.K., Stainback G.A. and Matta J.R. 2004. Agroforestry development: An environmental economic perspective. (This volume).

Alavalapatti J.R.R. and Nair P.K.R 2001. Socioeconomic and institutional perspectives of agroforestry: an overview. pp. 52–62. In: World Forests, Markets and Policies, Vol. 2 in the series World Forests, Kluwer, Dordrecht, The Netherlands.

Bass S., Dubois O., Mouracosta P., Pinard M., Tipper R. and Wilson C. 2000. Rural Livelihood and Carbon Management. IIED Natural Resources Paper No. 1. International Institute for Economic Development. London, UK.

Bateman I.J. and Willis K.G. (eds) 1999. Valuing Environmental Preferences: Theory and Practice of Contingent Valuation Method in the US, EU, and Developing Countries. Oxford University Press, New York.

Beer J., Bonnemann A., Chavez W., Fassbender H.W., Imbach A.C. and Martel I. 1990. Modelling agroforestry systems of cacao (*Theobroma cacao*) with laurel (*Cordia alliodora*) or poro (*Erythrina poeppigiana*) in Costa Rica. V. Productivity indices, organic material models and sustainability over ten years. Agroforest Syst 12: 229–249.

Brooks D.J. 1993. U.S. forests in a global context. USDA For. Serv. Gen. Tech. Rep. RM-228. US For. Serv. Rocky Mountain For. and Range Exp. Stn, Ft. Collins, CO.

Cairns M.A. and Meganck R.A. 1994. Carbon sequestration, biological diversity, and sustainable development: integrated forest management. Environ Manag 18: 13–22.

Clason T.R. and Sharrow S.H. 2000. Silvopastoral practices. pp. 119–147. In: Garrett H.E., Rietveld W.J. and Fisher R.F. (eds) North American Agroforestry: An Integrated Science and Practice. Am. Soc. Agronomy, Madison, WI.

Cooper J.C. and Kleim R.W. 1996. Incentive payments to encourage farmer adoption of water quality protection practices. Am J Agr Econ 78: 54–64.

Dixon R.K. (1995) Agroforestry systems: sources or sinks of greenhouse gases? Agroforest Syst 31: 99–116.

Dixon R.K., Winjum J.K., Andrasko K.J., Lee J.J. and Schroeder P.E. 1994. Integrated systems: assessment of promising agroforest and alternative land-use practices to enhance carbon conservation and sequestration. Climatic Change 30: 1–23.

Duguma B., Gockowski J. and Bakala J. 2001. Smallholder cacao (*Theobroma cacao* Linn.) cultivation in agroforestry systems of West and Central Africa: challenges and opportunities. Agroforest Syst 51: 177–188.

Evans J. 1999. Planted forests of the wet and dry tropics: their variety, nature, and significance. New Forestry 17: 25–36.

FAO 2000. Global Forest Resources Assessment 2000. Main Report. FAO, Rome, Italy 512 pp.
FAO 2001. State of the world's forests 2001. Food and Agriculture Organization of the United Nations. Rome, Italy, 181 pp.
FAO 2003. State of the World's Forests 2003. Food and Agriculture Organization of the United Nations. Rome, Italy, 126 pp.
Follett R.F., Kimble J.M. and Lal R. (eds) 2001. The Potential of U. S. Grazing Lands to Sequester Carbon and Mitigate the Greenhouse Effect. Lewis Publ., Boca Raton, FL.
Garrett H.E. and McGraw R.L. 2000. Alley cropping practices. pp. 149–188. In: H.E. Garrett, W.J. Rietveld and R.F. Fisher (eds) North American Agroforestry: An Integrated Science and Practice, Am. Soc. Agronomy, Madison, WI.
Harmon ME (2001) Carbon sequestration in forests: addressing the scale question. J Forest 99 (4): 24–29.
Houghton R.A., Davidson E.A. and Woodwell G.M. 1998. Missing sinks, feedbacks, and understanding the role of terrestrial ecosystems in the global carbon balance. Global Biogeochem Cy 12: 25–34.
ICRAF 1995. International Centre for Research in Agroforestry. Annual Report, 1995. ICRAF, Nairobi, 288 pp.
IPCC 2000. Special Report on Land Use, Land Use Change and Forestry. Summary for Policy Makers. Geneva, Switzerland. 20 pp.
Johnsen K.H., Wear D., Oren R.., Teskey R.O., Sanchez F., Will R., Butnor J., Markewitz D., Richter D., Rials T., Allen H.L., Seiler J., Ellsworth D., Maier C., Katul G. and Dougherty P.M. 2001. Meeting global policy commitments: Carbon sequestration and southern pine forests. J Forest 99 (4): 14–21.
Koskela J., Nygren P., Berninger F. and Luukkanen O. 2000. Implications of the Kyoto Protocol for tropical forest management and land use: prospects and pitfalls. Tropical Forestry Reports 22. University of Helsinki, Department of Forest Ecology. Helsinki. 103 pp.
Kursten E. 2000. Fuelwood production in agroforestry systems for sustainable land use and CO_2 mitigation. Ecol Eng 16: S69–S72.
Lal R., Kimble J.M., Follett R.F. and Cole C.V. (eds) 1999. The Potential of U. S. Cropland to Sequester Carbon and Mitigate the Greenhouse Effect. Lewis Publ, Boca Raton, FL.
Lott J.E., Howard S.B., Ong C.K. and Black C.R. 2000a. Long-term productivity of a *Grevillea robusta*-based overstorey agroforestry system in semi-arid Kenya. I. Tree growth. Forest Ecol Manag 139: 175–186.
Lott J.E., Howard S.B., Ong C.K. and Black C.R. 2000b. Long-term productivity of a *Grevillea robusta*-based overstorey agroforestry system in semi-arid Kenya. II. Crop growth and system performance. Forest Ecol Manag 139: 187–201.
Loveland T.R. and Belward A.S. 1997. The IGBP-DIS global 1 km land cover data set, DISCover ® first results. Int J Remote Sens 18: 3289–3295.
Montagnini F. 2001. Strategies for the recovery of degraded ecosystems: experiences from Latin America. Interciencia 26(10): 498–503.
Montagnini F., González E., Rheingans R. and Porras C 1995. Mixed and pure forest plantations in the humid neotropics: a comparison of early growth, pest damage and establishment costs. Commonw Forest Rev 74: 306–314.
Montagnini F. and Mendelsohn R. 1996. Managing forest fallows: improving the economics of swidden agriculture. Ambio 26: 118–123.
Montagnini F. and Porras C. 1998. Evaluating the role of plantations as carbon sinks: an example of an integrative approach from the humid tropics. Environ Manag 22: 459–470.
Myers N. 1996. The world's forests: problems and potential. Environ Conserv 23(2): 156–168.
Nair P.K.R. 2001. Agroforestry. pp. 375–393. In: Our Fragile World: Challenges and Opportunities for Sustainable Development, Forerunner to The Encyclopedia of Life Support Systems. UNESCO, Paris, France & EOLSS, UK.
Nair P.K.R., Buresh R.J., Mugendi D.N. and Latt C.R. 1999. Nutrient cycling in tropical agroforestry systems: Myths and science. p. 1–31. In: Buck L.E., Lassoie J.P. and Fernandes, E.C.M. (eds), Agroforestry in Sustainable Agricultural Systems. CRC Press, Boca Raton, FL, USA.
Nair P.K.R. and Nair V.D. 2003. Carbon Storage in North American Agroforestry Systems. pp. 333–346. In: J. Kimble, L.S. Heath, R.A. Birdsey and R. Lal (eds) The Potential of U.S. Forest Soils to Sequester Carbon and Mitigate the Greenhouse Effect. CRC Press, Boca Raton, FL, USA.
Noweg T.A. and Kurtz W.B. 1987. Eastern black walnut plantations: an economically viable option for conservation of reserve lands within the corn belt. North J Appl Forest 4: 158–160.
Oren R., Ellsworth D.S., Johnsen K.H., Phillips N., Ewers B., Maler C., Schaefer K.V.R., McCarthy H., Hendrey H., McNutty S.G. and Katul G.G. 2001. Soil fertility limits carbon sequestration by forest ecosystems in a CO_2-enriched atmosphere. Nature 411: 469–471.
Palm C.A. and 17 others 2000. Carbon sequestration and trace gas emissions in slash-and-burn and alternative land-uses in the humid tropics. Final Report, Alternatives to Slash and Burn (ASB) Climate Change Working Group, Phase II. ICRAF, Nairobi, Kenya.
Parrish J., Reitsma R. and Greensberg R. 2003. Cacao as crop and conservation tool. http://nationalzoo.si.edu/conservationandscience/migratorybirds/research/cacao/parrish.cfm.
Piotto D., Montagnini F., Ugalde L., and Kanninen M. 2003. Performance of forest plantations in small and medium sized farms in the Atlantic lowlands of Costa Rica. Forest Ecol Manag 175: 195–204.
Ruark G.A., Schoeneberger M.M. and Nair P.K.R. 2003. Agroforestry–Helping to Achieve Sustainable Forest Management. UNFF (United Nations Forum for Forests) Intersessional Experts Meeting on the Role of Planted Forests in Sustainable Forest Management, 24–30 March 2003, New Zealand. www.maf.govt.nz/unff-planted-forestry-meeting
Schlesinger W.H. and Lichter J. 2001. Limited carbon storage in soil and litter of experimental forest plots under increased atmospheric CO_2. Nature 411: 466–468.
Schroeder P. 1992. Carbon storage potential of short rotation tropical tree plantations. Forest Ecol Manag 50: 31–41.
Schroeder P. 1994 Carbon storage benefits of agroforestry systems. Agroforest Syst 27: 89–97.
Schroth G., D'Angelo S.A., Teixeira W.G., Haag D. and Lieberei R. 2002. Conversion of secondary forest into agroforestry and monoculture plantations in Amazonia: consequences for biomass, litter and soil carbon stocks after 7 years. Forest Ecol Manag 163: 131–150.
Schulze E.D., Wirth C. and Heimann M. 2000. Managing forests after Kyoto. Science 289: 2058–2059.
Sharrow S.H. and Ismail S. 2004. Carbon and nitrogen storage in agroforests, tree plantations, and pastures in western Oregon, USA. Agroforest Syst 60: 123–130.
Shepherd D. and Montagnini F. 2001. Carbon Sequestration Potential in mixed and pure tree plantations in the humid tropics. J Trop For Sci 13: 450–459.

Shivley G.E., Zelek C.A., Midmore D.J. and Nissen T.M. Carbon sequestration in a tropical landscape: An economic model to measure its incremental cost. Agroforest Syst (in press).

Shrestha R.K. and Alavalapati J.R.R. Valuing environmental benefits of silvopasture practices: A case study of the Lake Okeechobe Watershed in Florida. Ecol Econ (in press).

Smith J. and Scherr S.J. 2002. Forest Carbon and Local Livelihoods: Assessment of Opportunities and Policy Recommendations. CIFOR Occasional Paper 37, Centre for International Forestry Research, Jakarta, Indonesia.

Thevathasan N.V. and Gordon A.M. Enhancing greenhouse gas (GHG) sinks in agroecosystems through agroforestry based land-use practices in Canada. Forest Chron (in press).

Tipper R. 2002. Helping indigenous farmers to participate in the international market for carbon services: the case of Scolel Té pp. 222–233. In: Pagliola S., Bishop J. and Landell-Mills N. (eds), Selling Forest Environmental Services. Market-based Mechanisms for Conservation and Development. Earthscan, London.

Tomich T.P., de Foresta H., Dennis R., Ketterings Q., Murdiyarso D., Palm C., Stolle F., and van Noordwijk M. 2002. Carbon offsets for conservation and development in Indonesia? Am Alt Agr 17: 125–137.

USDA 1996. Grazing lands and people: A national program statement and guidelines for the cooperative extension service. USDA Ext. Serv., Dec. 1996. USDA, Washington, DC.

USDA–SCS 1989. Soil, Water, and Related resources on Nonfederal Land in the United States: Analysis of Conditions and Trends. Misc. Publ. 1482, Second RCA Appraisal, USDA Soil Cons. Serv., Washington DC.

US Dept of Energy 1999. Carbon Sequestration: State of Science. Draft Report. Ch. 4: Carbon Sequestration in Terrestrial Ecosystems. USDE, Washington, DC.

van Noordwijk M., Rahayu S., Hairiah K., Wulan Y.C., Farida A. and Verbist B. 2002. Carbon stock assessment for a forest-to-coffee conversion landscape in Sumber-Jaya (Lampug, Indonesia): from allometric equations to land use change analysis. Science in China Series C-Life Sciences 45: 75–86 Suppl. S OCT 2002. Science in China Press, Beijing.

Watson R.T., Noble I.R., Bolin B., Ravindranath N.H., Verardo D.J. and Dokken D.J. (eds) 2000. Land Use, Land-Use Change, and Forestry. Intergovernmental Panel on Climate Change (IPCC), Special report. Cambridge Univ. Press. New York.

Workman S.W., Bannister M.E. and Nair P.K.R. 2003. Agroforestry potential in the southeastern United States: Perceptions of landowners and extension professionals. Agroforest Syst 59: 73–83.

Young A. 1997. Agroforestry for Soil Management. 2nd Ed. C.A.B. International. Wallingford, UK, 320 pp.

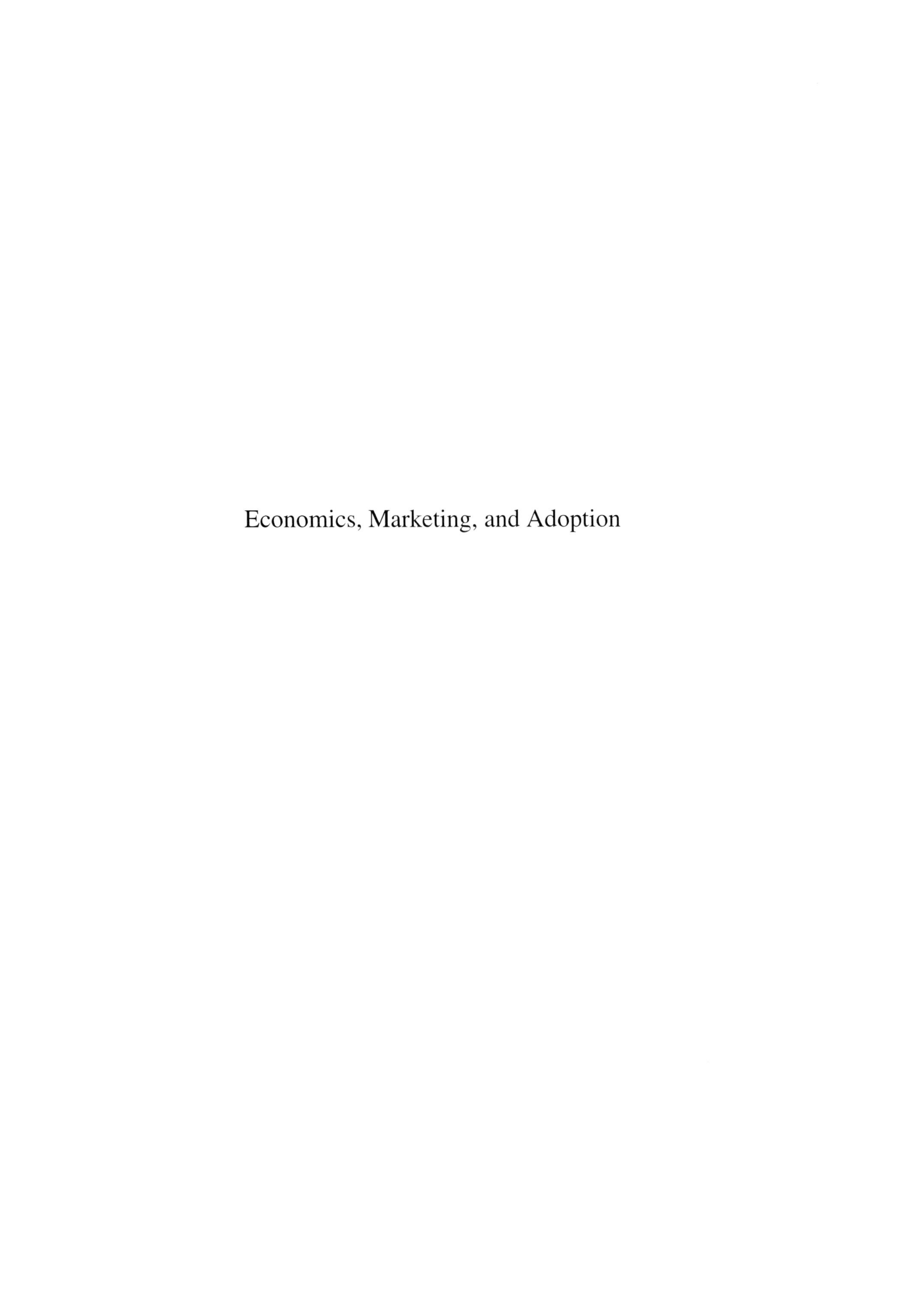

Economics, Marketing, and Adoption

Agroforestry Systems 61: 299–310, 2004.

Agroforestry development: An environmental economic perspective

J.R.R. Alavalapati*, R.K. Shrestha, G.A. Stainback and J.R. Matta
*School of Forest Resources and Conservation, University of Florida, P.O. Box 110410, Gainesville, Florida, 32611–0410 USA; *Author for correspondence (e-mail: janaki@ufl.edu)*

Key words: Contingent valuation, Externalities, Profitability, Silvopasture, Willingness to accept (WTA), Willingness to pay (WTP)

Abstract

Agroforestry systems (AFS) provide a mix of market goods and nonmarket goods and services. We postulate that if nonmarket goods and services can be internalized to the benefit of landowners, the adoption of AFS will increase. Using the theory of externality as a conceptual framework, this paper provides an environmental economic logic for developing incentive policies to internalize environmental services especially in the industrialized countries. Specifically, the paper addresses the following questions with focus on North America in general and southern United States in particular: What is the effect of environmental costs and benefits on the adoption of silvopasture? Do households care for carbon sequestration, water quality improvement, and biodiversity associated with silvopasture? Will they be willing to pay for them? If so, how much? Will ranchers adopt more silvopasture if incentives are provided? Which incentive policy, a price premium or a direct payment, is more effective? It has been found that the profitability of silvopasture would increase, relative to conventional ranching, if environmental services are included. Estimates of public willingness to pay for environmental services associated with silvopasture and estimates of ranchers' willingness to accept for the adoption of silvopasture will provide a scientific basis for policy development.

Introduction

Agroforestry is widely considered as a potential way of improving socioeconomic and environmental sustainability both in the tropics and in the temperate regions (Alavalapati and Nair 2001; Garrett et al., 2000; Nair 2001). Agroforestry systems (AFS) provide a mix of market goods such as food, wood products, and fodder to cattle, and nonmarket goods and services including soil conservation, water and air quality improvement, biodiversity conservation, and scenic beauty. As such, AFS contribute to the rural economy, employment, poverty alleviation, and environmental protection at a local, regional, and national level. One of the key factors in determining agroforestry adoption is the relative profitability of the practice in comparison with other land-use practices.

Determining AFS profitability can be complex. Profitability from a landowner perspective, generally called 'private profitability', can be different from that of a societal perspective, often referred to as 'social profitability.' The exclusion or inclusion of social benefits and costs and nonmarket values, often referred to as externalities, largely differentiates the private and social profitability.[1] In the context of AFS, private profitability analysis does not include nonmarket goods and services because these are public goods. These goods are generally indivisible and consumed by all at a level that is the same for all. Also, producers of these goods cannot exclude people from consuming these goods (Peterson 2001). Therefore, indivisibility and nonexclusivity attributes of public goods discourage landowners to consider them in their decision-making.

Research suggests that environmental services such as carbon sequestration, improvement in water quality, and biodiversity for example, associated with AFS are significant (Garrett et al. 2000). Furthermore,

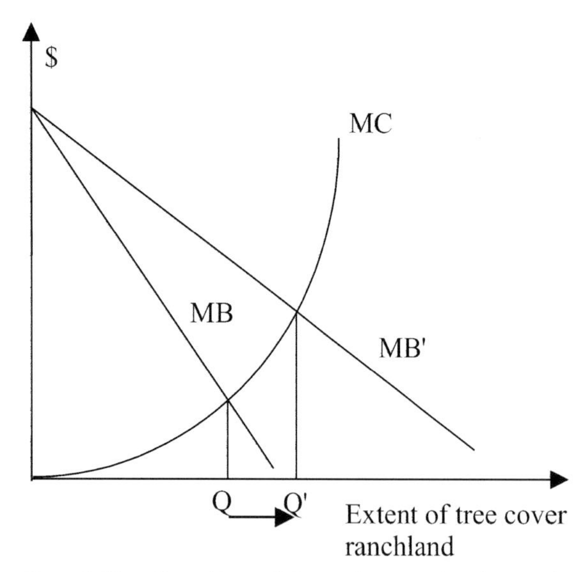

Figure 1. The effect of internalizing environmental services on the extent of tree cover on ranchlands.

there is growing appreciation for these services from scientific communities and the general public especially in the industrialized world (Montagnini and Nair 2004; Pimentel et al. 1995). Therefore, it is important and also rational to include them in ASF profitability analysis. Failure to account for them will result in gross under-valuation of AFS. Furthermore, if environmental costs of alternative land uses are greater than those of AFS[2] and if they do not enter into the analysis, it will result in over-valuation of alternative land-uses. Either way, the adoption of AFS will be less than optimal.

Figure 1 presents a graphical analysis to illustrate the effect of internalizing environmental services associated with temperate-zone silvopasture on its adoption. The horizontal axis measures the extent of trees and buffer strips on ranchlands, a proxy for silvopasture, while the vertical axis measures the costs/benefits of maintaining tree cover and buffer strips. Trees and buffer strips on ranch lands are considered beneficial to ranchers by providing shade to cattle and additional revenue from timber (MB). They also provide environmental benefits to the public by improving water quality and air quality. Research suggests that the trees' extensive root system and litter fall can reduce soil erosion; trees' uptake of nutrients can reduce fertilizer and pesticide runoff into nearby streams (Zinkhan and Mercer 1997); and growing trees sequester carbon by absorbing carbon dioxide from the atmosphere (see Montagnini and Nair 2004). The reduction in livestock numbers due to the addition of trees and buffer strips could result in further reduction of phosphorus and nitrogen wastes (Boggess et al. 1995). However, maintaining trees and buffer strips on ranchlands will cost ranchers (MC) but not the public. As forage production, a proxy for cattle production, is a function of tree basal area, an increase in the tree density will cause a reduction in the cattle output (see Stainback et al. 2004 for the relationship between tree density and cattle output).[3] Furthermore, the cost of fencing, tree seedlings, site preparation and pruning is expected to increase with an increase in the tree density.

In the absence of benefits from environmental services of silvopasture, ranchers will maintain only Q amount of tree cover on their ranches. This is determined by equating the private marginal cost of maintaining trees on ranchlands (MC) and private marginal benefit of trees (MB). However, if markets exist for environmental services and if ranchers can capitalize on water quality improvement and carbon sequestration for example, the marginal benefit of trees would be higher (MB'). This would motivate ranchers to maintain Q amount of tree cover on their ranchlands.

One of the main impediments for internalizing environmental services is limited information about public demand for these goods and the supply response of landowners. To our knowledge, very few efforts have been devoted to systematically investigate the following important questions. How do environmental costs and benefits influence the profitability of AFS and alternative land uses? Do households care for carbon sequestration, improvement in water quality, and biodiversity associated with AFS? If they do, will they be willing to pay for them and how much? Do landowners adopt AFS to improve environmental quality if incentives exist? Which incentive policy, a price premium or a direct payment, would be more effective?

The objective of this chapter is to illustrate the importance of environmental economic information in agroforestry development and suggest possible roles for government intervention. Specifically, we will attempt to address the above questions by drawing on the research that was conducted on sivlopasture in Florida. First, we present the theory of externalities to provide a framework for internalizing environmental costs and benefits of silvopasture. Second, the effect of environmental costs and benefits on the profitability of silvopasture and conventional ranching is assessed. Third, public demand for environmental services in terms of their willingness to pay is assessed. Fourth, ranchers' willingness to adopt silvopasture in the face of a price premium and a direct payment is analyzed. The final section discusses policy implications and the role of government.

Theory of externalities

An externality occurs when the activities of one agent affect the activities of another either positively or negatively and when these activities are not taken into account by the first agent in making his or her decision (van Kooten 1993). In the context of AFS, silvopasture practices on ranchlands, especially in Florida, may improve water quality in nearby streams and thus impact the utility of general public positively: chemical fertilizers used in pasture production can enter into nearby streams or groundwater and cause nonpoint source pollution (Nair and Graetz 2004) In the absence of incentives for the provision of positive externalities and penalties for generating negative externalities, a rational agent is less likely to produce them at societal optimum.

Institutional economics suggests that assigning property rights is an effective way to fix externality problems. Under exclusive, transferable, and enforceable property rights, people can be rewarded for positive externalities and penalized for negative ones. This would influence rational agents to produce externalities at optimum levels. Nobel Laureate Ronald Coase pioneered this principle in his famous article 'The Problem of Social Cost' published in 1960 (Coase 1960). He further states that under zero transaction costs, markets allocate resources such that externalities are produced at optimum levels regardless of the initial assignment of property rights to either party (Coase 1992; Felder 2001).

Consider, for example, a watershed with a group of ranchers in the upstream and a settlement of households in the downstream. Ranchers pollute a nearby stream through the application of chemical fertilizers and pesticides on pasture land. Households, who depend on the stream for clean water and other economic activities such as fishing, get affected negatively with water pollution. Ranchers, however, can withhold pollution by maintaining tree cover and buffer strips but it would cost them. On the other hand, households would benefit when ranchers can withhold pollution. In Figure 2, the horizontal axis represents the quantity of tree cover and buffer strips and the vertical axis represents costs or benefits of withholding pollution. The MC curve represents the marginal cost to ranchers and the MB curve represents the marginal benefits to households from withholding pollution.

Consider that ranchers do not have to maintain tree cover and buffer strips, i.e. ranchers have a right to pollute. However, they would be willing to maintain tree cover and buffer strips if households can cover the costs. In this scenario, households will have to pay ranchers an amount equal to area AED, the entire area under the MC curve, to enjoy clean water. Since the amount that households must pay ranchers beyond C is higher than the benefit they derive, households do not want tree cover and buffer strips beyond C. Alternatively, if households have a right to clean water, then ranchers have to pay households an amount equal to area ABD, the entire area under the MB curve to put up with the pollution. Given that the amount ranchers must pay households is initially much higher than the cost of withholding pollution, ranchers would want to maintain tree cover and buffer strips at C. At this point the amount that ranchers must pay equals the marginal cost of withholding pollution. The Coase theorem suggests that if property rights are defined, the parties will negotiate according to their benefits and costs and reach an optimal solution. Regardless of which one has the property rights, with zero transaction costs, the solution is reached at C.

A number of factors may prevent agents from realizing an optimum solution. In many cases property rights may not be clearly defined; negotiation between the parties often leads to litigation; benefits of environmental improvement are intangible; and those who benefit may have problems to get together and coordinate their actions (van Kooten 1993). In order to address these factors, a variety of information is needed including the environmental benefits and cost of silvopasture, respectively, from households' and ranchers' perspectives. Economists have recently developed a suite of environmental valuation tools to achieve this task. The following three sections will apply dynamic optimization, stated preference based choice experiment, and dichotomous contingent valuation techniques to conduct environmental economic analysis of sivlopasture in southern Florida.

Effect of environmental costs and benefits on silvopasture

Cattle ranching is an important agricultural enterprise in Florida. Ranchlands cover over 2.4 million hectares

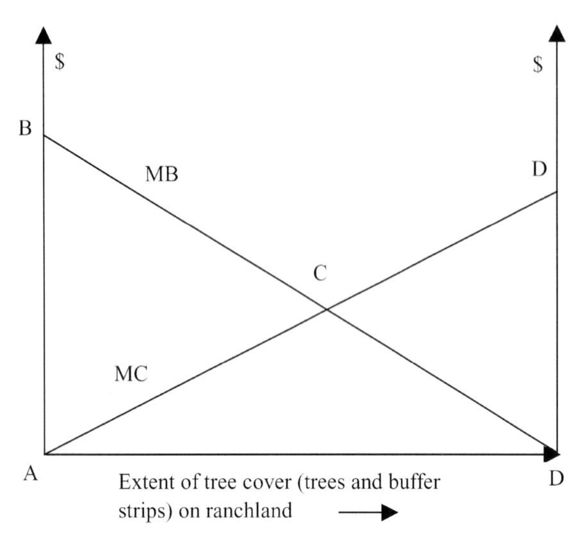

Figure 2. An illustration of optimal adoption of silvopasture with zero transaction costs.

and generate over $300 million each year with over 1.8 million cattle. More than 60% of cattle production in Florida occurs near Lake Okeechobee and the Everglades (south Florida), which is a large freshwater lake of 1891 square kilometers and a drainage basin of approximately 12 950 square kilometers (5000 square miles) (Harvey and Havens 1999). During the twentieth century, Lake Okeechobee and the Everglades have been subjected to massive water control projects to drain parts of the Everglades, to protect areas from flooding and to make land available for agriculture (including cattle ranching) and urban development. The resulting land-use changes have had a profoundly adverse impact on environmental quality in south Florida. In particular, the phosphorus content of Lake Okeechobee has more than doubled over the past century causing eutrophication and subsequent damage to its aquatic life (Harvey and Havens 1999). Since cattle ranching is one of the sources of phosphorus runoff, it has long been perceived as an environmental concern in this region.

Many studies considered silvopasture, a two-commodity system of producing trees and livestock forage, is an economically viable enterprise. Lundgren et al. (1983) found that pine based silvopasture systems in the southeastern United States could have as much as a 4.5% positive rate of return. Clason (1995) noted that silvopasture utilizing loblolly pine (*Pinus taeda*) in Louisiana could produce greater net returns than either pure pasture systems or pure timber systems. Grado et al. (2001) concluded that combining beef cattle and pine plantations can be profitable in southern Mississippi. Stainback and Alavalapati (2004) found combining longleaf pine production with cattle ranching is more profitable than conventional forestry or cattle ranching.

Stainback et al. (2004), however, found that combining slash pine with cattle production is not competitive with conventional ranching in Florida. The authors also found that silvopasture could be competitive with traditional ranching when environmental costs and benefits were factored into the analysis. Using a dynamic optimization model, they assessed the impact of a tax on phosphorus runoff and a payment for carbon sequestration on the profitability of traditional cattle ranching and silvopasture (see Stainback et al. 2004 for modeling details). The land values, which reflect the net present values of perpetual annual returns, under traditional ranching and silvopasture as a function of phosphorus runoff taxes and carbon payments are shown in Table 1. In the absence of pollution tax and carbon payments, land values under traditional ranching are shown to be greater than those of sivlopasture. As the pollution tax increases, land value decreases for both traditional ranching and silvopasture. If it is above $44 per kilogram or $20 per lb, then the land value for both traditional ranching and silvopasture would be negative. For all pollution tax rates with carbon payments of greater than $1 per ton ($10^6$ gram, Mg)land values under silvopasture are greater than those of traditional ranching. As the pollution tax increases, the carbon payment necessary for silvopasture declines. For instance, for a tax of $10 per 0.45 kg (lb), a carbon payment of $0.50 per 0.45 kg (lb) is not sufficient to make silvopasture competitive. On the other hand, with a pollution tax of $15 per 0.45 kg (lb), a carbon payment of $0.50 per 0.45 kg (lb) would make silvopasture financially competitive.

The optimal land management in terms of tree density and timber rotation age for various combinations of pollution taxes and carbon payments is also shown in Table 1. Results indicate that tree density and timber rotation age are more sensitive to carbon payments than to pollution taxes. For example, for a given pollution tax, the increase in carbon payments will result in significant changes in tree density. On the other hand, for a given carbon payment the increase in pollution tax will have modest impacts on tree density. For example, with a pollution tax of $20 per 0.45 kg (lb) doubling the carbon payment from $1 per ton to $2 per ton increases the optimal initial tree density from 200 stems per 0.4 hectare (acre) to 400 stems per 0.4 hectare (acre) and increases the optimal rotation age by 17 years. However, at a carbon price of $1.50 per ton doubling the pollution tax from $10 per 0.45 kg (lb) to $20 per 0.45 kg (lb) only increases the optimal initial tree density from 200 to 300 stems per 0.4 hectare (acre) and increases the optimal rotation age by only 10 years. Similar impacts are noticed with respect to carbon supply and phosphorus reduction.

Several conclusions can be drawn from this analysis. A pollution tax alone is not sufficient to induce landowners to adopt silvopasture. However, the profitability of silvopasture will be greater than that of traditional ranching when both costs of pollution and benefits of carbon sequestration are considered. In this case, payment for carbon is shown to be more effective in influencing ranchers to undertake silvopasture practices in Florida. Next, we will turn to questions – do households care for carbon sequestration, improvement in water quality, and biodiversity associated with silvopasture? Will they be willing to pay for these benefits? If so, how much?

Valuing environmental services of silvopasture

Welfare economics suggests that the value of a public good is the sum of what the individuals would be willing to pay for it (Varian 1992). This concept has been extensively used to value environmental services associated with forestry and agricultural practices (Cooper and Keim 1996; Lohr and Park 1994). Cameron et al. (2002) estimated nonmarket value of tree plantation on public lands accounting for the benefits of shade, windbreaks, and carbon sequestration. Loomis et al. (2000) assessed the value of buffer strips along streams in terms of erosion control, water quality improvement, and fish and wildlife habitat. Shrestha and Alavalapati (2004a) assessed the value of environmental services associated with silvopasture using a stated preference based choice experiment approach. They assume that silvopasture has a potential to reduce phosphorus run-off, sequester additional carbon dioxide, and improve the habitat for wildlife in the Lake Okeechobee watershed; silvopasture is compatible with the traditional ranching practices in this watershed; and information about public demand for environmental services is critical to formulate incentive schemes.

The theoretical construct of stated preference choice experiment (CE) approach stems from discrete choice analysis of consumer preferences. Random utility theory, which provides the basis for CE approach (McFadden 1974), considers each of the environmental attributes of silvopasture in the watershed as an alternative in a choice set.[4] The probability of choosing an alternative such that the utility of this alternative is greater than the utility of all other alternatives can be estimated using a multinomial logit (MNL) model or a random parameter logit (RPL) (McFadden 1974; Train 1998; Adamowicz et al. 1998; Louviere et al. 2000). Shrestha and Alavalapati (2004a) applied a random parameter logit (RPL) model to estimate household's willingness to pay (WTP) for environmental services.

A survey of households, who potentially benefit from silvopasture practices in the Lake Okeechobee watershed, was conducted. In designing the survey, each of three environmental attributes (water quality, carbon sequestration, and biodiversity) was identified and described at three levels (Table 2). The state utility tax that was specified at six levels chosen as a cost attribute to reflect environmental benefits of silvopasture because Florida prohibits state income tax, a commonly used variable in valuation surveys (Milon et al. 1999). Then, a choice experiment design was used to characterize these attributes and their levels in a survey questionnaire. An orthogonal main effect experimental design representing $3^3 \times 6 \times 2$ factorials was followed. Following the smallest orthogonal main effect plan 12 pairs of profiles in two blocks were identified using autocall macros of SAS version 8 (Kuhfeld et al. 1994). Each respondent was asked to value two options and a 'status quo' option in each scenario and continue with six scenarios.

The survey instrument included several sections. First, a map of the Lake Okeechobee watershed was presented and the rationale for selecting the study area was given. Second, a brief description of the traditional ranching, water pollution issues in the watershed, as well as potential environmental benefits and costs of silvopasture practices were outlined in the survey. Color photos and drawings were used to illustrate the silvopasture land-use changes. Third, the directions and an illustration of how to answer survey questions were provided. Fourth, the respondents were given six choice tasks, each with two options (A, B) and a status-quo option (C, and asked to choose one of the options. The valuation scenarios were designed in a referendum format (Figure 3). The choice question was followed by a rating question asking respondents to rate their confidence level in each choice task. Finally, the respondents were asked to provide their

Table 1. Estimated effect of payment for carbon sequestration and pollution tax on silvopasture profitability in southern Florida, USA.

Land values (U.S. $per 0.4 hectare (one acre)

	Silvopasture					Traditional ranch
	Carbon Payment ($/ ton)					
Phosphorus Tax ($/.45 kg or one lb)	0	0.5	1.0	1.5	2.0	
0	302	365	433	520	648	425
5	221	285	353	448	590	320
10	141	205	273	376	531	215
15	60	125	166	306	473	110
20	−20	45	125	242	415	5

Management and other variables

Carbon payment ($/ton)	Tree density	Timber rotation age	Carbon supply (tons per 0.4 hectare (acre)	Phosphorus reduction 0.45 kg per 0.4 ha (lb/acre)
Phosphorus tax = 0				
0.0	0	0	0	0
0.5	0	0	0	0
1.0	100	32	10	0.22
1.5	200	38	17	0.35
2.0	400	53	26	0.55
Phosphorus tax of $5 per 0.45 kg (lb)				
0.0	0	0	0	0
0.5	0	0	0	0
1.0	100	33	11	0.23
1.5	200	38	17	0.25
2.0	400	53	26	0.55
Phosphorus tax of $10 per 0.45 kg (lb)				
0.0	0	0	0	0
0.5	0	0	0	0
1.0	100	33	11	0.23
1.5	200	38	17	0.35
2.0	400	53	26	0.55
Phosphorus tax of $15 per 0.45 kg (lb)				
0.0	0	0	0	0
0.5	100	31	10	0.22
1.0	200	36	16	0.34
1.5	300	43	23	0.45
2.0	400	53	26	0.55
Phosphorus tax of $20 per 0.45 kg (lb)				
0.0	0	0	0	0
0.5	100	31	10	0.22
1.0	200	36	16	0.34
1.5	300	44	23	0.45
2.0	400	53	26	0.55

Source: Stainback et al. (2004).

Table 2. Attributes of silvopasture land-use in southern Florida, USA, used for the 'choice experiment'.

Environmental Attributes	Level of change in environmental attributes
1. Water quality	Reduce phosphorus runoff anywhere between 0 to 30%.
	Reduce phosphorus runoff anywhere between 31 to 60%.
	Reduce phosphorus runoff anywhere between 61 to 90%.
2. Carbon sequestration	Limited absorption of CO_2 from the atmosphere.
	Moderate absorption of CO_2 from the atmosphere.
	High absorption of CO_2 from the atmosphere.
3. Wildlife habitat	Limited improvement of habitats for wildlife.
	Moderate improvement of habitats for wildlife.
	High improvement of habitats for wildlife.
4. Increase in state utility tax (per year for 5 years) per household	\$0, \$10, \$20, \$40, \$80, \$120, \$140.

The first level of each attribute indicates the environmental quality without silvopasture.
Source: Shrestha and Alavalapati (2004a).

socioeconomic and demographic information. The survey was pre-tested with respondents from diverse educational and demographic backgrounds. Based on the pre-testing, the survey was revised by avoiding the use of technical terms and reducing the length of the description.

Survey sample was drawn randomly from households of ten counties in south-central Florida, namely, Glades, Hendry, Highlands, Martin, Okeechobee, Orange, Osceola, Palm Beach, Polk, and St. Lucie. A total of 504 packets were mailed, which contained a questionnaire, cover letter explaining the intent of the survey, and a magnet sticker with School of Forest Resources and Conservation and University of Florida logos. With second mailing and post card follow-ups a total of 152 survey responses were received.

The implicit values of the environmental quality improvement were estimated. The WTP estimates for moderate and high water quality improvement through reduced phosphorus runoff are, respectively, \$30.24 and \$71.17 per year for five years (Table 3). Corresponding estimates for moderate and high carbon sequestration levels are \$58.05 and \$62.72. Similarly, the WTP estimates for wildlife habitat improvements are \$49.68 and 41.06. These WTP estimates are within the range of values reported by Milon et al. (1999), who estimated household net WTP of \$59 and \$70 per year for hydrological and species restoration, respectively, in south Florida. The average WTP of household for a moderate level of improvement in all three environmental attributes is \$137.97 per year for five years. With 1.34 million households in the watershed, the total WTP for the environmental service would be \$924.40 million. This value reflects the households' total demand for environmental services associated with silvopasture in the Lake Okeechobee Watershed. Next we turn to the supply side of equation to assess ranchers' willingness to accept for sivopasture adoption and which policy incentive, a price premium or a direct payment, is more effective?

Ranchowners' willingness to adopt silvopasture

We stated before that the benefits of environmental services associated with silvopasture are external to cattle ranchers. Revenues from timber and additional hunting can offset the cost of sivlopasture only partially (Shrestha and Alavalapati 2004b). Therefore, cattle ranchers may have little or no motivation to adopt silvopasture unless policy incentives are provided. Literature suggests that landowners are responsive to policy incentives (Cooper and Keim 1996; Kingsbury and Boggess 1999; Purvis et al. 1989). Various federal and state programs such as Conservation Compliance, Sodbuster, Swampbuster, and the Conservation Reserve Program (CRP) in the United States are designed to encourage conservation practices on farmlands (Feather et al. 1999; Heimlich et al. 1998; Ribaudo et al. 1999; Westcott et al. 2002). Under these programs, conservation practices such as filter strips, riparian buffers, shelterbelts, windbreaks, and grass waterways, which are structurally and functionally similar to silvopasture, qualify for incentives.

Several studies have been conducted to assess the responsiveness of landowners to incentive programs. Cooper and Keim (1996) studied farmers' willingness to adopt water quality protection practices in the face of incentive payments. They conducted contingent valuation (CV) surveys in four watershed areas, i.e.,

Choice Scenario 1

Please vote for the plan that you prefer:

Environmental Response	Option A	Option B	Option C Current Condition
Water Quality Improvement (reduction of phosphorus runoff)	31 - 60%	61 - 90%	No change
Air Quality Improvement (absorption of CO_2)	No change	high	No change
Wildlife habitat Improvement (better habitat for wildlife)	Moderate	No change	No change
Annual Tax Increase (per year for 5 years)	$40	$120	$0

If the above plans are in a referendum, which one would you vote for?

☐ Option A ☐ Option B ☐ Option C

Please circle one number to indicate how certain you are with the choice you have just made:

Not certain — Very certain

1 ——— 2 ——— 3 ——— 4 ——— 5

Figure 3. An example of a choice set presented to the household to elicit the value of environmental change. Source: Shrestha and Alavalapati (2004a)

Eastern Iowa and Illinois Basin areas, the Albemarle-Pamlico Drainage area of Virginia and North Carolina, the Georgia-Florida Coastal Plain, and the Upper Snake River Basin area. Kingsbury and Boggess (1999) used a contingent valuation approach to study landowners' willingness to participate in conservation reserve enhancement program in Oregon. Shrestha and Alavalapati investigated Florida ranchers' willingness to adopt silvopasture[5]. Specifically, they assessed the effect of a premium on beef price and a direct payment on the adoption of silvopasture using a dichotomous choice CV approach (see note 5 for details of survey design). They apply a logit model for empirical estimation.

Table 4 presents the effect of price premium, direct payments, rancher's socioeconomic characteristics, and natural attributes of ranches on ranchers' likelihood of silvopasture adoption. The results show that the variable representing incentive payment offer has a positive and significant on the probability of adoption. The impact is shown to increase at a decreasing rate suggesting a nonlinear relationship. The variable representing access to the urban center (*ACCESS*), in the direct payment model, is shown to have a positive and significant effect on the likelihood of silvopasture adoption. This result is consistent with Kingsbury and Boggess (1999) who found that a lower opportunity cost of land increased the probability of

Table 3. Estimates of 'Marginal Willingness to Pay' for environmental improvement by landowners in relation to perceived benefits from silvopastoral practice in Florida, USA.

Environmental Attributes	Value for environmental improvement ($ per household per year for five years)	
	Moderate	High
Water quality	30.24 (19.63–40.86)	71.17 (53.41–88.93)
Carbon sequestration	58.05 (43.72–72.37)	62.72 (54.38–71.06)
Wildlife habitat	49.68 (38.08–61.28)	41.06 (35.41–46.72)

Numbers in parentheses are 95% confidence interval calculated from 1 000 draws from the distribution of coefficients in the model. Source: Shrestha and Alavalapati (2004a).

Table 4. Effect of policy incentives, socioeconomic, and natural attributes on ranchers' willingness to adopt silvopasture in Florida, USA.

Variable	Price support		Direct payment	
	Coefficient	SE	Coefficient	SE
Payment Offer and Opportunity Costs:				
Payments (PAYMT)	5.2159**	1.7901	0.2934**	0.0992
Square of payments (PAYMTS)	−4.9301**	2.0101	−0.0113**	0.0039
Plot size (ACRE)	−0.0001	0.0001	−1.83E-05	0.0001
Square of plot size (ACRES)	9.78E-07	1.22E-06	3.48E-07	1.05E-06
Northern Florida (NORTH)	0.2411	0.4042	−0.4486	0.3900
Central Florida (CENTRAL)	−0.2053	0.3535	−0.3893	0.3415
Distance to city (ACCESS)	0.0123	0.0082	0.0183**	0.0079
Natural Attributes:				
No of wildlife found (WNUM)	0.1001**	0.0519	0.1459**	0.0513
Creek and/or stream (CKST)	−0.2502	0.2680	0.4553*	0.2631
Marshland land (MARSH)	0.5935**	0.2886	−0.1990	0.2771
Forest cover (FOREST)	−0.2420	0.2692	−0.4102	0.2652
Longleaf pine (LLPINE)	0.6631**	0.2786	0.8586**	0.2680
Improved pasture (IMPPAST)	0.0024	0.0041	−0.0035	0.0040
Recreation Benefits:				
Wildlife hunting (HUNT)	0.6042**	0.2733	0.4164	0.2674
Fishing (FISH)	−0.1600	0.2701	−2.21E-01	0.2608
Horse riding (HBACK)	−0.5439	0.4736	0.0219	0.4597
Socioeconomic Characteristics:				
Household income (INC)	−0.0157	0.0107	−0.0230**	0.0095
Square of income (INCS)	0.0837	0.0660	0.0958*	0.0557
Respondents' age (AGE)	−0.0108	0.0080	−0.0091	0.0078
Respondents' education (EDU)	0.0467	0.0526	0.0438	0.0508
Member in environmental organization (MEMB)	−0.3484	0.3839	−0.2280	0.3824
Slope parameter (CONSTANT)	−1.2806	1.0485	−1.3255	1.0776
Log-L	−221.69		−231.47	
Chi-Square	52.03**		60.78**	
Correct prediction	64.45%		66.93%	
N	366		378	

*Coefficient significant at $p < 0.10$, **Coefficient/statistics significant at $P < 0.05$
Source: Shrestha and Alavalapati[5]

farmers' participation in conservation programs. The results also indicate that the existence of natural attributes will increase probability of silvopasture adoption. For example, variables representing wildlife presence (*WNUM*), existence of creeks and/or streams (*CKST*), marshlands (*MARSH*), and longleaf pine (*LLPINE*) are shown to have a positive impact on ranchers' adoption of silvopasture. The variable representing recreational hunting (*HUNT*) is positive and highly significant in the price support model suggesting that ranchers are more likely to adopt silvopasture if they currently use their ranches for recreational hunting.

Shrestha and Alavalapati estimated ranchers' mean WTA for the adoption of silvopasture[5].They found that on average, a price premium of $0.15 /lb. of beef or a direct payment of $9.32 /0.4 hectare (acre)/year was required for ranchers to adopt silvopasture practices. These estimates are much lower compared with previous studies on farmers' willingness to participate in conservation programs in the U.S. (e.g., Lohr and Park 1994; Cooper and Keim 1996). Lant (1991), for example, reported that average annual payment under Conservation Reserve Program was $48.93 /0.4 hectare (acre) at national average, while state averages ranged from $37.48 /0.4 hectare (acre) in Montana to $81.00 /0.4 hectare (acre) in Iowa. The lower WTA estimates of this study may be due to high compatibility of tree growing with cattle ranching. The authors also calculate the total annual incentive payments required for silvopasture adoption at a state level. With approximately 2.4 million hectares (6 million acres) of pasture and ranchlands (USDA 1997) and at the rate of $9.32 per 0.4 hectare (acre) per year, the direct payment policy would cost about $56 million annually. On the other hand, with nearly 2 million cattle resulting in annual sales of approximately 219 million kgs (482.84 million pounds) of beef (FAS, 2002) and with a premium of $0.15 per 0.45 kg (lb) beef, the price policy would cost about $72.43 million annually.

Although environmental economic valuation approaches that were presented here can be applied to various agroforestry practices, a caution must be taken in developing the questionnaire and collecting the data. The questionnaire must be customized with explicit consideration to locality and issue specific factors. For example, we used an increase in utility tax as a payment vehicle, as opposed to income tax, because Florida does not have state income tax. More caution must be taken in applying these methodologies in the tropics. Lack of familiarity with the issue under investigation, high illiteracy, and lack of experience in answering complex answers might pose additional challenges.

Conclusions

The importance of agroforestry systems in generating environmental services such as carbon sequestration, improvement in water quality, and biodiversity is being recognized increasingly the world over. If these services are not internalized to the benefit of landowners, the landowners (or producers)will have little incentive to produce them at a socially desirable level. Environmental economic analysis provides an effective framework for internalizing environmental services. Recent advancements in environmental valuation methodologies help quantify the potential demand for and supply of environmental services more accurately. Policy makers can use this information as a basis to formulate incentive or tax policies that further the economy and the environment.

Each policy may be unique in achieving the set objective and in bringing other desirable and undesirable effects. Policies must induce landowners search for land use innovations that reduce the cost of pollution and/or increase the profitability. For example, research findings reported in this paper indicate that phosphorus runoff taxes alone do induce desired changes in ranchland management. Payments for sequestering carbon along with pollution tax are found to be more effective in bringing desired changes. In the face of growing opposition to imposing additional taxes on farm or forest activities, the search for the least-cost incentive policies may be more desirable.

Several incentive mechanisms can be pursued including extension services, lump-sum payment on unit-area (hectare or acre) basis, price premium to reflect environmentally friendly practices, and cost-sharing programs. The opportunity cost, social acceptability, and administrative feasibility of each policy may be different. For example, results reported in this study indicate that a direct payment policy is less costly, relative to a price premium, to promote sivlopasture in Florida. It is desirable to assess these policies from a social acceptability and an administrative feasibility perspective. Institutional arrangements are critical for effective implementation of any incentive scheme. Government intervention is critical in developing institutions that help coordinate the actions of stakeholders and facilitate transactions at a minimum cost.

Acknowledgements

This publication was partially supported by U.S. Department of Agriculture, Cooperative State Research, Education, and Extension Service, Initiative for Future Agriculture and Food Systems (USDA/CSREES/IFAFS) grant number 00-52103-9702. Florida Agricultural Experiment Station Journal Series R-09892.

End Notes

1. The other factors include imperfect information, imperfect competition, ill-defined property rights, and presence of common-pool goods.
2. It is conceivable that environmental pollution such as phosphorous run-off for example, from conventional ranching would be higher than that under silvopasture system, where tree cover and buffer strips are maintained along with pasture.
3. Following Wolters (1973), the following equation specifies the forage produced, $f(t)$, in lbs of dry biomass per acre per year: $f(t) = 2048.9 - 14.7b(t)$, where $b(t)$ is tree basal area in square feet per acre. Forage production can be converted to cattle production using standard animal unit months (AUMs). One AUM is defined as the amount of forage required to support one adult cow weighing 1000 lbs for one month.
4. Alternative j represent one specific type of consumption bundle representing an improvement in the environmental quality of the watershed with its conditional indirect utility level V_j for a

household i and is expressed as $V_{ij} = v_{ij} + \varepsilon_{ij}$, where v_{ij} is the deterministic component of the model and ε_{ij} is the random component. Thus, selection of alternative j over alternative h implies that the utility of V_{ij} is greater than that of V_{ih}.

5. Shrestha R.K. and Alavalapati J.R.R. 2004. Florida ranchers' willingness to participate in silvopasture practices: an economic analysis and policy implications (In review).

References

Adamowicz W., Boxall P., Williams M. and Louviere J. 1998. Stated preference approach for measuring passive use values: Choice experiments and contingent valuation. Am J Agr Econ 80: 64–75.

Alavalapati J.R.R. and Nair P.K. 2001. Socioeconomic and institutional perspectives of agroforestry. pp. 71–81. In: Palo M. and Uusivuori J. (eds) World Forests, Society, and Environment – Markets and Policies. Kluwer, Dordrecht, The Netherlands.

Boggess C.F., Flaig E.G. and Fluck R.C. 1995. Phosphorus budget-basin relationships for Lake Okeechobee tributary basins. Ecol Eng 5: 143–162

Cameron T.A., Poe G.L., Ethier R.G. and Schulze W.D. 2002. Alternative non-market value-elicitation: Are the underlying preference the same? J Environ Econ Manag 44: 391–422.

Clason T.R. 1995. Economic implications of silvipastures on southern pine plantations. Agroforest Syst 29: 227–238.

Coase R.H. 1960. The problem of social cost. J Law Econ 3: 1–44.

Coase R.H. 1992. The institutional structure of production. Am Econ Rev 82(4): 713–719.

Cooper J.C. and Keim R.W. 1996. Incentive payments to encourage farmer adoption of water quality protection practices. Am J Agr Econ 78: 54–64.

FAS 2002. Beef cattle and calf inventory by county: Livestock, dairy, and poultry summary, Florida Agricultural Statistics Service. The USDA National Agricultural Statistics Service, http://www.nass.usda.gov/fl/.

Feather P., Hellerstein D. and Hansen.L 1999. Economic valuation of environmental benefits and the targeting of conservation programs: The case of the CRP. Washington DC, USDA Economic Research Service, Agricultural Economic Report No. 778.

Felder J. 2001. Coase Thorem 1-2-3. Am Econ 45: 54–60.

Garrett H.E., Rietveld W.J. and Fisher R.F. (eds) 2000. North American Agroforestry: An Integrated Science and Practice, American Society of Agronomy, Madison, Wisconsin, 402 pp.

Grado S.C., Hovermale C.H. and Louis D.J.S. 2001. A financial analysis of a silvopasture system in southern Mississippi. Agroforest Syst 53: 313–322.

Harvey R. and Havens K. 1999. Lake Okeechobee Action Plan. Lake Okeechobee Issue Team for the South Florida Ecosystem Restoration Working Group, South Florida Water Management District, West Palm Beach, Florida, 43 pp.

Heimlich R.E., Wiebe K.D., Claassen R., Gadsby D. and House R.M. 1998 Wetlands and agriculture: Private interests and public benefits. United States Department of Agriculture, Economic Research Service, Agricultural Economic Report No. AER765, Washington DC. 104 pp.

Kingsbury L. and Boggess W. 1999. An economic analysis of riparian landowners' willingness to participate in Oregon's Conservation Reserve Enhancement Program. Selected paper for the annual meeting of the Agricultural Economics Association, August 8-11, Nashville, Tennessee. 15 pp.

Kuhfeld W.F., Tobias R.D. and Garratt M. 1994. Efficient experimental design with marketing research applications. J Marketing Res 31: 545–557.

Lant C.L. 1991. Potential of the Conservation Reserve Program to control agricultural surface water pollution. Environ Manage 15: 507–518.

Lohr L., Park T. 1994. Discrete/continuous choices in contingent valuation surveys: Conservation decisions in Michigan. Rev Agri Econ 16: 1–15.

Loomis J.B., Kent P., Strange L., Fausch K. and Covich A. 2000. Measuring the total economic value of restoring ecosystem services in an impaired river basin: results from a contingent valuation survey. Ecol Econ 33: 103–117.

Louviere J.J., Hensher D.A. and Swait J.D. 2000. State Choice Methods: Analysis and Applications. Cambridge University Press, Cambridge, UK 402 pp.

Lundgren G.K., Conner J.R. and Pearson H.A. 1983. An economic analysis of forest grazing on four timber management situation. South J Appl For 7: 119–124.

McFadden D. 1974. Conditional logit analysis of qualitative choice behavior. pp. 105–142. In: Zarembka P. (ed.) Frontiers in Econometrics. Academic Press, New York.

Milon J.W., Hodges A.W., Rimal A., Kiker C.F. and Casey F. 1999. Public preferences and economic values for restoration of the Everglades/South Florida ecosystems. Economics Report 99–1, Food and Resource Economics Department, University of Florida, Gainesville. 124 pp.

Montagnini F. and Nair P.K.R. 2004. Carbon sequestration: An under-exploited environmental benefit of agroforestry systems. (This volume).

Nair P.K.R. 2001. Agroforestry. In Our Fragile World: Challenges and Opportunities for Sustainable Development, Forerunner to The Encyclopedia of Life Support Systems, Chapter 1.25: 375–393. UNESCO, Paris, France & EOLSS, UK.

Nair V.D. and Graetz D.A. 2004. Agroforestry as an approach to minimizing nutrient loss from heavily fertilized soils: The Florida experience (This volume).

Peterson E.W.F. 2001. The Political Economy of Agricultural, Natural Resource, and Environmental Policy Analysis. Iowa State University Press, Ames, Iowa, 373 pp.

Pimentel D., Harvey C., Resosudarmo P., Sinclair K., Kurz D., McNair M., Crist S., Shpritz L., Fitton L., Saffouri R. and Blair R. 1995. Environmental and economic costs of soil erosion and conservation benefits. Science 267: 1117–1123.

Purvis A., Hoehn J.P., Sorenson V.L. and Pierce F.J. 1989. Farmers' response to a filter strip program: Results from a contingent valuation survey. J Soil Water Conserv 44: 501–504.

Ribaudo M.O., Horan R.D. and Smith M.E. 1999. Economics of water quality protection from nonpoint sources. USDA/Economic Research Service, Washington, D.C.

Shrestha R.K. and Alavalapati J.R.R. 2004a. Valuing Environmental Benefits of Silvopasture Practice: A Case Study of the Lake Okeechobee Watershed in Florida. Ecol Econ (in press).

Shrestha R.K. and Alavalapati J.R.R. 2004b. Effect of Ranchland Attributes on Recreational Hunting in Florida: A Hedonic Price Analysis J Agr Applied Econ (in press).

Stainback G.A., Alavalapati J.R.R., Shrestha R.R., Larkin S. and Wong G. 2004. Environmental Economic Valuation of Silvopasture: A Dynamic Optimization Approach. J Agr Applied Econ (in press).

Stainback G.A. and Alavalapati J.R.R. 2004. An Economic Analysis Restoring Longleaf Pine on Ranchlands. Forest Policy Econ (in press).

Train K.E. 1998. Recreation demand models with taste differences over people. Land Econ 74: 230–239.
USDA 1997. Census of Agriculture: National, state, and county tables. United States Department of Agriculture, National Agricultural Statistics Service, Washington DC (http://www.nass.usda.gov/census/).
Varian H.R. 1992. Microeconomics Analysis, 3rd Edition, W.W. Norton & Company, New York, 364 pp.
van Kooten G.C. 1993. Land Resource Economics and Sustainable Development: Economic Policies and the Common Good, UBC Press, Vancouver, 450 pp.
Westcott P., Young C.E. and Price J.M. 2002. The 2002 Farm Act: Provision and implications for commodity markets. Washington DC, USDA Economic Research Service, Agricultural Information Bulletin No. AIB778. 67 pp.
Zinkhan F.C. and Mercer D.E. 1997. An assessment of agroforestry systems in the southern U.S.A. Agroforest Syst 35: 303–321.

Agroforestry Systems 204411: 311–328, 2004.

Adoption of agroforestry innovations in the tropics: A review

D.E. Mercer
Southern Research Station, USDA Forest Service, 3041 Cornwallis Road, P.O. Box 12254 Research Triangle Park, NC 27709; USA; e-mail: emercer@fs.fed.us

Key words: Decision-making, Diffusion, Risk, Tenure, Uncertainty

Abstract

The period since the early 1990s has witnessed an explosion of research on the adoption of agroforestry innovations in the tropics. Much of this work was motivated by a perceived gap between advances in agroforestry science and the success of agroforestry-based development programs and projects. Achieving the full promise of agroforestry requires a fundamental understanding of how and why farmers make long-term land-use decisions and applying this knowledge to the design, development, and 'marketing' of agroforestry innovations. This paper reviews the theoretical and empirical literature that has developed during the past decade analyzing agroforestry adoption from a variety of perspectives and identifies needed future research. Much progress has been made, especially in using binary choice regression models to assess influences of farm and household characteristics on adoption and in developing ex-ante participatory, on-farm research methods for analyzing the potential adoptability of agroforestry innovations. Additional research-needs that have been identified include developing a better understanding of the role of risk and uncertainty, insights into how and why farmers adapt and modify adopted systems, factors influencing the intensity of adoption, village-level and spatial analyses of adoption, the impacts of disease such as AIDS and malaria on adoption, and the temporal path of adoption.

Introduction

No matter how elegant, efficient, productive, and/or ecologically sustainable, agroforestry systems can contribute to sustainable land use only if they are adopted and maintained over long time periods (Raintree 1983; Scherr 1992; Sanchez 1995). Although there are some examples of significant adoption over the past two decades (Current et al. 1995; Buckles and Triomphe 1999; Barrett et al. 2002; Franzel and Scherr 2002), many have lamented the fact that adoption and diffusion have lagged behind the scientific and technological advances in agroforestry research reducing the potential impacts of agroforestry-based development projects (Adesina and Chianu 2002; Alavalapati et al. 1995; Bannister and Nair 2003; Lapar and Pandey 1999; Nair 1996; Sanchez 1995; Thacher et al. 1997). As a result, research on adoption of agroforestry innovations has attracted much attention and generated a relatively large literature during the past decade.

Approaches to analyzing agroforestry adoption tend to follow the vast literature on adoption of agricultural production technologies, most of which focus on new or improved production inputs (e.g. Green Revolution inputs) for conventional agricultural crops (Feder et al. 1985; Feder and Umali 1993). A number of features of agroforestry, however, make analysis of its adoption unique and deserving of its own review. Adoption of agroforestry is considerably more complex than traditional agriculture because it usually requires establishing a new input-output mix of annuals, perennials, green manure, fodder and other components, combined with new conservation techniques such as contour hedgerows, alleycropping, and enriched fallows (Rafiq et al. 2000). Unlike standard agriculture, there are few packaged agroforestry or farm-based, natural resource management (NRM) practices to deliver to farmers (Barrett et al. 2002). As a result, agroforestry and other NRM innovations are typically more knowledge-intensive than modern agricultural development packages based on improved seed, chemical, and/or mechanical inputs. Therefore, farmer education, experimentation, and modification are more important for agroforestry and NRM development than for conventional agriculture (Barrett et al. 2002).

This multicomponent, multiproduct nature of agroforestry may limit adoption due to the complex management requirements and the long period of testing and modification that is required compared to annual cropping technologies. An agroforestry system is likely to take three to six years before benefits begin to be fully realized compared to the few months needed to harvest and evaluate a new annual crop or method (Franzel and Scherr 2002). These characteristics can enhance opportunities for adoption by allowing more farmer experimentation and adaptation but can also complicate analysis of who adopts, what they adopt, and how they modify the system adopted (Vosti et al. 1998). The additional uncertainty inherent in these new input-output mixes is also an important reason for slower adoption rates and suggests that agroforestry projects will require longer time periods before becoming self-sustaining and self-diffusing than the earlier Green Revolution innovations (Amacher et al. 1993).

Most research supports the notion that decisions to adopt resource-conserving practices like agroforestry are largely driven by expected contributions to increased productivity, output stability through risk reduction, and enhanced economic viability compared to the alternatives (Arnold and Dewees 1995; Sain and Barreto 1996; Salam et al. 2000; Scherr 2000). Therefore, this review of the agroforestry adoption literature is primarily economics-based. Following on Pattanayak et al.'s (2003) meta-analysis of multiple regression based agroforestry studies, this paper examines the broader adoption literature to assess the current state of knowledge on agroforestry adoption.

Definitions and a brief history of adoption-diffusion research

Innovation, adoption, diffusion

From a sociological viewpoint, an innovation is an idea, practice, or object that an individual perceives as new. Since the focus is on the perception of the idea, the innovation need only be 'new' to the potential adopter. This suggests that adoption is the mental process from first hearing about an innovation to deciding to make full use of the new idea (Rogers and Shoemaker 1971; Rogers 1983; Evans 1988). Feder et al. (1985) argued, however, that sociological definitions of adoption are usually inadequate for 'rigorous theoretical and empirical analysis' due to their imprecision and failure to distinguish individual or farm-level adoption from aggregate adoption.

From an economic standpoint, an innovation is a technological factor of production with perceived and/or objective uncertainties about its impact on production. Farmers reduce uncertainty over time by acquiring experience, modifying the innovation, and becoming more efficient in its application. Therefore, economists have defined final adoption at the farm-level as 'the degree of use of a new technology in long-run equilibrium when the farmer has full information about the new technology and its potential' (Feder et al. 1985).

Diffusion of technological innovations has been defined as the spread of 'successful' innovations as they combine with or displace existing 'inferior' alternatives (Sarkar 1998). Thus, diffusion concerns the extent (spatially and temporally) to which the new innovation is put to productive use. Early adopters are often referred to as innovators and the diffusion process as the spread of the innovation to other members of the population (Feder and Umali 1993).

Adoption typically has been viewed from two perspectives. At the individual farm level, each household chooses whether or not to adopt and the intensity of adoption. Farm-level adoption studies, then, are concerned with the factors influencing the adoption decision either statically or dynamically by incorporating learning and experience. At a macro-level, diffusion studies examine how adoption evolves across a population or region. Since the objective is to identify specific trends in the diffusion cycle over space and time, diffusion models do not explicitly address the innovation process. Diffusion studies typically begin after the innovation is already in use and analyze the spread of the innovation as a dynamic, aggregative process over continuous time (Feder and Umali 1993).

A brief history

Although the adoption and diffusion of agricultural innovations depends on a combination of social, economic, and cultural factors, most theoretical and empirical adoption work has tended to be dominated by separate lines of research by sociologists, economists, and geographers. Economists historically emphasized profitability and investment risks, while sociologists concentrated on the social rewards and nature of communication channels associated with adoption. Geographers have studied the spatial differences in resource endowment and diffusion, and anthropologists have emphasized compatibility with the norms of society (Boahene et al. 1999).

Adoption-diffusion first emerged as an important research agenda in rural sociology in the 1940s and 1950s (Ruttan 1996). According to Rogers (1983), 'the research tradition that can claim major credit for initially forming the intellectual paradigm for diffusion research, and that has produced the largest number of diffusion studies, is rural sociology.' Sociologists have traditionally conceptualized the adoption-decision process by examining distinguishing characteristics of adopters and nonadopters and opinion leaders, farmers' perceptions of the attributes of the innovations, rates of adoption and diffusion, and the channels of communication during the various stages of the adoption decision process (Marra et al. 2003).

The number of adoption-diffusion studies by rural sociologists began to decline in the 1950s for U.S. and European studies, and in the mid-1960s for developing countries. By the mid 1970s, rural sociology began to lose its dominance.[1] Several reasons have been advanced for this decline including: i. the lack of attention to theory, ii. over dependence on 'search for variables' approaches to empirical analysis, iii. inadequate attention to how independent and dependent variables are specified, an issue referred to by economists as the identification problem, iv. noncritical use of the epidemic model and the assumed linear relationship between status and adoption, and v. a shift in social theory away from modernization and toward dependency and other class-based perspectives (Ruttan 1996). Ruttan (1996) concludes that the most important reason for the decline in sociological adoption research 'is that sociologists failed to embrace the more formal analytical methods introduced by geographers to understand the process of spatial diffusion or by the economists and technologists to understand the process of technological innovation, substitution, and replacement.'

Although agricultural economists' interest in adoption research began in the 1950s with Griliches' (1957) econometric study of hybrid corn (*Zea mays*) adoption, it was not until the 1970s that economists' work on adoption-diffusion studies began to rapidly expand. Ruttan (1996) provides three primary explanations for the increasing importance and dominance of economics in adoption-diffusion research. First, both domestic and international development agencies demand research that provides policy-relevant knowledge of the sources and diffusion of technical change. Second, economists typically remain skeptical of the sociologists' argument that technology is the problem rather than the solution. Finally, the arguments of agricultural economists that broader economic forces create many of the incentives to innovate have widespread appeal.

At least initially, economists drew heavily on the diffusion work by sociologists, but as the research evolved, economists embarked on an increasingly distinct theoretical and methodological path. The result was that neither sociologists nor economists influenced each other's research during the 1980s and 1990s (Ruttan 1996). Viewed from a multidisciplinary prospective, adoption is a multi-dimensional process dependent on a variety of factors such as perceived profitability, costs of establishment, compatibility with value systems, and the ability to communicate new knowledge and information between and among adopters and potential adopters (Boahene et al. 1999). As Kenneth Arrow stated:

> while *(the economists)* stress the profitability of the investment and risks involved, the sociologists are concerned with the nature of the channel connecting adopters of an innovation with potential followers....*(but)* there is nothing irreconcilable in the two viewpoints: in effect, the economists are studying the demand for information by potential innovators and sociologists the problems in the supply of communication channels' (1969, p.13).

Agroforestry adoption research has followed a similar historical path, with a 20-year lag. Initial efforts in the study of agroforestry adoption during the 1970s and 1980s tended to be descriptive and prescriptive and lacked formal theoretical development or rigorous empirical analysis (Raintree 1983; Fujisaka 1989; Allen 1990). For example, Swinkels and Scherr's (1991) bibliography of 230 studies of agroforestry economics lists only eight publications related to adoption, of which only three were in peer-reviewed journals, one in a conference proceedings, and four unpublished. By the early 1990s, agroforestry research was still primarily concerned with physical and biological interactions with little emphasis on economics or sociology (Swinkels and Scherr 1991; Current et al. 1995; Mercer and Miller 1998). Agroforestry adoption research blossomed in the 1990s beginning with the Current et al. (1995) case studies in Central America that relied primarily on nonstatistical analyses of project data, cost-benefit analyses, informal and formal surveys of farmers, focus group discussions and interviews with project staff. Since then, adoption studies have expanded considerably.

Adoption-diffusion theory

Technology adoption under uncertainty: general agricultural applications

The expected utility framework is the most common approach to modeling technology adoption under uncertainty (Feder et al. 1985; Feder and Umali 1993; Marra et al. 2003). Applying the expected utility framework to technology adoption under uncertainty was first proposed by Just and Zilberman (1983) to remedy the lack of a theoretical framework for explaining the stochastic relationship of production under new and old technologies[2]. The expected utility model assumes that adoption decisions are based on the maximization of expected utility or profit subject to land, credit, labor and other constraints. Since profit is a function of the farmer's choice of crops and technology in each time period, maximizing profit, or utility, depends on the farmer's discrete selection from a menu of alternatives, including traditional practices. Among the important results of the theory is that the correlation of outputs under alternative technologies is crucial to determining adoption rates. For example, if correlation between returns to the old and new technologies is high and if risk aversion decreases sufficiently with increased wealth, adoption of new profitable technologies may be constrained (Marra et al. 2003).

Most of the other adoption models have extended the expected utility framework. Portfolio models view the land allocation decision between new and old technologies as a decision process in which farmers maximize the expected utility of income by choosing a specific combination of crops or systems given their risk aversion levels, the stochastic interactions between variables, and the impacts of socioeconomic variables such as wealth, age and education (Feder and Umali 1993). The 'safety-first' models deviate from the usual assumption of a concave and well-behaved utility function, and instead assume that the utility of income is zero below a certain 'disaster' level and one above it (Feder et al. 1985). As a result, safety-first criteria for making choices between uncertain alternatives are concerned only with the risk of failing to achieve a certain minimum target or to secure prespecified safety margins such as subsistence (Bigman 1996).

Learning by doing and farmer experimentation models, as developed by Foster and Rosenzweig (1995), show that imperfect knowledge is a barrier to adoption. Although experience appears to initially augment the ability to make appropriate decisions about new technologies, the effect declines rapidly over time as experience increases. The impact of experience and experimentation can also have small but important spillover effects on neighbors. Those with neighbors experienced in the new technology tend to be significantly more profitable than those with inexperienced neighbors. Since the farmer's and neighbors' assets, net of the experience effect, have opposite effects on adoption, incentives exist for farmers to free-ride on the learning of others. For example, a farmer with neighbors experimenting with new technologies like agroforestry may curtail his own experimentation because he can realize higher short-term profits by using traditional methods and shifting to the new technology when there is sufficient experience from his neighbors to make adoption profitable (Foster and Rosenzweig 1995).

Choosing one specific model as a basis for empirical analysis of adoption, however, can produce estimation errors and biased estimators (Feder and Umali 1993). Perhaps the most comprehensive model of adoption was developed by Abadi Ghadim and Pannell (1999) in response to criticisms that most empirical studies of adoption suffer from omitted variable biases, poor model specification, failure to derive hypotheses from a sound theoretical framework, and/or failure to account for the dynamic learning process in adoption. The Abadi Ghadim and Pannell (2003) model depicts adoption as a multistage decision process that incorporates information acquisition and learning by doing by farmers who vary in their risk preferences and perceptions of the risks associated with the innovation.

As an alternative to the expected utility theory of choice, Gladwin (1976) applied hierarchical decision tree models derived from the work of cognitive anthropologists and psychologists to analyze fertilizer decisions in Mexico. The hierarchical decision tree approach assumes that choices are made in a decomposed, piecemeal basis (i.e. one dimension at a time) in contrast to the assumptions of expected utility theory in which people are assumed to examine all alternatives simultaneously, formulate separate subjective probabilities and pick the one with the largest expected utility (Gladwin et al. 2002a). During the first stage in the decision tree process, all alternatives with certain negative characteristics are immediately eliminated. The remaining alternatives are then ordered on one or more criteria and subjected to a series of disqualifying constraints. The highest ranked alternative that passes all constraints is chosen. Arrow (1963) referred to decision tree models as 'choice function(s) not built up from orderings.' Although the theory uses a discrete form of the standard neoclassical maximization subject to constraints approach, the ranking process is not connected and not necessarily transitive (Gladwin et al. 2002a).

Technology adoption under uncertainty: agroforestry applications

Theoretical models of agroforestry adoption, which began to appear in the literature in the mid-1990s, have primarily utilized a household production framework to model agroforestry adoption as an investment choice based on maximization of expected utility, or profit, subject to labor, capital, and income constraints (Amacher et al. 1993; Rafiq et al. 2000; Mercer and Pattanayak 2003). Amacher et al. (1993) were perhaps the first to apply household production theory and expected utility models to the problem of agroforestry adoption. Assuming decreasing absolute risk aversion, their model predicts positive impacts of all income sources (farm, off-farm, and forest based), household labor, capital, and land endowments, land tenure and price of tree products for adopting tree planting. In contrast, prices of nonforest consumption goods reduce adoption, while impacts of the variability (i.e. riskiness) of household production of forest products could not be determined by the model. Mercer and Pattanayak (2003) and Pattanayak et al. (2003) use a similar framework for a meta-analysis of multiple regression-based adoption studies to demonstrate that agroforestry adoption is a function of market incentives, biophysical conditions, resource endowments, risk and uncertainty, and household preferences.

Adding a household specific, safety-first constraint to the expected utility model allows the household to evaluate expected returns in terms of a probability distribution for minimum income, which depends on the household's income earning potential (Shively 1997). Shively uses this model to show that when the safety-first constraint is binding, adoption decisions depend on farm size, non-farm income, farm specific attributes such as soil quality and slope, the probability of a consumption shortfall, and, of course, a comparison of net benefits. Including the safety-first constraint shows that when adoption is costly, the probability of falling below the subsistence level for low-income households is crucial to decision-making.

Shively (2001) uses a dynamic expected utility model combined with an equation of motion for soil stocks that includes the probability of a catastrophic erosion event, to show how consumption risks and investment costs influence incentives to adopt contour hedgerows for soil conservation. Although unable to derive an analytical solution to the problem, Shively (2001) uses simulation and stochastic dominance to show that the household's valuation of soil conservation methods depends on investment costs, riskiness of the innovation, and capacity to bear risk. As the decision to invest in soil conservation depends on farm size, adoption is especially costly on small farms due to the increased short-run risk of consumption shortfall. Shively concludes that assuming risk-neutrality may lead to incorrect adoption predictions for low-income households whenever the risk of consumption shortfalls are high.

Besley (1995) examines the impact of property rights on investments in tree planting and other conservation methods in Ghana under three perspectives: security, collateral-based, and gains-from-trade. Besley's model maximizes returns to investment, which depend on the amount of capital invested at time t and property rights at time $t + 1$. The security case assumes the probability of land expropriation in period $t + 1$ is a decreasing function of tenure security. If investment costs are independent of tenure security, the results are analogous to a random tax on land, and investment increases with increasing tenure security. In the collateral-based scenario, when land is easier to collateralize, because individuals have better transfer rights, banks will charge lower interest rates. Since farmers equate the marginal return to investing in land to the interest rate, investments increase with increasing land tenure. The gains-from-trade model examines the relationship between land-tenure security, the costs of selling or renting land, and land conservation investment incentives. The implications are similar to those of the other two models: increased land-tenure security increases investment incentives.

Pannell (2003) utilizes a dynamic profit-function analysis, which allows the collection, integration and evaluation of new information to reduce uncertainty over time. His model demonstrates that uncertainty inhibits adoption, assuming farmers are risk averse, because uncertainty can lead to incorrect predictions of the expected benefits from adoption. As a result, farmers may be better off waiting to adopt in some cases. Pannell (2003) concludes that uncertainty has received inadequate attention as an impediment to adoption of innovative land conservation practices and that on-farm experimentation is indistinguishable from adoption because production systems are continually tested and modified as perceptions and expectations evolve.

Hierarchical decision tree models have been applied to adoption of agroforestry and natural resource management (NRM) by Swinkels and Franzel (1997), Gladwin et al. (2002a,b), and Swinkels et al. (2002). Correctly specified and empirically verified decision-tree models allow identification of constraints to adoption and detailed examination of the decision-making process by breaking the process into a series of sub-decisions which are mapped as branches of a tree (Franzel et al. 2002). Gladwin et al. (2002a,b) increase the rigor of decision tree analysis by subjecting hypotheses derived from the analysis to tests of statistical inference using logit and ordered probit analyses. This is a good example of combining rigorous scientific hypothesis-testing with participatory approaches to adoption analysis and how using cognitive decision models and econometric testing can improve both approaches.

Diffusion theories: General agricultural applications

Induced innovation

Although the basic idea behind the theory of 'induced innovation' can be traced to Hick's (1932) *Theory of Wages*, Boserup's (1965) analysis of agricultural growth was perhaps the major influence on the development of induced innovation theory. Boserup showed that as population densities rise and/or demand for agricultural products increases, the resulting land pressures induce adoption of technological and institutional innovations to intensify land use. Basically, the scarcity of land relative to labor and/or capital induces investment in additional labor/capital inputs to maintain or increase agricultural production. By the 1970s, induced innovation theory was applied to a wide range of new agricultural technologies to explain the impact of population and markets on the diffusion of innovations in both subsistence and commercial agriculture (Binswanger and Ruttan 1978; Ruttan and Hayami 1984; Pingali et al. 1987). More recently, induced innovation/ intensification analyses have used both micro- and macro-level data and incorporated additional determinants of technology diffusion such as environmental conditions, government policies, and land tenure (Goldman 1993; Humphries 1993; Turner and Ali 1996; Wiegers et al. 1999).

Epidemic or logistic models

The epidemic models of diffusion, first introduced in the 1950s, were based on analogies between the spread of contagious diseases and technological innovations. Since contact with other adopters and information leads to the spread of adoption across a population, diffusion is determined by the epidemic spread of information among potential adopters (Sarkar 1998). Hence, demonstration effects and learning from others' experiences underlie the epidemic model (Feder and Umali 1993). Epidemic models were first used to explain patterns discovered in empirical studies, such as time periods required for adoption within and across farms, varying diffusion rates, and the tendency for adoption to follow a logistic (sigmoid or S-shaped) time path beginning slowly initially, then speeding up, and finally tapering off. Epidemic models assume that adoption rates are a function of the product of the portion of the population already 'infected' with the new technology and the size of the 'uninfected' population (Sarkar 1998). Epidemic models have been criticized for their weak theoretical foundations, restrictive assumptions, and failing to establish theoretical links between decision-theoretic models of individual farmer behavior and the diffusion of innovations (Sarkar 1998).

Decision-theoretic models

In response to criticisms of the epidemic models researchers began to develop models that explain why adoption by different individuals varies over time, why individual households take time to switch between old and new technologies, and how diffusion patterns impact economic growth and employment (Sarkar 1998). Economists have developed decision-theoretic approaches along two lines: neo-classical equilibrium (NE) and evolutionary disequilibrium (ED) approaches.

Following dynamic neoclassical analysis, NE models assume that the diffusion process can be modeled as a sequence of shifting static equilibria among infinitely rational decision makers. NE approaches can be subdivided into the probit models, which assume that the temporal variation in adoption is due to heterogeneity among potential adopters, and the game-theoretic models in which strategic interactions among households rather than differences in household characteristics determine the diffusion path. All NE models, however, assume that adopters are infinitely rational and can evaluate and determine optimal strategies before any diffusion actually takes place (Sarkar 1998). The most important contributions of the NE models include demonstrating the importance of heterogeneity between adopters, interactions in the supply and demand for innovations and adoption rates, strategic interactions among adopters, and the importance of market structure.

Evolutionary disequilibrium (ED) models were developed in response to criticisms of the NE models' assumptions of perfect information, infinite rationality and that diffusion is a continuous equilibrium process. The fundamental features of evolutionary disequilibrium models are that i. adopters' rationality is bounded, ii. profit maximization may not be the only basis for adoption, and iii. diffusion is disequilibriating and endogenously driven and may not necessarily be continuous (Sarkar 1998). The spread of new technologies over time is determined by the competitive advantages of alternatives, behavioral attributes of farmers, and the economic and institutional environment.

In contrast to the NE models, the competitive selection process in ED models assumes that at least some households may not be able to calculate the relative advantages of alternative technologies due to information or cognitive limitations. Initial choices between alternatives are random and the diffusion process endogenously reveals the relative superiority of the alternatives. This view is rooted in path-dependency theory in which final outcomes are dependent on the sequence of prior states that derive from randomness. Hence the process is not necessarily optimal, incremental or cumulative. Decision-making differs between individuals due to differences in motivation, perceptions of possibilities, search behavior, enthusiasm for experimentation, and inferences from observations. Sarkar (1998) provides details on the diverse array of ED models that have evolved recently.

Spatial diffusion models

The role of infrastructure and supply in the diffusion process are the main contributions of the spatial diffusion models developed by social geographers to describe the aggregate spread of technological innovations (Marra et al. 2003). These models are based on the assumption that farmers are passive participants in the adoption process. Two major strands of research emphasize impacts of neighbors (the neighborhood effect) and the role of technology suppliers and innovators. As such, they have been primarily concerned with the spread of knowledge about the innovation rather than the rate of adoption. Since these models typically ignore the influence of adopters on non-adopters, information from extension agents, or the impacts of farmer experience and experimentation, some believe they have contributed little to understanding the adoption decision process (Marra et al. 2003). As Marra et al. (2003) state, spatial and temporal models 'have ignored the central determinant of the rate and pattern of adoption which is the decision process involved in moving from a state of awareness to actual adoption.'

Diffusion theories: agroforestry applications

Scant theoretical work examined the diffusion of agroforestry innovations prior to the 1990s. For example, Scherr (1992) stated that 'No comparable theoretical framework *(to induced innovation)* yet guides agroforestry policy' (italics added). Raintree and Warner (1986) were perhaps the first to apply induced innovation theory to agroforestry. They examined the potential pathways for intensification of shifting cultivation, emphasizing the adoption of improved fallows at various stages of intensification. When land is plentiful (the extensive stage) adoption potential for improved fallows is low because: i. investments are not seen as necessary for current or future soil fertility problems; ii. applying additional labor to fallows is not efficient since returns to labor are relatively low; and. iii. planted fallows may not be culturally acceptable. During the intermediate phase of intensification, soil fertility and fallow yields begin to decline and adoption of improved fallows may increase. However, when land pressures become acute during the final stages of intensification, Raintree and Warner (1986) predict that adoption of improved fallows will be rare due to decreasing farm size, abandonment of shifting cultivation for continuous cultivation, and reduction in available land for fallowing. Franzel (1999) shows how the recent widespread adoption of improved fallows in densely populated areas of Western Kenya requires modifying the Raintree and Warner interpretation of induced innovation theory because of access to off-farm income, bi-modal rainfall, and the use of single season fallows.

Scherr (1992) described some common patterns of agroforestry intensification. She suggested that four types of long-term pressures induce farmers to intensify tree growing: i. declining access to forests and increasing scarcity of wood products; ii. increasing demand for tree products due to population growth, new tree uses or products, and new markets for tree products; iii. increasing population density and declining farm size; and iv. declining land quality. She called for conceptual models of agroforestry intensification under alternative land use conditions, the development of which requires historical and comparative analyses of on-farm tree management under varying agroecological, socioeconomic, and policy conditions.

Applying induced innovation theory to agroforestry, Scherr hypothesized that 'historical increases in tree domestication and management intensity are a response to declining supplies of uncultivated tree resources, increased subsistence and commercial demand for tree products and perceived risks of ecological degradation... [and that] adoption of agroforestry interventions is most likely where consistent with underlying economic incentives for land use change' (1995, p. 788). However, she acknowledged that scarcity alone is insufficient to explain agroforestry diffusion as tree product scarcity may induce substitutions, increased trade, or reduced consumption.

Scherr (2000) applied induced innovation theory to analyze the linkages between poverty, agricultural production, environmental degradation, and adoption of resource conserving technologies such as agroforestry. Her model assumed that pressures from population growth, markets, new technology and other exogenous factors induced changes in local markets, prices and institutions. Community characteristics such as infrastructure, asset distribution, human and natural capital, market linkages and local knowledge and experience determine the local impacts, which induce a variety of household and community agricultural and natural resource management strategies. These strategies may include changes in land use investments and intensity, input and output mix, conservation practices and collective action. The responses may be path-dependent since they are conditioned by community characteristics and land use history. Hence, they may influence the environment, agricultural and tree production, and human welfare, which in turn feed back on local conditions, institutions and natural resource management decisions. The impact of policies and programs that promote agroforestry systems to reduce poverty and promote sustainable land use depends on the dynamics of local change and the relative importance of key factors influencing the poverty-environment relationship. After reviewing a variety of empirical studies, Scherr (2000) concluded that the key factors for increasing farmers' livelihood security while improving or conserving the local resource base were: i. local endowments, ii resource-conserving technologies (conditions for adopting conservation technology), and iii. local institutions supporting the poor.

Glendinning et al. (2001) applied Rogers' (1983) and Rogers and Shoemaker's (1971) sociological theories of innovation-diffusion to examine the relationship between social forestry extension approaches and farmer participation in farm forestry projects in eastern India. Assuming the adoption decision is primarily an 'information-seeking-and-processing activity' to reduce the uncertainty of the returns to adoption, Glendinning et al. (2001) concluded that the most important factor influencing adoption decisions was access to information.

Otsuka et al. (2001) developed a dynamic model of land use to derive the optimal timing of tree planting under different tenure rules. They assumed that households maximized the risk-adjusted, net expected value of land use constrained by the probability that the farmer retains rights to returns from the land. The adoption probability was a function of the time period of production and the land tenure institution. The model predicted two major results. First, shifting cultivation land use and declining land tenure security promoted early tree planting. Second, more secure land tenure would develop in response to increasing land scarcity relative to labor so that the landowner might reap the benefits from investing in land improvement technologies like agroforestry. By increasing the value of cleared land and decreasing the profitability of early tree planting, however, land policies that attempt to increase land rights for cleared forest are likely to be counter-productive for agroforestry adoption.

Empirical adoption studies

General agricultural studies

Feder et al. (1985) and Feder and Umali (1993) reviewed the extensive literature on adoption and diffusion of agricultural innovations. Focusing primarily on the initial stages of Green Revolution technology adoption and diffusion, Feder et al. (1985) concluded that farm size, risk and uncertainty, human capital, labor availability, credit constraints, and tenure were the most important factors determining adoption decisions. However, the impact of farm size depends on the characteristics of the technology and institutional setting and is often a surrogate for a large number of other factors such as access to credit, risk bearing capacity, wealth, and access to information. Since these factors differ spatially and temporally, the relationship between farm size and adoption also varies. In 1985, Feder et al. found that the empirical work on the role of subjective risk was not rigorous enough to draw conclusions concerning its impact on adoption, and this remains the case (Marra et al. 2003; Pannell 2003). Human capital received a good deal of attention in the early adoption literature, most of which suggests that farmers with better education are earlier and more efficient adopters. The direction of the effect of labor availability depends on whether the innovation is relatively labor saving or using. Concerning land tenure, Feder et al. (1985) found that empirical results often differed on the relation between land tenure and adoption and were 'in accordance with the unsettled debate in the theoretical literature over the relation between tenancy and adoption.'

Feder and Umali (1993) updated the Feder et al. (1985) survey emphasizing later stages of adoption-diffusion. As the adoption-diffusion process proceeds, many of the factors that Feder et al. (1985) cited as important for early adoption including farm size, tenure, education, extension, and credit are no longer significant. For example, in later stages of the diffusion process infrastructure variables such as population density; access to markets, roads, and fertilizers; and irrigation availability are more important determinants of diffusion rates of high yielding varieties of rice. However, the agroclimate appears to be the most significant determinant of locational differences in adoption rates (Feder and Umali 1993).

Feder and Umali (1993) also examined the literature on early adoption of soil conservation technologies, which included many agroforestry practices such as contour hedgerows. They concluded that younger, more educated, wealthier farmers who recognized erosion problems were more likely to adopt soil conservation technologies. As Feder et al. (1985) found for Green Revolution technologies, the impact of tenure on soil conservation investments varied among the empirical studies reviewed and requires more research.

Agroforestry studies

As in the more general adoption-diffusion literature since the 1970s, economists have dominated agroforesry adoption research (Mercer and Miller 1998). This work has two separate and distinct lines. Ex-ante studies of agroforestry adoption based on a 'farming systems' approach have emphasized the adoption potential of various agroforestry systems based on researcher-led and participatory on-farm research methods (Byerlee and Collinson 1980; Chambers et al. 1989; Scherr 1991a, 1991b). The goal is to define the 'boundary conditions' for adopting a particular system or practice based on the biophysical and socioeconomic circumstances that allow the innovation to be 'profitable, feasible and acceptable' to farmers (Franzel and Scherr 2002). Ex-post studies of agroforestry have focused primarily on explaining how characteristics of farmers, farms, projects, and other demographic and socio-economic variables are correlated with past adoption behavior. The empirical work has been based primarily on binary choice regression models estimated from cross-sectional household survey data. Many studies, however, have failed to link the empirical analysis to underlying theory and typically have not examined the full range of potential factors that may influence agroforestry adoption (Pattanayak et al. 2003).

Ex-ante studies

Ex-ante studies rely primarily on social and financial analyses of on-farm trials of agroforestry innovations to assess the adoption potential of and to improve the effectiveness and efficiency of developing, modifying and disseminating new agroforestry practices (Franzel and Scherr 2002). Typically, these studies provide information and data on financial and nonfinancial benefits, what works where and why, differential adoption behavior, intra-household distribution of benefits, and how and why farmers are using and modifying the technologies.

Donor agencies need this type of information and analysis to determine how/if the innovations contribute to household welfare and economic development as a basis for research and development allocation decisions. Researchers developing improved systems need this information to insure that their experimental systems are appropriate for farmers' needs, abilities, and circumstances. Finally, this type of information is invaluable to farmers as they attempt to make informed adoption decisions on agroforestry systems that typically require considerable resources, skills, and time to implement and manage (Franzel and Scherr 2002). Nevertheless, systematic ex-ante assessments of experimental systems during the early phases of adoption are rare, partly because some scientists believe they are too 'soft' or too 'subjective' to qualify as rigorous analyses (Franzel and Scherr 2002). The most important and major works in this area are the Current et al. (1995) collection of eight *ex-ante* evaluations of 21 agroforestry projects in Central America (Costa Rica, El Salvador, Guatemala, Honduras, Nicaraugua, and Panama) and the Caribbean (Dominican Republic and Haiti) and Franzel and Scherr's (2002) volume of five studies in Zambia and Kenya.

Franzel et al. (2002) develop a framework for assessing adoption potential. The basic approach uses a variety of participatory appraisals and surveys to identify farmers' problems and needs which are then used by researchers and farmers to develop and design systems for on-farm experimentation. The results of the trials and associated analyses are then used to evaluate the potential for widespread adoption of the systems. For further details on methods for *ex-ante* analysis see Current et al. (1995), Barrett et al. (2002), and Franzel and Scherr (2002).

In the first large-scale study of farm level profitability of agroforestry, the majority of the 56 agroforestry technologies in the Current et al. (1995) volume were labeled as potentially profitable, based on positive Net Present Values and assuming a 20% discount rate. In addition to the expected financial returns, the relation between the new technology and the total farm enterprise and the existing capital, labor, and land constraints were crucial to the adoptability of the systems. Local scarcities as reflected in the price of wood products appeared to be the key factor to profitability and adoption in Central America and the Caribbean. Negative NPVs were associated with poor yields on the annual crop component and/or low output prices. Based on these studies, Current et al. (1995) concluded that including trees in agricultural systems reduces sensitivity to annual crop yield and price variability and thereby improves overall profitability.

Likewise, in the five sub-Sahara African case studies in Franzel and Scherr (2002), agroforestry is shown to have potential to increase farm incomes and solve difficult environmental problems. In addition to the products and services provided, African farmers in Kenya and Zambia value the experimental agroforestry systems for their risk reducing impacts. These studies also confirm the impact of wealth on adoption with better-off families more likely to adopt improved fallows. However, Franzel and Scherr note the absence of absolute barriers to adoption by poor households, since about 20% of the poorest farmers in Zambia planted improved fallows during the first four years of on-farm experimentation.

Vosti et al. (1998) examine the adoption potential and related policy issues for adoption of five 'simple' agroforestry systems including cacao and/or coffee combined with bandarra and rubber and *cupuaçu/freijo*/black pepper combinations, in the western Brazilian Amazon. They found that high investment requirements, negative cash flows in early years, and uncertain local or international demand reduce adoption potential by small holders. Vosti et al. (1998) conclude that evaluating agroforestry adoption potential requires a thorough understanding of the physical and financial returns to all factors of production for all phases of the production process including establishment, maintenance, harvest, processing, and marketing/distribution of products. Although profitability of intermediate and final products sold off farm is crucial, other important factors in determining adoption potential are:

i. Scale of production: profitability and returns to factors of production change with scale of production (especially for crop mixes).
ii. Timing/Size of Investment: costs of under-investing or delaying investment can be quite high for agroforestry compared to annual cropping or pasturing.
iii. Maintenance Costs: can be high (especially labor) for agroforestry and may not vary with amount of product extracted.
iv. Costs of Abandonment: can be quite high, even higher than clearing primary forest.
v. Competing Supply Sources: market competition is key to long-range sustainability of agroforestry.
vi. Sources of Risk: markets, land tenure, weather, etc.

Ex-post studies

Pattanayak et al. (2003) reviewed 120 articles on adoption of agricultural and forestry technology by small holders and concluded that the following five categories of factors explain technology adoption: preferences, resource endowments, market incentives, biophysical factors, and risk and uncertainty. They then used this framework to perform a simple meta-analysis of 32 empirical, regression-based analyses of agroforestry and related soil conservation investments. Although household preferences and resource endowments were the most common variables included in the analyses, the most significant influences on agroforestry adoption behavior appeared to be uncertainty and risk, biophysical conditions, market incentives, and resource endowment factors. For example, when included in the analyses, risk and uncertainty variables were significant 71% of the time, market incentives 78%, biophysical factors 64%, and resource endowments 60‰. In contrast, household preference proxies were significant only 41% of the time (Pattanayak et al. 2003). Likewise in a comparison of tree planting between Brazil and Panama, Simmons et al. (2002) found that institutional variables were more important than household preference variables. The Pattanayak et al. (2003) framework and an expanded list of agroforestry adoption studies are used to take a closer look at the results of the ex-post empirical analyses of agroforestry adoption.

Risk and uncertainty Although risk and uncertainty has long been recognized as important in reducing adoption of a variety of agricultural innovations (Feder and Umali 1993; Smale and Heisey 1993), relatively little empirical research has directly addressed the issue (Abadi Ghadim and Pannell 2003; Pannell 2003). As Marra et al. (2003) point out, 'the issue of risk in adoption has rarely been addressed adequately, and strong empirical evidence to test the common view of its importance and impact has been rare and scattered.' For sustainable agricultural and land conservation systems like agroforestry, risk and uncertainty appear to be even more important to adoption decisions (Pannell 2003). Empirical evidence for the influence of risk and uncertainty on adoption is scarce because few studies include risk and uncertainty as explanatory variables, and the few that do have typically used crude proxies for farmers' risk perceptions and attitudes. Abadi Ghadim and Pannell (2003) suggest this omission has led to poor model specification.

In agroforestry adoption studies, risk has been proxied primarily through four types of variables: tenure, experience, extension and training, and membership in cooperatives and community organizations (Pattanayak et al. 2003). Tenure, experience and extension and training are far more likely than membership in organizations to be significant predictors of adoption of agroforestry, with tenure being significant 64%–72%, experience 90%, and extension 100% of the time when included as independent variables. Extension and training were included as risk proxies in 27% of studies analyzed by Pattanayak et al. (2003) and were always significant and positive. Although experience and extension are obviously important criteria for adoption, few studies have focused directly on them.

In 37 empirical studies of agroforestry adoption, (32 in Pattanayak et al 2003), more secure land tenure always had a positive impact on adoption when significant. In a few cases tenure was an insignificant predictor of adoption (Ayuk 1997; Pattanayak and Mercer 1998; Lapar and Pandey 1999; Adesina et al. 2000), but no studies exhibiting a negative relation between tenure and agroforestry adoption were found. Likewise, Simmons et al. (2002) found tree planting in the Brazilian Amazon to be 15.4 times more likely under secure land tenure, and Wiersum's (1994) study of contour hedgerow adoption in east Indonesia found that the landless comprised 31% of the population but only 11% of participants in *leucaena* based farming systems. This is in stark contrast with the conflicting empirical results that Feder et al. (1985) found for annual cropping. The longer term nature of agroforestry investments appears to inhibit adoption by farmers lacking secure tenure who might otherwise adopt risky innovations for short-term production goals. Thus, risk and uncertainty may be more important for agroforestry than for annual cropping decisions.

Using logistic regression analysis of household survey data in Sumatra, Otsuka et al. (2001) examined factors involved in the evolution of customary land tenure institutions and their impact on tree planting and agroforestry development. Higher population densities promoted individuation of land ownership through the pressures to convert primary forest and bush fallows to commercial tree plots, often using agroforestry systems. Both clearing forests and planting trees enhanced individual tenure rights, but tenure security acquired by clearing communal forests decreased over time when food crops were grown under shifting cultivation systems. Even though purchased bush fallow lands had the most secure tenure rights, tree planting rates were lower than for lands acquired by clearing primary forests because planting trees on cleared lands enhances tenure security on cleared primary forests but not on purchased bush-fallow lands. Otsuka et al. concluded that under these tenure institutions, policies promoting agroforestry may not only induce tree planting but also increase transformation of primary forests.

Only a few studies have concentrated explicitly on the impacts of various risks and uncertainties on agroforestry adoption. In a series of papers, Shively (1997, 1999a, 1999b, 2001) examined the impacts of yield, prices, and consumption risks on agroforestry adoption in the Philippines. Shively (1997) used a probit model to examine the roles of farm assets and relative consumption risk to explain patterns of contour hedgerows at the farm and plot level in the Philippines. A two-moment analysis elicited subjective yield distributions by asking farmers to guess the most likely, lowest, and highest harvest with and without hedgerows and calculating subjective means and variances of the estimated probability density function. Both mean and variance of farmers' estimates were larger for hedgerows, with the difference between true and subjective mean/variance estimates being higher/lower for adopters than non-adopters. Larger farm size, greater tenure security and higher labor availability were correlated with higher adoption probabilities. Shively posits that farm size may be a proxy for lower consumption risk exposure since larger farms have more productive capacity and greater liquidity. As adoption related consumption risk (the opportunity cost of adoption on a plot) increased, the probability of adoption dropped.

Combining stochastic efficiency analysis with a heteroskedastic regression model, Shively (1999a) found that hedgerow adoption was correlated with increased corn yields. Although hedgerows initially reduced effective and observed yields, over time hedgerows were positively correlated with yields and appeared to dampen or reverse the rate of yield reduction on farmers' fields[3]. Hedgerow intensities between 5%–10% of plot area were required to achieve yield increases. Although not as strong an effect, contour hedgerows were found to reduce corn yield variance by as much as 4% depending on the intensity of adoption. Using stochastic efficiency analysis, Shively found that hedgerows would be preferred by a risk-averse farmer only when the range for the coefficient of relative risk aversion (based on mean income) was rather high (3–5.5)[4].

Shively (2001) used the same Philippines data to perform a dynamic simulation to demonstrate how the probability of adoption of hedgerows for soil conservation depends on the opportunity costs of adoption and the household's ability to self-insure against low consumption. Break-even discount rates for risk-averse farmers depended on risk preferences, the opportunity costs of investment, and household exposure to consumption risk conditioned by farm size. Soil conservation was found to be a form of consumption insurance against declining yields and low-probability catastrophic soil erosion events. The pattern of under-investment in soil conservation on small farms implied that basing adoption predictions on risk-neutrality may be misleading when risk of food-insecurity is high. Shively found that small farmers' reluctance to adopt soil conservation measures was ameliorated by access to credit. By minimizing the impact of investment costs and consumption requirements, access to credit facilitated investments in risk-reducing and resource-conserving activities with relatively long payback periods, like contour hedgerows.

Household preferences Assuming that farm households are heterogeneous, farmers exhibit differing adoption patterns depending on their attitudes and preferences for a number of factors such as risk tolerance, conservation priorities, and intra-household homogeneity (Mercer and Pattanayak 2003). Since preferences are extremely difficult to measure directly, a variety of proxy variables are usually used to examine the impact of preferences on adoption. Education, age, gender, and socio-cultural status are the most frequently used proxies for household preferences and were included in almost half of the studies examined by Pattanayak et al. (2003). When included they were significant 40% of the time. Although only included in 36% of studies, gender is the most likely proxy for household preferences to be significant (in 63% of included studies). While education and age are more commonly included in the studies (77% and 64% of studies) they are much less likely to be significant (24% of the time for education and 29% of the time for age). Age, when significant was always positive, while education when significant was positive 75% of the time, and males were always more likely to adopt than females (Pattanayak et al. 2003). The lack of significance of education may be due to low variability in education variables among low-income farmers and correlations between education and other variables such age and wealth. The education of the head of the household may also be irrelevant if the head utilizes the knowledge of other more educated household members. Finally, since educational levels may affect livelihood choices of rural households, samples of only farmers may bias many preference variables like education due to self-selection bias (Barrett et al. 2002).

Several studies have taken a closer look at gender differences in adoption of agroforestry innovations. In a study of tree planting in Kenya, Scherr (1995) found significant gender differences, with male-headed households planting 50% more trees more intensely (more trees per hectare) than women. Men also tended to plant trees on crop land while women planted trees primarily for fuel wood. Using logit and ordered probit analyses, Gladwin et al. (2002a) found that women in female- headed households in Eastern Zambia were significantly more likely to adopt improved fallows than either women or men in male headed households. Adoption rates of improved fallows were almost equal for female headed (47%) and male headed (52%) households in Eastern Zambia (Gladwin et al. 2002a). Place et al. (2002) analyzed the relationship between wealth and gender in adoption of improved fallows in Kenya and Zambia and found that wealthier male headed households were more likely to adopt increased fertilizer and manure applications, while improved fallows, as the Gladwin studies found, were relatively wealth and gender neutral.

Resource endowments The assets and resources available to farmers for investing in new technologies such as labor, land, livestock, savings and credit are critical to adoption decisions. Both the theoretical and empirical literature tell us that early adopters tend to be the better-off households who are better situated to take advantage of new innovations with uncertain prospects. These households are more likely to have the necessary 'risk capital' such as larger incomes and more labor or land to facilitate risky investment in unproven technologies (Hyde et al. 2000; Patel et al., 1995; Scherr 1995). Resource endowments were significant in 60% of the studies in which it was included as an independent variable, with income being significant 50% of the time, assets 100%, labor 33%, and credit (although in only 5% of all studies) 100% of the time (Pattanayak et al. 2003). With the exception of income, depending on the relative importance of farm and off-farm income, and plot size (50% positive and 28% negatively correlated with adoption), resource endowments were unambiguously and consistently positive influences on adoption, i.e. better-off farmers are more likely to adopt (Pattanayak et al. 2003).

Market incentives Agricultural and tree-product prices are well known influences on land-use decisions (Godoy 1992; Hyde and Amacher 2000; Shively 1999b). Unfortunately, market incentives such as input and output prices, market availability, expected income changes, and transportation costs have rarely been included in agroforestry adoption studies (only 33% of studies) (Pattanayak et al. 2003). Market incentives perform well, however, being significant 55% of the time with overwhelmingly positive influences on adoption (Pattanayak et al. 2003).

Using farm-level data and wholesale agricultural price data, Shively (1999b) investigated how changes in the level and variability of market prices affect farmers' decisions to plant mango trees. A 1% increase in price volatility of rice and corn produced an equal increase in mango tree planting intensity in the Philippines. Tree planting was positively correlated with mango prices and negatively correlated with competing crop prices. However, the price of rice, a subsistence food crop, was more strongly correlated with tree planting than the price of the cash crop, corn. This suggests that the economic tradeoff between food and tree crops was more important than the tradeoff between two cash crops. Although it is not surprising that farmers are price sensitive, the fact that short-run price changes affect low-income farmers' decisions to plant mango trees suggests the need for further research on the impacts of short-run prices on tree planting behavior.

Biophysical factors Biophysical factors such as slope, soil quality, irrigation and others have rarely been included in agroforestry adoption studies. Pattanayak et al.'s (2003) meta-analysis found biophysical factors included in only 27% of agroforestry adoption studies. Nevertheless, they appear to be important predictors of adoption, being significant in 64% of studies when included. The direction of significance of many biophysical factors was inconsistent. Poorer quality soils were usually positively correlated with adoption; however, at some point soil quality can be so poor as to render investments pointless. In contrast, slope variables were positive in all but one previous study with steeper slopes providing incentives for farmers to adopt (Pattanayak et al. 2003).

Methodological issues The majority of agroforestry adoption studies have relied on logit or probit models to analyze dichotomous adoption decisions in which the dependent variable is binary (1 if adopts, 0 otherwise). The probit model is used when the error term is assumed to follow a normal distribution and the logit when a logistic cumulative distribution is assumed. Only in rare cases, such as when there are very few positive or negative responses, do the two models produce different results, and scant theoretical justification exists for choosing one model over the other (Greene 1997). Alavalapati et al. (1995), Sunderlin (1997), Thacher et al. (1997), Adesina et al. (2000), Salem et al. (2000), Otsuka et al. (2001), Owubah et al. (2001), Adesina and Chianu (2002), Mercer et al. (in press), and provide examples of the logit model, while applications of the probit model can be found in Shively (1997, 1999a, 2001), Pattanayak and Mercer (1998), and Lapar and Pandey (1999).

When analyzing the simultaneous decision of whether or not to adopt and the extent or intensity of use of the new technology, alternative specifications are required. Typically either the tobit (censored regression model) ordered multinomial logit, or two-stage Heckman models have been used in these situations. Unfortunately, these approaches are quite rare in the agroforestry adoption literature, as only a handful of studies have examined both the probability and extent of adoption.

The tobit model is used when the same independent variables influence both the probability and size of the dependent variable. The ordered tobit accounts for the dependent variable being truncated at either the upper or lower limits of its ranges by assuming the error term follows a truncated normal distribution. A major benefit of the tobit model is that it allows for elasticities measured at the means to be decomposed into an elasticity of adoption and the elasticity of effort when adoption occurs. Rajasekharan and Veeraputhran (2002) use a tobit model to analyze the extent of adoption, in terms of share of intercropped area in rubber plantations in Kerala, India. The ordered multinomial logit is used when the dependent variable is categorical, hierarchical and censored and when the same variables are assumed to influence both adoption and extent of adoption. Patel et al. (1995) apply the ordered multinomial logit model to analyzing tree planting on small farms in East Africa.

When different explanatory variables are assumed to affect the decision to adopt and the extent or intensity of adoption, a two-stage Heckman model is more appropriate. Generally the first stage consists of either a logit or probit analysis of the probability of adoption followed by an ordinary least squares (OLS) regression of the extent of adoption incorporating the sample selection control function (the inverse Mills ratio) from the first equation (Greene 1997). Caviglia and Kahn (2001) apply the Heckman model to analyze adoption of sustainable agriculture, including agroforestry systems, in Brazil.

Finally, several agroforestry adoption studies have made the common mistake of treating categorical independent variables as continuous. This is equivalent to assuming that the intercept shift is the same for each category of the independent variable in question. The regression would be $y = \beta_1 + \beta_2 x_2 + \delta q + \varepsilon$ where x is a continuous variable and q is a categorical variable with three or more categories and β and δ the respective estimated coefficients. The underlying model, however, is $y = \beta_1 + \beta_2 x_2 + \delta + \varepsilon$ for the first category and $y = \beta_1 + \beta_2 x_2 + 2\delta + \varepsilon$ for the second category, etc. This is a testable restriction on the correct formulation ($y = \beta_1 + \beta_2 x_2 + \delta_1 D_1 + \delta_2 D_2 + \delta_3 D_3 + \varepsilon$ where the D's represent dummy variables derived from the categorical variables) but it is unlikely to be appropriate (Greene 1997).

Conclusions

Although research on the adoption and diffusion of agricultural innovations has a long and rich history dating back to the 1950s, interest in understanding the process with the more complex agroforestry systems is relatively new, beginning in the early 1990s. A substantial literature on agroforestry adoption, however, has been developing over the past 10 years. While the majority of the work has emphasized *ex-post* econometric analysis of factors determining adoption behavior, substantial effort has also been made to understand the potential adoptability of systems *ex-ante*. These complementary efforts are critical to the success of agroforestry as an economically, socially, and environmentally sustainable land use system. *Ex-ante* studies of the profitability, feasibility and acceptability of experimental agroforestry systems are essential for researchers in helping design appropriate systems, for development agencies in determining how and where to allocate scarce program funds, and for farmers as they experiment and test new systems as part of the adoption process. *Ex-post* studies are equally important for predicting which segments of society will adopt at various times in the adoption cycle, for estimating the welfare and equity impacts of agroforestry projects, and for designing effective policies to encourage adoption by target populations.

As in traditional agricultural adoption, the major influences on adoption concern household preferences, resource endowments, market incentives, biophysical factors, and risk and uncertainty. Also similarly to adoption of agricultural production innovations, agroforestry adoption follows the predictions of economic theory. Farmers will invest in agroforestry when the expected gains from the new system are higher than the alternatives for the use of their land, labor and capital. Early adopters will tend to be those relatively better-off households who have more risk capital available in terms of higher incomes or more resource endowments (land, labor, capital, experience, education) to allow investments in uncertain and unproven technologies.

However, agroforestry adoption also differs from traditional agriculture, especially with regard to the role of risk and uncertainty. For example, although security of tenure has ambiguous impacts on adoption of annual cropping innovations, the review of *ex-post* agroforestry adoption studies found that tenure security has an unambiguous and positive impact on agroforestry adoption. This is likely due to the longer time periods required to begin to reap the benefits from agroforestry investments. It also highlights the relatively larger importance of risk and uncertainty in agroforestry adoption decisions compared to annual cropping innovations. Nevertheless, risk and uncertainty remain one of the most under studied aspects of adoption behavior. In only a few instances has risk been directly evaluated in agroforestry adoption studies. Typically, we just throw in a few risk proxies such as tenure, experience and extension, and conclude that risk is important. To remedy this situation, studies are required that directly measure risk preferences and perceptions and relate them to the adoption decision process. In particular, we need to understand farmers' perceptions of the impacts of the innovation on risk of production or consumption shortfalls; farmers' uncertainty about the innovation; and farmers' attitudes toward risk and uncertainty. This requires incorporating elicitation of risk preferences and subjective probabilities of the riskiness of alternative technologies into data collection efforts. Norris and Kramer (1990) provide a thorough review of approaches for eliciting subjective probabilities from farmers.

Furthermore, the *ex-ante* adoption literature suggests that in response to high perceived risks associated with new agroforestry systems, households tend to invest only incrementally in new agroforestry technologies and that substantial experimentation, testing, and modification occur before agroforestry innovations begin to diffuse through a community or region. The ability to adapt agroforestry practices over time to emphasize production of different goods and services depending on individual household circumstances and external forces such as markets, policies and weather, appears to be one of the risk-limiting advantages of some agroforestry systems and may be an important factor in differential adoption rates. Understanding this process and incorporating it into the development of dynamic agroforestry adoption models and analyses is essential to understanding the complex patterns of adoption. Other areas that require more research include factors influencing the variability in adoption intensities, the role of adoption of agroforestry practices as complements or substitutes for other farm operations, moving beyond household level analysis to village level and spatial analyses of adoption (e.g. Kristjanson et al. 2002), the impact of diseases like malaria and AIDS on adoption (e.g. Amacher et al. *in press*) and more emphasis in general on the cross disciplinary approaches being championed by the emerging natural resource management adoption literature (Barrett et al. 2002, Franzel and Scherr 2002).

Perhaps the largest deficit in agroforestry adoption research, however, concerns the temporal path of adoption. Adoption is a process that occurs over time. A few households adopt initially, then a few more, and so on, but seldom if ever do all households in a community or region adopt any technology, even over the longest time period. Current *ex-ante* adoption studies based on binary choice regressions have generally been useful only for increasing our understanding of who adopts first and for showing us which communities and which households within those communities to begin with when introducing new agroforestry systems, projects, or programs. If the goal is to assess the potential benefits of agroforestry, however, we need to understand the time path of adoption, the rates and intensity of adoption along that time path, who adopts at different times, and the final level of adoption. Yet, none of these temporal dimensions of agroforestry adoption have been studied empirically. Temporal issues have also rarely been examined in the more general agriculture, forestry, and natural resource management literature. It is time to begin to collect longitudinal data on agroforestry adoption that will allow analysis of the temporal nature of agroforestry adoption including the rate and time at which a technology will be abandoned. This will take us far in understanding the crucial linkage between micro-adoption and aggregate diffusion necessary for effective policy intervention.

End Notes

1. For example, Rogers (1983) reported that before 1964, 44% of the 950 adoption-diffusion publications were by rural sociologists, whereas between 1974–1983, only 8% (45 of 578 publications) were in rural sociology.
2. Prior to Just and Zilberman (1983), most rigorous adoption studies were forced to assume that the new technology produced stochastic yields due to the intractability of handling related random variables with prior theories of choice under uncertainty.
3. Although hedgerow plots outperformed conventional plots on both effective (net of hedgerow) and per hectare basis, yield difference may have partly been a result of higher average labor and fertilizer inputs on hedgerow plots (Shively 1999a).
4. The coefficient of relative risk averstion (CRRA)is defined as:
   $\mathrm{CRRA} = -wU''(w)/U'(w)$
   where w = wealth, U = utility function, $'$ = first derivative of U with respect to w, and $''$ = second derivative of U with respect to w.

References

Abadi Ghadim A.K. and Pannell D.J. 1999. A conceptual framework of adoption of an agricultural innovation. Agr Econ 21: 145–154.

Abadi Ghadim A.K. and Pannell D.J. 2003. Risk attitudes and risk perceptions of crop producers in Western Australia. pp. 114–133. In: Babcock B.A., Fraser R.W. and Lekakis J.N. (eds) Risk Management and the Environment: Agriculture in Perspective. Kluwer, Dordrecht.

Adesina A.A. and Chianu J. 2002. Determinants of farmers' adoption and adaptation of alley farming technology in Nigeria. Agroforest Syst 55: 99–112.

Adesina A.A., Mbila D., Nkamleu G.B. and Endamana D. 2000. Econometric analysis of the determinants of adoption of alley farming by farmers in the forest zone of southwest Cameroon. Agric Ecosyst Environ 80: 255–265.

Alavalapati J., Luckert M. and Gill D. 1995. Adoption of agroforestry practices: a case study from Andhra Pradesh, India. Agroforest Syst 32: 1–14.

Allen J. 1990. 'Homestead tree planting in two rural Swazi communities.' Agroforest Syst 11: 11–22.

Amacher G.S., Ersado L., Hyde W.F. and Haynes A. Tree planting in Tigray, Ethiopia: the importance of disease and water microdams. Agroforest Syst (In press).

Amacher G., Hyde W. and Rafiq M. 1993. Local adoption of new forestry technologies: an example from Pakistan's NW Frontier Province. World Dev 21: 445–453.

Arnold J.E.M. and Dewees P.A. 1995. Tree Management in Farmer Strategies: Responses to Agricultural Intensification. Oxford University Press, Oxford. 292 pp.

Arrow K. 1963. Social Choice and Individual Values. Wiley, New York. 124 pp.

Arrow K.J. 1969. Classification notes on the production and transmission of technological knowledge. Am Econ Rev 59: 29–35.

Arrow K.J. and Fisher A.C. 1974. Environmental preservation, uncertainty and irreversibility. Q J Econ 88: 312–319.

Ayuk E. 1997. Adoption of agroforestry technology: the case of live hedges in the Central Plateau of Burkina Faso. Agr Syst 54: 189–206.

Bannister M.E. and Nair P.K.R. 2003. Agroforestry adoption in Haiti: the importance of household and farm characteristics. Agroforest Syst 57: 149–157.

Barrett C.B., Place F., Aboudk A. and Brown D.R. 2002. The challenge of stimulating adoption of improved natural resource management practices in African agriculture. pp. 1–22. In: Barrett C.B., Place F. and Aboud A. (eds) Natural Resources Management in African Agriculture: Understanding and Improving Current Practices. CABI Publishing, Wallingford, UK.

Besley T. 1995. Property rights and investment incentives: theory and evidence from Ghana. J Polit Econ 103: 902–937.

Bigman D. 1996. Safety-first criteria and their measures of risk. Am J Agr Econ 78: 225–236.

Binswanger H. and Ruttan V. 1978. Induced innovation: technology, institutions, and development. Johns Hopkins University Press, Baltimore. 423 pp.

Boahene K., Snijders T.A.B. and Folmer H. 1999. An integrated socioeconomic analysis of innovation adoption: the case of hybrid cocoa in Ghana. J Policy Model 21:167–184.

Boserup E. 1965. The Conditions of Agricultural Growth. Aldine: Chicago. 124 pp.

Buckles D. and B. Triomphe. 1999. Adoption of mucuna in the farming systems of northern Hounduras. Agroforest Syst 47: 67–91

Byerlee, D. and Collinson M.P. 1980. Planning technologies appropriate to farmers. CIMMYT, Mexico. 71 pp.

Caveness, F.A. and Kurtz W.B. 1993. Agroforestry adoption and risk perception by farmers in Senegal. Agroforest Syst 21: 11–25.

Caviglia J. and Kahn J. (2001). Diffusion of sustainable agriculture in the Brazilian tropical rain forest: a discrete choice analysis. Econ Dev Cult Change 49: 311–334.

Clay D., Reardon T. and Kangasniemi J. 1998. Sustainable intensification in the highland tropics: Rwandan farmers' investments in land conservation and soil fertility. Econ Dev Cult Change 46: 351–378.

Chambers R., Pacey A. and Thrupp L.A. 1989. Farmer First: Farmer Innovation and Agricultural Research. Intermediate Technology Publications, London. 218 pp.

Current D., Lutz E. and Scherr S. 1995. Costs, benefits and farmer adoption of agroforestry. Chapter 1. pp. 1–27. In: Current D., Lutz E. and Scherr S. (eds) Costs, Benefits and Farmer Adoption of Agroforestry: Project Experience in Central America and the Caribbean. World Bank Environment Paper Number 14. The World Bank, Washington, DC.

Daru R.D. and Tips W.E.J. 1985. Farmers participation and socio-economic effects of a watershed management programme in Central Java (Solo river basin, Wiroko watershed). Agroforest Syst 3: 159–180.

Doss C.R. and Morris M.L. 2001. How does gender affect the adoption of agricultural innovations? The case of improved maize technology in Ghana. Agr Econ 25: 27–39.

Evans P.T. 1988. Designing agroforestry innovations to increase their adoptability: a case study from Paraguay. J Rural Stud 4: 45–55.

Feder G., Just R.E. and Zilberman D. 1985. Adoption of Agricultural Innovations in Developing Countries: A survey. Econ Dev Cult Change 33: 255–295.

Feder G. and Umali D.L. 1993. The adoption of agricultural innovations: A review. Technol Forecast Soc 43: 215–219.

Foster A.D. and Rosenzweig M.R. 1995. Learning by doing and learning from others: Human capital and technical change in agriculture. J Politi Econ 103:1176–1209.

Franzel S. 1999. Socioeconomic factors affecting the adoption potential of improved tree fallows in Africa. Agroforest Syst 47: 305–321.

Franzel S. and Scherr S.J. 2002. Introduction. p.1–11. In: Franzel S. and Scherr S.J. (eds) Trees on the Farm: Assessing the Adoption Potential of Agroforestry Practices in Africa. CABI, Wallingford.

Franzel S., Scherr S.J., Coe R, Cooper P.J.M. and Place F. 2002. Methods for assessing agroforestry adoption potential. pp. 37–64. In: Franzel S. and Scherr S.J. (eds) Trees on the Farm: Assessing the Adoption Potential of Agrforestry Practices in Africa. CABI, Wallingford.

Foster A.D. and Rosenzweig M.R. 1995. Learning by doing and learning from others: human capital and technical change in agriculture. J Politi Econ 103: 1176–1209.

Fujisaka S. 1989. The need to build upon farmer practice and knowledge: reminders from selected upland conservation projects and policies. Agroforest Syst 9: 141–153.

Gladwin C.H. 1976. A view of the Plan Puebla: an application of hierarchical decision models. Am J Agr Econ 58: 881–887.

Gladwin C.H., Peterson J.S. and Mwale A.C. 2002a. The quality of science in participatory research: a case study from Eastern Zambia. World Dev 30:523–543.

Gladwin C.H., Peterson J.S., Phiri D. and Uttaro R. 2002b. Agroforestry adoption decisions, structural adjustment and gender in Africa. pp. 115–128. In: Barrett C.B., Place F. and Aboud A. (eds) Natural Resources Management in African Agriculture: Understanding and Improving Current Practices. CABI Publishing, Wallingford, UK.

Glendinning A., Mahapatra A. and Mitchell C.P. 2001. Modes of communication and effectiveness of agroforestry extension in Eastern India. Hum Ecol 29: 283–305.

Godoy R.A. 1992. Determinants of smallholder commercial tree cultivation. World Dev 20:713–725.

Goldman A.C. 1993. Agricultural innovation in three areas of Kenya: Neo-Boserupian theories and regional characterization. Econ Geogr 69: 44–71.

Greene W.H. 1997. Econometric Analysis: Third Edition. Prentice Hall, Upper Saddle River, New Jersey. 1075 pp.

Griliches Z. 1957. Hybrid corn: an exploration in the economics of technological change. Econometrica 25: 501–523.

Hicks J.R. 1932. The Theory of Wages. Macmillan, London. 247 pp.

Humphries S.A. 1993. The intensification of traditional agriculture among Yucatec Mayan farmers: facing up the dilemma of livelihood sustainability. Human Ecol 21: 87–102.

Hyde W.F. and Amacher G.S. 2000. Economics of Forestry and Rural Development: An empirical introduction from Asia. University of Michigan Press, Ann Arbor. 287 pp.

Hyde W.F., Kohlin G. and Amacher G.S. 2000. Social Forestry Reconsidered. pp. 243–287. In: Hyde W.F. and Amacher G.S. Economics of Forestry and Rural Development: An empirical introduction from Asia. University of Michigan Press, Ann Arbor.

Just R.E. and Zilberman D. 1983. Stochastic structure, farm size, and technology adoption in developing agriculture. Oxford Econ Pap 35: 307–328.

Kristjanson P., Okike I., Tarawali S.A., Kruska R., Manyong V.M. and Singh B.B. 2002. Evaluating adoption of new crop-livestock-soil-management technologies using georeferenced village-level data: the case of cowpea in the dry savannahs of West Africa. pp. 169–180. In: Barrett C.B., Place F. and Aboud A. (eds) Natural Resources Management in African Agriculture: Understanding and Improving Current Practices. CABI Publishing, Wallingford, UK.

Lapar M.L. and Pandey S. 1999. Adoption of soil conservation: the case of the Philippines uplands. Agr Econ 21: 241–256.

Marra M., Pannell D.J. and Ghadim A.A. 2003. The economics of risk, uncertainty and learning in the adoption of new agricultural technologies: where are we on the learning curve? Agr Syst 75: 215–234

Mercer E., Haggar J., Snook A and Sosa M. Agroforestry adoption in the Calakmul Biosphere Reserve, Campeche, Mexico. J Crop Production (In press).

Mercer D.E. and Miller R.P. 1998. Socioeconomic research in agroforestry: progress, prospects, priorities. Agroforest Syst 38: 177–193.

Mercer D.E. and Pattanayak S. 2003. Agroforestry adoption by small holders. pp. 283–300. In: Sills E. and Abt K. (eds) Forests in a Market Economy. Kluwer, Dordrecht.

Nair P.K.R. 1996. Agroforestry directions and trends. pp. 74–95. In: McDonald P. and Lassoie J. (eds) The Literature of Forestry and Agroforestry. Cornell University Press, Ithaca, NY.

Norris P.E. and Kramer R.A. 1990. The elicitation of subjective probabilities with applications in agricultural economics. Rev Market Agri Econ 58: 127–147.

Otsuka K., Suyanto S., Sonobe T. and Tomich T. 2001. Evolution of land tenure institutions and development of agroforestry: evidence from customary land areas of Sumatra. Agr Econ 25: 85–101.

Owubah C.E., Le Master D.C., Bowker J.M. and Lee J.G. 2001. Forest tenure systems and sustainable forest management: the case of Ghana. Forest Ecol Manag 149: 253–264.

Patel S.H., Pinckney T.C. and Jaeger W.K. 1995. Smallholder wood production and population pressure in East Africa: evidence of an environmental Kuznets curve? Land Econ 71: 500–515.

Pattanayak S. and Mercer D.E. 1998. Valuing soil conservation benefits of agroforestry: contour hedgerows in the Eastern Visayas, Philippines. Agr Econ 18: 31–46.

Pattanayak S., Mercer D.E., Sills E. and Yang J. 2003. Taking stock of agroforestry adoption studies. Agroforest Syst 57: 173–186.

Pannell D.J. 2003. Uncertainty and adoption of sustainable farming systems. pp. 67–81. In: Babcock B.A., Fraser R.W. and Lekakis J.N. (eds) Risk Management and the Environment: Agriculture in Perspective. Kluwer, Dordrecht.

Perz S.G. 2003. Social determinants and land use correlates of agricultural technology adoption in a forest frontier: a case study in the Brazilian Amazon. Human Ecol 31: 133–165.

Pingali P., Bigot Y. and Binswanger H.P. 1987. Agricultural mechanization and the evolution of farming systems in sub-Saharan Africa. Johns Hopkins University Press, Baltimore. 216 pp.

Place F., Franzel S., DeWolf J., Rommelse R., Kweisga F., Niang A. and Jama B. 2002. Agroforestry for soil-fertility replenishment: evidence on adoption processes in Kenya and Zambia. pp. 155–168. In: Barrett C.B., Place F. and Aboud A. (eds) Natural Resources Management in African Agriculture: Understanding and Improving Current Practices. CABI Publishing, Wallingford, UK.

Rafiq M., Amacher G.S. and Hyde W.F. 2000. Innovation and adoption in Pakistan's Northwest Frontier Province. pp. 87–100. In: Hyde W.F. and Amacher G.S. (eds) Economics of Forestry and Rural Development: An Empirical Introduction from Asia. University of Michigan Press, Ann Arbor.

Raintree J.B. 1983. Strategies for enhancing the adoptability of agroforestry innovations. Agroforest Syst 1: 173–187.

Raintree J.B. and Warner K. 1986. Agroforestry pathways for the intensification of shifting cultivation. Agroforest Syst 4: 39–54.

Rajasekharan P. and Veeraputhran S. 2002. Adoption of intercropping in rubber smallholdings in Kerala, India: a tobit analysis. Agroforest Syst 56: 1–11.

Rogers E.M. 1983. Diffusion of Innovations, 3rd Edition. The Free Press, New York. 367 pp.

Rogers E.M. and Shoemaker F.F. 1971. Communication of Innovations: A Cross Cultural Approach. The Free Press, New York. 476 pp.

Rosenberg N. 1976. On technological expectations. Econ J 86: 523–535.

Ruttan V.W. and Hayami Y. 1984. Induced innovation model of agricultural development. pp. 59–74. In: Clark C.K. and Staatz J.M. (eds) Agricultural Development in the Third World. Johns Hopkins University Press, Baltimore.

Ruttan V.W. 1996. What happened to technology adoption–diffusion research? Sociologia Ruralis 36: 51–73.

Sain G. and Barreto H. 1996. The adoption of soil conservation technology in El Salvador: linking productivity and conservation. J Soil Water Conserv 51: 1495–1506.

Salam M.A., Noguchi T. and Koike M. 2000. Understanding why farmers plant trees in the homestead agroforestry in Bangladesh. Agroforest Syst 50: 77–93.

Sanchez P.A. 1995. Science in agroforestry. Agroforest Syst 30: 5–55.

Sarkar J. 1998. Technological diffusion: alternative theories and historical evidence. J Econ Surv 12: 131–176.

Scherr S.J. 1991a. On-farm research: the challenges of agroforestry. Agroforest Syst 15: 96–110.

Scherr S.J. 1991b. Methods for Participatory On-Farm Agroforestry Research: Summary Proceedings of an International Workshop. International Centre for Research in Agroforestry (ICRAF), Nairobi. 72 pp.

Scherr S.J. 1992. Not out of the woods yet: challenges for economics research on agroforestry. Am J Agr Econ 74: 802–808.

Scherr S.J. 1995. Economic factors in farmer adoption: patterns observed in Western Kenya. World Dev 23: 787–804.

Scherr S.J. 2000. A downward spiral? Research evidence on the relationship between poverty and natural resource degradation. Food Policy 25: 479–498.

Shively G. 1997. Consumption risk, farm characteristics, and soil conservation adoption among low-income farmers in the Philippines. Agr Econ 17: 165–177.

Shively G. 1999a. Risks and returns from soil conservation: evidence from low-income farms in the Philippines. Agr Econ 21: 53–67.

Shively G. 1999b. Prices and tree planting on hillside farms in Palawan. World Dev 27: 937–949.

Shively G. 2001. Poverty, consumption risk, and soil conservation. J Dev Econ 65: 267–290.

Simmons C.S., Walker R.T. and Wood C.H. 2002. Tree planting by small producers in the tropics: a comparative study of Brazil and Panama. Agroforest Syst 56: 89–105.

Sunderlin W.D. 1997. An ex-post methodology for measuring poor people's participation in social forestry: an example from Java. Agroforest Syst 37: 297–310.

Swinkels R. and Franzel S. 1997. Adoption potential of hedgerow intercropping systems in the highlands of Western Kenya: II. Economic and farmers' evaluation. Exp Agr 33: 211–223.

Swinkels R. and Scherr S.J. 1991. Economic Analysis of Agroforestry Technologies: An Annotated Bibliography. ICRAF. Nairobi, Kenya. 215 pp.

Swinkels R., Shepherd K.D., Franzel S., Ndufa J.K., Ohlsson E. and Sjogren H. 2002. Assessing the adoption potential of hedgerow intercropping for improving soil fertility, Western Kenya. pp. 89–110. In: Franzel S. and Scherr S.J. (eds) Trees on the Farm: Assessing the Adoption Potential of Agrforestry Practices in Africa. CABI, Wallingford.

Smale M. and Heisey P.W. 1993. Simultaneous estimation of seed-fertilizer adoption decisions. Technol Forecast Soc 43: 353–368.

Thacher T., Lee D. and Schellas J. 1997. Farmer Participation in reforestation incentive programs in Costa Rica. Agroforest Syst 35: 269–289.

Turner B.L., II and Ali A.M.S. 1996. Induced intensification: agricultural change in Bangladesh with implications for Malthus and Boserup. P Natl Acad Sci, USA 93: 14984–14991.

Vosti S.A., Witcover J., Oliveira S. and Faminow M. 1998. Policy issues in agroforestry: technology adoption and regional integration in the western Brazil Amazon. Agroforest Syst 38: 195–222.

Wiegers E.S., Hijmans R.J., Hervé D. and Fresco L.O. 1999. Land use intensification and disintensification in the upper Cañete Valley, Peru. Human Ecol 27: 319–339.

Wiersum K.F. 1994. Farmer adoption of contour hedgerow intercropping, a case study from east Indonesia. Agroforest Syst 27: 163–182.

Agroforestry Systems **61:** 329–344, 2004.

Scaling up the impact of agroforestry: Lessons from three sites in Africa and Asia

S. Franzel*, G.L. Denning, J.P.B. Lillesø and A.R. Mercado, Jr.
*World Agroforestry Centre, Box 30677, Nairobi, Kenya; *Author for correspondence: e-mail: S.franzel@cgiar.org*

Key words: Extension, Farmer-centered research, Fodder shrubs, Landcare, Natural vegetative strips, Tree fallows

Abstract

This paper assesses recent lessons in scaling up agroforestry benefits, drawing on three case studies: fodder shrubs in Kenya, improved tree fallows in Zambia and natural vegetative strips coupled with the Landcare Movement in the Philippines. Currently more than 15 000 farmers use each of these innovations. Based on an examination of the main factors facilitating their spread, 10 key elements of scaling up are presented. The key elements contributing to impact were a farmer-centered research and extension approach, a range of technical options developed by farmers and researchers, the building of local institutional capacity, the sharing of knowledge and information, learning from successes and failures, and strategic partnerships and facilitation. Three other elements are critical for scaling up: marketing, germplasm production and distribution systems, and policy options. But the performance of the three case-study projects on these was, at best, mixed. As different as the strategies for scaling up are in the three case studies, they face similar challenges. Facilitators need to develop exit strategies, find ways to maintain bottom-up approaches in scaling up as innovations spread, assess whether and how successful strategies can be adapted to different sites and countries, examine under which circumstances they should scale up innovations and under which circumstances they should scale up processes, and determine how the costs of scaling up may be reduced.

Introduction

During the past two decades, researchers have worked with farmers throughout the tropics to identify and develop improved agroforestry practices that build on local indigenous practices and offer substantial benefits to households and to the environment (Cooper et al. 1996; Franzel and Scherr 2002; Place et al. 2002a; Sanchez 1995). Research and development projects have demonstrated in many instances that agroforestry increases household incomes, generates environmental benefits, and is particularly well suited to poor and female farmers. But in most cases these success stories have been confined to localized sites, often with unusually concentrated institutional support from research and development organizations. Binswanger's (2000, p. 2173) comments concerning successful HIV/AIDS programs are equally relevant to successful agroforestry projects: 'Like expensive boutiques, they are available to the lucky few'.

As a consequence, considerable attention has been devoted in recent years to 'scaling up' the benefits of research, that is, 'bringing more quality benefits to more people over a wider geographical area, more quickly, more equitably, and more lastingly (IIRR 2000)'. The issue of scaling up is particularly important to agroforestry and natural resources management innovations, because they are relatively 'knowledge intensive', and, unlike Green Revolution technologies, may not spread easily on their own. Moreover, tree seed is often scarce, as many tree species do not seed very quickly after planting. Drawing on a range of expertise, Cooper and Denning (2000) identified 10 essential elements for scaling up agroforestry innovations: building local capacity, facilitation, farmer-centered research and extension approaches, germplasm, knowledge and information sharing, learning from successes and failures, market options, policy options, strategic partnerships, and technology options (Figure 1). Franzel et al. (2001a)

Figure 1. Essential elements for scaling up agroforestry innovations (from Cooper and Denning 2000).

reviewed nine case studies of scaling up across the tropics and drew lessons, highlighting the varying approaches that different organizations took in different sites.

The objective of this paper is to assess recent lessons learned in scaling up agroforestry benefits, drawing on three case studies in Kenya, Zambia, and the Philippines. Two of these, from Kenya and the Philippines, were reported on in Franzel et al. (2001a), but this paper will show important developments in them since then. First, concepts and definitions of scaling up are reviewed. Next, the case studies are presented and their use of 10 fundamental elements of scaling up is discussed. Finally, conclusions are drawn and research challenges are discussed.

Scaling up: Definitions and concepts

Scaling up is claimed to be reminiscent of the Loch Ness monster – many have sighted it but its description is as varied as the people who have written about it (Uvin and Miller 1994[1]; Gündel et al. 2001). The proliferation of terms describing scaling up has caused some confusion; but as Uvin and Miller note the diversity is important in that it allows for an analysis from a range of perspectives. Their typology involves 17 different kinds of scaling up, focusing on structure, when a program expands its size; strategy or degree of political involvement; and resource base, referring to organizational strength. IIRR (International Institute of Rural Reconstruction) title their book 'going to scale', but use 'scaling up' and 'going to scale' synonymously (IIRR 2000). Others consider scaling out, expanding geographically or involving more farmers, as different from scaling up, involving more organizations and functions. In practice these are difficult to distinguish, since involving more farmers is often associated with involving more organizations and broadening functional objectives. Others use the term 'up-scaling'. In this paper we follow Gündel et al. (2001), who adopt the IIRR definition of scaling up, above, and distinguish between two types, horizontal and vertical. As they note, the IIRR definition 'stresses the importance of a people-centred vision… introduces the quality dimension … without neglecting the quantitative dimension, and highlights the importance of time, equity, and sustainability' (p. 9). Horizontal scaling up is the spread across geographical areas and to more people while vertical scaling up is institutional in nature, involving different types of organizations with different functions from grass roots farmer groups to extension services, policy makers, private companies, and national and international organizations. In fact, horizontal and vertical scaling up often take place simultaneously.

IIRR (2000) notes that the 'scaled up state' can occur either spontaneously or because of the deliberate, planned efforts of governments, NGOs (nongovernmental organizations) or other change agents. There are two important implications. First, much can be learned from studying how spontaneous dissemination of innovations takes place, and in particular the role of farmer-to-farmer dissemination. Second, scaling up is primarily a communication process and change agents have to understand how farmers receive, analyse, and disseminate information in order to facilitate the scaling up process. Of particular relevance is the emerging literature on agricultural knowledge and information systems, that is, how organizations and individuals involved in the generation of agricultural knowledge acquire, transmit and exchange information (Garforth 2001).

Gündel et al. (2001) note that it is wrong to wait until the scaling up phase to consider how to scale up. Rather, they highlight the importance of developing key strategies for scaling up in the design and implementation phase of a research project. These strategies include carefully defining target groups, engaging policy makers, mobilizing dissemination partners, and ensuring that research outputs are simplified, adaptive and flexible and that markets are available for projected increases in output.

Case studies from Kenya, Zambia and The Philippines

Fodder shrubs, Kenya

The site, the farming system, and the farmers' problem being addressed

The low quality and quantity of feed resources is a major constraint to dairy farming in central Kenya, where farm size averages one to two hectares and about 80% of households have one or two dairy cows each. The dairy zone ranges in altitude from 1300 m to 2000 m and rainfall occurs in two seasons, varying from 1200 mm to 1500 mm annually. Soils, primarily Nitosols (Alfisols), are deep and of moderate to high fertility. The main crops are coffee (*Coffea* sp.) produced for cash, and maize (*Zea mays*) and beans (*Phaseolus* spp. and *Vigna* spp.), produced for food. Most farmers also grow napier grass (*Pennisetum purpureum*) as fodder (cut and fed to cows). But napier grass is insufficient in protein so milk yields of animals fed on it are low, about 8 kg cow^{-1} day^{-1}. Commercial dairy meal is available but farmers consider it expensive and most do not use it (Franzel et al. 2003).

The innovation and how it was developed

Researchers (of the Kenya Agricultural Research Institute – KARI , the Kenya Forestry Research Institute – KEFRI, and the World Agroforestry Centre) and farmers tested several fodder shrubs around Embu, Kenya, in the early 1990s and *Calliandra calothyrsus* (hereafter referred to as calliandra) emerged as the best performing and most preferred by farmers. Most of the trials were farmer-designed and managed. For example, farmers tested the feasibility of growing calliandra in a range of 'neglected niches' on their farm. They found that the shrub could be successfully planted in hedges along internal and external boundaries, around the homestead, along the contour for controlling soil erosion, or intercropped with napier grass. When pruned at a height of 1 m, the shrubs did not compete with adjacent crops. Farmers were able to plant 500 shrubs, at a spacing of 50 cm, around their farms, and to begin pruning them within a year after planting. Five hundred shrubs are required to feed a cow throughout the year with 2 kg dry matter per day, adding about 0.6 kg crude protein. On-farm feeding trials confirmed that the farmers could use the shrubs as a substitute for dairy meal or as a supplement to increase milk production. Growing 500 shrubs increased farmers' incomes by between \$98 and \$124 per year. By the late 1990s, two other shrub species, *Morus alba* (mulberry) and *Leucaena trichandra,* were introduced to farmers following successful on-farm testing (Franzel et al. 2003).

The scaling up process

By 1999, eight years after the introduction of fodder shrubs, about 1 000 farmers around the Embu on-farm research sites had planted them. But the scope for reaching the 625 000 dairy farmers in Kenya was limited. Seed was scarce and the farmers, extension staff and NGOs were not aware of fodder shrubs except in areas around the on-farm trials. During 1999–2001, KARI, the World Agroforestry Centre, and the International Livestock Research Institute collaborated in a project to scale up the use of fodder shrubs in central Kenya. An extension facilitator, working with a range of government and NGO partners, assisted 180 farmer-development groups comprising 3200 farmers across seven districts to establish nurseries and plant fodder shrubs. The approach proved to be very effective for facilitating the spread of the practice. By 2002, each farmer had an average of 340 shrubs and each had given information and planting material, seeds or seedlings, to an average of six other farmers. Sixty percent of participating farmers were women. Research projects aimed at understanding the scaling up process examined factors affecting the performance of nursery groups and factors affecting whether adopters disseminated planting material and germplasm to other farmers (Place et al. 2002b[2]).

Beginning in 2002, a project financed by the Forestry Research Programme of DFID (UK Department for International Development) and implemented by the Oxford Forestry Institute and the World Agroforestry Centre is helping a range of partner organizations to increase the adoption of fodder shrubs in Kenya and four other countries. By early 2003, there were about 22 000 farmers planting fodder shrubs in Kenya and several thousand in the other countries. Facilitators are helping to train the extension staff of a range of different organizations, including government, NGOs, churches, community-based organizations, farmer groups and private sector firms. The project is also helping to facilitate the emergence of private seed producers and dealers, and to help link them to buyers in areas where seed demand is high (Franzel et al. 2003; Wambugu et al. 2001).

Improved tree fallows, Zambia

The site, the farming system, and the farmers' problem being addressed

The plateau area of eastern Zambia is characterized by a flat to gently rolling landscape and altitudes ranging from 900 m to 1200 m. Rainfall averages about 1000 mm per year with about 85% falling in four months, December-March. The main soil types are loamy sand or sand Alfisols interspersed with clay and loam. About half of the farmers practice ox cultivation, the others cultivate by hand hoe. Cropped land per farm is 1 ha to 1.6 ha for hand-hoe cultivators and 2 ha to 4 ha for ox cultivators. Maize is the most important crop accounting for 60% of cultivated area; other crops include sunflower (*Helianthus annuus*), groundnuts (*Arachis hypogaea*), cotton (*Gossypium* spp.), and tobacco (*Nicotiana tabacum*).

Surveys identified soil fertility as the farmers' main problem; fertilizer use had been common during the 1980s but the collapse of the parastatal marketing system and the cessation of subsidies caused fertilizer use to decline by 70% between 1987 and 1995 (Howard et al. 1997[3]). Farmers had a strong-felt need for fertilizer but lacked cash for purchasing it (Franzel et al. 2002; Kwesiga et al. 1999).

The innovation and how it was developed

In 1987, the Zambia/ICRAF Agroforestry Research Project began on-station research on improved fallows, using *Sesbania sesban* (hereafter referred to as sesbania).Results were encouraging and on-farm trials began in 1992. By 1995, several hundred farmers were involved in a range of different trials, testing and comparing different options. In researcher-led trials, farmers chose among three different species and two different management options, intercropping with maize vs. growing the trees in pure stands, and compared their improved fallows with plots of continuously cropped maize with and without fertilizer. In farmer-led trials, farmers planted and managed the improved fallows as they wished. Most farmers opted for a two-year fallow and planted their main food crop maize, for two seasons following the fallow. *Tephrosia vogelii, Cajanus cajan*, and *Gliricidia sepium* (hereafter referred to as tephrosia, cajanus, and gliricidia, respectively) were the main species used. Maize yields following improved fallows averaged 3.6 tons, almost as high as for continuously cropped maize with fertilizer (4.4 Mg ha^{-1}) and much higher than maize planted without fertilizer (1.0 Mg ha^{-1}). Returns to land were higher for the fertilized option than for improved fallows but returns to labor were about the same. Both gave much higher returns to land and to labor than maize without fertilizer (Franzel et al. 2002; Kwesiga et al. 1999).

The scaling up process

Extension activities began in earnest in 1996 when an extension specialist in the Zambia/ICRAF project set up demonstrations, facilitated farmer-to-farmer visits, and trained Ministry of Agriculture extension staff and staff of several NGOs and development projects in Eastern Province. The project helped launch an adaptive research and dissemination network, consisting of representatives from several organizations, farmer associations, and projects. (Katanga et al. 2002). The extension effort received a big boost with the start of a USAID (U.S. Agency for International Development)-financed agroforestry project in 1999, covering five districts. ICRAF also facilitated the visits of farmers from Malawi to their counterparts across the border in Zambia; this helped launch the practice in Malawi (Böhringer et al. 1998). Scaling-up objectives include sensitization, building grassroots capacities, developing effective partnerships, promoting policies more conducive to adoption, monitoring and evaluation, and conducting research on the scaling up process (Böhringer et al. 2003). By 2001, over 20 000 farmers in eastern Zambia had planted improved fallows (Kwesiga et al. in press[4]).

Natural vegetation strips (NVS) and landcare, The Philippines

The site, the farming system, and the farmers' problem being addressed

The upland municipality of Claveria is located in Northern Mindanao, the Philippines. With rainfall of 2200 mm yr^{-1}, distributed relatively evenly over nine to 10 months, a farming system of two maize crops per year was widely practiced in the decades following forest clearing by migrants, mainly from the Visayas region of the central Philippines. However, with this high rainfall, cultivation of sloping fields, and use of animal tillage up and down the slope, soil loss through erosion had degraded lands and led to declining maize yields. In most severely eroded areas, farmers have abandoned their fields to cogon grass (*Imperata cyl-*

indrica) and other low-value species. Typical of much of the cultivated uplands in Southeast Asia, the Oxisols of Claveria are now mostly acidic, low in P, and generally unproductive

Building on experience elsewhere in Mindanao, applied research began in 1985 on contour hedgerow systems using nitrogen-fixing trees to minimize erosion, restore soil fertility and improve crop productivity. But adoption of this system was slow and many hedgerows were abandoned due to the high labor requirement to maintain hedgerows, poor adaptation of leguminous trees to acid soils, and competition between the trees and the maize crop. Alternative approaches were clearly needed if these degraded soils were to become productive and a sustainable source of rural livelihood (Mercado et al. 2001).

The innovation and how it was developed

Through participatory on-farm experiments, ICRAF researchers concluded that the *concept* of contour hedgerows remained popular and that farmers were indeed concerned about soil erosion and loss of productivity. Researchers observed that farmers appreciated the value of plowing along contour lines and leaving crop residues and/or natural vegetation in strips between plowed fields. The latter innovation evolved into natural vegetative strips (NVS) and emerged as a crucial entry point for reversing land degradation on sloping fields.

Over several years, the NVS technology, coupled with contour plowing, spread spontaneously among farmers. This innovative farmer-based system and its components were the subjects of intensive on-farm research. Farmer innovations such as the 'cow's back method' for establishing contours were identified as acceptable alternatives to the more technical 'A-frame' technique (ICRAF 1997). In the cow's back method, farmers use the view of the ox's backbone when plowing to maintain a reasonable trajectory for laying out contour lines. Researchers developed a variety of systems of bund width and inter-bund spacing to accommodate a wider range of individual farmer preferences. Some farmers demonstrated interest in cash crops, such as fruit, timber, and coffee; others preferred improved fodder grasses and legumes. In all cases, these innovations built on and enriched the foundation of the NVS. The resulting diversification of enterprises helped increase and stabilize income, and coupled with conservation farming practices, demonstrated that upland farming systems could be both sustainable and profitable (Mercado et al. 2001).

The scaling up process

With the spontaneous visible spread of NVS in and around ICRAF's applied research sites, considerable interest emerged from communities, local and provincial government agencies, and NGOs to learn more about this low-cost approach to sustainable farming in the uplands. In 1996, ICRAF responded to requests by communities for technical support and training. ICRAF facilitators introduced and tested the appropriateness of Landcare, a participatory, community-based approach from Australia involving the development of community groups in partnership with local government to promote conservation-farming practices (Campbell and Siepen 1994; Catacutan et al. 2001[5]; Mercado et al. 2001). Farmers' interest led to the formation of the Claveria Landcare Association, which has emerged as the platform for widespread dissemination of conservation farming based on NVS.

Farmers' incentives for joining Landcare groups include access to training and information, material benefits provided by government agencies, and social benefits of belonging to a group. Community activities have extended to a wide range of functions including waste disposal, livestock raising, and mortuary funds. In 1999, Landcare was extended to another ICRAF research site in nearby Lantapan municipality. By 2002, there were an estimated 500 Landcare groups, involving more than 15 000 farmers in Claveria and other parts of the Philippines. This impressive expansion has taken place as a result of donor-financed facilitation, the support of local government, and farmer – farmer communication. There are four defining characteristics of these Landcare groups (Garrity 2000).

i. They develop their own research and development agenda according to priorities of the group.

ii. They are organized at the sub-village level.

iii The impetus for formation and continuation comes from the community.

iv. Technical and financial support from local government, NGOs, and other agencies enhances continued momentum and sustainability of effort.

Based on the experience from Claveria and nearby municipalities, the Landcare approach has attracted the attention of the national government, donors and international researchers. The Philippines government was in the process of decentralizing in the late 1990s and saw Landcare as an effective partner, providing information and services through groups was more efficient than providing them to individuals. The national watershed management strategy is now based on Landcare. As a result, Landcare principles and experiences have been scaled out to other parts of the country. In addition, ICRAF scientists and national partners are critically examining Landcare's relevance to rural development in Africa

Comparing how the key elements of scaling up were managed in the case studies

Farmer-centered research and extension

Participatory research, in which farmers play a critical role in the design, implementation, and evaluation of research, has been shown to improve the effectiveness of research and to reduce the time between initial testing and uptake (CGIAR/PRGA 1999[6]). In Kenya and Zambia, participatory diagnostic surveys helped identify farmers' critical problems, that is, lack of quality livestock-feed and poor soil fertility, respectively. In both cases, the new practices built on farmers' indigenous practices, that is, the lopping of branches of indigenous fodder trees in Kenya and farmers' bush fallows in Zambia. In the Philippines, there was a clear awareness by farmers of the loss of productivity caused by soil erosion. This led initially to the interest in the use of tree hedgerows and subsequently to the identification of NVS as a more appropriate solution for farmers.

Farmers were involved in the early stages of the development of technologies at all three sites. In Zambia, once on-station trials demonstrated that improved fallows could increase crop yields, researchers began researcher-led and farmer-led on-farm trials (Kwesiga et al. 1999). In the Embu area of Kenya, the first trials testing fodder shrubs were on farmers' fields. At both sites, researcher-led and farmer-led trials were conducted simultaneously, the former primarily to assess biophysical response and the latter, for socioeconomic assessment (Franzel et al. 2001b). Encouraging farmers to experiment with the practice as they wished had important consequences: farmer innovations greatly improved the practices at both sites, reducing costs, promoting adoption and making scaling up more rapid. For example in Zambia, a female farmer obtained potted seedlings at a farmer training center but to reduce the cost of transporting them to her farm, she removed them from the pots and carried them 'bare-rooted' to her farm. After she successfully planted the seedlings, researchers conducted trials, which confirmed that seedlings could be raised in beds instead of pots and transplanted 'bare-rooted', reducing establishment costs by about 60% (Franzel et al. 2002). In Kenya, farmers tested different seed pre-treatment methods and found that soaking the calliandra seed for 48 to 64 hours instead of the recommended 24 hours resulted in higher germination rates. Researchers then confirmed the farmers' findings and the recommendations were revised (Wambugu et al. 2001).

In the Philippines, researchers introduced contour hedgerows of pruned leguminous trees in the 1980s but there was little adoption, due to high labor requirements and competition between the hedges and crops. But a farmer innovation proved very popular, leaving crop residues along the contour, which revegetated forming NVSs. Researchers later found that these strips were effective in controlling soil erosion and required little maintenance. The NVSs spread rapidly and farmers continued to innovate, planting fruit and timber trees or fodder grasses along the strips (Mercado et al. 2001).

The farmer-centered approach taken in Claveria provided the stimulus for innovation and evolution of the NVS system. Engaging directly with farmers revealed the need for low-cost, low-labor innovations. Also, establishing a long-term field presence in Claveria enabled researchers to identify and validate farmers' innovations such as the 'cow's back' method, mentioned above. The long-term field presence also ensured that researchers were able to assist farmers to adjust the NVS system to better reflect their interests, in particular, by introducing cash-generating enterprises such as timber and fruit.

There was some variation in extension strategies among the three case studies. In Kenya, extension facilitators provided training to government extension and NGO staff and representatives of village-based farmer development groups, which are common throughout central Kenya. This approach resulted in a significant amount of farmer-to-farmer extension. A similar strategy was implemented in eastern Zambia, except that facilitators established a network of farmer trainers, who are selected by their communities and receive training and some token support from NGOs and development projects. Lead farmers help train 50 to 100 farmers each season.

In the Philippines, partnership with farmers in on-farm research paved the way for active farmer participation in the scaling up of both the technical innovations and the Landcare approach. As in the Kenya and Zambia cases, early adopters became the focus of cross-site visits and farmer-to-farmer training and extension. In addition, the more articulate and experienced farmers partnered with government extension technicians and outside technical facilitators to form 'local conservation teams' in building awareness and skills for conservation farming (Mercado et al. 2001).

Technical options

Offering a range of options to farmers rather than a specific recommendation is important for several reasons (Franzel et al. 2001c).
i. Most farmers prefer to diversify in order to minimize production and market risks.
ii. Different farmers may have different preferences.
iii. Different options are likely to perform differently as environments change.

In all three sites, researchers and farmers succeeded in quickly developing a range of options for the technologies in question. In Kenya, farmers choose among a range of three fodder shrubs and also a herbaceous legume, *Desmodium intortum*, which is easier to propagate and faster growing than the shrubs but does not perform well during the dry season. Farmers can also plant fodder shrubs in a range of different niches and arrangements on their farms, intercropped, on boundaries, or along contours. Moreover, they can feed the leaves fresh or dry and store them, for use during the dry season.

In Zambia, farmers choose from among four different species and a range of management options for their improved fallows, depending on their preferences and available labor. Sesbania, the most effective species, requires substantial labor because the seedlings need to be raised in nurseries while tephrosia and cajanus can be directly seeded in the field. All species can be intercropped with annual crops during the first year following establishment; intercropping reduces crop yields and tree growth but many farmers prefer it because it economizes on land and labor use relative to planting in pure stands. Thus whereas intercropping with tephrosia requires only about three days per ha of extra labor, as compared to cultivating maize, planting sesbania in pure stand requires about 36 extra days of labor, but with a substantial additional payoff. Farmers can choose from a range of options between these two extremes. Moreover, the most recently introduced species, gliricidia, does not need to be removed before cropping; it can be cut back and then it re-sprouts. Its advantage is that it never needs to be replanted; its disadvantage is that it needs to be cut back during the cropping phase or it may compete with crops. In the mid-1990s, most farmers planted sesbania, the first species to be tested, but tephrosia became a more popular species by 2000. By 2002, gliricidia had become more popular than sesbania and tephrosia in areas where it was available. Cajanus has never been widely used because being palatable to livestock, it is often consumed by free-ranging livestock. Gliricidia and sesbania are also palatable but have not been consumed by livestock in the study area.

In the Philippines in the early 1990s, farmers and researchers began with a single innovation, the NVS. But by the end of the decade, farmers had introduced 31 different perennials on the NVSs, on their own initiative or with advice from facilitators. These different options included fruits (e.g., mango – *Mangifera indica*, rambutan – *Nephelium lappaceum*, and durian – *Durio zibethinus*), timber trees (e.g., *Gmelina arborea*, *Eucalyptus* spp., and *Sweitienia* spp.) as well as fodder grasses (e.g., *Setaria* spp., *Pennisetum purpureum* and *Panicum maximum*) and legumes (*Flamingia congesta* and *Desmodium rensonii*). Most of the fruit- and timber trees were planted with the intention of earning cash (Mercado et al. 2001).

Expanding the range of technical options acknowledges the diversity and interests of farmers. For example, wide spacing of NVS in Claveria was preferred by many farmers wanting to establish timber and fruit trees. This practice delayed the closure of tree canopies and enabled farmers to continue planting annual crops for a longer period while the trees approached maturity. In contrast, farmers with more land preferred closer spacing of NVS as they had the option of planting annual crops on separate parcels once the tree canopy closed (Mercado et al. 2001).

In both Kenya and Zambia, efforts were made to promote and assess involvement of poor and female farmers. An important step in this process was identifying technical options that were particularly suited to these groups, that is, options that required less labor. In Zambia, direct seeding and intercropping, as mentioned above, were important labor-saving practices of special interest to women and the poor. Phiri et al. (2004) conducted a census of all 218 households in four villages and found that 32% of male-headed households and 23% of female-headed households had planted improved fallows, but there was no significant difference between the two proportions. In contrast, fertilizer and manure use was strongly associated with male-headed households (Place et al. 2002a). In Zambia, considerable attention was also given to assessing the participation of the poor in planting improved fallows. Phiri et al. (2004) found that while higher income farmers planted more frequently than poor farmers, 22% of poor households and 16% of the very poor households planted, demonstrating there were no absolute barriers preventing them from planting. But designing appropriate technical options was not the only means of reaching the poor and women households. In Zambia, facilitators encouraged partners to ensure that 50% of beneficiaries were women (Böhringer 2002). In central Kenya, many of the farmer groups that received assistance were women's groups, hence it was not surprising that 60% of beneficiaries in the Systemwide Livestock Project were women.

Local institutional capacity

Empowering local communities to plan their own development and mobilize resources is fundamental to any successful development strategy (Binswanger 2000). The three case studies used different approaches to building local institutional capacity. In central Kenya and eastern Zambia, extension facilitators provided training to village-based groups on the technologies they were promoting but there were few direct efforts to otherwise build the capacities of these groups, except in individual circumstances. For example, in central Kenya, facilitators on a number of occasions helped farmer groups to plan development projects, submit proposals to donor organizations, and link with NGOs and others providing capacity-building services.

In Eastern Province, Zambia, in the mid-1990s, ICRAF assisted partner organizations to form an adaptive research and dissemination network, composed of NGOs, government extension staff, community-based organizations, traditional authorities, and farmer associations and groups. The network functions are to plan, implement and evaluate on-farm research, training, and dissemination activities. Chaired by the provincial research officer, the network brings together representatives of several dozen organizations two times per year to review and plan development activities. Other functions include feeding problems back to research, organizing training, sourcing and exchanging seed, and evaluating different extension and development approaches. The network facilitates the involvement of local groups in the plans and activities of research and development organizations. The involvement of local organizations in the decision making of the network enhances their capacity and feelings of ownership of the network and practices (Katanga et al. 2002).

Landcare in the Philippines has gone further in building local institutional capacity than the organizations promoting improved fallows and fodder trees. Landcare is recognized as an innovative approach to empowering rural communities and local government. The emergence and federation of Landcare chapters in Claveria provided an institutional platform for accessing new information and resources relevant to conservation farming and broader aspects of livelihood development. Landcare enabled communities to share knowledge and experience, influence the agenda of researchers and local policy makers, and mobilize financial resources for technical training and community development. Mercado et al. (2001) noted that the 'greatest success of Landcare' was the attitudinal change of farmers, policy makers, local government and landowners with respect to land use and environmental management.

Landcare has also empowered local government units (LGUs). LGUs have emerged as important partners in raising awareness and in contributing financial resources to Landcare initiatives. Many local government officials speak with a sense of enthusiasm and pride in areas where Landcare has taken root. Indeed some local politicians have used Landcare as a base for political advancement. While local government support can be a powerful force in scaling up, there is a downside risk that the original ethics and principles of Landcare – community leadership and ownership – will be diminished. Indeed, in the rush to expand Landcare, some evidence has emerged of counter-productive LGU directives aimed at 'non-adopters' (J. Tanui, pers. comm. 2003; D. Russell pers. comm. 2003).

Germplasm

The lack of planting material is repeatedly identified as one of the most important constraints to the wider adoption of agroforestry innovations (Simons 1997). National Tree Seed Centres have been unable to deliver seed to large numbers of smallholders and, as with crop seed, 'the seed demand-supply relationship in a large proportion of Africa's smallholder farming systems appears to represent a situation of market failure' (DeVries and Toenniessen 2001).

Successful production and distribution of quality tree seed to resource-poor farmers depends on a number of factors, some of which are biophysical, such as identifying adapted provenances and seed sources, and ensuring sufficient genetic variation and germination capacity. Other factors are economical, organizational and institutional such as protection of and ownership of seed sources, cost-efficient production and distribution networks, institutional support to extension and information networks. A well-functioning seed system is 'one that uses the appropriate combination of formal, informal, market and non-market channels to stimulate and efficiently meet farmers' evolving demand for quality seeds' (Maredia et al. 1999[7]).

The calliandra case study typifies the dilemma of seed systems for new trees: farmers unfamiliar with the practice cannot be expected to buy seed, yet provision of free seed discourages farmers from harvesting seed and undermines the emergence of private sector marketing systems. In the worst scenario, farmers receiving free seed sell it back to the NGO, which thinks it is buying seed from growers! In western Kenya, a few private seed dealers harvest and buy seed from their neighbors and sell to governmental and nongovernmental organizations in central Kenya, where most of the demand is. These organizations distribute seed to farmers for free or at subsidized rates. ICRAF and KARI are trying to improve the situation in four ways.

i. Helping link dealers in western Kenya to buyers in other parts of Kenya.

ii Assisting fodder shrub growers and private nurseries in central Kenya to produce high-quality seed and seedlings and become seed dealers.

iii Working with an NGO, Farm Input Provision Services, to help private dealers to package and sell seed through stockists. However, the legal status of selling through stockists is precarious, as the Kenya Plant Health Inspectorate Service (KEPHIS) requires registration of seed sellers. The fees and documentation requirements make it very difficult for the stockists to market fodder shrub seed profitably. ICRAF is currently seeking permission from KEPHIS for an exemption for fodder shrubs, since the scale of shrub seed sales is much lower than for seed of most annual crops.

iv. Promoting private firms to produce fodder shrub seed or buy seed from seed dealers. One firm has started planting fodder shrubs to explore the feasibility of commercial seed production and another has started buying seed from private dealers in western Kenya, for sale in central Kenya.

The calliandra case study presents two further challenges for scaling up seed-distribution: sufficient seed sources of good genetic quality need to be in production by the time farmers begin adopting, and the species choices for a particular technology should be widened early in the process.

The situation of improved fallows in Zambia has many similarities. Scaling up was constrained by the lack of seed and in a single year, 1996, seven tons of sesbania seed were imported, for free distribution to farmers. The distribution of free seed has discouraged farmers from harvesting it. Later, the USAID-financed project initiated a scheme in which farmers were loaned seed, and promised to give back double the amount they took. Large quantities of seed have been returned for re-distribution to new farmers, and these are managed locally by community seed management committees. The sustainability of this system is uncertain. No private seed dealers have yet emerged, despite the wide-scale adoption of improved fallows and the large potential for expansion.

In the expansion of Landcare, hundreds of communal and private, individual tree nurseries have been established to provide seedlings for fruit and timber species. An aggressive government-led program to promote the use of *Gmelina arborea* has been overtaken by farmers' strong interest in diversifying to more profitable species. There is anecdotal evidence that local entrepreneurs are responding to market opportunities for quality seed and seedlings for high value tree species. In Lantapan, farmers organized themselves to create the Agroforestry Tree Seed Association of Lantapan (ATSAL). ATSAL is a farmer-operated seed collection, production, processing, and marketing association. Founded in 1998 with 15 members, there are now more than 60 active ATSAL members in Lantapan. The organization has trained more than a thousand farmers in seed collection, handling and marketing for both exotic and indigenous tree species and has extended its operations to other areas of the country.

Marketing

Linking farmers to markets and adding value to raw products have great potential for improving the incomes of smallholders and facilitating the scaling up process (Dewees and Scherr 1996[8]). All three of the main practices promoted in the case studies produce inputs: fodder for increased milk production and soil erosion control and soil fertility for crop production.

That all three of the products are intermediate and that only one of them, fodder, can be sold explains the relatively low emphasis given to marketing and product transformation in the case studies. Nevertheless, the uptake of the practices in the three case studies depends on the availability of markets for the final products: milk from fodder, crops following improved fallows, and timber and fruits in NVS systems.

Efforts are needed in all three case studies discussed here to promote the marketing of seed and seedlings. In the case of fodder shrubs in Kenya, a seed marketing study showed that a few private seed producers produced most of the available seed but that they were only able to reach smallholder farmers by selling to government services and NGOs. As mentioned above, facilitators are helping to link the seed producers to seed marketers such as a private seed companies. They are also helping a range of different entities to produce and market seed to smallholders, including a milk processor, a seed company, a nut-marketing firm, individual entrepreneurs, smallholder farmers, farmer groups, and private nurseries.

Moreover, there are also options for increasing the marketing of fodder from shrubs. Fodder may be promoted as a cash crop for farmers who do not own livestock but can sell to their neighbors who do own them. In Kenya, there is also great potential for selling leaf meal as a protein source to millers producing dairy concentrates, who currently import protein in the form of fish meal, soybeans, and cottonseed cake. Leaf meal is a component of dairy meal in India and Central America and it could probably become a cash crop for smallholders in Kenya as well.

For thousands of low-income farmers in the Philippines, the NVS system has evolved as a means to graduate from subsistence maize farming to cash cropping. Claveria is well connected by road to the large port city of Cagayan de Oro opening up potential markets for a range of agroforestry products. Initially there was great enthusiasm to plant *Gmelina arborea* to meet a seemingly insatiable demand: 'Plant gmelina and become a millionaire'. However, prices have since declined and the income expectations of most farmers who rushed to plant gmelina have not been met. This experience has highlighted the value of diversification, which is now increasingly accepted by farmers. NVS adopters in Claveria are now observed to be growing a wider range of timber and fruit trees and are increasingly expressing interest in backyard livestock enterprises to diversify and stabilize their incomes. Market access has been critical for the intensification and diversification of the NVS system.

Policy options

An enabling policy environment is critical for successful scaling up of natural resource management innovations. Policy interfaces with scaling up in several different ways: policy constraints may limit adoption of new practices, policy incentives may be drawn upon for promoting adoption, and policy makers may be engaged to promote or even finance scaling up activities. Whereas there is much attention given to how policy affects the uptake of new practices, less attention is given to directly engaging policy makers in the scaling up process. Raussen et al. (2001) argue that policymakers, at both the local and national level, are an often-untapped resource in scaling-up strategies.

In Zambia, facilitators engaged local leaders to promote agroforestry practices. Numerous workshops were conducted for local leaders, both traditional authorities and government administrators. Local leaders played important roles in promoting improved fallows in two ways. First they helped sensitize and mobilize their constituents to plant improved fallows. Second, they passed and in some cases promoted the enforcement of bylaws to remove two of the main constraints to agroforestry adoption: the setting of uncontrolled fires and free grazing of livestock. Bylaws against free grazing in the Ngoni areas were very effective, primarily because of buy-in by local leaders, because livestock owners were relatively few as compared to Chewa communities, and because the rules were set out with clearly-stated investigative procedures and penalties. Bylaws against setting fires were ineffective, largely because of the difficulty in identifying the persons who set the fires (Ajayi et al. 2002).

The Landcare movement has benefited from and, in turn, reinforced the Philippine government policy of decentralization and devolution of responsibilities to local government. The LGUs are now seen as important partners in local natural resource management initiatives (Catacutan and Duque 2002[9]). Local government units provide policy support for institutionalizing Landcare and conservation farming practices, training staff, and financing Landcare activities (Catacutan et al. 2001[10]). National recognition of the Claveria Landcare experience led to the incorporation of the Landcare approach into the Philippines National Strategy for Improved Watershed Resources Management (Mercado et al. 2001).

As in many countries, land tenure is a constraint to scaling up the establishment of agroforestry enterprises. In situations where improvements are associated with an ownership claim by tenant farmers, there is reluctance by landowners to permit investments in NVS and related tree enterprises. Land tenure was not an important issue in the Kenyan and Zambian case studies, where private rights to land are fairly well established.

One policy risk that has been recognized is the 'projectization' of the Landcare movement. With growing recognition of the Landcare approach and its achievements, it is not difficult to attract external resources, both government and donor, to support replication in other regions of the Philippines and other countries. However, without a full appreciation of Landcare's fundamental concept and principles, sustainability of effort is unlikely.

Knowledge, information sharing, and learning from successes and failures

The dissemination of knowledge and information among stakeholders in scaling up is necessary for ensuring that they are able to make effective decisions. Monitoring and evaluation systems, both formal and informal, are needed to ensure the generation of such information at a range of different scales and from the perspectives of different stakeholders (Cooper and Denning 2000).

In Kenya and Zambia, monitoring and evaluation have been conducted in several different ways. A first step was to find out farmers' assessments and expectations of impacts from technologies they are testing and using. Village workshops were conducted to get this information and farmers broke into focus groups to give their assessments, as well as to identify critical constraints to adoption. These workshops generated many of the same indicators that researchers and extension staff had expected, but there were some important exceptions. For example, in Zambia, many farmers were not using improved fallows to increase their maize production but rather to reduce the area needed to supply their families with maize, so that they could devote more area to cash crops, such as cotton, tobacco, and sunflower (Place 1997; Kristjanson et al. 2002).

In both Zambia and Kenya, ICRAF staff and partners engage in collaborative monitoring and evaluation, in which partners have agreed on a minimum data set to collect. In Zambia, the adaptive research and dissemination network meetings provide a forum for sharing the data. In Kenya, partners jointly developed a simple monitoring form that each fills out every other year. In addition, each organization collects other data that it requires. All organizations take interest in the detailed studies that researchers and students conduct, which most of the partners lack the necessary resources to collect. These studies include economic analyses, impact assessments, and assessing factors affecting adoption. The system in both countries is not without problems, not all organizations involved in scaling up participate and some are unable to collect even the minimum data required. But the collaborative mechanism gives partners a greater sense of ownership and buy-in as well as access to more information and feedback (Nanok 2003[11]).

Knowledge sharing and learning are priorities at all three sites. As highlighted earlier, Landcare groups have proven to be an effective vehicle for knowledge sharing in areas of conservation farming and livelihood improvement. Training in new technologies and sharing of information and experience are among the major benefits of Landcare (Mercado et al. 2001). This institutional platform for knowledge sharing is especially valuable in heterogeneous biophysical environments and dynamic local economies where farming systems are constantly evolving.

Strategic partnerships and facilitation

Partnerships in scaling up offer high potential benefits: organizations with complementary strengths, resources and 'reach' can improve the efficiency and effectiveness of their scaling up efforts (Cooper and Denning 2000). However, these need to be weighted against potential risks: high costs in terms of time and resources, compromised impact, and loss of identity (Jacquet de Haveskercke et al. 2003). Landcare in the Philippines is based on a strategic partnership of farmers, local government units, and technical facilitators (such as ICRAF and government line agencies). They form local conservation teams who help organize Landcare chapters at village and sub-village levels. Federations of village-level Landcare chapters may occur at the municipal level, incorporating a number of relevant line agencies. These alliances are an important aspect of the Landcare movement as they strengthen the influence of farming communities on policy formulation and resource allocation decisions.

An innovative strategic partnership with schools has emerged in Claveria and Lantapan. The Land-

Table 1. Agroforestry practices, extension strategy, policy options, and institutional innovations promoted in the three scaling up case studies.

Technology		Extension strategy	Policy options				Institutional innovations		
Type	Origin	Facilitating extension services, NGOs, farmer groups	Engaging local government in facilitative role	Obtaining local government financing	Lobbying for local policy changes	Lobbying for national policy changes	Capacity building of farmer groups	Facilitating the creation of federations of farmer groups	Promoting consortiums of partners
Fodder shrubs, Kenya	Researcher- and farmer-led trials	X					X		
Improved fallows, Zambia	Researcher- and farmer-led trials	X	X		X		X		X
Natural Vegetative Strips/ Landcare, Philippines	Farmer innovation with research inputs	X	X	X	X	X	X	X	X

care in Schools program started with an information and education campaign that included the training of teachers on technical issues and facilitation skills. The program progressed to include formation of clubs, establishment of school nurseries, and demonstrations of conservation farming and agroforestry. Parents and teachers associations, local communities, and technical facilitators are involved (Catacutan and Colonia 2000[5]). The success of this initiative along with other pilot cases in Africa inspired ICRAF to initiate a multi-region 'Farmers of the Future' program to facilitate and contribute to the integration of agroforestry and natural resource management in basic education (Vandenbosch et al. 2002). After seven years of Landcare in the Philippines, the role of facilitators is emerging as one of the most crucial elements of success. A facilitator works to bring about the various alliances and partnerships needed to develop and support Landcare groups and helps communities develop their potential through advice on organizational structure, training, organization of study tours, and identification and nurturing leadership (Mercado et al. 2001). One concern is that ICRAF continues to play a seemingly indispensable role in facilitation. More effort is needed to develop local facilitation skills, preferably in the community or with NGOs that have a long-term commitment to the local communities. Only then will Landcare be fully recognized as a sustainable institutional innovation.

Conclusions: Research challenges

The review of the three case studies highlights the fact that there is no single recipe to scaling up (Table 1). Different approaches to scaling up can be successful, depending on the innovation, the environment, and the resources at hand. The key elements contributing to impact in the case studies were a farmer-centered research and extension approach, a range of technical options developed by farmers and researchers, the building of local institutional capacity, the sharing of knowledge and information, learning from successes and failures, and strategic partnerships and facilitation. Three other elements are critical for scaling up: marketing, germplasm production and distribution systems, and policy options. But the performance of the cases on these was, at best, mixed.

Although the facilitators in the three case studies used many of the same elements, their scaling up strategies were very different. The case of fodder shrubs in Kenya offers the simplest approach, 'scaling-up-light', in which a focus on on-farm research and facilitating extension services, NGOs and farmer groups reaped surprisingly high benefits. Certainly the limited scope of the innovation – it being one among several important management practices in the dairy enterprise – prevents certain scaling up approaches from being used. For example, facilitators declined to try to develop a network of partners or seek local government financing for promoting fodder shrubs. Interestingly, they probably have involved the private sector in scaling up to a greater extent than facilitators in the other two case studies.

The case of improved fallows in Zambia provides an intermediate case, in which a more complex management practice relevant to several enterprises is being scaled up. Facilitators are using several strategies in addition to those used by fodder shrub facilitators, including engaging local government in a facilitative role, lobbying for policy changes, and promoting a network of partners. These have greatly added to the success of the innovation and to its spread across eastern Zambia.

The case of NVS/Landcare in the Philippines presents the most far-reaching set of innovations, a technical one accompanied by an institutional one. The technical innovation is simple, yet serves as a platform for a multiplicity of other technical innovations and, indeed, a transformation of the farming system. The institutional innovation, Landcare, has had far-reaching ramifications, as federations of farmer groups can wield not only increased economic power but political power as well. In addition to the strategies used by those promoting fodder shrubs and improved fallows, facilitators in the Philippines have obtained local government financing and have facilitated the establishment of federations of groups. Moreover, they have persuaded policy makers to incorporate the Landcare approach into local and national policy.

But as different as the case studies are, they face five similar challenges. First, in none have the facilitators yet articulated a clear exit strategy. At what point can farmers continue to implement the innovations on their own, with limited local backstopping? At what point will the innovations be able to spread to new areas on their own, using farmer-to-farmer dissemination mechanisms and locally available facilitation? What types of local backstopping will be required, e.g., research to respond to shocks (e.g., pests), seed production and marketing mechanisms, policy support, monitoring and evaluation, and extension services? Can some of these tasks be devolved to the private sector or to farmer organizations? What steps need to be taken to phase out external facilitation without sacrificing project success?

Second, as innovations are introduced to new areas and new partners, how can the bottom-up, participatory nature of the scaling up process be maintained? Many government services and NGOs have top-down approaches; they may be interested in promoting the superficial aspects of the innovation without the underlying philosophy needed for it to be successfully adapted to new environments. This problem is as critical for technical innovations, such as improved fallows, as it is for institutional innovations, such as Landcare.

Third, under what circumstances can scaling up innovations and processes at one site or country be adapted for use at another site or country? There are certainly many cases of scaling-up-processes and innovations being successfully extended and adapted to new areas and to new institutions; for example, the Landcare approach in Claveria, northern Mindanao has been successfully adapted to sites and institutions in other parts of Mindanao (Catacutan et al. 2001[10]). But, as mentioned above, the adaptation of the approach in Visayas region has proved more difficult, in terms of the appropriateness of both technical and institutional innovations. Can the evolution of Landcare in the Philippines, including the challenges of its adaptation beyond Claveria, provide valuable lessons for rural development in Africa? In the search for new models for community-driven development, the Landcare experience presents an exciting opportunity to learn and adapt. The basic preconditions for Landcare appear to exist in much of rural Africa: communities' recognition of land degradation as a critical problem; a range of technical options available to address degradation; decentralization and devolution of government authority in some countries; national commitment to reduce rural poverty in many countries; and existing community organization/group structures, at the village level if not at higher levels.

Fourth, and related to the above, under what circumstances should facilitators seek to scale up technologies and under what circumstances should they seek to scale up the process by which adoption and adaptation have taken place? Certainly there are examples where one or the other or both can be successfully scaled up but what guidelines can practitioners use to decide? To what degree is any scaling up strategy useful: only for a particular technology (a seed treatment method), for several innovations (a nursery), for any type of agricultural innovation (a farmer group) or for agriculture in general (improving market access). These questions are crucial, with major implications for the resource demands of efforts to scale up.

Fifth, what are the costs of scaling up relative to the benefits, and how can costs be reduced? How can informal farmer-to-farmer information systems be promoted or formalized, and can these replace expensive, government or NGO-financed systems? Can farmer organizations such as Landcare take on some of the functions of these systems, such as extension and local testing and adaptation of new innovations? What role can the private sector play and how can facilitators promote the involvement of the private sector in scaling up?

All of the above issues are at least to some extent researchable. For example, careful assessments of the costs and benefits and advantages and disadvantages of different strategies can be made. Simple planned comparisons of different scaling up mechanisms can be undertaken. Just as learning and knowledge sharing are critical functions in the scaling up of innovations, they are critical for identifying effective and efficient scaling up strategies. Investment in understanding scaling up processes will reap important rewards leading to improved livelihoods of beneficiaries.

Acknowledgements

The comments and suggestions of Chris Garforth and Robert Tripp on earlier drafts are gratefully acknowledged.

End Notes

1. Uvin P. and Miller D. 1994. Scaling up: Thinking through the issues. World Hunger Program. Brown University, Providence, RI, USA. www.globalpolicy.org/ngos/role/intro/imp/2000/1204.htm
2. Place F., Kariuki G., Wangila J., Kristjanson P., Makauki A. and Ndubi J. 2002b. Assessing the factors underlying differences in group performance: Methodological issues and empirical findings from the highlands of central Kenya. CAPRI Working Paper no. 25. CGIAR Systemwide Program on Collective Action and Property Rights, International Food Policy Research Institute, Washington, D.C.
3. Howard J.A., Rubey L. and Crawford E.W. 1997. Getting technology and the technology environment right: Lessons from maize development in Southern Africa. Paper presented at the Meeting of the International Association of Agricultural Economics. 10-16 August, (1997), Sacramento, CA, USA.

4. Kwesiga F., Franzel S., Mafongoya P., Ajayi O., Phiri D., Katanga R., Kuntashula E., Place F. and Chirwa T. (in press) Improved fallows in Eastern Zambia: History, Farmer practice, and impacts. IFPRI discussion paper.
5. Catacutan D.C. and Colonia G. 2000. Landcare in school: the Lantapan experience. Farmers of the Future Workshop, 2-3 October 2000, Nairobi, Kenya.
6. CGIAR/PRGA (Consultative Group on International Agricultural Research/Program on Participatory Research and Gender Analysis). 1999. Crossing perspectives: Farmers and scientists in participatory Plant Breeding. Cali, Colombia.
7. Maredia M., Howard J. and Boughton D. 1999. Increasing Seed System Efficiency in Africa: Concepts, Strategies and Issues. MSU International Development Working Paper No. 77. Department of Agricultural Economics. Department of Economics. Michigan State University, USA.
8. Dewees P.A. and Scherr S.J. 1996. Policies and markets for non-timber tree products. Environment and Production Technology Division Discussion Paper No. 16. International Food Policy Research Institute, Washington, D.C.
9. Catacutan D.C. and Duque C.E. 2002. Locally-led Natural Resource Management. Proceedings of a regional workshop, 8-9 November 2001, Valencia City, Bukidnon, Philippines.
10. Catacutan D.C., Mercado A.R. and Patindol M. 2001. Scaling up the Landcare and NRM planning process in Mindanao, Philippines. LEISA Magazine on Low External Input and Sustainable Agriculture. 17:3, 31-34.
11. Nanok T. 2003 Assessment of the adoption and dissemination of improved fallows and biomass transfer technologies in western Kenya. M.Sc. thesis. Department of Agricultural Economics. Egerton University, Njoro, Kenya.

References

Ajayi O.C., Katanga R., Kuntashula E., and Ayuk E.T. 2002. Effectiveness of local policies in enhancing adoption of agroforestry technologies: the case of by-laws on grazing and fire in eastern Zambia. pp. 93–99. In: Kwesiga F., Ayuk E. and Agumya A. (eds) Proceedings of the 14th Southern Africa Regional Review and Planning Workshop, 3–7 September 2001, Harare, Zimbabwe, SADC-ICRAF Regional Programme, Harare, Zimbabwe.

Binswanger H.P. 2000. Scaling up HIV/AIDS programs to national coverage, Science 288: 2173–2176.

Böhringer A. 2002. Facilitating the wider use of agroforestry for development in southern Africa. pp. 434–448. In: Franzel S., Cooper P., Denning G.L. (eds) Development and Agroforestry: Scaling up the Impacts of Research. Development in Practice Readers Series. Oxfam Publishing, Oxford, UK.

Böhringer A., Ayuk E.T., Katanga R. and Ruvuga S. 2003. Farmers nurseries as a catalyst for developing sustainable land use systems in southern Africa. Part A: Nursery productivity and organization. Agr Syst 77: 187–201.

Böhringer A., Moyo N. and Katanga R. 1998. Farmers as impact accelerators: Sharing agroforestry knowledge across borders. Agroforestry Today 10 (2): 9–11.

Campbell A. and Siepen G. 1994. Landcare: Communities Shaping the Land and the Future, Allen and Unwin, Sydney. 344 pp.

Cooper P.J.M. and Denning G.L. 2000. Scaling Up the Impact of Agroforestry Research. ICRAF, Nairobi. pp. 43.

Cooper P.J., Leakey R.R.B., Rao, M.R. and Reynolds L. 1996. Agroforestry and the mitigation of land degradation in the humid and sub-humid tropics of Africa. Exp Agr 32: 235–290.

DeVries J and Toenniessen G. 2001. Securing the harvest: Biotechnology, breeding, and seed systems for Africa's crops. CABI Publishing, Wallingford U.K. 266 pp.

Franzel S., Cooper P., Denning G.L. and Eade D. (eds) 2001a. Development and Agroforestry: Scaling up the Impacts of Research. Development in Practice (Special volume) 11: 405–534.

Franzel S., Coe R. Cooper P., Place F. and Scherr S.J. 2001b. Assessing the adoption potential of agroforestry practices in sub-Saharan Africa. Agr Syst 69: 37–62.

Franzel S., Cooper P. and Denning G.L. (eds) 2001c. Scaling up the benefits of agroforestry research: lessons learned and research challenges. Development in Practice 11: 524–534.

Franzel S., Phiri D. and Kwesiga F.R. 2002. Assessing the adoption potential of improved tree fallows in Eastern Zambia. pp. 37–64. In: Franzel S. and Scherr S.J. (eds) Trees on the Farm: Assessing the Adoption Potential of Agroforestry Practices in Africa. CABI, Wallingford, U.K.

Franzel S. and Scherr S.J. (eds) 2002. Trees on the Farm: Assessing the Adoption Potential of Agroforestry Practices in Africa. CABI, Wallingford, U.K. 197 pp.

Franzel S., Wambugu C. and Tuwei P. 2003. The adoption and dissemination of fodder shrubs in central Kenya, Agricultural Research and Network (AGREN) Series Paper No. 131. Overseas Development Institute, London, UK. 2003. 10 pp.

Garforth C. 2001. Agricultural knowledge and information systems in Eritrea: a study in sub zoba Hagaz. The University of Reading, Reading UK. 59 pp.

Garrity D. 2000. The farmer-driven Landcare Movement: an institutional innovation with implications for extension and research. pp. 7–9. In: Cooper P.J.M. and Denning G.L. (eds) Scaling Up the Impact of Agroforestry Research. ICRAF, Nairobi, Kenya.

Gündel S., Hancock J. and Anderson S. 2001. Scaling up strategies for Research in Natural Resources Management: A comparative review. Natural Resources Institute, Chatham, UK. 61 pp.

ICRAF. 1997. Annual Report 1996, International Centre for Research in Agroforestry, Nairobi. 340 pp.

IIRR (International Institute of Rural Reconstruction) 2000. Going to Scale: Can We Bring More Benefits to More People More Quickly? IIRR, Silang, Cavite, Philippines. 114 pp.

Jacquet de Haveskercke C., Böhringer A., Katanga R., Matarirano L., Ruvuga S. and Russell D. 2003. Scaling out agroforestry through partnerships and networking: the experience of the World Agroforestry Centre in southern Africa. Southern Africa Agroforestry Development Series No. 2. World Agroforestry Centre, Zomba, Malawi.

Katanga R., Phiri D., Böhringer A. and Mafongoya P. 2002. The Adaptive Research and Dissemination Network for Agroforestry: a synthesis of the adaptive workshops. pp. 93–99 In: F. Kwesiga, E. Ayuk and A. Agumya (eds) Proceedings of the 14th Southern Africa Regional Review and Planning Workshop, 3–7 September 2001, Harare, Zimbabwe, SADC-ICRAF Regional Programme, Harare, Zimbabwe.

Kristjanson P., Place F., Franzel S. and Thornton P.K. 2002. Assessing research impact on poverty: the importance of farmers' perspectives. Agr Syst. 72: 73–92.

Kwesiga F.R., Franzel S., Place F., Phiri D. and Simwanza, C.P. 1999. *Sesbania sesban* improved fallows in eastern Zambia: their inception, development, and farmer enthusiasm. Agroforest Syst 47: 49–66.

Mercado A.R., Marcelino P. and Garrity D.P. 2001. The Landcare experience in the Philippines: technical and institutional innovations for conservation farming. Development in Practice 11 (4): 495–508.

Phiri D., Franzel S., Mafongoya P., Jere I., Katanga R. and Phiri S. (2004). Who is using the new technology? A case study of the association of wealth status and gender with the planting of improved tree fallows in Eastern Province, Zambia. Agr Syst. 79: 131–144.

Place F. 1997. A methodology for a participatory approach to impact assessment: implications from village workshops in Eastern Province, Zambia. ICRAF. 14 pp.

Place F., Franzel S., DeWolf J., Rommelse R., Kwesiga F., Niang A. and Jama B. 2002a. Agroforestry for soil fertility replenishment: Evidence on adoption processes in Kenya and Zambia. pp. 155–168. In: Barrett C.B., Place F. and Aboud A.A. (eds) Natural Resources Management in African Agriculture: Understanding and Improving Current Practices. CAB International, Wallingford, UK. 2002.

Raussen T., Ebong G. and Musiime J. 2001. More effective natural resources management through democratically elected, decentralised government structures in Uganda. Development in Practice 11 (4): 460–470.

Sanchez P.A. 1995. Science in agroforestry. Agroforest Syst 9: 259–274.

Simons A.J. 1997. Delivery of improvement for agroforestry trees. In: Dieters M.J. and Matheson C.A. (eds) Tree Improvement for Sustainable Tropical Forestry. Queensland Forestry Research Institute, Gympie, Australia. 261 pp.

Vandenbosch T., Taylor P., Beniest J. and Bekele-Tesemma A. 2002. Farmers of the Future – a strategy for action. World Agroforestry Centre (ICRAF), Nairobi, Kenya. 37 pp.

Wambugu C., Franzel S., Tuwei P. and Karanja. G. 2001. Scaling up the use of fodder shrubs in central Kenya. Development in Practice 11: 487–494.

Agroforestry Systems **61:** 345–355, 2004.

Trees of prosperity: Agroforestry, markets and the African smallholder

D. Russell* and S. Franzel
*World Agroforestry Centre, Box 30677, Nairobi, Kenya; *Author for correspondence: e-mail: d.russell@cgiar.org*

Key words: Charcoal, Enterprise development, Nurseries, Smallholders, Tree-crops, Tree seed

Abstract

In many developing countries, especially in Africa, farmers have been introduced to agroforestry with little consideration for the markets for trees and tree products aside from potential productivity gains to food crops. It is now being recognized that expanding market opportunities for smallholders particularly in niche markets and high value products is critical to the success of agroforestry innovations. Some recent work presented in this paper on marketing agroforestry products in Africa, linking farmers to markets and assisting farmer organizations, shows how constraints are tied to both long-standing market structures as well as shifting market imperatives. Forest policy, physical and social barriers to smallholder participation in markets, the overall lack of information at all levels on markets for agroforestry products, and the challenges to outgrowing schemes and contract farming inhibit the growth of the smallholder tree product sector in Africa outside of traditional products. Notwithstanding these constraints, there are promising developments including contract fuelwood schemes, small-scale nursery enterprises, charcoal policy reform, novel market information systems, facilitating and capacity building of farmer and farm forest associations, and collaboration between the private sector, research and extension.

Introduction

Since the advent of trade networks in Africa, smallholder farmers with trees on farm or in their common areas have been drawn into markets for benefit and survival. Trade in certain forest and tree products such as timber, fuelwood, charcoal, palm wine, medicinal products, honey, nuts and fruit has gone on for centuries (Bohannan and Dalton 1968; Terray 1974). Overall, however, 'colonial interventions in Africa left a profound legacy of neglect of smallholder farmers in favor of estate farm producers (in eastern and southern Africa) and European export trading companies (in West Africa)' (Kherallah et al. 2000).

Agroforestry extension in industrialized countries, including South Africa, is now focused on the market potential for trees and tree products such as Christmas trees and non-timber forest products (Gold et al. 2004). Even the environmental services of agroforestry systems such as for windbreaks or soil stability are fostered largely because of their economic value to farmers and landowners (e.g., Thomas and Schumann 1993; Small Woodlands Program of British Columbia 1994; Harrison 2002). Yet in Africa, agroforestry often is portrayed and promoted, still, as a conservation or 'natural resource management' activity.[1] It is surprising that many farmers have been introduced to agroforestry with little consideration for the markets for trees and tree products aside from potential productivity gains to staple crops.

The interface among private sector actors, researchers and extensionists has been quite weak in agroforestry. This gap became evident, for example, at a workshop on agroforestry and markets in western Kenya in 2002.[2] In the workshop, which was held after more than a decade of agroforestry extension in the area, it became clear that many participants including farmers, researchers and extensionists 'met the market' for key tree products for the first time. In part the problem lies with the dearth of marketing expertise in extension agencies and other institutions interfacing with farmers. In part it reflects the predominant paradigm of agroforestry as a 'sustainable land use system', in contrast to monocropping and plantations, rather than a moneymaker[3].

If agroforestry systems and innovations are to contribute to the fight against poverty, we must explore, enable, and expand market and value addition opportunities for smallholders particularly in niche markets and high value products that present opportunities beyond staple crops (Hellin and Higman 2002). Indeed, despite its environmental attractions, agroforestry has a dim future if it supplies few direct monetary benefits to farmers and landowners.[4] However, we need to question initiatives that focus only or largely on high-end products for Western consumers and on certification schemes that involve very few farmers; instead we advocate concentrating on products that are important for Africa's development, for example in the energy, construction, and locally valued food product sectors.

Many threats and bottlenecks hamper smallholders' access to markets and to value addition. Gabre-Madhim (2002) of the International Food Policy Research Institute (IFPRI), working on smallholders' access to markets, noted that post-reform markets in Africa continued to be characterized by high transaction costs, limited and asymmetric market information, lack of coordination, missing markets for storage and finance, lack of smallholder market power and increased risk. In particular, challenges facing asset-poor small producers are the lack of sufficient scale to access global markets, the high penalties faced for logistical bottlenecks, problems of standardization and limited access to post-harvest technology and market information.

These constraints are compounded in the agroforestry/forestry sector by policies inhibiting tree use, even poorer market information systems than for staple crops, a diminishing resource base, weak internal consumer demand due to poverty and food habits, monopolization by elites, strong commodity chains for exports that neglect 'minor forest products', poor infrastructure and greater risk from longer production cycles (Kaimowitz 2002). A key research and development question, then, is how can research and NGO (non-government organization) partners working together with market actors bring about change in the structure and operation of markets? Can farmer groups really be empowered in the sub-Saharan Africa to develop replicable enterprise strategies that can be adapted for a range of tree products? Is the demand for these products relatively elastic or inelastic – where is it located and how could farmers better understand it?

We present case examples of products with high local demand that the World Agroforestry Centre (ICRAF) and partners have been researching including charcoal, smallholder timber, peri-urban nurseries, indigenous fruits, and calliandra (*Calliandra calothyrsus*, a fodder shrub) seed. These cases illustrate the added problems associated with 'non-traditional' smallholder agroforestry products that must be tackled to improve the prospects for this sector and increase the economic value of agroforestry systems.

Case examples

Charcoal marketing[5]

Charcoal-making has occurred in forests and woodlands for hundreds of years and has caused conflicts with forest authorities. Until cheap electricity comes to all households, it is going to be an important source of energy—perhaps even after that time as it is often the preferred cooking fuel even in homes with electricity. ICRAF's Regional Land Management Unit (RELMA) in East Africa has been working on charcoal policy since 2001. This work concludes that it is possible to produce charcoal from planted trees and market it sustainably. A major hindrance is the forest/woodland conservation mindset of the policy-makers in some countries and of some development partners (cf. Girard 2002, which focuses on technical constraints). Below we compare the charcoal markets and institutions in Kenya and Uganda (that do not have an organized charcoal market) to that of Sudan, which has an organized market and a fuelwood policy and strategy centered in one institution.

In Kenya, policymakers have been reluctant to discuss charcoal let alone make a decision about its sustainable development. Producing and transporting charcoal from private farms is legal while doing the same from forests is illegal. The main problem faced by producers and traders is that they are unable to transport or trade charcoal because police and other law enforcers assume that it is from forests, even when they have documentation showing otherwise.

Kenya has no oil resources and depends on imported oil. It is therefore necessary for Kenya to plan to produce charcoal sustainably by focusing on species and production methods that are less environmentally damaging than those that currently prevail (see Utama et al. 1999 for an example in Indonesia). Although it is technically possible, Kenya has no long-term production plans. Similarly in Uganda, 92% of the energy consumed is from woodfuel. Despite the fact that 'modern' energy provides only 8% of the national energy needs, the energy policy in Uganda emphasizes the development of modern energy (electricity and petroleum) and there is insignificant mention of charcoal in the policy.

Kenya lacks an institution that assumes full responsibility for charcoal development.[6] In Kenya three ministries handle charcoal: the Ministries of Energy, Environment and Natural Resources, and Local Government and Provincial Administration. The overlap causes confusion. Similarly, the Ugandan charcoal industry is disorganized: in Uganda, laws relating to wood are scattered across different sectors, making enforcement difficult. Currently, the Forest Department in Uganda deals with supply while the Energy Department is responsible for managing demand.

In contrast to Kenya and Uganda, Sudan emphasizes woodfuel development despite its oil reserves. In Sudan, the Forest National Corporation plans and implements production and regulates the industry. While the government of Sudan obtains substantial revenue from the charcoal industry, the government of Kenya loses revenue because of the illegal status of the commodity. Hence there is no capital for re-investment. Charcoal producer and traders' associations are registered and legal in Sudan. They were initiated to assist members in negotiating with the government and with related interests. They follow government rules in all their production and marketing activities but they also have their own rules and by-laws. They expel members who do not obey the rules, for example those who do not pay taxes and fees and those who are corrupt. Although the charcoal enterprises are profitable, the Sudan markets could be improved: merchants face high taxes, unclear boundaries and conflicts due to animal routes through contracted land. They would like the government to give them forestland, which they could harvest rotationally and manage as their own.[7]

For charcoal to be produced and marketed sustainably, studies by ICRAF, RELMA and partners that have compared systems come to several conclusions.
1. African governments should recognize the importance of charcoal as a source of energy and allocate adequate human and financial resources for the development of the industry.
2. Charcoal should be legalized and measures to ensure its sustainable supply put in place.
3. Administration of the industry should be decentralized to allow for faster enforcement of the law.
4. The charcoal policy and institutions should have adequate provision for the participation of the private sector.
5. Lessons from countries with reformed charcoal policies can be applied and sounder charcoal production technologies can be adopted when national charcoal institutions and markets are reformed.

Smallholder timber in Kenya[8]

During 1996–2000, there has been a 1% per capita increase in wood demand worldwide (Scherr, 2004). The wood supply crisis is severe in several tropical countries. There are logging bans in natural forests and plantations, which make timber from these sources illegal. Large plantations are difficult to site in many countries and they incur risks, for example fire. In Europe and the United States, there is a consumer preference for sustainably grown timber and there is also developing country experience in establishing outgrower group schemes, for example, poplars in India, eucalypts in South Africa, and hardwoods in Ghana.

ICRAF's smallholder timber program is an outcome of a joint Kenya Agricultural Research Institute (KARI)/Kenya Forest Research Institute (KEFRI)/ICRAF initiative in Embu, Kenya, in 1998 on high value trees in the Kenya Central Highlands. Preliminary appraisals revealed that the main farm enterprises include coffee (*Coffea* spp), dairy, maize (*Zea mays*) and the tree *Grevillea robusta*. Men are the decision makers on choice of species, marketing and utilization. Farmers are unable to estimate the quantities of wood and value of their trees and there is poor knowledge of tree management including silvicultural techniques for high-quality timber production (Holding-Anyonge and Roshetko 2003). Farmers could not pinpoint a definite market for timber: they did not perceive organized timber markets. They received low prices for tree and tree products and also pointed to poor conditions of roads, and lack of capital hindered tree selling. Tree competition with crops was another problem mentioned by farmers.

Additional surveys revealed that farmers used their timber domestically: they sell trees to neighbors or traders who come to the farms, often when coffee prices are low (S. Carsan, pers. comm. 2003).

Trees are sold to any willing buyer at any agreed price. The surveys reinforced the findings of the initial appraisal on poor knowledge on marketing, silvicultural techniques, mensuration, and valuation. The farmers did not consider trees on their farms to be income-generating enterprises. This perception remained among farmers around Imenti Forest in Meru even after the forest was closed to use and fenced off and new market opportunities from tea factories arose in Meru (R. Barr, pers. comm. 2003).

A survey of sawmillers found that mobile sawmillers were strategically placed in the marketing of on-farm timber. The sawmillers were more knowledgeable than farmers but had problems acquiring timber and logs from farmers. They needed more information on better tree use and on existing supply of trees and logs on farm. The sawmillers also needed to improve output efficiency and to find methods of adding value to timber.

In sum, the research seeks to outline how the local market chain is organized and seeks to project demand from smallscale farms, then to build capacity of farmers and other stakeholders to improve the market access and added value for smallholders through market information and intelligence systems and technical skills. The research and development approach was holistic: looking at institutional, technical, economic, and environmental/resource angles. Lessons emerging from this project include the fact that trees on the farm act as buffers to the poor in times of hardship, however, the poor lack the necessary knowledge, capital or legal rights to exploit market opportunities.[9] The results of the pilot project show there are many opportunities to enhance strategies for farmer-led market analysis and development given current trends of market liberalization and decentralization (Holding et al. 2001; Holding-Anyonge et al. 2002). To achieve the desired impact, however, we need to recognize that the relative success of research support to farmers' initiatives in smallholder timber production depends to a great extent on political and economic forces well beyond the control of research and development program. Communities are heterogeneous and tree products interests may fragment within a given community and may even conflict along gender issues, wealth class, and the political elite. These issues in the smallholder timber sector are not unique to Africa (See ASB 2001 and Holding-Anyonge and Roshetko 2003 for experiences from Asia and Murray and Bannister 2004 for experience from Haiti).

Tree nursery enterprises[10]

In the 1980s and 1990s, governments promoted centralized, publicly-funded nurseries to provide seedlings to farmers. These programs have been expensive to operate, have reached very few farmers and therefore are considered largely unsuccessful. Research by ICRAF and others has indicated that to establish 'sustainable seedling supply systems', small-scale private producers are the most appropriate. Group nurseries, typically used by agroforestry extension, produced significantly fewer seedlings than individual nurseries, although the performance of these varied greatly (Böhringer et al. 2003). Small-scale individual enterprises are flexible and able to meet changing demands; they are local and have good contacts with their customers. The enterprises can be profitable and thereby sustainable. In this context ICRAF decided to focus on improving small-scale tree nursery enterprises.

In ICRAF's research in central Kenya, which has not been published yet, the problems identified by nursery operators have all been market related. From surveys, the nursery operators identified the main problems as few customers or poor markets for seedlings. Probing these problems further, several areas of intervention can be identified:

i. lack of market intelligence systems,
ii. minimal business management skills,
iii. few links with other tree nursery operators and seed suppliers,
iv. weak links with research and extension systems, and
v. distorted markets through state or NGO subsidies.

ICRAF's staff has carried out market analysis and development activities to establish different market chains from seed procurement to final sale and planting, to make detailed analysis of the investment and returns on the nursery enterprises, to identify factors that affect nursery performance, and to focus on building networks and providing training and assistance in financial management and market intelligence.

The simple act of getting nursery operators together to share issues and develop capacity in business skills has had a positive effect in the peri-urban area around Nairobi. Operators have formed a group that meets every month at a tree nursery where they discuss new ideas, markets and species. This cooperation seems to be working well as the group is now expanding, bringing in new members from further afield. Financial management and market intelligence were taught in a workshop environment that got the operators to analyze their own enterprises and discuss their conclusions with their peers. This enabled them to recognize and analyze simple concepts such as profit margins, choice of species, competition and collaboration. Few operators are keeping good records of their businesses but many have made market related decisions to improve their enterprise, such as to open up stalls in urban areas, share big orders with their neighbors, adjust their pricing and stock new tree species.

Experiences with rural operators in Central Kenya have shown that the techniques used for peri-urban tree nursery enterprises are not effective in rural areas. In rural situations the tree nursery enterprise is usually not the main income earner, market orientation is not as strong, education levels are not as high and geographical distances among the operators are larger. In addition, the market has been distorted through past government interventions in seedling supply. Key priorities include adapting methods for training peri-urban operators in enterprise development to suit the rural nursery operators, investigating why some groups of tree nursery operators are effective and creating or demonstrating incentives for the formation of new groups. ICRAF's research into nurseries and development activities in training entrepreneurs in business management lead to the conclusion that the state and NGO sector should adopt a unified approach to seedling supply that favors smallscale enterprises. Indeed, these entrepreneurs could be seen as the 'tree extension agents' of the future with proper training on tree growth, management and use.

Indigenous fruits in southern and central Africa[11]

Research and development institutions in southern Africa have neglected indigenous fruits, yet they represent a wealth of resources in forests and on farms. There are viable national, regional and international markets for some tree products. ICRAF and partners' tree domestication experience and knowledge is being used to develop these products with smallholder farmers. There are also constraints in the development of indigenous fruits because farmers earn low returns from fruits and culinary products. These constraints stem from seasonal gluts, limited and conflicting market knowledge, lack of networks and associations, policies that constrain use of trees and tree products, and inadequacy in processing and storage methods.

Marketing research began in the late 1990s. Kaaria (1998) assessed the production and marketing potential for *Ziziphus mauritiana* in Malawi. Mithofer (2003) examined the economics of collection, use, and sale of three indigenous fruits (*Uapaca kirkiana, Strychnos* spp. and *Parinarium curatellaefolium*) in Zimbabwe. Labor productivity from collecting and selling was relatively high, suggesting that growing the fruits on farm may be profitable, especially since wild fruits are becoming scarce. Ramadhani (2002) found significant barriers to trade in indigenous fruits in Zimbabwe; local policies often prohibited their sale, in the name of protecting a natural resource. But the bans had the additional effect of discouraging farmers from planting the trees on their farms. The establishment of farmer marketing groups, improved price information, standardization of measurements, washing and grading were among the recommendations for improved marketing.

As in southern Africa, ICRAF is carrying out market studies of indigenous fruits and non-timber forest products (NTFPs) in Cameroon, Nigeria, Gabon, and Equatorial Guinea following in the footsteps of the extensive studies of NTFP markets carried out by the Centre for International Forestry Research (CIFOR) over the years (e.g., Ndoye et al. 1997; Awono et al. 2002; Ruiz Pérez 2000). Trade in indigenous fruit products in the Congo Basin area is wide but dispersed. Demand is likely to increase with wild populations diminishing as forest recedes. Processing is done at the village level. The most important areas for improvement are market information for farmers, especially those involved in domestication and nursery management, grading and quality control, and processing improvements.

Calliandra seed

Calliandra calothyrsus is a fast growing, nitrogen-fixing leguminous shrub principally used for animal fodder, soil stabilization and amelioration, stakes for agricultural crops such as climbing beans and fuelwood. In Kenya, its main use is for fodder and since dissemination began in the late 1990s over 20 000 dairy farmers have planted the shrub (Franzel et al. 2003). Adequate and sustained supply of planting materials is a problem wherever calliandra planting is expanding. A study carried out by Technoserve/Kenya and ICRAF explored the development of sustainable planting material distribution and marketing systems for calliandra.[12] Forty-four interviews were conducted of dairy farmers, institutional key informants, commercial seed farmers, seed vendors, seedling sellers and farmer groups in central and western Kenya.

Most of Kenya's calliandra seed is produced in western Kenya, while most of the demand is in central Kenya, where most smallholder dairy farmers reside. The private sector in western Kenya was effective in providing seed for sale to institutional buyers but not to farmers. In central Kenya these institutional buyers supplied farmer groups with free seed. The groups then established group nurseries to supply seedlings for their members. Farmers were encouraged to leave some trees uncut for seed production in their farms. This practice may ensure that three years following calliandra adoption the area could be self-sustaining with respect to seed. But a key problem is that land is scarce and few farmers are willing to leave 30 trees for seed, the number required to ensure maintenance of genetic diversity. Group members are thus being encouraged to pool and mix their seed.

However, it will be much more difficult for the private sector to provide seed to new areas, where calliandra is not found. For example, at times seed vendors in western Kenya have been unable to sell all of their seed, and yet there is a lack of seed in many areas of central Kenya. Currently, there is little private sector involvement in producing and marketing tree seeds in central Kenya, where most of the demand is located. There appears to be insufficient incentives for the private sector to undertake calliandra seed distribution, especially given the poor performance of the milk marketing system, which discourages farmers from testing new technologies. Demand for calliandra seed needs to be developed by raising awareness and training of farmers on calliandra management and use. A subsidized seed distribution and marketing system is still necessary and the following marketing options are being explored:

1. Link seed marketing to other economically attractive activities like milk collection, bulking and processing and marketing of other forage. ICRAF has a partnership with a major milk processor to promote fodder shrubs.
2. Develop and/or strengthen partnerships with organizations and institutions in dairy development and seed distribution. One private company has planted fodder shrubs with the intention of exploring the profitability of seed marketing.
3. Explore and initiate pilot private sector seed production and distribution systems in areas where calliandra planting is expanding. ICRAF and KARI are currently assisting private nurseries and seed vendors in central Kenya, to develop their business skills and to link with small-scale farmers and institutions interested in buying seed. An NGO is making calliandra seed available in small packets for sale to agricultural extension agents in Kiambu District, who in turn sell the seed, and provide information on fodder trees to farmers. A large agricultural seed retailer is currently assessing the feasibility of marketing calliandra seed.

The most important lesson learned is that the main reason behind the lack of available seed is a lack of information about calliandra, both the potential of the shrub as a livestock feed as well as how to harvest, store, and market seed and seedlings to farmers who want to buy it. Organizations such as ICRAF can play a critical, facilitative role linking seed producers and vendors to farmers and providing information to private companies and individuals interested in exploring business opportunities in seed production and marketing.

Common problems and new prospects

A common misstep of enterprise projects is to begin by producing more of a commodity or product with the idea that markets will emerge on their own. This is the flip-side of projects to promote tree-planting that do not consider market factors. The strategy often leads to surpluses that act as *disincentives* to farmers' investment. Another mistake is to focus exclusively on one commodity that seems to be in high demand with a ready market. Markets are by their nature risky and changeable, particularly in Africa and particularly in forest and tree products. Dewees and Saxena (1995) find a weak understanding of markets in tree planting projects: 'When markets have turned out to provide significant incentives for tree growing, this has usually been understood only after-the-fact. It has been more commonly the case that forestry planners have recognized the importance of markets after the project investments have been made, usually in tree planting options which were intended for vastly different purposes.'

Diversification is essential for stability of revenue streams, equity of opportunity within households and communities, and to maintain ecological processes. Yet diversification strategies confront commodity chains that demand numbers of uniform-quality products easily transported and bulked—even with certified and organic products. Can diversification be supported with an eco-certification approach that focuses on certifying the integrity of the farming system, including fair returns to the farmer, rather than on one commodity? This idea stems from the process of 'appellation' in the French agricultural system.[13]

Quality issues and competition have to be taken into account when considering encouraging farmers to enter the market. For example, Kenya, Uganda and Tanzania compete with South Africa in fruit production for export, and even for internal consumption. South Africa puts more investment into infrastructure, improvement and advertising quality. Kenyan on-farm timber producers compete with unregulated forest timber coming from the Democratic Republic of Congo that is used by furniture manufacturers in East Africa. Lack of policy or enforcement in one country creates externalities in a neighboring country if material flows across borders.

These situations can be highly fluid as demand and supply for a commodity shift, as forests close and other policies change. How can groups working with smallholders assist them to deal with quality concerns and changing markets?

One way is through market oriented agroforestry workshops. Trade fairs and market days can also make links especially if held in rural areas. Another avenue is market information systems appropriate to different types of farmers, products and situations. Donor-funded market information systems in East Africa include only scattered bits of information on charcoal and citrus (*Citrus* spp.) fruit.[14] Market information systems are developing rapidly however. One example involves a mobile phone innovation in Kenya.[15] There is potential also to link the various agroforestry and rural development efforts to share market information, which is now proposed for western Kenya through the Consortium for Scaling up Options for Increased Farm Productivity (www.ugunja.org/cosofap).

Government and NGO extension agencies lack up-to-date information and links with the private sector. They are not trained in market analysis. This is particularly true with respect to tree products. Training for those directly involved in extension and agroforestry development, including farmer groups, can help but direct contact with market actors is likely to produce better results.

Farmer organizations are evolving but seem to be stuck between cooperative models that were inefficient, top-down, and often corrupt and new models that may not provide the necessary structure or information for farmers to deal with the private sector. The latter may include 'common interest groups' that are forming across Africa as a result of NGO or government intervention. What is the proper role for international organizations in helping farmers to organize for production and link with the private sector? How can networks and associations be supported and encouraged so that they garner market information, lobby and access resources but continue to support local members effectively? FAO (Food and Agriculture Organization of the United Nations) and RELMA for example have programs to help to strengthen farmers' associations and also to link them to associations and farmer-support organizations in Europe and the United States (http://www.relma.org/Activities_FarmerOrganization.htm). Landcare International, based in Australia, is another group concerned with helping farmers associate and federate for better services and access to markets (www.landcareweb.com, Franzel et al. 2004). Landcare initiatives have started up in South Africa and are now moving to east Africa, building on many small initiatives in the region.[16] The nursery associations that ICRAF has nurtured show promise to address some key problems in the sector, at least in the peri-urban situation.

The long-term nature of production makes contract tree farming and outgrowing schemes more risky than annual crops and thus more attractive to larger and absentee landowners.[17] Contract law enforcement is weak to non-existent in many areas of Africa with respect to smallholders. What models could be adapted or systems put into place to facilitate better market linkages? Coulter et al. (2001) provide guidance for farmers on contract farming in Africa. Homalime, a lime enterprise in western Kenya, is now working with ICRAF and other partners to develop a fuelwood outgrowing scheme. Tea plantations are also trying these schemes in Western Kenya. Thuiya Enterprises was created to link smallholder charcoal producers to the market in Western Kenya.[18] Harrison (2002) describes 10 community-based farm forestry-outgrower efforts in South Africa covering over 5000 ha of land.

FAO and ICRAF held a workshop on farm agribusiness linkages that highlighted ways forward (FAO/ICRAF 2003) and offered some relevant recommendations.[19]

1. Promote links between farmers and agribusinesses and at the same time assist farmers to start their own businesses and cottage industries.
2. Create an enabling environment for business development, which includes providing macro-economic stability, investment-friendly policies, and infrastructure development.

3. Forge partnerships between the private sector, NGOs, and governments to promote farmer-agribusiness linkages. Governments and NGOs should focus on facilitative roles, not on running businesses.
4. Stimulate agricultural development through effective investment promotion centers: services include one-stop licensing, market research and feasibility studies, and a national reference library for sector studies and marketing information.
5. Discourage NGOs and governments from creating unfair competition by offering subsidized (and hence unsustainable) incentives and services that undercut private sector enterprises.

Market information services need to be expanded from common price information services to include more general market information such as trend analysis, identification of key players, and provision of trade information. Market information services need to be tailored to specific target groups. High transaction costs in dealing with smallholders, for example, can be reduced by strengthening farmer groups to take on assembling functions and by adopting appropriate management systems. And building trust and mutual accountability between firms and smallholder farmers is far more important than establishing formal mechanisms, such as contracts.

Tree protection policies dating from the colonial era, which prohibit cutting and/or transporting trees and tree products, significantly inhibit the development of market linkages between farmers and agribusiness for tree products such as charcoal and firewood. How can we learn from the experience of countries such as Namibia and Sudan in liberalizing these sectors? What, if any, are the conservation trade-offs? Policy workshops on charcoal continue in Kenya and elsewhere as part of comprehensive reviews of forest and energy policies. We must avoid setting up top-heavy marketing boards and creating new and conflicting legislation, and focus instead on streamlining and communicating regulations, working directly with affected communities and stakeholders to craft locally-relevant regulations. We can use the example of ASB (Alternatives to Slash-and-Burn) program in Indonesia to identify appropriate species for exploitation (ASB 2001).

Similarly, land and tree tenure policies and traditions deeply affect how smallholders can benefit from trees and tree products. These are in flux all over Africa with changes in laws, practices, migration and private sector initiatives. Even in areas with relatively secure smallholder tenure such as Kenya, absentee landlordism on the one hand impedes development of tree industries by discouraging long-term investment by leasers and other users. On the other hand it may represent the only non-industrial sector that can afford the risks of investment in producing trees for market. In areas with insecure tenure based on 'customary use' such as much of francophone Africa, small home-gardens and diversification of cash crop plantations are likely to be the way forward to avoid conflicts in common lands.

Finally, how can agroforestry researchers fully integrate knowledge of market demand for agroforestry products into research at all levels? How can we engage the private sector more actively in research without losing the focus on smallholder farmers? NEPAD (New Partnership for African Development) and other regional and global initiatives demand that we develop natural resource management initiatives that contribute to poverty reduction (www.nepad.org). Developing a major thematic thrust on 'trees and markets' is a good start within ICRAF (World Agroforestry Centre 2003). Studying examples of agroforestry product marketing from Australia, Europe and the United States is instructive but if the structural constraints that we have discussed are not addressed, marketing efforts will fall short of the aim of poverty reduction.

Conclusion

We propose a set of research and development efforts to help bring about change in the structure and operation of markets in Africa that could foster smallholder tree-based enterprises. First, we hypothesize that, as elsewhere in the world, performance of markets in Africa is linked to overall growth in rural economies. As rural economies improve, trees and their products will be increasingly purchased and diversity can be encouraged. If a tree and its products are economically valuable, people will plant and nurture it. If people plant and nurture trees, society at large will benefit. Consequently, agroforestry should put emphasis on rural development and not on tree planting or technologies alone. Efforts to strengthen rural development by building and supporting rural industries should be prioritized in research and development. Efforts should focus on developing products for local and regional markets such as local fruits, timber and charcoal with the assumption that products for international markets and schemes such as certification, although privileged by donors and international NGOs, benefit only a few and are highly volatile. Research and development should also center on removing barriers to smallholder use and transformation of tree products.

To accomplish this, we need to analyze consumer and local industry needs for tree crops by expanding contacts between researchers and local industries through Rotary Clubs, Chambers of Commerce, trade associations and other small business forums and sponsoring more venues for multiple actors in a subsector market chain such as was done for smallholder timber in central Kenya.

To create and promote market intelligence systems, we can build on mobile phone networks that are growing throughout Africa and better integrate market information into extension programs. We will have to create new cultivars that meet needs for rural development (including construction, energy, nutrition, and beautification), create models for tree diversity in farms, plantations and woodlots and carry out biophysical and economic analyses of tree crops in farming systems.

Finally we need to help farmer associations to share technologies, lessons and policy advocacy, find and disseminate best practices in contracting and outgrowing trees/products, tree seed production and dissemination. We can promote business development units within agroforestry extension programs. Let us for example look to nursery entrepreneurs as tree crop extension agents of the future.[xx]

Acknowledgements

Many thanks to FAO, especially Alexandra Rottger and Doyle Baker, for funding and co-organizing a workshop on farm-agribusiness linkages in Africa at ICRAF where an earlier version of this paper was presented. We would also like to thank those who helped to collect data in the field and write case studies, particularly Fridah Mugo, Sammy Carsan, Charly Facheux, Will Frost, Christine Holding-Anyonge, Pauline Mwangi, and Tony Simons. Thanks to Rebecca Ashley and Anand Aithal for documentation and finally to the farmers who are taking risks to grow trees for markets and merchants developing these subsectors.

End Notes

1. For example, see topics covered in the World Agroforestry Centre's (ICRAF) annual report 2001-2 (World Agroforestry Centre 2003). For an early attempt to promote marketing in agroforestry to smallholders see FAO 1996, however when searching the FAO database it is interesting to note that there are only 11 documents under 'marketing and agroforestry.'
2. Comments from evaluation of 'Market oriented Agroforestry Workshop' in Kisumu, Kenya (Russell 2002) included such statements as:
   a. Even agroforestry products I thought could not be sold or marketed I have learned have value and can be marketed.
   b. I learned a new dimension of agroforestry as a business rather than environmental concern.
   c. It made me look at forestry from a commercial point of view. It opens avenues on profit-making areas in forest industry in addition to social-ecological contribution of the industries. Market opportunities motivate farmers to grow trees more than anything else.
   See: Russell, D. 2002. Moving ahead with market-oriented agroforestry in Western Kenya: Report of a workshop in February 2002, Kisumu, Kenya. ICRAF, Nairobi.
3. There is also a longstanding element of distrust between the private sector and NGOs and government involved in forestry and agroforestry based on perceived extortion of farmers by traders and merchants and concern about deforestation.
4. This is not to say, simplistically, that monetary benefit is the only incentive to plant trees or maintain agroforestry systems. Arnold (1995: see page 12 of the book edited by Arnold and Dewees, listed against Dewees and Saxena 1995 in the references section) clarifies that 'tree growing can be explained as one or more of four categories of response to dynamic change,' including maintaining supplies of tree products, meeting demands for tree products, maintaining agricultural productivity and reducing risk.
5. This material was provided primarily by Fridah Mugo, Thuiya Enterprises and the following unpublished material:
   Mugo, F. 1999. Charcoal trade in Kenya. RELMA working paper No 5. Regional Land Management Unit, Nairobi.
   Kenya, Republic of. 2003. The National Wood Energy Policy. First Draft. May 2003. Nairobi: Ministry of Energy, Republic of Kenya.
   RELMA 2002. Regional Workshop on Charcoal Policy and Legislation, March 4-6, 2002 at ICRAF, Nairobi, Kenya: Regional Land Management Unit (RELMA) in collaboration with Forest Action Network. Nairobi: RELMA.
6. The Draft National Wood Energy Policy of Kenya (May 2003) calls for the creation of a Wood Energy Board. Given the poor performance of agricultural marketing boards in Africa, this may not be the best solution unless there is high participation of communities and the private sector (Bates 1981).
7. However, Dewees and Saxena (1995: 209) point out that Sudan has had a thriving charcoal market for a long time based on clearance of acacia woodlands for mechanized agriculture. They claim that this system produces cheaper charcoal than 'sustainable wood supply' on plantations even taking transport costs into account.
8. Material for this section comes from Sammy Carson of ICRAF and Christine Holding-Anyonge from FAO (formerly from ICRAF). See Holding et al. 2001, Holding–Anyonge et al. 2002 and Holding-Anyonge and Roshetko 2003 as well as the following unpublished material:

Carsan S. 2001. Proceedings of the second Meru timber marketing stakeholders workshop. Partners: ICRAF, Forest Action Network (FAN) and Ministry of Agriculture and Rural Development (MoARD), 25-26 June, ICRAF House, Nairobi Kenya. Nairobi: World Agroforestry Centre.
Kenya, Republic of. 2001. Marketing of timber and other tree products in Meru Central district, Kenya: Proceedings of pilot phase review meeting (7-9 March 2001). Meru Central, Kenya: Secretariat for District agricultural office.
Mwangi, E. 2002. Smallholder farmers' marketing channels for on-farm timber in Kenya: A case of Embu District. MSc thesis for Master of Environmental Studies at Kenyatta University, Kenya.
Opanga, P. S. 2000. Emerging markets for *Grevillea robusta* farm timber in Igoki and Ntakira Locations of Meru Central District. Forestry Action Network and District Agriculture and Livestock Office, Meru.

9. One reviewer noted that in absence of market opportunities, trees could become a liability for farmers and cause them to have negative attitudes toward agroforestry.
10. Material for this case was provided by Will Frost, ICRAF and the following unpublished materials:
Basweti, C., Lengkeek, A., Prytz L. and Jaenicke, H. 2001. Tree nursery trade in urban and peri-urban areas: a survey in Nairobi and Kiambu districts in Kenya. RELMA Working Paper No. 13. Regional Land Management Unit, Nairobi.
Njenga A. and Frost W. 2001. Economic and market condition analysis of peri-urban nurseries: Results of a PFSA workshop conducted in Kiambu District, Central Kenya. Nairobi: World Agroforestry Centre.
11. Congo Basin information from Charly Facheaux of ICRAF's Cameroon office.
12. See Mwangi P., Franzel S. and Wambugu C. 2003. *Calliandra Calothyrsus*: Sustainable planting material distribution and marketing systems: report and recommendations for action. Technoserve/Kenya and ICRAF. Nairobi: World Agroforestry Centre. The study was financed by a DFID (Department for International Development, U.K.)-financed project managed by the Oxford Forestry Institute and ICRAF that is promoting the adoption of fodder trees throughout East Africa.
13. This idea comes from conversations with colleagues from CIRAD (Centre de coopération internationale en recherche agronomique pour le développement).
14. FoodNet (IITA) 2001. Personal communication at workshop for participants in market information systems in east Africa. Nairobi, Kenya. (See www.foodnet.cgiar.org/) [Accessed Nov 2003]
15. The Kenya Agricultural Commodity Exchange (KACE) for example offers a Guide to Accessing Daily Market Price Information through Your Safaricom Mobile Phone, and provides names and telephone numbers for persons who can assist.
16. See Tanui, J. 2002. Incorporating a Landcare approach into community land management efforts in Africa: A case study of the Mount Kenya region. AGILE Working Paper No. 2. African Grassroots Innovations for Livelihood and Environment. ICRAF and African Highlands Initiative, Kampala, Uganda.
17. Baumann notes that private investors 'are increasingly concentrated on crop processing and marketing [as opposed to smallholder production schemes] where there are clear margins and quick profits' (Baumann 2000: 45). She finds that government support is crucial for the tree-contracting sector.
Baumann, P. 2000. Equity and efficiency in contract farming schemes: The experience of agricultural tree crops. Working Paper 139. Overseas Development Institute, London.
18. Thuiya Enterprises is a pilot project devised by Fridah Mugo to assess farmers' responses to tree planting for a guaranteed market in Kenya. The idea involves facilitating contract outgrowing of wood for charcoal on farmers' land that would otherwise remain uncropped. The project facilitates seedling production, efficient processing of charcoal, market research, extension services and helping farmer groups to become legal charcoal producers.
19. See FAO and ICRAF 2003. Expert Consultation on Strengthening Farm-Agribusiness Linkages in Africa. Food and Agricultural Organization of the United Nations and the World Agroforestry Centre. 24-27 March 2003. Rome: Food and Agricultural Organization (draft proceedings available).
20. This idea came from Will Frost.

References

ASB. 2001. Policy Brief No. 3: Deregulating agroforestry timber to fight poverty and protect the environment. ICRAF, Nairobi, Kenya: Alternatives to Slash and Burn. http://www.asb.cgiar.org/PDFwebdocs/PolicyBrief3.pdf

Awono A., Ndoye O., Schreckenberg K., Tabuna H., Isseri F. and Temple L. 2002. Production and marketing of safou (*Dacryodes edulis*) in Cameroon and internationally: market development issues. Forests, Trees and Livelihoods 12: 125–147.

Bates R.H. 1981. Markets and states in tropic Africa: The political basis of agricultural policies. University of California Press, Berkeley, pp. 178 pages.

Bohannan P.J. and Dalton G. (eds) 1968. Markets in Africa: University of Illinois Press, Evanston, IL, 762 pages.

Böhringer A., Ayuk E.T., Katanga R. and Ruvuga S. 2003. Farmer nurseries as a catalyst for developing sustainable land use systems in southern Africa: Part A: Nursery productivity and organization. Part B: Support systems, early impact and policy issues. Agricultural Systems 77: 187–217.

Coulter J., Goodland A., Tallontire A. and Stringfellow R. 2001. Marrying farmer co-operation and contract farming for agricultural service provision in sub-Saharan Africa. From: The Guide to Developing Agricultural Markets and Agro-Enterprises. World Bank, Washington DC, www.worldbank.org/essd/essd.nsf/Agroenterprise/marrying

Dewees P.A. and Saxena N.C. 1995. Wood product markets as incentives for farmer tree growing. pp. 198–241. In: Arnold J.E.M. and Dewees P.A. (eds) Tree Management in Farmer Strategies: Responses to Agricultural Intensitification, Oxford University Press, Oxford, U.K.

FAO, Forestry Department. 1996. Marketing in forestry and agroforestry by rural people. Food and Agriculture Organization, Rome. 76 pages.

Franzel S., Wambugu C. and Tuwei P. 2003. The adoption and dissemination of fodder shrubs in central Kenya, Agricultural Research and Network (AGREN) Series Paper No. 131. Overseas Development Institute, London, 11 pp.

Gabre-Madhin E.Z. 2002. Getting markets right in Africa: Opportunities and challenges. AICHA Seminar Presentation. AICHA, Africa, African agriculture, agriculture, agricultural growth. Washington DC: International Food Policy Research Institute. http://www.ifpri.org/themes/aicha/gabremadhin0402.pdf

Girard, P. 2002. Charcoal production and use in Africa: What future? Unasylva 211/53: 30–35.

Gold M.A., Godsey L.D. and Josiah S.J. 2004. Markets and marketing strategies for agroforestry products in North America. (This volume)

Harrison G. 2002. Making a difference in South African forestry. Chapter 12 in proceedings of a conference on equitable partnerships between corporate and smallholder partners: relating partnerships to social, economic and environmental indicators sponsored by FAO and CIFOR. 21–23 May 2002, Bogor, Indonesia. Centre for International Forestry Research, Bogor, Indonesia, www.fao.org/DOCREP/005/Y4803E/ Y4803E00.HTM [accessed: 24 Nov 2003]

Hellin J. and Higman S. 2002. Smallholders and niche markets: Lessons from the Andes. Agren Network Paper No. 118, January 2002. ODI Agricultural Research and Extension Network, London. 16 pp.

Holding C., Njuguna P. and Gatundu C. 2001. Farm sourced timber: The restructuring of the timber industry in Kenya – Opportunities and challenges. In: Race D. and Reid D. (eds) Proceedings of the international conference on Forestry extension: Assisting forest owner, farmer and stakeholder decision-making Lorne, Australia. 29 October–2 November, 2001. Joint Venture Agroforestry Program and Agriculture, Fisheries and Forestry-Australia. http://iufro.boku.ac.at/iufro/iufronet/d6/wu60603/proc2001/ holding.htm [accessed December 6, 2004].

Holding-Anyonge C., Carsan S., Mwaura M., Muchai D. and Gatundu C. 2002. Farm sourced timber: the changing structure of the timber industry in Kenya. Linking research, extension and advocacy to address farmers' needs. Agricultural Research and Extension Network (AgREN) Newsletter 46: 11. http: //www.fao.org/forestry/foris/webview/forestry2/index.jsp? siteId=2141&langId=1 [Accessed Nov 2003]

Holding-Anyonge C.H. and Roshetko J. M. 2003. Farmer-level timber production: Orienting farmers toward the market. Unasylva 212/54: 48–55.

Kaaria S. 1998. The economic potential of wild fruits in Malawi. Ph.D. Thesis, University of Minnesota, U.S.A.

Kaimowitz D. 2002. Not by bread alone...Forests and rural livelihoods in sub-saharan Africa. pp 45-64 In: Tapani Oksanen, Brita Pajari and Tomi Tuomasjukka (eds) Forests in Poverty Reduction Strategies: Capturing the Potential. Proceedings 47. European Forest Institute, Finland. http://www.efi.fi/publications/Proceedings/47.html

Kherallah M., Delgado C., Gabre-Mahdin E., Minot N. and Johnson M. 2000. Agricultural market reforms in sub-Saharan Africa: A synthesis of research findings. International Food Policy Research Institute, Washington DC. 104 pp.

Mithofer D. 2003. Socioeconomic factors in the collection and use of indigenous fruits in Zimbabwe. Ph.D. Thesis, Department of Economics, University of Hannover, Germany.

Murray G.F. and Bannister M.E. 2004. Peasants, agroforesters, and anthropologists: A 20-year venture in income-generating trees and hedgerows in Haiti. (This volume).

Ndoye O., Ruiz Pérez M. and Eyebe A. 1997. The markets of non-timber forest products in the humid forest zone of Cameroon. ODI Rural Development Forestry Network Paper No. 22c. Overseas Development Institute, London.

Ramdhani T. 2002. Marketing analysis of *Uapaca Kirkiana* indigenous fruits in Zimbabwe. Ph.D. Thesis, Department of Economis, University of Hannover, Germany.

Ruiz Pérez M., Ndoye O., Eyebe A. and Puntodewo A. 2000. Spatial characterization of non-timber forest products markets in the humid zone of Cameroon. Int For Rev 2/2: 71–83.

Scherr S. 2004. Building opportunities for small-farm agroforestry to supply domestic wood markets in developing countries (This volume).

Small Woodlands Program of British Columbia 1994. Guide to Agroforestry, Chapter 12: Marketing Agroforestry Products. See: http://www.swp.bc.ca/ Accessed November 15, 2003.

Terray E. 1974. Long distance exchange and the formation of the State: The case of the Abron Kingdom of Gyman. Economy and Society 3: 315–345.

Thomas M.G. and Schumann D.R. 1993. Income opportunities in special forest products: Self-help suggestions for rural entrepreneurs. Department of Agriculture, Agriculture Information Bulletin AIB-666, May 1993. Washington DC.

Utama R., Rantan D., de Jong W. and Budhia S. 1999. Income generation through rehabilitation of imperata grasslands: production of vitex pubescens as a source of charcoal. pp. 175–184. In: Roshetko J. and Evans D.O. (eds), Domestication of agroforestry trees in Southeast Asia: Proceedings of a regional workshop in Yogyakarta Indonesia November 4–7 1997. Forest, Farm and Community Tree Research Reports Special Issue. Winrock International Institute for Agricultural Development, Morrilton AK, USA.

World Agroforestry Centre 2003. World Agroforestry Centre 2001-2, Annual Report: Transforming Lives and Landscapes. ICRAF, Nairobi, 111 pp.

World Agroforestry Centre, 2003. World Agroforestry Center 2004-6, Medium-Term Plan. ICRAF, Nairobi, 147 pp.

Agroforestry Systems **61:** 357–370, 2004.

Building opportunities for small-farm agroforestry to supply domestic wood markets in developing countries

S.J. Scherr
Forest Trends, 1050 Potomac St., N.W., Washington, D.C. 20036, USA; e-mail: SScherr@forest-trends.org

Key words: Domestic markets, Market analysis, Market policies, Smallholder wood production, Wood markets

Abstract

The fastest-growing demand for wood products is in domestic markets of developing countries. These markets could offer significant economic opportunities for hundreds of millions of small-scale agroforestry producers, in market niches where they can offer competitive advantages such as control over commercially valuable tree resources, lower cost structure, better monitoring and protection or branding for socially responsible markets. The most promising opportunities for small-scale farmers to sell high-value timber are as outgrowers for industrial buyers or by selling to intermediaries wood grown in agroforestry systems. Farmers located in forest-scarce regions near pulp mills may benefit from outgrower arrangements for pulpwood. Farmers located near inland urban markets may be competitive in some commodity wood and woodfuel markets. Opportunities in processed wood products are mainly in pre-processing, milling to supply low-end products, niches that cannot be efficiently served by industrial-scale producers, and through contracts for selected operations in vertically-integrated industries. To develop viable wood market enterprises, producers must improve their market position, strengthen their organizations, and forge strategic business partnerships. Forest market institutions must adapt by providing business services to small-scale farm producers, investing in regional forest enterprise development to fill gaps in the value chain for wood products, and targeting research, education and training. It is essential to remove policy barriers to small-farm participation in markets, by removing excessive regulations, creating fair and open competitive market environment, and involving farmers' organizations in forest policy negotiations.

Introduction

An estimated 500 million to 1 billion smallholder farmers grow farm trees or manage remnant forests for subsistence and income. Yet most forest-product markets, and market institutions and policy, are structured to serve large-scale natural forest and plantation producers. A major challenge for rural development in developing countries in the twenty-first century will be to reshape efficient markets that also serve small-farm producers.

The first wave of modern agroforestry in the 1970s and 1980s focused attention of policymakers and researchers on enhancing the subsistence, farming input and conservation roles of trees on farms. Some agroforestry projects did promote production for local and national markets, and in the 1990s much attention was paid to developing methods for local farmers and communities to determine short-term market potentials (e.g., Lecup and Nicholson 2000). The search for commercial opportunities has most commonly focused on higher-value niche export markets in the industrialized countries, even though these account for a small fraction of total demand. Few countries have systematically analyzed, at the sector and sub-sector levels, the potential medium- to long-term competitive advantages for farm-grown tree products in domestic markets.[1] In part because such market analyses and strategies do not exist, agroforestry rarely appears on the 'radar screen' of policymakers, much less in national or regional strategies for rural economic development and poverty reduction.

Yet small-farm participation in commercial wood markets could potentially make a major contribution to rural economic growth, poverty reduction and ecosystem objectives. This paper builds on the findings of an international review[2] by Forest Trends and the Center for International Forestry Research (CIFOR) of forest product markets for low-income producers in developing countries, to analyze the implications for small-scale agroforestry. It presents a broad analysis of trends in domestic demand for wood products in the developing countries; identifies promising areas where small-farm agroforestry producers have, or could readily develop, a competitive market position; then discusses key actions needed to develop more efficient and lucrative markets for small-farm producers.

Rapid growth in domestic wood demand

Global wood demand grew by more than 50% from the 1960s to the mid-1990s, although per capita consumption was fairly stable on average, even declining in the late 1990s (Gardner-Outlaw and Engelman 1999). Industrialized countries presently consume about 75% of industrial roundwood production (solid wood and panels), but demand in these countries grew by only 0.6% per year during 1961 to1997. By contrast, consumption grew by 3.2% per year in developing countries during the same period (Bazett 2000). Forest resources play an important role in economic development: to earn foreign currency, to build urban centers and infrastructure, and to provide fuel for industrial production. Looking forward, domestic demand for forest products in developing countries is projected to continue rising in the next few decades, driven mainly by income and population growth. Nonindustrial demand – for products such as fuelwood, construction materials, and rough furniture – is expected to be especially high in those countries in the early stages of economic growth. Urbanization, income growth and new preferences drawn from cross-cultural contact have greatly diversified forest products. New technologies are increasing demand for wood products that are especially attractive financially and feasible for low-income producers to supply. Modern sawmills can utilize a much wider range of tree species than was historically the case. New processing technologies allow commercial use of small-diameter, 'low-quality' wood for many higher-value products.

In most developing countries, domestic consumption of wood absorbs the largest share of total production. The proportion is nearly 100% for fuelwood and charcoal production. For industrial roundwood, the proportion is generally over 95%, even in importing exporting countries like Bolivia and Brazil. Over two-thirds of all paper products and pulp for paper produced in most developing countries go for domestic consumption, and the proportion in countries like Angola, China, India, and Mexico is over 90% (FAO 2003). To illustrate the scale of this demand: Brazil's domestic consumption of tropical timber, estimated at 34 million cubic meters of logs in 1997, exceeded timber consumption in all of the Western European countries combined (Smeraldo and Verissimo 1999).

Timber imports are projected to triple or quadruple in India, China and other forest-scarce developing countries over the next two decades (FAO 2001). Indeed, after trending slightly downward from the 1960s to 1990, the ratio of wood to grain prices has been rising since then (FAO 1998). This shift in relative prices has been one of the basic drivers behind the expansion of agroforestry.

Competitive advantages of small-farm agroforestry

There is thus a promising potential demand for agroforestry products. However, large-scale natural forest producers and plantations, both foreign and domestic, can more economically supply some parts of the market. A large share of forest product demand is 'commodities,' that is, highly standardized products purchased in large volumes. Moreover, internal transport costs for wood to major urban markets from many agroforestry areas are high relative to product value. Globalization is encouraging large-scale buyers to seek very-low-cost, high-volume producers. Prices are also kept low by the availability of nonwood substitutes. Continued land-clearing and illegal extraction in many regions supplies wood and NTFPs (nontimber forest products) to the market at a lower cost than can sustainable agroforestry (Scherr et al. 2003). To identify those niches where small farm producers can realistically compete requires assessing their competitive advantages in each particular setting. Four factors may provide such a competitive advantage.

Control of commercially valuable forest resources

During the past 15 years, smallholders' ownership and control over commercially available tree resources has grown significantly in developing countries, as a result of increased on-farm planting of timber trees (reflecting local scarcities), and devolution of forests previously controlled by governments to farmers and communities (White and Martin 2002). This control over the primary resource may give smallholders a competitive advantage. Ownership greatly improves their negotiating position with buyers of high-value wood and NTFPs. Producers located near centers of growing consumer or industrial demand, particularly inland cities far from commercial ports, may be competitive with imports or distant suppliers of lower-value products, due to the high cost of internal transport. Smallholders who have secure land tenure may have an advantage over forest communities with unclear ownership rights, and even over large-scale concessionaries or forest owners whose rights may be in dispute from local communities.

Lower cost structure for some products

Some small-scale farm producers may be able and willing to supply products at a lower cost than large-scale or corporate suppliers, because of lower opportunity costs for land and labor, lower production costs from intercropping, or because they value collateral benefits such as local employment, environmental services or local lifestyle. Timber trees can be managed and harvested during periods when labor demands for other activities are low. Farm trees can increase agricultural productivity when grown as windbreaks, fodder banks, live fences, or nurse trees for perennial cash crops. Local producers may be more familiar with local product and processing preferences, more flexible in supplying small quantities as needed to local traders, or providing fresh supplies of perishable NTFPs (Current et al. 1995).

Better monitoring and protection

Local people may have a greater ability than outside companies or agencies to protect forest resources from risks like encroachment, illegal harvest, fire and social unrest, because of superior capacity for monitoring and community interest in forest protection. Insurance companies consider this to be a critical factor in assessing forestry risk and insurability.[3]

Branding in socially responsible markets

Small-farm producers may be able to secure an advantage in marketing their products to consumers or investors in socially and environmentally responsible market niches, and to companies that are sensitive to reputation. For example, Fair Trade markets in Europe favor smallholder producers.

Market strategies to manage livelihood risk

Forest markets present two major types of risk to producers. First, prices are often highly erratic – as with many commodities – as a result of cycles of seasonal and year-to-year changes in global supply and demand. Second, the forest processing industry has tended to follow a ‘boom and bust’ cycle: overexploiting cheap forest resources and then moving on. This strategy may make sense for large-scale industrial product buyers or short-term natural forest concessionaires. But for farm producers relying on their own tree resources, it makes more sense to treat them as a long-term capital asset – composed of particular tree species mix and spatial pattern – capable of producing multiple streams of income. Those streams may derive from harvesting different products from a multi-purpose tree, by harvesting at different ages, or harvesting from a different mix of species. Low-income producers need a ‘portfolio’ of products in different income/risk categories, including agricultural and non-farm enterprises. Thus, standing timber trees serve as a ‘savings bank’ for farmers in central Kenya, who can sell them in times of great need. That portfolio will reflect cultural, social and aesthetic values important to local people. Finding at least one product that provides a reliable source of annual income is essential, if such an income flow is not provided by farming or off-farm employment. Non-timber forest products or payments for ecosystem services may also serve this function, though not discussed in this paper (but see Alavalapati et al. 2004).

Risks may be reduced by planting only a portion of the farm in promising market species, using agroforestry configurations, and managing resources in such a way that a variety of different products could potentially be marketed in response to changing market conditions. Establishing small-scale plantations or agroforestry may be done gradually over time, using farm or unemployed community labor, rather than in large parcels all at once, which would require credit to hire labor.

Cash-poor producers need to develop enterprises that require low cash investment, at least initially. External advisers and business-service providers must understand that *ex-ante* analyses based on business models from large-scale commercial production may not be suitable for small-farm agroforestry systems. Farmers may focus on a more diverse set of products, with more outputs of short rotation, may use assisted natural regeneration more than seedlings, or may use household labor at times when its opportunity cost is lower than the wage rate (Scherr 1995).

Growers can use multipurpose trees for small-diameter wood products and also for food in a lean year. (For example, fruit tree prunings can supply fuelwood and stakes.) Such strategies are more likely than large-scale industrial enterprises to result in landscape patterns and management practices that protect ecosystem services. Land use mosaics may also benefit commercial production through reduced disease incidence, lower monitoring costs, or higher densities of valued 'edge' NTFP species.

Promising market niches for wood products from small-farm agroforestry

The analysis below, summarized in Table 1, assesses which segments of existing and emerging markets are likely to offer real opportunities to small-scale agroforestry producers, and which do not.[4] Good market analyses are essential to identify the opportunities in specific countries and sub-regions, and these must clearly assess and develop strategies for overcoming existing institutional constraints to linking smallholders with market buyers. It is important to distinguish between markets in which producers are likely to earn only supplemental or 'safety net' income, from those that could raise incomes significantly, with strong development multipliers. Policies should seek to encourage the latter, and not restrict low-income producers to economically stagnant market niches.

Commodity wood

Low-income producers are unlikely to be competitive in export markets for commodity-grade[5] timber, which require large volumes and high product consistency. When new industrial plantation-grown wood comes into the market beginning in 2005, and a second wave around 2010, sharp downward pressure on prices is predicted (Leslie 1999). By contrast, there is a large potential market for low-income producers in commodity-grade products for segments of domestic markets that do not trade in very large volumes. Urbanization, rural housing and infrastructure construction all demand large quantities of commodity-grade wood; intensification of agriculture demands wood for fencing, storage structures, crop and tree supports, and packing crates. Iron and steel production depends heavily on wood energy. More than half the total roundwood harvested in developing countries is burned directly as fuelwood or charcoal, and woodfuel demand rises in the early stages of economic growth, even as growth in use of substitute fuels accelerates. Biomass fuel markets are also increasing in some countries.

These markets could benefit tens of millions of smallholder farmers near rapidly growing inland population centers. Farmers can compete only in markets where buyers need greater supply flexibility, or where transport costs make alternative supplies too costly, such as countries or sub-regions with poor port and transport facilities linking them to international markets or low-cost natural forest supplies. Commodity wood production may be more profitable and lower risk for small-scale farmers if integrated with other components of livelihood strategies, as through agroforestry, by-products from managing timber or tree crop stands, or wood from fallow stands.[6] Much of total supply has been, and will continue to be, generated as a by-product of the agricultural cycle, e.g., production of woodfuels where land clearance is taking place. Farmers may simply sell stumpage to outside loggers, or sell logs themselves if they can acquire the necessary equipment. Mechanisms for bundling products from small-volume producers, as well as for grading and sorting, are often essential to negotiate reasonable prices from buyers, as are grading and sorting.

Successful smallholder agroforestry and farm forestry for commodity-grade construction wood has been established in a number of forest-scarce countries. Most notable has been the degree of farm forestry in India, based on fast-growing species. Construction wood for national markets is produced by thousands of farmers in Karnataka, India (Dewees and Saxena 1995). A farm forestry scheme developed by a match company works with 30 000 farmers on 40 000 hectares in Uttar Pradesh (Desmond and Race 2000). Dynamic domestic-wood demand provided incentives for small-scale farmers to produce commodity wood in agroforestry systems in Mindanao, the Philippines (ICRAF 2001), and in Bangladesh and Nepal (Arnold and Dewees 1995).

Table 1. Market characteristics that enable small-scale producers to compete.

Enabling conditions	How conditions benefit small-scale agroforest producers	Comments
Supply factors		
Low-cost processing technology exists	Can benefit from higher-value segment of market value chain	Economics of 'value-added' location-specific
Production technologies locally known	Reduces adoption risks, maintenance costs	Training and extension programs can provide
Neutral or declining returns to scale for production	No economic advantage for large-scale producers	Especially where labor-intensive management
Limited direct competition from very low-cost producers	Greater potentials distant from ports, distant for agricultural land-clearing	
Environmental services can be produced together with forest or agricultural production	Environmental service payments supplement, rather than replace, production income	May require change in landscape design, location of production, management
Demand factors		
Large number of buyers (transporters, wholesalers, processors, service users)	More competitive prices and terms of sale for sellers; more interest by buyers in negotiating long-term relationships	Monopsony currently characterizes a majority of forest product and environmental service markets
Products with growing demand	Greater opportunity for new entrants	
Niche market buyers interested in supporting rural development	Potential to 'brand' product or access higher-paying consumers or investors	Limited scale of market
Demand for natural species that are difficult to domesticate, replace	Creates asset value for natural forests, 'volunteer' farm trees	Most species have domesticated or synthetic substitutes
Flexible quality standards	Can use greater variety and quality of wood species	Difficult to reliably supply raw materials for international markets
Long-term supply contracts offered	Provides more stable income source, reducing livelihood risks	Usually offered by high capital cost processing firms (e.g., pulpwood) for steady supply
Low capital costs of market entry	Existing or low-cost capital equipment for production or processing; low costs to find buyers (e.g., advertising)	Often low-value products; many low-cost technologies exist but not known locally
Small and variable volumes are purchased	Producers can move in and out of the market easily; Cases where no economic advantage for large-volume producers	For example, in direct retailing of medicinal plants, local fuelwood markets
Open, transparent and unrestricted bidding processes	Avoids discrimination against small-scale suppliers or raw material purchasers	
Marketing intermediary established for small-scale producers	Provides 'bundling', technical support, financing; achieves economies of scale in marketing, production	Established by producer cooperatives, NGO's, parastatals, buyer company
Market regulation		
Low regulatory costs of market entry	No registration fees; competitive bidding for small timber volumes; low-cost management plans; no bribes required	
No producer/consumer subsidies	Greater competitiveness for small-scale producers	Large producers or buyers most benefit from subsidies
Low-cost regulatory environment	Few harvest, transport, sales permits required; reduced risk and corruption	
Secure local rights for forest products, environmental services	Reduces risk of 'forest grab' by more powerful actors	Especially for long-term product, service contracts

Source: Scherr et al. (2003).

Table 2. Promising market opportunities and business models for small-scale agroforestry producers.

Main opportunities will be found where:	Scale of market opportunity for poor	Business models	Potential to raise incomes	Examples
Commodity wood				
Forest-scarce inland regions with rapid income or population growth; humid/ sub-humid areas	***	Farm forestry, products sold to local traders	**	Eucalyptus farming in India (Dewees and Saxena 1995)
		Farm forestry or outgrower schemes that directly link producers with large-scale sawmills, commodity wholesalers or final users	***	Match Company farm forestry scheme with 30,000 farmers on 40,000 hectares in Uttar Pradesh, India; Kolombangara Forest Products, Ltd. Informal sawlog grower scheme with 100 growers (Desmond and Race 2000)
		Farm forestry, with cooperative wood marketing organization	***	Widespread in India, Philippines, Bangladesh, Nepal
High-quality timber				
Mainly in forest-scarce regions with growing incomes and demand for high-value products good market access; areas with secure tenure; mainly in humid/sub-humid areas	**	Small farms or communities participate in outgrower or crop-share schemes with private companies to establish plantations of improved high-value timber	**	Prima Woods project for teak production in Ghana (Mayers and Vermeulen 2002)
		Farmers grow timber at low densities in agroforestry systems and remnant forest to sell cooperatively	*(*)	Philippines Agroforestry Cooperatives (ICRAF 2001)
Certified wood				
Farmer groups, mainly in humid/sub-humid regions, with high capacity for natural forest management and marketing, that can achieve low certification costs	*	Farm producer groups with established contracts or agreements with certified wood users or market intermediaries	**	Klabin pulp and paper company of Brazil assists outgrowers to obtain certification and to supply local furniture company demand (Mayers and Vermeulen 2002)
Industrial pulpwood				
Densely settled, forest-scarce countries with large pulp and paper or engineered wood industry, and limited foreign exchange; farmers located near pulp mills; humid/sub-humid areas	**	Outgrower arrangements: industry assists farmers to establish and manage pulpwood plantations, in guaranteed supply contracts	***	Aracruz Cellulose 'timber partner program' in Brazil (Desmond and Race 2000; Saigal, Arora and Rizvi. 2002)

Table 2 continued.

Main opportunities will be found where:	Scale of market opportunity for poor	Business models	Potential to raise incomes	Examples
		Farm forestry: farmers establish plantations with technical support from industry; sell output without purchase contracts	**	ITC Bhadrachalam Paperboards, Ltd., integrated pulp and paper mill in Andhra Pradesh State, India (Lal 2000; Saigal et al. forthcoming)
		Land leasing by farmers to private companies for pulpwood production	**	Jant limited wood chipping operation in Madang, Papua New Guinea (Mayers and Vermeulen 2002)
Forest product processing				
Simple tools, furniture, other basic commodities for poor consumers in growing rural or urban areas	**	Community or group enterprise	**	Small-scale processing firms in Africa (Arnold et al. 1994)
Sawmilling, in markets where large-scale, high efficiency mills do not compete (humid/sub-humid forest regions)	*	Cooperative community, farmer or group sawmill enterprise with identified buyers	**	Small-scale logging in the Amazon (Padoch and Pinedo-Vasquez 1996)
Finished or semi-finished processing, where commercial links can be forged with businesses serving higher-income consumers	*	Forest community or farmer cooperative for sale direct to wholesalers/retailers	***	

High-value wood

Wood demand in high and middle-income countries and urban centers is diversifying into higher-value and specialty products such as finished furniture and home improvement products. Appearance-grade wood for solid wood and veneer may retail for a price three to four times higher than low-quality construction grade timber, and much higher than for low-value products such as fuelwood. Long-term wood-price increases are projected only for these higher-end segments of the market, as a result of the scarcity of large-diameter timber and the greater opportunity to differentiate products (FAO 2001). For example, retail prices for mahogany (*Swietenia* spp.) are 25% higher today than a decade ago (Robbins 2000).

The most valued woods are those grown primarily in natural forests – such as mahogany, red cedar (*Acrocarpus* sp.), and rosewood (*Dalbergia* spp.) – that have become scarce through over-exploitation. But a few appearance-grade timbers, such as teak (*Tectona grandis*), have been domesticated, and fast-growing cultivars can be grown in plantations and on farms. Some mahogany is now grown in plantations (largely outside its native habitat, where the species is susceptible to more pests), though the wood quality is considered inferior (Robbins 2000). Demand is rising for non-traditional appearance-grade wood species, as knowledge about their processing and use characteristics develops (e.g., Vlosky and Aguirre 2001). Research and development on production systems for small-scale farmers working with diverse species has been weak in the past; there may be significant potential for increasing productivity and marketability (Leslie 1999). Only a minority of low-income farmers will have the stand quality, quantity or market contacts to supply higher-value appearance-grade wood markets. But this still means that millions could benefit[7], as illustrated by projections by Beer et al. (2000) for Central America. Producers will benefit most economically where they develop long-term partnerships with buyers to produce higher-value products, or where there is active competition among buyers. Models include both timber grown in agroforestry systems, and outgrower farm forestry schemes with private companies. In Ghana for example, the Prima Woods company buys teak from smallholder plantations.

Certified wood

Without major changes in certification processes, few farming communities will directly participate in certification markets. Certification schemes are not presently targeted to small-farm wood producers. Moreover, constraints include very high economies of scale in certification processes, exclusion from the certified 'chain of custody' processes set up by major buyers to ensure that source certification is accurate, dependence on external professional technicians, difficulty identifying potential buyers, and the limited price premium for certified wood. In wood markets where certification is becoming a standard, this is inadvertently erecting another market barrier for small farm producers.

For small-scale farmers to participate in certified wood markets, the costs of achieving certification must be low; location, cost of access and quality must meet market criteria; communities must have direct links to wholesale or retail buyers, to establish chain of custody and ensure access to higher-value markets to justify costs; and producer groups will need partners who are willing to underwrite certification costs and facilitate the process (Rametsteiner and Simula 2001).[8] For example, Klabin pulp and paper company of Brazil assists outgrowers to obtain certification and supply local furniture company demand (Mayers and Vermeulen 2002).

Greater efforts are needed to facilitate group certification for smallholder farmers, who could benefit significantly in middle-income countries where domestic demand for certified wood is growing. As international agricultural-product buyers become more sensitive to the need for environmentally and socially sustainable supply chains, interesting new markets are emerging for a variety of farm products specially branded as produced in 'biodiversity-friendly' production systems, as championed in Latin America by the Rainforest Alliance.

Processed wood products

Because the cost of raw material is such a small proportion of the final value of many forest products, local producers often seek to find ways to add value to their product and secure greater market certainty through processing. Exports offer few opportunities for small-scale farmers. While international trade in furniture, moldings, builders' woodwork and other processed tropical wood products grew more than 250% since 1990, from $1.8 to $6.5 billion, nearly all tropical exports are from industrial plantations in Indonesia, Malaysia, Thailand, Brazil and the Philippines (ITTO 2003).[9]

By far the largest market for processed wood products of developing country producers will be domestic consumers. In newly industrialized and developing countries, per capita consumption of furniture is typically low, and demand is growing rapidly. Much of this demand will be met by large-scale manufacturing facilities, often in vertically integrated industries and imports; indeed, developing countries account for 10 out of the world's 15 largest net importers (Poschen and Lougren 2001). But there may be significant scope for communities and small-scale producers to manufacture low-end products for local and domestic urban markets, and to supply niche markets that cannot be efficiently served by industrial-scale processors. Millions of small-scale producers could participate profitably in value-added wood processing enterprises, particularly through pre-processing, milling for local markets, contracts for selected operations in vertically integrated industries, and high-value artesanal production near urban centers or exports for specialty markets. Producer groups could subcontract with larger companies for labor-intensive operations, such as upholstering, and production of low-value furniture and wood products (Poschen and Lougren 2001). Production of handicrafts that have economies of small scale and no mechanically produced substitutes may be a low-cost strategy to add value, as may woodworking products that do not have pronounced economies of scale.[10] Small-scale manufacturing enterprises may be able to sub-contract with large-scale companies. Small-scale forest product processing is already one of the largest and fastest-growing sources of rural non-farm employment. In eastern and southern Africa, woodworking for urban and rural markets grew 10 times as fast as other forest product enterprises (Arnold et al. 1994).

Processing involves additional management complexity and investment. If producers face markets where low cost and high volumes of consistent quality is at a premium, industrial-scale operations will be more competitive. Even where markets are promising, investment to improve milling efficiency and add value will be essential. Fortunately, new technologies have been developed for small-scale wood processing.[11] Advantages of on- or near-site processing include returning residues such as bark, sawdust and trimmings back to the forest, flexibility in meeting market demands (i.e., being able to supply as the opportunity arises); reduction of transport costs by reducing logs to commercially recoverable timber on-site; and the ability to harvest and mill the timber in both small and inaccessible areas. Farmers in parts of the Amazon are actively involved in small-scale milling activities (Padoch and Piñedo-Vasquez 1996). From construction-grade wood, lumber of various diameters and grades can be produced, including appearance-grade wood, certified wood furniture, flooring and decorative wood products. Drying kilns can further increase the quality, and thus price, of wood products.

Industrial pulpwood

Demand for industrial pulpwood (chemically-treated wood products) has grown faster than any other market segment in recent decades, and now accounts for more than a quarter of industrial wood consumption. Although technological improvements and recycling have greatly reduced the volume of pulp required in production[12], pulp consumption in developing countries is rising by five percent annually. Continued demand growth is expected, as average paper consumption is only 15 kg per capita per year, as compared to 200 kg in the EU and 300 in the United States. International tropical trade of reconstituted panels, pulp and paper grew by more than 200% during 1990–2002, from $1.5 to $5 billion. Most exports are based on wood produced in industrial plantations, with only four countries – Indonesia, Brazil, Thailand and Malaysia – accounting for 94% of export production (ITTO 2003).[9]

Due to improved industrial efficiency, increased availability of low-cost wood, and entry of plantation-grown wood, pulpwood prices have, however, been declining despite the rapid growth in paper demand. In general, small-scale farmers will not be able to compete with large-scale industrial plantations in most international pulpwood markets on price, scale or reliability. However, pulp production is highly capital-intensive; a single plant may require 1 million to 2 million cubic meter raw material each year (Bazett 2000). Though the raw material represents a small share of total costs, it is essential to have a reliable supply to ensure continuous use of equipment. Thus, in countries with large domestic markets for pulp, and limited scope for large-scale harvest from natural forests, millions of farmers may find commercial opportunities through out-grower arrangements, farm forestry cooperatives, joint ventures with industry or land leasing to private companies. Already, 60% of firms producing wood pulp in developing countries source at least some of their supply from farmers.[13] In coastal Andhra Pradesh, India, for example, over 40 000 hectares of small-holder farmland are estimated to be under trees suitable for pulp and the district is supplying 700 000 tons of pulpwood annually to different wood-based industries (Mayers and Vermeulen 2002). Furthermore, technological innovations that utilize wood chips to process boards have increased demand for small-diameter wood that can be grown on farms in shorter rotations and for lower quality wood.

Opportunities are geographically limited to areas in close proximity to major pulp mills (within 100 km), with good transportation infrastructure, good growing conditions and uncompetitive agricultural alternatives. For low-income farmers to participate in this market, they require a role in selecting the species to be planted, clear tree rights, financial support while trees mature, good prices; adequate returns on investment, and diversified markets (IIED 1996). For pulp companies to participate, they may need assistance in identifying partners and must adapt operations to the small size of farms and woodlots, and overcome the lack of mutual trust to sign contracts between companies and farmers. Political insecurity may discourage long-term investments/contracts (Mayers and Vermeulen (2002). To safeguard local livelihoods and environment, pulpwood plantations need to be developed in ways that respect conservation guidelines, planting in mosaic patterns that retain areas for natural forest and farming.

Developing markets for farm-grown tree products

Whether producers are able to capitalize on *potential* market advantage depends on the characteristics of the market environment in which they must operate. Experience in both smallholder agriculture and community forestry suggests some of the characteristics of markets that are likely to favor low-income producers. These are identified in Table 2. Small-scale farmers will benefit where production and processing systems have low capital costs, no economies of scale, and where provision of ecosystem services is compatible with economic activities. They benefit from more competitive and open markets (where they do not compete directly with very low-cost producers), in market niches that prefer small-scale suppliers, where local farmers' tree resources provide valued species or ecosystem services, and where costs of market entry are low. Small-scale producers flourish most in regulatory environments with low cost of entry and operation, few burdensome regulations, few subsidies to large industry and secure forest rights. Moreover, where new types of products are being marketed, key actors and functions in the 'value chain' from producer to consumer may also be missing.

Thus, in addition to helping local forest enterprises to develop, which has been the principal focus of smallholder market interventions to date, it will be essential to strengthen forest market institutions to serve small-farm producers and remove policy barriers to local market participation, as articulated also by Puri and Nair (2004) based on experience in India. Lessons from case study analyses and experience of small-farm crop and livestock production suggest some strategies to achieve these.

Develop local forest enterprises

Targeted support is needed to develop local forest enterprises: to improve the market positions of those already operating, strengthen producer organizations to carry out marketing and processing activities, and forge strategic partnerships with large-scale businesses.

Improve market position

To raise incomes significantly, farmers need to analyze the value chain, define the market and establish a competitive position. Sale of low-value wood products with mainly local or *stagnant* regional demand (for example, rough-finished furniture) are important mainly for the role they play in livelihood security of the poor. These are products for which prices are low and quite competitive because the cost of market entry is low. Low-cost substitutes are often available in these markets, such that earnings and profits are also generally low.

Significant longer-term income growth will depend upon selling products for which there is growing domestic demand, such as construction-grade timber, packing materials, or inexpensive semi-finished furniture. Product differentiation does not exist to any significant extent and prices are competitive, so the main opportunities are likely to be in forest-scarce regions with large populations, especially around urban areas distant from major ports. Success requires building supply networks that link producers to markets and increasing production efficiency.[14] These markets should, in most cases, be the main targets for rural development forestry investment with low-income producers.

A small number of farmers may be able to access high-value specialty markets, mainly high-value timbers and processed products. To be successful in these markets, producers must be highly responsive to consumer preferences and have good marketing strategies. Volumes required are typically lower, but products are highly differentiated and rely on branding or direct linkages with buyers, so marketing costs can be high (Scherr et al. 2003). Success requires close attention to product design, promotion and marketing, and product quality. Producers will typically need to improve production and marketing technology, product quality or reliability of supply, and reduce unit costs (Hyman 1996). As enterprises develop, producer groups will need to seek supplemental investment financing, through grants, public subsidies, rural credit, development banks, or from socially responsible firms. In cases where market-oriented agroforestry development can contribute to forest ecosystem or watershed restoration, conservation finance may be a source.

Strengthen producer organizations

Strong farmer organizations are needed for successful commercial forestry enterprises, to capture the potential gains from vertical and horizontal integration. Farmer organizations may play diverse roles: to mobilize capital for critical investments (for example, through micro-finance), undertake joint processing activities, organize marketing deals or establish product quality or conservation controls. Groups can contract with intermediaries to assure minimum supplies to a large-scale buyer. In regions with underdeveloped market institutions, groups of producers can work together to overcome value chain 'gaps,' by setting up reliable transport services, recruiting regional traders, establishing log sorting yards or agreeing to quality standards. In the Philippines, for example, farmers in Agroforestry Cooperatives grow timber in agroforestry systems and remnant forests to sell cooperatively (ICRAF 2001). Successful models for capacity-building exist (e.g., Ford 1998, Fisher 2001, Colchester et al. 2003) but will need to be scaled up (Franzel et al. 2004).

Forge strategic business partnership

Strategic business partnerships between private industry and local farm producers can benefit both. Industrial firms can access wood fiber at a competitive cost, along with forest asset production, local ecosystem expertise and social branding opportunities. Local partners can benefit from high-quality planting materials, technical assistance, quality control, investment resources for expansion and marketing and business expertise. Comparative studies of operating forestry partnerships around the world found that effective partnership requires a long-term perspective for business development, flexible contract terms, special attention to reducing business risks (such as spreading sources of supply among different producer groups) and mechanisms to reduce transaction costs (Desmond and Race 2000; Mayer and Vermeulen 2002). Industrial partners, accustomed to specialization, need to respect the diversified livelihood strategies of their lower-income partners.

Strengthen forest market institutions to serve small-farm producers

Historically most commercial forest products flowed through privileged large-scale public forest industry monopolies or multinational companies. As these were often highly vertically integrated, a private sector network of business, financial, technical and marketing services and market infrastructure did not develop to serve smaller-scale, independent producers. Development of the small-farm agroforestry sector requires that such services become available to a large number of producers or producer organizations.

Encourage agroforestry business-service providers

Local business success will often depend on access to essential business services, tailored to meet the special requirements of low-income farm producers. Key elements include management services, organizational support, technical assistance for production and processing, market information, insurance, marketing assistance and financing. In the early stages of local forest-market development, such services rarely exist in most rural communities. They must be provided by nonprofit, public- or civic agencies, or a private business partner. Agricultural extension services need to be strengthened and their scope expanded to include commercial agroforestry. Producer networks may play a role in providing some of these services (Colchester et al. 2003). As local capacity and scale of production expand, the private sector may find profitable opportunities.

Invest in regional forest enterprise development to fill gaps in the value chain

In regions with high potential for commercial small-farm agroforestry, public or quasi-public agencies can serve to accelerate development of the necessary physical market infrastructure and institutions at a regional scale. Such programs may facilitate business partnerships, financing for local forestry businesses, training, developing of grading systems, and other services.[15]

Target research, education and training

Forming a commercially viable small-farm agroforestry sector will require developing, disseminating and adapting to new production, processing and management systems. Education and training programs must foster this new expertise, integrating sustainable forest management, business and marketing skills with skills in community facilitation. Research efforts need to focus on technical, economic, institutional and policy problems relevant to this sub-sector.

Remove policy barriers to local market participation

Most developing countries have begun the process of dismantling and reforming forest policies that were originally put in place to tightly control commercial production by privileged, subsidized suppliers, and are moving to a more market-based system. But existing policy frameworks still greatly disadvantage small-scale producers, and aggressive market policy reform is essential for development of a thriving, legal and sustainably managed small-farm agroforestry system. Reform should include three key elements.

Reduce the excessive regulatory burden

Forest market activity in most developing countries is choked by excessive state regulation. Complex, poorly understood and contradictory regulations from various agencies make compliance difficult and expensive, and encourage selective enforcement (Puri and Nair 2004). There are a plethora of rules that distort markets and burden small-scale producers, maintain product standards biased against producers (such as over-dimensioning of lumber), and set excessive taxes and forest agency service charges. This drives millions of people to operate illegally, and induces farmers to destroy wildlings of tree species whose harvest is regulated by the state (Kaimowitz 2003). Reducing this regulatory burden is essential for small-farm wood producers to develop sustainable enterprises. For example, it should be possible to deregulate markets very quickly for many tree species that are not grown at all in natural forests (ASB 2001).

Create a fair and open competitive market environment for local producers

Forest market policies still widely favor dominant large-scale producers, distributors and buyers, often even establishing official monopoly buyers. For example, rules for bidding on wood from public forests may set minimum lot sizes that effectively exclude small-scale buyers, while eligibility criteria for receiving public subsidies may set a minimum size of farm plantation that excludes most small-scale agroforesters. Lower-income agroforestry producers benefit most from a fair and open competitive market environment consisting of markets with many buyers and sellers, few limitations on market entry or operation, flexible quality and volume requirements, and no subsidies or regulations that favor large-scale actors (Scherr et al. 2003). Certification systems will need to be adapted specifically for small-scale agroforestry producers (Atyi and Simula 2002).

Involve farmers in forest policy negotiations

Local farm producers' active involvement in forest market policy negotiations is needed to promote more practical, realistic and lower-cost laws, market regulations and development plans. In some countries, democratization has enabled greater participation, and forced greater transparency in forest markets. Forest rights and regulatory reforms have been achieved in some areas, such as southern Mexico (De Walt et al. 2000), through political alliances, involving local producers networks, private industry, government agencies and/or environmental groups that will benefit from sustainable market development.

Conclusions

Trends in forest resource scarcity, ownership, demand for forest products, governance, and forest productivity are opening up unprecedented opportunities for small-scale farms to benefit from commercial markets for wood products from agroforestry systems. A more strategic use of overseas development assistance, non-governmental organization and public funds could leverage these private flows and incentives to transform forest markets and new instruments for ecosystem services into positive contributors to both ecosystem conservation and poverty alleviation.

Wood from small-farm agroforestry and farm forestry systems is likely to supply only a small share of exports of commodity wood or high-value timber, though the benefits farmers receive from this trade could increase. But their role in domestic wood supply could potentially be far larger than it is today, with aggressive efforts to develop the necessary market infrastructure, services and institutions. The impacts on rural-poverty reduction from such a strategy are hard to calculate at this time because they are so dependent on achieving the necessary policy reforms and capacity building. But those impacts could be large during the next 25 to 50 years – benefiting hundreds of millions of low-income people.

End Notes

1. The World Agroforestry Centre, the Center for International Forestry Research, FAO and local collaborators have such studies underway in several African countries.
2. Scherr S.J., White A. and Kaimowitz D. 2003. A New Agenda for Forest Conservation and Poverty Alleviation: Making Markets Work for Low-Income Producers. Forest Trends and CIFOR: Washington, D.C. www.forest-trends.org
3. ARM and Mundy J. 2000. Policy Risks for the Forest Sector Associated with the Implementation of Kyoto Protocol. Report to Forest Trends. Washington, DC. May.
4. There is little rigorous data documenting small-scale forest producers' experience in these markets in aggregate, or the impacts on their livelihoods. Our analysis, adapted from Scherr, White

and Kaimowitz (2003), is based largely on case study evidence of 'successes' and 'failures,' and personal observation of the structure and function of existing and emerging forestry markets, and extension of lessons learned from promoting smallholder agriculture in developing countries, which is much better documented than forestry.

5. The term 'commodity' is used in this paper to refer to products characterized by a high level of standardization (tree species, size, quality), which can be traded interchangeably with similar products from other sources, in large-scale national or international markets. This is distinguished from 'niche' or 'specialty' products, which may have unique properties, special uses, or are traded in small lots.
6. Field analysis of 56 agroforestry systems in eight countries of Central America and the Caribbean found that the payback period for most systems other than woodlots was one to six years. The ratio of benefits to costs was over one in most cases, and over two in eight cases. The most profitable systems for farmers were *taungya*, various types of intercropping, and homegardens (Current et al. 1995).
7. Where high-value commercial timber is found in public forests, the bulk of revenues is likely to accrue to other actors.
8. Rametsteiner E. and Simula M. 2001. Forging Novel Incentives for Environment and Sustainable Forest Management. For the European Commission DG Environment. Workshop on Forest Certification, Brussels, Sept. 6-7, pp. 13-72.
9. ITTO. 2003. Tropical timber trade and production trends. International Tropical Timber Organization, Yokohama, Japan.
10. For example, an assessment of value-added opportunities for aboriginal forest producers in North America identified: joinery stock, door and window frames, cabinets, flooring, housing components for specialty markets, edge-glued panels for shelving and furniture, finger-jointed products; moldings, garden furniture; canoe paddles, chopsticks, and log homes.
11. Diverse types of small-scale mills are now available, including chainsaw mills, horizontal band saws, single and double circular saws, and one-person bench sawmills (FFAQI 2000).
12. The worldwide shift from chemical to mechanical pulping has cut the wood required for a ton of pulp by half (Imhoff 1999). Consumers now recycle a global average of 40% of their paper. Some 30% of global wood fiber for paper now comes from manufacturing residues.
13. Two dozen examples were documented in a review of community-industry forestry partnerships (Mayers and Vermeulen 2002).
14. The first wave of farm forestry in India inspired high farmer participation, but led to market saturation and much lower prices than anticipated. During the second wave of farm forestry, market dynamics were widely recognized and incomes have been more reliable (Saigal et al. 2002).
15. A highly successful example in an impoverished region of Appalachia in the United States is described in Poffenberger M. and Selin S. (eds.) 1998. Communities and Forest Management in Canada and the United States. A Regional Profile of the Working Group on Community Involvement in Forest Management. Berkeley, CA: Forests, People and Policies.

References

Alavalapati J.R.R., Shrestha R.K., Stainback G.A. and Matta J.R. 2004 Agroforestry development: An environmental economics perspective (This volume).

ARM and Mundy J. 2000. Policy Risks for the Forest Sector Associated with the Implementation of the Kyoto Protocol. Report to Forest Trends. Washington, DC. May.

Arnold J.E.M., Liedholm C., Mead D. and Townson I.M. 1994. Structure and growth of small enterprises using forest products in southern and eastern Africa. OFI Occasional Paper No. 47. Oxford Forestry Institute, Oxford and GEMIN Working Paper No. 48. Growth equity through microenterprise investments and institutions. Gemini Project: Bethesda, MD, USA, 34 pp.

ASB 2001. Deregulating agroforestry timber to fight poverty and protect the environment. ASB Policy Brief No. 3. Alternatives to Slash-and-Burn Programme, World Agroforestry Centre, Nairobi, Kenya, 4 pp.

Atyi R.E. and Simula M. 2002. Forest certification: pending challenges for tropical timber. ITTO Technical Series No. 19. International Tropical Timber Organization, Yokohama, Japan, 59 pp.

Bazett M. 2000. Long-term changes in the location and structure of forest industries. Global Vision 2050 for Forestry. World Bank/World Wildlife Fund Project, Washington, DC, 23 pp.

Beer J., Ibrahim M. and Schlonvoigt A. 2000. Timber production in tropical agroforestry systems of Central America. pp. 777–786. In: Krishnapillay B., Soepadmo E., Lotfy Arshad N., Wong A., Appanah S., Wan Chik S., Manokaran N., Lay Tong H. and Kean Choon K. (eds), Forests and Society: The Role of Research. Sub-plenary Session, Vol. 1, XXI IUFRO World Congress, 7–12 August 2000, Kuala Lumpur. International Union of Forest Research Organizations Secretariat, Vienna, Austria and Forest Research Institute Malaysia, Kuala Lumpur, Malaysia.

Colchester M., Apte T., Laforge M., Mandondo A. and Pathak N. 2003. Bridging the gap: communities, forests and international networks. CIFOR Occasional Paper No. 41. Synthesis report of the project 'Learning Lessons from International Community Forestry Networks.' Center for International Forestry Research, Bogor, Indonesia, 72 pp.

Current D., Lutz E. and Scherr S.J. (eds) 1995. Costs, benefits, and farmer adoption of agroforestry: project experience in Central America and the Caribbean. World Bank Environment Paper No. 14. World Bank, Washington, DC, 212 pp.

Daley-Harris S. (ed.) 2002. Pathways out of Poverty: Innovations in Microfinance for the Poorest Families. Kumarian Press, Bloomfield, CT, 400 pp.

Desmond H. and Race D. 2000. Global survey and analytical framework for forestry out-grower arrangements. Final Report submitted to the Food and Agriculture Organization (FAO) of the United Nations, Rome, Italy. Australian National University Forestry, Canberra, Australia, 54 pp.

DeWalt B.R., Olivera F.G. and Correa J.L.B. 2000. Mid-term evaluation of the Mexico Community Forestry Project. World Bank, Washington, DC, 18 pp.

Dewees P.A. and Saxena N.C. 1995. Wood product markets as incentives for farmer tree growing, pp. 198–241. In: Arnold J.E.M. and Dewees P.A. (eds) Tree Management in Farmer Strategies: Responses to Agricultural Intensification. Oxford University Press, Oxford, UK.

FAO 2003. State of the World's Forests. Food and Agriculture Organization, Rome, Italy, 152 pp.

FAO 2001. FAOSTAT. Food and Agriculture Organization, Rome, Italy.

FAO 1998. FAOSTAT. Food and Agriculture Organization, Rome, Italy.

FFAQI 2000. Portable sawmills: weighing up the potential for on-farm processing. RIRDS Short Report No. 38. Forest Farmers Association of Queensland Inc., Highgate Hill, Australia.

Fisher R.J. 2001. Poverty alleviation and forests: experiences from Asia. Regional Community Forestry Training Center for Asia and the Pacific, Kasetsart University, Bangkok, Thailand.

Ford Foundation 1998. Forestry for Sustainable Rural Development. A Review of Ford Foundation-Supported Community Forestry Programs in Asia. Ford Foundation, New York, NY, 58 pp.

Franzel S. and Scherr S.J. 2002. Trees on the Farm: Assessing the Adoption Potential of Agroforestry Practices in Africa. CAB International, Wallingford, UK, 208 pp.

Franzel S., Denning G.L., Lillesø J.P.B. and Mercado A.R. Jr. 2004. Scaling up the impact of agroforestry: Lessons from three sites in Africa and Asia (This volume).

Gardner-Outlaw T. and Engelman R. 1999. Forest Futures: Population, Consumption and Wood Resources. Population Action International, Washington, DC, 68 pp.

Hellin J. and Higman S. 2003. Feeding the Market: South American Farmers, Trade and Globalization. Kumarian Press, Bloomfield, CT, 240 pp.

Hyman E. 1996. Technology and the organisation of production, processing, and marketing of non-timber forest products. p. 197-218. In: Ruiz M. Perez M. Ruiz and Arnold J.E.M. (eds). Current Issues in Non-Timber Forest Products Research. Proceedings of the Workshop 'Research on NTFP', Hot Springs, Zimbabwe, 28 August–2 September 1995. Center for International Forestry Research, Bogor, Indonesia.

ICRAF, World Agroforestry Centre 2001. Negotiation support system for natural resource conflict resolution to enhance environmental services in Southeast Asia. EAPEI proposal. World Agroforestry Centre (ICRAF), Nairobi, Kenya.

IIED 1996. Towards a Sustainable Paper Cycle: An Independent Study on the Sustainability of the Pulp and Paper Industry. Report prepared for the World Business Council for Sustainable Development. International Institute for Environment and Development, London, UK, 258 pp.

Imhoff D. 1999. The Simple Life Guide to Tree-Free, Recycled and Certified Papers. Simplelife: Philo, California.

Kaimowitz D. 2003. Forest law enforcement and rural livelihoods. Int. For. Rev. 5(3): 199–210.

Lecup I. and Nicholson K. 2000. Community-Based Tree and Forest Product Enterprises: Market Analysis and Development Field Manual. Food and Agriculture Organization, Rome, Italy, 195 pp.

Leslie A.J. 1999. For whom the bell tolls: what is the future of the tropical timber trade in the face of a probable glut of plantation timber? In: Tropical Forest Update 9(4). Publication of the International Tropical Timber Organization, Yokohama, Japan. (Available at http://www.itto.or.jp/newsletter/v9n4/7.html).

Mayers J. and Vermeulen S. 2002. Company-Community Forestry Partnerships: From Raw Deals to Mutual Gains? Instruments for Sustainable Private Sector Forestry Series. International Institute for Environment and Development, London, UK, 176 pp.

Piñedo-Vasquez M., Zarin D1.J., Coffey K., Padoch C. and Rabelo F. 2001. Post-boom logging in Amazonia. Hum Ecol 29(2): 5.

Poffenberger M. (ed.) 1996. Communities and forest management. A report of the IUCN Working Group on Community Involvement in Forest Management, with recommendations to the Intergovernmental Panel on Forests. IUCN-The World Conservation Union, Washington DC, 44 pp.

Poschen P. and Lougren M. 2001. Globalization and sustainability: the forestry and wood industries on the move. Report for discussion at the Tripartite Meeting on the Social and Labour Dimensions of the Forestry and Wood Industries on the Move. International Labour Organization, Geneva, Switzerland, 125 pp.

Puri S. and Nair P.K.R. 2004. Agroforestry research for development in India: 25 years of experiences of a national program (this volume).

Raintree J.B. and Francisco H.A. (eds) 1994. Marketing of multipurpose tree products in Asia. Proceedings of an international workshop held in Baguio City, Philippines, 6–9 December 1993. Winrock International, Bangkok, Thailand, pp. 289–300.

Robbins C.S. 2000. Mahogany matters: the U.S. market for big-leafed mahogany and its implications for the conservation of the species. TRAFFIC North America, Washington, DC, 58 pp.

Saigal S., Arora H. and Rizvi S. 2002. The New Foresters: The Role of Private Enterprise in the Indian Forestry Sector. Instruments for Sustainable Private Sector Forestry Series. Ecotech Services (New Delhi, India), and International Institute for Environment and Development, London, UK, 192 pp.

Scherr S.J. 1995. Meeting household needs: farmer tree-growing strategies in western Kenya, pp. 141–173. In: Arnold J.E.M. and Dewees P.A. (eds). Tree Management in Farmer Strategies: Responses to Agricultural Intensification. Oxford University Press, Oxford, UK.

Scherr S.J., White A. and Kaimowitz D. 2003. A New Agenda for Forest Conservation and Poverty Alleviation: Making Markets Work for Low-Income Producers. Forest Trends and CIFOR, Washington, DC, 90 pp. Available at: http://www.forest-trends.org/resources/pdf/A%20New%20Agenda.pdf

Simons A.J., Jaenicke H., Tchoundjeu Z., Dawson I., Kindt R., Oginosako Z., Lengkeek A. and De Grande A. 2000. The future of trees is on farm: tree domestication in Africa, pp. 752–760. In: Krishnapillay B., Soepadmo E., Lotfy Arshad N., Wong A., Appanah S., Wan Chik S., Manokaran N., Lay Tong H. and Kean Choon K. (eds). Forests and Society: The Role of Research. Sub-plenary Session, Vol. 1, XXI IUFRO World Congress, 7-12 August 2000, Kuala Lumpur. International Union of Forest Research Organizations Secretariat, Vienna, Austria and Forest Research Institute Malaysia, Kuala Lumpur, Malaysia.

Smeraldi R. and Verissimo A. 1999. Hitting the target: timber consumption in the Brazilian domestic market and promotion of forest certification. Amigos da Terra – Programa Amazonia, São Paulo, Brazil, 41 pp.

Vlosky R.P. and Aguirre J.A. 2001. Increasing marketing opportunities of lesser known wood species and secondary wood products in tropical Central America and Mexico. Louisiana Forest Products Development Center, Louisiana State University, Baton Rouge. August, 6 pp.

White A. and Martin A. 2002. Who Owns the World's Forests? Forest Trends, Washington, DC, 32 pp.

Agroforestry Systems **61:** 371–382, 2004.

Markets and marketing strategies for agroforestry specialty products in North America

M.A. Gold[1,*], L.D. Godsey[1] and S.J. Josiah[2]
[1]*University of Missouri Center for Agroforestry, 203 ABNR Building, Columbia, MO 65211 USA;* [2]*School of Natural Resources, UNL East Campus Lincoln, NE, USA;* [*]*Author for correspondence: e-mail: goldm@missouri.edu*

Key words: Alternative crops, Market value chain, Niche markets, Porter Five Forces Model, Woody florals, Eastern red cedar

Abstract

In agroforestry, marketing is unique for several reasons: many products typically lack established marketing institutions, market information, and grade or quality standards. All that is known about the market for many agroforestry specialty products is that someone is growing the product and consumers are buying it. What happens to the product as it moves through the value chain from producer to consumer is unknown, the 'black box' of agroforestry markets. To shed light on the black box and stimulate adoption of agroforestry practices, successful marketing strategies must be developed. Porter's 'Five Forces Model' is presented as a viable approach to explore the competitive environment of unexplored markets. The Five Forces Model looks at the underlying fundamentals of competition that are independent of the specific strategies used by market competitors including: (1) potential entrants; (2) suppliers; (3) buyers; (4) substitutes; and (5) industry competitors. By understanding the competitive forces within an industry, market strengths, weaknesses, opportunities and threats can be identified and successful strategies can be developed. Standard marketing methods are highlighted to guide the development of specialty product market strategies. Woody florals research is described to illustrate a successful approach to agroforestry niche market development. Results of eastern red cedar market research are presented to illustrate use of the Five Forces Model as a framework for producers to look critically at potential markets and successfully identify opportunities, develop strategies, and avoid pitfalls. The information presented attempts to go beyond identification of the market participants (who, what and where) and seeks to identify the forces (how and why) of agroforestry markets.

Introduction

Widespread adoption of agroforestry in North America is lagging. This is due, in part, to risk-averse producers' understandable reluctance to establish agroforestry practices in the absence of readily available market information. Understanding niche markets and marketing is a key ingredient in the success of profitable agroforestry enterprises that produce commercially valuable specialty products.

Unlike other types of conservation practices where land is taken out of production, agroforestry is 'productive conservation'. Agroforestry practices enable landowners to generate income from the production of a wide range of conventional and specialty products while simultaneously protecting and conserving soil, water and other natural resources. Many observers have examined the potential of these dual-purpose market-driven conservation systems in North America, including Smith (1950), Kurtz et al. (1984), Campbell (1991), Kurtz et al. (1991), Garrett et al. (1994), Kurtz et al. (1996), Garrett and Harper (1999), Kays (1999) and Josiah et al. (2004a, 2004b).

Specialty products, also referred to as specialty forest products, nontimber forest products and nonwood forest products, are produced from trees, within forests, or in myriad combinations with trees or shrubs in a variety of agroforestry practices (Garrett et al.

2000; Teel and Buck 1999). Many of these products have proven economic value but have been ignored by, or were unknown to, agricultural and forest landowners. Specialty products in North America are divided into groups based on product use: edibles (e.g., mushrooms, nuts, berries); herbal medicinals (e.g., ginseng, goldenseal, witch hazel); specialty wood products (e.g., diamond willow canes, red cedar closet liners, walnut gunstock blanks); floral and greenery products (e.g., curly and pussy willow, ferns, salal); fiber and mulch (e.g., cedar pet bedding, pine straw); and recreation (e.g., agritourism, fee hunting) (Hammett and Chamberlain 1998).

In the tropics, there is an enormous range of indigenous and exotic species producing commercially valuable products that are traditional to or have been introduced into agroforestry systems. In Haiti, for example, pineapple (*Ananas comosus*), fruit trees (e.g., mango – *Mangifera indica*), vetiver grass (*Vetiveria zizanioides*), and many other plant species that produce valuable products are used in hillside alleycropping configurations, arresting soil erosion, enhancing soil fertility, and generating new sources of income (Bannister and Josiah 1993). This diversity creates numerous opportunities for landowners to earn supplemental income or, in some cases, generate more income per unit area than traditional agricultural commodities currently provide. In the community of Tomé-Açu, Para, Brazil, over 55 different crops grown in intensively managed, highly diverse agroforestry systems are commercially marketed (Yamada and Gholz 2002). Shade grown coffee (*Coffea* spp.) is produced in the highland tropics worldwide for the international marketplace in two- and three-tier agroforestry systems (Hull 1999). Koppell (1995) has written a hands-on manual for developing market-oriented production in tropical agroforestry systems. Clay (1992) described eleven broad principles and strategies to consider when marketing tropical specialty forest products in temperate industrialized countries.

Marketing diverse specialty products from agroforestry practices presents a unique set of challenges for the producer. Diversification is a well-accepted strategy for reducing risk from price and output uncertainty. It can increase profitability by making more efficient use of labor and other resources. Yet, diversification also can increase the complexity of production and marketing decisions. Agroforestry enterprises that produce multiple niche product and commodity goods in complex integrated production systems require meticulous planning and coordination in order to grow, harvest, and profitably market those goods. Producers are faced with the dual challenge of marketing agricultural commodities and niche market specialty products.

Because there is limited experience using market research methodologies to examine specialty product markets, and since detailed market information is lacking for many specialty products, the remainder of this paper focuses on: 1) A discussion and comparison of commodity and specialty product markets; 2) An introduction to the Five Forces Model (Porter 1980) as a tool to explore the competitive environment of specialty product niche markets; 3) A quick review of basic approaches to the development of marketing strategies; and 4) A presentation of two real-world examples to show the application of marketing strategy and the Five Forces Model to specialty product niche markets. The information presented attempts to go beyond identification of the market participants (who, what and where) and seeks to identify the forces (how and why) of agroforestry markets. The tools and methods presented should assist forest and agricultural landowners who want to participate in unique markets by helping them identify critical market characteristics and marketing strategies grounded in a solid understanding of underlying market forces.

Commodity vs. specialty-product niche markets

Marketing involves the process of planning and implementing a strategy for exchanging goods. This includes idea development, pricing, promotion, and distribution right through to the exchange of a product for cash. More than just selling products, marketing is focused on solving problems for customers whose needs are at the center of the business's activities. As mentioned, agroforestry practices often combine production of both commodity and specialty niche products. Producers must understand the differences between commodity and niche product markets and take advantage of how both market categories meet the needs of their customers.

Commodity markets

The nature of modern farming and forestry has created an environment for marketing commodities requiring a minimum amount of effort from the producer. For example, marketing #2 yellow corn (= maize, *Zea mays*) can be done by selling to the local grain elevator. In commodity markets, the buyer wants the cheapest price and lowest cost. Success in commodity markets requires high-volume production and is reliant on advantages of size and scale. Examples of commodity markets include major grain and fruit crops, beef and dairy, lumber and veneer.

In commodity markets, an infrastructure has been established that reduces risk inherent in marketing by simplifying the processes of: 1) search and information gathering; 2) bargaining and decision making; and 3) supervision and enforcement. For agricultural commodities, established markets such as livestock auctions, grain elevators, and Boards of Trade (Chicago or Kansas City) minimize the time and risk involved in the search and information gathering process. The Chicago and Kansas City Boards of Trade along with the United States Department of Agriculture (USDA) provide objective market information that minimizes the risks associated with the bargaining and decision-making processes. Further, accepted standards for agriculture commodities have been established that are easily enforced by marketing institutions. Timber markets are similar to agricultural markets. Grading standards are established, market prices are publicly available, and market participants are often brought together by an established bidding process.

Specialty product niche markets

Markets for specialty products often differ from markets for more common agricultural products. They are usually characterized as 'niche' or 'product' markets – small, volatile, specialized, and with relatively few buyers. The scale of product markets is normally much smaller than commodity markets. Product markets differ from commodity markets in that they rely upon the concept of value-added, represent unique niche products 'with a face and a place' and seek to increase benefits to the customer. Customers purchasing niche products are attracted not only by quality and consistency but by the individual(s) or company behind the product(s) and their story along with the location (origin) of the product. In other words, specialty-product niche markets are based on trust and authenticity. Large firms cannot manufacture trust and authenticity, a competitive advantage that goes to smaller firms. Further, the unique, personalized, customer-oriented approach presents an inherent competitive advantage for those who are in niche product markets compared to mass markets.

In contrast to commodity markets, specialty products often lack established marketing institutions, lack accessible market information, and lack established grade or quality standards. The resultant lack of market information creates understandable disincentives to individuals considering specialty-product-based enterprises. Thus, producers often continue to produce commodities with known markets at a loss instead of pursuing other novel but potentially lucrative opportunities where markets are not well known or established. Further, because knowledge of the value chain is so valuable in niche markets, producers may be reluctant to share their sources of information, methods of production, or current or potential markets. This can make it difficult for newcomers to enter the niche marketplace (Vollmers and Vollmers 1999).

The 'black box' of agroforestry enterprises

As stated, agroforestry enterprises often produce specialty products for markets about which little is known. All that may be known about the market is that someone is growing the product and consumers are buying the product (Thomas and Schumann 1993). What happens to the product between producer and consumer is unknown, i.e., a 'black box'. From a producer's perspective, the list of unanswered questions can be long. How many times does the product change hands before it reaches the final consumer? How do I get into the market? What are my costs and potential returns? Who are my customers? Who are my competitors? What strategy should I use in order to be successful in this market? These and many other questions complicate the decision to produce and market specialty products (Bogash 1998; Sullivan and Greer 2000). Application of the Porter Five Forces Model and standard market research approaches to specialty products provides answers to these questions and enables development of sound marketing strategies.

The five forces model and the value chain

Porter's 'Five Forces Model' (Oster 1994; Porter 1980) of competitive market forces provides a framework for examining competition that transcends industries, particular technologies, or management approaches. The Five Forces Model looks at the underlying fundamentals of competition that are independent of the specific strategies used by market competitors. These areas include (1) potential entrants (barriers to entry); (2) suppliers (bargaining power of suppliers); (3) buyers (bargaining power of buyers); (4) substitutes (substitute products); and, (5) industry competitors (rivalry among existing firms). Another force that is added to the Porter model is the influence of governmental policies on the market (Figure 1). The framework of the Five Forces Model can be applied to both the macro and micro levels in order to develop market characteristics for specialty products, and provide a map to trace how raw product moves across the value chain from producer to the consumer (Gold and Godsey 2001).

The market value chain represents a link between all steps a product moves through from producer to consumer, from raw material to final retail product including all stages where value is added. Value chains represent market-focused collaboration in which different business enterprises work together to produce and market products and services in an effective and efficient manner. A value chain can be thought of an as extended enterprise. When the chain's products and processes are difficult for others to copy, the created value chain has a competitive advantage. Value chains can improve access to a market and reduce the time it takes to respond to changing customer needs (Parker et al. 2002). In the case of eastern red cedar (*Juniperus virginiana*), the value chain links raw material producers (loggers), primary manufacturers (sawmills), secondary manufacturers, wholesalers, retailers, and final consumers.

Potential entrants (barriers to entry)

For many individuals considering entry into new specialty product markets, risks and uncertainties may deter them from looking at the markets' possibilities. Porter (1980) identified six major entry barriers including (1) economies of scale, (2) capital requirements, (3) product differentiation, (4) switching costs, (5) access to distribution channels, and (6) cost advantages independent of scale. Not all of these barriers will be evident in all markets. However, every market will have some entry barrier, either real or perceived, that must be overcome. These market risks and uncertainties can be overcome by understanding what critical resources are needed and what entry barriers exist. Often, entry barriers and critical resources are the same.

Incumbent advantages (Oster 1994) appear to be the most common barrier to entry for specialty products. Incumbent advantages can range from the existence of established relationships to 'learning-curve' knowledge. To maintain competitive advantage, existing producers may be reluctant to share their sources of product, methods of production, or potential markets. Another common specialty product barrier to entry is access to distribution channels (i.e. reaching customers). Reaching customers is commonly achieved through advertising, which can be costly. Successful specialty product niche market entrepreneurs are able to generate substantial amounts of free publicity to reach their customers. Specialty-products producers can take full competitive advantage of 'face' and 'place'. These businesses are growing and marketing products that are unique. If the producers are enthusiastic about what they do, they must get the word out by attending trade shows, giving presentations and bringing products for sale, holding field days and contacting journalists. Face and place stories are often compelling and attract new customers. As a result, magazines write articles about the producer and their business.

Paul Easley owns Oak Leaf Wood and Supplies, a 'stump to store' niche market business in central Illinois. Easley combines vision and experience with patience, taking the time to cut wood to maximize its value. He puts the needs of his customers, crafters and other high-end users at the center of his business activities. Easley does not actually make any finished products in the retail business. He recognized that there are many different facets to woodworking and many market niches to be filled. The products that he manufactures and sells include cabinet grade stock, carving stock, ballpoint pen blanks. The Easley's market area includes the entire United States and seven foreign countries. This occurred through tremendous free publicity and making a product that people wanted. Easley's business has been in close to a hundred magazines.

Suppliers and buyers (bargaining power)

The goal of any entrepreneur is to have the bargaining power in the market. Bargaining power fluctuates over time as market participants enter and exit the market and consumer demands change. However, all market participants are both buyers and sellers in a market. Therefore, it is important to understand how to minimize pressure on profits that can be exerted through

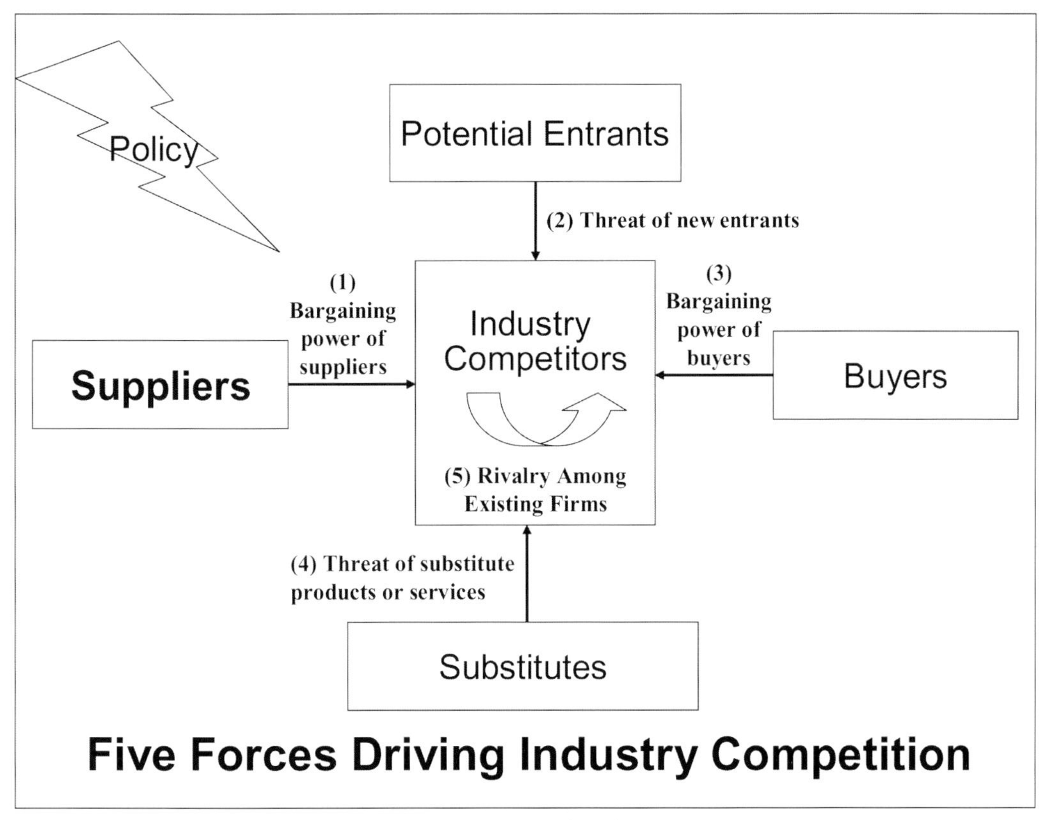

Figure 1. Five Forces Driving Industry Competition: Porter's Five Forces Model including the influence of governmental policy. Source: Adapted from Porter (1980).

bargaining power. Suppliers with bargaining power can extract excess profit by increasing input costs. Likewise, buyers with bargaining power can extract excess profit by putting downward pressure on output prices. Walmart is an example of overwhelming bargaining power in the hands of the buyer, dictating price to suppliers and switching suppliers whenever a better price can be extracted. Another example of buyer bargaining power can be found in U.S. agricultural commodity producers. The U.S. farmer has a limited number of large firms supplying farm inputs and a limited number of large firms buying the commodities. As a result, the farmer's profits are continually being eroded by higher input costs and lower market prices. Specialty niche market producers must focus on products that retain value to buyers through uniqueness, relative scarcity and value added approaches. New products and new markets must be continually kept in mind. By taking these steps, niche producers can retain bargaining power at the supply level and retain more profit through value added production, thereby controlling more of the value chain and retaining more of the profit.

Substitutes (substitute products)

Substitute products can replace other products with little or no lost value to the consumer. If the market perceives product 'B' to be a close substitute to product 'A', then small price increases in product 'A' will increase demand for product 'B'. As a result, close substitutes can have an effect on market prices by providing an option to consumers when prices fluctuate.

Substitute products can play an important role in new markets where demand is increasing rapidly or there are few direct competitors. When supply cannot keep up with demand, substitutes can be used to 'fill the gap'. Yet, in the process of 'filling the gap', these substitutes limit the profit potential of the original product. However, in periods of excess capacity or in highly competitive markets, substitutes have little or no effect on market prices.

Godsey et al. (2004) examined the question of potential substitutes for eastern red cedar (*Juniperus virginiana*) as part of a comprehensive market analysis. In order to understand the potential threat from substitutes, phone survey respondents were asked to describe substitutes for red cedar in terms of quality, availability, and price. Most respondents noted that there were not any direct substitutes. Western red cedar (*Thuja plicata*) is often sold as 'red cedar', but it does not have the same marketable properties as the 'aromatic' eastern red cedar. Different uses of eastern red cedar have different substitutes. For example, western red cedar or chromated copper arsenate (CCA)-treated lumber could be used as substitutes for eastern red cedar for home construction or dimensional lumber. For novelties, other cypress species are often substituted. Many pine species (*Pinus* spp.) are used for pet bedding and mulch in place of red cedar. However, eastern red cedar is considered to be cheaper, more available, and unique in quality when compared to those substitutes. Information regarding potential substitutes in the red cedar market can help develop strategies for interaction with other market participants and identify marketing niches.

Industry competitors (rivalry among existing firms)

Rivalry within an industry is a function of the number of competitors, the size distribution of the competitors, the homogeneity of the products, the level of fixed investments, and the volatility of market demand. In markets where rivalry is intense, business strategies are often implemented that focus on competitive posturing and less on profit generation. For example, a producer may cut prices for their goods in order to undercut the prices of a competitor (Oster 1994). In other markets, e.g., hazelnuts (*Corylus avelana*) or nonnative chestnut (*Castanea* spp.) in the Pacific northwest, rivalry may not be a factor and market participants actually work together to further the demand for their products and increase profitability for all participants.

Government (policy impacts)

Although governmental policies are not specifically listed as one of the five competitive forces in the Five Forces Model, they do have a dramatic impact on the market. Governmental policies that impact markets include: intellectual property rights – patents, trademarks, and copyrights can be used as entry barriers or as methods of product differentiation; regulations – specific laws dictating how a market must operate can reduce rivalry among market participants, since all firms must follow the same guidelines; and quotas and tariffs – price fixing strategies such as quotas and tariffs can increase the number of market participants by allowing some inefficient producers enough income incentive to stay in the market.

Governmental policies affect all aspects of the market, from rivalry to entry barriers. Specialty products will typically have fewer governmental policies impacting the market than agricultural commodities, because of the scale of the market. Specialty product markets will often feel the effect of policies that were not written for that market but have an indirect application. CCA-treated lumber widely used due to its rot resistance, is being phased out by order of the U.S. Environmental Protection Agency because of public health concerns about arsenic. This phase-out has the potential to stimulate market demand for eastern red cedar lumber due to its natural rot-resistance.

Marketing strategy

Market Research: 1. documenting the markets

In order to exploit specialty forest product markets, entrepreneurs must develop a marketing plan that opens the 'black box' and overcomes the information asymmetry inherent in these niche markets. Market research is the starting point in developing a marketing plan, providing a broad yet detailed overview of the industry that is being examined. Market research identifies markets, inventories of raw materials, and answers specific questions about product(s). As markets are constantly changing, market research is an ongoing process that influences all aspects of a marketing strategy and continues throughout the life of a business and is essential to survive and thrive in a changing marketplace.

Two major categories of market research can be identified: primary and secondary. Primary market research is custom-made to answer specific questions about a business and to help operators make more informed decisions. Primary market research is based on direct consultations with existing or prospective customers, observations of and conversations with competitors, discussions with other businesses outside the potential market area, observations of how and where potential product(s) are currently being marketed, discussions with brokers, retailers, and distributors. Market research firms can help obtain this information, although this can be costly. Secondary market research consists of information obtained from other sources (library, government, professional associations, university, Internet). Specifically, this includes published reports and market studies, trade magazines and journals, newspapers, books, literature from competitors, and business directories.

Some producers choose to start with market research to identify product needs and then determine if they have the necessary raw materials. Others inventory what they have and then work to identify or develop markets. Both approaches are valid, and are often conducted simultaneously, but markets must exist if an enterprise is to be successful (Freed 1999).

Market Research: 2. Analysis of market data

Once primary and secondary market data have been collected, they must be organized and analyzed in a way that reveals the most about the market environment. Market research methods vary from industry to industry based on the availability of information (Frigstad 1995; Hester 1996). Several approaches have been used to describe agricultural commodity and forest product markets (Mater et al. 1992; Sinclair 1992). There is limited experience using market research methodologies to examine specialty product markets.

Agricultural commodity market studies are approached at least two ways: analyzing the marketing functions (functional approach), or analyzing the various institutions and firms that are involved in the marketing process (institutional approach) (Kohls and Uhl 1998). The functional approach identifies areas where value is added to the product. The institutional approach identifies and describes the value chain (producers, processors, wholesalers, and distribution centers). Both of these approaches are typically macro level analyses and are difficult to apply to specialty products, which are often micro-level, niche markets.

Conventional forest product market studies have focused on macro-level industry development (Vlosky et al. 1998; Jensen and Pompelli 2000; Mangun and Phelps 2000), competitive advantage analysis (Hansen and Punches 1999) and competitive strategies (Hansen et al. 2002). These studies all incorporate some form of competitive analysis, whether it is identifying a market's strengths, weaknesses, opportunities and threats (SWOT analysis) or describing specific market strategies. Greene (1998) conducted an exploratory and qualitative study of nontimber forest products in southwestern Virginia, United States, focusing on value addition and market outlets, pricing, promotion, and the distribution and marketing chain *Nontimber forest product marketing systems and market players in southwest Virginia: A case study of craft, medicinal and herbal, specialty wood, and edible forest products.* M.S. thesis, Virginia Polytechnic Institute and State University) (Greene S.M. 1998). DeCourley (1992) identified potential market opportunities by understanding timing, placement, and people.

Numerous guides have been published through university extension services and elsewhere that are designed to assist producers in marketing products (Grudens-Schuck and Green 1991; Salatin 1998). Usually these are general marketing guides that typically focus on the four 'P's' – product, promotion, placement, and pricing – and are not specific to any product.

All of these efforts identify key elements of a market that can be used to generally describe market characteristics. Given the niche nature of specialty products, detailed studies that focus on specific aspects of the competitive market environment are needed to understand the market characteristics and develop successful product marketing strategies. Competitive market analysis helps to identify the factors that coordinate and control each unique product market (Porter 1980).

Market research: 3. Developing a marketing strategy

Once market data have been collected and analyzed, a market strategy can be developed (Crawford and Di Benedetto 2000). Two common marketing strategies for specialty products are (1) direct marketing, and (2) adding value. In addition, new value chains can be established to build specialty product markets. Several different approaches are used for direct marketing specialty products. Direct marketing generally provides greater financial returns per unit of product than through wholesale commodity marketing, but require greater marketing time and skills (Freed 1999).

Outlets for direct marketing specialty products to consumers include U-Pick (mushrooms, fruit, nuts, herbs, and flowers/florals), farmers markets, festivals, garden shows, trade events, County/State Fairs, WWW/catalog sales and retail outlets. Direct marketing is growing in popularity because consumers now demand safer, high quality products. Buyers place a value on coming face-to-face with the producer and their production location (farm, farmers' market location, business location, etc). Customers seek more information about the products produced. Directly marketed products can be promoted through the farmers' markets, Internet, radio, weekly newspapers, and local access television (Freed 1999).

Other market outlets include wholesalers and retailers, small or ethnic groceries, restaurants, florists, mass merchandisers including grocery store chains, and local or regional wholesalers. One of the challenges in the wholesale marketplace is the need to provide a constant, dependable supply of large quantities of quality goods to the wholesaler. Wholesalers (particularly with woody florals) may only accept a few weeks worth of product at any one time, forcing the producer to incur storage and multiple delivery costs.

A value-added strategy denotes a deliberate game plan to achieve or increase profitability. Customer value equates to perceived benefits as a function of price. Value can be added by determining the customers purchasing criteria and increasing the bundle of benefits to the consumer. Benefits can be bundled by improving quality (remove defects, consistent high standards), increasing functionality, enhancing the consumer's perception of value (e.g., by incorporating woody floral stems in floral arrangements to increase height and width), changing product form (kiln dry), place (deliver), time, or ease of possession.

New value chain configurations can be created through the establishment of new partnerships among farmers, processors and retailers. Specialty-product producers must strive to preserve their products identity and their unique story as the product moves across the value chain from producer to consumer (farm to table). Producers must take full advantage of face and place by marketing the farm. Higher prices are paid for identity preserved niche products that come with a story (Hartman 2003).

Putting the five forces model and marketing strategy development to work

Two real-world examples of marketing show the Five Force Model's broad application and effectiveness as a market analysis tool. The first example, woody florals, demonstrates a successful approach to specialty product niche market development. The second example, eastern red cedar, applies the Five Forces Model and reveals the model's effectiveness in understanding the competitive forces underlying red cedar markets.

Example #1: Woody floral production systems and markets in Nebraska, USA

In 1999, the University of Nebraska – Lincoln (UNL) and a number of partners embarked on a research program to test the concept of 'market-driven' or 'productive' conservation. The research approach included both primary and secondary market research to identify woody plants that could produce commercially valuable specialty products and be integrated into agroforestry configurations such as riparian buffers and hillside and flatland alley cropping arrangements. Replicated field trials of woody plants, established on five locations in Nebraska, tested more than 30 species and cultivars of trees and shrubs. Expected specialty products included small fruits, nuts, woody floral stems, and herbal medicinals.

Unique to this effort, the researchers participated directly in the woody floral marketplace, and due in part to their direct participation, were able to describe the market value chain. In addition to growing the woody floral products, researchers conducted several local and regional market assessments to determine the nature and characteristics of the markets, including product prices, for each crop class. A telephone census of nearly all retail and wholesale florists in Nebraska ($n = 93$) was conducted to determine numbers of stems used by cultivar or species. A second telephone survey that sampled wholesale florists across the United States (except Pacific coast states) documented an approximately $8 million market for fresh woody floral stems. UNL researchers also sold woody stems produced in field trials in the marketplace whenever possible. This allowed the acquisition of real-world price, quality criteria, and local market data. Beginning two growing seasons after establishment, woody floral stems (curly willow – *Salix matsudana* , pussy willow – *Salix caprea*, scarlet curls willow – *Salix 'Scarcuzam'*, and red – *Cornus sericea (stolonifera)* and yellow – *Cornus sericea 'Flaviramea'* stemmed dogwoods) were harvested. In order to familiarize local florists with and build demand for locally produced woody florals, researchers distributed this first harvest (worth several thousand dollars) to the Nebraska Florist Society, a professional organization of wholesale and retail florists. In return, researchers obtained valuable feedback on quality and packaging criteria, and market trends. Based on this information, and on subsequently strengthened personal relationships with florists, the next two years' crops were sold to four wholesale florists in Lincoln and Omaha, NE. Project staff interviewed wholesaler employees during each delivery to gather market-, price- and buyer-requirement details. Labor and associated costs required for production, harvesting, processing and marketing activities were recorded, enabling the development of enterprise budgets.

Simultaneously, a major extension effort was launched to educate landowners about the value of integrating woody florals and other specialty woody crops into agroforestry practices. Despite nearly two years of extension programming (1999–2001), it was not until UNL began to actually sell woody florals in the marketplace that any landowners adopted this practice. Since late 2001, (when the first sales occurred), at least 24 landowners in Nebraska have integrated woody florals and other specialty products into their farming and land use systems, independent of any federal or state subsidy programs. UNL also sold harvest rights to woody florals produced in one of its trials to several landowners, further moving the enterprise into the private sector. Landowners are now in the process of forming a Nebraska Woody Florals Grower Association to attain better prices, pool production, secure larger, longer term contracts, reduce risk, and to jointly explore other markets (e.g., dried/preserved woody florals). Floral design students at UNL and local community colleges are now trained in the use of woody stems in floral arrangements, and practicing floral designers receive similar training through the Nebraska Florist Society. These educational efforts increase demand for woody floral stems at the consumer level.

The key to stimulating landowner adoption of this new approach to agroforestry was UNL's provision of a comprehensive package that directly addressed all major landowner concerns. This package contained practical, straightforward, research-based information on species/cultivar selection, sources and prices of planting stock, production, harvesting, processing and packaging guidelines, labor requirements, and real-world cost, price and market information.

Example #2: Applying the Five Forces Model to the eastern red cedar industry, USA

Often considered a 'trash or nuisance' tree, eastern red cedar (*Juniperus virginiana*) is a prime example of a 'black box' market. Due to a lack of knowledge and understanding of the markets for red cedar, the species has been subject to management practices that call for its eradication. Oklahoma and Missouri natural resource agencies have adopted management polices that pay outside contractors to eradicate eastern red cedar from State lands by means of burning or herbicide application.

The policy of paying someone to destroy a resource that has a clear market value (evidenced by the quarterly timber report published in Missouri by the same agency that pays to have eastern red cedar removed), seems irrational. This inconsistency indicated that traditional approaches to understanding markets failed to describe important aspects of small or niche markets. Several aspects of the red cedar market were unanswered by traditional market-analysis approaches. First, there was no aggregated information about who the market participants were or what the scale of the industry was. It appeared that the red cedar market had developed as a disjointed group of producers who operated within the boundaries of very narrow local market areas, each oblivious to the others' activities. Second, information about the factors that impact profitability and the competitive environment, such as output prices and bargaining power of suppliers or buyers, was not known. Finally, because eastern red cedar was not considered a major commercial timber species, even though it has great potential for certain lumber applications, information regarding its future growth potential and current availability was not known.

Porter's Five Forces Model of competitive market forces was used to develop and analyze these and other market characteristics of eastern red cedar (Godsey et al. 2004). Primary data regarding market participants were collected through meetings with producer groups, Internet searches, retail market visits, and personal contacts. This information identified market participants and different products that were bought and sold along the eastern red cedar value chain. From the list of market participants collected, a mail survey ($n = 187$) was developed and sent to businesses and individuals from all over the United States who were actively participating in the eastern red cedar market. The survey was designed to create a basic understanding of the size, location, position, and production within the eastern red cedar market, and to obtain a basic understanding of supply and demand perceptions.

One output from this market documentation was a directory that included all those who were surveyed and wished to be included. From this eastern red cedar market directory, networks of buyers and sellers were created. For example, a producer of eastern red cedar cants in southwest Missouri was connected with a fencing company in St. Louis and a working relationship was developed. From these survey respondents, a subsample of respondents (n=25) was selected to participate in follow up phone surveys. The phone survey consisted of a directed series of questions based on competitive forces within the eastern red cedar market. Results from both surveys were analyzed using the Five Forces Model. Data from both surveys identified the level of production throughout the country and provided valuable decision making information for developing market strategies. From these surveys it was determined that the eastern red cedar industry generated nearly $60 million in gross revenue annually. More importantly the surveys identified the potential for future growth in the market.

By focusing survey questions on areas that were perceived to be barriers to entry, critical resources for successful entry into the red cedar market at all levels of the value chain were highlighted. These critical resources of access to labor, market knowledge, access to the raw material resource, financial resources, equipment and the cultivation of personal relationships among players in the market value chain may seem obvious. In typical market analysis approaches only a few of these important resources would be considered and assumptions about availability would have been implied.

A closer look at the bargaining power of suppliers and buyers led to the need for further analysis drawn from U.S. Forest Service Forest Inventory and Analysis data. The data revealed that four States, Arkansas, Tennessee, Kentucky, and Missouri, possessed nearly half of the nation's current and future growing stock for eastern red cedar. The inventory data coincides precisely with the primary locations of the eastern red cedar industry. This information is critical to any current or potential industry participant concerned about the location, cost and availability of current and future supply of eastern red cedar.

Other important observations that resulted from the application of the Five Forces Model stemmed from the investigation of governmental policies, such as the ban on CCA treated lumber and eradication as a management practice, which were framed from an industry perspective. Impacts of these policies were shown to affect the demand for red cedar products or the supply of raw red cedar resources. By looking at the red cedar market through the lens of the Five Forces Model, key issues that impact the market were identified. The opening of the 'black box' of the eastern red cedar market permits industry participants to pursue pragmatic strategies that are based on solid information regarding a broad range of market impact. Weaknesses in the market, such as a lack of market networking, were identified and strategies including the development of a market directory were used to overcome these problems. The Five Forces Model proved to be a valuable tool in market analysis.

Conclusions

Agroforestry enterprises are multiproduct, multiyear activities often requiring three to 12 years before multiple income streams are derived. Market assessment and strategic marketing are essential for agroforestry enterprise success. Although agroforestry practices that produce commercial specialty products may increase management and marketing complexity, uncertainty and risk can be reduced by good market research. The woody florals study reveals new methods that can be used to document actual markets and stimulate landowner adoption. The red cedar example demonstrates how the Five Forces Model can serve as a framework for producers to look critically at potential markets and successfully identify opportunities, develop strategies, and avoid pitfalls.

Acknowledgements

Case study research in this paper was funded, in part, through the University of Missouri Center for Agroforestry under cooperative agreement AG-02100251 with the USDA Agriculture Research Service, and the Cooperative State Research, Education, and Extension Service, U.S. Department of Agriculture, under Agreement No. 99-34292-7388. Any opinions, findings, conclusions, or recommendations expressed in this publication are those of the authors and do not necessarily reflect the view of the U.S. Department of Agriculture, and may not represent the policies or positions of the ARS.

References

Bannister M. and Josiah S.J. 1993 Agroforestry training and extension: The experience from Haiti. Agroforest Syst 23: 239–251.

Bogash S.M. 1998. Marketing of agricultural and natural resource income enterprises: Learning from and sharing with entrepreneurs, pp. 21–25. In: Kays J.S., Goff G.R., Smallidge P.J., Grafton W.N. and Parkhurst J.A. (eds). Natural Resources Income Opportunities on Private Lands Conference. Hagerstown, MD, April 5–7, 1998. University of Maryland Cooperative Extension Service.

Campbell R.D. 1991. High value windbreaks for cash crop lands, pp. 177–178. In: Finch S. and Baldwin C.S. (eds). Third International Windbreak and Agroforestry Symposium. Ridgetown College, Ontario, Canada.

Clay J. 1992. Some general principles and strategies for developing markets in North America and Europe for nontimber forest products, pp. 302–309. In: Plotkin M. and Famolare L. (eds). Sustainable Harvest and Marketing of Rain Forest Products. Island Press. Washington, D.C.

Crawford C.M. and Di Benedetto C.A. 2000. New Products Management. 6th ed. McGraw Hill, Boston. 534 pp.

DeCourley C. 1992. Opportunity marketing: Another way of looking at marketing. Small Farm Today, August 1992: 14–15.

Freed J. 1999. Developing Special Forest Products Markets for Non-Industrial Private Forest Land owners, pp. 34–39. In: Josiah S. (ed.) North American Conference on Enterprise Development Through Agroforestry: Farming the Agroforest for Specialty Products. Minneapolis, MN. Oct 4–7, 1998.

Frigstad D.B. 1995. Know Your Markets. Oasis Press. Grants Pass., 177 pp.

Garrett H.E. and Harper L.S. 1999. The science and practice of black walnut agroforestry in Missouri, USA; A temperate zone assessment, pp. 97–110. In: Buck L.E., Lassoie J.P. and Fernandes E.C.M. (eds). Agroforestry in Sustainable Agricultural Systems. CRC Press and Lewis Publishers, New York.

Garrett H.E., Buck L.E., Gold M.A., Hardesty L.H., Kurtz W.B., Lassoie J.P., Pearson H.A. and Slusher J.P. 1994. Agroforestry: An integrated land-use management system for production and farmland conservation. Resource Conservation Act Appraisal Document. Washington DC: USDA – SCS, 58 pp.

Garrett H.E., Rietveld W.J and Fisher R.F. (eds.) 2000. North American Agroforestry: An Integrated Science and Practice. Agronomy Society of America, Madison, WI, 402 pp.

Godsey L.D., Gold M.A. and Cernusca M.M. 2004. Eastern red cedar: A competitive market analysis, pp. 128–141. In: Sharrow S. (ed.) Eighth Biennial Conference on Agroforestry in North America, June 22–25, 2003, Corvallis, OR, USA.

Gold M.A. and Godsey L.D. 2001. A framework to analyze markets for traditional and nontraditional Midwestern agroforestry products: The case of eastern red cedar, pp. 179–184. In: Schroeder W. and Kort J. (eds). Seventh Biennial Conference on Agroforestry in North America and Sixth Annual Conference of the Plains and Prairie Forestry Association, Regina, Saskatchewan, Canada. August 13–15, 2001.

Grudens-Schuck N. and Green J. 1991. Farming alternatives: A guide to evaluating the feasibility of new farm-based enterprises. Northeast Regional Agricultural Engineering Service, Cornell University, Ithaca, NY. 88 pp.

Hammett A.L. and Chamberlain J.L. 1998. Sustainable use of non-timber forest products: alternative forest-based income opportunities, pp. 141–147. In: Kays J.S., Goff G.R., Smallidge P.J., Grafton W.N. and Parkhurst J.A. (eds). Natural Resources Income Opportunities on Private Lands Conference, Hagerstown, MD, April 5–7. University of Maryland Cooperative Extension Service, College Park, MD.

Hansen E., Seppälä J. and Juslin H. 2002. Marketing strategies of softwood sawmills in western North America. Forest Products Journal 52(10): 19–25.

Hansen E. and Punches J. 1999. Developing markets for certified forest products: A case study of Collins Pine Company. Forest Products Journal 49(1): 30–35.

Hartman H. 2003. Reflections on a Cultural Brand: Connecting with Lifestyles. The Hartman Group, 93 pp.

Hester E.L. 1996. Successful Marketing Research. Wiley, New York. 240 pp.

Hull J.B. 1999. Can coffee drinkers save the rain forest? Atlantic Monthly 284(2): 19–21.

Jensen K. and Pompelli G. 2000. Marketing and business assistance needs perceived by Tennessee forest products firms. Forest Products Journal 50(7/8): 48–54.

Josiah S.J., Brott H. and Brandle J. 2004a. Producing woody floral products in an alleycropping system in Nebraska. J Hort Technology (in press).

Josiah S.J., St. Pierre R., Brott H. and Brandle J. 2004b. Productive conservation: Diversifying farm enterprises by producing specialty woody products in agroforestry systems. J Sustainable Agr (in press).

Kays J.S. 1999. Improving the success of natural-resource based enterprises, pp. 171–174. In: Josiah S. (ed.). North American Conference on Enterprise Development Through Agroforestry: Farming the Agroforest for Specialty Products, Minneapolis, MN. Oct 4–7, 1998.

Kohls R.L. and Uhl J.N. 1998. Marketing agricultural products. 8th ed. Prentice Hall Publishing, Upper Saddle River, NJ. 560 pp.

Koppell C. 1995. Marketing information systems for non-timber forest products, K.S. Schoonmaker Freudenberger, ed. FAO Community Forestry Field Manual No. 6, 115 pp.

Kurtz W.B., Garrett H.E and Kincaid W.H. 1984. Investment alternatives for black walnut management. J Forestry 82: 604–608.

Kurtz W.B., Thurman S.E., Monson J.J. and Garrett H.E. 1991. The use of agroforestry to control erosion – financial aspects. Forest Chron 67: 254–257.

Kurtz W.B., Garrett H.E., Slusher J.P. and Osburn D.B. 1996. Economics of Agroforestry. MU Guidesheet # G5021. University of Missouri, Columbia, MO. 2 pp.

Mangun J.C. and Phelps J.E. 2000. Profiling existing markets: The Illinois secondary solid wood products industry. Forest Products Journal 50(5): 55–60.

Mater J., Mater M.S. and Mater C.M. 1992. Marketing Forest Products. Miller Freeman, Inc. San Francisco, 290 pp.

Oster S.M. 1994. Modern Competitive Analysis, 2nd ed. Oxford University Press, New York. 411 pp.

Parker D., Billings L.L., Thiessen M. and Colero L. 2002. Value chain handbook: New strategies to create more rewarding positions in the marketplace. Alberta Agriculture, Food and Rural Environment, Alberta, Canada, 38 pp.

Porter M.E. 1980. Competitive Strategy: Techniques for Analyzing Industries and Competitors. The Free Press, New York. 396 pp.

Salatin J. 1998. You Can Farm: The entrepreneur's guide to start and succeed in a farm enterprise. Polyface, Swoope, VA, 480 pp.

Sinclair S.A. 1992. Forest Products Marketing. McGraw-Hill, Inc. New York, 403 pp.

Smith J.R. 1950. Tree Crops: A Permanent Agriculture. Harper and Row, New York. 408 pp.

Sullivan P. and Greer L. 2000. Evaluating a Rural Enterprise – Marketing and Business Guide. ATTRA. Fayetteville, AR, 14 pp.

Teel W.S. and Buck L.F. 1999. From wildcrafting to intentional cultivation: The potential for producing specialty forest products in agroforestry systems in temperate North America, pp. 7–24.

In: Josiah S. (ed.). North American Conference on Enterprise Development Through Agroforestry: Farming the Agroforest for Specialty Products, Minneapolis, MN. Oct 4–7, 1998.

Thomas M.G. and Schumann D.R. 1993. Income Opportunities in Special Forest Products: Self-help suggestions for rural entrepreneurs. USDA Forest Service Agriculture Information Bulletin 666. Washington, D.C., 206 pp.

Vlosky R.P., Chance N.P., Monroe P.A., Hughes D.W. and Blalock L.B. 1998. An integrated market-based methodology for value-added solid wood products sector economic development. Forest Products Journal 48(11/12): 29–35.

Vollmers C. and Vollmers S. 1999. Designing Marketing Plans for Specialty Forest Products, pp. 175–182. In: Josiah S. (ed.). North American Conference on Enterprise Development Through Agroforestry: Farming the Agroforest for Specialty Products, Minneapolis, MN. Oct. 4–7, 1998.

Yamada M. and Gholz H.L. 2002. An evaluation of agroforestry systems as a rural development option for the Brazilian Amazon. Agroforest Syst 55: 81–87.

Agroforestry Systems **61:** 383–397, 2004.

Peasants, agroforesters, and anthropologists: A 20-year venture in income-generating trees and hedgerows in Haiti

G.F. Murray[1,*] and M.E. Bannister[2]
[1]*Department of Anthropology, University of Florida, Gainesville, Florida 32611;* [2]*School of Forest Resources and Conservation, University of Florida, Gainesville, Florida 32611;* [*]*Author for correspondence: e-mail: murray@anthro.ufl.edu*

Key words: Anthropology, Domestication, Land tenure, Tree tenure, Project design

Abstract

This chapter examines the evolving trajectory and emerging lessons from twenty years of agroforestry project activities in Haiti that made it possible for more than 300 000 Haitian peasant households – over a third of the entire rural population of Haiti – to plant wood trees as a domesticated, income-generating crop on their holdings. Unusual popular enthusiasm for the project derived from several anthropological and technical design factors: the adaptation of the project to pre-existing Haitian land tenure, tree tenure, and market systems; the elevation of micro-economic over macro-ecological themes; the decision to bypass the Haitian government and operate the project through local NGOs (non-government organizations); the use of a joint-venture mode in which smallholders supplied land and labor and the project supplied capital in the form of seedlings; the use of professionally managed small-container seedling technology rather than backyard nurseries; and a project management policy that *encouraged farmer-induced deviations* from project assumptions in matters of tree deployment and harvesting schedules. Issues of secure tree tenure were central to farmer planting decisions. The article discusses how secure tree tenure was possible under the heterogeneous informal arrangements that characterize Haitian peasant land tenure. The approach generated the birth of several creative Haitian peasant agroforestry configurations described in the chapter. In discussing lessons learned, the authors argue that long-term environmental payoffs should be viewed, not as the principal project goal, but as secondary side effects of smallholder tree planting decisions made for short-term micro-economic reasons.

Introduction

This chapter examines the evolving trajectory and emerging insights of 20 years of program efforts to promote tree planting and sustainable hillside farming practices in rural Haiti. The Agroforestry Outreach Project, which was launched in 1981 and continued in modified form and under different names until 2000, made it possible for over 300 000 Haitian smallholders voluntarily to plant several hundred fast-growing wood trees and/or to install hedgerows on their land.

One of the authors (Murray, an anthropologist who had lived and worked in rural Haiti) was heavily involved in the initial theoretical conceptualization and programmatic design of the project and was the project 'chief of party' for the first 18 months of implementation. The other author (Bannister, an agroforester) first came to Haiti as a regional coordinator for the initial project and stayed with the implementing agency, the Pan American Development Foundation (PADF), for most of the two decades in which the project continued. An unusually heavy level of interdisciplinary collaboration between agroforesters and anthropologists was a key feature of the project in its initial years.

The protagonists in the tree saga, however, are neither anthropologists nor agroforesters but the smallholders of Haiti, inhabitants of a nation with troubling economic, political, epidemiological, educational, and ecological indicators. The doom-and-gloom tone that dominates international discussions of Haiti began long ago. Indeed the entire Caribbean island of Hispaniola has had a tragic demographic history, beginning with the death of as many as one million Amerindians, in the early sixteenth century, largely from European diseases, and followed by their replacement with half a million African slaves coerced into what was the most prosperous but the harshest of New-World slave regimes. In 1804, after 13 years of revolutionary bloodshed, the French colony of Saint-Domingue was renamed Haiti, as the first independent nation of rebel ex-slaves was born, (Rogozinsky 2000).

Two centuries later more than 8 million Creole-speaking descendants of these rebel slaves are crowded into the mountainous western third of Hispaniola, tensely sharing the island with 8 million more prosperous Spanish-speaking Dominican neighbors to the east. Ecologically Haitians inherited the 'wrong side' of the island. Haiti's surface area of 27 750 sq. km. has few fertile lowland plains. Most of the country is mountainous and 75% of the country would be classified as sloping highlands (Weil et al. 1973). Limestone substrate underlies 80% of the land area; the rest is basaltic or alluvial (Ehrlich et al. 1985). The country's thin subtropical topsoils, vulnerable to start with, long ago succumbed to erosion under land-use systems based not on the practices of unknown African ancestors but on the extractive technologies of a market-oriented colonial plantation system.

Tree cover has virtually disappeared. An application of Holdridge's (1947) classification showed a country whose 'life zones' consisted of subtropical moist forest and subtropical dry forest, subtropical wet and rain forest zones being common in the middle and upper elevations. Half a century later, however, the few forests that remained in the 1940s have now virtually disappeared. The few autochthonous tree-planting traditions that emerged in post-revolutionary Haiti tended to focus on fruit trees. Wood trees, in contrast, were viewed as natural goods supplied by nature – or rather by *Bon-Dye*, the Creole version of the French word for God – for human extraction. Though some individual wood trees were considered sacred and left standing, protective folk religious traditions were ecologically impotent in the face of a growing rural population that needed to clear land for farming and a growing urban population that needed charcoal as cooking fuel. In short, its troubling economic and social statistics, its political chaos, and its denuded hillsides make Haiti an unlikely setting for a happy tree story.

The earliest reforestation attempts were emphatically not happy stories. They began under foreign prodding with the arrival of development agencies after World War II. These early projects were largely based on the theme of *reboisement,* protectionist and conservationist reforestation premises inappropriate to a virtually treeless but densely inhabited country. Furthermore international donor funds were routinely entrusted to the fiscal management of predatory and mistrusted state bureaucracies whose authoritarian commands to plant and protect trees were routinely ignored by villagers and whose foreign-funded seedlings therefore died in nurseries for want of interested planters.

Some three decades of tree planting failures led frustrated expatriate donors, by the late 1970s, to be open to new paradigms. It was at this period that the United States Agency for International Development (USAID) contracted anthropologists to propose new models. The result was the Agroforestry Outreach Project, and its successor descendants, to be described here. The approach was based on several factors, including (1) the adaptation of the project to pre-existing Haitian land tenure, tree tenure, and market systems, (2) the elevation of micro-economic over macro-ecological themes, (3) the decision to bypass the Haitian government and operate the project through local NGOs (non-government organizations), (4) the use of a joint-venture mode in which smallholders supplied land and labor and the project supplied capital in the form of seedlings, (5) the use of professionally managed small-container seedling technology rather than backyard nurseries, and (6) a project management policy that *encouraged farmer-induced deviations* from project assumptions in matters of tree deployment and harvesting schedules.

Information sources

To go beyond personal anecdotes, we will base our description of the approach and our analysis of its results on a now voluminous literature from Haiti, which includes over a dozen empirical studies of Haitian smallholders who planted wood-trees on their holdings in the course of the project. The research scrutiny given to one project has been quite unusual. Pre-project feasibility investigations include Murray (1979[1]; 1981[2]) and Smucker (1981[3]). Two years after the project started Murray published the first description of the project (Murray 1984), followed by an analysis focusing on anthropological issues (Murray 1987). A project agroforester returned to villages where he had delivered trees three years earlier to examine their fate (Buffum 1985[4]; Buffum and King 1985[5]). One anthropologist wrote his doctoral dissertation on the project (Balzano 1989). Another anthropologist examined decision-making processes in a community of early tree planters (Conway 1986a[6]) and synthesized the results of five additional studies of tree-planting communities done under the auspices of either PADF or the University of Maine, a project research partner (Conway 1986b[7]; Lauwerysen 1985[8]). An economist calculated monetary returns to tree planters and documented higher-than-predicted internal rates of return (Grosenick 1986[9]). Another economist examined the charcoal and pole markets (McGowan 1986[10]), and Smucker (1988[11]) analyzed six years of tree planting in several communities. In the early 1990s, Bannister and Nair (1990) discussed the soon-to-be expanded hedgerow component of the project, and Bannister and Josiah (1993) examined extension and training issues. An anthropologist/forester team (Smucker and Timyan 1995[12]) did case studies that included harvest information. The following year Timyan (1996) published a volume on the trees of Haiti. Land-tenure issues were analyzed by Smucker et al. (2002). Most recently Bannister and Nair (2003) analyzed data from 1540 households and 2295 plots that had received project interventions. There is, in short, a substantial body of empirical information provided by anthropologists, agroforesters, and economists from which we will draw.

In terms of secondary literature, the project has received more attention in the professional circles of anthropology than those of forestry or agroforestry. Two years after its onset, it won the international Anthropological Praxis Award, a competitive annual prize for applied anthropology. The project is now one of the most frequently cited cases of applied anthropology in recent college cultural anthropology textbooks (e.g., Robbins 1993, Nanda and Warms 1998, Peoples and Bailey 1997, Ferraro 1998, Harris and Johnson 2000). A description of the project has been reprinted in four successive editions of a widely circulated reader in applied anthropology (Podolefsky and Brown 1989). We have come across only one anthropological critique of the project (Escobar 1991), whose author had not been to Haiti but who excoriated this and several other anthropological projects for the sin of 'commodification', i.e. opening the 'natural systems' of peasants to the 'penetration of capital' and exposing peasants to the perilous world of markets and money. (This romantic but poorly conceived desire to protect smallholders from access to money would meet with little sympathy among intended peasant protégés anywhere in the world).

The project has received much less attention, however, in what may be its most proper professional habitat, the world of agroforestry. In this article we will therefore focus on issues germane to agroforesters.

Projects as evolving systems

We will proceed systemically rather than anecdotally. Agroforestry configurations are best viewed, not as a collage of discrete practices, but as dynamic, evolving, integrated systems. By the same token, *externally funded projects themselves are best viewed as evolving, problem-solving systems.* We propose that, whatever culture-specific idiosyncratic arrangements may be instituted in different countries, the typical agroforestry project nonetheless has four universal or quasi-universal underlying problem sets to solve: technical planning, benefit-flow planning, fund management, and village outreach strategies. We will examine the Haiti project through the lens of these four broad systemic components.

Technical base

For market-related reasons discussed in Murray (1987) the planting of wood was a better income-generating venture in Haiti than the planting of fruit. The project therefore focused on the distribution of wood tree seedlings. Because of the limited inventory of existing small-container tree nurseries in Port-au-Prince, the project began with six exotic fast-growing hardwood species: *Acacia auriculiformis*, *Azadirachta indica*, *Casuarina equisetifolia*, *Eucalyptus camaldulensis*, *Leucaena leucocephala*, and *Senna siamea*. By the end of the project twenty years later farmers were planting 74 species, many of them local trees. As we shall see below, the farmers supplied land and labor. The project supplied several hundred seedlings free of charge to each participant.

Both for quality control and economies of scale, the project opted for professionally run nurseries rather than backyard nurseries or direct-seeding techniques. At the apogee of the tree planting component in the late 1980s the project was being supplied by 36 nurseries producing nearly 10 million seedlings a year all over Haiti. The nurseries were not owned by the project. NGOs (non-governmental organizations) operated the nurseries as profit-generating microenterprises. Villagers who had been contracted as tree extensionists would elicit orders for quantities and species of wood seedlings among their kin and neighbors. The nursery would produce and sell the seedlings to PADF at an agreed on price. PADF would then organize the distribution of seedlings. To facilitate transportation the nurseries used small containers rather than polyurethane bags. Backyard nurseries with polyurethane bags were experimentally introduced in the late 1980s, and increased in the 1990s.

To enhance the soil conservation element, the second author began experimenting with *Leucaena leucocephala* hedgerows in 1984. Extension of tree-based soil conservation structures was officially adopted by the project in 1987, but after 1991 it became the dominant project element. Because of marketing weaknesses, fruit trees had been a minor component of the project in its earliest years. This component, however, increased dramatically in the project's second decade, including top-grafting of adult fruit trees. A new component, improved food crop germplasm, was introduced in the 1990s. The logic of this new agricultural component was that plots protected by some form of soil conservation would eventually become more fertile, and therefore would merit a larger investment in crop production.

Benefit flow arrangement

Agroforestry projects cannot limit attention to technical matters. The resolution of a second problem set, the guaranteeing of satisfactory *benefit flows* is equally essential to project acceptance. We assured benefit flows by sharing costs and risks with farmers and by guaranteeing their total control of harvesting and marketing. We established a joint venture arrangement in which participants supplied two of the factors of production – land and labor – and the project supplied capital in the form of seedlings. Participants would risk covering part of their land with a new crop – the wood tree – and plant enough to make a measurable economic difference down the road. As for labor, we broke militantly with the 'Food-for-Work' subsidy that had been used in most previous tree-planting projects. Participants had to supply all of the ground preparation and planting labor, either by doing it themselves or by using one of the many labor-mobilizing arrangements (e.g., exchange labor with neighbors) found in traditional Haiti.

As for capital, the project supplied it in the form of several hundred free-of-cost wood seedlings to each participant. The original minimum of 500 seedlings was quickly dropped to 200 seedlings and eventually even less. The project was having difficulty supplying the unexpected demand which it provoked, and labor constraints (more so than land constraints) made it difficult for many farmers to plant 500 seedlings at one fell swoop, since the planting of seedlings had to be timed with rains, which was the trigger for planting other crops as well. Each seedling cost about 10 U.S. cents to produce. The project thus made a modest average contribution of about $20 in seedlings to each participant, who in turn allocated land and labor. We recommended different planting strategies, but planters made the final seedling-deployment decisions. Of major importance, we guaranteed tree tenure and full harvest rights. We repeated regularly that planters would be sole owners of seedlings and trees. Villagers needed no permission to harvest trees. In the second decade, other benefit flow arrangements were instituted for hedgerow and improved food-seed components.

Fund management arrangements

Besides a solid technical base and well-designed benefit flows, projects have a third challenge: fund management. Had funds been entrusted to the Haitian government, we would be doing a post-mortem on why the project failed. The money was managed, instead, by PADF, the implementing NGO, utilizing an ad-hoc 'umbrella strategy' of grant management that was devised specifically for the project. USAID made a macro-grant of $4 million – i.e. $1 million per year – to PADF. PADF then entered into agreements with local NGOs all over Haiti to hire and train village tree agents who in turn would invite their kin and neighbors to plant trees on their own land. PADF's support to these groups was in the form of contracts that included in-kind seedlings and small amounts of cash for overhead, salaries of promoters, and other minor expenses. This umbrella strategy permitted one central macro-grant to feed hundreds of localized minicontracts. It was a welcome buffering mechanism appreciated by all parties. USAID staff had to manage only one grant. Local NGOs were shielded from stringent USAID accounting requirements.

Outreach structures

A fourth universal agroforestry problem-set is outreach. Local NGOs selected and hired villagers to serve as tree extensionists. These villagers, trained by the project and compensated on a part-time basis by tasks performed, would explain the project to their kin and neighbors and invite them to allocate part of their land and labor to wood tree seedlings over which, once planted on their land, they were guaranteed total ownership and harvest rights. As nursery seedlings matured and rains arrived, the seedlings, which were grown in commercial organic potting mixes that guaranteed good root development, would be removed from the containers, placed in boxes, and loaded on pickup trucks. Villagers were pre-alerted and plots were prepared. Participants gathered at designated drop-off sites and received their seedlings. Seedlings were generally in the ground 48 hours after pickup. Similar outreach structures were employed in the second decade for hedgerow and improved crop promotion. During the second ten years of the project (beginning in 1992) there was a shift from larger NGOs to smaller community-based organizations (CBOs) and the strengthening of farmer groups became an important goal.

The project measures just described are presented, not as a cookbook recipe to be applied elsewhere, but as an abstract project paradigm that may be generalizable beyond Haiti. The four problem sets addressed in the Haiti project will have to be addressed in many and perhaps most projects around the world: (1) technical decisions, (2) benefit flow arrangement, (3) fund management, and (4) outreach. Different specific solutions will apply in other world regions. But the problem-sets themselves are widespread if not universal.[13]

Qualitative results: evolving agroforestry micro-systems

We have more research information on the results of the tree component than on the later hedgerow component. The PADF component alone[14] tripled its projected seedling output in the first four years. And the 65 million seedling figure for two decades, all voluntarily planted in small lots of several hundred or fewer by over 300 000 peasants on their own holdings, leaves absolutely no doubt as to the enthusiasm generated by the project for the planting of wood trees in a country where they were formerly extracted from nature.

These figures by themselves, however, reveal little about the character of the agroforestry systems that emerged. Enthusiasm seemed homogeneous, but not specific tree-deployment decisions. The creativity of diverse local responses produced a rich heterogeneity of agroforestry configurations. They have been descriptively documented in several of the studies cited in the Introduction section. We will describe the systems as a series of questions.

Who planted project wood trees, who did not plant, and why?

Several studies (among them Conway1986a[6]; Balzano 1989; Bannister and Nair 2003) explore statistical links between tree planting as the dependent variable and personal or household variables as independent variables governing tree-planting decisions. Socioeconomic differentials and gender are two particularly interesting clusters of independent variables.

Haitian villages are not homogeneous; modest socioeconomic differentials exist. A 5-hectare holding would be 'large'. The national holding mean is closer to 1.5 ha (Zuvekas 1979). Balzano (1986[15]) documented a slight statistical tendency for tree-planters to be older, to have slightly more land under secure tenure, and to be in a better position to hire labor. Similar findings were found to hold as well more than a decade later in Bannister and Nair (2003). Though statistically significant, however, the cross tabs and correlations using household and plot characteristics as predictors explain only a small fraction of the differences between planters and non-planters, or between heavier and lighter participation. The project hope of reaching even land-poorer sectors was attained. (Total landlessness is rare in rural Haiti.)

As for gender, trees were offered to females as well as males. In the project's final year, 135 436 males and 41 121 females (23%) were listed as participants. These figures warrant explanation. The gender skewing reflects local customs of formal household headship. In households with a conjugal couple the male traditionally presents himself as household head and was listed as the project participant. Female participants are generally from households headed by a female. Even in households with a conjugal couple, however, case studies (e.g. Smucker and Timyan 1995[12]) have shown that wives of participants were heavily involved in household decisions whether to plant or not to plant and in the subsequent management of the plantings.

For what purposes did peasants plant trees?

The stated goals of tree planters in all studies were overwhelmingly economic. The dominant goal was harvest of the wood, precisely as had been predicted in the anthropological 'social soundness analysis' (Murray 1981[2]) that formed the conceptual backbone of the original USAID project document. The definition of 'economic', however, had to be broadened. Murray (1984) alluded to the themes of 'cash cropping' and the 'domestication of energy' (e.g., of charcoal) as central to the project in the first article published on the project a year and a half into its implementation. The prediction was that the project would permit a shift from the then prevalent extractive mode of charcoal procurement. Balzano (1989) and Smucker and Timyan (1995[12]) give case study evidence of farmers in some regions behaving exactly as predicted, planting *jaden chabon*, 'charcoal gardens'. They would prepare a field, plant it with annuals, intercrop wood tree seedlings, continue cropping until shade competition from the trees no longer permitted, clear cut the growing trees, convert the wood into charcoal, and begin again.

But these were the exceptions. More farmers planted with a view to more valuable saw-timber further down the road. And few harvested all of their trees at once. They used the tree rather to store value, less vulnerable to drought than annual crops (cf. Conway 1986a[6]; Smucker and Timyan 1995[12]). In some regions farmers calculate a 50% probability of losing one's annual food crop to drought. Under such conditions of agrarian peril the domestic tree stand becomes an economic safety net that is protected until absolutely needed, and even then cut only selectively. Balzano[16] revisited his dissertation community 10 years later and found large number of project trees still standing, some of them over 15 m high, being protected as vehicles of savings for times of emergency.

But farmers' harvest decisions reminded us further to broaden the term 'economic' beyond money. We have observed tree planters for the first time in Haitian history 'growing' parts of their houses, particularly posts and rafters. No cash is generated in such self-use; but money is saved. Self-use is more frequent with poles and beams than with charcoal and timber, probably because of clearer price-setting dynamics for low-valued charcoal and high-valued planks (McGowan1986[10]; Smucker and Timyan 1995[12]). But whether marketed or used for one's own house, the motivation is economic. The trees did provide ecological benefits to the fields as well (Grosenick 1986[9]). Exactly as was predicted in pre-project documents, however, ecological benefits from trees came as secondary side effects of behaviors in which smallholders engaged for economic reasons.

The primacy of economic motives came out dramatically even in the most heavily ecological component of the project, contour hedgerows. In theory, peasants were to top-prune hedgerows to about 50 cm, use the leafy material as a soil amendment in the alleyways, and the cut woody material as a soil-retention barrier uphill from the hedgerow. In actuality, peasants were more prone to using the leafy material as fodder for livestock rather than as fertilizer in the soil. In other words economic payoffs from well-fed goats were viewed as more attractive than the ecological advantages of well-fertilized land.

But farmers' economic maneuvers went even further. They invented a new type of hedgerow consisting of perennial food crops (especially plantains (*Musa* spp.), sugar cane (*Sacharum officinarum*), cassava (*Manihot esculenta*), pigeon peas (*Cajanus cajan*), and pineapple (*Ananas comosus*)) as the structural component, holding the soil, in combination with annual crops (particularly sweet potato (*Ipomoea batatas*), yam (*Diascorea* spp.), and others) grown underneath them in a wider contour band whose width could reach 2 or 3 m up and down slope (the width of the typical hedgerow is a meter). The name of this new invention was *bann manjé* – a play on words that can mean 'a band of food' or 'a bunch of food'. The bulk of the food on the *bann manje* was destined for sale in markets. What the peasants of market-oriented Haiti were doing was in effect converting what technicians had intended to be a conservation and fertility-enriching strategy – i.e. an ecological strategy into a microeconomic income-enhancing strategy (cf. Ashby et al. (1996) for similar behaviors in Colombia. Bunch (1999) also found Central American farmers independently modifying project-promoted conservation structures. Garcia et al. (2002) discuss evolving hedgerow technologies in the Philippines.)

Site management strategies: Where did they plant trees?

The question can be subdivided into holding-management and plot-management issues. On which plots, within the typical Haitian multi-plot holding, were trees planted? And where were trees deployed within the selected plots? As for the former, several research documents identify three variables that governed plot selection: land tenure, distance from the home, and the edaphic/ topographic characteristics of the plot itself.

With respect to land tenure, pre-project predictions hypothesized that farmers would plant principally on the two types of plots over which Haitian villagers have reasonably secure control: plots which they have purchased, and plots that they have inherited and subdivided among siblings. Rarely is the subdivision done with a fully legal surveyor's chain. It is almost universally done with ropes in the presence of community witnesses. The separation is informal rather than legal. But once that subdivision occurs with community witnesses, the recipient is the de-facto owner and can safely plant trees. It was conversely predicted that smallholders would be less inclined to plant trees on sharecropped land, on rented land, and on undivided inheritance land over which they had no exclusive control. These predictions were borne out strongly in the studies of several tree-planting villages done in the mid 1980s, several years after project launching, by Buffum (1985[4]), Lauwerysen (1985[8]), Balzano (1986[15]), and Conway (1986a[6]). Smucker and Timyan (1995[17]) also pointed out that the functionally important variable is not legally deeded ownership of the plot (which is extremely rare) but rather secure control over the trees, which can be acquired through informal land ownership. Purchased and subdivided inheritance plots are recognized by the community as 'owned' by farmers even in the absence of surveyed deeds. Such de-facto control is required for secure ownership of the trees planted. The ownership of the tree is, in fact, the key variable predicting willingness to plant. The statistical analysis done by Bannister and Nair (2003) confirms the same tendency for more trees to be planted on plots under more secure tenure.

Exceptional cases were found on which even sharecroppers or managers could work out tree planting arrangements with the owners. But the pre-project predictions about the importance of land tenure as a determinant of tree planting were fully borne out. The project was therefore possible in Haiti, where smallholdings may be the norm but landlessness is rare. In countries with large landless sectors special project measures will be required to avoid favoring only the well-off with privately owned trees.

A second variable, distance from the home, also played a role in plot selection. The result was the planting of project trees closer to home. Soil-quality and slope variables also played a role (Bannister and Nair 2003). Trees were rarely planted in any numbers on precisely the plots that outsiders would deem the ideal site, heavily sloped and eroded agriculturally marginal plots where crops cannot grow but where wood trees could, albeit slowly. Trees tended to be planted on agriculturally better land in conjunction with food crops.

This tendency to avoid distant and degraded plots is fully logical but economically and ecologically unfortunate. The logic is that farmers are aware, much more than planners, that trees in Haiti are vulnerable to three dangerous predators that can wreak havoc on distant plots or agriculturally marginal plots: (1) free ranging livestock, (2) nocturnal thieves, and (3) one's own kin. On plots distant from one's home, young seedlings are more vulnerable to free browsing livestock, and mature trees can be and were cut by thieves. And as for the agriculturally marginal denuded hillsides all over Haiti that would be prime candidates for hardy trees, it is precisely such land that kin groups keep in common for grazing purposes. There is no individually recognized owner of any particular plot.

If seedlings are planted on this collective land, and if the seedlings miraculously survived the free-ranging goats, a distant cousin of the planter could harvest the wood and would not have to do so under cover of darkness. The issue is tree tenure. The tree is not safely yours unless the land on which it stands has been subdivided by common informal agreement. And even when the tree is on your land, you will come under pressure from kin to give them permission to cut some of your trees. Wood was traditionally a free good and, though now planted, retains its earlier 'communal' aura. Relatives might never dream of asking you to let them harvest part of your bean crop. But they can and do ask you to let them cut some of your planted wood.

These three predatory actors – livestock, thieves, and relatives – generate an economically and ecologically formidable barrier to the expansion of tree planting onto precisely the underutilized plots where the competition from trees would be lowest and the economic increments from wood would be greatest. The problems will all be solvable at a more advanced stage of the transition to planting wood. Kin groups and communities can make arrangements to neutralize each of the three predators and to make tree planting possible on these marginal landscapes. The present project, however, merely observed the barriers, without being able to circumvent them.

The result was a strong tendency to plant project trees in conjunction with one's plots, which is precisely the arrangement envisioned in classical agroforestry. The agroforestry configurations that did emerge on agricultural land and on house sites were rich in their diversity. The preferred strategy was to convert the border of the plot into the principal locus of the trees. And no clear-cutting is generally performed on these trees. They are kept rather as a permanent source of wood, to be harvested selectively as needed. Many trees were also planted near the homestead.

In some cases plots were allowed to evolve into permanent woodlots. On agricultural land more distant from homesteads, farmers often place agrarian risks on sharecroppers. As a general rule, however, the sharecropping arrangement impedes tree-planting. Barring exceptional arrangements, the tenant will not plant trees as he will probably not be the one to harvest them. The landowner will not plant, as the sharecropper will take subtle measures to eliminate them. One arrangement observed was for the landowner to designate a block of land on one edge of the plot for a permanent woodlot whose vegetation belongs to him, not to the tenant. Since the tenant cannot plant there, there is no incentive to help the seedlings die. In one reported case (Smucker and Timyan 1995[12]) the household derived more benefits from the woodlot in the form of charcoal and pole wood than it did from the portion of the meager harvest that they received on the sharecropped part of the plot.

In a small number of cases even entire plots with at least some agricultural potential have been allowed to evolve into permanent woodlots. But this is done for exceptional reasons – the owner may be aging and may have less energy for agrarian pursuits, or may migrate temporarily to the Dominican Republic and can get more from the land by turning it over to low-maintenance trees. This is not done with enough frequency to cause a cut in local food supplies.

How many seedlings survived the first year?

The entire undertaking, of course, is an exercise in futility if the seedlings die. Survival- monitoring procedures were instituted by PADF. An increasing 12-month tree survival rate was achieved as the project progressed in its first decade from an early project average hovering near 30% to a later average closer to 50%. Data are not available for seedling survival during the more intensive 1990s phase. Because of closer interactions between project and participants, the survival rates were probably higher. A reasonable generalization, at least for Haiti, would be that in a project of this type, in which seedlings are transplanted onto farmer-controlled rather than project-controlled plots, approximately two seedlings will be required for every mature tree.

When were the trees harvested?

The timing of tree harvest is determined less by agronomic perceptions on the part of the peasant, e.g., the mean annual growth increment, than by special crises or special occasions in which expenditures are required: funerals, illnesses, weddings, school tuition. In this sense peasant treatment of the tree is quite different from what is done in developed industrial plantations. This cutting-in-crisis is not erratic short-sighted behavior on the part of the peasant. It is part of a long-term strategy in which the wood tree becomes a partial surrogate for the savings that used to be achieved principally through livestock raising.

A negative theme that recurs in the case studies is the obligation that farmers felt to accelerate wood harvest. The initial objective was often the harvest of timber, the highest value of wood trees. In some communities studied (Conway1986a[6]) few if any participants reported planting with a view to the charcoal market. But economic pressures often lead to faster cutting of the trees for charcoal.

The harvest schedule of wood trees, because it is so discretionary, is much more vulnerable to political events than is true of other crops. Rice (*Oryza sativa*) and beans (*Phaseolus* spp.) have to be harvested on schedule, whatever happens in national and international politics. Not so trees. But the political chaos following the ousting of a dictator in 1986 increased lawlessness and thievery. A U.S. embargo of the 1990s, instituted after an elected president was ousted in a coup, created hardships that led to premature harvest of the trees. Political events thus accelerated tree harvesting, as tree-owners switched from timber to charcoal goals either to meet urgent cash needs or to protect against the clandestine cutting of trees by thieves.

What happened to the plots after tree harvest?

There are no case studies that document a post-harvest 'good riddance' attitude on the part of farmers. What the case studies, particularly those of Smucker and Timyan (1995[12]), show is the emergence, as a result of project participation, of a transformed land orientation in which the wood tree is now seen to be a *danre*, an income-generating crop that can be planted like other crops. The post-harvest replacement strategies entail careful managing of coppice and transplanting of wildlings. If seedlings are made available farmers will replant and even expand tree-planting into new areas of their holdings. Internal attitudes and external behaviors toward the wood tree have been profoundly modified, not through educational messages and not through ecological homilies, but through planting and harvesting.

Quantitative indicators

Tree seedlings

If the project had reached only 10 communities and 300 families, the preceding descriptions of emergent agroforestry orientations would still make it humanly and scientifically interesting. But it reached hundreds of thousands of families. During the first 10 years (1982–1991), when large-scale tree distribution was the main focus, about 48 million wood tree seedlings were distributed to farmers by PADF alone. In the project's second decade, from 1992 to 2000, the PADF component of the project delivered 14 million seedlings. An additional three million seedlings were produced during this second decade by farmer-operated, small, 'plastic bag' nurseries.

In spite of the dwindling in average annual seedling output during the second decade (for extraneous reasons to be discussed later), a total of 65 million seedlings, over 95% of them wood tree seedlings of 74 different autochthonous and exotic species, were voluntarily integrated by Haitian peasants onto their own holdings during the project's 20 years through the PADF project. CARE (Cooperative for Assistance and Relief Everywhere, Inc.) was also active in tree-planting activities in Haiti's arid northwestern region during this period under the overall USAID project. Although we lack precise data on such CARE-sponsored tree-planting, a conservative estimate of the total project-facilitated out-planting is over 100 million trees on Haitian peasant land during the 20-year period.

Hedgerows, gully-plugs, and other soil conservation interventions

During the first decade of the project, soil conservation interventions were minimal. Six hundred fifty (650) km of hedgerows were installed, and 2200 gully plugs were built. In the second decade, with its shift into a greater emphasis on conservation practices, farmers installed over 12 000 km of hedgerows, 3000 km of rock walls, and 94 000 gully plugs. In addition, the 'agro' component of agroforestry was increased, and improved food-crop germplasm was distributed to cover over 11 000 ha of land, and 30 000 vegetable gardens were planted.

Numbers of participating smallholders

Such aggregate tree and hedgerow statistics by themselves meant nothing to project implementers. If the 65 million trees planted had all ended up on State land or on the land of 500 wealthy landowners, the project would have been seen as a failure. As Nair (1993) points out, agroforestry is not merely about biomass. It attempts to benefit the rural poor as well. For us, the key statistic in the Haiti project is not number of seedlings distributed but number of households participating.

During the first 10 years, two factors confounded the counting of real participants. About a third of the participants in any season may have been 'repeat-planters', not new ones. But there was even stronger skewing in the opposite direction in the form of 'non-registered participants'. It is known that many farmers receiving 200 or 300 seedlings would distribute a substantial but impossible-to-quantify number to relatives or friends, who thus became de-facto – but uncounted – project beneficiaries. A conservative estimate of numbers of distinct households planting the 48 million seedlings distributed during the first decade would be 190 000 households, or about 250 seedlings per household. During the final eight years, 1992–2000, when the project concentrated in smaller geographic areas, the data are more precise. A total of 176 557 farmers are known to have participated through 83 local organizations that employed about 1000 extensionists. To accommodate the possibility of up to 15% of non-registered planters, a recent article by Bannister and Nair (2003) raise the figure to 200 000 participants for the second decade. That means a total of 390 000 households for the two periods. Conservatively, we can state with confidence that during its 20-year life the project involved a minimum of 350 000 Haitian farm families.

When placed in the context of Haitian demography, these figures startle even the most enthusiastic proponent of the approach used. The population average for the 20-year period can be set at 6 million, about 70% (or 4.2 million) of whom were living in rural areas. Survey data carried out by the second author yielded an average of 5.7 persons per rural household. If extrapolated to all rural areas, there are about 737 000 rural households in Haiti. With the project reaching a minimum of 350 000 households as stated above, more than 40% and perhaps nearly 50% of the households of rural Haiti may have received seedlings or otherwise participated in the project at one point or another during the two decades. Even if these national participation figures are dropped by 10 or even 20 percentage points for 'safety's sake', the level of nationwide involvement in and enthusiasm for a tree planting project must still be seen as unprecedented in the annals of agroforestry.

We cite these figures, not to tout the 'success' of the project. Success must be evaluated on criteria that go beyond crude number crunching. We consider the qualitative descriptions of emergent agroforestry systems presented in the preceding section to be better indications of the effectiveness of the approach. The national statistics reveal less about the project than about the peasants, or rather about the potential for the emergence, certainly in Haiti and possibly elsewhere, of an evolutionary restructuring of the relationship between the smallholder and the wood-tree. If the poorest farmers, with the smallest holdings, of the Western Hemisphere, can be moved by anthropologically and technically creative project design – and *by an abundant supply of high quality seedlings* – to incorporate wood trees into their agrarian inventory, the potential for smallholder tree planting must be even greater in economically less-stressed settings. Stated differently, there was a latent readiness in Haiti for a shift into massive wood-tree planting, a type of subterranean 'potential energy' waiting to be released. A well designed project that deals with issues of tree tenure, harvest rights, and (above all) *seedling supply* can act as a catalyst to convert this potential energy into kinetic energy – to convert interest in trees into the planting of trees.

Generalizable lessons?

We have learned lessons in Haiti. Their generalizability to other settings, however, is a matter for professional debate. Let us simply conclude by briefly ventilating several controversial issues, each of which by itself warrants an article. Four of them, we believe, have been settled, at least for Haiti. Two are still contested.

Economic vs. ecological goals: The Neolithic analogy

The project entailed a paradigm shift for both planner and peasant in their view of the wood tree. Earlier project planners, with themes of reforestation and conservation, viewed the wood tree as a natural resource to be protected. Haitian villagers also treated it as a natural resource, but one to be exploited, not protected. The project moved militantly away from any such 'natural resource' construal of the tree at all. We presented the wood tree instead as a slow-growing crop that could be planted, harvested, and sold or used as any other crop. Villagers made the shift in their own way, usually treating their trees not as a crop to clear-cut in one fell swoop, but as a store of value to slowly harvest when needed. But the shift into a domesticated mode of wood production was made by a substantial percentage of the population of rural Haiti.

As Murray (1987) pointed out, there were ancient anthropological precedents for this shift into domestication. Archeologists and cultural anthropologists have studied that ancient food crisis in the Fertile Crescent that led some 12 millennia ago into the shift away from dependence on the gathering of wild vegetation and the hunting of wild animals into the domestication of crops and livestock, a process referred to as the Neolithic transition. The shift was provoked by food shortages, by 'Paleolithic overkill' of wild animals. This food crisis was *not* solved, however, by 'natural resource management', by better stewardship of nature's resources, or by a shift to 'sustained yield' hunting and gathering. It was solved instead by *domestication*, by a shift from an extractive mode to a productive mode of resource procurement. Gathering of wild vegetation yielded to crop cultivation, hunting to livestock. (Murray 1987; cf. Simons and Leakey 2004).

Parallel problems evoke parallel solutions. The shift into a domesticated mode of wood production is a replay in the domain of wood of the process that led to the domestication of food. The evolutionary readiness of Haitian villagers to this alternative emerged because two of three conditions were already present in local economic repertoires. (1) The cash-cropping farmers of Haiti had been planting for markets for nearly two centuries. (2) Increasing wood scarcity and burgeoning construction and charcoal markets endowed wood with more commercial value than most annuals. The missing connection was the *planting* of the wood to be cut and sold. Once a seedling supply had been established, however, and tree tenure had been guaranteed, the shift from extracting wood for sale to planting and harvesting it for sale was a logical, gentle step, and an attractive alternative to explore (Murray 1987).

Rational farmers dealing with real problems easily make the shift into a new paradigm. Whether planners and intellectuals can make the shift – a focus on messages of domestication and production – or whether they will cling to archaic protect-Mother-Nature paradigms, is another question. If the Neolithic analogy holds, the wood crisis on planet earth is more likely to be solved by the production of planted wood, as Haitian villagers have begun doing, than by the protection of nature's wood.

Let us not push the analogy too far; Amazonian and Orinoco rain forests can and should be protected. But let us conversely desist from infecting tree programs in settings like Haiti with inappropriate protect-Mother-Nature conservationist themes. The woodlot planted by a farmer is no longer a 'natural resource'. It is a crop. The natural tree stands of Amazonia should be hugged and protected. The domesticated tree stands of rural Haiti should be hugged and then harvested.

Professional nurseries vs. backyard nurseries

When the anti-subsidy policy in USAID led to the closing of NGO nurseries in 1991, the tree component of the project had to shift to backyard nurseries operated by individual farmers or small groups. The subsidies were just as great; polyurethane sack containers, seeds, and watering devices were supplied by the project. But they were masked under a camouflage of grassroots thatched-shed 'peasant initiative' that the anti-subsidy vigilantes lacked either the talent or inclination to penetrate. These peasant-managed backyard nurseries, however, should not be romanticized as more sustainable or developmentally superior. Their volume of output was lethargic compared to that of professionally run small-container nurseries. And the 80% subsidy that nurtured them is no more sustainable than the 100% dependence by NGO nurseries on project purchases.

Several arguments can be made in favor of the professionally run nursery. Sixty five million seedlings distributed to Haitian farmers were incorporated as crops into their farming systems. A responsibility comes with this level of output – to ensure a sustainable source of the best quality tree seeds possible. The NGO nurseries established with project assistance purchased most of their tree seeds from PADF, who imported some of them from overseas suppliers and collected the others locally using local entrepreneurs who were given some training in what characteristics to look for. The nurseries also collected seeds themselves from their regions.

The difficulty with seed supply in Haiti is that trees with the most desirable characteristics were harvested long ago, leaving the inferior individuals to supply seeds, with the resulting drop in production. Even more troubling is that seed of some of the popular exotic species, such as *Senna siamea*, *Azadirachta indica*, and *Swietenia macrophylla*, were collected from a very small number of parent trees planted near government buildings. This makes for a very narrow genetic base for a large population of trees. Since resistance to pests and diseases is a highly heritable trait in trees, this genetically narrow population could be at risk. In any case, a broader genetic base of tree germplasm is necessary to support future selection for increased production, ability to grow in marginal sites, and improved growth rate (Zobel and Talbert 1984). A tree seed selection program was begun by the project in the late 1980s with large private landowners to establish selection trials that would eventually be converted to seed orchards (J. Timyan, pers. comm. September 2003). In short, professionally managed seed and seedling operations have clear advantages over the backyard user-managed nursery in these matters. And in terms of seedling quality, volume of production, and the ability to excite farmer enthusiasm, the professionally run NGO nursery was a better seedling-supply option than the backyard nursery in Haiti.

Public vs. private fund management

A third matter has been 'settled', at least for Haiti. In the absence of functioning government institutions, the NGO route was selected for project administration. Cordial working relationships were eventually established with local government agronomists. But at no time did the Haitian government control any project funds.

The NGO mode of project implementation does not sit well in all international development circles. One concern is the issue of local State sovereignty in development matters. Another is the fly-by-night, predatory behavior of at least some NGOs. The respect-sovereignty anthem, however, rang flat in

Duvalierist Haiti, whose bureaucrats had for decades exercised their sovereign right to plunder donor funds intended for their people. Even the most ardent government-to-government 'institution builders' had to back down and invite NGOs.

We agree, however, with the legitimacy of the second concern against premature beatification of the NGO-in-shining-armor. One of the authors (Murray), in an unpublished report for USAID/El Salvador, proposed a dichotomy between two types of NGOs: the ONGO, or 'operational NGO', with a bona-fide service track record, and the FONGO, the 'foraging NGO', that worldwide genus of parasitical profit-oriented 'non-profit' group dedicated first and foremost to the foraging and capture of international donor funds. In the early 1980s, there were dozens of bona-fide service-providing NGOs with whom we could work. In view of current (2003) international sanctions against the Haitian government, the NGO route continues to be the preferred project mode.

Tree programs in other political and cultural settings may eschew the NGO option. Few governments could have worked as efficiently as the NGOs of Haiti. But we will avoid doctrinaire generalizations. We simply pose a question for further professional debate. Under what conditions should agroforestry project funds be entrusted to a local government? And under what conditions should planners use every measure that is legally and politically feasible, as was done in Haiti, to protect a project against government interference?

Pedagogues vs. partners: Assigning education its proper role

Some would be puzzled that our project paradigm has no separate education component; education is instead a minor component under 'outreach'. The demotion is intentional. In the early years the project was almost 'anti-education'. In its emphasis on seedling supply, tree tenure, and harvest rights, the approach was a militant philosophical rejoinder to questionable pedagogical theories that viewed peasant knowledge deficits as the cause of Haiti's ecological problem and environmental education as the major solution. The educational component of the project was upgraded in later years with Creole-language manuals and training sessions in administration for CBOs. But we continue to surround education with two caveats. First, educational message-flows must be linked to material flows. They are analogous to user manuals that come with computers or printers. A manual is useless without access to the hardware. Equally useless are environmental education projects whose budgets finance the educational manuals but expect smallholders then to go forth and obtain seedlings on their own. Secondly, educational flows should be bi-directional, moving as frequently from farmer to project staff as vice-versa. While technical information regarding nursery and agronomic practices was valuable and necessary, the knowledge that staff in Haiti carried in their brains was, after 20 years of field immersion, more heavily influenced by farmer inputs than by lessons learned long ago in school. Stated differently, we abandoned the podium and the pulpit. We were neither pedagogues nor ecological preachers, but partners in a long-term joint venture, encapsulated by the Creole slogan used by the field staff '*Plantè se kolèg*' (planters are colleagues). We suspect that in many other agroforestry settings as well, issues of seedling supply, tree tenure, and harvest rights are more critical than the mission of remedying presumed peasant knowledge deficits by itinerant environmental educators.

Santa vs. Scrooge: The issue of seedling subsidy

The preceding issues have been comfortably resolved in Haiti. The 'Santa vs. Scrooge' tensions – subsidy-and-sustainability issues – have not. Seedling supply is the most hotly debated subsidy question: to gift or not to gift. Villagers in the 32 Nigerian settlements studied by Osembo (1987) stated that they would plant trees under three conditions: tree seedlings would be free of cost, farmers could interplant trees and food crops without losing crop yield, and that it would be possible to earn income from the trees. That verdict could have come straight from the mouth of Haitians. Hwang et al. (1994), Leakey and Tomich (1999), and Herrador and Dimas (2000), take similar stands justifying subsidies and incentives. But other observers have misgivings about the impacts of subsidies (Arnold and Dewees 1999; Bunch 1999; Napier and Bridges 2002).

Anti-subsidy voices in Haiti were varied. Hard-liners wanted farmers to pay full price for seedlings. Soft-liners wanted a symbolic penny per-seedling, about 10% of production costs. (Soft-liners often alluded to character-building themes. People would be 'spoiled' if they get free handouts.). Had hard-liners prevailed, no trees would have been planted. Had soft-liners prevailed perhaps 1000 farmers per year all over Haiti would have bought 10 or 20 wood trees each, yielding a 20-year total of 20 000 households planting

400 000 trees (and that is probably a speculative overestimation). With our joint venture mode, in which farmers supplied land and labor and the project supplied a modest average $20-per-participant seedling subsidy, 350 000 households planted 65 million trees. In retrospect we now know with certitude that a 'character building' decision on our part to exact at least a penny per seedling would have been an exercise in institutional idiocy. It would have suppressed over 99% of the tree planting that occurred with absolutely no sustainability benefits. A project in which peasants pay for 10% of the seedlings is no more sustainable than one in which the project simply donates seedlings.

We can go further and quantify the damage done by anti-subsidy interventions. Some 10 years into the project a USAID director opposed to subsidies ordered PADF, under threat of de-funding, to desist from purchasing and delivering free seedlings to farmers. If they want wood trees, let them pay full price. No farmers did, of course. This led to the shutting of the 36 NGO nurseries that had emerged during the previous decade with PADF as their sole or principal purchaser, and which had been pumping out about ten million seedlings a year onto peasant land. (Such anti-subsidy sentiment is often selective and whimsical. The same director who withheld free trees from villages continued to instruct his subordinates to endow the same villages with abundant flows of free contraceptives).

The seedling flow was reactivated under his successor, who reversed his order, but under hastily constructed new nurseries the volume never recuperated. The original nurseries had lost their investment in infrastructure and no longer trusted USAID or PADF. We can calculate that, had those nurseries not been closed, they would have distributed during the decade of the 1990s as many as 80 million, rather than 14 million, seedlings. At 200 trees per participating household, this 66 million-seedling gap translates into more than 300 000 rural Haitian households who were blocked from initiating or repeating the planting of tree seedlings. An annual multi-million-seedling flow that had taken ten years to create and nurture was stopped by the dogmatic flick of an anti-subsidy pen.

Philosophical dogmatism divorced from empirical reality but linked to administrative power can indeed be destructive. One powerful empirical fact that should have been honestly confronted by an American agency before cutting off the free seedling flow to impoverished Haitians is the *tree-subsidy policy of the U.S. government toward its own farmers*. The U.S. Department of Agriculture has a national cost-share program in the 2002 Farm Bill called the Forest Land Enhancement Program (FLEP) that provides up to $10 000 per landowner per year for tree plantation and management practices (http://www.fl-dof.com/Help/FLEP.html). FLEP is not unique, and was preceded by other similar programs, such as the Forest Incentive Program (FIP) and the Conservation Reserve Program, among others. If a U.S. farmer can receive $10 000 per year in tree subsidies without lethal damage to the American economy or to the moral character of the recipient, we would appreciate arguments as to why the same is not true of a one-shot $20 seedling subsidy to a Haitian villager.

The issue of sustainability

Will farmers continue agroforestry once project support has ceased? The question must be broken down into two components. (1) Will they protect and manage coppice and volunteer seedlings? (2) Will they henceforth purchase wood seedlings commercially or begin producing them in their backyards? The answer to the first question is an empirically solid yes. We have already indicated not only that farmers have left some trees standing as a bank of value, but also that they have creatively managed coppice regrowth and volunteer seedlings long after project-termination.

We have no evidence, however, that they will on their own, without container subsidies, establish backyard wood-seedling nurseries and we doubt that they will purchase seedlings commercially. To our knowledge neither of these behaviors has occurred (in Haiti), nor have any studies addressed these specific questions. We know well that coppice and wildling management will sustain only a modest percentage of the trees for several years at most. Without new inputs of germplasm, plots that received trees are in danger of reverting eventually to their former treeless condition. No matter how attractive they are, the returns from tree planting are still too far down the road to compete with more immediate cash needs. This is as true for the prosperous American farmer who will not plant trees without a $10 000 per-annum subsidy program, as it is for the Haitian farmer who will perform amazing agroforestry feats with the modest $20 seedling grant which our project supplied.

Are we declaring wood-tree planting to be impossible without subsidies? Of course not. Georgia Pacific, Weyerhaeuser, and similar companies have a long record of planting, managing, and harvesting wood trees without any public support. We can, in sanctimonious avoidance of subsidies, leave the re-greening of the landscape to these powerful economic giants. If the goal, however, is to involve impoverished tropical smallholders in the transition to domesticated wood, in our view there is no magical mantra, no quick-fix educational or motivational gimmicks, that can circumvent the need for sustained public or philanthropic seedling support in the foreseeable future.

To conclude, in arguing for the approach used in Haiti, we are not peddling panaceas. The approach adopted here cannot by itself protect natural forests and cannot by itself provide total coverage of a watershed. It was not meant to. What it will do can be stated with simplicity and focus: It will enable local shifts into the domesticated production of the wood that was formerly scavenged from nature. We now know with certitude that a latent readiness to make this shift existed in Haiti. We believe that that same readiness for wood-tree planting exists in other world regions and other cultural settings as well, able to be activated on as massive a scale as was activated in Haiti. The catalyzing impact will come only if conservationist homilies and protectionist penalties are replaced with a rich supply of seedlings made available to villagers *under the same tenure and market assumptions that govern other crops*. The Haitian peasants have taught us a major lesson. We now know that it is possible, through anthropologically and technically sound project planning, to create the conditions by which even impoverished smallholders can participate in the exciting transition now occurring, as humans replicate the food-domestication achievements of their Neolithic ancestors and now bring even the wood tree itself into that subset of flora that humans plant, tend, and harvest.

Acknowledgements

The authors sincerely appreciate the candid comments and suggestions made by the reviewers, which greatly improved this chapter. Florida Agricultural Experiment Station Journal Series Number R-09895.

End Notes

1. Murray G. 1979. Terraces, trees, and the Haitian peasant: An assessment of 25 years of erosion control in rural Haiti. USAID, Port-au-Prince.
2. Murray G. 1981. Peasant tree planting in Haiti: A social soundness analysis. USAID, Port-au-Prince.
3. Smucker G. 1981. Trees and charcoal in Haitian peasant economy: A feasibility study of reforestation. USAID, Haiti.
4. Buffum W. 1985. Three years of tree planting in a Haitian mountain village: A socio-economic analysis. Pan American Development Foundation, Port-au-Prince.
5. Buffum W. and King W. 1985. Small farmer tree planting and decision making: Agroforestry extension recommendations. Pan American Development Foundation, Port-au-Prince.
6. Conway F.J. 1986a. The decision-making framework for tree planting in the Agroforestry Outreach Project. University of Maine Agroforestry Outreach Research Project, Port-au-Prince.
7. Conway F.J. 1986b. Synthesis of socioeconomic findings about participants in the USAID/Haiti Agroforestry Outreach Project. University of Maine Agroforestry Outreach Research Project, Port-au-Prince.
8. Lauwerysen H. 1985. Socioeconomic study in two tree-planting communities (St. Michel de L'Atalaye and Bainet). Pan American Development Foundation, Port-au-Prince.
9. Grosenick G. 1986. Economic evaluation of the Agroforestry Outreach Project. University of Maine Agroforestry Outreach Research Project, Port-au-Prince.
10. McGowan L.A. 1986. Potential marketability of wood products, rural charcoal consumption, peasant risk aversion strategies, and the harvest of AOP trees. University of Maine Agroforestry Outreach Research Project, Port-au-Prince.
11. Smucker G. 1988. Decisions and motivations in peasant tree farming: Morne-Franck and the PADF cycle of village studies. Pan American Development Foundation, Port-au-Prince.
12. Smucker G. and Timyan J. 1995. Impact of tree planting in Haiti: 1982-1995. Haiti Productive Land Use Systems Project, Southeast Consortium for International Development and Auburn University, Petionville, Haiti.
13. Some readers may wonder at the absence of a separate educational rubric as a key project component. This will be discussed below.
14. The project was also implemented by CARE in a separate region of Haiti, in the semi-arid Northwest.
15. Balzano A. 1986. Socio-economic aspects of forestry in rural Haiti. University of Maine Agroforestry Outreach Research Project, Port-au-Prince.
16. Balzano A. 1997. A Haitian community: ten years after. Presented to The Society for Anthropology in Community Colleges, 1997 Annual Meeting, Toronto, Canada, April 16-19, 1997.

References

Arnold J.E.M. and Dewees P.A.A 1999. Trees in managed landscapes: factors in farmer decision making. In: Buck L.E., Lassoie J.P. and Fernandes E.C. (eds). Agroforestry in Sustainable Agricultural Systems. CRC Press, Lewis Publishers, Boca Raton Florida, USA.

Ashby J.A., Alonso Beltrán J., del Pilar Guerrero M. and Fabio Ramos H. 1996. Improving the acceptability to farmers of soil conservation practices. J. Soil Water Conserv. 51: 309–312.

Balzano A. 1989. Tree-planting in Haiti: Agroforestry and rural development in a local context. Ph.D. dissertation, Rutgers University, USA.

Bannister M.E. and Josiah S.J. 1993. Agroforestry training and extension: the experience from Haiti. Agroforest Syst 23: 239–251.

Bannister M.E. and Nair P.K.R. 2003. Agroforestry adoption in Haiti: the importance of household and farm characteristics. Agroforest Syst 57: 149–157.

Bannister M.E. and Nair P.K.R. 1990. Alley cropping as a sustainable technology for the hillsides of Haiti: experience of an agroforestry outreach project. Am J Alt Agr 5: 51–57.

Bunch R. 1999. Reasons for non-adoption of soil conservation technologies and how to overcome them. Mt Res Dev 19: 213–220.

Ehrlich M., Conway F., Adrien N., LeBeau F., Lewis L., Lauwerysen H., Lowenthal I., Mayda Y., Paryski P., Smucker G., Talbot J. and Wilcox E. 1985. Haiti Country Environmental Profile: a field study. USAID, Port-au-Prince, Haiti.

Escobar A. 1991. Anthropology and the development encounter: The making and marketing of development anthropology. Am Ethnol 18: 658–682.

Ferraro G. 1998. Cultural Anthropology: An Applied Perspective (third edition). West/Wadsworth, New York. 397 pp.

Garcia J.N., Gerrits R.V. and Cramb R.A. 2002. Adoption and maintenance of contour bunds and hedgerows in a dynamic environment. Experience in the Philippine uplands. Mt Res Dev 22: 10–13.

Harris M. and Johnson O. 2000. Cultural Anthropology (5th Edition). Allyn and Bacon, Boston, 374 pp.

Herrador D. and Dimas L. 2000. Payment for environmental services in El Salvador. Mt Res Dev 20: 306–309.

Holdridge L.R. 1947. Determination of world plant formations from simple climatic data. Science 105: 367–368.

Hwang S.W., Alwang J. and Norton G.W. 1994. Soil conservation practices and farm income in the Dominican Republic. Agr Syst 46: 59–77.

Leakey R.B. and Tomich T.P. 1999. Domestication of tropical trees: from biology to economics and policy, pp. 319–338. In: Buck L.E., Lassoie J.P. and Fernandes E.C. (eds). Agroforestry in Sustainable Agricultural Systems. CRC Press, Lewis Publishers, Boca Raton Florida, USA.

Murray G. 1984. The wood tree as a peasant cash crop: An anthropological strategy for the domestication of energy, pp. 141–160. In: Foster C.R. and Valdman A. (eds). Haiti – Today and Tomorrow. University Press of America, Lanham, MD.

Murray G. 1987. The domestication of wood in Haiti: A case study in applied evolution, pp. 223–242. In: Wulff R.M. and Fiske S.J. (eds). Anthropological Praxis: Translating Knowledge into Action. Westview Press, Boulder CO.

Nair P.K.R. 1993. An Introduction to Agroforestry. Kluwer, Dordrecht, The Netherlands, 499 pp.

Nanda, S. and Warms R. 1998. Cultural Anthropology. West/Wadsworth, New York, 429 pp.

Napier T.L. and Bridges T. 2002. Adoption of conservation production systems in two Ohio watersheds: a comparative study. J Soil Water Conserv 57: 229–235.

Osembo G.J. 1987. Smallholder farmers and forestry development: a study of rural land-use in Bendel, Nigeria. Agr Syst 24: 31–51.

Peoples J. and Bailey G. 1997. Humanity: An Introduction to Cultural Anthropology. West/Wadsworth, New York, 446 pp.

Podolefsky A. and Brown P.J. (eds) 1989. Applying Anthropology: An Introductory Reader. Mountain View, Mayfield, CA, 324 pp.

Robbins R. 1993. Cultural Anthropology: A Problem-based Approach. F.E. Peacock Publisher, Itasca, IL, 222 pp.

Rogozinsky, J. 2000. A Brief History of the Caribbean: From the Arawak and Carib to the Present (Revised Edition). Penguin Putnam (Plume), New York. 415 pp.

Simons A.J. and Leakey R. 2004. Tree domestication in agroforestry in the tropics (this volume).

Smucker G.R., White T.A. and Bannister M.E. 2002. Land tenure and the adoption of agricultural technology in Haiti, pp. 119–146. In: Meinzen-Dick R., Knox A., Place F. and Swallow B. (eds). Innovation in Natural Resource Management: The Role of Property Rights and Collective Action in Developing Countries. Johns Hopkins University Press for the International Food Policy Research Institute, Baltimore, MD.

Timyan J. 1996 Bwa yo: Important Trees of Haiti. South-East Consortium for International Development, Washington, D.C., 418 pp.

Weil T.E., Black J.K., Blutstein H.I., Johnston K.T., McMorris D.S. and Munson F.P. 1973. Area Handbook for Haiti. U.S. Government Printing Office, Washington, DC. 189 pp.

Zobel B.J. and Talbert J.T. 1984. Applied Forest Tree Improvement. John Wiley & Sons, Inc., New York, 505 pp.

Zuvekas C. 1979. Land tenure in Haiti and its policy implications: A survey of the literature. Soc Econ Stud 28: 1–30.

Knowledge Integration

Agroforestry Systems **61:** 401–421, 2004.

Computer-based tools for decision support in agroforestry: Current state and future needs

E.A. Ellis[1,*], G. Bentrup[2] and M.M. Schoeneberger[2]
[1]*School of Forest Resources and Conservation, University of Florida, PO Box 110410, Gainesville, FL 32611-0410, USA;* [2]*USDA National Agroforestry Center, USFS Rocky Mountain Research Station, UNL-East Campus, Lincoln, Nebraska 68583–0822, USA;* **Author for correspondence: e-mail: eaellis@ufl.edu*

Key words: Databases, Decision Support Systems, Geographical Information Systems, Models

Abstract

Successful design of agroforestry practices hinges on the ability to pull together very diverse and sometimes large sets of information (i.e., biophysical, economic and social factors), and then implementing the synthesis of this information across several spatial scales from site to landscape. Agroforestry, by its very nature, creates complex systems with impacts ranging from the site or practice level up to the landscape and beyond. Computer-based Decision Support Tools (DST) help to integrate information to facilitate the decision-making process that directs development, acceptance, adoption, and management aspects in agroforestry. Computer-based DSTs include databases, geographical information systems, models, knowledge-base or expert systems, and 'hybrid' decision support systems. These different DSTs and their applications in agroforestry research and development are described in this paper. Although agroforestry lacks the large research foundation of its agriculture and forestry counterparts, the development and use of computer-based tools in agroforestry have been substantial and are projected to increase as the recognition of the productive and protective (service) roles of these tree-based practices expands. The utility of these and future tools for decision-support in agroforestry must take into account the limits of our current scientific information, the diversity of aspects (i.e. economic, social, and biophysical) that must be incorporated into the planning and design process, and, most importantly, who the end-user of the tools will be. Incorporating these tools into the design and planning process will enhance the capability of agroforestry to simultaneously achieve environmental protection and agricultural production goals.

Introduction

> *'Few things disappoint a landowner more than spending money, time, and effort on a project that fails ... especially one like agroforestry, where it can be years before problems become apparent' (Dosskey and Wells 2000).*

Agroforestry, the deliberate integration of trees into crop and livestock operations, has the potential to achieve many of the environmental, economic, and social objectives being demanded from working landscapes by landowners and society. By adding structural and functional diversity to the landscape, these tree-based plantings can perform ecological functions that have significance far greater than the relatively small amount of land they occupy (Guo 2000; Nair 2001; Ruark et al. 2003). Realizing this potential is, however, a complex task of determining what opportunities, limitations, and trade-offs exist in each situation, and of designing an agroforestry practice that achieves the best balance among them. There are numerous impacts created by agroforestry plantings, ranging from intended to nonintended and, therefore, ranging from detrimental to advantageous, occurring both on- and off-site, and varying over time. Consequently, if agroforestry is to be a viable strategy in promoting agroecosystem sustainability, the decision-making process must incorporate many considerations, not only at the practice scale but also at the larger scales of farm, landscape, and watershed (Schoeneberger et al. 1994). Simply put, agroforestry creates a complex system of interactions that must be managed for multiple objectives, multiple alternatives and multiple social interests and preferences, while being applied over a wide range of landscapes and landscape features.

The decision-making process involved in agroforestry research, development and application is composed of several components: the person or group making the decision, the problem, the approach or method to solve the problem, and the decision. Decision support tools (DST) are a wide variety of technologies that can be used to help integrate diverse and large sets of information. DSTs do not replace the decision-making by the landowner or natural resource manager, but they do facilitate the decision-making process by making the planning process more informed and more objective (Grabaum and Meyer 1998). Although agroforestry, like most natural-resource management sciences, is characterized by high complexity of which we have limited understanding and data (Sanchez 1995; Nair 1998), the science and application of agroforestry can be greatly enhanced through the use of these tools.

Computer-based decision support tools in natural resource management

Computers now play an integral role for information management and decision-making in all disciplines related to natural resource management. Constant accretion of data and information on agriculture, forestry, agroforestry, and natural resource management has created the need to synthesize, organize and manipulate this growing knowledge base and facilitate its accessibility and use for education, research, and decision-making (Davis 1988; Schmoldt and Rauscher 1996). The complexity of natural resource management, considering the diversity of resources, interests, objectives, constraints, and stakeholders involved, adds to the difficulty of making sound management decisions (Schmoldt and Rauscher 1996). Computer-based DSTs provide an effective means to compile and sort out the medley of variables, information and knowledge (quantitative, qualitative, spatial, and heuristic) that managers must consider when making informed management decisions. In other words, DSTs synthesize the wide array of information and offer a holistic approach for evaluating land and resource management problems and finding appropriate solutions (Schmoldt and Rauscher 1996). There are five major categories of computer-based technologies used for decision support: databases, geographical information systems (GIS), computer-based models, knowledge-based or expert systems (KBS), and hybrid systems (Table 1).

In the past decade the use of computer-based technologies in agriculture, forestry, and natural resource management has been impressive. In the field of agriculture, the use of computers can be considered part of the agricultural revolution of the 20th century, advancing scientific research, facilitating farm management, and improving production (Paarlberg and Paarlberg 2000). The development of crop models, expert systems for agricultural management, and precision farming that incorporates GIS has advanced considerably and are being used in many farming operations (NRC 1997; Zazueta and Xin 1998; Ahuja et al. 2002). Forestry, compared to production agriculture, is often faced with more complex and multiple objective management scenarios. The adoption and use of computers for decision support in forestry have had to evolve from simpler mathematic models used for harvesting scheduling to more complex computer DSTs used to help make management decisions where timber production must be balanced with wildlife conservation, water quality, recreation, and other objectives, often involving policy and social issues (Rauscher 1999). Table 2 lists some of the more recognized DSTs used in agricultural and forest management and describes their degree of complexities and integration of the major computer-based technologies.

Databases

Databases are computer-based tools used to access and query large quantities of data and information. They are often key components within other DSTs. The database DST consists of a database (a logically coherent collection of data) and a database management component (the software system), which allows a user to define, create, and maintain a database (Mata-Toledo and Cushman 2000). Computer databases are often implemented as a Relational Database Management System (RDBMS), designed around the mathematical concepts of relational algebra linking together two-dimensional data tables (Sanders 2000; Sunderic and Woodhead 2001). Query statements can be developed, allowing users to search and analyze data as well as extract specific information from huge datasets. This ability to extract pieces of informa-

Table 1. Major categories of computer-based decision support technologies.

Category	Description
Databases	Organizes and facilitates the management and querying of large quantities of data and information
Geographical Information Systems (GIS)	Brings in a geographic or spatial component to a database; manages, manipulates and analyzes spatial data
Computer-Based Models	Mathematical computer models that represent real world processes and predict outcomes based on input scenarios
Knowledge-Based or Expert Systems	Adopts 'Artificial Intelligence' in the form of organizing, manipulating and obtaining solutions using knowledge in the form qualitative statements, expert rules (i.e. rules of thumb) and a computer language representation system for storing and manipulating knowledge.
Hybrid Systems	Integrates two or more of the above computer-based technologies (e.g. (GIS, KBS and Models) for more versatile, efficient and comprehensive DSTs.

Table 2. Computer-based decision support tools used in agricultural and forestry management.

Decision support tool	Description	Reference
GOSSYM-COMAX	Used for management of water, nitrogen, herbicide and growth regulator in cotton	Reddy et al. 2002; Gertsis and Whisler 1998;
GLYCM	Soybean production model	Timlin et al. 2002; Acock et al. 1997; Manning 1996
CERES	Production model for crops in the tropics and subtropics	Ritchie and Otter 1985; Ahuja et al. 2002
CROPGRO	Production model for crops in the tropics and subtropics	Boote et al. 1998a; Boote et al. 1998b; Ahuja et al. 2002
DSSAT	Package of crop-soil models to facilitate the evaluation and application of different cropping systems and the input and organization of relevant scientific data	Jones et al. 1998; Jones et al. 2003
FORPLAN/SPECTRUM	Optimization models for forest management to evaluate financial efficiency, land allocation and resource scheduling	Field 1984; Kent et al. 1991; Rauscher 1999
STEWPLAN	Knowledge-based computer tool to assist landowners develop stewardship plans based on forest stand descriptions.	Knopp and Twery 2003
NED	Hybrid decision support tool integrating forest models, GIS, graphic visualization and a knowledge base for multi-use forest management.	Twery et al. 2000; Twery et al., 2003
EMDS	Landscape scale tool that integrates GIS, knowledge-based reasoning and decision modeling technologies for ecosystem management decision support.	Reynolds et al. 2002; Rauscher 1999
EPIC	Erosion-Productivity Impact Calculator determines relationships between soil erosion and crop productivity	Jones et al. 1991; Easterling et al. 1997
CO2FIX	Estimates and evaluates dynamics of carbon in forest management and afforestation projects	Masera et al. 2003

tion based on user-specified criteria makes a RDBMS an excellent computer-based technology for decision support. A multitude of natural-resource-related databases (i.e., ecosystems, flora, fauna, soils, hydrology, etc.) are now widely available and used in management decisions.

Geographical information systems

A GIS can be defined as a data management system designed to input, store, retrieve, manipulate, analyze, and display spatial data for the purposes of research and decision-making (De Mers 1997). In a GIS, a database is associated with map features, and data values are geographically referenced, so resource managers can spatially represent information such as soil types or plant communities. Since land use and a diversity of related disciplines (i.e., agriculture, forestry, rural planning, and conservation) all deal with spatial characteristics of landscapes (Lacher 1998), GIS has gained considerable use in land use planning and natural-resource management, providing a spatial framework to aid in the decision-making process (Zeiler 1999).

Additional technologies are often associated with GIS, such as Global Positioning Systems (GPS) and remote sensing. GPS is a means for inputting spatial data with real world coordinates into a GIS and has become an important tool for researchers locating and recording information in the field. Remote sensing involves using spatial data from photographic and satellite images, and software tools to analyze and interpret these data.

Computer-based models

For the most part, computer-based models refer to the translation of data and information into a mathematical form using algorithms to represent a real world 'process' or 'system' and to forecast outcomes of different scenarios. In the realm of environmental and natural-resource related fields, these models try to mathematically represent ecological processes (Skidmore 2002). Environmental models mathematically define ecological interactions and processes between biotic and abiotic components based on current or past conditions or states of these components. The goal of these models is to quantitatively predict future states of these components, serving a valuable role in defining the key processes in agroforestry practices (Peng et al. 2002; Skidmore 2002).

Knowledge-based systems

Knowledge-Based Systems (KBS) or expert systems are part of the broad field of Artificial Intelligence (AI) involving the creation of computer programs that attempt to mimic human intelligence or reasoning, 'learn' new information and tasks, and draw useful conclusions about the world around us (Patterson 1990). In a KBS, knowledge is defined as a 'body of facts and principles accumulated by humankind or the act, fact or state of knowing' (Patterson 1990). KBS are used to acquire, organize and manipulate knowledge, often using heuristic rules, analogous to 'rules of thumb' or 'good judgments,' to help make sound deductions (Nikolopoulos 1997). Often, experts in the subject are used to define these rules; however, knowledge for a KBS can be acquired from the literature, databases or other sources. A 'knowledge engineer' extracts knowledge, information and data from experts and other sources and translates it into programming languages so a computer can utilize and reason with it (Nikolopoulos 1997; Patterson 1990; Schmoldt and Rauscher 1996). With an appropriate user interface, the user can input problem scenarios and make enquiries to find solutions (Nikolopoulos 1997).

Hybrid systems

Many DSTs today integrate a variety of computer-decision support technologies such as RDBMS, GIS, Models and KBS (Davis 1988; Liebowitz 1990). Increasingly, land-use planning and natural-resource management DSTs are merging GIS and KBS to develop very effective and efficient spatial planning tools (Loh and Rykiel 1992). Nowadays, application development programs, modeling tools and GIS software are designed to be compatible with other systems and allow a relatively easy integration of the different computer-based technologies.

Applications of computer-based DSTs for agroforestry

Considerable advances have been made in research, planning and development for a variety of agroforestry systems in a wide range of agroecological regions, from tropical to temperate. Prior to 1991, computer use in agroforestry research began with the development of databases as aids in guiding plant selection (Nair 1998). As the use of agroforestry has broadened to address such issues as climate change and crop growth, carbon sequestration, biodiversity and even green infrastructure, so has the need to simulate agroforestry's longer-term effects across larger scales, further necessitating use of computer-based DSTs. These early DSTs used in agroforestry were generally those already in place in the fields of agriculture and forestry. For instance, the effect of shelterbelts on maize productivity under hypothesized climate change scenarios was examined using the Erosion-Productivity Impact Calculator (EPIC) crop model, an agricultural model originally developed to determine the relationship between soil erosion and crop productivity (Jones et al. 1991; Easterling et al. 1997). Even today, many of these models developed for agriculture or forestry are still a first choice for use in agroforestry exercises. CO2FIX, a user-friendly model for dynamically estimating the carbon sequestration potential of forest management and afforestation projects, is readily adaptable for agroforestry (Masera et al. 2003).

Today we have several DSTs developed exclusively for agroforestry applications for the purposes of selecting suitable species, identifying suitable lands, modeling different systems and predicting outcomes of different scenarios. Several different types of computer-based DSTs that have been applied or are strongly applicable to agroforestry research, planning and development are listed in Table 3 and are discussed in further detail below. Additionally the intended uses, targeted end-users and current status of these major agroforestry DSTs are summarized in Table 4.

Agroforestry databases

An initial effort to use computers to manage agroforestry data began in the late 1980s with the Agroforestry Systems Inventory Database (AFSI) developed by the International Centre for Research in Agroforestry (ICRAF), now the World Agroforestry Centre. AFSI involved a global collection of data or information on agroforestry systems using a questionnaire as the survey instrument. Data and information collected and entered into the database included general description, geographical location, biophysical characteristics, socioeconomic aspects, system evaluation, components of the system and their uses, and identification of research gaps. With AFSI, the user is able to query the database, extracting information such as geographical locations of different agroforestry systems and the species found within these systems in different locations (Nair 1987; Oduol et al. 1988). AFSI was apparently developed as a research and information tool for researchers, particularly in ICRAF. No documentation could be found about current versions or availability of AFSI. Unfortunately, many early DSTs like AFSI often fail to be maintained and upgraded and therefore fade with time.

Another early agroforestry database was the Multi-Purpose Tree and Shrub Database (MPTS), also developed by ICRAF in 1991 (von Carlowitz et al. 1991). The MPTS database, developed for researchers and extension agents, helped to select the right tree or shrub species for agroforestry practices, primarily for the tropics and subtropics. MPTS Database Version 1.0 contained information for 1093 species including site-specific requirements (e.g., soils), morphological and phenological descriptions, management characteristics and environmental responses (Schroder and Jaenicke 1994). A simple climate model was included to predict climatic conditions based on the input of geographical coordinates, and tree and shrub species were selected via a database search or query that matched the descriptors that the user selected. The descriptors included 19 different criteria covering aspects of location, climate and soil conditions, products, environmental services, management and cultivation. The user was also able to use boolean operators (i.e., and, or, and not) to fine-tune the search to their specific needs. References are also included to provide further information on selected agroforestry species (Schroder and Jaenicke 1994).

The current and revised version of MPTS is now the AgroforesTree Database (AFT). Unlike its predecessor this database is more widely accessible and available on the Internet and as a CD-ROM. It is a database management system intended for use by researchers and fieldworkers to select agroforestry trees that are being deliberately grown and managed for more than one output and expected to make significant economic and/or ecological impacts (Salim et al. 1998; World Agroforestry Centre 2003a). More than 300 species are incorporated into AFT with information on ecology and distribution, propagation and management, functional uses and pest and diseases (Salim et al. 1998). With AFT, users are able to search and select trees according to geographical location, biophysical limits and other management criteria selected. In addition, there are references, research contacts, a seed supplier's directory, images of trees, and a glossary of terms to help agroforesters obtain vital information and make wise decisions concern-

Table 3. Computer-based decision support tools used in agroforestry.

Decision support tool	Type	Description	Reference
AFSI (Agroforestry Systems Inventory Database)	Database	Agroforestry system inventory database describing geographic location and biophysical, socioeconomic and species characteristics	Nair 1987; Oduol et al. 1988
MPTS (Multipurpose Tree and Shrub Database)	Database	Multi-purpose tree and shrub database used for tree selection and species information	Schroder and Jaenicke 1994
AgroforesTree Database	Database	Internet and CD-Rom application for reference and selection guide of agroforestry trees.	Salim et al. 1998; World Agroforestry Centre 2003a
Subtropical Tree and Shrub Database	Database	On-line database on potential agroforestry tree and shrub species for the American subtropics.	Ellis et al. 2003
Forestry Compendium	Database	Compilation of knowledge on forestry, agroforestry and plantations and information on trees for management decision-making and species selection	CABI 1998; Kleine et al. 2003
Agroforestry System Suitability in Africa	GIS	Spatial analysis using climate, soil land use and other spatial data alongside plant species data to determine species and agroforestry suitability	Booth et al. 1989; Booth et al. 1990; Unruh and Lefebvre 1995
Agroforestry System Suitability in Ecuador	GIS	Spatial analysis to determine suitable areas of *Annona cherimola* agroforestry systems in Southern Ecuador.	Bydekerke et al. 1998
Agroforestry System Assessment in Nebraska	GIS	Spatial suitability assessment for willow and forest farming agroforestry systems in a Nebraska watershed	Bentrup and Leininger 2002
Agroforestry Parklands in Burkina Faso	GIS	Spatial analysis of dynamics of agroforestry parklands and species distribution due to human impacts	Bernard and Depommier 1997
Historical Transformation of Agroforestry Landscape in Canada	GIS	Spatial analysis of census and geomorphologic data to explore dynamics of agroforestry in 19th century Canadian landscape	Paquette and Domon 1997
Field-level spatial analysis of temperate agroforestry system	GIS	Spatial analysis using ground penetrating radar (GPR) to evaluate root biomass and distribution and soil nutrient crop-tree interactions in temperate alley cropping	Jose et al. 2001
AME (Agroforestry Modeling Environment)	Modeling Tool	Object-oriented tool to graphically visualize, construct, integrate and exchange agroforestry models	Muetzelfeldt and Taylor 1997
HyPAR	Model	Biophysical model combining crop and forest models and integrating climate, hydrology, light interception, water and nutrient competition, and carbon allocation processes in agroforestry systems	Mobbs et al. 2001
HyCAS	Model	Biophysical model for agroforestry systems with cassava simulating competition for light, water and nutrients including phosphorus cycles	Matthews and Lawson 1997

Table 3. Continued

Decision support tool	Type	Description	Reference
WaNuLCAS (Water, Nutrient and Light Capture in AgroforestySystems)	Model	Biophysical model of tree-crop interactions based on above and below-ground resource capture and competition of water, nutrients and light under different management scenarios in agroforestry systems	Van Noordwijk and Lusiana 1999; World Agroforestry Centre 2003b
SCUAF (Soil Changes under agroforestry)	Model	Nutrient cycling model predicts changes in soil conditions under different agroforestry systems based on parameters of biophysical environment, land use and management, plant growth, and plant–soil processes	Young and Muraya 1990; Vermeulen et al. 1993; Menz et al. 1997; Macadong et al. 1998; Nelson et al. 1997
FALLOW (Forest, Agroforest, Low-value Landscape or Wasteland?)	Model and GIS	Model to evaluate impacts of shifting cultivation and fallow rotations at a landscape-scale evaluating transitions in soil fertility, crop productivity, biodiversity and carbon stocks	World Agroforestry Centre 2003c; Van Noordwijk 2002
BEAM (Bio-economic Agroforestry Model)	Model	Bioeconomic model to assess physical and financial performance of agroforestry systems based on tree and crop biometric and economic models	Willis et al. 1993; Willis and Thomas 1997
AEM (Agroforestry Estate Model)	Model	Economic model to evaluate agroforestry in combination with other farm activities assessing effects of tree production and physical and financial resources on-farm	Middlemiss and Knowles 1996
DESSAP (Agroforestry Planning Model)	Model	Multi-objective linear programming model to assess feasible agroforestry alternatives based on land, labor and cash constraints	Garcia-de Ceca and Gebremedhin 1991
ATK (Agroforestry Knowledge Toolkit)	KBS	KBS to store, manipulate and analyze a variety of information and knowledge acquired on agroforestry systems	Walker et al. 1995
AES (Agroforestry Expert System)	KBS	KBS used heuristic knowledge or expert 'rules of thumb' to determine optimal species and spacing for alley cropping systems in the tropics	Warkentin et al. 1990
AGFADOPT (Agroforestry Adoption Evaluation Tool)	Decision Tree KBS	KBS based on decision trees used to assess adoption of agroforestry based on economic and social factors faced by small-scale farmers	Robotham 1998
Agroforestry Planning Tool in China	Hybrid GIS, Models and KBS	Hybrid DST integrating GIS data, regression models plus expert knowledge to assess biophysical, social and economic suitability of *Paulownia* intercropping agroforestry systems	Liu et al. 1999
PLANTGRO (Plantation and Agroforestry Species Selection Tool)	Hybrid GIS/KBS	Plantation and agroforestry species selection tool integrates GIS and expert system on plant growth	Booth 1996; Hackett and Vanclay 2003
SEADSS (Southeastern Agroforestry Decision Support System)	Hybrid Database, GIS, KBS	Landscape and site-scale agroforestry planning and species selection DST for landowners and extension agents of Southeast US that integrates GIS, tree and shrub database and expert knowledge	Ellis et al. 2003

Table 3. Continued

Decision support tool	Type	Description	Reference
Conservation Buffer Planning Tools for Western Cornbelt Region, USA	Hybrid GIS/Models/ Visualization	Suite of GIS, economic models and visualization tool for landowners and resource managers to evaluate agroforestry strategies in Midwest Cornbelt region of the USA	Bentrup et al. 2003

ing the use and selection of agroforestry trees (Salim et al. 1998). Even though the AgroforesTree Database is recognized and linked to a variety of rural and agricultural development Websites, its specific use and impact on agroforestry research and development projects are difficult to assess at this stage.

Although not solely for agroforestry, the Forestry Compendium is an extremely useful database for agroforestry research and planning. The development of the Forestry Compendium was undertaken by both the Commonwealth Agricultural Bureau International (CABI) and International Union of Forestry Research Organizations (IUFRO) and consists of a compilation of knowledge on multipurpose forestry, including agroforestry, plantations, and natural forest management (CABI 1998; Kleine et al. 2003). The compendium gives information about what trees could be planted in a particular environment and for what purposes, how they will perform, how they should be managed, and provides current documents available regarding each species (Kleine et al. 2003). A Species Selection Module aids in decision-making for selecting suitable species according to a variety of criteria including geographical location, climate, type of use, and other management options (Kleine et al. 2003).

GIS applications in agroforestry

Considering that GIS technology is widely available and affordable today and the fact that agroforestry is directly dependent upon spatial characteristics, it is logical to expect to have several agroforestry-specific GIS DSTs; but the reality is that only a few are available. An early GIS application compiled information on 173 species including their descriptions, soil and climate preferences, and management characteristics for Africa (Booth et al. 1989). This application allowed users to query the database and generate maps showing the climatic suitability for different species. At a regional scale, Booth et al. (1990) created a similar application for Zimbabwe, demonstrating how GIS applications can be done at many scales. Unruh and Lefebvre (1995) performed a similar GIS application for sub-Saharan Africa to determine areas suitable for different agroforestry systems. Integrating ICRAF's agroforestry database with spatial data on geographic regions, climate and land uses in the region, their application was able to map out potential regions for 21 specific types of agroforestry systems.

Most of the past agroforestry GIS applications mentioned above have been research-oriented. The Southeastern Agroforestry Decision Support System (SEADSS), developed recently by the Center for Subtropical Agroforestry (CSTAF) at the University of Florida brings on-line GIS capabilities directly to extension agents and landowners; it offers county soils, land use and other spatial data for selecting suitable tree and shrub species in a specified location (Ellis et al. 2003). The USDA National Agroforestry Center (NAC) is currently using GIS to facilitate conservation buffer planning in the Western Corn Belt ecoregion in the central United States (Bentrup et al. 2000). GIS-guided assessments, derived from publicly available datasets, are being used to evaluate four key issues of the Western Cornbelt: biodiversity, soil protection, water quality, and agroforestry products. By combining these assessments, information is generated for use in identifying opportunities and constraints on the landscape where multiple benefits from conservation buffers, especially agroforestry plantings, can be achieved (Bentrup et al. 2000). Utilizing the agroforestry product assessments (Bentrup and Leininger 2002) in conjunction with the riparian buffer connectivity assessments, areas were identified where riparian forest buffers could be located to improve habitat connectivity while offering landowners the option to grow woody florals for profit (G. Bentrup and T. Kellerman, presentation to 8th North American Agroforestry Conference, June 2003).

GIS-guided agroforestry suitability analysis will only improve as spatial data and computer resources become more accessible. Many states and countries already are assembling internet-accessible GIS data

Table 4. Uses, targeted end-users and current status of major decision support tools used in Agroforestry.

Decision support tool	Intended use	Targeted end-users	Current status and availability
AFSI (Agroforestry Systems Inventory Database)	General agroforestry research & planning for ICRAF	Researchers and ICRAF	No current versions or availability
MPTS (Multipurpose Tree and Shrub Database)	Species selection for agroforestry research & planning	Researchers & Extension Agents, Foresters	Upgraded to AgroforesTree
AgroforesTree Database	Species selection for agroforestry research & planning (World-wide)	Researchers & Fieldworkers	CD-ROM 1998 and currently available on-line from World Agroforestry Centre http://www.worldagroforestrycentre.org/Sites/TreeDBS/AFT/AFT.htm
Subtropical Tree and Shrub Database	Species selection and information for agroforestry extension, planning & development (American Subtropics & Caribbean)	Landowners, Extension Agents, Researchers	Currently under development and evaluation. Available on-line from Center for Subtropical Agroforestry http://cstaf.ifas.ufl.edu/tree&shrubdb.asp
Forestry Compendium Database	Species selection and information for forestry, agroforestry and plantation research and planning and development (World-wide)	Foresters, Policy Makers, Conservationists, Consultants, Extensionists	CD-ROM 2003 and internet version available through CABI International http://www.cabi.org/compendia/fc/index.asp
AME (Agroforestry Modeling Environment)	Development of agroforestry models for research	Researchers	Now SIMILE for building general ecology models available from Simulistics http://simulistics.com/
HyPAR Model	Research on biophysical process and interactions in agroforestry systems	Researchers	HyPAR v 4.5 available through Center for Ecology and Hydrology, Edinburgh, UK http://www.nbu.ac.uk/hypar/
HyCAS Model	Research on biophysical process and interactions in cassava agroforestry systems	Researchers	HyCAS available through Cranfield University, UK http://www.silsoe.cranfield.ac.uk/iwe/research/hycas.htm
WaNuLCAS Model (Water, Nutrient and Light Capture in AgroforestrySystems)	Research on biophysical processes and interactions in agroforestry systems	Researchers	WaNuLCAS v2.11 Available through World Agroforestry Centre http://www.worldagroforestrycentre.org/sea/Products/AFModels/WaNulCAS/index.htm
SCUAF Model (Soil Changes under agroforestry)	Environmental evaluation of agroforestry systems used for research and development projects	Researchers	SCUAF v4.0 available through Centre for Resource and Environmental Studies, Australian National University http://incres.anu.edu.au/imperata/imp-mods.htm
FALLOW Model (Forest, Agroforest, Low-value Landscape or Wasteland?)	Impact assessment on landscape dynamics due to socioeconomic and land-use changes	Researchers	FALLOW available through World Agroforestry Centre http://www.worldagroforestrycentre.org/sea/Products/AFModels/FALLOW/Fallow.htm

Table 4. Continued

Decision support tool	Intended use	Targeted end-users	Current status and availability
BEAM (Bio-economic Agroforestry Model)	Bio-economic assessment of agroforestry systems used for research and development projects	Researchers	Available through University of Wales, Bangor, UK http://www.safs.bangor.ac.uk/
AEM (Agroforestry Estate Model)	Evaluation of agroforestry physical and financial yields for planning & development projects	Consultants	Available through Forest Research, Rotorua, New Zealand http://www.forestresearch.co.nz
DESSAP (Agroforestry Planning Model)	Evaluation of feasibility of agroforestry systems planning & development	Planners, Managers, Extensionists	No current versions and availability unknown
ATK (Agroforestry Knowledge Toolkit)	Build agroforestry knowledge bases for research, planning and development	Development professionals	ATK 5 available through University of Wales, Bangor, UK http://www.bangor.ac.uk/~afs40c/afforum/akt5/akt5_frame.htm
AES (Agroforestry Expert System)	Alley cropping planning & Development	Land use officials, Researchers, Extensionists	No current versions and availability unknown
AGFADOPT (Agroforestry Adoption Evaluation Tool)	Assess Agroforestry adoption used in Dominican Republic	Researchers, Planners	No current versions and availability unknown
PLANTGRO (Plantation and Agroforestry Species Selection Tool)	Species selection and land use planning used for plantation forestry planning in Indonesia	Planners, Development professionals	Available through CIFOR TROPIS http://www.cifor.cgiar.org/scripts/default.asp?ref=research_tools/tropis/plantgro-infer.htm
SEADSS (Southeastern Agroforestry Decision Support System)	Site evaluation and species selection for agroforestry planning and development	Extensionists, farmers, land-owners	Currently under development and available on-line from Center for Subtropical Agroforestry http://cstaf.ifas.ufl.edu/seadss.htm
Conservation Buffer Planning Tools for Western Cornbelt Region, USA	Facilitate planning and designing conservation buffers for multiple objectives	Planners, Landowners	Various tools available through National Agroforestry Centre http://www.unl.edu/nac/conservation/index.html

clearinghouses to facilitate the use of spatial information.

Agroforestry models

Computer-based agroforestry models are useful for efficient handling of the many social, economic, and ecological variables that must be considered when dealing with the highly complex systems created by agroforestry. Output from these models can assist in evaluating agroforestry alternatives, testing research hypothesis, and understanding the processes and interactions in these systems, potentially saving time and money (Jagtap and Ong 1997; Muetzelfeldt and Taylor 1997). A concerted effort to start developing and implementing agroforestry models began in the mid 1990s with the Agroforestry Modeling Project (AMP). AMP was funded by the Forestry Research Programme of the UK Department of International Development and undertaken by the University of Edinburgh in association with other universities (Nottingham, Reading, North Wales and Cranfield), ICRAF and the International Institute of Tropical Agriculture (IITA). The main objectives of this project were: 1) promote liaisons between agroforestry modelers and researchers in order to add value and rigor to information obtained from experiments and models, and to improve advice given to farmers; 2) promote the integration of information obtained from agroforestry models and experiments; 3) develop process-based agroforestry models which address tree, crop and soil interaction (C, N and water cycles); 4) use the models to test hypotheses regarding competition between trees and crops for light, water and nutrients; and 5) define optimal agroforestry practices for different regions (Lawson and Cannell 1997).

One of the most fundamental products that came from AMP was the Agroforestry Modeling Environment (AME). Developed by Edinburgh University, AME is a tool to help visualize, construct, integrate and exchange agroforestry models. It uses a user-friendly, object-oriented environment for model construction where users can select, characterize and link predefined components and run mathematical processes (Muetzelfeldt and Taylor 1997). Users can construct models without programming by drawing model diagrams and using easily understood concepts such as 'sub-model,' 'compartment' and 'flows.' The objective of AME is to stimulate rapid model development, re-use and standardize model components developed in the past, increase the ease of building and comprehending models, and increase the efficiency and effectiveness of the modeling process (Muetzelfeldt and Taylor 1997).

Another important product that came out of AMP is the HyPAR model. It was recognized that agroforestry models needed to synthesize experimental and empirical data on tree and crop interactions; pinpoint and prioritize knowledge gaps; extrapolate research results to new combinations of soil, climate, species and management conditions; and to provide decision support to policy makers, researchers and extension staff (Mobbs et al. 2001). HyPAR combines two models: 'Hybrid,' a forest canopy model developed by the Centre for Ecology and Hydrology, and PARCH, a crop growth model developed by the University of Nottingham for application in the dry tropics. HyPAR incorporates biophysical processes to calculate light interception through a disaggregated canopy of individual types of trees in known positions along with water competition, nutrient competition, daily carbon allocation, hydrology and the impacts of different management scenarios in crop-tree agroforestry systems (Mobbs et al. 2001). A friendly, graphical user-interface guides the user input and running process with a set of menus and dialogs, allowing rapid editing of input parameters and simulation settings. In a test using a site in Ghana, the model gave validated outputs for net primary productivity in natural forest/woodland vegetation and potential sorghum grain yield. It has been shown to be useful in exploring opportunities for complementarity of light and water use by trees and sorghum in a wide variety of climates (Cannell et al. 1998; Mobbs et al. 1998).

A similar model is WaNuLCAS (Water, Nutrient and Light Capture in Agroforestry Systems), designed to model tree-soil-crop interactions for a wide range of agroforestry practices (van Noordwijk and Lusiana 1999; World Agroforestry Centre 2003b). WaNuLCAS incorporates a plant-plant interaction model focusing on above- and below-ground resource capture in a competitive situation (van Noordwijk and Lusiana 1999). The model links mulch production, its effect on soil fertility and shading effects of trees on crop yields (van Noordwijk and Lusiana 2000). The model allows different management options such as plot size, tree spacing and choice of tree species, cropping cycles, pruning, organic and inorganic fertilizer inputs and crop residue removal (van Noordwijk and Lusiana 2000).

SCUAF (Soil Changes Under Agroforestry), a nutrient cycling model that predicts annual changes in soil conditions (e.g. soil erosion, carbon, nitrogen, phosphorus and organic matter) and the effect of soil changes upon plant growth and harvest, has been used for agroforestry research since the early 1990s (Young and Muraya 1990). Research of miombo woodlands in Zimbabwe for agroforestry purposes utilized SCUAF (Vermeulen et al. 1993), as did a more recent project sponsored by the Australian Centre for International Agricultural Research (ACIAR) and the Center for International Forestry Research (CIFOR) entitled, 'Improving smallholder farming systems in *Imperata* areas of Southeast Asia: a bioeconomic modeling approach,' (Menz et al. 1998). For this project, SCUAF was used in the evaluation of replacing *Imperata* fallows with the use of *Gliricidia* fallows as a means to increase soil nutrient concentrations and reduce erosion (Grist et al. 1998). Magcale-Macadong et al. (1998) used SCUAF to demonstrate the use of napier grass (*Pennisetum purpureum*) to control soil erosion, aid productivity of hedgerow agroforestry systems by providing mulch, and improve livestock systems by using it as feed.

FALLOW (Forest, Agroforest, Low-value Landscape or Wasteland) is a model that scales-up the assessment of land-use systems by evaluating the impacts of shifting cultivation or crop-fallow rotations at a landscape scale (van Noordwijk 2002; World Agroforestry Centre 2003c). It can take into account a mosaic of land-use plots (100 fields) within the landscape and investigate transitions in soil fertility, crop productivity, biodiversity and carbon stocks based on the dynamics of different land-use scenarios (van Noordwijk 2002). FALLOW is unique in comparison to many other models because it considers the roles of stakeholders in transforming landscapes, as well as stakeholders' feedbacks caused by a changing landscape (van Noordwijk 2002).

The Bio-Economic Agroforestry Model (BEAM), developed by the University of Wales, predicts the physical and financial performance of agricultural monocultures and silvopastoral and agri-silvopastoral polycultures under different scenarios. Originally designed to only address poplar (*Populus* spp.) systems in conditions present in the United Kingdom (Willis et al. 1993), BEAM is now sufficiently generalized to allow users to predict performance under a variety of managerial, silvicultural and economic conditions (Willis and Thomas 1997). BEAM was incorporated into the ACIAR and CIFOR project mentioned earlier as a means to evaluate the bio-economic impact and interaction of rubber plantations on *Imperata* grass. Purnamasari et al. (2002), using a modified version of BEAM to assess the impact of Indonesian rubber production under uncertainties of prices and climate, concluded that as a risk aversion strategy, it was better to use lower planting densities, undertake longer rotations and start tapping later in the life of the trees.

Many of the models presented above, and as noted in Table 4, are complex, predominantly used by researchers, and not very friendly to the layperson. Although some have been applied outside research (for example, BEAM and SCUAF), there is little evidence of use by decision makers, planners, extension agents and landowners. While many current versions of these models can still be obtained, there has not been a notable effort of their application toward extension and planning purposes.

Agroforestry knowledge-based or expert systems

The Agroforestry Knowledge Tool Kit (AKT) is perhaps the best initial attempt so far to construct and develop a 'true' KBS applicable to agroforestry (Walker et al. 1995). Recognizing the dearth of quantitative data in agroforestry and the need to consider a variety of information from a multidisciplinary range of sources (including farmers), AKT provides a framework to synthesize heuristic knowledge related to agroforestry systems and their ecological dynamics. AKT applies a formal language representation system to store and link together knowledge collected from different sources, which users, by using diagrams or a text-based interface, can access and infer the knowledge base in a flexible manner. Inferences on the knowledge base are especially useful in detecting gaps in the knowledge base and in extracting information for extension and planning purposes. Its application in representing temporal and spatial aspects of agroforestry, however, is limited (Walker et al. 1995).

AKT's use by researchers and development professionals includes the development of knowledge bases for agroforestry systems in South Asia, Southeast Asia and Africa, such as tree-fodder systems in Nepal, fermented tea production in the hills of Thailand, rangeland management with trees in Tanzania, and the Kandy forestry gardens in Sri Lanka. During 2001, ICRAF and the University of Wales conducted workshops for research institutions and NGOs (non-governmental organizations) in Thailand on the value of local ecological knowledge using the current AKT-5 (World Agroforestry Centre 2003d). Currently ATK is being used for various projects funded by the United Kingdom Department for International Development (DFID) (Dixon et al. 2001).

Hybrids and combination decision-support tools

Many DSTs used in agroforestry involve the use and integration of several different types of computer-based technologies. In China, a knowledge-based model developed for regional agroforestry planning of *Paulownia*–crop intercropping (PCI) integrates GIS, regression models and expert knowledge in order to spatially determine the biophysical, social and economic suitability of PCI within a planning area landscape (Liu et al. 1999).

CSTAF's SEADSS mentioned earlier is developed to address agroforestry problems at various scales (site

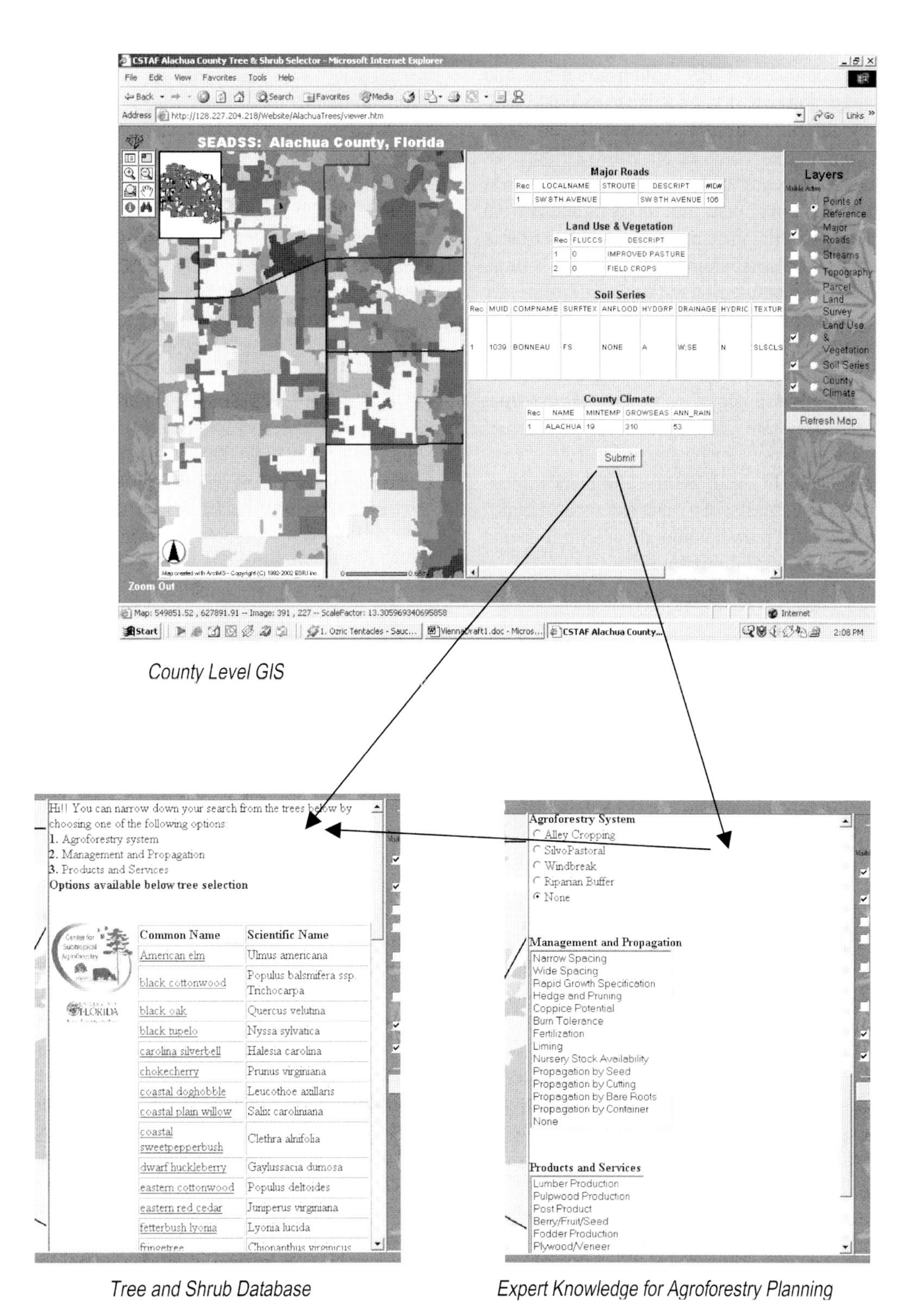

Figure 1. Hybrid decision support tool integrating GIS, sub-tropical tree and shrub database, and expert knowledge for agroforestry species selection in southeastern counties. Source: Ellis et al. 2003.

and landscape) and for different purposes (farm production and rural sustainability). It is an on-line hybrid DST linking county-level GIS, a subtropical tree and shrub database and expert knowledge to evaluate suitable tree or shrub components for specific sites and agroforestry practices (Ellis et al. 2003) (Figure 1). SEADSS selects trees from the database using queries that take into account climate and soil parameters from a selected geographic site and management criteria. Expert knowledge incorporated into SEADSS's database queries was obtained by surveying agroforestry professional using questionnaires to determine the following: 1) desired morphological and habit characteristics of species for each agroforestry practice, 2) economic and environmental services obtained from an agroforestry system, and 3) optimal plant characteristics for different environmental services (e.g. erosion control) (Ellis et al. 2000). SEADSS's intended end-users are landowners and extension agents, and their inputs and participation are essential during the validation and evaluation process.

Researchers and technology transfer experts at the USDA National Agroforestry Center and the University of Missouri recognized the need for a diverse range of DSTs to accommodate each unique setting in which agroforestry may be applied. Their Conservation Buffer Planning Project is developing a variety of DSTs (e.g., GIS-guided assessments, visual simulations, and cost-benefit models) to address the biophysical, social and economic issues critical in the planning and designing of systems that can better balance production with environmental stewardship (Figure 2) (Bentrup et al. 2000). The diversity of DSTs incorporated into this project reflects the need to meld concerns and objectives that occur at various spatial scales, the variability of information that may be available, and the recognition that people are differentially influenced in a decision-making process.

Challenges and opportunities for agroforestry decision-support tools

The major challenge in the development of effective computer-based DSTs for agroforestry is dealing with its complex nature. This challenge is further exacerbated by the limited knowledge base we have to work with currently and the many unresolved issues, making the decision process in agroforestry planning even more complex. Regardless of this challenge, the need for these tools is becoming more imperative, especially for extension and planning activities.

From databases to GIS and knowledge bases

DSTs incorporating databases are extremely valuable in organizing and accessing data and information. A database inventorying and describing agroforestry systems, such as AFSI, is extremely valuable and has great potential to be integrated into a GIS to evaluate the spatial distribution, characteristics and species components of agroforestry systems throughout the world. To ensure usability, databases need to be developed, tested and evaluated with end-user involvement. Up-front participation by end-users will result in better matching the tool to the end-user's needs and capabilities, and in increasing the awareness and therefore subsequent use of the tool by the end-user groups. The greater the utility for and use by end-users, the greater the support to maintain its utility and use over time. For example, AFSI has not been upgraded and does not appear to be used.

Descriptive and biophysical data for a majority of species are still unknown, making it difficult to develop useful databases. For instance, much of the early research on agroforestry species has concentrated on the aboveground productivity and has paid little attention to belowground issues such as, root characteristics and root responses to management practices (Sinclair 1996). As new pertinent information becomes available, databases will need to be updated to maintain their value and future utility. Database developers will therefore need to consider potential mechanisms for maintaining and upgrading their database. For example, AgroforesTree is both maintained and available on-line, making it more popular than its predecessor (MPTS).

Although the adoption of GIS applications has been slow in agroforestry research and planning, this trend is changing as a greater variety of spatial data is becoming accessible and affordable. Data on soils and land use are often free, and remotely sensed images (aerial photographs and satellite imagery) are also now available free or at very low costs. As GIS technology continues to become cheaper and easier to use, integrating agroforestry databases with geo-referenced data will occur due to the spatial nature and landscape issues involved with agroforestry planning and development. Users of this technology may not even be required to have GIS software and hardware with the advent of on-line GIS applications like SEADSS,

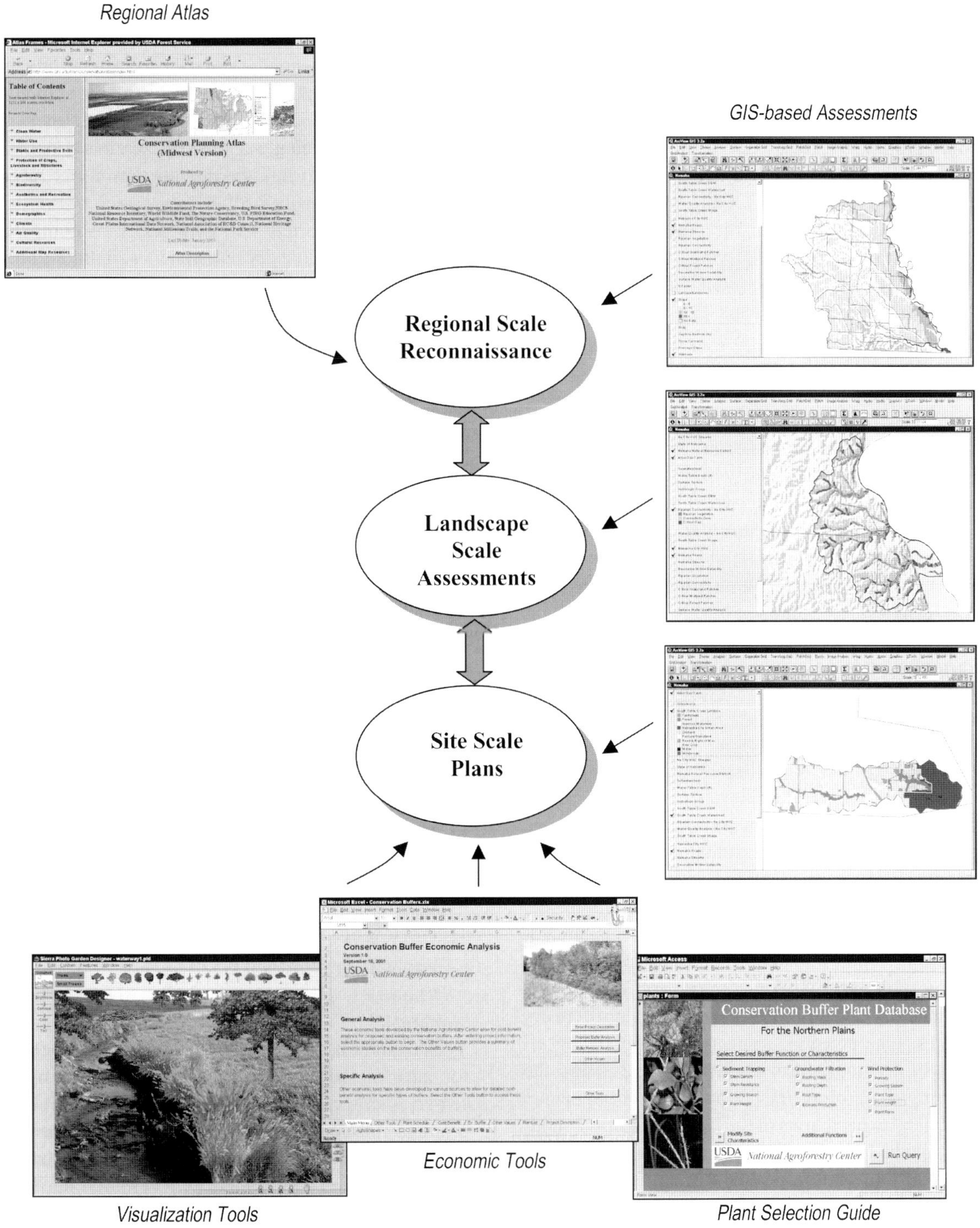
Regional Atlas
Table of Contents
Conservation Planning Atlas
(Midwest Version)
GIS-based Assessments
Regional Scale
Reconnaissance
Landscape
Scale
Assessments
Site Scale
Plans
Conservation Buffer Economic Analysis
Conservation Buffer Plant Database
For the Northern Plains
Economic Tools
Visualization Tools
Plant Selection Guide

Figure 2. Suite of computer-based tools being developed for a multi-scale and multi-issue conservation planning process for agroforestry buffers in the Western Corn-belt Ecoregion (Bentrup et al. 2000).

for example, which offers county-level GIS to extension agents and landowners within a user-friendly interface.

Knowledge-Based Systems have value because effective decision-making in agroforestry must involve the use of all available scientific, professional, and traditional information and knowledge. Currently much of our agroforestry knowledge is incomplete, contentious, observational, and qualitative. KBS overcomes some of these problems by tying these diverse sets of information together for the purposes of evaluating and synthesizing qualitative and traditional knowledge in addition to conventional research on agroforestry practices and systems (Sinclair and Walker 1998; Walker and Sinclair 1998; Dixon et al. 2001). Dixon et al. (2001) list some examples of reasoning tasks that knowledge bases could provide, including 1) generating synthesized reports on the current state of knowledge for a specific topic, 2) exploring the knowledge base to identify discrepancies between scientifically and locally derived knowledge, 3) correlating scientific information with professional and traditional knowledge, and 4) identifying key gaps in understanding agroforestry topics and practices.

From models to hybrid and combined decision-support tools

Much effort has focused on building sophisticated agroforestry models for research. Jagtap and Ong (1997) suggest that the ultimate goal of modeling should be to increase the relevance and efficiency of research by integrating the major social, economic and biophysical driving forces influencing agroforestry practices. They have identified several areas still in need of research that they feel are necessary for the development of efficient and successful models, including below-ground interactions, root architecture and dynamics, organic matter turnovers, nutrient and water competition, and livestock grazing on agroforestry practices. One potential result of these modeling efforts will be whole-farm evaluations that assess the suitability of agroforestry options to meet landowner and societal expectations of the land. In addition, models should be linked to GIS to predict where and how agroforestry technologies might be used; facilitating research and development, technology transfer, and policy and program activities (Bentrup and Leininger 2002).

Many of the models developed to-date have proved to be difficult to use for many practitioners, requiring large sets of specific parameter information that are frequently unavailable. As a result, some models have been hard to use for environments different than those in which they were originally developed, creating frustration among potential users of the models. While this highlights the need to collect additional and standardized data on crop, tree, soil and climate parameters for different geographical regions, it also clearly demonstrates the need to link the models with users' needs and resources.

Due to the variety and complexity of computer-based tools, agroforestry researchers and practitioners will need to partner with experts in computer science to develop effective and efficient DSTs (Walker and Lowes 1997). Through the development and evaluation of DSTs, knowledge gaps and future research needs can be determined. As with any innovative technology, agroforesters must be careful, however, not to view the DSTs as the ends but as a means to assist decision-making. As Wu and Hobbs (2002) state, 'We need to avoid having powerful methodologies in search of meaningful questions to answer; rather we need to seek the right techniques to answer pressing questions.'

Future direction for DST development in agroforestry

To place this review into the larger context of decision support, we need to step back from the specifics of the technology. Critical up-front questions need to be addressed as computer-based DSTs are being developed:

What is the primary purpose of DSTs?
What issues will the DSTs address?
Who will be the end-users of DSTs?
Who will develop the DSTs?
Who will implement/maintain the DSTs after development?

As mentioned earlier, many of the DSTs developed for agroforestry systems to-date are for testing research hypotheses. The tools primarily explore and evaluate the multitude of biophysical parameters for determining the key interrelationships in agroforestry practices. Developed primarily for research purposes, these tools play a valuable role in building the scientific foundation of agroforestry and will continue to be an important area of focus. On the other hand, development of DSTs for the adoption, planning, and implementation of agroforestry practices is vastly lacking in comparison to research-based applications. Developers of research-based tools often make the assumption that their applications can be easily used in the planning and design process of agroforestry systems. The answers to the critical questions guiding tool development, specifically, what is needed to influence end-user decision making, are more often than not different in research and planning-driven applications. The primary purpose for tools in research is to address the *why* of agroforestry practices, whereas in the planning and design process, the focus is on the *what*, *where* and *how*.

Because of the time and cost involved in the development of DSTs, they should be targeted to match end-user's needs and resources (Robinson 1996). Users of agroforestry tools for planning and design are primarily landowners and resource professionals working together in partnership to develop agroforestry plans. In this case, when the end-users are not directly involved in the development process, the result will be ineffective tools that do not respond to users' problems and needs, creating a waste of project funds and bitter feelings between developers and users (Hoag et al. 2000, Turner and Church 1995).

DSTs for research by necessity tend to focus on a limited number of issues while landowners must necessarily incorporate numerous and diverse issues in their decision-making process on whether or not to adopt agroforestry practices (Walker and Lowes 1997). Due to each individual's unique situation, resources, and personal value system, these biophysical, economic, and social issues are weighted differently in every potential application of agroforestry. The DSTs we create for agroforestry planning and design must be flexible enough to accommodate the range of potential issues and each individual unique decision-making process. Again, due to the nature of these issues, they need to be analyzed and synthesized at a variety of spatial scales. For instance, Helenius (1995) points out the advantages of being able to plan for ecological pest management at the landscape level where 'the benefits of improved logistics and economy of scale may provide sufficient incentive for the necessary local cooperation between farmers.'

Because one DST will not satisfy all of these demands, a suite of tools must be created to address the variety of issues at multiple scales. Ideally, these should be loosely coupled rather than intricately woven together. This approach allows users to select the DSTs appropriate for their situation and it facilitates the integration of new tools into the planning process.

From this brief discussion, several important principles are listed that can serve as a starting point for developers embarking on tool development. The DSTs should be:

- Focused on the what, where and how
- Capable of addressing multiple issues
- Developed with user participation
- Suitable for multiple spatial scales
- A loosely coupled suite of tools, and be
- Amenable to standardization of data formats.

Several projects illustrate some of these principles. SEADSS, for example, is working with landowners and extension agents to develop an online GIS application for plant selection at site and landscape scales (Ellis et al. 2003). The Comprehensive Conservation Buffer Planning Project offers a suite of loosely coupled DSTs for simultaneously addressing multiple issues across multiple scales (Bentrup et al. 2000). This project is also expanding the traditional definition of DSTs used in agroforestry by broadly considering the landowner adoption process. One such category of tools is computer-based visual simulations that can graphically depict future agroforestry scenarios at various spatial and temporal scales. This enables landowners and other stakeholders to better conceptualize and understand the agroforestry alternatives and it seems to be more influential in landowner/stakeholder adoption than information generated by ecological or economic DSTs (Al-Kodmany 1999; Nassauer et al. 2001).

Successful application of agroforestry systems depends upon pulling together diverse sources of information, in a manner that responds to users' needs and resources. Computer-based DSTs that accommodate these tasks can greatly facilitate the decision-making process that seeks to simultaneously balance environmental and production goals that meet landowner and societal needs. As Nassauer et al. (2001) state, we must go beyond providing tools that only address the ecological and economic aspects of sustainability and provide those that also enhance the cultural sustainability of agroforestry systems; that is, it must elicit sustained human attention over time or else the benefits may be compromised as land ownership changes, as development pressure increases, or as different political viewpoints arise.

Acknowledgements

Florida Agricultural Experiment Station Journal Series Number R-09865. This publication was partially supported by U.S. Department of Agriculture, Cooperative State Research, Education, and Extension Service, Initiative for Future Agriculture and Food Systems (USDA/CSREES/IFAFS) grant number 00-52103-9702. This paper was written and prepared by a U.S. government employee on official time, and therefore it is in the public domain and not subject to copyright.

References

Acock B., Pachepsky Y.A., Acock M.C., Reddy V.R. and Whisler F.D. 1997. Modeling soybean cultivar development rates using field data from the Mississippi Valley. Agron J 89: 994–1002.

Ahuja L.R., Ma L., Howell T.A. 2002. Agricultural System Models in Field Research and Technology Transfer. Lewis Publishers. Boca Raton, FL, 357 pp.

Al-Kodmany K. 1999. Using visualization techniques for enhancing public participation in planning and design: process, implementation, and evaluation. Landscape Urban Plan 45: 37–45.

Bentrup G., Dosskey M., Schoeneberger M., Wells M., Leininger T. and Klenke K. 2000. Planning for multi-purpose riparian management. pp. 423–426. In: Wigington P.J. and Beschta R.L. (eds), Riparian Ecology and Management in Multi-Land Use Watersheds. American Water Resources Association, Middleburg, VA.

Bentrup G. and Leininger T. 2002 Agroforestry mapping: the way GIS. J Soil Water Conserv 57: 148–153.

Bernard C. and Depommier D. 1997. The systematic approach and the role of GIS in the characterization and monitoring of agroforestry parks. XI World Forestry Congress, Antalya, Turkey, 13 to 22 October 1997, Vol. 1., p. 87.

Boote K.J., Jones J.W. and Hoogenboom G. 1998a. Simulation of crop growth: CROPGRO Model. pp. 651–692. In: Peart R.M. and Curry R.B. (eds), Agricultural Systems Modeling and Simulation. Marcel Dekker, New York, NY.

Boote K.J., Jones J.W., Hoogenboom G. and Pickering N.B. 1998b. The CROPGRO model for grain legumes. pp. 99–128. In: Tsuji G.Y., Hoogenboom G. and Thornton P.K. (eds), Understanding Options for Agricultural Production. Kluwer Academic Publishers, Dordrecht, The Netherlands.

Booth T.H. 1996. Tree selection and growth improvement. pp. 311–316. In: Dieters M.J., Matheson A.C., Nikles D.G., Hardwood C.E. and Walker S.M. (eds), Tree Improvement for Sustainable Tropical Forestry: Proceedings of OFRI-IUFRO Conference. Queensland Forestry Research Institute. Gypie, Qld, Australia.

Booth T.H., Stein J.A., Nix H.A. and Hutchinson M.F. 1989. Mapping regions climatically suitable for particular species: an example using Africa. Forest Ecol Manag 28: 19–31.

Booth T.H., Stein J.A., Hutchinson M.F. and Nix H.A. 1990. Identifying areas within a country climatically suitable for particular tree species: an example using Zimbabwe. Int Tree Crops J 6: 1–16.

Bydekerke L., Van Ranst E., Vanmechelen L. and Groenemans R. 1998. Land suitability assessment for cherimoya in southern Ecuador using expert knowledge and GIS. Agric Ecosys Environ 69: 89–98.

Cannell M.G.R., Mobbs D.S. and Lawson G.J. 1998. Complementarity of light and water use in tropical agroforests II. Modelled theoretical tree production and potential crop yield in arid to humid climates. Forest Ecol Manag 102: 275–282.

CABI (Commonwealth Agricultural Bureau International) 1998. The Forestry Compendium: A Silvicultural Reference. CABI, Oxford, United Kingdom. CD-ROM.

Davis M.W. 1988. Applied Decision Support. Prentice-Hall, Englewood Cliffs, NJ. 256 pp.

De Mers M.N. 1997. Fundamental of Geographic Information Systems. Wiley, New York, 486 pp.

Dixon H.J., Doores J.W., Joshi L. and Sinclair F.L. 2001. Agroecological Knowledge Toolkit For Windows: Methodological Guidelines, Computer Software and Manual for AKT5. School of Agricultural and Forestry Sciences, University of Wales, Bangor UK, 171 pp.

Dosskey M. and Wells G. 2000. Planning agroforestry practices. Agroforestry Notes (AF) #20. USDA National Agroforestry Center, 4 pp.

Easterling W.F., Hays C.J. Easterling M.M. and J.R. Brandle J.R. 1997. Modeling the effect of shelterbelts on maize productivity under climate change: an application of the EPIC model. Agric Ecosys Environ 61: 163–176.

Ellis E.A., Nair P.K.R., Linehan P.E., Beck H.W., Blanche C.A. 2000. A GIS-based database management application for agroforestry and tree selection. Computers and Electronics in Agriculture. 27: 41–55.

Ellis E.A., Nair P.K.R. and Jeswani S.D. 2003. The southeastern agroforestry decision support system (SEADSS): a Web-based application for agroforestry planning and tree selection. pp. 1–12. In: Vacik H., Lexer M.J., Rauscher M.H., Reynolds, K.M. and Brooks R.T. (eds), Decision Support for Multiple Purpose Forestry. A Transdisciplinary Conference on the Development and Application of Decision Support Tools for Forest Management. 23–25 April 2003, University of Natural Resources and Applied Life Sciences, Vienna Austria, CD-ROM.

Field R.C. 1984. National forest planning is promoting US Forest Service acceptance of operations research. Interfaces 14: 67–76.

Garcia-de-Ceca J.L. and Gebremedhin K.G. 1991. A decision support system for planning agroforestry systems. Forest Ecol Manag 45: 199–206.

Gertsis A.C. and Whisler F.D. 1998. GOSSYM: a cotton crop simulation model as a tool for the farmer. Acta Horticulturae. 476: 213–217.

Grabaum R. and Meyer B. 1998. Multicriteria optimization of landscapes using GIS-based functional assessments. Landscape Urban Plan 43: 21–34.

Grist P., Menz K. and Nelson R. 1998. *Gliricidia* as improved fallow. pp. 133–148. In: Menz K.M., Magcale-Macandog D. and Rusastra W.I. (eds), Improving Smallholder Farming systems in Imperata Areas of Southeast Asia: Alternatives to Shifting Cultivation. ACIAR Monograph No. 52.

Guo Q. 2000. Climate change and biodiversity conservation in the Great Plains. Global Environmental Change 10: 289–298.

Hackett C. and Vanclay J.K. 2003. Mobilizing expert knowledge of tree growth with the PLANTGRO and INFER systems. Plantsoft Services, Stirling Australia and CIFOR, Jakarta, Indonesia, 18 pp.

Helenius J. 1995. Regional crop rotation for ecological pest management (EPM) at landscape level. pp. 255–260. In: Integrated Crop Protection: Towards Sustainability? Proceedings of the British

Crop Protection, 1995, September 11–14, Edinburgh, Scotland. BCPC Registered Office, Surrey, UK.

Hoag D.L., Ascough II J.C. and Frasier W.M. 2000. Will farmers use computers for resource and environmental management? J Soil Water Conserv 55: 456–462.

Jagtap S. and Ong C.K. 1997. Perspective on issues, needs and opportunities for agroforestry models. pp. 110–112. In: Agroforestry Modelling and Research Coordination. Annual Report to ODA July 1996–June 1997. Institute of Terrestrial Ecology. Natural Environment Research Council. Edinburgh, UK.

Jones C.A., Dyke P.T., Williams J.R., Kiniry J.R., Benson V.W., Griggs R.H. 1991. EPIC: and operational model for evaluation of agricultural sustainability. Agr Syst 37: 341–350.

Jones J.W., Tsuji G.Y., Hoogenboom G., Hunt L.A., Thornton P.K., Wilkens P.W., Imamura D.T., Bowen W.T. and Singh U. 1998. Decision support system for agrotechnology transfer; DSSAT v3. pp. 157–177. In: Tsuji G.Y., Hoogenboom G. and Thornton P.K. (eds), Understanding Options for Agricultural Production. Kluwer Academic Publishers, Dordrecht, The Netherlands.

Jones J.W., Hoogenboom G., Poter C.H., Boote K.J., Batchelor W.D., Hunt L.A., Wilkens P.W., Singh U., Gijsman A.J. and Ritchie J.T. 2003. The DSSAT cropping system model. Eur J Agron 18: 235–265.

Jose S., Gillespie A.R., Seifert J.R. and Pop P.E. 2001. Comparison of minirhizotron and soil core methods for quantifying root biomass in a temperate alley cropping system. Agroforest Syst 52: 161–168.

Kent B., Bare B.B., Field R.C. and Bradley G.A. 1991. Natural resource land management planning using large scale linear programs: the USDA Forest Service experience with FORPLAN. Oper Res 39: 13–27.

Kleine M., Scott P., de Neergard N.B., Pasiecznik N. and Becker K. 2003. New technologies to support decision making and global collaboration in multiple-purpose forest management-the Forestry Compendium and the Global Forest Information System. pp. 1–10. In: Vacik H., Lexer M.J., Rauscher M.H., Reynolds K.M. and Brooks R.T. (eds), Decision Support for Multiple Purpose Forestry. A Transdisciplinary Conference on the Development and Application of Decision Support Tools for Forest Management. April 23–25, 2003, University of Natural Resources and Applied Life Sciences, Vienna Austria, CD-ROM.

Knopp P.D. and Twery J. 2003. Stewplan: software for creating forest stewardship plans (Version 1.3). Gen. Tech. Rep. NE-301. U.S. Department of Agriculture, Forest Service, Northeastern Research Station, 12 pp.

Lacher T.E. 1998. The spatial nature of conservation and development. pp. 3–12. In: Savitsky B.G. and Lacher T.E. (eds), GIS Methodologies for Developing Conservation Strategies: Tropical Forest Recovery and Wildlife Management in Costa Rica. Columbia University Press, New York, NY.

Lawson G.J. and Cannell M.G.R. 1997. Introduction and Formal DFID Annual Progress Report. pp. 1–9. In: Modelling and Research Coordination. Annual Report to ODA July 1996–June 1997. Institute of Terrestrial Ecology. Natural Environment Research Council. Edinburgh, UK.

Liebowitz J. 1990. The Dynamics of Decision Support Systems and Expert Systems. Dryden Press, Chicago. 211 pp.

Liu J., Shao G. and Li W. 1999. Expert knowledge-based model for regional agroforestry planning. pp. 1–8. In: Li B Geoinformatics and Socioinformatics: The Proceeding of Geoinformatics'99 Conference. Ann Arbor, Michigan, 19–21 June. University of California, Berkely, CA.

Loh D.K. and Rykiel E.J. 1992. Integrated resource management systems: coupling expert systems with data-base management and geographical information systems. Environ Manage 16: 167–177.

Magcale-Macadong D.B., Predo C.D., Menz K.M. and Calub A.D. 1998. Napier grass strips and livestock: a bioeconomic analysis. Agroforest Syst 40: 41–58.

Manning E. 1996. GLYCIM: crop model for soybeans. Progressive Farmer 11: 33.

Masera O.R., Garza-Caligaris, J.F., Kanninen M., Karjalainen T., Liski J., Nabuurs G.J., Pussinen A., de Jong B.H.J., Mohren G.M.J 2003. Modeling carbon sequestration in afforestation, agroforestry, and forest management projects: the CO2FIX V.2 approach. Ecol Mod 164: 177–199.

Mata-Toledo R.A. and Cushman P.K. 2000. Schaum's Outline of Fundamentals of Relational Database. McGraw-Hill Professional. New York, NY. 249 pp.

Matthews, R.B. and Lawson G.J. 1997. Structure and applications of the HyCAS model. Agroforest Forum 8(2): 14–17.

Menz K.M., Magcale-Macandog D. and Rusastra W.I. (eds) 1998. Improving Smallholder Farming Systems in Imperata areas of Southeast Asia: Alternatives to Shifting Cultivation. ACIAR Monograph No. 52. 280 pp.

Middlemiss P. and Knowles L. 1998. AEM Agroforestry Model User Guide for v.4.0g. Forest Research Institute, Rotorua, New Zealand. 121 pp.

Mobbs D.C., Cannell M.G.R., Crout N.M.J., Lawson G.J., Friend A.D., Arah J. 1998. Complementarity of light and water use in tropical agroforests I. Theoretical model outline, performance and sensitivity. Forest Ecol Manag 102: 259–274.

Mobbs D.C., Lawson G.J., Brown T.A.W. 2001. Model for Agroforestry Systems User Guide: HyPAR Version 4.1. Centre for Ecology & Hydrology, Forestry Research Programme. Edinburgh, UK. 52 pp.

Muetzelfeldt R.B. and Taylor J. 1997. The Agroforestry Modelling Environment. pp. 10–20. In: Agroforestry Modelling and Research Coordination. Annual Report to ODA July 1996-June 1997. Institute of Terrestrial Ecology. Natural Environment Research Council. Edinburgh, UK.

Nair P.K.R. 1987. Agroforestry System Inventory. Agroforest Syst 5: 319–338.

Nair P.K.R. 1998. Directions in tropical agroforestry research: past, present and future. Agroforest Syst 38: 223–245.

Nair P.K.R. 2001. Agroforestry. pp. 375–393. In: Our Fragile World: Challenges and Opportunities for Sustainable Development, Forerunner to The Encyclopedia of Life Support Systems, Chapter 1.25, vol. I. UNESCO, Paris, France, & EOLSS, UK.

Nassauer JI, Kosek SE and Corry RC. 2001. Meeting public expectations with ecological innovation in riparian landscapes. J American Water Resources Association 37: 1439–1443.

NRC (National Research Council). 1997. Precision Agriculture in the 21st Century: Geospatial and Information Technologies in Crop Management. National Academy Press, Washington, D.C. 149 pp.

Nelson R.A., Grist P.G., Menz K.M., Cramb R.A., Paningbatan E.P. and Mamicpic M.A. 1997. A cost-benefit analysis of hedgerow intercropping in the Philippine uplands using the SCUAF model. Agroforest Syst 35: 203–220.

Nikolopoulos C. 1997. Expert Systems: Introduction to First and Second Generation and Hybrid Knowledge Based Systems. Marcel Dekker, New York. 331 pp.

Oduol .P.A., Muraya P., Fernandes E.C.M. and Nair P.K.R. 1988. The agroforestry systems database at ICRAF. Agroforest Syst 6: 253–270.

Paarlberg D. and Paarlberg P. 2000. Agricultural Revolution of the 20th Century. Iowa State University Press, Ames, IA. 154 pp.

Paquette S. and Domon G. 1997. The transformation of the agroforestry landscape in the nineteenth century: a case study in southern Quebec, Canada. Landscape Urban Plan 37: 197–209.

Patterson D.W. 1990. Introduction to Artificial Intelligence and Expert Systems. Prentice-Hall, Englewood Cliffs, NJ. 448 pp.

Peng G., Leslie L.M., Shao Y. 2002. Environmental Modelling and Prediction. Springer-Verlag. Berlin, Germany. 480 pp.

Purnamasari R., Cahco O. and Simmons P. 2002. Management strategies for Indonesian rubber production under yield and price uncertainty: a bioeconomic analysis. Agroforest Syst 54: 121–135.

Rauscher M.H. 1999. Ecosystem management decision support for federal forests in the United States: a review. Forest Ecol Manag 114: 173–197.

Reddy K.R., Kakani V.G., McKinion J.M. and Baker D.N. 2002. Applications of a cotton simulation model, GOSSYM, for crop management, economic and policy decisions. pp. 33–54. In: Ahuja L.R., Ma L. and Howell T.A. (eds), Agricultural System Models in Field Research and Technology Transfer. Lewis Publishers. Boca Raton, FL.

Reynolds K.M., Rodriguez S. and Bevans K. 2002. Ecosystem Management Decision Support. USDA Forest Services, Pacific Northwest Station. Corvallis, OR. 80 pp.

Ritchie J.T. and Otter S. 1985 Description and performance of CERES-Wheat: a user oriented wheat yield model. pp. 159–175. In: ARS Wheat Yield Project. ARS 38. National Technical Information Service, Springfield, VA.

Robinson B. 1996. Expert systems in agriculture and long-term research. Canadian J Plant Science 76: 611–617.

Robotham M.P. 1998. AGFADOPT: a decision support system for agroforestry project planning and implementation: pp. 167–176. In: El-Swaify S.A. and Yakowitz D.S. (eds), Multiple Objective Decision Making for Land, Water and Environmental Management: Proceedings of the First International Conference on Multiple Objective Decision Support Systems for Land, Water and Environmental Management: Concepts, Approaches, and Applications. Lewis, Boca Raton, FL.

Ruark G.A., Schoeneberger M.M. and Nair P.K.R. 2003. Agroforestry – Helping to Achieve Sustainable Forest Management. UNFF (United Nations Forum for Forests) Intersessional Experts Meeting on the Role of Planted Forests in Sustainable Forest Management, 24–30 March 2003, New Zealand. pp. 1–13.

Salim A.S., Simons A.J., Waruhiu A., Orwa C. and Anyango C. 1998. Agroforestry Database: A Tree Reference and Selection Guide. Version 1.0. ICRAF. Nairobi, Kenya. CD-ROM.

Sanchez P.A. 1995 Science in agroforestry. Agroforest Syst 30: 5–55.

Sanders R.E. (2000) DB2 Universal Database: Application Programming Interface (API) Developer's Guide. McGraw-Hill Professional. New York, NY. 640 pp.

Schmoldt D.L. and Rauscher H.M. 1996. Building Knowledge-Based Systems for Natural Resource Management. Chapman & Hall, New York. 386 pp.

Schoeneberger M., Dix M. and Dosskey M. 1994. Agroforestry-enhanced biodiversity: the good, the bad and the unknown. pp. 207–215. In: Rietveld W. (ed.), Agroforestry and Sustainable Systems: Symposium Proceedings; 1994, August-10: Fort Collins, CO. General Technical Report RM-GTR-261. Fort Collins, CO: US Department of Agriculture, Forest Service, Rocky Mountain Forest & Range Experiment Station.

Schroder J.M. and Jaenicke H. 1994. A computerized database as decision support tool for the selection of agroforestry tree species. Agroforest Syst 26: 65–70.

Sinclair F.L. 1996. The emergence of associative tree ideotypes from ecophysiological research and farmers' knowledge. Agroforest Forum 7: 17–19.

Sinclair F.L. and Walker D.H. 1998. Acquiring qualitative knowledge about complex agroecosystems. Part 1: representation as natural language. Agr Syst 56: 341–363.

Skidmore A. (ed.) 2002. Environmental Modelling with GIS and Remote Sensing. Taylor & Francis Inc. London, UK. 268 pp.

Sunderic D. and Woodhead T. 2001. SQL Server 2000: Stored Procedure Programming. McGraw-Hill Professional. Berkeley, CA. 732 pp.

Timlin D.J., Pachepsky Y.A., Whisler F.D. and Reddy V.R. 2002. Experience with onfarm applications of GLYCIM/GUICS. pp. 55–69. In: Ahuja L.R., Ma L. and Howell T.A. (eds), Agricultural System Models in Field Research and Technology Transfer. Lewis Publishers. Boca Raton, FL.

Turner B.J. and Church R. 1995. Review of the use of the FORPLAN (FORest PLANning) model. Department of Conservation and Natural Resources, Victoria, Australia. 25 pp.

Twery M.J., Rauscher M.H., Bennett D.J., Thomasma S.A., Stout S.L., Palmer J.F., Hoffman R.E., DeCalesta D.S., Gustafson E., Cleveland H., Grove J.M., Nute D., Kim G. and Kollasch P.R. 2000. NED-1: integrated analyses for forest stewardship decisions. Computers and Electronics in Agriculture 27: 167–193.

Twery M.J., Rauscher M., Knopp P.D., Thomas S.A., Nute D.E., Potter W.D., Maier F., Wang J., Dass M., Uchiyama H. and Glende A. 2003. NED-2: An integrated forest ecosystem management decision support system. pp. 1–11. In: Vacik H., Lexer M.J., Rauscher M.H., Reynolds K.M. and Brooks R.T. (eds), Decision Support for Multiple Purpose Forestry. A Transdisciplinary Conference on the Development and Application of Decision Support Tools for Forest Management. 23–25 April 2003, University of Natural Resources and Applied Life Sciences, Vienna Austria, CD-ROM.

Unruh J.D. and Lefebvre P.A. 1995. A spatial database approach for estimating areas suitable for agroforestry in sub-Saharan Africa: aggregation and use of agroforestry case studies. Agroforest Syst 32: 81–96.

van Noordwijk M. 2002 Scaling trade-offs between crop productivity, carbon stocks and biodiversity in shifting cultivation landscape mosaics: the FALLOW model. Ecol Model 149: 113–126.

van Noordwijk M. and Lusiana B. 1999. WaNuLCAS, a model of nutrient and light capture in agroforestry systems. pp. 217–242. In: Auclair D. and Dupraz C. (eds), Agroforestry for Sustainable Land-Use Fundamental Research and Modelling with Emphasis on Temperate and Mediterranean Applications. Kluwer Academic Publishers. Dordrecht, The Netherlands.

van Noordwijk M. and Lusiana B. 2000. WaNuLCAS version 2.0, Background on a Model of Water Nutrient and Light Capture in Agroforestry Systems. International Centre for Research in Agroforestry (ICRAF). Bogor, Indonesia. 186 pp.

Vermeulen S.J., Woomer P., Campbell B.M., Kamukondiwa W., Swift M.J., Frost P.G.H., Chivaura C., Murwira H.K., Mutambanengwe F. and Nyathi P. 1993. Use of the SCUAF model to simulate natural miombo woodland and maize monoculture ecosystems in Zimbabwe. Agroforest Syst 22: 259–271.

von Carlowitz P.G., Wolf G.V. and Kemperman R.E.M. 1991. Multipurpose Tree and Shrub Database: An Information and Decision Support System. ICRAF, Nairobi, Kenya & GTZ, Eschborn, Germany. 104 pp.

World Agroforestry Centre 2003a. Agroforestry Database. http://www.worldagroforestrycentre.org/Sites/TreeDBS/AFT/AFT.htm Accessed June 15, 2003.

World Agroforestry Centre 2003b. WaNuLCAS: A model of Water Nutrient and Light Capture in Agroforestry Systems. http://www.worldagroforestrycentre.org/sea/agromodels/WaNuLCAS/index.htm Accessed June 15, 2003.

World Agroforestry Centre 2003c. Forest, Agroforest, Low-value Landscape or Wasteland? http://www. worldagroforestrycentre.org/sea/agromodels/FALLOW/fallow.htm Accessed June 15, 2003

World Agroforestry Centre 2003d. Training on Local Ecological knowledge (LEK) and Knowledge-based systems (KBS). Chiang Mai, Thailand: 8–12 June 2001. http://www. worldagroforestrycentre.org/sea/Products/training/

GroupTra/LEK/#Objectives Accessed September 12, 2003.

Walker D.H. and Lowes D. 1997. Natural resource management: opportunities and challenges in the application of decision support systems. AI Applications 11(2): 41–51.

Walker D.H. and Sinclair F.L. 1998. Acquiring qualitative knowledge about complex agroecosystems. Part 2: formal representation. Agr Syst 56: 365–386.

Walker D.H., Sinclair F.L. and Kendon G. 1995. A knowledge-based system approach to agroforestry research and extension. AI Applications 9: 61–72.

Warkentin M.E., Nair P.K.R., Ruth S.R. and Sprague K. 1990. A knowledge-based expert system for planning and design of agroforestry systems. Agroforest Syst 11: 71–83.

Willis R.W. and Thomas T.H. 1997. POPMOD: A Bioeconomic Agroforestry Model. The BEAM Project. School of Agricultural and Forest Sciences, University of Wales, Bangor, UK. 13 pp.

Willis R.W., Thomas T.H., van Slycken J. 1993. Poplar agroforestry: a re-evaluation of its economic potential on arable land in the United Kingdom. Forest Ecol Manag 57: 85–97.

Wu J. and Hobbs R. 2002. Key issues and research priorities in landscape ecology: an idiosyncratic synthesis. Landscape Ecol 17: 355–365.

Young A. and Muraya P. 1990. SCUAF: Soil Changes Under Agroforestry. ICRAF, Nairobi. 124 pp.

Zazueta S.P. and Xin J. 1998. Computers in Agriculture: Proceedings of the 7th International Conference. Orlando, Florida, October 26–30. American Society of Agricultural Engineers, St. Joseph, MI. 999 pp.

Zeiler M. 1999. Modeling our World: the ESRI guide to geodatabase design. ESRI Press. Redlands, CA. 199 pp.

Agroforestry Systems **61:** 423–435, 2004.

Anthropogenic grasslands in Southeast Asia: Sociology of knowledge and implications for agroforestry

M.R. Dove
School of Forestry and Environmental Studies, Yale University, 205 Prospect Street New Haven, Connecticut 06511-2189, U.S.A.; e-mail: michael.dove@yale.edu

Key words: Development failure, Grassland myths, Grassland research, *Imperata cylindrica*, Scientific paradigms, Western environmental thought.

Abstract

Anthropogenic grasslands, meaning grasslands that have been influenced and modified by humans, are one of the most important land covers of the tropics, but their management is dominated by conflicted and contested views, which is reflected in the problematic record of grassland development intervention. This chapter analyzes the historic, cultural, political, and institutional factors that affect the way grasslands are viewed, drawing largely on data from Southeast Asia. These data suggest that perceptions of grasslands are colored in part by the marginal place that they occupy in the cosmology of western industrialized societies, which idealize forest covers. Consequently, national and international agencies view grasslands not as a common land cover but as a development problem. The agendas of government and development agencies are often not grounded in a proper understanding of the local human and bio-physical ecology of grasslands or of successful local agroforestry practices; and research on many of the most important dimensions of grassland management is poorly conducted and/or utilized. The recent rise in scientific interest in indigenous knowledge, environmental history, and non-equilibrium systems, has opened up new possibilities for the study of grasslands. Agroforestry, given its inherent bridging of nature and culture, is ideally suited to benefit from these possibilities, by focusing research attention on the bio-social factors that determine the appearance, disappearance, and maintenance of anthropogenic grasslands.

Introduction

The importance to the successful conservation and development of natural resources of understanding human social, economic, and political behavior has become widely accepted over the past generation, including within the discipline of agroforestry. Less accepted is the importance of understanding the more fundamental aspects of our own knowledge of natural resources and their use. Attention to how knowledge is produced, attention to the sociology of knowledge, is especially important in fields such as agroforestry that manifestly cut across more orthodox ways of scientifically and institutionally dividing up the world. It is even more important when agroforestry is practiced in environments that are themselves considered to be somewhat anomalous, like grasslands. The production of knowledge in grassland agroforestry simultaneously presents stiff new challenges and opens up possibilities for theoretical and also practical innovation.

Tropical grasslands were one of the earliest ecosystems to be exploited by humans and they remain one of the most important (Harlan 1982; Solbrig 1993). Tropical grasslands and savannas occupy one-third of the earth's land surface (40% of the tropics) and are inhabited by one-fifth of the earth's human population (Werner et al. 1990; Solbrig 1996). In some (semi)arid and arctic regions, such grasslands may form the climax vegetation; but in (sub)humid tropical regions, grasslands are mostly 'anthropogenic' in origin and character, meaning 'having its origin in the activities of man' (Oxford English Dictionary 1989). A decade ago in Southeast Asia, grasslands dominated by one species, *Imperata cylindrica* (L.) Beauv., were alone estimated to cover 20 million ha (Vandenbeldt 1993). Given the depth and breadth of their place on the human landscape, it is remarkable how conflicted our understanding of such grasslands is. Tropical, anthropogenic grasslands are one of the most misunderstood human modification of the earth's natural environments.

Discussions of research and development in tropical anthropogenic grasslands have long been characterized by paradox and contest. Local communities and extra-local development agencies often hold diametrically opposing views of grasslands, with the former often seeing them as an integral component of their agro-ecology and the latter seeing them as wastelands. Villagers may see grasslands as fragile, whereas officials see them as tenacious. Scholars themselves have vacillated over the past two generations between seeing grasslands as models of equilibrium versus disequilibrium. As might be predicted from these conflicted views, development interventions intended to convert grasslands to other covers or uses have been quite problematic.

This pattern of conflicted intellectual and developmental engagement with grasslands is not new: its roots extend back into the colonial administrations of the nineteenth century and, in particular, to their concern for estate crop production and biases against extensive, subsistence agriculture. Although contests over the interpretation and management of grasslands have thus dominated their 'ecology' [in Bateson's (1972) sense of the word] for a century and a half, research on tropical grasslands has focused and continues to focus largely on their bio-physical dimensions, as noted by a number of the contributors to Agroforestry Systems 36 (1997) special issue on *Imperata*. All of this suggests that we need both to better understand tropical anthropogenic grasslands and local agroforestry practices therein and to understand much better the way this same understanding is constrained by historical, cultural, and institutional factors.

The purpose of the current analysis is to outline some of these constraints, drawing largely on data (both primary and secondary) from Southeast Asia, which has some of the most extensive and most intensively debated anthropogenic grasslands in the world.[1] I will first show how anomalous grasslands are within the environmental views of the western, industrialized countries. Next I will examine how an anomalous view of grasslands contributes to their conceptualization as an object for development intervention and, similarly, how the imperatives of development contribute to the misunderstanding of grasslands. I will then examine how state agendas contribute to the misunderstanding of grasslands and how new scientific interest in indigenous knowledge, environmental history, and nonequilibrium systems is opening up new possibilities for transcending this misunderstanding. I will conclude by summarizing the different historical stages of research on anthropogenic grasslands and discuss the appropriateness of agroforestry approaches for future research on them.

The forest bias

Grasslands hold an anomalous place within western environmental thinking, which has been prejudicial to the treatment of anthropogenic grassland within conservation and development planning. This anomaly stems from the relative emphasis on trees versus grasses in western world views, and the related focus of western environmentalism on forests to the exclusion of other habitats.

Trees versus grasses in Western and Non-Western thought

The tree and the forest occupy a powerful place in western thought. Their role as 'root metaphors' (Ortner 1973), or central organizing images, is evident in pioneering works like Frazer's 'Golden Bough' (1951), wherein tree beliefs are used as the central optic through which Western and non-Western religion are compared and contrasted. A recent contribution is Davies' (1988) 'The Evocative Symbolism of Trees.' In analyzing the centrality of the tree to western thought, he writes, 'In scaling the tree of knowledge without getting too far out on any limb, in exploring the many branches of thought, and in attempting to get at the root of the matter, we pursue a branching task' (Davies 1988). Davies (1988) attributes this centrality not to cultural or historical factors but to the 'intrinsic suggestivity and the inherent attractiveness of trees as the basis for and evocative of symbolic responses.' He argues that by comparison, 'After all is said it remains true that grass, the most universal and successful of plants, has seldom fed the flames of creative thought to any marked extent. Trees have done so because they possess not only a variety of parts but because they stand over and against human generations in a way which demands acknowledgment.' Whereas the longevity of trees does seem to impress itself on human thought in many societies, both Western and non-Western[2], this naturalization of tree symbols has come in for increasing critique among scholars.

This critique has been based, in part, on the comparative study of tree metaphors in non-western societies (Dove 1998; Rival 1998). The critique also has been driven by a more critical study of Western tree metaphors (Bourdieu 1977; Bouquet 1995). The impact of this scholarship has been to de-naturalize the tree metaphor and to alert us to the possibility of its cultural and even political loading. This latter point has been pushed perhaps furthest by Deleuze and Guattari (1987), who compare arboreally-oriented western thought unfavorably with 'rhizome-oriented' Eastern thinking.

Such a sweeping dichotomization of western tree-based thought and eastern root-based thought is untenable. The central role of tree metaphors documented by Frazer in European folklore and religion is easily matched, as Frazer also found, by the tree-of-life, which is still a central cosmological symbol all over Southeast Asia [see Schärer (1963) on Borneo].[3] This ritual role parallels the centrality of trees in human economic and ecological history [especially in (sub)humid biomes]. The oldest trade-good in the world may well be tree exudates. Camphor (*Dryobalanops aromatica* Gaetn. F.), 'dragon's blood' (*Daemonorhops Blume, spp.*), gum benjamin (*Styrax spp.*), and various pine resins (especially from *Pinus merkusii*) are the oldest trade products of Southeast Asia; and they fit into a trade niche that was originally created for the even more ancient traffic in the tree saps of the Middle East, the fabled frankincense (*Boswellia spp.*) and myrrh (*Commiphora spp.*). Even more important perhaps, one of the oldest and most successful forms of agriculture, swidden cultivation, is based on the cyclic clearing, burning, and sowing of forest plots, with trees thus providing the essential environmental conditions for crop cultivation. Swidden agriculture was prevalent in Europe and North America through the nineteenth century and is still ubiquitous in Southeast Asia and other parts of the tropics. Estimates suggest that swidden agriculture supports as many as one billion people today, representing 22% of the population of the developing world in tropical and subtropical zones (Thrupp et al. 1997). Wherever and whenever swidden agriculture was replaced by more intensive, permanent-field forms of agriculture, the clearing of forest assumed a central role in cultural conceptions of social development and evolution. For example, a common Javanese term for development, *babad alas*, literally means 'to clear the forest' (Horne 1974).

This long and intimate association of people and trees has contributed to the ubiquitous importance of tree symbols in the region. Since this association is most often based not simply on management of forests, however, but on their clearance and re-growth or conversion, involving relationships between forests and other sorts of vegetative covers as well, it should come as no surprise to find that grass symbols are equally ubiquitous. Grasses in fact have played an important role in traditional ritual throughout Southeast Asia (e.g., Dove 1986b; Wessing 1992), as we find in Hadiwidjojo's (1956) notes on the prominence of *Imperata cylindrica* in traditional Hindu Javanese ritual. He quotes the traditional Javanese saying that 'God lies in the tip of a stalk of Imperata,' which refers to the fact that stalks of *Imperata* were used to sprinkle holy water during traditional Hindu ceremonies (ibid. 6). To this day, *Imperata* is used in most domestic rituals in central Java, as in marital ceremonies, when it is placed under the mat on which bride and groom kneel (Carpenter 1987).

The traditional symbolic loading of grasses in Southeast Asia is attested to by their role in myths pertaining to the origin of culture, for example among the aboriginal Batek of the Malay Peninsula (Tuck-Po 2004):

> In the beginning, all *bangsa'* 'races' were the same. All were Batek. One day, Adam, Tohan's younger brother, set fire to the *lalang* 'grass' where the Batek were living ... One family fled into the forest. They left behind their *surat kitab* 'religious books'. ... Another family was near the riverbank. They jumped into the river and took the *surat kitab* with them down river towards the sea. These became the Malays. The Batek looked for their *surat kitab* but the Malays had hidden them. This is why the Malays pray now. So the *surat kitab* belonged to the Batek first.

The prominence of grasses in the landscape of all-important cultural myths both reflects and attests to their role as part of the basic cultural idiom for thinking about the world. Although this does not validate Deleuze and Guattari's 'Eastern rhizome' thesis, it does help to make it apparent that the Western idealization of trees and forests is also cultural and historical and, thus, subjective.

Forest focus of Western environmentalism

This discussion of the prominent role of trees and forests in Western versus non-Western thought is of more than academic interest. It seems probable that this prominence has influenced the agenda of academic ecological and botanical research and that this in turn has influenced the agenda of international environmental policy-making and intervention. For more than a century, anthropogenic grasslands and forests have fallen on either side of an academic *cordon sanitaire*, with the latter being seen as a more legitimate and the former as a less legitimate subject of research. The distinction is based on two, related characteristics: the perceived difference between natural and nonnatural vegetation, and the perceived difference between forests and everything else.

The pioneering ethno-botanist Edgar Anderson was one of the first to identify the bias against purportedly non-natural vegetation one-half century ago and argue for a different path. He wrote:

> Once he is in the field, the average taxonomist is an incurable romantic. Watch him take a group of students on a field trip. The nearest fragments of the original flora may be miles away and difficult of access but that is no barrier. With truck, bus, train, jeep, or car, on foot if need be, the class is rushed past the domesticated and semi-domesticated flora among which they spend their lives to the cliff side or peat bog or woodland which most nearly reflects nature in pre-human times. (Anderson 1971 pp. 45–46)

Whereas Anderson's point was that the theoretical questions pertaining to human-influenced vegetation are just as interesting and important as those pertaining to so-called 'pre-human' or natural vegetation, it is now recognized that the perceived dichotomy between the former and the latter is much more problematic that was once supposed. Work over the past decade in historical ecology and allied fields has demonstrated that what we once thought to be 'natural' or 'virgin' forest has in most cases been modified by human behavior, however subtly (Denevan 1992; McDade 1993; Balée 1994; Corlett 1995). The past conception of virgin and unchanging forest is now seen to reflect the premises of a modern, essentializing, equilibrium-based model of the environment, which has been largely discredited. As Worster says, the forest is now seen as just 'an erratic, shifting mosaic of trees and other plants.' (1990, p. 8). Considerable academic interest has shifted from traditional studies of climax vegetation to fallow dynamics and secondary forests (e.g., Denevan and Padoch, 1987; Nyerges, 1989; Lambert, 1996; Lawrence and Mogea, 1996; Coomes et al. 2000; Chokkalingam et al., 2001).

The past valorization of natural vegetation has been associated in most policy as well as academic contexts with an emphasis on forests (and also tree plantations) as opposed to other vegetative covers. Buttel (1992) memorably termed this emphasis, in a critique of the focus of international environmentalism on Third World forests, 'forest fundamentalism.' Stott (1991) similarly wrote that 'With a few notable exceptions ... the 'green' debate in the tropics nearly always focuses on the forests'. This forest bias is not confined to the exercise of international policy in the tropics; it periodically surfaces in Western industrialized nations as well. As the environmental historian Cronon writes, '...to this day there is no national park in the grasslands [of the United States]'. (1996, p. 73) This bias has political implications when it is exercised in less-developed non-Western countries, however, that it does not have when exercised in North America. As Stott writes (cf. Buttel 1992):

> The dominant forest ideology is most unfortunate because it blinds us to the facts that the rain forests are not, and never were, the most widespread formations in the tropics, and that the majority of people eking out a living in the tropical world do so in lands derived from other, non-forest, formations, above all the savannas'. (1991, p. 18)

In short, the forest bias is deprivileging for other land covers like savannas and grasslands and for the peoples who live in them[4]. The lack of attention to these nonforest biotas in policy and regulatory circles, both national and international, can be inimical to their continued existence. The most obvious example of this involves the ubiquitous policy 'solution' to anthropogenic grasslands, namely reforestation.

Grassland myths and development interventions

This cultural-historical view of grasslands as 'other' is associated with a particularly hostile developmental stance toward them. From the colonial to the postcolonial era, most government policies have problematized tropical anthropogenic grasslands and sought to replace them with other land-uses, albeit with very limited success. The persistence of such unsuccessful policies can be attributed to the opportunities that they create for development intervention in local resource-use systems.

Interventions and failures

The basis for development interventions in grasslands is laid in the way that they are classified. During colonial as well as post-colonial times throughout Southeast Asia, anthropogenic grasslands have been labeled 'wastelands' or something similar, a term that is inherently supportive of alternate land-uses (as perhaps is also the biological classification of them as 'dysclimaxes' and the anthropological classification of them as 'green deserts')[5]. The implication in such labels of an unchanging and degraded ecosystem is prejudicial both to recognition of any existing active management of the resource and to any local claim to the resource. This classification helps to construct an official landscape, which can be contrasted with what Jackson (1984) has called the 'vernacular' landscape, which he defines as local, customary, fluid, and adaptive. Jackson writes (1984, p. 148), 'A landscape, like a language, is the field of perpetual conflict and compromise between what is established by authority and what the vernacular insists upon preferring.' The gap between the former and the latter has been characteristically marked in the case of Southeast Asia's grasslands.

Beginning in colonial times and extending throughout Southeast Asia, national governments have typically viewed grasslands (especially those dominated by *Imperata*) as an unproductive and undesirable land-use that should be replaced with something thought to be more productive and more desirable, such as forests, plantations, or permanent agricultural fields. A generation of research before World War II and another following it has been devoted to facilitating this replacement. These efforts have been characterized by an emphasis on technological innovation with limited consideration for economic costs (except on plantations) and/or social costs (cf. Whyte 1962). The perceived need to reclaim grasslands that has driven these efforts has persisted to the present day. For example, in the early 1990s the Indonesian Ministry of Forests designated grasslands as one of five priority land-use types for conversion to industrial tree plantations (Hamzah 1993).

Modern efforts to 'reclaim' or 'convert' grasslands have been remarkably problematic. They have seldom succeeded at any large scale and without subsidies, beyond the test plots and pilot projects, especially where there are proximate farming communities. In this respect, development interventions in grasslands resemble those that have been made in many other dimensions of rural resource-use. The failure of modern rural development planning has been so consistent as to prompt scholars over the past decade to re-think the concept of 'failure' itself. That is, if interventions *always* fail, then what meaning does 'failure' still hold? The development community itself has typically blamed failure on recalcitrant rural peoples, sometimes on poor implementation, and rarely on poor development policy or planning. But even the explanation that hits closest to home, poor policy and planning, is naive, because it treats this and other factors as isolated and atypical phenomena as opposed to normal phenomena that are deeply 'embedded' in wider social and historical processes. As Hecht and Cockburn (1989, p. 99) write of development failures in the Amazon: 'To take 'poor policy' as the precipitating factor in deforestation is to see 'policy' as something conceived and executed by technocrats secluded from a country's political economy, and is an extraordinarily naive assumption' (cf. Ascher 1993). Such assumptions result in an inability to correct, and thus a tendency to perpetuate, development failures (Esteva 1987).

Some light is shed on what failure means in the context of development interventions in Southeast Asia's grasslands by examining the occasional examples of success. The historic development of the famous Deli tobacco cultivation system in Sumatra in the second half of the nineteenth century is one such example (cf. Bartlett 1956; Pelzer 1978). During the early decades of the industry, *Imperata* spread like wildfire. It consumed the tobacco fields after the initial harvest, but the planters little cared. But then the planters' attitude changed, and the *Imperata* simply disappeared, as summarized by Pelzer (1978, pp. 29–30):

> So long as the planters believed that their land could produce only *one* tobacco crop, they did nothing to combat the spread of grasses. Once they realized, however, that they had been far too pessimistic and that tobacco could be planted repeatedly provided the land lay fallow under the second-growth forest, or *blukar*, for not less than seven or eight years, they took measures to prevent the burning of the grasses and the concomitant killing of young trees.... These actions greatly altered the physiognomy of the tobacco plantations, as second-growth forest smothered the

grasses and spread steadily at the expense of the savannas.'

In short, then as now, the appearance and disappearance of *Imperata* is much more a function of human wishes than the outcome of a fierce battle between people and nature.[6] This is drolly attested to by Bartlett's (1956) characteristically astute observation that *Imperata* was easily-enough suppressed, by mowing, whenever Europeans wanted to establish golf courses in the tropics (cf. Bartlett 1955,I).

Development logic

Recognition of the problematic history of development interventions in grasslands leads to a question regarding the motives of such intervention. The Deli tobacco example demonstrates that successful management of *Imperata* grasslands is not difficult when it is clearly in the managers' self-interest. This is also obvious from a number of studies that have been conducted of traditional peasant management of *Imperata*, which reveal systems even more sophisticated and successful than that of the Deli tobacco planters (Conklin 1959; Clarke 1966; Seavoy 1975; Sherman 1980; Dove 1981, 1986a, 1986b; Potter 1987, 1997). With few if any exceptions, however, these local successes have been ignored by developers, who treat local communities as incapable of managing grasslands on their own. This raises the question: What are the implications of portraying skilled, indigenous resource managers as needy victims, as people confronted with resource degradation with which they cannot cope? In short, what are the implications of asking, 'How can we [outsiders] 'help'?'[7]

The representation of a situation as one in which help is needed is empowering of the one doing the helping and dis-empowering of the one being helped (Edelman 1974). It is critical to its own self-interest for any development agency to be able to publicly portray potential development subjects as needy, in particular with respect to the resources that the agency offers (Escobar 1995). It is critically important for any development agency, in short, to create a conceptual welcoming niche for itself, as Ferguson (1990) demonstrated in his pioneering analysis of the development programs of the World Bank and the U.S. Agency for International Development (USAID) in Lesotho. The larger the potential niche, the greater the self-interest of the development agency in making this effort. Estimates in Indonesia, for example, have suggested that *Imperata* grasslands alone cover over 10 million ha (Tjitrosoedirdjo 1993; Vandenbeldt 1993), which is an area so large as to give to its conversion to alternate uses profound economic, social, and political implications. There is abundant historical precedent for state authorities viewing askance indigenous regimes of resource use and tenure simply because of their magnitude [cf. Cronon (1983) and Bryant (1994) on colonial policies in seventeenth century North America and nineteenth century Burma, respectively].

The creation of a niche for development intervention can involve great misrepresentation. For example, Ferguson (1990) states that USAID and the World Bank, in order to create a niche for themselves in Lesotho, portrayed a nation that was essentially as a labor reserve for the South African mines as an autonomous, subsistence-oriented agrarian society. Reality, in short, was upended. Similarly, Fairhead and Leach (1996) show how a process of anthropogenic afforestation of grassland zones in Guinea has been mis-represented as a process of anthropogenic forest-grassland succession, in order to construct the picture of environmental crisis needed to justify national governmental intervention and international donor funding. In research that I have done on grassland management among the Banjarese of South Kalimantan, Indonesia, I also identified a reversal of the reality of grassland ecology. Whereas the government saw the *Imperata* grasslands as a problem, the Banjarese saw them as an important solution to many of life's problems; and whereas the imperative in government planning was to get rid of the grasslands, the imperative in the peasant system of grassland management was to maintain them (Dove 1981, 1986a, 1986b).[8] This coexistence of vigorous anthropogenic grasslands and equally vigorous antigrassland policies and interventions exemplifies Jackson's landscape (1984) of conflict and compromise.

Anthropogenic grasslands are one of a complex of perceived resource development problems – including swidden cultivation, wild land fire, etc. – that appear to be insoluble; but this very insolubility raises questions about the validity of the problem (Thompson et al. 1986). There often is no grassland problem in the sense as conceived by the development agency, which is thus in the position of offering a solution in search of a problem (Stone 1989). Often the only real problem in grassland landscapes is the conflict between the local community and extra-local policy-makers, which is exacerbated by the common tendency of state entities to view *Imperata* grassland as *prima facie* evidence of the absence of local tenurial claims. For example, Javanese transmigrants in southern Sumatra say that *'Alang-alang hanya menjadi persoalan kalau dilihat petugas proyek' (Imperata* only becomes a problem [for the transmigrants] if it is seen by the transmigration project officials) (Dove 1986b)[9]. This local view of the 'problem' of grasslands is neither recognized nor addressed in most current grassland management interventions.

Understanding and mis-understanding anthropogenic grasslands

In order to understand and move beyond the bleak history of development interventions in anthropogenic grasslands, there is a need to better understand how knowledge of such grasslands is produced and managed. Systemic misunderstanding of grassland dynamics must be seen not as accidental but as a function of particular political-economic institutions. This relationship is reflected in patterns of grassland research that systematically avoid critical topics like local land uses and rights. Nonequilibrium approaches to society and environment are making it more possible to understand anthropogenic grasslands as successful versus failed resource use regimes. In turn, this suggests new avenues for agroforestry research and development in grassland regions.

Institutional dimensions of misunderstanding

The misunderstanding of anthropogenic grasslands in developmental as well as scientific circles is integral to the practical ecology of grasslands. It is not a simple function of the challenge of understanding Southeast Asia's (e.g.) natural environment; it is a function of past and present structures of political and economic power in the region. This is reflected in the fact that the misunderstandings are not random: they are consistent in both content and implications. The dominant views of grasslands in policy circles in Southeast Asia tend to privilege the science, policy, and resource-use regimes of the region's states and their allies, as opposed to the knowledge and resource-use systems of local communities, many of which are politically and socially marginal.

The self-privileging discourses of states are generally supported by the government institutions in charge of the construction of knowledge about rural areas. This linkage has been analyzed by Koppel and Oasa (1987) for the green revolution and by Taylor (1997) for desertification. Ascher (1993) analyzes this linkage in the context of the rapid degradation of Indonesia's tropical forests under former President Soeharto's 'New Order.' He attributes this degradation to a rent transfer strategy, the persistence of which was dependent upon "embarrassment minimization".[10] According to Ascher the construction of knowledge versus ignorance and of understanding versus misunderstanding, is a social product (cf. Agrawal 1995). The social generation and imposition of uncertainty or misunderstanding tends to be an implicit not explicit product of policy (cf. Leach and Fairhead 1994).

Grassland studies

A critical element in the production of such policy is the distinctive patterns and norms that have governed research on anthropogenic grasslands. It is not true, for example, that little or no research has been done on Southeast Asia's grasslands; but what has been done has rarely been so directed as to challenge the dominant developmental myths. The most common research topic is the elimination of grasslands. Abundant experiments have been done on the eradication of grasslands through the planting of replacement grasses, mechanical tillage, and especially the use of herbicides (Bagnall-Oakeley et al. 1997). Some more recent research has focused on the conversion of grasslands to agroforestry systems (e.g., Garrity 1997); but the value of these latter studies has been limited by the lack of additional study of the economics of such conversions. As Tomich et al. (1997) write, 'There has been almost no economic analysis of agroforestry approaches to conversion of *Imperata* grasslands.'[11]

The indifference to the realities of local systems of resource-use that is reflected in this neglect is also seen in the absence of study of local grassland management systems (and ways of building on them). As Bagnall-Oakeley et al. (1997, p. 100) write: 'There has been relatively little research into integrated *Imperata* management practices for Smallholders ... There has been even less participatory on-farm research that takes account of Smallholders' views in the development of technologies for their use.' Aspects of these smallholder systems that are inconsistent with the dominant grassland myths are almost never empirically examined.[12] This applies to the myths pertaining to grasslands's purported (1) infertility, (2) lack of conservation of soil and water, (3) uselessness for agriculture and animal husbandry, and (4) rationality of local management practices, especially those involving the use of fire. Bagnall-Oakeley et al. (1997, p. 100) write: 'The relationship between *Imperata* in smallholdings, its control, and the risk of fire have not been researched at all as far as we are aware... Its neglect can probably be explained by the fact that fire-related issues are not normally addressed by agricultural research programmes. Consequently, fire is not recognised by researchers as a research issue; not recognised by research institutes as being within their mandate...' As Bagnall-Oakeley et al. (1997) also note, the research lacunae extend to some of the most obvious agroforestry topics. They write, 'It is well known that *Imperata* competition can have a severe retarding effect on rubber trees during the first few years after planting, although there has been only one published piece of research on this subject.' (p. 85) This lack of research on subjects of such importance is remarkable. As Bagnall-Oakeley et al. (1997, p. 101) themselves conclude, 'One would expect that this subject would have been well-researched. It appears, however, that it has not.'

The lack of study is especially marked as regards the role of policymakers in the study of, and intervention in, grasslands. Policymakers characteristically problematize the grassland landscape and proximate communities; but they do not question their own premises that these grasslands are unmanaged and unproductive or that their rehabilitation is economically feasible. Yet Tomich et al. (1997) maintain that given current technologies, markets, and government policies, it is unprofitable or at best marginally profitable to convert large areas of existing grasslands in Southeast Asia to other uses (cf. Vandenbeldt 1993). The general lack of critical reflection regarding development views and interventions is clearly at least partly responsible for the failure of so many of these same interventions.

The fact that needful grassland research has not been done, or has been done in error, or has been done correctly but without impact, is sociologically meaningful. As Thompson et al. (1986) have written, repeated developmental mistakes are, in effect, development 'signposts,' which point us toward the most important development truths. The truth being pointed toward in this case is that the misunderstanding of grassland ecology merits our attention as much as any other dimension of this ecology.

Over the past generation considerable scholarly work has been devoted to the objectification of scientific research. This encompasses broad studies of the sociology of scientific knowledge (e.g., Barnes 1974), lab-based ethnographic studies (e.g., Traweek 1988), actor-centered analyses (Latour 1988), post-structural analyses of nature and culture (Haraway 1991), and most recently and especially studies of the production and practice of scientific knowledge (Pickering 1992, 1995). One of the pioneering contributions to the study of scientific practice was Kuhn's (1962), thesis of paradigm change. According to this thesis, scientists spend most of their time doing 'normal science' within an (often unconscious) paradigm, which is a pre-theoretical ordering of reality. Over time, anomalies that cannot be accounted for within the paradigm build up, confidence in the paradigm weakens, normal science breaks down, and the old paradigm is challenged by a new one in what potentially amounts to a scientific revolution. Because of differences in worldview and conceptual language, there is no communication between followers of different paradigms, which helps to explain the complete dysjuncture between those who view Southeast Asia's grasslands as wastelands and those who view them as a productive peasant landscape.

New approaches to the study of grasslands

As suggested earlier, outside observers once saw grasslands as telling us something not just about environmental relations but specifically about *failed* environmental relations. As Potter (1988, p. 129) writes of Dutch colonial views of the extensive grasslands of Southeast Kalimantan: 'The very existence of the Hulu Sungai... with its *sawah* fields in close proximity to grass-covered uplands dominated by *Imperata cylindrica* and followed at a distance by swidden sites in secondary forest, was felt to be an *object lesson* [added] to all.' Colonial officials ignorant of and unreceptive to local traditions of grassland management did not read this landscape as an intentionally 'patchy' one. Instead, they read the proximity of grassland and non-grassland as a moral story (cf. Cronon, 1992) about the unwanted spread of grasslands, about environmental decline.

The moral character of such stories is reflected in their directionality. The presence of grasslands was and is commonly interpreted as a 'lesson' about what *could happen* to non-grassland covers if sufficient care is not taken; but the reverse and ecologically equally likely lesson is rarely if ever drawn. That is, observers almost never interpret non-grassland covers as a lesson about what could happen to grasslands. Vandenbeldt

(1993, p. 3) states the conventional wisdom as follows: '*Imperata cylindrica* grasslands are a fire-climax vegetation type derived from cleared forest lands.' On the other hand, it is equally ecologically valid, yet unheard-of, to describe *Imperata* as 'an unstable vegetation type often terminating in closed forest' (or to describe tropical rainforest as a 'climax vegetation type derived from unmanaged grasslands'). Indeed, the mere possibility of a spontaneous succession from grassland to non-grassland, which is not directed by state authorities, can be unwelcome. A dramatic example of this is given in Fairhead and Leach's (1996) analysis of how a historic progression from grassland to forest in Guinea is interpreted by the government as just the opposite, in order to attract more funding for afforestation.

Increasingly, however, grassland history is being interpreted not in terms of an equilibrium model, which assumes a static condition that is only periodically and unnaturally upset, but in terms of a non-equilibrium model in which ongoing perturbation is assumed to be the natural state of affairs. The former model, which is particularly associated with Clements' (1916) work on vegetative succession and Odum's (1971) work on homeostasis, has been supplanted by the latter over the past two decades in much of the natural and social sciences (Holling 1978). Worster (1990 pp. 3,11) writes that 'Odum's ecosystem, with its stress on cooperation, social organization, and environmentalism' was replaced with an image of 'nature characterized by highly individualistic associations, constant disturbance, and incessant change' (cf. Cronon 1983). Worster (1990) summarizes the change as one from 'a study of equilibrium, harmony, and order' to a study of 'disturbance, disharmony, and chaos.' Our understanding of grassland ecology has dramatically changed as a result of this shift in explanatory paradigms. Whereas most scholars once saw grasslands as exemplars of ecological stability, they now see them as just the opposite, as models of instability and disturbance (Worster 1990). The older belief prevails to this day, however, in much of the policy and development community, where it is still assumed that grasslands are a stable and tenacious climax community, which will not disappear unless dramatic steps are taken to make them disappear.

The perceived instability of grasslands is associated with a more prominent role for humans in grassland ecology: scholars now agree that many grasslands depend for their existence upon *continued* human perturbation of the environment, without which they will in fact disappear. In my study of the Banjarese, for example, I found that continuous management inputs were critical to the maintenance of grasslands. This is reflected in the fact that if grassland is left alone for so many years that the signs of such inputs begin to disappear and a natural process of afforestation commences, tenure to that patch of grassland lapses and it is considered to be a free good (Dove 1981). To understand grassland ecology in such a system it is necessary to study not a single, episodic interaction between people and environment but rather *the ongoing* relations between society and environment. It is necessary to ask what reasons local communities may have to continue or discontinue the burning that typically accounts for the appearance versus disappearance, respectively, of the grasslands. This is the key question that developers do *not* ask, because the local 'agency' implied by this question is so inimical to the development niche discussed earlier.

The shift in thinking can be illustrated by my own experience in my research on the Banjarese grasslands. I initially traveled to Southeast Kalimantan as part of an international project to study the 'breakdown' in Banjarese swidden agriculture that was thought to be responsible for the grasslands and to assess the prospects for a state afforestation program. But what I found was a system of active local management of the grasslands, of which people said 'We could not live without them.' The Banjarese were cultivating the grasslands using brush-sword, fire, and cattle-drawn plough, and they reaped an average crop of 2200 kilograms of rice per hectare, for up to seven years in a row followed by a three-year fallow (Dove 1981). They also were burning the grasslands to create graze for their livestock. Such a dramatic gap between indigenous and exogenous views of anthropogenic grasslands shows how important it is to study local grassland management knowledge and technology. Prior failure to study extant local management systems is the signal failure of most development interventions.

Implications for agroforestry research and development

The most obvious insight for agroforestry from this review of anthropogenic grasslands in Southeast Asia is that such lands are commonly integrated into and valued within local systems of agriculture and animal husbandry. Because of this value, such grasslands tend to be managed and maintained; they are not simply unmanaged climaxes. The corollary point is that these grasslands are almost always valued far more highly in the eyes of local peoples than extra-local observers and administrators, which has great significance for development interventions.

In any development plan to eradicate or replace anthropogenic grasslands, it is vitally important to ask if local valuation of these lands and local roles in their dynamics have been given due consideration. In particular, simplistic schemes to reforest such grasslands, based on exogenous views of their utility, should be critically examined. The prospects for tree cultivation in such lands needs to be assessed in the context of wider agro-ecological, economic, and political dynamics.

One of the most important but overlooked dimensions of anthropogenic grasslands, and one that should be at the forefront of planning of any development intervention, is their political character: these grasslands are typically contested landscapes – contested in terms of how existing and alternative land-uses are evaluated. As a result, this gives a political inflection to any act of tree cultivation or reforestation in anthropogenic grasslands. The cultivation of trees in such lands typically has serious implications for local tenurial relations and government control, which should be assessed before any such intervention proceeds.

Some of the research priorities for agroforestry in anthropogenic grasslands have already been mentioned, including on-farm research of such topics as the economics of grassland improvement or conversion and the dynamics of interactions such as that between *Imperata* and rubber (*Hevea brasiliensis*) trees. Fire management is a particularly important research topic for these grasslands, which can be called a 'pyrophytic' habitat (cf. D'Antonio and Vitousek 1992, p. 73). In any anthropogenic grassland, fire tolerance is the foremost criteria for successful tree growth. The study of the possible role of fire-tolerant trees within (not outside of) local schemes of grassland management is a greatly understudied topic.

Summary and conclusions

The earliest stage of research on tropical anthropogenic grasslands consisted of brief commentaries in the early colonial, ethnographic, and travel literature.[13] These commentaries were often more insightful than might be expected, because the observers' vision was less influenced by the orthodox beliefs concerning grasslands that accompanied later specialists to the field. The second stage of studies consisted of the work of these same specialists, working mostly within the estate crop sector and focusing on how to eliminate grasses such as *Imperata*. The third stage of study consisted of efforts by both natural and social scientists to interpret particular systems of grassland management and ecology in light of the actual social and biological evidence as opposed to prevailing orthodoxy. The fourth and still ongoing stage consists of efforts to analyze this orthodoxy and to identify and try to overcome the gaps that it has left in our knowledge. According to the 1997 *Agroforestry Systems* special issue on *Imperata*, these gaps are greatest with respect to local smallholder grassland management systems and the extralocal institutional forces that impact them.

These gaps are a function of the past research emphasis on natural versus social dynamics in anthropogenic grasslands, a curious emphasis given the fact – obvious to all – that the grasslands in question are a human product.[14] This emphasis reflects a still persisting difficulty in science in bridging natural and cultural arenas. Ironically, the success of local, smallholder grassland managers rests precisely upon such a bridge – articulating field and forest, culture and nature. Agroforestry, with its conceptual mandate to cross this same divide, is uniquely suited to studying grasslands and offering ways to improve their management. Even if grasslands do not contain woody species, as a transitional stage between field and forest they in effect represent a key stage in a larger agroforestry system, and agroforestry research has a key role to play in examining the role of grasslands within this larger system. How are the dynamics of succession of ligneous and herbaceous species balanced and managed in the grasslands; how do the economics and ecology of grassland management fit into wider land-use systems often comprising both annuals and perennials and both more- and less-intensive cultivation methods; and how and why do different parties – local as well as extralocal – see the grasslands in different ways? These questions, and others like them, are well-suited to agroforestry perspectives and approaches.

This analysis of tropical anthropogenic grasslands also raises a number of points of relevance to the wider natural resource management issues of the day. It shows that how we draw the line between natural processes on the one hand and cultural processes on the other is highly subjective and self-interested. It shows that whether we view the environment as dynamic or static has enormous implications for both environmental and social relations. It shows that where we locate agency, in both society and environment, is fraught with implications for state projects of intervention. It shows that the subjective use of scientific concepts by policy-makers and scientists alike is an integral part of ecology and a part that we need to study more.

End Notes

1. Anthropogenic grasslands are the center of debates all over the world, however the specific issues involved in these debates differ. Thus, in South America the debates focus on conversion of rain forest to pasture, and in Africa they focus on the purported savannization of forests. (I am grateful to one of the compendium's anonymous reviewers for this observation.)
2. For example, among the Kantu' of West Kalimantan, Indonesia, there was traditionally a pantang (proscription) against the use of Borneo's foremost timber, from the ironwood tree (Eusideroxylon zwageri), in house construction. The Kantu' rationale for this proscription was that the durability of this wood and anything constructed from it surpassed and thus challenged the life-span of the people using it.
3. See McDonald (2002) on the history of the 'tree of life' in the Middle East.
4. 'Savannas' are distinguished from 'grasslands' by the presence of scattered trees in a continuous grass cover (Scholes and Archer 1997).
5. The strength of this long-entrenched view of Imperata is reflected in the fact that one of the conclusions of a recent review of the field simply focused on its rebuttal: 'Imperata grasslands are not 'wastelands' (Garrity 1997).
6. This does not mean that grassland management is not onerous: the Banjarese of South Kalimantan report that the annual labor costs of cultivating grassland fields by hoe and plough are about twice that of making swiddens in the forest using axe and fire (Dove 1986a).
7. Cf. Tomich et al.'s (1997) pointed distinction between creating positive distortions and removing distortions, between the unlikely benefits of creating new incentives for grassland development and the much more likely benefits of simply removing existing dis-incentives.
8. There is great and systematic variation in local attitudes toward anthropogenic grasslands, however. In a river valley adjacent to the one in which I studied, Potter (1987) found that cattle-using villages liked it whereas swiddening villages did not. Similarly, in a comparative study of a number of different sites around Indonesia, I found that appreciation of Imperata varied inversely with the intensity of cultivation (Dove 1986b). Conklin's (1959) pioneering analysis showed that attitudes toward Imperata may even vary depending upon where the grass is found within the territory of a single village.
9. Transmigrant attitudes toward Imperata vary with their possession of cattle, however: with cattle, it is easy to cultivate Imperata grassland, upon which the cattle also can feed. Without cattle, the grasslands are harder to cultivate and of less economic use. (I am grateful to one of the compendium's anonymous reviewers for this observation.)
10. The consequences are reflected in Gillis' comment (1987; cited in Ascher 1993) that 'The quality of technical information on African and most Latin American tropical forests is superior to that available for the dense Dipterocarp forests of Indonesian Kalimantan, where most harvesting takes place.'
11. Tomich et al. (1997) add that Turvey's 'definitive survey of the literature for Southeast Asia [in 1994] does not contain a single reference to any study of the financial or economic feasibility of agroforestry applications for Imperata grassland conversion.'
12. As Sherman (1980) notes with regard to the myth of barren grassland soils: 'It should come as no surprise that in all the time it was assumed that forest-covered soil regenerated its former fertility while grassland caused erosion and leaching, no tests were done on the possibility of increased fertility levels under grassland conditions.'
13. These are thoroughly reviewed in Bartlett's bibliographies of 1955, 1957, and 1961.
14. I have used the term 'anthropogenic' throughout this analysis for want of a better term, although I recognize that its earlier-cited definition of 'having its origin in the activities of man' oversimplifies a complex and varied range of interactions between society and environment.

References

Agrawal A. 1995. Dismantling the divide between indigenous and scientific knowledge. Dev Change 26: 413–439.

Anderson E. 1971/1952. Plants, Man and Life. University of California Press, Berkeley. 251 pp.

Ascher W. 1993. Political Economy and Problematic Forestry Policies in Indonesia: Obstacles to Incorporating Sound Economics and Science. Ms.

Bagnall-Oakeley H., Conroy C., Faiz A., Gunawan A. Gouyon A., Penot E., Liangsutthissagaon S., Nguyen H.D. and Anwar C. 1997. *Imperata* management strategies used in a smallholder rubber-based farming system. Agroforest Syst 36(1–3): 83–104.

Balée W. 1994. Footprints of the Forest: Ka'apor Ethnobotany – The Historical Ecology of Plant Utilization by an Amazonian People. Columbia University Press, New York. 396 pp.

Barnes, B. 1974. Scientific Knowledge and Sociological Theory. Routledge, London.

Bartlett H.H. 1955. Fire in Relation to Primitive Agriculture and Grazing in the Tropics: An Annotated Bibliography. Vol. I. University of Michigan Botanical Gardens, Ann Arbor, MI. 568 pp.

Bartlett H.H. 1956. Fire, primitive agriculture, and grazing in the tropics. pp. 692–720. In: Thomas W.L. (ed.) Man's Role in Changing the Face of the Earth. University of Chicago Press, Chicago.

Bartlett H.H. 1957. Fire in Relation to Primitive Agriculture and Grazing in the Tropics: An Annotated Bibliography. Vol. II. University of Michigan Department of Botany, Ann Arbor, MI. 873 pp.

Bartlett H.H. 1961. Fire in Relation to Primitive Agriculture and Grazing in the Tropics: An Annotated Bibliography. Vol. III. The University of Michigan Department of Botany, Ann Arbor, MI. 216 pp.

Bateson G. 1972. Steps to an Ecology of Mind. Ballantine Books, New York. 541 pp.

Bouquet M. 1995. Exhibiting knowledge: The trees of Dubois, Haeckel, Jesse and Rivers at the Pithecanthropus centennial exhibition. pp. 31–55. In: Strathern M. (ed.). Shifting Contexts:

Transformations in Anthropological Knowledge, Routledge, London.
Bourdieu P. 1977. Outline of a Theory of Practice, Richard Nice (trans.), Cambridge University Press, Cambridge [Original, Esquisse d'une théorie de la practique; Précédé de trois études de'ethnologie Kabyle, 1972, Librairie Droz S.A. Switzerland]. 248 pp.
Bryant R.L. 1994. Shifting the cultivator: The politics of teak regeneration in colonial Burma. Modern Asian Studies 28(2): 225–250.
Buttel F.H. 1992. Environmentalization: Origins, processes, and implications for rural social change. Rural Sociol 57(1): 1–27.
Carpenter C. 1987. Brides and Bride-Dressers in Contemporary Java. Ph.D. dissertation, Cornell University.
Chokkalingam U., De Jong W., Smith J. and Sabogal C. (eds) 2001. Secondary Forests in Asia: Their Diversity, Importance, and Role in Future Environmental Management. J Trop For Sc 13(4). Special Issue.
Clarke W.C. 1966. From extensive to intensive shifting cultivation: A succession from New Guinea. Ethnology 5: 347–359.
Clements F. 1916. Plant Succession: Analysis of the Development of Vegetation. Carnegie Institute, Washington D.C.
Conklin H.C. 1959. Shifting cultivation and succession to grassland climax. Proceedings of the Ninth Pacific Science Congress 7: 60–62.
Coomes O.T., Grimard F. and Burt G.J. 2000. Tropical forests and shifting cultivation: Secondary forest fallow dynamics among traditional farmers of the Peruvian Amazon. Ecol Econ 32: 109–124.
Corlett R.T. 1995. Tropical secondary forests. Prog Phys Geog 19: 159–172.
Cronon W. 1983. Changes in the Land: Indians, Colonists, and the Ecology of New England. Hill and Wang, New York. 241 pp.
Cronon W. 1992. A place for stories: Nature, history and narrative. J Am Hist 78: 347–376.
Cronon W. 1996. The trouble with wilderness, or getting back to the wrong nature. pp. 69–90. In: Cronon W. (ed.). Uncommon Ground: Rethinking the Human Place In Nature. W.W. Norton, New York.
D'Antonio C.M. and Vitousek P.M. 1992 Biological Invasions by Exotic Grasses, The Grass/Fire Cycle, and Global Change. Annu Rev Ecol Syst 23: 63–87.
Davies D. 1988. The evocative symbolism of trees. pp. 32–42. In: Cosgrove D. and Daniels S. (eds) The Iconography of Landscape. Cambridge University Press, Cambridge.
Deleuze G. and Guattari F. 1987. A Thousand Plateaus: Capitalism and Schizophrenia. Trans. Brian Massumi. (Orig. Mille Plateaux, vol. 2 of Capitalisme et Schizophrénie, 1980, Les Editions de Minuit, Paris). University of Minnesota Press, Minneapolis. 610 pp.
Denevan W.M. 1992. The pristine myth: The landscape of the Americas in 1492. The American Geographer 82(3): 369–385.
Denevan W.M. and Padoch C. (eds) 1987. Swidden-Fallow Agroforestry in the Peruvian Amazon. The New York Botanical Gardens, New York. 107 pp.
Dove M.R. 1981. Symbiotic relationships between human populations and *Imperata cylindrica:* The question of ecosystemic succession and persistence in South Kalimantan. pp. 187–200. In: Nordin M., Latiff A., Mahani M.C. and Tan S.C. (eds) Conservation: Inputs from Life Sciences. Universiti Kebangsaan Malaysia, Bangi, Malaysia.
Dove M.R. 1986a. Peasant versus government perception and use of the environment: A case study of Banjarese ecology and river basin development in South Kalimantan. J South Asian Stud 17: 113–136.
Dove M.R. 1986b. The practical reason of weeds in Indonesia: Peasant versus state views of *Imperata* and *Chromolaena*. Hum Ecol 14: 163–190.
Dove M.R. 1998. Living rubber, dead land, and persisting systems in Borneo: Indigenous representations of sustainability. Bijdragen tot de Taal-, Land-en Volkenkunde 154(1): 20–54.
Edelman M. 1974. The political language of the helping professions. Polit Soc 4: 295–310.
Escobar A. 1995. Encountering Development: The Making and Unmaking of the Third World. Princeton University Press, Princeton, NJ. 290 pp.
Esteva G. 1987. Regenerating people's space. Alternatives 12(1): 125–152.
Fairhead J. and Leach M. 1996. Misreading the African Landscape: Society and Ecology in Forest-Savanna Mosaic. Cambridge University Press, Cambridge, UK. 354 pp.
Ferguson J. 1990. The Anti-Politics Machine: 'Development,' Depoliticization and Bureaucratic Power in Lesotho. Cambridge University Press, Cambridge, UK.
Frazer Sir J.G. 1951. The Golden Bough: A Study in Magic and Religion, abr. ed. MacMillan, New York.
Garrity D.P. 1997. Agroforestry innovations for *Imperata* grassland rehabilitation: Workshop recommendations. Agroforest Syst 36: 263–274.
Gillis M. 1987. Multinational enterprises and environmental and resource management issues in the Indonesian tropical forest sector. pp. 64–89. In: Pearson C. (ed.) Multinational Corporations, Environment, and the Third World: Business Matters, pp. 64–89. Duke University Press, Durham, NC.
Hadiwidjojo GPH. 1956. Alang-Alang Kumitir. Paper read at the Radyapustaka (Court library). Solo (Central Java, Indonesia), December 28.
Hamzah Z. 1993. Pembinaan HTI di Tanah Kosong dan Tanah Alang-Alang (I) (Regulation of Tree Plantations in Empty Lands and *Imperata* Lands (I)). Mangala Wanabakti 9(3): 1–6.
Haraway D. 1991. Simians, Cyborgs, and Women: The Reinvention of Nature. Routledge, New York. 287 pp.
Harlan J.R. 1982. Human interference with grass systematics. pp. 37–50. In: Estes J.R., Tryl R.J. and Brunken J.N. (eds) Grasses and Grasslands: Systematics and Ecology. University of Oklahoma Press, Norman, Oklahoma.
Hecht S. and Cockburn A. 1989. The Fate of the Forest: Developers, Destroyers and Defenders of the Amazon. Verso, London.
Holling C.S. 1978. Myths of ecological stability: Resilience and the problem of failure. pp. 97–109. In: Smart C.F. and Stanbury W.T. (eds) Studies on Crisis Management. Institute for Research on Public Policy, Toronto.
Horne E.C. 1974. Javanese-English Dictionary. Yale University Press, New Haven, CT.
Jackson J.B. 1984. Discovering the Vernacular Landscape. Yale University Press, New Haven, CT. 728 pp.
Koppel B.M. and Oasa E. 1987. Induced innovation theory and Asia's green revolution: A case study of an ideology of neutrality. Dev Change 18: 29–67.
Kuhn T.S. 1962. The Structure of Scientific Revolutions. University of Chicago Press, Chicago. 172 pp.
Lambert D.P. 1996. Crop diversity and fallow management in a tropical deciduous forest shifting cultivation system. Hum Ecol 24: 427–453.
Latour B. 1988. The Pasteurization of France. Harvard University Press, Cambridge, MA. 157 pp.

Lawrence D.C. and Mogea J.P. 1996. A preliminary analysis of tree diversity in the landscape under shifting cultivation north of Gunung Palung National Park. Tropical Biodiversity 3: 297–319.

Leach M. and Fairhead J. 1994. Natural resource management: The reproduction and use of environmental misinformation in Guinea's forest-savanna transition zone. Institute of Development Studies Bulletin 25(2): 81–87.

McDade L. (ed.) 1993. La Selva: Ecology and Natural History of a Neotropical Rainforest. University of Chicago Press, Chicago.

McDonald J.A. 2002. Botanical determination of the Middle Eastern tree of life. Econ Bot 56: 113–129.

Nyerges A.E. 1989. Coppice swidden fallows in tropical deciduous forests: Biological, technological, and sociocultural determinants of secondary forest succession. Hum Ecol 17: 379–400.

Odum E.P. 1971. Fundamentals of Ecology, 3rd ed. Saunders College Publishing, Philadelphia. 574 pp.

Ortner S.B. 1973. On key symbols. Am Anthropol 75: 1338–1346.

Oxford English Dictionary. 1989. 2nd ed. Clarendon Press, Oxford.

Pelzer K.J. 1978. Planter and Peasant: Colonial Policy and the Agrarian Struggle in East Sumatra, 1863–1947. Verhandelingen 84. Martinus Nijhoff, 's-Gravenhage. 163 pp.

Pickering A. (ed.) 1992. Science as Practice and Culture. University of Chicago Press, Chicago. 474 pp.

Pickering A. 1995. The Mangle of Practice: Time, Agency, and Science. University of Chicago Press, Chicago.

Potter L. 1987. Degradation, innovation and social welfare in the Riam Kiwa Valley, Kalimantan, Indonesia. pp. 164–176. In: Blaikie P. and Brookfield H. (eds). Land Degradation and Society. Methuen, London.

Potter L. 1997. The Dynamics of *Imperata*: Historical overview and current farmer perspectives, with special reference to South Kalimantan, Indonesia. Agroforest Syst 36: 31–51.

Rival L. (ed.) 1998. The Social Life of Trees: Anthropological Perspectives on Tree Symbolism. Berg, Oxford.

Schärer H. 1963. Ngaju Religion: The Conception of God Among a South Borneo People, R. Needham (trans.). Koninklijk Instituut voor Taal-, Land- en Volkenkunde, Translation Series 6. Martinus Nijhoff, The Hague. 229 pp.

Scholes R.J. and Archer S.R. 1997. Tree-grass interactions in savannas. Annu Rev Ecol Syst 28: 517–544.

Seavoy R.E. 1975. The origin of tropical grasslands in Kalimantan, Indonesia. J Trop Geogr 40: 48–52.

Sherman D.G. 1980. What 'green desert'? The ecology of Batak grassland farming. *Indonesia* 29: 113–148.

Solbrig O.T. 1993. Ecological constraints to savanna use. pp. 21–47. In: Young M.D. and Solbrig O.T. (eds). The World's Savannas: Economic Driving Forces, Ecological Constraints and Policy Options for Sustainable Land Use. UNESCO, Paris and Parthenon, Carnforth, U.K.

Solbrig O.T. 1996. The diversity of the savanna ecosystem. pp. 1–30. In: O.T. Solbrig O.T., Medina E. and Silva J.F. (eds). Biodiversity and Savanna Ecosystem Processes: A Global Perspective. Springer, Berlin.

Stone D.A. 1989. Causal stories and the formation of policy agendas. Polit Sci Quart 104: 281–300.

Stott P. 1991. Recent trends in the ecology and management of the world's savanna formations. Prog Phys Geog 15: 18–28.

Taylor P.J. 1997. Re/constructing socio-ecologies: System dynamics modeling of nomadic pastoralists in sub-Saharan Africa. pp. 115–148. In: A. Clarke and J. Fujimura (eds). The Rights Tools for the Job: At Work in Twentieth Century Life Sciences. Princeton University Press, Princeton, NJ.

Thompson M., Warburton M. and Hatley T. 1986. Uncertainty on a Himalayan Scale: An Institutional Theory of Environmental Perception and a Strategic Framework for the Sustainable Development of the Himalaya. Ethnographica, London. 162 pp.

Thrupp L.A., Hecht S. and Browder J. 1997. The Diversity and Dynamics of Shifting Cultivation: Myths, Realities, and Policy Implications. World Resources Institute, Washington, D.C.

Tjitrosoedirdjo S. 1993. *Imperata* cylindrica grasslands in Indonesia. pp. 31–52. In: Vandenbeldt R.J. (ed.) Imperata Grasslands in Southeast Asia: Summary Reports from the Philippines, Malaysia, and Indonesia. The Forestry/Fuelwood Research and Development Project, Report #18. Winrock International, Washington, D.C.

Tomich T.P., Kuusipalo J., Menz K. and Byron N. 1997. *Imperata* economics and policy. Agroforest Syst 36: 233–261.

Traweek, S. 1988. Beamtimes and Lifetimes: The World of High Energy Physicists. Harvard University Press, Cambridge, MA.

Tuck-Po L. 2004. Changing Pathways: Forest Degradation and the Batek of Pahang, Malaysia. Lexington Books, Lanham, Maryland.

Vandenbeldt R.J. 1993. *Imperata* grasslands in Southeast Asia: Executive summary. pp. 1–5. In: Vandenbeldt R.J. (ed.). *Imperata* Grasslands in Southeast Asia: Summary Reports from the Philippines, Malaysia, and Indonesia. The Forestry/Fuelwood Research and Development Project, Report #18. Winrock International, Washington, D.C.

Werner P.A., Walker B.H. and Stott P.A. 1990. Introduction. Special issue on: Savanna ecology and management: Australian Perspective and Intercontinental Comparisons. J Biogeography 17: 343–344.

Wessing R. 1992. A tiger in the heart: The Javanese rampok macan. Bijdragen tot de Taal-, Land-en Volkenkunde 148: 287–308.

Whyte R.O. 1962. The myth of tropical grasslands. Trop Agr 39: 1–11.

Williams R. 1980. Problems in Materialism and Culture: Selected Essays. NLB, London.

Worster D. 1990. The ecology of order and chaos. Environ Hist Rev 14: 1–18.

Agroforestry Systems **61:** 437–452, 2004.

Agroforestry research for development in India: 25 years of experiences of a national program

S. Puri[1,*] and P.K.R. Nair[2]
[1]*Department of Forestry, Indira Gandhi Agricultural University, Raipur 492 006 India;* [2]*School of Forest Resources and Conservation, University of Florida, 118 Newins-Ziegler Hall, P.O. Box 110410, Gainesville, FL 32611-0410, USA;* **Author for correspondence: e-mail: sunilpuri56@hotmail.com; spuri56@yahoo.com*

Key words: Competition, Evergreen revolution, Policy issues, Productivity, Smallholder tree farms, Soil amelioration

Abstract

India has been in the forefront of agroforestry research ever since organized research in agroforestry started worldwide about 25 years ago. Considering the country's unique land-use, demographic, political, and sociocultural characteristics as well as its strong record in agricultural and forestry research, India's experience in agroforestry research is important to agroforestry development, especially in developing nations. Agroforestry has received much attention in India from researchers, policymakers and others for its perceived ability to contribute significantly to economic growth, poverty alleviation and environmental quality, so that today agroforestry is an important part of the 'evergreen revolution' movement in the country. Twenty-five years of investments in research have clearly demonstrated the potential of agroforestry in many parts of the country, and some practices have been widely adopted. But the vast potential remains largely underexploited, and many technologies have not been widely adopted. This situation is a result of the interplay of several complex factors. The understanding of the biophysical issues related to productivity, water-resource sharing, soil fertility, and plant interactions in mixed communities is incomplete and insufficient, mainly because research has mostly been observational in nature rather than process oriented. Methods to value and assess the social, cultural and economic benefits of various tangible and nontangible benefits of agroforestry are not available, and the socioeconomic processes involved in the success and failure of agroforestry have not been investigated. On the other hand, the success stories of wasteland reclamation, and poplar-based agroforestry show that the technologies are widely adopted when their scientific principles are understood and socioeconomic benefits are convincing. An examination of the impact of agroforestry technology generation and adoption in different parts of the country highlights the major role of smallholders as agroforestry producers of the future. It is crucial that progressive legal and institutional policies are created to eschew the historical dichotomy between agriculture and forestry and encourage integrated land-use systems. Government policies hold the key to agroforestry adoption.

Introduction

Agroforestry has been a way of life in India for thousands of years. That is perhaps no wonder in a country with over 5000 years of civilization and culture of agriculture. The two great Indian epics *Ramayana* and *Mahabharata* (based on events that have supposed to have happened in 7000 B.C. and 4000 B.C. respectively) contain an illustration of *Ashok Vatika*, a form of what are called homegardens today. It is comprised of plants encircling a *sarovar* (water-tank) in three rings, with trees such as *ashok* (*Saraca indica*), neem (*Azadirachta indica*), *pipal* (*Ficus religiosa*), and *bargad* (*Ficus bengalensis*) in the outer ring, fruit trees such as jackfruit (*Artocarpus heterophyllus*), mango (*Mangifera indica*), banana (*Musa* spp.) and guava (*Psidium guajava*) in the middle ring, and flowers such as *champa* (*Michelia champaka*), *chameli* (*Jasminum grandiflorum*), *juhi* (*Ervatamia divaricata*), and *bela* (*Jasminum arborescens*) in the inner ring. The tribal dances and other rituals that are performed even today around the trees are symbolic of the reverence with which trees are held in the society.

In spite of this long-standing social, ethnic, and religious significance of trees in India and the practice of various traditional forms of agroforestry in different parts of the country (Tejwani 1994), recognition of trees as components of farming systems (agroforestry, as perceived today: Nair 2001) has been rather limited (with the notable exception of interest in farm woodlots during the past two to three decade). This is partly because of the traditional divide between agriculture and forestry in government business – just as it is elsewhere in the world. The several traditional land-use practices that have been in existence in India for a long time (Nair 1989; Tejwani 1994) were ignored or bypassed while organized research endeavors in agriculture and forestry were developed along strict disciplinary lines – again, a situation not unique to India. India ranks, however, among a select few countries where organized research in agroforestry started almost at the same time when such initiatives started on the global scene in the late 1970s. [The reference here is to research under the name 'Agroforestry,' not to belittle the importance of research of this nature conducted during and before the 1970s; although that was then not known as agroforestry, today it is (a notable example is studies on crop combinations with coconuts and other perennial plantation crops: Nair 1979)]. Under the newly found banner of agroforestry, a National Seminar on Agroforestry organized at Imphal, Manipur, by the Indian Council of Agricultural Research (ICAR) in northeastern India in 1979 (ICAR 1981) is one of the first – if not the first – such national initiatives. Following that, an All India Coordinated Research Project on Agroforestry was launched by ICAR in 1983 in 20 centers all over the country – the number of centers has risen to 37 in 2003. This is, again, the first major national initiative of this magnitude anywhere in the world. Further, a National Research Centre for Agroforestry was established by ICAR in 1988 at Jhansi in central India, and agroforestry research and education has been taken up in a substantial manner in most of the 32 state agricultural universities. The Indian Council of Forestry Research and Education (ICFRE) that was set up along the lines of ICAR to coordinate and promote forestry research at the national level also has agroforestry as a major program area in most of its eight national research institutes and three advance centers. In addition to these public-sector undertakings, some private and nongovernmental institutions are involved in research that is of relevance to agroforestry. Notable among them is the Tata Energy Research Institute, Delhi, which has a major research interest in biomass-productivity enhancement and large-scale multiplication of elite trees. Today, about 2000 scientific and technical personnel are involved in agroforestry research and education in India. Mark Twain, who once said 'India is the cradle of the human race, the birthplace of human speech, the mother of history, the grandmother of legend, and the great grand mother of tradition,' would perhaps add agroforestry to that list of virtues today. Given the geographic, ecological, demographic, social, cultural, political, and economic complexities of a country like India, its rather unparalleled experience in agroforestry research and development will be of considerable interest to the international community. With India's much acclaimed track record of green revolution and the success story of creating a well-organized infrastructure in agricultural research, India's experiences in agroforestry research for development present a unique case study.

Despite the considerable efforts that have been expended toward promoting and implementing agroforestry practices in India (National Agriculture Policy of 2000; http://www.incg.org.in/Agriculture/Policies/National Agriculture Policy.htm), 'modern' agroforestry technologies[1] have not been widely adopted in the country, except perhaps for the poplar (*Populus* spp.)-based system in the northwestern parts. Moreover, during recent years, agroforestry practices have generated some controversies[2]. Critics argue that such activities are taking over land that previously provided them with the means to feed themselves and earn a living. They claim that planting large areas with eucalypts or eucalyptus (*Eucalyptus* spp.), acacias (*Acacia* spp.), pines (*Pinus* spp.) and poplars have adversely affected the environment and ruined smallholder farmers (FAO 1995; Nambiar et al. 1998). Others say the practices will help protect natural forests and provide economic growth. Understandably there is much confusion. Part of the problem can, arguably, be related to the location-specificity of agroforestry systems. But agricultural systems are site specific too; yet modern agricultural technologies have gained widespread adoption in India. The problem could not (only) be site-specificity of agroforestry, but (also) the lack of a science base in agroforestry. It could well be that the scientific principles of successful indigenous systems have not been adequately understood and/or recognized (see Kumar and Nair 2004), nor are the so-called 'modern' agroforestry technologies based on sound scientifc principles. It could be that there are serious disincentives to agroforestry adoption in terms of social, cultural, economic, and policy issues. In this paper, we will try to assess the extent to which 25 years of research and experience have contributed to the knowledge base of agroforestry using a subset of some examples of agroforestry experience in India, and highlight some issues that have emerged consequent to recent adoption of tree planting on farms. We wish to make it clear that the objective of the paper is not to present a review of agroforestry research in India.

Major research topics

Systems inventory

Although the general perception about agroforestry in India is one of tree planting outside designated forestland, a wide variety of tree planting activities (known by different terms such as social forestry, farm forestry, community forestry, urban forestry, etc.) and a bewildering array of associations involving trees, crops, and animals are found outside forests throughout the country. Several reports describing the major agroforestry systems in the country are available (Nair 1989; Tejwani 1994; Kumar 1999; Pathak and Solanki 2002). An interesting outcome of this effort is the several fascinating and imaginative terms that have been coined by various workers to describe specific agroforestry systems, and arguments and counterarguments on the relevance and validity of these various terms. The different terms are, however, only indicative of the fact that each system is a specific local example of the association or combination of the components, characterized by geography, environment, plant species, and their arrangement, management and socioeconomic functioning (Nair 1985). The existence of a plethora of such terms – many of them poorly defined – has, nevertheless, caused some confusion; but that can be taken as a passing phase in the development of the subject. The major agroforestry practices in India are listed in Table 1; each of these practices has an infinite number of variations.

Plant-to-plant interactions

The farming community in India, as well as elsewhere, believes that, in general, agricultural crops grow poorly beneath or near trees. This is considered an expression of competitive interaction between trees and crops grown together (Rao et al. 1998) and is viewed as a negative aspect of agroforestry. Indeed, this view has led to the argument, mostly outside India, that one should shy away altogether from simultaneous agroforestry systems, especially in drier climatic zones and focus on fallow rotations (Sanchez 1995). Such arguments do not, however, consider the system's productivity and value from a holistic viewpoint. For example, farmers in the subhumid central parts of India have traditionally followed a sort of parkland agroforestry system that includes a fast growing native tree species, *Acacia nilotica,* with rice (*Oryza sativa*) (Viswanath et al. 2000). The tree provides a wide range of products for household needs including a termite-resistant wood and a variety of nontimber products (gum, seeds); and its leaves and fruits provide fodder for goats, sheep and cattle. On the other hand, farmers of the Indo-Gangetic plains of northern India dislike this tree on their farms, because they perceive it to be competing with agricultural crops causing 40% to 60% reductions in crop yields compared to sole crops (Chaturvedi and Das 2002; Puri et al. 1994c). For farmers in the central parts of India, the tree is quite valuable because of the additional benefits and cash income generated from the tree-based systems even at the cost of reduced crop yields, whereas for the relatively more prosperous farmers in northern plains, any practice that may cause even slight reduction in crop yields is not acceptable. This is a good example how the criteria of judging the suitability of tree species and the outlook on competition can vary depending upon socioeconomic conditions and farmer perceptions.

Following the major emphasis that complementary sharing of resources between crops and trees has received in agroforestry research worldwide (e.g., Ong and Huxley 1996), this topic has been an important one in India too. Early studies began with the general assumption that trees are more deeply rooted than crops and therefore complementarity in utilization of belowground resources could be expected (Nair 1984). Subsequent studies have shown, however, that spatial complementarity (i.e., tree roots capturing nutrients and other belowground resources that are not taken up by crops) may not always be a reality. Many tree species could indeed be shallow rooted and could compete with crops for soil resources; e.g., *Populus deltoides* (Puri et al. 1994d), *Gmelina arborea* (Swamy et al. 2003a). As generalized by Schroth (1999), the distribution of tree-root system in space

Table 1. Major agroforestry practices followed in India.

Practice (Listed in alphabetical order)	Agroecological region / states
Boundary planting and live hedges	In all regions
Farm woodlots	Throughout the country
Homegardens	Mainly tropical west coast region, especially Kerala, southern Karnataka, Andaman & Nicobar Islands
Industrial plantations with crops	Intensively cropped areas in northern and northwestern regions: Haryana, Himachal Pradesh, Punjab, Uttar Pradesh; also in southern states (Andhra Pradesh, Karnataka, Kerala, and Tamil Nadu)
Scattered trees on farms, parklands	All regions, especially semiarid and arid regions
Shaded perennial systems with plantation crops	Mainly humid tropical region in the southern region; also Assam, W. Bengal
Shelterbelts and windbreaks	In wind-prone areas, esp. coastal, arid, and alpine regions
Silvopasture:	
Fodder trees,	Throughout the country
Intercropping/grasses with fruit trees	Subtropical and tropical; orchards in hilly regions
Seasonal forest grazing	Semiarid and mountainous ecosystem
Taungya	Eastern region
Tree planting for reclamation of saline soils and wastelands	Semiarid and canal-irrigated regions, mostly in the northern and northwestern regions
Woodlots for soil conservation	In hilly areas, along sea coast and ravine lands

and time is usually influenced by both the genetic character of the species and localized soil conditions.

Quite a few studies related to tree–crop interactions have been reported from India; the major topics of research and the salient results are summarized below.

Crop yields

One of the most widely reported systems is the *Prosopis cineraria* parklands in the arid and semi-arid regions, mainly in Rajasthan and other western parts of the country (Shankarnarayan et al. 1987; Puri et al. 1994a). Millet (*Lennisetum* spp.) and sorghum (*Sorghum bicolor*) crops growing within a 5-to-10 m radius around mature *P. cineraria* trees have been reported to yield two to three times more than crops growing adjacently but away from trees. This yield increase is attributed to soil fertility improvement under the trees (Shankarnarayan et al. 1987), just as in the case of the *Faidherbia albida* system in West Africa (Boffa 1999; Nair 1989). The animals like the tree's pods and fodder; because of that, and also for shade, they congregate under the trees; moreover, birds perch on tree branches. These activities result in considerable accumulation of animal and bird excreta, adding nutrient inputs to the soil under the tree canopy. *P. cineraria* also fixes nitrogen with nodules located on roots several meters below the soil surface near the water table (Tejwani 1994). Although the merits of this traditional agroforestry practice have been well appreciated, our understanding of the system is not satisfactory (Yadav and Blyth 1996), and several practical questions remain unanswered: for example, the optimum planting density of *Prosopis cineraria* trees in crop fields, and the length of time it takes for such positive effects of trees become recognizable. It has been postulated that the soil-related beneficial effects of *Prosopis* on crop yields will start only after the trees are more than 10 years old (Puri and Kumar 1995). On the other hand, the negative effect of trees because of competition for growth resources can be noted within a much shorter time span.

The *P. cineraria* example needs to be taken as one end of the general spectrum of trees' effect on crop yields. The slow-growing trees such as *Prosopis* may not influence crop yields for many years after their establishment (Puri and Bangarwa 1992), but the fast-growing trees such as eucalypts reduce crop yields within two to three years after their establishment. At Hissar, in western India, 3.5-year-old *Eucalyptus tereticornis* on field boundaries reduced crop yields by 41% up to 10 m from the tree row (Malik and Sharma 1990). The biophysical bottom line is how to manage the interaction for natural resources (light, water and nutrients) between the tree component and the crop and/or livestock components for the benefit of the farmer (Rao et al. 1998). Agroforestry literature on tree–crop interactions is largely based on studies with young trees, with the exception of some studies of the parkland system and shaded perennial systems. Although considerable evidence exists that mature scattered trees in such environments are associated with improved soil nutrient supply in parkland systems, it is uncertain whether trees actually increase the total pool of nutrients in the system or simply re-distribute existing soil nutrients more effectively across the landscape (Kumar et al. 1999; Nair et al. 1999). If higher nutrient status under trees is due to nutrient redistribution, no net gain in production at the landscape scale can be expected. Unfortunately, not too many studies have sampled soil nutrients randomly over sufficiently large areas to test this hypothesis. The extent of shading caused by trees and its impact on crops are also not adequately understood. Crop yield declines – relative to yields in open, treeless fields – have been reported under canopies of large, evergreen unmanaged trees. Under *Acacia nilotica* trees in northern India, wheat (*Triticum* sp.) yields were reduced by up to 60% (Puri and Bangarwa 1992) and mustard (*Brassica* sp.) yields by up to 65% (Yadav et al. 1993). In environments where seasonal rainfall (above ca. 800 mm during the cropping season) or irrigation provides adequate water for crop growth, reduction of crop yields under trees could mainly be due to reduced light availability to the understory crop. Light availability can be altered through management interventions such as pruning or lopping the tree canopy, altering spatial geometries of tree planting, and growing crops when tree becomes deciduous (e.g. poplars). Several studies on tree management practices (lopping, pollarding, thinning, fertilizer application, and so on) and their effects on subcanopy light availability and yields of interplanted and overstory species are available (Kumar et al. 2001; Shujauddin and Kumar 2003). Planting of shade-tolerant crops such as ginger (*Zingiber officinale*) and turmeric (*Curcuma longa*), and medicinal plants (such as *Pogostemon cablin*, and *Kaempferia galanga*) is another option to adjust to reduced light conditions (Kumar 1999; Kumar et al. 1999; Rao et al. 2004).

It is important to have a realistic understanding of the extent of crop-yield changes on an overall-area basis. In spite of the seemingly large yield variations under or near trees, the overall effect of trees on crop yields could be small because only a small proportion of the area is subjected to tree/crop interactions (Rao et al. 1998). Considering a scattered-tree system that contains 20 trees ha^{-1} with each tree affecting crops over a 100 m^2 area, if the trees increase or decrease crop yields by 50% over that area, the overall yield of the system will be only 10% higher or lower than that of sole crop.

Water relations

The relation between trees and water continues to be a 'hot topic' of public debate in India. The fact that young tree plantations, especially of evergreen species, tend to use more water than the older, deciduous forests or agricultural (nonirrigated) lands, has gained attention in the form of the much-publicized 'Eucalyptus Debate.' In southern India, eucalypt plantations have been alleged to not only extract *all* the rainfall that enters the soil, but also to utilize an additional 100 mm of water from each meter depth of soil that the roots penetrate (Kallarackal and Soman 1997). This is a matter of concern as studies in Karnataka in southern India have shown that eucalypt roots can reach to a depth of 8 m in three years (Calder et al. 1997). Consequently, when planting trees in association with crops, it is essential to consider the implications of increased water use on the medium- and longer-term water budgets. Special consideration should be paid to the source of water used by trees, the rate of water depletion below the crop rooting zone, and the prospects for deep recharge; using water at faster than replacement rates demonstrates a serious ecological consequence of not mimicking the resource use patterns of locally adapted species (Ong and Leakey 1999). While there is no reason to single out *Eucalyptus* species in this regard, the high water use of fast growing, evergreen trees can be a concern in areas with a shortage of groundwater or subsurface flows of water. In other areas such interception of subsurface flows can be seen as an environmental service function, where it prevents salt movement in groundwater flows. The environmental service perceptions will thus depend on the local agroecosystem, much the same way as the perception to the *Acacia nilotica* system described earlier.

Competition for water between trees and crops can occur both within the rainy season and through the buildup of a soil water deficit caused by tree transpiration during the dry season. Interseasonal water deficit can be particularly severe when there is a succession of droughts and when recharge is small and infrequent (McIntyre et al. 1996). Thus, temporal complementarity is only possible if residual water is available after the crop's harvest or if rainfall occurs when there is no scope for cropping. For example, in Hyderabad in the northern part of peninsular India, 20% (152 mm) of the annual rainfall occurs outside the normal cropping season and even the traditional intercrop of sorghum/pigeonpea (*Cajanus cajan*) utilizes only 41% of the annual rainfall. The remainder of the rain is lost as runoff (26%) or deep percolation (33%) (Ong et al. 1992). In such environments, agroforestry practices increase rainfall utilization by extending the growing season.

It is advocated that trees harvest resources from their surroundings and accumulate them under the trees – the so-called 'hydraulic lift.' The hypothesis is that water taken up by roots from moist zones of soil is transported through the root system and released into drier soil. Rainfall captured through stem flow, especially by a woody canopy, can be stored deep in the soil close to the roots and be returned to the topsoil beneath the canopy by hydraulic lift for later use with associated benefits for the understory species (Ong and Leakey 1999). An intriguing possibility is the prospect of capitalizing on this phenomenon of resource capture at a distance through strategic use of hydraulic lift. Considering the vast areas of arid and semiarid lands in India, this is an area of agroforestry with potentially wide application.

In the context of an on-going environmental debate and the emergence of a powerful environmental lobby in India, the hydrological effects of agroforestry and other tree-based systems, particularly the effect of trees on water yields, have been the subject of much discussion and more speculation. Almost every instance of a change in weather pattern or water cycles – and they are of common occurrence – is alleged to have been either caused or aggravated by deforestation. And this is generally accompanied by the claim that the obvious way to prevent its recurrence is to plant trees in the water catchment, the idea being that trees will 'soak up' excess water. Rigorous scientific data are lacking, however. Planting trees in agroforestry can be justified for a variety of reasons; but it is doubtful if trees can guarantee a steady flow of water under all climatic conditions.

Soil conservation and amelioration

Keeping the soil resource productive and in place is one of the major sustainability issues. Many agroforestry systems help keep the soil resource in place by biological means. Closely spaced trees on slopes reduce soil erosion by water through two main processes: first, as a physical barrier of stems, low branches, superficial roots and leaf litter against running surface water; and, secondly, as sites where water infiltrates faster because of generally better soil structure under trees than on adjacent land. Scientifically acceptable evidence to support these is overwhelming (Rao et al. 1998). Trees do not, however, provide these functions until they are well established and have developed a litter layer. Simultaneous agroforestry systems that include crops or ground cover while the trees are small may overcome this limitation.

If the beneficial effects of parkland trees on soil properties are linked to trees with a high proportion of woody aboveground structures, then it would take a long time before the beneficial effects can be realized. Therefore it is not surprising that the positive interactions between mature, deep-rooted and widely spaced *P. cineraria* trees would require 10 to 15 years. Such long time scales are well beyond the planning horizon of many farmers for the relatively small benefit in crop productivity and may well explain why farmers rarely plant these trees, even though they are aware of tree species that improve soil and climate.

The use of trees and shrubs for soil amelioration, especially in the reclamation of salt-affected soils, is an aspect that has received considerable attention in India. An estimated 9 million ha of semiarid lands in the states of Andhra Pradesh, Gujarat, Haryana Karnataka, Maharashtra, Punjab, Rajasthan, and Uttar Pradesh are affected by salinity/alkalinity (Government of India 1992). The problem could extend – if it is not already there – to all the more than 20 million ha of canal-irrigated lands (Tomar 1997). Essentially two approaches have been followed for the rehabilitation of such lands: improving the soil properties through suitable chemical amendments, and growing plant genotypes that can withstand sodic soil conditions (Gill and Abrol 1991; Singh et al. 1992; Tomar 1997). In the latter option, which is relatively far less expensive, biomass addition by the species and its consequent decomposition and mineralization will improve the soil in the long run (Garg and Jain 1992). Salt-tolerant tree species such as *Acacia nilotica*, *Dalbergia sissoo*, *Prosopis juliflora*, and *Terminalia arjuna* are now planted to reclaim large tracts of such salt-affected soils; the trees may be underplanted with suitable fodder grass species, and the land will be reverted to agricultural crops after the salinity levels of the top soil drop down (Singh et al. 1992; Garg 1998). Screening of germplasm of salt-tolerant species and development of appropriate management strategies for their large-scale multiplication and planting are being pursed vigorously in a number of locations in the country (Gill and Abrol 1991; Singh et al. 1988; Singh et al. 1992; Tomar 1997).

Other topics of biological research

Tree selection and improvement

Systematic research on genetic improvement of tree species started in India during the 1950s, at the Forest Research Institute, Dehra Dun. The efforts were concentrated on relatively few species, mostly of the genera of commercial significance (*Tectona*, *Eucalyptus* and *Pinus*), usually of industrial plantation species (Kedharnath 1984). But it was during the late 1970s that tree improvement began to attain recognition as an integral part of Indian forestry (Kedharnath 1986), and genetic improvement of agroforestry trees did not gain any attention until the mid 1980s (Khosla 1985). Although some efforts have subsequently been made in outlining the breeding and selection principles of agroforestry species for specific agroforestry practices (Puri 1998) and the importance of the task has been recognized (Bangarwa 2002), the activities are yet to gather momentum.

A major deficiency in this area is the lack of a legislative mechanism for certification of seeds of forestry species or registration of clones and certification of clonal planting stock. In spite of that, some private initiatives have been made in forestry research and development. Notable among them are the promotion of poplar (*Populus deltoides*) plantations by WIMCO Seedlings Limited, and clonal technology research and promotion of large-scale eucalyptus plantations on marginal farmlands by ITC Bhadrachalam Paperboards Limited (Jones and Lal 1991; Nair et al. 1996). Vegetative propagation of some indigenous tree species has also been attempted, with the main objective of productivity gains and quality improvement; e.g., *Dalbergia sissoo* (Puri and Swamy 1999), and neem (*Azadirachta indica*) (Palanisamy et al. 1998; Puri and Swamy 1999). Field trials for germplasm selections have also been carried out for many indigenous agroforestry species. For example, one of the earliest attempts to evaluate genetic variability in neem indicated significant provenance related variations in seed weight, leaf length and height growth (Mishra 1995). Chaturvedi and Pandey (2001) evaluated 30 germplasm lines of *Bombax ceiba* for genetic divergence. Based on mean performance, genetic distance and clustering of germplasm lines, progenies with greater plant height, stem diameter and improved timber quality were selected. Many indigenous agroforestry tree species are also reported to have significant provenance variations in growth and biomass production e.g., *P. cineraria* (Arya et al. 1999) and *D. sissoo* (Bangarwa 1996). These provenance differences are now being exploited through selection of the best provenances for their use in plantations.

Germplasm collection and testing of *Leucaena* species was initiated in India in 1970 with the objectives of selecting new provenances which were straight-stemmed, late flowering with low seed production, and having high ratio of edible to woody parts, good coppicing ability, high leaf retention in dry periods and low mimosine content in forage. Gupta et al. (1996) tested a total of 496 accessions from 14 species and found *L. leucocephala* ssp. *glabrata* as most promising in northern India that is free of the devastating psyllid, *Heteropsylla cubana* that has devastated large areas of *Leucaena* in many other regions of the world. Other promising varieties were K8, K217, K601, K397, K340, IGFRI 23-1, Sil 4, S-24, S-10, and S-14.

Naina et al. (1989) were successful in inducing genetic transformation in neem using *Agrobacterium tumefaciens* strains K 12 × 562 E and K 12 × 167. The transgenic plants were successfully regenerated *in-vitro*. However, this is still a developing field of science in India, and achievements in most cases are at the best at laboratory level.

In order to improve trees for agroforestry, it is important to know the natural breeding systems of the species. Although the breeding systems are well known for the major industrial species (pine, spruce, eucalypt, teak) and for some multipurpose trees, particularly legumes, the natural systems are not known for many of the species promising for agroforestry (Burley et al. 1986). The mating system was investigated in a natural population of the tetraploid taxon *Acacia nilotica* sp. *leiocarpa* using open-pollinated seeds from 15 families (Mandal et al. 1994). Estimates of single-locus ($t_s = 0.358$) and multilocus ($t_m = 0.384$) population out crossing rates were homogeneous indicating substantial selfing in the population. This enables the breeder to develop the strategy for seed collection for breeding and conservation purposes. Phenological, morphological and anatomical studies are also required and artificial pollination methods have to be developed for many agroforestry tree species.

Multipurpose trees

The need for selection and characterization of woody perennials that can provide multiple products and services in agroforestry systems has long been recognized in India. Although many traditional multipurpose trees such as *P. cineraria* have been recognized as valuable, efforts in selection and improvement of such species have been generally lacking. Toky et al. (1992) listed more than 250 indigenous multipurpose trees present in different ecoregions of India, including many lesser-known species that need domestication. Many such species (e.g., *Albizia lebbek*, *A. procera*, *Azadirachta indica*, *Casuarina equisetifolia*, *Dalbergia sissoo*, *Dendrocalamus* spp., *Gmelina arborea*, *Grewia optiva*) are in a process of domestication and/or commercialization (Naugraiya and Puri 2001; Tewari 2001). While identifying such plants, emphasis is on the species and provenances with wide adaptability to sites (including those with problem soils), ease of establishment, resistance to biotic and abiotic forces, fast growth and biomass production potential, shorter harvest rotations, etc. (Pathak 2002). In India, woody tree species constitute the main energy source, and this dependence on woody vegetation is unlikely to change for many years. Ravindranath et al. (1997) reported for a Karnataka village that 79% of all the energy used came mainly from trees and shrubs. With rising demand for fuels, tree planting has undergone both expansion and change. Efforts to increase wood resources were initiated in the late 1960s, and were intensified with the World Bank aided Social Forestry program in the early 1980s, with emphasis on planting of multipurpose trees (both indigenous and exotic). Research into the efficiency of the trees as an energy source and fuelwood characteristics of some Indian trees and shrubs have been documented (Bhatt and Todaria 1990; Singh 1988). Wood density is known to positively correlate with calorific value (Bhatt and Todaria 1990). The fuelwood potential of indigenous (*Acacia nilotica*, *Azadirachta indica*, *Casuarina equisetifolia*, *Dalbergia sissoo*, *P. cineraria* and *Ziziphus mauritiana*) and exotic (*Acacia auriculiformis*, *A. tortilis*, *E. camaldulensis* and *E. tereticornis*) trees was studied (Puri et al. 1994b). The calorific values ranged from 18.7 to 20.8 MJ kg^{-1} for indigenous tree species and 17.3 to 19.3 MJ kg^{-1} for exotics. Another measure used, the fuelwood value index (FVI), was reported to be high for indigenous tree species but was highest for *C. equisetifolia* (FVI 2815); however, since the parameters used to calculate FVI (heat of combustion, specific gravity, and water content) are autocorrelated, these values have little importance. Pathak (2002) opined that species such as *C. equisetifolia*, *P. juliflora*, *L. leucocephala* and *Calliandra calothyrsus* have become prominent due to their potential for providing wood energy at the highest efficiency, shorter rotation and also their high adaptability to diverse habitats and climates.

Tree fodder and silvopastoral systems

In India, an estimated 196 million cattle and 80 million buffaloes – which account for 15% and 52% of world totals of these animals – are used for smallholder farm situations (Singh and Upadhyaya 2001). Trees and shrubs often contribute a substantial amount of leaf fodder for these animals during 'lean' periods of fodder availability, through lopping/pruning of trees (Singh 1982), popularly known as top feeds. Tree fodder and browsing systems are more common in drier areas, degraded lands and community lands. In the semiarid regions, silvopastoral systems involving native tree species (e.g., *Albizia procera*, *A. lebbek*, *Acacia* spp., *Azadirachta indica*, *Dalbergiasissoo*, *Morus alba*, and *Pongamia pinnata*) have been practiced for many years (Singh and Roy 1993). Similarly, in arid regions, fodder tree species such as *A. nilotica*, *Hardwickia binata*, *P. cineraria*, *Tecomella undulata*, and *Ziziphus nummularia*, form an important component of land use (Tejwani 1994). Complementarity between tree and pasture species grown in association is important for increased growth and biomass production.

The synergies of tree–pasture association and their effects on livestock productivity have been evaluated to only a limited extent in India (Pathak 2002). Rai et al. (1999, 2001) have reported increase in land productivity through silvopastoral systems in the shallow red gravely soils under semiarid condition at Jhansi in central India, using trees such as *Acacia nilotica* var. *cupressiformis*, *Albizia lebbek*, *A. procera*, and *Hardwickia binata*. Results of studies on performance of grass (*Cenchrus* sp.) and legume (*Stylosanthes* sp.) species grown under various tree species (*Albizia amara*, *A. lebbek*, *Acacia tortilis*, *Dalbergia sissoo*, *L. leucocephala*, and *P. cineraria*) have also been reported from a number of places in India (Rao and Osman 1994; Sharma et al. 1996; Pathak 2002). Considerable information on silvopastoral systems under arid conditions is also available. In addition to indigenous tree species (e.g., *Prosopis cineraria* and *Z. nummularia*), the introduced species such as *A. tortilis*, *Dichrostactys cinerea*, and *Colophospermum mopane* have showed good performance in silvopastoral systems under such arid conditions (Ahuja 1980). Biomass productivity of arid lands increased two to three times by replacing natural grass cover with *Cenchrus ciliaris* and introducing top feed (tree) species such as *Z. nummularia* and *Grewia tanax* under silvopastoral system (Sharma et al. 1994).

The information on the performance of livestock production under silvopastoral grazing conditions is meager (Rai et al. 1999). Other topics of research in this area include economics (mostly benefit/cost calculations) (Kareemulla et al. 2002; Pathak 2002), and the impact of rotational grazing by sheep and goats on plant composition and dominance of natural grassland and established silvopastoral systems (Sharma and Ogra 1990; Rai et al. 1999). Most of these studies are, understandably, location specific, and their results have little transferability to other locales. The results could, however, be quite useful for developing comprehensive simulation models on silvopastoral system performances in a broad context, which is an aspect that is worth pursuing in future research.

Other ecosystem functions

Most of the agroforestry research in India to date has been limited to plot- or field-level studies, with practically no attention given to landscape and ecosystem level studies. One of the emerging areas of interest related to environmental services at both farm and ecosystem levels in India, following the global interest in the subject, is carbon sequestration potential of agroforestry. The science of carbon sequestration is relatively simple: trees and crops convert carbon dioxide into carbon compounds, in the form of biomass (see Montagnini and Nair 2004). Carbon cannot be stored in crops for long durations. Much of the opportunity to store carbon through afforestation in India will occur on agricultural lands due to the vast land area devoted to agriculture. Obviously, the nature of trees (fast or slow growing) and the purpose for which the timber is used will determine whether it is beneficial from an environmental and social point of view. Critics of carbon sequestration deals fear that sequestering carbon in one area by, for example, planting trees, could lead to pressure on forests elsewhere, thus neutralizing the benefits of the scheme. They also suggest that carbon sequestering projects involving trees will have a very short-term impact on carbon budgets. As in other places, research related to carbon sequestration in agroforestry practices is relatively new in India. Swamy et al. (2003b) estimated that a six-year-old *Gmelina arborea*-based agrisilviculture system in Raipur, eastern India, sequestered 31.4 Mg ha^{-1} carbon. In a silvopastoral system in Kurukshetra, northern India, carbon flux in net primary productivity increased due to integration of *Prosopis juliflora* and *Dalbergia sissoo* with grasses (Kaur et al. 2002). Such sporadic studies are uncoordinated in nature and they mostly report a derivative of biomass productivity rather than C sequestration in soils; their value from the sequestration point of view depends on the end-use of the wood and other products, as mentioned before. Nevertheless, the studies are indicative of the nascent interest in the topic.

Socioeconomic issues

During the last quarter of the 20th century, India has had some unique and remarkable experiences, including controversies, surrounding tree planting on agricultural fields. Most of these were concentrated in the north and northwestern parts of the country; the following discussion is therefore focused on that region and that particular practice. It needs to be recognized that most of the issues discussed in this section are not ‘lessons learned’ from research, but are ‘spin-off’ issues that have surfaced in the recent past following farmers’ initiatives in tree planting on farms. Research is needed to address such ‘unanticipated’ and ‘nonconventional’ issues.

Recognizable impacts

Although agroforestry has been a way of life in India for a very long time, it was the planting of *Eucalyptus* sp. in agricultural fields during the mid-1960s that heralded a sort of 'revolution' in the economic activity related to tree planting (Pathak and Pateria 1999). During 1970–1975, 7% of the farmers in Punjab state in northwestern India had taken up planting *Eucalyptus* and earned substantial financial gains. That 'success' led to widespread adoption of the practice: 46% of farmers had adopted it by 1980–1985. But, during the late 1980s to early 1990s, when these plantations became harvestable, the prices crashed due to a glut in the market for *Eucalyptus* timber and low consumption by paper/pulp mills (Negi et al. 1996). That resulted in a sharp decline in the percentage of *Eucalyptus*-growing farmers (5% in 1990–1991), and diverted farmers' attention to planting poplars (in locations above 28°N latitude because of climatic adaptability of the species). The number of farms adopting poplar planting increased from 3% in 1970–1975 to 48% in 1990–1995 in northwestern India (Joshi 1996; Dhanda and Kaur 2000). During the late 1990s, a few clones of the species were also introduced into the central parts of India (Puri et al. 2002). Presently poplar is grown on a large scale and has replaced *Eucalyptus* in the expansive irrigated agricultural belt of the Indo-Gangetic plains. Easy marketability of the timber, rapid growth and high productivity (average 10–30 m^3 ha^{-1} yr^{-1}; 46.92 m^3 ha^{-1} yr^{-1} under intensive management), deciduous nature allowing cultivation of shade-intolerant winter crops, soil fertility maintenance through litterfall (deciduous nature), and service role as a windbreak in fruit orchards and thus as a boundary marker, are some of the characteristics of the species that endear it to farmers. Plantations of other tree species such as *Dalbergia sissoo*, *G. arborea*, and bamboos have also been undertaken to a limited scale. Farmers have also developed agrisilvicultural systems involving many fruit trees (Chaturvedi and Jha 1998; Kaushik et al. 2002).

Assessment of the impact of agroforestry is hampered by the lack of rigorous procedures and methodologies (Nair 1999). Nevertheless, some clear experiences are available that can be taken as indicators of impacts of the efforts.

1. Farmers of Punjab, Haryana and western Uttar Pradesh (particularly in the *Tarai* belt) have adopted the poplar- and eucalyptus-based agroforestry in a big way (approximately 48% farms of the region covering an estimated 11.48 million ha of land) (Pathak and Pateria 1999). Today, poplar trees planted on agricultural fields and field boundaries are harvested in 6-to-8 year rotations, and the average economic return per hectare of poplar-agroforestry system is three to five times more than that from sole crop agriculture (Joshi 1996).
2. Agroforestry played a major role in the recent past in the rehabilitation of wastelands such as deserts and lands that have been degraded by salinization and ravines, gullies, and other forms of water and wind erosion hazards (section 2. 2. 3) . For example, by using the agroforestry technologies developed at Central Soil Salinity Research Institute, Karnal, the state forest departments, nongovernmental agencies (NGOs), National Wastelands Development Board (NWDB), and other developmental agencies have rehabilitated more than 1 million hectares (out of a total of 7.1 million hectares) of salt-affected soils, particularly the village community lands, government lands, absentee lands, areas along road sides, canals and railway tracks (Tomar 1997).
3. Indian Farm Forestry Development Co-operative Limited has implemented agroforestry in the states of Uttar Pradesh, Madhya Pradesh, Chhattisgarh and Rajasthan. This has been done by people's participation through the formation of cooperative societies (Singh 1999). Similarly, the Tree Growers Cooperative in Gujarat has come in a big way to implement agroforestry on farmlands and wastelands (Misra 2002).
4. Agroforestry got a major boost in northern India through the farmer-friendly policies of an industrial plywood company, WIMCO Ltd. This company has not only motivated the farmers to plant trees, such as poplar, on their farm lands but also created a big impact on the economy of these states by creating assured market for poplar wood (Joshi 1996; Chandra 2001).
5. Increased supply of wood in the market has triggered a substantial increase in the number of small-scale industries dealing with wood and wood-based ventures in the near past. Such industries have promoted agroforestry and contributed significantly to increasing area under farm forestry (Lal et al. 1997).
6. Recognizing agroforestry as a viable venture, many business corporations, limited companies have entered into this business and initiated agroforestry activities in collaboration with farmers on a large scale (Kinhal 1995; Misra 2002).

Quantitative data on the impact of these efforts are not yet available; but changes are clear and visible.

Plantations vs smallholder tree husbandry

In India the division between forestry and agriculture is particularly pronounced. The Indian constitution places the control of natural resources in the hands of the state governments and stipulates that the forests must be managed for the benefit of people. Foresters have experimented for more than 150 years, starting with the introduction of the *Taungya* system in the 1850s in the then British India (King 1987), with ways to get local farmers' participation in their efforts to plant and manage trees, in various management schemes that are forerunners to what are called agroforestry, farm forestry, and social forestry today. Most of these efforts have been a hard way to learn a simple lesson: unless farmers share substantially in the long-term benefits of forest plantation efforts, the interaction between the 'agro' and the 'forestry' components will remain competitive rather than complementary. Because of land scarcity, large-scale plantations and smallholder development programs tend to be mutually exclusive. Although some successes have been attained in augmenting wood production through plantation forestry, the merits of such industrial or government-sponsored schemes are being questioned for lack of people's participation (Maikhuri et al. 1997).

In contrast, the loss of local forest resources often leads to increased incentives for spontaneous expansion of smallholder tree husbandry. The spread of the poplar-agroforestry program in northwestern India described earlier is a case in point. In such cases, farmers plant and protect trees on their own farms to provide tree products for household needs and sale for cash income. The trees they plant represent a conscious investment for which other options have been forfeited. Farmers generally integrate *tree growing* with their crop and animal production activities. The management practices undertaken to assure good food-crop yields – cultivation, weed control, and fertilization – also benefit their trees. Because landholdings are small, farmers select the farm niches most appropriate for tree production.

Usually smallholders start timber production systems by planting short-rotation species to meet household and local market needs. As more farmers begin producing timber, supply meets or exceeds demand and prices decline. At this point, farmers either stop further investment in tree planting or diversify into long-rotation, premium-quality timbers (Saxena 1995). Farmers and policymakers alike have poorly understood the dynamics of supply and demand of tree products and marketing at the level of smallholder farmers (Alavalapati et al. 1995).

Economies of scale

Many pulp mills are finding it increasingly difficult to gain access to land where they can establish plantations, and finding it harder to source wood from natural forests. As a result, many of them are entering into contracts with local communities and small farmers to grow wood. 'Outgrower' or 'joint-venture schemes,' as they are known, are becoming a popular way of large-scale tree planting in India (Saxena 1995; Hobley 1996). One popular form of the scheme involves plantation companies providing local people with all the planting materials they need and the inputs required to maintain the plantations. At the time of harvest, the company buys the wood (Lal et al. 1997; Chandra 2001). Company/smallholder relationships have rapidly evolved, with wood being traded on the open market, and companies competing to meet their wood requirements. Outgrower schemes have benefited from a government policy that limits the area of land that can be held by a private owner. Such statutory ceilings limit the agricultural land holdings both for individual farmers and corporate sector to extremely low level in all states, e.g. up to 22 ha of the worst category of land in Andhra Pradesh (Reddy and Reddy 1995). This has forced the wood-based industries to buy most of their supplies from smallholders, and small-scale tree growing is increasingly seen as a viable land use. Companies rely on farmers for their wood procurement and the farmers, in turn, depend on companies to provide services such as research and development. There is a strong demand from participating farmers for improved planting stock, and several companies have understood that the best, and often the only, thing they have to do to support smallholders is to focus on tree improvement work and the commercial production, in local nurseries, of high-yielding clonal seedlings (Joshi 1996).

What all these schemes have in common is their capacity to provide wood fiber, which would otherwise be unavailable, while still allowing the industry to achieve economies of scale in situations where land is scarce. Local involvement may mean there is less likelihood of conflict, which is a common problem for many plantation owners. As for local communities and landholders, they benefit in a variety of ways. It provides employment and income for some, and profits, as well as a way to spread risks between agricultural and timber crops, for others. From an environmental point of view, a large numbers of small wood lots scattered in agricultural fields may be preferable to a few vast blocks of trees. In short, this is a 'win-win' situation in which both the farmers and the companies benefit from the other partner's 'economies of scale.'

Marketing of farm-grown wood

Marketing is recognized as an important aspect of tree cropping in India. The contrast between tree marketing and agricultural crop marketing is sharp: for marketing agricultural products, the state governments have set up a vast infrastructure; remunerative procurement prices are announced several months in advance of the crop; generous subsidies are built into the system of government purchase, which absorbs surplus and stabilizes prices; market yards have been set up to reduce exploitation; and the middleman's commission has been fixed. Tree marketing does not have any of these 'privileges.' On the contrary, markets for forest products are dominated by bureaucratic regulations, shrouded in secrecy and impeded by legal bottlenecks (Saxena 1995). Compared with markets for other agricultural products, wood markets are still not fully geared to receiving farm production. Transport of wood is regulated by legal restrictions that were designed to prevent illicit felling from government forests. But, today they act against the interests of producers, forming a barrier between the producer and the market and bringing uncertainty in the operation of sale transactions (Negi et al. 1996). Farmers are usually unaware of buyers and the prevailing market imperfections, and the middlemen's margins in some areas are seemingly large; these lead to high price differentials between what producers get and the consumers pay. One of the largest risks reported by middlemen is unreliable quality and quantity of smallholder products.

Private forestry requires security of land and tree tenure, and secure access to markets. A great source of market imperfections in tree product markets is the existing, antiquated, legal and procedural framework, which makes cutting and selling privately owned trees so difficult, irksome, complicated, and nonremunerative.

Other institutional and policy issues

It is well known that the success of every agricultural project or program will be affected by the prevailing governmental policies. A serious deficiency that limits the realization of the benefits of agroforestry is the lack of adequate policy and institutional infrastructure. Experience from past forestry and agricultural development programs indicates that institutional support, exerted through equitable land tenure legislation, research and extension, market support and development, and credit availability, is as important to farmer response as the biological performance of the promoted technologies.

Currently, almost all forestlands that are deforested and degraded through inappropriate wood extraction and cultivation – which, interestingly enough, are prime candidates for rehabilitation through agroforestry systems – still largely remain under the jurisdiction and control of state forest departments. These governmental bodies invariably cling to the classical forestry concept of perpetual dedication of these lands to long-term, large-scale, and industry-oriented forestry, and continue to regard agricultural-forestry crop integration as completely incompatible with, and alien to, accepted forestry practices. Furthermore, they believe that farmer participation in forest development is neither suited nor needed, and they reinforce this view by depriving farmers of forestland tenure and of equitable share in forest benefits (Follis and Nair 1994).

Yet another problem is that the policy measures aimed at protecting forests also are applied to agroforests that are managed sustainably by small-scale farmers (Lakhmipathy 1992). The unintended result of treating all timber alike, regardless of its origin in forests or on farms, is that smallholders who plant and tend trees are unfairly penalized. Except for a few tree-species (e.g., poplar and eucalypts) there are restrictions on the felling of trees standing on private land (Mohanan et al. 2001). The laws make it mandatory to seek written permission from the administration before felling certain trees in certain areas. The legal provision, due to historical reasons, are scattered in revenue laws, forest laws and laws specially created for the purpose. The implicit assumption of the law is that the rural people cannot be expected to act responsibly in managing their trees and, therefore, the state should impose restrictions on the felling of trees on private land for the benefit of the rural people and preservation of the environment. The restrictions apply selectively to species, regions, and nature of landholding. It is impossible for the administration to enforce such restrictions on so many holdings, and the law becomes a ritual in clearance seeking, leading to harassment and corruption (Sen 1995). For private forestry and agroforestry to survive, the current laws must be repealed and replaced by more forward-looking, producer-friendly ones.

Conclusions

Admittedly, the analysis presented here is not based on a comprehensive treatment of all aspects of agroforestry research that has happened in all parts of India during the past 25 years. But it gives a clear indication of the advances made in understanding and appreciating the potential of agroforestry, the efforts made in realizing the potential, and the constraints and impediments that have caused setbacks in achievements. Some of these experiences and tribulations could be an eye-opener for other nations as well.

India's efforts in agroforestry research have largely been self-supported. Although some international support has been received at various stages, they do not constitute a significant contribution to the overall agroforestry research enterprise in the country. This approach of self-reliance in research is commendable in spite of its shortcomings. A major shortcoming is the apparent lack of clear goals, objectives, and political commitment to agroforestry. Agroforestry is perhaps somewhere at the bottom ebb of research priorities even in agricultural and forestry sectors. Agroforestry may not be the proverbial 'rocket science,' but India's advancement in rocket science is a clear example of how the long-term, sustained, determination and commitment at the highest levels of government can pay rich dividends. Agroforestry research in India has been spasmodic. It has had its highs and lows. A more coordinated, focused, determined effort will, undoubtedly, provide more and better results.

A disturbing aspect of all research efforts in agroforestry in India during the past quarter century is that research has mostly been conducted on research stations in relatively small plots or/and laboratories according to the norms and procedures of conventional agricultural/forestry research. Little or no research has been done on an ecosystem or landscape level. A second feature is the relative short-term nature of studies (less than three years). This is understandable, given the restrictions of government procedures and budget cycles. Yet another feature is that all research is of the so-called 'applied' (as opposed to 'basic') nature. Most research was designed for 'solving' some local problems. Again, this is a reflection of the accountability attached to utilization of public funds. A serious problem with most of these research efforts is that the research has been limited mostly to technology testing and gathering of observations (of what happens when something is done) with very little investigations on the reasons for the observed results. In other words, the 'how' and 'why' questions (i.e., the scientific reasons and processes underlying the observed behavior) have been vastly neglected. Because of that, the research results are of limited applicability in situations or locations other than the study sites, and even under different soil and other ecological conditions of the same study region. An argument could be made that since agroforestry research has been state-financed by local taxpayers' money, it is imperative that the research is focused on solving local problems. Admittedly, this argument has some merit. Nevertheless, the fact remains that any research that is not based on strong scientific foundation will not produce reproducible results. Technology testing has its value only if the technology itself is based on solid scientific foundations. It has also been amply illustrated with the successful examples of reclamation of salt-affected soils and other so-called 'wastelands' and the poplar-based agroforestry that when the scientific principles of a technology are well understood, the technology will 'take off.'

India's experience in tree farming on agricultural land also clearly shows how the existing agricultural and forestry laws and procedures offer serious limitations and impediments to agroforestry development. Many of the policies that are in place were created to keep agriculture and forestry separate and to empower bureaucracy to facilitate state control and are therefore serious disincentives to tree planting on farms. These experiences have brought out one thing very clearly: enabling policies and supportive political will at the highest levels of government are essential for the success of agroforestry. The analysis also shows that new paradigms such as farmers' initiatives on tree planting on farms will bring to surface new problems; socioeconomic and policy research will need to be geared to address such new challenges.

End Notes

1. Here, the term 'agroforestry technology' refers to agroforestry practices that have been developed recently with research backing – involving either improvement of traditional practices or introduction of new ones (Nair 1993).
2. The term agroforestry is used in a broader context to include all forms of tree planting outside forests, be they known as social forestry or farm forestry or whatever else.

References

Ahuja L.D. 1980. Grass production under Khejri, pp. 28–30. In: Mann H.S. and Saxena S.K. (eds), Khejri (*Prosopis cineraria*) in Indian Desert – Its Role in Agroforestry. Monograph No. 11. CAZRI, Jodhpur, India.

Alavalapati J.R.R., Luckert M.K. and Gill D.S. 1995. Adoption of agroforestry practices: a case study from Andhra Pradesh, India. Agroforest Syst 32: 1–14.

Arya S., Toky O.P., Bisht R.P., Tomer R. and Harris P.J.C. 1999. Variation in growth and biomass production of one-year seedling of 30 provenances of *Prosopis cineraria* (L.) Druce in arid India. Indian J Forestry 22: 169–173.

Bangarwa K.S. 1996. Sissoo Breeding. Agriculture & Forestry Information Centre, Hisar, India. 156 pp.

Bangarwa K.S. 2002 Plus tree selection and progeny testing for establishment of first generation seed orchard in *Dalbergia sissoo* Roxb. Indian J Agroforest 4: 122–131.

Bhatt B.P. and Todaria N.P. 1990. Fuelwood characteristics of some mountain trees and shrubs. Biomass 21: 233–238.

Boffa J.M. 1999. Agroforestry parklands in sub-Saharan Africa. FAO Conservation Guide 34. Food and Agriculture Organization of the United Nations, Rome, Italy.

Burley J., Hughes C.E. and Styles B.T. 1986. Genetic systems of tree species for arid and semiarid lands. Forest Ecol Manag 16: 317–344.

Calder I.R., Rosier P.T.W., Prasanna K.T. and Parameswarappa S. 1997. Eucalyptus water use greater than rainfall input – a possible explanation from southern India. Hydrology and Earth System Science 1: 249–256.

Chandra J.P. 2001. Scope of poplar cultivation. Indian Forester 127: 51–61.

Chaturvedi O.P. and Das D.K. 2002. Studies on rooting patterns of 5-year-old important agroforestry tree species in north Bihar, India. Forests, Trees and Livelihood 12: 329–339.

Chaturvedi O.P. and Jha M.K. 1998. Crop production and economics under *Litchi chinensis* Sonn. Plantation across 1 to 9 year age series in north Bihar, India. Int Tree Crops J 9: 159–168.

Chaturvedi O.P. and Pandey N. 2001. Genetic divergence in *Bombax ceiba* L. germplasms. Silvae Genetica 50: 99–102.

Dhanda R.S. and Kaur I. 2000. Production of poplar (*Populus deltoides* Bartr.) in agri-silviculture system in Punjab. Indian J Agroforestry 2: 93–97.

FAO 1995 Plantations in tropical and subtropical regions – mixed and pure. FAO of the UN, Rome, Italy.

Follis M.B. and Nair P.K.R. 1994 Policy and institutional support for agroforestry: An analysis of two Ecuadorian case studies. Agroforest Syst 27: 223–240.

Garg V.K. 1998. Interaction of tree crops with a sodic soil environment: potential for rehabilitation of degraded environments. Land Degradation and Development 9: 81–93.

Garg V.K. and Jain R.K. 1992. Influence of fuelwood trees on sodic soil. Can J Forestry Res 22: 729–735.

Gill H.S. and Abrol I.P. 1991. Salt affected soils, their afforestation and its ameliorating influence. Int Tree Crops J 6: 239–260.

Government of India 1992. Indian Agriculture in Brief. Directorate Economic Statistics, Department of Agriculture, Ministry of Agriculture, New Delhi. 460 pp.

Gupta V.K., Pathak P. and Solanki K.R. 1996 Genetic improvement of genus *Leucaena* in India, pp. 455–458. In: Dieters M.J., Matheson A.C., Nikles D.G., Harwood C.E. and Walker S.M. (eds). Tree Improvement of Sustainable Tropical Forestry. Proc. QFRI – IUFRO Conference, Caloundra, Queensland, Australia.

Hobley M. 1996. Participatory Forestry: The Process of Change in India and Nepal. Rural Development Forestry Study Guide 3, Rural Development Forestry Network, ODI, London. 337 pp.

ICAR 1981. Proceedings of the Agroforestry Seminar, Imphal, 1979. Indian Council of Agricultural Research, New Delhi. 268 pp.

Jones N. and Lal P. 1991. Commercial poplar planting in India under agro-forestry system. Commonw Forest Rev 68: 19–26.

Joshi B.C. 1996. Poplar cultivation under agroforestry, pp. 409–423. In: Bach I. (ed.). Environmental and Social Issues in Poplar and Willow Cultivation and Utilization. FAO Proceedings 20th Session of the International Poplar Commission. Budapest, Hungary.

Kallarackal J. and Somen C.K. 1997 Water use by *Eucalyptus tereticornis* stands of differing density in southern India. Tree Physiol 17: 195–203.

Kareemulla K., Rai P., Rao G.R. and Solanki K.R. 2002. Economic analysis of a silvopastoral system for degraded lands under rainfed condition. Indian Forester 128: 1346–1350.

Kaur B., Gupta S.R. and Singh G. 2002. Carbon storage and nitrogen cycling in silvipastoral system on a sodic soil northwestern India. Agroforest Syst 54: 21–29.

Kaushik W., Kaushik R.A., Saini R.S. and Deswal R.P.S. 2002. Performance of agri-silvi-horticulture systems in arid India. Indian J Agroforest 4: 31–34.

Kedharnath S. 1984. Forest tree improvement in India. Proceedings of Indian Academy of Sciences (Plant Sci.) 93: 401–412.

Kedharnath S. 1986. Genetics and improvement of forest trees. Indian J Genetics 46 (Suppl.): 172–180.

Khosla P.K. 1985. Genetic improvement of agroforestry trees. In: Khosla P.K. and Puri S. (eds). Agroforestry Systems – A New Challenge, pp. 151–160. Indian Society of Tree Scientists, Solan (HP), India.

King K.F.S. 1987. The history of agroforestry, pp. 1–11. In: Steppler H.A. and Nair P.K.R. (eds) Agroforestry: A Decade of Development. ICRAF, Nairobi, Kenya.

Kinhal G.A. 1995. Technical and financial evaluation of green equities. Indian Forester 121: 566–572.

Kumar B.M. 1999. Agroforestry in the Indian tropics. Indian J Agroforest 1: 47–62.

Kumar B.M. and Nair P.K.R. 2004. The enigma of tropical homegardens. (This volume).

Kumar B.M., Thomas J. and Fisher R.F. 2001. *Ailanthus triphysa* at different density and fertiliser levels in Kerala, India: tree growth, light transmittance and understorey ginger yield. Agroforest Syst 52: 133–144.

Kumar S.S., Kumar B.M., Wahid P.A., Kamalam N.V. and Fisher R.F. 1999. Root competition for phosphorus between coconut, multipurpose trees and kacholam (*Kaempferia galanga* L) in Kerala, India. Agroforest Syst 46: 131–146.

Lakhmipathy C.H. 1992. Tree Farming, pp. 159–168. In: Khosla P.K. (ed.) Status of Indian Forestry Problems and Perspectives. ISTS, Solan, India.

Lal P., Kulkarni H.D., Srinivas K., Venkatesh K.R. and Santakumar P. 1997. Genetically improved clonal planting stock of *Eucalyptus* – a success story from India. Indian Forester 123: 1117–1138.

Maikhuri R.K., Semwal R.L., Rao K.S. and Saxena K.G. 1997. Rehabilitation of degraded community lands for sustainable development in Himalaya: a case study in Garhwal Himalaya, India. Int J Sust Dev World 4: 192–203.

Malik R.S. and Sharma S.K. 1990. Moisture extraction and crop yield as a function of distance from a row of *Eucalyptus tereticornis*. Agroforest Syst 12: 187–195.

Mandal A.K., Ennos R.A. and Fagg C.W. 1994. Mating system analysis in a natural population of *Acacia nilotica* subspecies *leiocarpa*. Theoretical Applied Genetics 89: 931–935.

McIntyre B.D., Riha S.J. and Ong C.K. 1996. Light interception and evapotranspiration in hedgerow agroforestry systems. Agr Forest Meteorol 81: 31–40.

Mishra R.N. 1995. Neem improvement research at Arid Forest Research Institute, Jodhpur. Indian Forester 121: 997–1002.

Misra V.K. 2002 Greening the wastelands: experiences from the Tree Grower's Cooperatives Project, pp. 334–354. In: Marothia D.K. (ed.) Institutionalizing Common Pool Resources. Concept Publishing Company, New Delhi, India.

Mohanan C., Mathew G., Krishnankutty C.N., Seethalakshmi K.K. and Renuka C. 2001. Policy and Legal Issues in Cultivation & Utilization of Bamboo, Rattan and Forest Trees in Private & Community Lands. KFRI, Peechi, India. 175 pp.

Montagnini F. and Nair P.K.R. 2004. Carbon sequestration: An underexploited environmental benefit of agroforestry systems. (This volume).

Naina N.S., Gupta P.K. and Mascarenhas A.F. 1989. Genetic transformation and regeneration of transgenic neem (*Azadirachta indica*) plants using *Agrobacterium tumefaciens*. Current Science 58: 164–187.

Nair K.S.S., Chundamannil M., Chacko K.C., Ganapathy P.M. and Iyer C.S. 1996. Characteristics of private sector forestry research in India, pp. 65–80 In: Enters T., Nair C.T.S. and Apichart K. (eds) Emerging Institutional Arrangements for Forestry Research. FORSPA Publication No. 20/11998. Bangkok, Thailand.

Nair P.K.R. 1979. Intensive multiple cropping with coconuts in India. Advances Agro Crop Sci 6: 1–12.

Nair P.K.R. 1984. Soil Productivity Aspects of Agroforestry. Science and Practice of Agroforestry 1. ICRAF, Nairobi, Kenya. 85 pp.

Nair P.K.R. 1985. Classification of agroforestry systems. Agroforest Syst 3: 97–128.

Nair P.K.R. (ed) 1989. Agroforestry Systems in the Tropics. Kluwer, Dordrecht, The Netherlands. 664 pp.

Nair P.K.R. 1993. An Introduction to Agroforestry. Kluwer, Dordrecht, The Netherlands. 499 pp.

Nair P.K.R 1999. Agroforestry research at a crossroads. Ann Arid Zone 38: 415–430.

Nair P.K.R. 2001. Agroforestry. In Our Fragile World: Challenges and Opportunities for Sustainable Development, Forerunner to The Encyclopedia of Life Support Systems, Chapter 1.25, pp. 375–393, vol. I. UNESCO, Paris, France, & EOLSS, UK.

Nair P.K.R., Buresh R.J., Mugendi D.N. and Latt C.R. 1999. Nutrient cycling in tropical agroforestry systems: myths and science, pp. 1–31. In: Buck L.E., Lassoie J.P. and Fernandes E.C.M. (eds). Agroforestry in Sustainable Agricultural Systems. CRC Press, Boca Raton, FL.

Nambiar E.K.S., Cossalter C. and Tiarks A. (eds) 1998. Site management and productivity in tropical plantation forests. CIFOR, Indonesia. 76 pp.

Naugraiya M.N. and Puri S. 2001. Performance of multipurpose tree species under agroforestry systems on entisols of Chhattisgarh plains. Range Manag Agroforest 22: 164–172.

Negi Y.S., Tewari S.C. and Kumar J. 1996. Eucalyptus marketing in Punjab – a comparative inter-market analysis. Indian Forester 122: 1127–1135.

Ong C.K. and Huxley P. 1996. Tree–Crop Interactions: A Physiological Approach. CAB International UK. 386 pp.

Ong C.K. and Leakey R.R.B. 1999. Why tree crop interactions in agroforestry appear at odds with tree-grass interactions in tropical savannahs. Agroforest Syst 45: 109–129.

Ong C.K., Odongo J.C.W., Marshall F. and Black C.R. 1992. Water use of agroforestry systems in semiarid India, pp. 347–358. In: Calder I.R., Hall L.R. and Adlard P.G. (eds) Growth and Water Use of Plantations. Proceedings of the International Symposium on the Growth and Water Use of Forest Plantations, Bangalore. Wiley, Chichester, UK.

Palanisamy K., Anasari S.A., Kumar P. and Gupta B.N. 1998. Adventitious rooting in shoot cuttings of *Azadirachta indica* and *Pongamia pinnata*. New Forests 16: 81–88.

Pathak P.S. 2002. Common pool degraded lands: technological and institutional options, pp. 402–433. In: Marothia D.K. (ed.) Institutionalizing Common Pool Resources. Concept Publishing Co., New Delhi, India.

Pathak P.S. and Pateria H.M. 1999. Agroforestry in the Indo-Gangetic Plains: an analysis. Indian J Agroforestry 1: 15–36.

Pathak P.S. and Solanki K.S. 2002. Agroforestry Technologies for Different Agroclimatic Regions of India. National Research Centre for Agroforestry, ICAR, New Delhi. 42 pp.

Puri S. 1998. Tree Improvement: Applied Research and Technology Transfer. Science Publishers, New Hampshire, USA. 390 pp.

Puri S. and Bangarwa K.S. 1992 Effects of trees on the yield of irrigated wheat crop in semiarid regions. Agroforest Syst 20: 229–241.

Puri S. and Kumar K. 1995. Establishment of *Prosopis cineraria* (L.) Druce in the hot desert of India. New Forests 9: 21–33.

Puri S. and Swamy S.L. 1999. Geographical variation in rooting ability of stem cuttings of *Azadirachta indica* and *Dalbergia sissoo*. Genetic Resource and Crop Evolution 46: 29–36.

Puri S., Kumar A. and Singh S. 1994a. Productivity of *Cicer arietinum* (chickpea) under *Prosopis cineraria* based agroforestry system in arid regions of India. J Arid Environ 27: 85–98.

Puri S., Singh S. and Bhushan B. 1994b. Evaluation of fuelwood quality of indigenous and exotic tree species of India's semiarid region. Agroforest Syst 26: 123–130.

Puri S., Singh S. and Kumar A. 1994c. Growth and productivity of crops in association with an *Acacia nilotica* tree belt. J Arid Environ 27: 37–48.

Puri S., Singh V., Bhusan B. and Singh S. 1994d. Biomass production and distribution of roots in three stands of *Populus deltoides*. Forest Ecol Manag 65: 135–147.

Puri S., Swamy S.L. and Jaiswal A.K. 2002. Evaluation of *Populus deltoides* clones under nursery, field and agrisilviculture system in subhumid tropics of Central India. New Forests 23: 45–61.

Rai P., Solanki K.R. and Rao G.R. 1999. Silvipasture research in India – a review. Indian J Agroforest 1: 107–120.

Rai P., Yadav R.S., Solanki K.R., Rao G.R. and Singh R. 2001 Growth and pruned production of multipurpose tree species in silvo-pastoral systems on degraded lands in semi-arid region of Uttar Pradesh, India. Forests, Trees and Livelihoods 11: 347–364.

Rao M.R., Palada M.C. and Becker B.N. 2004. Medicinal and aromatic plants in agroforestry systems (This volume).

Rao J.P. and Osman M. 1994. Studies on silvipastoral systems in non-arable drylands, pp. 755–760. In: Singh P., Pathak P.S. and Roy M. (eds). Agroforestry Systems for Degraded Lands. Oxford & IBH, New Delhi.

Rao M.R., Nair P.K.R. and Ong C.K. 1998. Biophysical interactions in tropical agroforestry systems. Agroforest Syst 38: 3–50.

Ravindranath N.H., Chanakya H.N. and Kerr J.M. 1997. An ecosystem approach for analysis of biomass energy, pp. 561–587. In: Ker J.M., Marothia D.K., Singh K., Ramaswamy C. and Bentley W.R. (eds) Natural Resource Economics Theory and Application in India. Oxford & IBH, New Delhi.

Reddy P.R. and Reddy P.S. 1995. Commentaries on Land Reforms Laws in Andhra Pradesh. 3rd edition, Asia Law House, Hyderabad, India. 100 pp.

Sanchez P.A. 1995. Science in agroforestry. Agroforest Syst 30: 5–55.

Saxena N.C. 1995. Wood markets for farm eucalyptus in north-west India, pp. 351–378. In: Saxena N.C. and Ballabh V. (eds) Farm Forestry in South Asia. Sage Publications India, New Delhi.

Schroth G. 1999. A review of belowground interaction in agroforestry focusing on mechanisms and management options. Agroforest Syst 43: 5–34.

Sen D. 1995. The management of transfer of farm forestry technologies – emerging trends, pp. 197–218. In: Saxena N.C. and Ballabh V. (eds). Farm Forestry in South Asia. Sage Publications India, New Delhi.

Shankaranarayan K.A., Harsh L.N. and Kathju S. 1987. Agroforestry in the arid zones of India. Agroforest Syst 9: 259–274.

Sharma K. and Ogra J.L. 1990. Influence of continuous grazing by goats and sheep under three tier pasture on their performance and constituent vegetation. Range Manag & Agroforest 11: 99–102.

Sharma S.K., Datta B.K. and Tiwari J.C. 1996. *Prosopis cineraia* (L) Druce in silvipastoral system in arid regions of Western Rajasthan. Range Manag & Agroforest 17: 81–85.

Sharma S.K., Singh R.S., Tiwari J.C. and Burman U. 1994. Silvopastoral studies in arid and semiarid degraded lands of western Rajasthan, pp. 749–754. In: Singh P., Pathak P.S. and Roy M. (eds). Agroforestry Systems for Degraded Lands. Oxford & IBH, New Delhi.

Shujauddin N. and Kumar B.M. 2003. *Ailanthus triphysa* at different densities and fertiliser regimes in Kerala, India: Biomass productivity, nutrient export and nutrient use efficiency. For Ecol Manag 180: 135–151.

Singh B. 1988. Biomass potentials of some firewood shrubs of North India. Biomass 4: 235–238.

Singh G., Abrol I.P. and Cheema S.S. 1988. Agroforestry in alkali soils – effect of planting methods and amendments in initial growth, biomass accumulation and chemical composition of mesquite grown with and without Karnal grass (*Leotochloa fusca*) in inter-space. Agroforest Syst 7: 135–160.

Singh G.B. 1999. Agroforestry research in India – issues and strategies. Indian J Agroforest 1: 1–14.

Singh K., Yadav J.S.P. and Singh B. 1992. Tolerance of trees to soil sodicity. J Indian Soc Soil Sci 40: 173–179.

Singh P. and Roy M.M. 1993. Silvopastoral systems for ameliorating productivity of degraded lands in India. Annals Forestry 1: 61–73.

Singh P. and Upadhyaya S.D. 2001 Biological interaction in tropical grassland ecosystem, pp. 113–143. In: Shiyomi M. and Koizumi H. (eds) Structure and Function in Agroecosystem Design and Management. CRC Press, USA.

Singh R.V. 1982. Fodder Trees of India. Oxford & IBH, New Delhi. 285 pp.

Swamy S.L., Mishra A. and Puri S. 2003a. Biomass production and root distribution of *Gmelina arborea* under an agrisilviculture system in subhumid tropics of central India. New Forests 26: 167–186.

Swamy S.L., Puri S. and Singh A.K. 2003b. Growth, biomass, carbon storage and nutrient distribution in *Gmelina arborea* Roxb. stands on red lateritic soils in Central India. Bioresource Technol 90: 109–126.

Tejwani K.G. 1994. Agroforestry in India. Oxford & IBH, New Delhi. 233 pp.

Tewari D.D. 2001. Domestication of non-timber forest products (NTFP) – a case of bamboo farming in Kheda district, Gujarat, India. Indian Forester 127: 788–798.

Toky O.P., Bisht R.P. and Singh R.R. 1992. Potential multipurpose trees of Indian forests vegetation, pp. 359–378. In: Khosla P.K. (ed.) Status of Indian Forestry Problems and Perspectives. ISTS, Solan, India.

Tomar O.S. 1997. Technologies of afforestation of salt-affected soils. Int Tree Crops J 9: 131–158.

Viswanath S., Nair P.K.R., Kaushik P.K. and Praksasam U. 2000. *Acacia nilotica* trees in rice fields: a traditional agroforestry system in central India. Agroforest Syst 50: 157–177.

Yadav J.P. and Blyth J.B. 1996. Combined productivity of *Prosopis cineraria* – mustard agrosilviculture practice. Int Tree Crops J 9: 47–58.

Yadav J.P., Sharma K.K. and Khanna P. 1993. Effect of *Acacia nilotica* on mustard crop. Agroforest Syst 21: 91–98.

Agroforestry Systems 61: 453–462, 2004.

Public/private partnerships in agroforestry: the example of working together to improve cocoa sustainability

H-Y Shapiro[1,*] and E.M. Rosenquist[2]
[1]*Mars Incorporated, 800 High Street, Hackettstown, NJ 07840, USA;* [2] *International Programs, USDA–ARS, 5601 Sunnyside Avenue, Beltsville, MD 20705–5139, USA;* [*]*Author for correspondence: e-mail: howard.shapiro@effem.com*

Abstract

As information on the economic, environmental and social benefits of cocoa has grown, so has the understanding that only a coordinated effort by all stakeholders can ensure cocoa sustainability. This chapter describes how challenges to cocoa supplies brought seemingly disparate – if not competitive – groups together in unique public/private partnerships. While it is not meant to be an exhaustive listing of every initiative that has been developed, it provides an overview of how working across sectors has benefited all of those involved in the cocoa industry – corporations, governments, nongovernment organizations and individual farmers. The progress they have made and the lessons learned from these partnerships will help frame policies and practices aimed at ensuring a healthy future for all involved in the cocoa industry, and be a model for such initiatives for the development of other shaded perennial crops in agroforestry systems.

Introduction

Although cocoa (*Theobroma cacao* L.) has a history dating back some 2000 years, the full extent of its contributions to the global economy, its vital role to its surrounding environment, and its social impact on its farmers and their communities are only now being recognized. Approximately 3 million tons (3×10^6 Mg) of cocoa beans are produced annually, with an average market value of approximately $4 billion (www.chocolateandcocoa.org), but the supply is fragile.

Editors' Notes
Unlike the other chapters in this volume, this chapter is not a scientific review, but is a summary of reflections and experiences of leaders of research and development in industry and public institutions. Such information cannot be obtained from scientific literature. The rationale for including such an account in this volume is that public/private partnerships of this nature could be a promising strategy for development of other perennial cash crops such as coffee (*Coffea* spp.), black pepper (*Piper nigrum*), and tree spices that are grown in shaded-perennial agroforestry systems.

Most scientific literature uses the term 'cacao' to refer to the plant *Theobroma cacao* L., but the product of commerce from the plant is known as 'cocoa'. In order to avoid the confusion that may arise from repeated use of these two similar terms, the 'cocoa' is used in this chapter to refer to both the plant and its products.

Native to the upper Amazon basin, cocoa trees, the source of cocoa beans, require constant warmth and rainfall to survive – they grow only in tropical regions within 20 degrees of the equator. Because they need protection from sun and wind, they grow well as part of a multilayered agroforest, rather than in plantation-style cropping. When grown in the open, the full-sun environment initially increases yields but in the long term creates stress on the trees and makes them more susceptible to pests and diseases. Small family farms are the heart of the cocoa industry, with 5 million to 6 million smallholder farmers providing more than 85% of the world's cocoa bean crop (World Cocoa Foundation, www.chocolateandcocoa.org). Typically, each cocoa farmer owns less than 2 ha of land and may grow approximately 1000 cocoa trees. In ideal conditions, the trees can produce fruit for 75 to 100 years.

An estimated one-third of the world's cocoa crop is lost to pests and diseases every year, having potentially devastating impact on small-scale farmers whose livelihoods depend on healthy crops. Economic conditions and political environments of cocoa producing countries are sometimes uncertain and chaotic. And many of the small-scale farmers lack the training and resources that can make them more successful.

Many groups have an interest in the issues facing cocoa: the chocolate industry needs a stable supply of raw ingredients, environmental groups seek to preserve the wildlife habitats that cocoa creates, development groups aim to raise rural incomes, and governments look to support domestic agricultures. Most efforts by these groups prior to late 1980s were limited in scope and not coordinated in any strategic or cohesive way. During the past decade, however, these varied interests have begun to unite. As cocoa stakeholders recognized that a sustainable agricultural system could meet current economic, social, and environmental needs as well as those of the future, they also began to recognize that a truly sustainable cocoa supply would require coordinated efforts of all interested parties. 'In the last several years, relationships have emerged, frameworks were created, and common goals established. Unique, successful public/private partnerships among industry, governments, international donor and development organizations, nongovernment organizations, and cocoa farmers now share knowledge and resources to build a sustainable cocoa supply chain. These are significant developments that have aided efforts to improve the economic and environmental sustainability associated with the global cocoa supply' (J. Lunde, pers. comm., December 2003)[1].

These cooperative partnerships were established with the following objectives:

- Raise the standard of living for small-scale cocoa farmers,
- Create stability in cocoa-producing communities,
- Improve surrounding ecosystems,
- Create jobs globally for cocoa producers and farmers of associated products, and
- Provide quality raw materials to satisfy consumer demand for chocolate and chocolate products.

Collectively, these efforts have shaped consensus around promoting the farming and marketing of quality cocoa, improving market access and income for small-scale producers, and creating systems that are environmentally friendly, socially responsible, and economically sustainable.

Early efforts to improve cocoa production

For almost a century, processors of cocoa and manufacturers of chocolate products in the developed world have formed cocoa- and confectionary-related trade associations, most of which funded and administered research, promoted chocolate to the general public, and advocated for the industry in dealings with government agencies. Several of these associations were tied to tropical research stations in cocoa-growing countries, which had ties with western countries dating from colonial periods.

In 1930, the International Office of Cocoa, Chocolate, and Sugar Confections (IOCCC) was created for efficient sharing of scientific, technical, and market research. Early members included most of the major manufacturing trade associations and represented the global reach of cocoa, such as the Chocolate Manufacturers Association of America (CMA); the Confectionery Manufacturers of Australasia; the Brazilian Chocolate, Cocoa & Confectionery Manufacturers Association (ABICAB); the Association of the Chocolate, Biscuit and Confectionery Industries of Europe (CAOBISCO); and the United Kingdom's Biscuit, Chocolate, Cake and Confectionary Alliance (BCCCA). Today, the IOCCC represents more than 2000 companies in 23 countries.

Coordinated efforts to improve cocoa cultivation began in the late 1940s. The American Cocoa Research Institute (ACRI), created by the CMA, helped establish a cocoa research and training center in Turrialba, Costa Rica, to teach farmers modern production methods. The center was associated with the Inter-American Institute for Cooperation in Agriculture (IICA), an agency of the Organization of American States (OAS). In the mid-1960s, ACRI began awarding annual grants to the Centre for International Cooperation for Agricultural Research and Development (CEPLAC), a cocoa research center in Brazil, to assist that country's efforts to improve cocoa production efficiency. As an ancillary effort, CEPLAC examined the social benefits that resulted from industrialization of cocoa. Also during this period, ACRI began participating in numerous cooperative activities with the U.S. Department of Agriculture (USDA), including establishing a cocoa varietal collection and protection station in Puerto Rico. In 1973, the International Cocoa Organization (ICCO) was established under the aegis of the United Nations to maintain cocoa price stability through buffer stocks. It has since grown to become a global forum for gathering and disseminating information on cocoa, as well as for promoting the findings of cocoa research, including economic studies of cocoa production, consumption and distribution.

Some of the international agricultural institutions that were established in the 1960s also had an interest in, if not a mandate for, improving cocoa farming. For example, the International Institute of Tropical Agriculture (IITA) that was founded in 1967 at Ibadan, Nigeria, to conduct research to improve food production in the humid tropics and to develop sustainable production systems, had interest in improving cocoa farming systems.

In the early 1980s, as emphasis on genetics research increased, the Cocoa Research Unit (CRU) in Trinidad, aided by the BCCCA, obtained funding from the European Development Fund for a program focused on cocoa germplasm conservation. This became the International Cocoa Genebank, Trinidad (ICG,T), where nearly 3000 cocoa germplasm samples from around the world are stored. CRU and ICG,T are now supported by the governments of Trinidad and Tobago, as well as the BCCCA.

In the mid-1980s, another important partnership was formed among governments of cocoa-producing countries, national research centers, and IOCCC. This partnership set out to identify the best economic management systems for cocoa production in the presence of witches' broom fungus (*Crinipellis pernicosa*), a cocoa disease that can have devastating effects on the crop[2]. South American countries were the primary participants in the project, which successfully identified more stringent husbandry requirements for producing cocoa in a witches' broom area. In 2000, the World Cocoa Foundation (WCF) (www.chocolateandcocoa.org) was created by cocoa industry members to focus on cocoa sustainability issues, such as training smallholder farmers and stimulating global investment in crop production. ACRI continues to be devoted to research in all scientific areas related to cocoa and chocolate.

Global threats spur broader efforts at partnerships

These efforts proved insufficient, however, when in 1989, witches' broom broke out in Bahia, Brazil, the largest South American cocoa producer at the time. Damage to the crops was so severe that production dropped by more than 70% in less than 10 years, changing the status of Brazil from a major cocoa exporter to a cocoa importer. According to Lunde[1], 'No one had seen devastation like this before. We were thinking that if this spreads to Africa, the results would be catastrophic. Brazil's experience opened everyone's eyes to the fact that a global, cooperative effort was needed to protect cocoa.'

Other threats around the world emphasized that need. In Malaysia, cocoa production grew rapidly in the 1970s and 1980s, but dropped off significantly through the 1990s, mainly due to the insect pest, cocoa pod borer (*Conopomorpha cramerella*)[3]. Damage to the crops was so severe that the increased work required to complete a harvest nearly devastated the industry.

In West African countries, privatization was presenting its own challenges. The governments of the region's major cocoa producers, Cote d'Ivoire and Ghana, previously controlled the entire cocoa market from farmer to port, but state commodity boards were starting to give way to liberalized markets due to pressure from industrial countries, including the G-7 and the International Monetary Fund (IMF). The short-term effect was a general decline in the character of the cocoa market, creating undesirable effects such as the degradation of cocoa quality and increased risks for banks and international traders, which in turn reduced financing available for smallholder farmers.

On a worldwide scale, the decline in Brazilian production was temporarily mitigated by the rapid rise in Indonesian cocoa production, which grew steadily through the 1980s and 1990s. Indonesia enjoyed good yields, an efficient supply chain, and, significantly, farmers received a fair proportion of the world market price. Unfortunately, the cocoa pod borer (CPD) arrived in 1993, threatening to decimate the Indonesian cocoa sector just as it did in Malaysia. In response, ACRI, the United States Agency for International Development (USAID), and the Indonesian Cocoa Association (ASKINDO), organized a workshop in Indonesia to review research and approaches to controlling the CPB. As a result, another partnership was formed – ASKINDO, ACRI and the United Kingdom's Biscuit, Chocolate, Cake and Confectionary Alliance (BCCCA) joined together to establish the Cocoa Pod Borer Management Project (CPBMP) to verify the effectiveness of the workshop recommendations of pruning, frequent harvesting, and targeted spraying. Implementation of farmer-training and technology-transfer was underwritten through United States' food assistance funds, which are tied to loans made in the currency of the recipient country and to be used for food and agricultural assistance programs. To date, these funds have been used successfully to promote sustainable tree crop efforts, including cocoa, in Indonesia and Bolivia. 'Decision- and policy makers in the cocoa and confectionary industry began to realize that the threats to cocoa were enormous – they were not going to wipe out one company, they were going to wipe out the industry.' 'By the mid-1990s, people started changing their vision. Their approach was moving from short-term individual projects to more long-term projects based on a common vision for the industry' (P. Petithuguenin, pers. comm. November 2003)[4].

As chocolate manufacturers worried about supply and governments worried about rural incomes in the developing world, environmental groups were taking notice of the effect of rainforest destruction on wildlife. For example, the Smithsonian Institute determined that migratory bird populations in North America had been declining for a decade, mostly due to farmers harvesting canopy trees for income and in related efforts to increase agricultural production (Smithsonian Tropical Research Institute, www.stri.org). The seemingly unlikely link between environmentalists and the cocoa industry was forged when it was reported that a newly discovered species of bird, the pink-legged graveterio (*Acrobatornis fonsecai)*, faced extinction, largely due to opening up the birds' canopy tree habitat to make up for cocoa production losses from witches' broom. 'Scientists at our Institute (Smithsonian Tropical Research Institute – STRI) have long been interested in understanding how tropical forests function – how the vast diversity of the tropics originated and is maintained. We were intrigued by research on traditional, small-scale, shade-grown cocoa plantations that implied that such farms could be effective in conserving much of the forest's natural diversity' (L. Barnett, pers. comm. November 2003)[5].

At the same time, breakthrough research on rainforest destruction was being conducted at STRI (T. Lovejoy, pers. comm. November 2003)[6]. Lovejoy believed that one way to save the rainforest was to encourage sustainable income-producing activities – such as growing cocoa. He introduced representatives from Mars, Incorporated, one of the world's largest chocolate producers, to researchers from STRI and the Smithsonian Migratory Bird Center, which had then conducted a workshop on sustainable coffee (*Coffea* spp.) production and its role in protecting migratory bird habitats. These introductions led to the so-called Panama Conference, which would later become recognized as an important event to date in the sustainability movement.

The cocoa community gathers at the Panama Conference

On April 2, 1998, the Smithsonian Migratory Bird Center and the Smithsonian Tropical Research Institute convened a conference in Panama at which 85 representatives from industry, environmental groups, foundations, universities, and agricultural research centers gathered to discuss sustainable cocoa production. 'The cocoa industry's need to protect cocoa trees aligned well with our research into ways in which naturally occurring, potentially beneficial fungi may contribute to plant growth and survival and specifically, may help keep plant diseases in check,' said Barnett of STRI (L. Barnett, pers. comm. December 2003)[5]. Officially named the First International Workshop on Sustainable Cocoa Growing, but referred to as the Panama Conference, participants embraced the idea that cocoa grown within a biologically diverse and environmentally sustainable agricultural system is capable of providing long-term economic, social, and environmental benefits to the millions of small-holder farmers who are uniquely better placed to cultivate cocoa.

The consensus statement of the Panama Conference established five principles, which currently guide the public/private efforts. Specifically, it said that sustainable production of cocoa will:

- Be based on cocoa grown under a diverse shade canopy in a manner that sustains as much biological diversity as is consistent with economically viable yields of cocoa and other products for farmers;
- Use constructive partnerships that are developed to involve all stakeholders with special emphasis on small-scale farmers;
- Build effective policy frameworks to support these partnerships and address the particular needs of smallholder farmers for generations to come;
- Encourage future cocoa production that rehabilitates agricultural lands and forms part of a strategy to preserve remnant forests and develop habitat corridors; and
- Maximize the judicious use of biological control, techniques of integrated management of pests, disease, and other low-input management systems.

Conference participants concluded that cocoa was a low-input small-farm crop, not a plantation crop; that cocoa could be a source of biodiversification, not a cause of rainforest destruction; that cocoa was essentially an 'orphan crop' that had not received the public and private sector support as other major agricultural commodities such as corn (maize, *Zea mays*), wheat (*Triticum* spp.) and soybeans (*Glycine max*); and that a focus on improved cocoa cultivation required a multifaceted effort that included training, technology transfer, credits, and improved genetic materials.

'The public/private partnership got off the ground in a substantive way starting with industry engaging governments, research institutions, development institutions and conservation agencies. The goal was to develop a holistic, integrated approach for not only the environmental sustainability of cocoa bean production, but more importantly the economic sustainability of private sector-led growth in the rural areas, both on the farms and in the local markets' (J. Lunde, pers. comm. December 2003)[1].

As the cocoa industry, governments, NGOs and environmentalists discovered synergies, a number of participants from USAID, USDA, the World Bank and foundations met in 1999 to develop a comprehensive, integrated approach to cocoa research. Similar to the consensus achieved in Panama, the delegation concluded that cocoa was not adequately supported in terms of extension, farmer organization, research, technology transfer, and market access. The delegation ended with a major industry and government review in Paris that led to the 'Paris Declaration,' through which all stakeholders pledged to work toward a sustainable cocoa economy. 'The entire cocoa supply chain began recognizing the multi-dimensional benefits of cocoa and coming together in ways that we never imagined' (B. Guyton, pers. comm. November 2003)[7].

Other examples of partnership

The West Africa Sustainable Tree Crops Program (STCP) is another example of the industry coming together, in this case, to address the needs of cocoa farmers to improve the economic, social and environmental conditions of small-scale farmers (www.treecrops.org). Managed under the leadership of IITA, the STCP is guided by a steering group of stakeholders representing the cocoa-growing regions, research, government, NGOs, and the global chocolate industry. The four basic program components of STCP are research and technology transfer, grower and business support services, policy change and implementation, and market and information systems. The original focus of STCP was to increase incomes and wellness, as well as promote environmental protection. More recently, it has also been addressing the needs of children in cocoa-farming communities. STCP pilot projects are in progress to compare, test, and validate different approaches and interventions to develop sustainable and integrated agricultural production systems in Cote d'Ivoire, Cameroon, Ghana, Guinea, and Nigeria. The overall goal of STCP is to improve the livelihood of rural cocoa producers in West Africa by improving their ability to respond to the demands of global markets. 'Suddenly we had an integrated model that went beyond any single issue. By working with private industry, we were able to raise visibility of our efforts and become much more credible in terms of gaining support from the cocoa farmers, other interested parties, and from the general public' (S. Weise, pers. comm. October 2003)[8]. In one respect, the role and experience of the IITA in the STCP effort underlines the policy deficiencies in the current international agricultural research scheme, particularly in regard to sustainable tree crops traded as commodities. Although the IITA was given responsibility for the coordination of cocoa improvement in Africa in the absence of a dedicated international center, its role is limited to Africa in the short to intermediate term, and does not address global problems related to germplasm exchange, exploitation of genetic resources, and the provision of long-term credits. Additionally, the STCP initially suffered from a limited integration of existing public and private sector cocoa research programs. By working together, however, public and private stakeholders have addressed many of these issues.

A parallel effort, initiated in 1994 by the USDA, the (United States) Department of State, focused on the cash crop requirements of farmers in the Andean Region. Although initially oriented toward reducing cultivation of illicit *coca* (*Erythroxylum coca*), from which cocaine is derived, funding for this effort provided an impetus for Integrated Pest Management of cocoa diseases and a coordinated international genetic improvement effort. Originally focused on U.S. security concerns in South America, in fiscal year 2000, the U. S. Congress enlarged USDA's appropriations language to address the needs of smallholders in West Africa, which demonstrated a growing, broader recognition of the global importance of cocoa cultivation.

Goals of the public/private partnership

Because agriculture will likely continue to be the main source of economic growth for many developing countries in several decades to come, cocoa industry stakeholders are beginning to recognize that cross-sector partnerships are the only way to effectively improve all facets of the cocoa supply chain, including breeding, pest and disease research, productivity and quality, technology transfer, and marketing and market access. Specific roles for the private sector, governments, and donor community are envisaged in these partnerships.

Private sector

The private sector in any industry is concerned with ensuring adequate raw material supplies to produce the products for which it has created demand. This is certainly true in the cocoa industry, but the sustainability of public/private partnerships also has made chocolate manufacturers more sensitive to the broader issues surrounding their business. Consequently, the industry is embracing the notion that sustainability has economic, social and environmental dimensions that it must be aware of and actively address.

The private sector also brings a global perspective and knowledge of the efficient operation of worldwide cocoa markets, including a sense for the key quality drivers that differentiate chocolate brands and trends in consumer demand. The industry is also in a position to educate the consumer on issues faced by rural small-scale farmers. 'By tapping into the knowledge and experience of the private sector, we created much stronger efforts across the board. There was a new global realization that governments alone cannot solve all problems, and neither can institutions' (C.L. Brookins, pers. comm. November 2003)[9].

Public sector

In the public sector, governments are creating environments in which other stakeholders can cooperate. By acting as facilitators and organizers, governments and parastatal organizations, nongovernment organization (NGOs), and the private sector can work together to support community development activities and assist disadvantaged groups in gaining greater access to resources and markets. In a recent example, USAID acknowledged valuable lessons learned working with other commodities and created a framework for smallholder cocoa farmers to benefit from globalization, to access technological innovations, and become more competitive. The goal was not to create a new institution; rather, it was to create a network that allowed existing groups to communicate more effectively (J. Hill, pers. comm. September 2003)[10]. 'The U.S. government was able to work as an inter-country facilitator and bring together resources from a broad variety of groups. U.S. agencies brought fundamental knowledge and understanding of how to research the crop's diseases and implement disease control measures and we were able to learn from their experiences' (J. Lunde, pers. comm. November 2003)[1].

At the USDA, focus was on domestic production of commodity crops such as peanuts (*Arachis hypogaea*), cotton (*Gossypium* spp.), soybeans, and wheat. But as the global importance of cocoa sustainability became more apparent, the USDA adopted a more global view regarding cocoa that linked its production with maintaining economic stability for U.S. chocolate producers and producers of related commodities – dairy, sugar, and peanuts.

The USDA conducts its research through its Agricultural Research Service (ARS), which has more than 130 locations in the United States, Puerto Rico, and five other countries. Originally ARS tropical crops division targeted coffee, bananas (*Musa* spp.), and oil palm (*Elaeis guineensis*), with programs oriented toward the U.S. foreign aid effort in Central and South America. ARS has since expanded its tropical agriculture program to include a range of activities for pest management, genetics, and breeding, as well as noncommodity programs such as nutrition and endangered plant species.

Mars, Incorporated, the USDA/ARS and their international partners are currently involved in a cooperative effort to study ways to use integrated pest management systems that combat cocoa pests and diseases through natural bio-control agents. The effort is being conducted under a Cooperative Research and Development Agreement (CRADA), which allows the Federal government and non-Federal partners to optimize their resources, share technical expertise in a protected environment, share intellectual property emerging from the effort, and speed the commercialization of federally developed technology.

ARS was chosen 'because they have the research capabilities and mindset of a university and the problem-solving mission of the private sector' (J. Lunde, pers. comm. November 2003)[1].

Development/donor organizations

Until recently, many development organizations funded agricultural projects as a way to fulfill their goals of improving security, eliminating poverty, and protecting the environment in developing countries, but none had programs specific to cocoa. An example is the Consultative Group on International Agricultural Research (CGIAR) centers; IITA is one of the several such 'CG' centers (www.cgiar.org). CGIAR is an informal association of more than 40 governments and 15 international organizations and private foundations, created and committed to address food security needs of the poor in developing countries. As mentioned earlier, cocoa was not viewed as a priority in CGIAR programs, because the major emphasis was on commodities that are locally produced and consumed in developing countries, emphasizing a supply-oriented concern and effort to raise the availability of food.

That view began to change after an October 1999 meeting convened by the World Bank. Participants included officials from the United States, United Kingdom, France, Japan, the European Union, the United Nations, the Ford Foundation, and the Rockefeller Foundation. In a workshop on public/private partnerships in Africa, chocolate industry representatives explained their concern for needing a stable long-term supply of cocoa, which meant focusing on supporting small-scale farmers and stabilizing the tropical farm. The meeting resonated with the donor organizations, which saw an opportunity to improve lives of rural people in the tropics by improving cocoa production. 'There is a wide body of research from the World Bank and from ACRI demonstrating how integrated the agricultural sector is with other aspects of the world economy, so that any improvement in revenues going into improving agriculture efficiencies is going to have a multiplier effect in terms of reducing poverty and giving people choice and opportunity.' said Brookins (C. L. Brookins, pers. comm. November 2003)[9].

Early successes of private/public partnerships

In the five years since the Panama Conference, cocoa sustainability efforts have generated remarkable results.

Breeding programs

A number of breeding programs are underway around the world to develop cocoa varieties with resistance to various pests and diseases, such as the CRADA between Mars and the USDA centered at the National Germplasm Repository in Miami, Florida. Research and germplasm management continues at International Cocoa Genebank, Trinidad (ICG,T); the Tropical Agronomic Centre for Research and Education (CATIE), Costa Rica; the USDA National Plant Germplasm System; the International Cocoa Germplasm Database (ICGD) and facilities such as University of Reading and the CIRAD facility at Montpellier, France, where potential transplants are quarantined for safety precautions before being delivered.

Worldwide cocoa breeding also has been strengthened by projects such as the 'Project on Cocoa Conservation and Utilization: A Global Approach,' which was designed to help smallholder farmers achieve sustainable production of cocoa and to reduce their need for expensive inputs by developing better varieties and through more efficient conservation and use of cocoa genetic resources. The project was developed by the International Plant Genetic Resources Institute (IPGRI) and the ICCO and is supported by CIRAD, ACRI, BCCCA, CFC, and the University of Reading, UK. The cocoa producing countries involved in the project are Brazil, Cameroon, Côte d'Ivoire, Ecuador, Ghana, Malaysia, Nigeria, Papua New Guinea, Trinidad and Tobago, and Venezuela.

Also, to ensure that quality characteristics are not lost in the drive for disease resistance, research is being conducted to identify genetic markers for flavor and fat content, two factors that are essential to the marketability of cocoa on the world market.

Pest and disease management

Because cocoa is always under the threat of catastrophic losses from pests and diseases, developing pest resistant materials, and management strategies focused on increasing yields, improved quality, and benefits to the smallholder farmer will continue to be a priority. Primary efforts are being directed toward breeding, biological control, responsible chemical control, and good agronomic practices. One of the most successful projects to date was supported by Association of the Chocolate, Biscuit & Confectionery Industries of the EU, involving researchers in France (CIRAD), Trinidad (CRU), Côte d'Ivoire (CNRA) and Cameroon (IRAD) to develop genetic markers for black pod[11] resistance. These were the first disease-resistance markers developed for cocoa and will accelerate the development of more resistant cocoa varieties for growers. Another example: the USDA, working under a formal cooperative research agreement with Mars, Incorporated, is currently studying ways to use Integrated Pest Management Systems that combat cocoa pests and diseases through natural bio-control agents. Recent field tests in Peru, while still preliminary in their findings, have shown a reduction of up to 50% in Witches' Broom[2] symptoms, and a corresponding increase in crop yields of up to 20%.

Farmer organization

While cocoa prices are presently high, farmers in certain growing regions often receive scarcely half of the world price for their crops. Using knowledge gained around the world, the industry is working to encourage cooperation among cocoa farmers in order to improve their participation in the world economy and receive a greater percentage of the price of cocoa. This effort has shown great success in Indonesia, where, by reducing production costs, raising productivity, and removing market and policy inefficiencies, farmers now earn more than 80 % of the world price. As incomes increase, so will opportunities for education, health care, nutrition and other social needs.

Farmer groups also facilitate the delivery of information, technology, and financial credit to farmers. In addition, cooperatives already are proven effective at communicating methods for improving quality, marketing cocoa better, and providing safer farm environments. 'By strengthening farmer organizations, we're creating leaders, bargaining power, and the capacity to really be partners in cocoa farming. Once farmers work together on something like managing their cocoa purchasing and marketing, they can build their own confidence and become more effective in taking hold of the destiny of their communities.' (C. Brookins, pers. comm. October 2003)[9].

Farmer training

After little change in the past 100 years, cocoa farming techniques are improving rapidly as a consequence of training programs developed to produce a more sustainable cocoa crop. These programs focus on improving soil nutrient supply, trimming tree canopies to increase light and nutrients, rehabilitating and rejuvenating older trees, developing integrated pest management programs that reduce use of pesticides, reducing post-harvest losses through processing and storage improvements, improving communication and cooperation among farmers, and diversifying farms to include other cropping systems such as coconut, rubber, oil palm, coffee, fruit trees, and timber trees.

New training techniques include farmer participatory approaches, which aim to give farmers the agroecological knowledge and the confidence to make their own crop-management decisions. Under the supervision of extension officers, farmers conduct their own experiments to evaluate or adapt new technologies, based on their individual needs and circumstances.

One example of public/private training is the Sustainable Cocoa Extension Services for Smallholders (SUCCESS) Project, a large-scale training program on Integrated Pest Management and pesticide-free control of the cocoa pod borer. Developed in conjunction with ACRI and BCCCA, Agricultural Cooperative Development International and Volunteers in Overseas Cooperative Assistance (ACDI/VOCA) is expected to train more than 26 000 farmers in Indonesia alone. 'When the productivity of cocoa does not increase for 30 years on a smallholder farm, something is wrong. The stream of innovations – technical, institutional, marketing, and policy – that were needed to stimulate productivity and incomes were not reaching farmers until very recently. Now that they are, farmers have the opportunity to see rapid improvements in a relatively short period of time. The emerging challenge is to scale these efforts up and ensure stability in the cocoa supply chain' (J. Hill, pers. comm. September 2003)[10].

Community development

The core purpose of the West African Sustainable Tree Crops Program (STCP) has always been to improve the well being of smallholder tree crop farming communities and to protect the tropical environment. Recently, new African government ministerial partners and major multilateral agencies have become involved in STCP programs, bringing an increased focus to the development of social services. Future programs are being developed that will increase vocational learning opportunities for young people beyond cocoa farming. 'It's profoundly important because we are teaching producers not just the institutional structures of coming together and coalescing, we are improving their abilities to manage their own money, and how to negotiate. These are the drivers of private-sector led growth – the small enterprises,' said Brookins[9].

Building trust among partners

As the cocoa stakeholders began to roll out programs, building trust among the partners was one of the early challenges. The decision on how to handle intellectual property issues not only demonstrated the way to clear that hurdle, but also became perhaps the most unique aspect of the cocoa sustainability public/private partnership: All of the activities conducted on behalf of the cocoa industry are shared so that all countries, manufacturers, and government agencies operate in an open, even environment.

Lunde states, 'Once you agree that you're not going to hold intellectual property over the head of someone that needs that material, then most of the issues of mistrust are over. Mars did not want to own the intellectual property rights on cocoa, just like it does not own cocoa farms. Information needs to be shared – whether it's a rich country, a poor, developing nation or a semiindustrialized nation, everyone competes on the same, even playing field. And that's the virtue of this public/private partnership. It is not giving anyone an advantage over anyone else, but together we all receive a benefit' (J. Lunde, pers. comm. November 2003)[1].

With information on research, development, marketing and cooperative extension activities available to all stakeholders, cocoa sustainability programs are currently underway in Indonesia, Vietnam, and Papua New Guinea in Southeast Asia; in Côte d'Ivoire, Nigeria, Cameroon, Ghana, and GuineaConakry in West Africa; and in Brazil, Ecuador, Peru, Panama, Mexico, Honduras, Guatemala and Costa Rica in Latin America. 'The common vision and commitment has been solid enough that it has helped all parties to grow together. The industry has learned a lot more about how we do business and we've learned a lot more about how business does business' (J. Hill, pers. comm. November 2003)[10].

The road ahead

Pests, diseases, and civil unrest remain legitimate threats to the global cocoa supply, but considerable progress has been made to recognize and develop programs to sustain the economic, environmental, and social aspects of cocoa in a relatively very short period of time (a little more than a decade). Thanks to unprecedented cooperation among diverse stakeholders, integrated, holistic programs now exist that not only work toward ensuring cocoa sustainability and its power to be an engine of economic development for farmers and farming communities, but can be used as a model for other tree crops.

As Brookins (C. Brookins, pers. comm. October 2003)[9] states, 'The cocoa public/private partnerships in place now can set the standard for other similar types of activities that combine science, public policy, finance, and business best practices. By lifting up the economic viability of cocoa producers – and ultimately other tree crop producers in some of the poorest developing countries – it is possible to build the kind of supply chain and institutional frameworks that engage all parties in providing any proven economic, environmental and social infrastructure.'

End Notes

1. John Lunde, Director of International Environmental Programs for Mars, Incorporated, USA.
2. Witches' Broom (*Crinipellis perniciosa*): witches' broom is a fungal disease that infects the trees and is spread by spores. It causes broom-like stems that grow from branches. The infected branches turn brown and die from the tip back toward the tree. Finally, small mushrooms grow on the dead brooms, releasing spores that infect other trees. Broom growth uses much of the tree's energy, causing production of lower number of pods as well as pods with inferior-quality beans.
3. Cocoa Pod Borer (*Conopomorpha cramerella*): Cocoa pod borer is an insect approximately 1 cm long that flies like a mosquito. It is common in Southeast Asia, especially in Malaysia and Indonesia. The female lays a tiny egg on the furrowed surface of the pod. After a few days the egg hatches, a larva emerges and burrows into the pod, spoiling the beans inside. The pod dries up after the larva has fed on the pulp and its entry hole allows infections to rot the pod. Approximately two weeks after hatching, the larva leaves the pod, usually producing a silk thread with which to reach the ground.
4. Philippe Petithuguenin, Director of Cocoa Programme, Centre for International Agricultural Research and Development, Montpellier, France.
5. L. Barnett, Development Officer, Smithsonian Tropical Research Institute, Washington DC, USA.
6. Tom Lovejoy, then Counselor to the Secretary for Biodiversity and Environmental Affairs at the Smithsonian Institution, Washington DC, USA.
7. Bill Guyton, President of the World Cocoa Foundation (WCF) and past Vice-President of cocoa research with ACRI.
8. Stephan Weise, Program Manager for West Africa Sustainable Tree Crops Program (STCP), US-AID, Washington DC, USA.
9. Carole L. Brookins, United States Executive Director for The World Bank
10. Jeff Hill, Senior Agricultural Advisor for USAID's Africa Bureau, USAID, Washington DC, USA.
11. Black Pod (*Phytophthora* spp.): black pod is a fungal disease affecting trees grown in humid conditions. There are two strains of the disease: (1) *Phytophthora megakyrya* that is the faster moving, and thereby the more dangerous, and is currently restricted to Cameroon, Nigeria and Ghana, and (2) *Phytophthora*

palmivora that acts more slowly and is thereby more easily controlled. Both strains attack all parts of the plant but this is most pronounced on the pods, which develop dark brown lesions, later becoming dusted with white spores. It is further spread by rain. Both strains may be controlled by selective pruning of diseased pods together with the use of copper fungicides.

Frosty pod is another fungal disease, which is caused by *Moniliophthora rereri* that attacks only young growing pods. It is difficult to detect in its early stages but once infected, the pods become irregularly swollen, then discolored and then grow spores on the surface which are released after rains for up to 10 months and can travel great distances on clothes and shoes. The spores are much smaller than those of witches' broom and are far more resistant to dry heat and intense, direct sunlight.

List of Reviewers

Alavalapati, J., IFAS, Univ. of Florida, Gainesville, FL, USA
Bannister, M.E., IFAS, Univ. of Florida, Gainesville, FL, USA
Bishaw, B., Oregon State Univ., Corvallis, OR, USA
Buck, L.E., Cornell Univ., Ithaca, NY, USA
Clement, C., INPA, Manaus, Brazil
Current, D., Forest Resources, Univ. of Minnesota, St Paul, MN, USA
De Forresta, H., ENGREF, Montpellier, France
Dove, M., Yale Univ., New Haven, CT, USA
Franzel, S., World AF Cent. (ICRAF), Nairobi, Kenya
Garforth, C., Univ. of Reading, UK
Gold, M.A., Univ. of Missouri Cent of Agroforestry, MO, USA
Gordon, A.M., Univ. of Guelph, Guelph, ON, Canada
Harwood, C., CSIRO For. & For. Products, Kinston ACT 2604, Australia
Hildebrand, P.E., IFAS, Univ. of Florida, Gainesville, FL, USA
Hyde, W.F., Grand Junction, CO, USA
Jama, B., World AF Cent. (ICRAF), Nairobi, Kenya
Buresh, R.J., Int. Rice Res. Inst., Los Baños, The Philippines
Jones, J.W., IFAS, Univ. of Florida, Gainesville, FL, USA
Jose, S., IFAS, Univ. Of Florida, Milton, FL, USA
Josiah, S.J., Univ. Nebraska, UNL East campus, Lincoln, NE, USA
Kalmbacher, R.S., IFAS, Univ. of Florida, Ona, FL, USA
Kumar, B.M., Kerala Agri. Univ., Thrissur, Kerala, India
Leakey, R.R.B., James Cook Univ., Cairns, Australia
Long, A.J., IFAS, Univ. of Florida, Gainesville, FL, USA
Mercer, E., US For. Service Southern Res. Stn, RTP, NC, USA
Nair, V.D., Soil & Water Sci., IFAS, Univ. of Florida, Gainesville, FL, USA
O'Neill, M., New Mexico State Univ., Farmington, NM, USA
Palada, M.C., Univ. of the Virgin Islands, St Croix, VI, USA
Puri, S., I.G. Agri Univ., Raipur, India
Rao, M.R., Secunderabad, Andhra Pradesh, India
Rauscher, H.M., USDA For. Service, Asheville, NC, USA
Sanchez, P.A., Colombia Univ., New York, NY, USA
Scherr, S.J., Forest Trend, Washington, DC, USA
Schroth, G., CIFOR, Santarem, Brazil
Schultz, R.C., Iowa State Univ., Ames, IO, USA
Shannon, D., Auburn Univ., AL, USA
Torquebiau, E., CIRAD, Montpellier, France
Tripp, R., Overseas Dev. Inst., London, UK
Udawatta, R.P., Univ. of Missouri Cent. for Agroforestry, MO, USA
Van Noordwijk, M., World AF Cent., Bogor, Indonesia
Walker, D., CSIRO Sust. Ecosystems, Townsville, Queensland, Australia
Wiersum, K.F., Agri. Univ., Wageningen, The Netherlands
Wight, B., NRCS, Nat'l AF Cent., Lincoln, NE, USA
One anonymous reviewer, whose help was solicited by a reviewer.

Agroforestry Systems **61:** 465–480, 2004.

Subject index Volume 61–62 2004